Author Index

G. G. Agrawal 83
B. Aktaş 323
M. A. Alam 729
S. K. J. Al-Ani 293
K. R. Allakhverdiev K139
V. K. Arora K171
M. Asche 607

S. S. Babaev K139
D. Backs 343
N. A. Bakhyshov K139
M. Balkanski K113
R. Baltramiejūnas K135
G. Bambakidis K67
S. Bandyopadhyay K117
F. Bechstedt 575
V. Bethkenhagen K71
A. Blacha 11
M. D. Blokh 487
W. Böhm 381
I. Bójko 371
L. Borisova K155
G. Borstel 617
E.-H. Böttcher K165
H. Boudriot K149
W. Brauer 629
M. D. Brown 361
R. K. Bull 361

S. G. Canagaratna 141
E. Canessa K1
M. Cardona 11
K. Čermák 273
S. Chattopadhyay 53
R. Chen 361
E. Chilla K91
D. R. Chopra 279
Chuan-zhang Yang K125
W. Cooke 687
J. W. Corbett 157

Dang Dinh Thong K159
G. L. M. Dassen 213
A. Daud 495
S. K. Deb K117
D. Della Sala 125
K. Deus K149
A. M. El-Dib 587
T. N. Dimov 261
J. Dominec K87
M. P. Dos Santos K113
M. Doyama 453
F. Dujardin 329
M. Dupont 323

V. Dvořák 459
A. A. Dyshekov 37

I. K. Efendieva K139
Zh. Egemberdiev 407, K63
R. Enderlein 575
P. Enders K177
M. V. Entin 487
B. Esser 393
J. Estel 121
L. R. Evangelista K29
F. Evangelisti 125

J. L. Farvacque 351
H. Fehske 235
R. Francini 305
D. Fröhlich 305

K. P. Ghatak K41, K47
M. N. Ghatikar 279
C. Giovanella 125
U. H. Gläser 741
O. Goede K159
A. V. Goltsev 537
H. S. Greenside 511
V. P. Gribkovskii 415
V. I. Gridnev 565
R. Gründler K149
H.-L. Günter 733

D. Hajduković K33, K37
O. P. Hansen 721
H. F. Hassan 587
T. K. Hatwar 279
R. Hayn 335
W. Heimbrodt K159
O. Heinrich 575
K. Helming 43
A. Hertrampf 343
C. Hirlimann K113
Hoang Ngoc Cam 247
C. A. Hogarth 293
B. Holland 511
G. Hüber K101
K. Hübner K71

I. A. Iliev 261
T. Irie K145
T. Itoh 71
N. B. Ivanov K131
K. Iwahana 693

M. S. Jahan 687
S. Jalaj 669

E. Jarocki 191
M. Jirsa 547
R. N. Joarder 443
W. John 335
S. K. Joshi 669
Juh Tzeng Lue 713

E. Kaden 733
H.-Matsuo Kagaya . . . 71, K5, K9, K109
H. G. Kahle 381
J. Kalus 121
V. Kamberský 547
A. A. Kaminskii 659
V. V. Kaplin 565
D. V. Kapor K175
W. Karaś 191
J. P. Karins 157
I. N. Karnaukhov 207
P. K. Khabibullaev 311
Z. M. Khakimov 311
N. M. Khamidullina 253
Yu. P. Khapachev 37
A. K. Khitrin 481
L. E. Kholodenkov 659
G. G. Kirakosian K83
D. S. Kiselev 37
M. Korkmaz 323
D. Korn 425
E. R. Korutcheva K19
R. Kotzott 223
A. I. Kozlov K59

V. V. Laiko 481
P. T. Landsberg 141
H. Lange 261
M. Lazaridou K123
J. Lech 371
A. Lehmeyer 223
A. A. Levin 311
H. Löffler 741
B. Lorenz 235
R. Luzzi 63

S. Mahajan 467
A. G. Makhanek 659
A. Sh. Makhmudov 311
A. Z. Maksymowicz 191
T. G. Mamedov K139
C. Mande 53
M. J. Marcinkowski 527
D. F. Mariani K1
K. Masuda-Jindo 643
V. K. Mathur 361
S. Matthies 43
A. P. Mayer 91
S. W. McKeever 361
W. T. KcKeown K67
A. O. O. Mekhrabov 453
A. I. Melker 133, 517
L. Merten 223
M. Mijatović K33
I. F. I. Mikhail 721
S. Mochizuki 105
M. Mondal K41, K47
A. M. Monti 163
M. A. de Moura 213
G. Munschy 329

V. Nagirnyi 407, 653, K63
H. Nakanishi K145
J. Narain 83
D. K. Neb K15
I. M. Nepomnyashi K53
Nguyen Ai Viet 247
Nguyen Hong Shon K105
Nguyen Van Hieu 247
V. N. Nishanov K53
S. B. Nurmagambetov 565

B. A. Orłowski 285

B. D. Padalia 279
S. Pak 565
S. Palchaudhuri 443
W. Pastusiak K77
U. Pietsch 595
L. Pintschovius 121
P. Plenkiewicz 285
D. Ploumbidis K101
S. Prakash 467
H. Presting 11
B. N. Provotorov 481

J. P. S. Rana 495
R. V. Gopala Rao 443
R. P. S. Rathore 87
J. F. Rhodes 361
S. N. Romanov 133, 517
M. Rösler 629
E. I. Rozum 565
I. Rückmann K135
V. N. Rudko 703

V. L. Safonov 197
S. N. Sahu 157
E. Yu. Salaev K139
O. G. Sarbei 607
I. M. Savić K37
V. K. Saxena K29
W. Schiller K91
M. Schlüter 511
W. F. Schmidt K165
H. A. Schneider K149
J. P. Šetrajčić K175

J. Shanker 77, 83
K. C. Sharma 59
G. M. Shmelev K105
I. Singh 669
R. K. Singh K15
R. P. Singh 495
S. P. Singh 87
M. J. Škrinjar K175
A. V. Skripov 557
A. Ślęzak 371
I. S. Smirnova 177
F. Smutný 459
V. V. Sobolev K59
T. Soma 71, K5, K9, K109
T. Soovik 407, 653, K63
H. N. Spector K171
B. Stébé 329
A. P. Stepanov 557
T. Stobiecki 191
T. Stöckert K149
V. I. Sugakov 703
M. Suhrke 393
V. K. Syal 59

G. Tamulaitis K135
G. Thörner 617
M. Tomak 677, 729
V. I. Torgashev 459
M. I. Török K97
S. Trimper K23
G. I. Tsurkan K105
C. Tuncay 677

S. R. Uglov 565
H.-G. Unruh 115
D. I. Uzunov K19

B. A. Vaid 59
R. A. Vardanian K83
G. B. Varshney 495
R. K. Varshney 77
A. R. Vasconcellos 63
D. Vignaud 351
J. Vignolo K1
S. V. Voitikov 415
S. A. Vorobiev 565

U. Walzer 752
H.-J. Weber 305
R. K. Wehner 91
K. Wesser 157
J. M. Wesselinowa 229
R. Wille 305
W. Wüchner 381

V. M. Yartsev 501
M. S. Yunusov 311

S. Zazubovich 407, 653, K63
Zhen-ya Li K125
A. F. Zhuravlev 207
A. Žukauskas K135

physica status solidi (b)

ISSN 0370-1972 * VOL. 126 · NO. 1 · NOVEMBER 1984

Classification Scheme

1. Structure of Crystalline Solids
   1.1 Perfectly Periodic Structure
   1.2 Solid-State Phase Transformations
   1.3 Alloys–Metallurgy
   1.4 Microstructure (Magnetic Domains See 18; Ferroelectric Domains See 14.4.1)
   1.5 Films
   1.6 Surfaces
2. Non-Crystalline State
3. Crystal Growth
4. Bonding Properties
5. Mössbauer Spectroscopy
6. Lattice Dynamics. Phonons
7. Acoustic Properties
8. Thermal Properties
9. Diffusion
10. Defect Properties (Irradiation Defects See 11)
    10.1 Metals
    10.2 Non-Metals
11. Irradiation Effects (X-Ray Diffraction Investigations See 1 and 10)
12. Mechanical Properties (Plastic Deformations See 10)
    12.1 Metals
    12.2 Non-Metals
13. Electron States
    13.1 Band Structure
    13.2 Fermi Surfaces
    13.3 Surface and Interface States
    13.4 Impurity and Defect States
    13.5 Elementary Excitations (Phonons See 6)
        13.5.1 Excitons
        13.5.2 Plasmons
        13.5.3 Polarons
        13.5.4 Magnons
14. Electrical Properties. Transport Phenomena
    14.1 Metals. Semi-Metals
    14.2 Superconductivity. Superconducting Materials and Devices
    14.3 Semiconductors
        14.3.1 Films
        14.3.2 Surfaces and Interfaces
        14.3.3 Devices. Junctions (Contact Problems See 14.3.4)
        14.3.4 High-Field Phenomena, Space-Charge Effects, Inhomogeneities, Injected Carriers (Electroluminescence See 20.3; Junctions See 14.3.3)
    14.4 Dielectrics
        14.4.1 Ferroelectrics
15. Thermoelectric and Thermomagnetic Properties
16. Photoconductivity. Photovoltaic Effects
17. Emission of Electrons and Ions
    17.1 Field Emission Microscope Investigations
18. Magnetic Properties
    18.1 Paramagnetic Properties
    18.2 Ferromagnetic Properties
        18.2.1 Ferromagnetic Films
    18.3 Ferrimagnetic Properties
    18.4 Antiferromagnetic Properties

(Continued on cover three)

physica status solidi (b)

basic research

Volume 126 · Number 1 · Pages 1 to 432, K1 to K108, and A1 to A8

November 1, 1984

PSSB 126(1) 1—432, K1—K108, A1—A8 (1984)

ISSN 0370-1972

AKADEMIE-VERLAG · BERLIN

and

VERLAG CHEMIE INTERNATIONAL INC.
DEERFIELD BEACH/FLORIDA

Enquiries, subscriptions, and orders for single copies from customers in the USA and CANADA should be directed to:

Verlag Chemie International Inc., 303 N.W. 12th Avenue, Deerfield Beach, FL 33441, USA;

from all other countries to:

Akademie-Verlag, Leipziger Str. 3–4, DDR-1086 Berlin.

Application to mail at second class postage rate is pending at Jamaica, N.Y. 11431.
Air freighted and mailed in the United States by Publications Expediting, Inc., 200 Meacham Avenue, Elmont, N.Y. 11003.
USA Postmasters: Send address changes to physica status solidi (b) to Verlag Chemie International Inc., 303 N.W. 12th Avenue, Deerfield Beach, FL 33441, USA.

Published 6 volumes with 2 issues each per year by Akademie-Verlag, Berlin, GDR. Annual subscription price: US $ 695.00 including postage.

Editorial Note:

"physica status solidi (b)" undertakes that an original paper *accepted for publication* before the 8th of any month will be published within 50 days of this date unless the author requests a postponement. In special cases there may be some delay between *receipt* and *acceptance* of a paper due to the review and, if necessary, revision of the paper.

Schriftleiter und verantwortlich für den Inhalt:
Professor Dr. Dr. h. c. P. Görlich, DDR-1086 Berlin, Leipziger Str. 3–4, Postfach 1233 bzw. DDR-6900 Jena, Schillbachstr. 24.

Verlag:
Akademie-Verlag, DDR-1086 Berlin, Leipziger Straße 3–4; Fernruf: 2236221 und 2236229; Telex-Nr.: 114420; Bank: Staatsbank der DDR, Berlin, Kto.-Nr.: 6836-26-20712.

Chefredakteur:
Dr. H.-J. Hänsch.

Redaktionskollegium:
Prof. Dr. E. Gutsche, Dr. H.-J. Hänsch, Dr. H. Lange, Dr. S. Oberländer.

Anschrift der Redaktion:
DDR-1086 Berlin, Leipziger Str. 3–4; Postfach 1233; Fernruf: 2236279.

Veröffentlicht unter der Lizenznummer 1310 des Presseamtes beim Vorsitzenden des Ministerrates der Deutschen Demokratischen Republik.

Gesamtherstellung:
VEB Druckerei „Thomas Müntzer", DDR-5820 Bad Langensalza.

Erscheinungsweise:
Die Zeitschrift „physica status solidi (b)" erscheint jeweils am 1. eines jeden Monats. Jährlich erscheinen 6 Bände zu je 2 Heften.

Bestellnummer dieses Bandes: 1068/126.

Printed in the German Democratic Republic.

Contents

Review Article

A. Blacha, H. Presting, and M. Cardona
Deformation Potentials of $\boldsymbol{k} = \boldsymbol{0}$ States of Tetrahedral Semiconductors 11

Original Papers and Short Notes

Structure

Yu. P. Khapachev, A. A. Dyshekov, and D. S. Kiselev
The Theory of X-Ray Diffraction Analysis of Elastic-Strain States in Epitaxial Films 37

K. Helming and S. Matthies
On the Interpretation of Orientation Distributions and Qualitative Ghost Corrections for Hexagonal-Orthorhombic Textures 43

S. Chattopadhyay and C. Mande
XANES and EXAFS Study of the Compound CoAsSe 53

E. Canessa, D. F. Mariani, and J. Vignolo
Liquid Structure Factor of Sb: Application of the Canessa-Mariani-Vignolo Model K1

Lattice properties

B. A. Vaid, K. C. Sharma, and V. K. Syal
A Disorder Model of Melting (II) 59

A. R. Vasconcellos and R. Luzzi
On Laser-Induced Lattice Heating in a Polar Semiconductor 63

T. Soma, T. Itoh, and H.-Matsuo Kagaya
Pressure Dependence of the Melting Point of Al 71

R. K. Varshney and J. Shanker
Temperature Dependence of Third-Order Elastic Constants of CsCl, CsBr, and CsI Crystals 77

J. Narain, G. G. Agrawal, and J. Shanker
Compressibility of Rare-Gas Solids under the Effect of High Pressure . . . 83

S. P. Singh and R. P. S. Rathore
Phonons in Face-Centred Cubic Calcium and Strontium 87

A. P. Mayer and R. K. Wehner
Calculation of Grüneisen Constants in Si 91

S. Mochizuki Radiative Optical Modes of Vibration in NiO Thin Film 105

H.-G. Unruh Dielectric Dispersion and Dynamics of Nitrite Ions in $K_2Ba(NO_2)_4$ 115

J. Estel, J. Kalus, and L. Pintschovius
Measurement of Phonon Energies in Silicon under Uniaxial Stress 121

D. Della Sala, C. Giovannella, and F. Evangelisti
Study of Bonding Configurations in Amorphous Ge_xSi_{1-x}:H Alloys. . . . 125

T. Soma and H.-Matsuo Kagaya
Compression Effect on Mode Grüneisen Parameters of Alkali Metals . . . K5

H.-Matsuo Kagaya and T. Soma
Dynamical Elastic Constants of Covalent Compounds K9

R. K. Singh and D. K. Neb
Mean Square Displacements of Rare Gas Solids K15

Defects, atomistic aspects

A. I. Melker and S. N. Romanov
Group Representation of the Formation Probabilities of Athermal Vacancy Clusters in Atomic Collision Cascades 133

P. T. Landsberg and S. G. Canagaratna
The Grand Partition Function in Defect Statistics 141

K. Weser, S. N. Sahu, J. P. Karins, and J. W. Corbett
Comparison of an Extended Interstitial with a Split Interstitial in Silicon 157

A. M. Monti Divacancy in H.C.P. Metals 163

I. S. Smirnova Dislocations in $A^{II}B^{VI}$ and $A^{III}B^{V}$ Piezoelectric Crystals 177

E. R. Korutcheva and D. I. Uzunov
On the Long-Range Random Critical Behaviour K19

Magnetism

A. Z. Maksymowicz, T. Stobiecki, E. Jarocki, and W. Karaś
Charge Transfer and Concentration Dependence of Magnetization in Amorphous Fe–B Alloys 191

V. L. Safonov Interactions of Magnons in the Six-Sublattice Antiferromagnet 197

I. N. Karnaukhov and A. F. Zhuravlev
Magnetic State of Crystals with Dislocations 207

G. L. M. Dassen and M. A. de Moura
Soliton Solution for the Excited State in the Spin-Peierls $S = 1/2$ Antiferromagnetic Chain 213

L. Merten, R. Kotzott, and A. Lehmeyer
Dispersion Relations for Magnon-Polaritons in Uniaxial Ferrimagnetic Crystals 223

J. M. Wessilinowa
Some Critical Properties of the s–d Model 229

B. Lorenz and H. Fehske
Critical Study of the Static Functional Integral Method in the Hubbard Model 235

S. Trimper Anisotropic Interactions in $2 + \varepsilon$ Dimensions K23

L. R. Evangelista and V. K. Saxena
Random-Field Ising Model with a Zero Temperature Multicritical Point K29

D. Hajduković and M. Mijatović
On the Ground-State Properties of the Ising Antiferromagnet with General Spin on the Honeycomb Lattice K33

D. Hajduković and I. M. Savić
On the Critical Properties of the Potts Antiferromagnets of the Bond-Decorated B.C.C. and F.C.C. Lattice K37

Extended electronic states and transitions

Hoang Ngoc Cam, Nguyen Van Hieu, and Nguyen Ai Viet
On the Damping of the Excitonic Polaritons 247

N. M. Khamidullina
Calculation of Some Characteristics of an Intrinsic Random Field in Hydrogenated Amorphous Silicon 253

T. N. Dimov, I. A. Iliev, and H. Lange
Peculiarities of Polarization Properties of CdS Single Crystals around the Isotropic Point . . . 261

K. Čermák The Width of the Energy Gap of $Tl_3^IB^VC_4^{VI}$-Type Compounds . . . 273

T. K. Hatwar, B. D. Padalia, M. N. Ghatikar, and D. R. Chopra
Extended X-Ray Absorption Fine Structure in RAl_2Ga_2 Intermetallics 279

B. A. Orłowski and P. Plenkiewicz
Electronic Band Structure of CdF_2. Photoemission Experiment and Pseudopotential Calculations . . . 285

S. K. J. Al-Ani and C. A. Hogarth
The Spin Density and Optical Properties of Amorphous Thin Arsenic Silicate Films . . . 293

D. Fröhlich, H.-J. Weber, R. Wille, and R. Francini
Temperature Dependence of the Exciton Lineshape of GeO_2 . . . 305

M. Mondal and K. P. Ghatak
Magnetic Susceptibility of Electrons in Ultrathin Films of Semiconductors in the Presence of a Quantizing Magnetic Field . . . K41

M. Mondal and K. P. Ghatak
On the Effective Mass in Tetragonal Semiconductors in the Presence of an Arbitrarily Oriented Quantizing Magnetic Field . . . K47

I. M. Nepomnyashi and V. N. Nishanov
The Localization Energy of Wannier-Mott Excitons at the Interface . . . K53

V. V. Sobolev and A. I. Kozlov
Free Exciton and Exciton–Impurity Complexes of Zinc Diphosphide and Zinc Diarsenide . . . K59

Localized electronic states and transitions

A. Sh. Makhmudov, Z. M. Khakimov, A. A. Levin, M. S. Yunusov, and P. K. Khabibullaev
Green's Function Method in the Equivalent Orbital Basis. Point Defects and Complexes of Two Point Defects in Silicon . . . 311

M. Korkmaz, B. Aktas et M. Dupont
Le spectre de RPE de l'ion Mn^{2+} dans le tartrate de cadmium ($CdC_4H_4O_4 \cdot 5\,H_2O$) . . . 323

F. Dujardin, B. Stébé, and G. Munschy
Neutral Bound Excitons in a Low Magnetic Field . . . 329

R. Hayn and W. John
Non-Exponential Localization in an Inhomogeneously Disordered Chain 335

D. BACKS and A. HERTRAMPF
Optically Induced EPR of Conduction and Donor Band Electrons in Zinc Sulphide 343

D. VIGNAUD and J. L. FARVACQUE
Free-Carrier Optical Absorption Induced by Dislocation Scattering Mechanisms in III–V Compounds 351

R. CHEN, V. K. MATHUR, J. F. RHODES, M. D. BROWN, S. W. MCKEEVER, and R. K. BULL
Thermoluminescence Governed by Simultaneous Thermal Stimulation of Electrons and Holes 361

J. LECH, A. ŚLĘZAK, and I. BÓJKO
EPR Powder Spectra of Low-Symmetric Mn^{2+} Complexes in Aragonite Lattice 371

W. BÖHM, H. G. KAHLE, and W. WÜCHNER
Spectroscopic Study of the Crystal Field Splittings in $TbPO_4$ 381

M. SUHRKE and B. ESSER
Time-Resolved Luminescence of Charge Carriers in Localized States in Amorphous Semiconductors 393

ZH. EGEMBERDIEV, V. NAGIRNYI, T. SOOVIK, and S. ZAZUBOVICH
Decay Kinetics and Polarization of the A_T Emission of Pb^{2+} Centres of Different Structure in $KBr:PbBr_2$ 407

ZH. EGEMBERDIEV, V. NAGIRNYI, T. SOOVIK, and S. ZAZUBOVICH
Slow Decay Component of the A_T Emission in KBr:Pb Crystals K63

G. BAMBAKIDIS and W. T. MCKEOWN
C_5H_9 Cluster as a Model for the (111) 1×1 Diamond Surface K67

V. BETHKENHAGEN and K. HÜBNER
Cluster-Bethe-Lattice Approach to the Electronic Structure of a Twofold Coordinated Silicon Atom as a New Intrinsic Defect in v-SiO_2 . . . . . K71

W. PASTUSIAK Dynamical Relativistic Crystal Field Contributions to the Zero Field Splitting of S-State $4f^7$ Ions K77

Electric transport

S. V. VOITIKOV and V. P. GRIBKOVSKII
On the Theory of Crystal Excitation by Electric Discharges 415

D. KORN Electrical Resistivity of Amorphous SnMn 425

R. A. VARDANIAN and G. G. KIRAKOSIAN
Electron Scattering by a Charged Dislocation K83

J. Dominec — Semiconductor–Semimetal Transition in Bi–Sb Alloys Observed in Ultrasound Attenuation K87

E. Chilla and W. Schiller
The Influence of Electron–Phonon Interaction on the Thermoelectric Power at High Temperatures K91

M. I. Török — On the Poole-Frenkel Constant K97

G. Hüber and D. Ploumbidis
Electrical Resistivity and NMR Investigations of the V–Co System at High Temperatures K101

G. M. Shmelev, Nguyen Hong Shon, and G. I. Tsurkan
Magnetic Oscillations of Acoustoelectrical Current of Photoelectrons in p-GaAs Type Semiconductors K105

Pre-Printed Titles

of papers to be published in the next issues of physica status solidi (b) and physica status solidi (a) A1

Systematic List

Subject classification:	Corresponding papers begin on the following pages (pages given in italics refer to the principle subject classification):
1	*53*
1.2	*115*, *381*, *K19*, K23
1.3	*43*
1.4	*43*
1.5	*37*
2	*59*, *135*, 191, *253*, *293*, *335*, 393, *425*, *K1*
4	*83*, 125, *273*, *K15*
6	*11*, 63, *71*, *87*, *91*, 105, 115, *121*, *K5*, K9, *K15*, *K105*
7	*K87*
8	59, *63*, 71, 235, *K5*, K15
10	*141*, *K19*
10.1	*163*, *207*
10.2	*157*, *177*, 351, K71, *K83*
11	*63*, *133*
12	37
12.1	87, K5
12.2	*77*, *83*, *K9*
13	*191*, *235*, *335*, *K47*
13.1	*11*, *273*, *285*, *293*, *311*, *351*, *K41*
13.3	*K53*, *K67*
13.4	*157*, *311*, *323*, *329*, *343*, *361*, *371*, *381*, 393, *407*, *K63*, *K71*, *K77*
13.5	105, *247*
13.5.1	*305*, *329*, *K53*, *K59*
13.5.4	*197*, *223*
14.1	*425*, K87, *K91*, *K101*
14.3	*177*, *K83*, *K87*
14.3.4	*415*, *K97*
14.4	*115*
15	*K91*
16	*K105*
18	*191*, *207*, *235*, *K29*, *K47*
18.2	*229*, *K23*
18.3	223
18.4	*197*, *213*, 381, *K33*, *K37*
19	207, 293, *323*, *343*, *371*, *K77*, *K101*
20	53, *279*
20.1	*105*, 125, *223*, 253, *261*, 273, 293, *305*, *351*, 381, *K59*
20.3	*361*, *393*, *407*, *K63*

21	43, 71, 163, 425, K91
21.1	53, K101
21.1.1	191
21.2	K5
21.3	87
21.4	279
21.7	K1
22	63, 361, 393, K41
22.1	11, K67
22.1.1	125, K83
22.1.2	91, 121, 125, 157, 253, 311
22.2	11, 177, 329, 351, K9
22.2.1	63, K47, K105
22.3	K59
22.4	11, 177, 329, K9
22.4.1	261, 343
22.5	197, 285
22.5.2	59, 77, K63
22.5.3	K77
22.5.4	407
22.6	105, 293, 305, K71
22.7	K87
22.8	229, 273, 371
22.8.1	115, 323, 381
23	83, K15

Contents of Volume 126 Continued on Page 435

Review Article

phys. stat. sol. (b) **126**, 11 (1984)

Subject classification: 6 and 13.1; 22.1; 22.2; 22.4

Max-Planck-Institut für Festkörperforschung, Stuttgart[1])

Deformation Potentials of $k = 0$ States of Tetrahedral Semiconductors

By

A. Blacha, H. Presting, and M. Cardona

Contents

1. Introduction

2. Definitions of deformation potentials for the Γ_{15}^{v}, Γ_{15}^{c}, and Γ_{1}^{c} states

3. Pseudopotential calculations

3.1 The pseudopotential method
3.2 Calculation of d_0
3.3 Calculation of a, b, and d'

4. LCAO calculation

4.1 LCAO Hamiltonian
4.2 sp^3-bonded semiconductors
4.3 Copper halides

5. Discussion

References

1. Introduction

The concept of deformation potential was introduced by Shockley and Bardeen [1] to describe the effect of uniform strains on the electronic energies of semiconductors. These deformation potentials are simply related to the electron-phonon coupling constants for long-wavelength acoustic phonons. More recently this concept has been generalized to include the effect of other deformations, such as those which correspond to the eigenvectors of the optical phonons at the center of the Brillouin zone (BZ) [2, 3]. The deformation potentials play an important role in phenomena as varied as the electrical conductivity [4], Raman [5] and Brillouin scattering [6], hyper-Raman scattering [7], phonon renormalizations due to free carriers [8], splittings, and shifts of excitons and band edges under stress [9], temperature dependence of band edges [10], displacive ferroelectricity [11], stress dependence of effective masses [12], etc. There is a great deal of experimental information about deformation potentials of germanium and zincblende-type semiconductors, especially those which correspond to uniform strains [2, 4, 9 to 11, 13]. Theoretical studies are less abundant. They fall into two categories: first principles and semiempirical [2, 12, 14 to 16]. The former have the disadvantage of lack of physical transparency in their lengthy numerical calculations.

---

[1]) Heisenbergstr. 1, D-7000 Stuttgart 80, FRG.

The latter, which use either the empirical pseudopotential method (EPM) or the linear combination of atomic orbitals (LCAO) method, can sometimes be reduced to simple analytic expressions which enable one to track the influence of the various parameters on the calculated pseudopotentials [2, 12, 17]. Semiempirical calculations are not suitable to obtain the *absolute* hydrostatic deformation potentials of band states [18]. They do yield, however, reasonable values for the dependence of energy differences (e.g., optical gaps) on hydrostatic strain.

In this paper we present EPM and LCAO calculations of deformation potentials for uniform stresses (both hydrostatic and uniaxial) and for Γ phonons of the group IV, III–V, and II–VI tetrahedral semiconductors. For the isomorphic I–VII semiconductors (e.g. CuCl) the valence states consist of hybridized copper 3d and halogen p states and the pseudopotential technique becomes inadequate. Calculations are, however, performed with the LCAO method [19]. The present calculations are restricted to the most important band states at the Γ point of the BZ, namely the top of the valence band, which we designate in the zincblende notation as Γ_{15}^{v}, the s-like (Γ_1^c), and the p-like (Γ_{15}^c) conduction band states. We do not only evaluate the intraband deformation potentials which couple these states with themselves and lead to splittings and shifts. We also calculate the interband deformation potentials which couple different states with each other and about which little information is hitherto available [14, 20]. Calculations for other points of the Brillouin zone will be presented elsewhere. The main purpose of the present paper is to display the systematics of the various deformation potentials for the materials under consideration and, when possible, to give simple analytic expressions for their evaluation.

The outline of the paper is as follows. In Section 2 we define the deformation potentials to be calculated and give the equivalence to other notations in use. In Section 3 the pseudopotential calculations are described: The results are listed in Tables 1 to 4. Section 4 describes the LCAO calculations with the results for the group IV, III–V, and II–VI materials listed in Tables 6 to 8. The results calculated with the LCAO method for the copper halides are listed in Tables 10 and 11. In Section 5 we discuss the results of both methods and compare them with experimental data.

2. Definitions of Deformation Potentials for the Γ_{15}^{v}, Γ_{15}^{c}, and Γ_1^c States

Under the action of a hydrostatic strain all electronic states shift but do not split. The components of these shifts linear in strain are represented by the deformation potentials a_i in (eV) defined as:

$$a_i = \frac{\mathrm{d}E_i}{\mathrm{d}\ln V} = \frac{\mathrm{d}E_i}{3\varepsilon_{jj}} = C_1 + \frac{2}{3}\,C_2\,, \tag{2.1}$$

where E_i represents the energy of state i, V the volume of the crystal, and ε_{jj} the diagonal component of the hydrostatic strain, C_1 and C_2 are found in [58]. Sometimes one finds in the literature instead of these deformation potentials the corresponding pressure coefficients $\mathrm{d}E_i/\mathrm{d}p$ which are related to the a_i through the bulk modulus B,

$$\frac{\mathrm{d}E_i}{\mathrm{d}p} = a_i B^{-1}\,. \tag{2.2}$$

Uniaxial (i.e. shear) strains and also a displacement of the two sublattices without changing the translation vector (i.e., an optical phonon at $\boldsymbol{k} = \boldsymbol{0}$) split degenerate levels such as the Γ_{15}^{v} top of the valence band. There are two irreducible components of pure shear strains, representing elongations along [100] (Γ_{12} symmetry) and [111] (Γ_{15} symmetry), respectively. Different deformation potentials are required to represent the effect of these strains on a given electronic state at Γ. The sublattice displacement, i.e. an optical phonon at Γ, has the same symmetry properties as the [111] strain (Γ_{15} symmetry). Hence the effects of such deformations are isomorphic to those of [111] strains in the sense of the Wigner-Eckart theorem, although the deformation potentials are different. A wide variety of notations are possible for these deformation potentials, differing from each other by factors of the order of unity. We use here the b and d notation of Pikus and Bir [3]. A more systematic notation, based on irreducible strain components, was introduced by Kane [21]. We give the equivalence below.

The deformation potential d_0 is defined by the splitting of a Γ_{15} state produced, *in the absence of spin–orbit interaction,* by a phonon along [111],

$$\delta\omega_0 = d_0 u/a_0 = d_{5,0}(u/a_0)\sqrt{2}\,, \tag{2.3}$$

where $\delta\omega_0$ is the shift of the singlet component of Γ_{15} polarized along [111], a_0 the lattice constant, and u the change of bond length in the [111] direction due to the displacement of the two sublattices $\left(\boldsymbol{u} = u(1, 1, 1)/\sqrt{3} = a_0\delta(1, 1, 1)/4\right)$. $d_{5,0}$ is the deformation potential defined in [2]. The inclusion of spin–orbit splitting requires the introduction of an additional deformation potential $d_{2,0}$ to represent the effect of the deformation on the spin–orbit coupling [22].[2]) It is supposed to be small but no quantitative information was available about its value until recently [22a]. Hence we shall not discuss it any further. Experimental values of d_0 are obtained from resonant Raman scattering [5, 23], phonon self-energies [24, 25], and exciton broadening with increasing temperature [24].

The effect of the Γ_{15} component of a pure shear is represented by the deformation potential d (d_5 in Kane's notation). For the shear along [111],

$$\boldsymbol{\epsilon} = \varepsilon(\Gamma_{15})\begin{pmatrix} 0 & 1 & 1 \\ 1 & 0 & 1 \\ 1 & 1 & 0 \end{pmatrix}, \tag{2.4}$$

the shift of the component of a Γ_{15} state polarized along [111] is given by

$$\delta\omega = 2\sqrt{3}\, d\varepsilon = \sqrt{6}\, d_5\varepsilon = -\,4\, D_{u'}\varepsilon\,, \tag{2.5}$$

where we have also defined the equivalent deformation potential $D_{u'}$, sometimes found in the literature.

The effect of a Γ_{12} component of a pure shear is represented by the deformation potential b (d_3 in Kane's notation) which for a shear along [100],

$$\boldsymbol{\epsilon} = \varepsilon(\Gamma_{12})\begin{pmatrix} 2 & & \\ & -1 & \\ & & -1 \end{pmatrix}, \tag{2.6}$$

gives the shift of the component of Γ_{15} states polarized along [100] as

$$\delta\omega = 6b\varepsilon = 2\sqrt{3}\, d_3\varepsilon = -4D_u\varepsilon = 2C_2\varepsilon\,. \tag{2.7}$$

Equation (2.7) also defines d_3 (Kane's notation), D_u, the counterpart of $D_{u'}$ and C_2 [58].

---

[2]) Reliable calculations of d_2 and $d_{2,0}$ for Si have recently been performed by Christensen [22a]. He finds $d_2 = 0.05$ eV.

Spin–orbit coupling requires the introduction of two additional deformation potentials b_2 and d_2 about which little theoretical information is available [22a]. In any case they are small ($\leqq 0.1$ eV) and we shall neglect them here.

Because of the fact that the Γ_{15} component of the strain and the sublattice displacement have the same symmetry they can "mix". The effect of a deformation of this symmetry is then given by the affine deformation

$$\boldsymbol{r} = (1 + \boldsymbol{\epsilon}) \cdot \boldsymbol{r} \tag{2.8}$$

plus the relative sublattice displacement [23]

$$\boldsymbol{u}_i = \sum_{j,k} \zeta e_{ijk} \varepsilon_{jk} a_0/4 \,, \tag{2.9}$$

where $e_{ijk} = 1$ if all subindices are not equal, zero otherwise. Correspondingly, the deformation potentials d can be decomposed into a component d' which represents the deformation of (2.8) and a component proportional to d_0 [23, 27],

$$d = d' - \tfrac{1}{4} \zeta d_0 \,, \tag{2.10}$$

where ζ is the internal strain parameter [28, 29]. Note that the definitions of b, d, and d' given here differ by a factor of 3 from those in [12].

The deformation potentials just defined are responsible for first-order perturbations of the energies of the Γ_{15}^{v}, Γ_{15}^{c}, and Γ_1^{c} states. It is also possible to define and calculate deformation potentials for the coupling between two of these states induced by either the optical phonon or a uniaxial stress. Such deformation potentials are of interest in problems of phonon renormalization [25], hyper-Raman effect [7], and changes in effective masses induced by strain [12].

We are interested in the coupling between a valence (Γ_{15}^{v}) and a conduction band state (either Γ_{15}^{c} or Γ_1^{c}). In the germanium structure these states have opposite parities. They thus cannot be coupled by the perturbations under consideration. In the zincblende structure, however, parity is lifted by the antisymmetric component of the potential. A Γ_{15}-like perturbation ([111] stress, optical phonon) can couple Γ_{15}^{v} with either Γ_1^{c} or Γ_{15}^{c} while a Γ_{12} perturbation ([100] stress) can only couple Γ_{15}^{c} with Γ_{15}^{v}. We define the corresponding coupling constants (deformation potentials) in a way analogous to that used for d_0, d, and b ((2.3), (2.5), (2.7)). The deformation potentials $d_0^{\mathrm{c,c}}$ and $d^{\mathrm{c,c}}$ for the coupling by optical phonons and by trigonal strains between Γ_1^{c} and Γ_{15}^{c}, allowed for the germanium structure, are also of interest [12].

$$\langle \Gamma_{15,z}^{\mathrm{v}} |\, \delta H_{\mathrm{phonon}} \,| \Gamma_1^{\mathrm{c}} \rangle = d_0^{\mathrm{v,cs}} \frac{u}{a_0}, \tag{2.11a}$$

$$\langle \Gamma_{15,z}^{\mathrm{v}} |\, \delta H_{\mathrm{phonon}} \,| \Gamma_{15,z}^{\mathrm{c}} \rangle = d_0^{\mathrm{v,cp}} \frac{u}{a_0}, \tag{2.11b}$$

$$\langle \Gamma_{15,z}^{\mathrm{c}} |\, \delta H_{\mathrm{phonon}} \,| \Gamma_1^{\mathrm{c}} \rangle = d_0^{\mathrm{c,c}} \frac{u}{a_0}, \tag{2.11c}$$

$$\langle \Gamma_{15,z}^{\mathrm{v}} |\, \delta H_{\mathrm{strain}} \,| \Gamma_1^{\mathrm{c}} \rangle = 2\sqrt{3}\, d^{\mathrm{v,cs}} \varepsilon \,, \tag{2.11d}$$

$$\langle \Gamma_{15,z}^{\mathrm{v}} |\, \delta H_{\mathrm{strain}} \,| \Gamma_{15,z}^{\mathrm{c}} \rangle = 2\sqrt{3}\, d^{\mathrm{v,cp}} \varepsilon \,, \tag{2.11e}$$

$$\langle \Gamma_{15,z}^{\mathrm{c}} |\, \delta H_{\mathrm{strain}} \,| \Gamma_1^{\mathrm{c}} \rangle = 2\sqrt{3}\, d^{\mathrm{c,c}} \varepsilon \,, \tag{2.11f}$$

$$\langle \Gamma_{15,x}^{\mathrm{v}} |\, \delta H_{\mathrm{strain}} \,| \Gamma_{15,x}^{\mathrm{c}} \rangle = 6 b^{\mathrm{v,cp}} \varepsilon \,. \tag{2.11g}$$

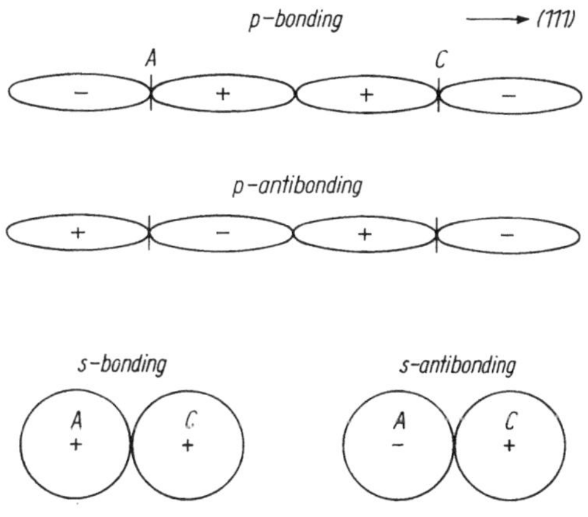

Fig. 1. Schematic diagram of the signs (phases) of the wave functions used in this work. This diagram is essential for understanding the signs of the interband deformation potentials

In (2.11 d, e, f), the strain is along [111], while in (2.11g), it is along [100]. The index $\bar{z}$ corresponds to polarization along the [111] direction, x to polarization along [100]. The deformation potentials of (2.11) are affected by some phase arbitrariness. These phases are well defined once the phases of the wave functions are chosen. They are important in interference phenomena. In the following calculations we have chosen the wave functions to be real, with the sign sketched in Fig. 1. The same convention will be made for the LCAO calculations in Section 4. The deformation potentials $d^{\mathrm{v,c}}$ can also be decomposed, in a manner similar to that of (2.10), into a part $d'^{\mathrm{v,c}}$ corresponding to an affine deformation, and an internal deformation equivalent to an optical phonon.

3. Pseudopotential Calculations

3.1 *The pseudopotential method*

We use the local empirical pseudopotential method without spin–orbit interaction. As can be seen in plane wave representation the kinetic and potential energy part of the Hamiltonian has the form

$$H_{\boldsymbol{G}\boldsymbol{G}'} = T_{\boldsymbol{G}}\delta_{\boldsymbol{G}\boldsymbol{G}'} + V(\boldsymbol{G} - \boldsymbol{G}');$$

$$T_{\boldsymbol{G}} = \frac{\hbar^2}{2m_0}(\boldsymbol{k} + \boldsymbol{G})^2\,, \tag{3.1}$$

$$V(\boldsymbol{G}'') = V_{\mathrm{s}}(|\boldsymbol{G}''|)\cos \boldsymbol{G}''\cdot\boldsymbol{\tau} - iV_{\mathrm{as}}(|\boldsymbol{G}''|)\sin \boldsymbol{G}''\cdot\boldsymbol{\tau}\,,$$

with

$$\boldsymbol{G}'' = \boldsymbol{G} - \boldsymbol{G}'\,, \qquad V_{\mathrm{s}} = (V_{\mathrm{A}} + V_{\mathrm{C}})/2\,, \qquad V_{\mathrm{as}} = (V_{\mathrm{A}} - V_{\mathrm{C}})/2\,, \tag{3.2}$$

while the pseudo-wave function has the form

$$\langle \boldsymbol{r} \mid n, \boldsymbol{k}\rangle = \psi_{n,\boldsymbol{k}}(\boldsymbol{r}) = \frac{1}{\sqrt{\Omega_0}}\sum_{\boldsymbol{G}} C_{n,\boldsymbol{k}}(\boldsymbol{G})\,\mathrm{e}^{i(\boldsymbol{k}+\boldsymbol{G})\cdot\boldsymbol{r}}\,. \tag{3.3}$$

Here $\boldsymbol{G} = (2\pi/a_0)\,(j, k, l)$ is a vector of the reciprocal lattice, $\boldsymbol{k}$ is a wave vector in the first BZ, a_0 the lattice constant, and Ω_0 means the volume to which the pseudo-wave functions are normalized, V_{A} and V_{C} are the Fourier transformed and appropriately screened atomic potentials of the anion and the cation from which the symmetric and antisymmetric pseudopotentials V_{s} and V_{as} are obtained. We have chosen the origin of direct space between the two atoms so that $\boldsymbol{\tau} = (a_0/8)\,(1, 1, 1)$ is the position vector of the cation and $-\boldsymbol{\tau}$ that of the anion.

The coefficients $C_{n,k}(G)$ of (3.3) as well as the band structure $E_n(k)$ are determined by solving Schrödinger's equation

$$(\sum_G H_{GG'} - E_n(k)\,\delta_{GG'})\,C_{n,k}(G) = 0\,. \tag{3.4}$$

We solved (3.4) numerically taking into account either 89 or 15 plane waves (see remarks in Sections 3.2 and 3.3). We used the pseudopotential form factors given in [30] and [31].

3.2 Calculation of d_0

We assume an optical phonon along the [111] direction with a sublattice displacement $u = (u/\sqrt{3})\,(1, 1, 1)$ (see (2.3)). The potential in the presence of the phonon has the form

$$V(G, u) = V_s \cos\left[G\cdot\left(\tau + \frac{u}{2}\right)\right] - i V_{as} \sin\left[G\cdot\left(\tau + \frac{u}{2}\right)\right].$$

Hence in plane wave representation the matrix elements of the electron–phonon perturbation Hamiltonian become

$$\langle n, k = 0|\,\delta H_{ep}(u)\,|n, k = 0\rangle =$$
$$= \sum_{\substack{r,s,t\\ j,k,l}} C^{n*}_{r+j,\,s+k,\,t+l} C^{n}_{r,s,t} \frac{\pi u}{\sqrt{3}} \frac{j+k+l}{a_0} \{V_s \sin G\cdot\tau + i V_{as} \cos G\cdot\tau\}\,. \tag{3.5}$$

The matrix elements of $\delta H_{ep}(u)$ lead to d_0 defined in (2.3). We have evaluated two intraband terms (that for the p valence band $(\Gamma_{25'}, \Gamma_{15}) \equiv d_0$ and that for the p conduction band $(\Gamma_{15}) \equiv d_0^c$, and three interband terms, namely p valence to s conduction band $(\Gamma_{2'}, \Gamma_1) \equiv d_c^{v,cs}$, p valence–p conduction band $\equiv d_0^{v,cp}$, and p conduction–s conduction band $d_0^{c,c}$. The results are listed in Table 1. By comparing the results given here for d_0 with those of [2] we conclude that nonlocality corrections do not seem to be important.

We would like to point out that in diamond structure crystals the matrix elements (3.5) can be calculated analytically using the symmetrized representation of 15 plane waves since there are only two sets of states of Γ_{15} symmetry. One finds for d_0 of the Γ_{15}^v band the simple expression [12, 17]

$$d_0 = \frac{-4\pi}{\sqrt{3}} [\beta^2(V_4 - V_8) + \gamma\beta\sqrt{2}\,(V_3 - V_{11})]\,, \tag{3.6}$$

where V_3, V_4, V_8, and V_{11} are the pseudopotential form factors for $j^2 + k^2 + l^2 = 3, 4, 8$, and 11, respectively [17, 30, 31]. The eigenvectors $\beta \approx 0.83$ and $\gamma \approx 0.55$ give the linear combinations of symmetrized $G = 2\pi a_0^{-1}(111)$ and $G = 2\pi a_0^{-1}(200)$ waves which make up the Γ_{15}^v state [17]. With (3.6) one finds for Ge $d_0 \approx 33$ eV, in rather good agreement with the results of Table 1 obtained with 89 plane waves. An equation similar to (3.6) can also be written for the corresponding deformation potential d_0^c of the Γ_{15}^c state.

While 15 and 89 plane waves give nearly the same results for Ge, the lower symmetry of the zincblende materials leads to errors in d_0 as high as 50% if only 15 plane waves are used. Hence the results for d_0 and d_0^c of Table 1 were obtained with 89 plane waves. For the interband coupling terms we found reasonably good convergence with 15 plane waves. Hence we restricted the listing of $d_0^{v,c}$ in Table 1 to the values obtained with 15 plane waves. We note that the values of $d_0^{v,cs}$ are particularly small as a result of a fortuitous cancellation of the contributions of the symmetric and the antisymmetric components of the perturbing potential [20]. For comparison, experimental values of d_0 are also listed in Table 1 [2, 13, 32].

Table 1

Deformation potentials (in eV) for optical phonons calculated by pseudopotential theory

material	d_0	d_0^c	$d_0^{v, cs}$	$d_0^{v, cp}$	$d_0^{c, c}$	d_0 (exp.)
C	90.4	−43.4	0.	0.	18.0	90
Si	35.0	−16.9	0.	0.	3.4	40
Ge	34.7	−15.4	0.	0.	4.6	34
α-Sn	29.9	−11.6	0.	0.	6.5	
BN	28.2	3.7	3.2	3.9	10.5	
BP	60.3	−26.4	5.5	23.4	10.3	
AlN	12.7	1.6	2.2	0.1	0.1	
AlSb	34.0	−15.1	2.8	6.3	4.9	37
GaP	28.9	−11.4	1.9	−0.3	1.9	44
GaAs	36.4	−12.0	0.3	2.4	4.5	48
GaSb	32.3	−13.1	1.2	9.1	6.1	32
InP	35.6	−15.1	1.1	5.9	5.3	35
InAs	33.3	−13.9	3.0	3.3	4.7	42
InSb	26.8	−10.3	0.3	4.2	5.7	39
ZnS	24.4	− 7.4	0.3	1.8	0.6	4*)
ZnSe	25.2	− 9.4	2.1	7.1	1.6	12*)
ZnTe	26.3	− 6.4	1.3	2.4	2.8	23
CdTe	22.8	− 6.8	1.3	0.2	0.6	22
CdS*)	6.9	− 1.0	5.0	1.9	0.7	
CdSe*)	8.9	− 2.76	8.7	2.0	0.1	

The deformation potentials d_0 for the Γ_{15}^v state and d_0^c (for Γ_{15}^c) were calculated with 89 plane waves while the coupling potentials $d_0^{v, cs}$, $d_0^{v, cp}$, and $d_0^{c, c}$ were calculated with 15 plane waves. Experimental values from [2, 13, 32, and 60] of d_0 are listed for comparison.

*) Zincblende modification.
**) Lower limit.

3.3 Calculation of a, b, and d′

Upon application of a uniaxial or hydrostatic stress, the reciprocal vectors of the perturbed lattice become

$$\boldsymbol{G}' = \boldsymbol{G} \cdot (1 - \boldsymbol{\epsilon}); \qquad \boldsymbol{G} = \frac{2\pi}{a_0} (j, k, l) . \tag{3.7}$$

The matrix elements of the corresponding perturbation Hamiltonians are (in Ryd units, $e = \sqrt{2}$, $m_0 = 1/2$, $\hbar = 1$)

$$\langle n', \boldsymbol{k} = \boldsymbol{0} | \, \delta H(\Gamma_1) \, | n, \boldsymbol{k} = \boldsymbol{0} \rangle = - \sum_{\substack{r, s, t \\ j, k, l}} C^{*n'}_{r+j, s+k, t+l} C^n_{r, s, t} \times$$
$$\times \, \varepsilon(\Gamma_1) \, (r^2 + s^2 + t^2) \, A_0\big(V(G), \boldsymbol{G}\big) \, \delta_{r+j, r} \delta_{s+k, s} \delta_{t+l, t} \, , \tag{3.8a}$$

$$\langle n', \boldsymbol{k} = \boldsymbol{0} | \, \delta H(\Gamma_{12}) \, | n, \boldsymbol{k} = \boldsymbol{0} \rangle = - \sum_{\substack{r, s, t \\ j, k, l}} C^{*n'}_{r+j, s+k, t+l} C^n_{r, s, t} \times$$
$$\times \, \varepsilon(\Gamma_{12}) \, (2j^2 - k^2 - l^2) \, A\big(V(G), \boldsymbol{G}\big) \, , \tag{3.8b}$$

$$\langle n', \boldsymbol{k} = \boldsymbol{0} | \, \delta H(\Gamma_{15}) \, | n, \boldsymbol{k} = \boldsymbol{0} \rangle = - \sum_{\substack{r, s, t \\ j, k, l}} C^{*n'}_{r+j, s+k, t+l} C^n_{r, s, t} \times$$
$$\times \, \varepsilon(\Gamma_{15}) \, [j(k + l) + k(j + l) + l(j + k)] \, A\big(V(G), \boldsymbol{G}\big) \, , \tag{3.8c}$$

where

$$A_0(V(G), \boldsymbol{G}) = = 2\Omega\delta_{r+j,r}\delta_{s+k,s}\delta_{t+l,t} + \left[\left(\frac{1}{3}\frac{\partial V_s(G)}{\partial \ln G} + V_s(G)\right)\cos \boldsymbol{G}\cdot\boldsymbol{\tau} - - i\left(\frac{1}{3}\frac{\partial V_{as}(G)}{\partial \ln G} + V_{as}(G)\right)\sin \boldsymbol{G}\cdot\boldsymbol{\tau}\right].$$

$$A(V(G), \boldsymbol{G}) = = 2\Omega\delta_{r+j,r}\delta_{s+k,s}\delta_{t+l,t} + \left[\frac{\partial V_s(G)}{\partial \ln G}\cos \boldsymbol{G}\cdot\boldsymbol{\tau} - i\,\frac{\partial V_{as}(G)}{\partial \ln G}\sin \boldsymbol{G}\cdot\boldsymbol{\tau}\right],$$

and

$$\Omega = (2\pi/a_0)^2$$

In (3.8) we have separated the three irreducible components of the strain: hydrostatic $(\varepsilon(\Gamma_1))$, tetrahedral shear $(\varepsilon(\Gamma_{15}))$, and tetragonal shear $(\varepsilon(\Gamma_{12}))$. The difference between A_0 and A comes from the fact that hydrostatic strain changes the volume of the crystal in contrast to uniaxial strain. As usual, we assume that the screening is not changed by the strain [17].

Table 2

The hydrostatic deformation potentials a (in eV) calculated by pseudopotential theory. The experimental values are taken from [13] unless otherwise indicated

material	$-a(\Gamma_{15}^v)$	$-a(\Gamma_1^c)$	$-a(\Gamma_{15}^c)$	$-a = a(\Gamma_{15}^v) - a(\Gamma_1^c)$	exp.	
					$-a$	$\lvert\dot{a}(\Gamma_1^c)\rvert$
C	36.1	30.7	16.3	−19.8[a]		
Si	10.2	15.3	10.0	5.1		
Ge*)	12.4	3.8	8.5	8.8	12.7	19.6[b]
α-Sn	3.3	13.7	5.8		—	—
BN	7.3	36.5	26.7	29.2	—	—
BP	−4.2	43.2	13.3	17.5[a]		
AlN	7.5	19.3	14.7	11.8	—	—
AlSb	4.4	14.3	5.5	9.9	5.9	—
GaP	6.8	16.1	7.8	9.3	9.3	19.9[b]
GaAs	4.8	18.3	7.3	13.5	9.8	17.5[b]
GaSb	2.6	15.4	6.1	12.8	8.3	—
InP	5.1	13.7	6.8	8.6	6.4	18.0[b]
InAs	5.2	11.7	6.1	6.5	6.0	—
InSb	2.6	12.9	5.5	10.3	7.7	14.6[b]
ZnS	9.2	13.1	6.8	3.9	4.0[f]	14.5[c]
ZnSe	8.6	14.8	5.6	6.2	5.4[f]	11.5[c]
ZnTe	3.0	12.6	6.3	9.6	5.8[d]	9.5[c]
CdTe	4.8	9.9	5.6	5.1	3.4	9.5[c]
CdS[e]	17.5	27.1	8.0	9.6	3.1	—
CdSe[e]	8.9	11.0	8.6	2.1	3.0	—

[a]) In this case $-a = (\Gamma_{15}^v - \Gamma_{15}^c)$, i.e., the deformation potential of the lowest direct gap at Γ has been listed.
[b]) [18].
[c]) [54].
[d]) [55].
[e]) Zincblende modification.
[f]) [58].
*) For Ge we used the pseudopotentials of [2] which lead to more reasonable values than the ones from [30].

We have determined the matrix elements of (3.8), and accordingly a, b, and d' (see (2.1), (2.5), and (2.7)). The derivatives of the pseudopotentials which appear in (3.2) were determined by numerical fitting of the pseudopotential form factors for $j^2 + k^2 + l^2 = 3, 4, 8$, and 11 to a cubic spline. The hydrostatic deformation potentials were also determined by performing a small change ($\approx 1\%$) of the lattice constant in the band structure calculation and calculating the corresponding change in the energy eigenvalues. The results were the same as those obtained with perturbation theory.

The hydrostatic deformation potentials a (2.1) calculated by this procedure for Γ_{15}^{v}, Γ_{1}^{c}, and Γ_{15}^{c} are listed in Table 2. We note that in the calculation the change in the average pseudopotential V_0 with volume has been neglected, hence the separate a's of the three states Γ_{1}^{c}, Γ_{15}^{v}, Γ_{15}^{c} may be in error by an additive constant, the same for all states [18]. This should not affect the difference between two a's, i.e., the volume dependence of an energy gap. Thus we have also listed in Table 2 this difference for the lowest direct gaps, usually $a(\Gamma_{15}^{v}) - a(\Gamma_{1}^{c})$.

While the volume dependence of V_0 is, in principle, not negligible, it has been shown by means of self-consistent, first-principles LMTO calculations in [18] that it is small for Ge and Si. Comparison with existing experimental results [18] (from transport

Table 3

The acoustic deformation potentials b and d' (in eV) calculated with the empirical pseudopotential method. The experimental values of b are from [13] unless otherwise stated

material	b	b c)	$b^{v,cp}$	b exp.	d'	d' c)	$d'^{v,cs}$	$d'^{v,cp}$	$d'^{cs,cp}$
C	−11.1	−1.6	0.		− 2.8	−27.2	0.	0.	5.4
Si	3.0	0.5	0.	−2.2	− 2.3 e)	−11.0	0.	0.	5.7
Ge	− 3.1	0.3	0.	−2.3	− 0.8	−10.7	0.	0.	4.7
α-Sn	− 2.1	1.0	0.		0.6	− 6.9	0.	0.	3.8
BN	−18.1	10.5	−9.5		−21.5	−18.8	3.3	−22.0	4.2
BP	− 3.8	1.7	0.6		1.4	−15.2	1.9	− 7.4	3.7
AlN	2.3	0.8	4.0		2.5	−12.2	0.3	− 7.0	0.8
AlSb	− 1.5	0.7	1.5	−1.4	0.8	− 7.2	0.3	− 0.3	3.5
GaP	− 0.1	−1.5	4.2	−1.8	− 1.1	− 9.0	0.4	− 3.5	1.2
GaAs	− 2.2	1.6	1.4	−2.0	2.7	− 7.8	0.6	− 4.1	3.6
GaSb	− 1.9	1.7	1.0	−1.8	0.1	− 6.6	0.9	− 2.7	3.7
InP	− 1.7	0.9	2.1	−2.0	0.6	− 7.4	1.3	− 3.1	2.8
InAs	− 1.0	0.1	2.6	−1.8	1.2	− 7.4	0.2	− 0.8	2.3
InSb	− 1.2	1.1	1.2	−2.0	0.5	− 5.6	1.3	− 2.7	3.1
ZnS	1.1	−3.3	5.9	−0.53 a) −0.7 c)	− 0.9	− 8.0	0.2	− 3.6	0.1
ZnSe	1.2	−3.4	8.2	−1.2 c) f)	− 0.3	− 8.1	1.1	− 2.1	2.1
ZnTe	− 1.0	0.3	2.1	−1.8 b)	− 0.3	− 5.6	0.5	− 2.5	3.3
CdTe	1.2	−2.9	5.0	−1.2	2.0	− 6.3	0.4	− 0.8	0.4
CdS d)	1.6	−3.2	4.1		− 3.1	− 9.6	0.7	− 6.3	0.3
CdSe d)	0.7	−1.8	2.8		− 2.7	− 8.7	0.2	− 5.6	0.2

a) [56].
b) [57].
c) [58].
d) Zincblende modification.
e) A value of $d' = -2.46$ eV has recently been calculated self-consistently by Christensen [22a].
f) [64].

measurements) also listed in Table 2 suggests that V_0 can also be assumed to be independent of volume for the other materials of the family.

The values of b and d' for the splitting of the Γ_{15}^c and Γ_{15}^v bands calculated by perturbation theory using 89 plane waves are listed in Table 3 together with experimental values of b. This table also includes the deformation potentials $d'^{v,cs}$, $d'^{v,cp}$, and $b^{v,cp}$ for the coupling between valence and conduction bands by the strain. The deformation potentials d' cannot be directly compared with the experimentally measured d. In order to compare d with d' we must replace in (2.10) the experimental or theoretical values of d_0 and the internal strain parameters ζ. The parameter ζ has only been measured for Ge [33], Si [34], and GaAs [35]. These values of ζ are listed in Table 4. For the other materials in this table we have estimated ζ through extrapolation and interpolation using periodic table systematics and the results of [36]. We have left out of this table diamond and the large-gap III–V compounds as for these materials estimates of ζ are too uncertain ($\zeta \approx 0.21$ is given in [36] for diamond) and no experimental values are known.

We should mention that within the 15 × 15 plane wave Hamiltonian analytic expressions for a, b, d', and d_0 similar to (3.6) but involving the $\mathrm{d}V/\mathrm{d}G$'s are possible for the group IV materials. They are given in [12] and [17].

Table 4
Comparison of experimental and calculated values of d (in eV).
The experimental values are from [13] unless otherwise indicated

material	a)		b)		d_{exp}
	ζ	d_{calc}	ζ	d_{calc}	
C	—	—	0.21	−7.5	—
Si	0.73	−8.7	0.56	−7.2	−5.1
Ge	0.74	−7.2	0.55	−5.6	−5.0
α-Sn	0.8	−5.4	—		
BN	—	—	—		—
BP	—	—			—
AlN	—	—	—		—
AlSb	0.8	−6.1	0.65	−4.7	−4.3
GaP	0.75	−6.5	0.59	−5.4	−4.5
GaAs	0.76	−4.2	0.60	−2.8	−5.4
GaSb	0.77	−6.1	0.61	−4.8	−4.6
InP	0.86	−7.1	0.70	−5.6	−5.0
InAs	0.84	−5.8	0.68	−4.5	−3.6
InSb	0.85	−5.2	0.70	−4.2	−5.0
ZnS	0.9	−6.4	0.74	−5.4	−3.7[c]
ZnSe	0.89	−5.9	0.72	−4.8	−3.8[d], −4.9[h]
ZnTe	0.87	−6.0	0.71	−5.0	−4.6[e]
CdTe	0.86	−2.9	0.79	−2.5	−5.4[c]
					−4.8[g]

The calculated values were obtained with (2.10) using for the internal stress parameter ζ the values listed here. a) ζ interpolated as described in Section 3, b) ζ from [36].

a) Obtained from the data of [33, 34] for Ge and Si and extrapolation to other materials using [36].
b) [36].
c) [56].
d) [58].
e) [57].
f) [62].
g) [63].
h) [64].

4. LCAO Calculations

In this section we describe a simple semi-empirical tight-binding (LCAO) calculation of the deformation potentials of tetrahedrally bonded semiconductors. A similar method has been applied by Pötz and Vogl [2] to the calculation of d_0 for some group IV, III–V, and II–VI materials. The LCAO approach yields analytic expressions for the deformation potentials as a function of the energy levels of the free atoms and a few interatomic matrix elements. As shown by Harrison [37, 38], the interatomic matrix elements can be described by universal parameters which are independent of the chemical nature of the atoms, and the interatomic distances. Hence the LCAO formalism is particularly useful to predict trends in the deformation potentials. Moreover, in the case of sp^3-bonded compounds, the tight-binding theory yields simple relations between the different types of deformation potentials: The deformation potentials b and d differ only by prefactors which do not depend on material parameters. Hence one can estimate, for example, the deformation potential d if one knows the optical deformation potential d_0 and the internal strain parameter ζ without any knowledge about the parameter d' which cannot be obtained experimentally (see (2.10)).

The tight-binding theory is also quite helpful for the treatment of the zincblende modifications of the copper halides. The deformation potentials of these materials are known to be anomalously small. Their valence band wave functions are admixtures of Cu-3d and halogen-p states. In the tight-binding approach the d states can be taken into account in a straightforward way, in contrast to techniques such as the EPM. One can easily show that the p–d admixture is responsible for the small values of d_0, b, and d in these materials.

In the following we describe first the LCAO formalism used (Section 4.1). Afterwards we discuss the LCAO calculation of the deformation potentials for the sp^3-bonded semiconductors (Section 4.2) and the copper halides (Section 4.3) as model compounds for semiconductors with p–d hybridized valence bands.

4.1 LCAO Hamiltonian

The Hamiltonian H of the crystal is represented in a minimal basis set $\{\psi_l\}$ of normalized linear combinations of atomic orbitals assuming only nearest-neighbor interatomic interaction. The basis functions ψ_l have the form

$$\psi_l(\boldsymbol{k}) = \frac{1}{\sqrt{N}} \sum_j \mathrm{e}^{i\boldsymbol{k}\cdot\boldsymbol{r}_j} \varphi_l(\boldsymbol{r}_j) , \tag{4.1}$$

where $\boldsymbol{k}$ is the wave vector, $\varphi_l(\boldsymbol{r}_j)$ an atomic orbital localized at the place $\boldsymbol{r}_j$, N the number of atoms in the crystal, and l denotes the different electronic states of the atoms. With respect to the basis $\{\psi_l\}$ the Hamiltonian matrix elements can be written as a function of the matrix elements for the orbitals φ_l, $\varphi_{l'}$,

$$H_{ll'}^{jj'} = \langle \varphi_l(r_j)| \, H \, |\varphi_{l'}(r_{j'})\rangle \, . \tag{4.2}$$

The $H_{ll'}^{jj'}$ are assumed to be parameters which can be obtained by a fit to the band structure at high symmetry points [37, 38]. For the intraatomic matrix elements $(j = j')$ one writes

$$H_{ll'}^{jj} = E_l^j \delta_{ll'} , \tag{4.3}$$

where E_l^j is the energy of the state $|\varphi_l(\boldsymbol{r}_j)\rangle$ of the free atom.

Equation (4.3) may not represent the effect of lattice deformations. Lattice deformations may lead to changes of the kinetic energy and therefore the diagonal elements of the Hamiltonian matrix may be functions of the lattice parameters: This has been shown to be an important effect for hydrostatic deformations [18] which cause a volume change of the lattice and therefore a large change of the kinetic energy. However, for lattice deformations due either to optical phonons or to a traceless shear the volume of the crystal remains constant. One may thus accept as a working hypothesis (4.3) with matrix elements independent of strain. This was shown to work for the calculation of d_0 in [2]. The validity of the method for b and d' will be demonstrated by the agreement of the results with experiments.

The interatomic matrix elements $H_{ll'}^{jj'}$ $(j \neq j')$ can be written as a function of two two-center integrals $V_{ll'\sigma}^{jj'}$ and $V_{ll'\pi}^{jj'}$ and the direction cosines $\cos(\theta_{j'x})$, $\cos(\theta_{j'y})$, $\cos(\theta_{j'z})$ of the vector $\boldsymbol{r} = \boldsymbol{r}_{j'} - \boldsymbol{r}_i$ referred to the x-, y-, z-axis, respectively. Recently, Harrison [37, 38] found that the two-center integrals $V_{ll'\sigma}^{jj'}$ and $V_{ll'\pi}^{jj'}$ are given for all compounds by the universal function

$$V_{ll'm}^{jj'} = \eta_{ll'm} \frac{\hbar^2}{m_0} d_{jj'}^{-\xi}; \qquad m = \sigma, \pi\,, \tag{4.4}$$

if one assumes a model which takes into account only nearest-neighbor interatomic interaction. The $\eta_{ll'm}$ are universal coefficients independent of the atoms, m_0 is the free electron mass, $d_{jj'}$ is the distance between atoms j and j', and the exponent ξ a number which depends on the angular momenta of the interacting atomic states.

In the following we make use of (4.4). The coefficients $\eta_{ll'm}$ and ξ published by Harrison [37, 38] can be found in Table 5 of this paper. Note that there are two different sets of parameters $\eta_{sp\sigma}$, $\eta_{pp\sigma}$, $\eta_{pp\pi}$ for interacting s and p states. These sets have been obtained from different models which take into account a different number of basis functions (see next section).

To calculate the deformation potentials of the energy bands of a solid with the LCAO method, we define a minimal basis set which can describe the properties of the upper valence band and the lowest conduction band, and we calculate the Hamiltonian matrix for different deformations of the crystal lattice. Linear expansion of the eigenvalues or the Hamiltonian as a function of the deformation yields the deformation potentials.

Table 5

Parameters $\eta_{ll'm}$ and ξ defined in (4.4) for interatomic matrix elements between s, p, and d orbitals ($m = \sigma, \pi$)

l	l'	sp^3 model [37]		sp^3s^* model [38]		ξ
		$\eta_{ll'\sigma}$	$\eta_{ll'\pi}$	$\eta_{ll'\sigma}$	$\eta_{ll'\pi}$	
s	s	−1.4	—	−1.32	—	2
s	p	1.84	—	1.42	—	2
p	p	3.24	−0.81	2.22	−0.63	2
p	d	−2.95	1.36	—	—	7/2

4.2 sp³-bonded semiconductors

For the minimal basis set we choose one s and three p orbitals for each of the two atoms of the unit cell, a total of eight orbitals. For the atomic orbitals we choose the notation

$$|s\rangle_a^j, |p_x\rangle_a^j, |p_y\rangle_a^j, |p_z\rangle_a^j; \qquad |s\rangle_c^{j'}, |p_x\rangle_c^{j'}, |p_y\rangle_c^{j'}, |p_z\rangle_c^{j'} . \tag{4.5}$$

The subscripts a, c distinguish between anion and cation in the case of a zincblende lattice. The superscripts j, j' indicate the positions $\boldsymbol{r}_j$, $\boldsymbol{r}_{j'}$ of the atoms and x, y, z the directions of the p orbitals referred to the cubic x, y, z axes.

The wave functions (4.1) are linear combinations of the wave functions φ of (4.5)

$$\psi_{l,a} = \frac{1}{\sqrt{N}} \sum_{j=1}^{N} \varphi_{l,a}^{j}, \qquad \psi_{l,c} = \frac{1}{\sqrt{N}} \sum_{j'=1}^{N} \varphi_{l,c}^{j'}, \tag{4.6}$$

where the index l indicates the angular momentum of the atomic orbital. In order to obtain the deformation potentials we evaluate the 8×8 Hamiltonian matrix with respect to the basis functions (4.6). We assume only nearest-neighbor interaction. In this approximation the Hamiltonian matrix elements $H_{ll'} = \langle \psi_{l,a} | H | \psi_{l',c} \rangle$ are

$$H_{ll'} = \sum_{j'=1}^{4} \langle \varphi_{l,a}^{j} | H | \varphi_{l',c}^{j'} \rangle = \sum_{j'=1}^{4} H_{ll'}^{jj'}, \tag{4.7}$$

where the superscript j' denotes the nearest neighbors to atom j.

The matrix elements $H_{ll'}^{jj'}$ in (4.7) can be taken from the Koster and Slater tables of interatomic matrix elements [39],

$$\begin{aligned}
H_{ss}^{jj'} &= V_{ss\sigma}^{jj'}, \\
H_{sx}^{jj'} &= \cos(\theta_{j'x})\, V_{sp\sigma}^{jj'}, \\
H_{xx}^{jj'} &= \cos^2(\theta_{j'x})\, V_{pp\sigma}^{jj'} + [1 - \cos^2(\theta_{j'x})]\, V_{pp\pi}^{jj'}, \\
H_{xy} &= \cos(\theta_{j'x}) \cos(\theta_{j'y})\, [V_{pp\sigma}^{jj'} - V_{pp\pi}^{jj'}] .
\end{aligned} \tag{4.8}$$

The superscripts l, l' have been replaced by s for an s orbital and x, y, z for a p_x, p_y, p_z orbital, respectively. In (4.8) $\cos(\theta_{j'x})$, $\cos(\theta_{j'y})$, $\cos(\theta_{j'z})$ are the direction cosines of the vector $\boldsymbol{r} = \boldsymbol{r}_{j'} - \boldsymbol{r}_j$ with respect to the cubic axes x, y, z. Interatomic matrix elements missing in (4.8) can be generated by cyclic permutations of x, y, z. The quantities $V_{ss\sigma}^{jj'}$, $V_{sp\sigma}^{jj'}$, $V_{pp\sigma}^{jj'}$, $V_{pp\pi}^{jj'}$ have been introduced in Section 4.1.

The top of the valence bands in tetrahedrally bonded semiconductors is purely p-like. Therefore to first order the effect of a deformation of the valence band is due to a coupling between the p states. Consequently for the calculation of the intra-valence-band deformation potentials we can neglect the strain-induced coupling between the p and s states. The s states give rise only to interband coupling.

First we deal with stress-induced intra-valence-band coupling. We diagonalize the 6×6 Hamiltonian with respect to the p states for a) a sublattice shift $\boldsymbol{u} = u(111)/\sqrt{3} = a_0\delta(111)/4$, b) a rhombohedral ($\Gamma_{15}$) macroscopic strain with zero internal strain ($\zeta = 0$), and c) a tetrahedral (Γ_{12}) strain.

The diagonal Hamiltonian matrix elements are the atomic energies E_{pa} and E_{pc} of the anion and the cation. E_{pa} and E_{pc} can be taken from [40].

The off-diagonal matrix elements (4.7) can be obtained from (4.8). Using the symbols $V_{pp\sigma}$, $V_{pp\pi}$ for the two-center integrals (4.4) of the undeformed lattice, the result is a) for a sublattice shift $\boldsymbol{u}(\delta)$ to first order in $\boldsymbol{u}$ (i.e., first order in δ),

$$\begin{aligned} H_{xx}(\delta) &= \tfrac{4}{3} V_{pp\sigma} + \tfrac{8}{3} V_{pp\pi} + O(\delta^2)\,, \\ H_{xy}(\delta) &= -\tfrac{16}{9} [V_{pp\sigma} - V_{pp\pi}]\,\delta = V_{xy}\delta; \end{aligned} \tag{4.9a}$$

b) for a $\varepsilon(\Gamma_{15})$ strain (2.4) with zero internal strain $\varepsilon(\Gamma_{15})$,

$$\begin{aligned} H_{xx}(\varepsilon(\Gamma_{15})) &= \tfrac{4}{3} V_{pp\sigma} + \tfrac{8}{3} V_{pp\pi} + O(\varepsilon^2(\Gamma_{15}))\,, \\ H_{xy}(\varepsilon(\Gamma_{15})) &= -\tfrac{8}{9} [V_{pp\sigma} - V_{pp\pi}]\,\varepsilon(\Gamma_{15}); \end{aligned} \tag{4.9b}$$

c) for a $\varepsilon(\Gamma_{12})$ strain (2.6),

$$\begin{aligned} H_{xx} &= \tfrac{4}{3} V_{pp\sigma} + \tfrac{8}{3} V_{pp\pi} + \tfrac{16}{3} (V_{pp\sigma} - V_{pp\pi})\,\varepsilon(\Gamma_{12})\,, \\ H_{yy} &= H_{zz} = \tfrac{4}{3} V_{pp\sigma} + \tfrac{8}{3} V_{pp\pi} - \tfrac{8}{3} (V_{pp\sigma} - V_{pp\pi})\,\varepsilon(\Gamma_{12})\,, \\ H_{yz} &= 0\,. \end{aligned} \tag{4.9c}$$

Cyclic permutations of the subscripts x, y, z yield the complete set of matrix elements. In the derivation of (4.9 a, b, c), we have used (4.4) with $\xi = 2$. It is trivial to extend these equations to other values of ξ.

One can see from (4.9 a to c) that the first-order coefficients of the expansion of the matrix elements with respect to the parameters δ, $\varepsilon(\Gamma_{15})$, $\varepsilon(\Gamma_{12})$ are equal to $V_{pp\sigma} - V_{pp\pi}$ times a number which depends only on the type of deformation. Therefore, the derivatives of the energy of the valence band with respect to δ, $\varepsilon(\Gamma_{15})$, and $\varepsilon(\Gamma_{12})$ differ only by numerical factors: the intraband deformation potentials d_0, d', b are proportional to each other.

In order to obtain the optical intra-valence-band deformation potential d_0 we have to diagonalize the Hamiltonian matrix (4.9a). From the splitting of the valence band as a function of δ one obtains [2]

$$d_0 = \frac{1}{\sqrt{3}} \frac{32}{3} \frac{H_{xx}(\delta = 0)\, V_{xy}}{[((E_{pa} - E_{pc})/2)^2 + H_{xx}^2(\delta = 0)]^{1/2}}. \tag{4.10}$$

The diagonalization of (4.9b) and (4.9c) yields d' and b, respectively. The result is

$$d' = \frac{1}{16} d_0\,, \tag{4.11}$$

$$b = \frac{-\sqrt{3}}{16} d_0\,. \tag{4.12}$$

Note that this result is only valid if in (4.4) $\xi = 2$ (see Table 5). In Tables 6 and 7 we present the intra-valence-band deformation potentials obtained from (4.10) to (4.12) for group IV, III–V, and II–VI semiconductors. The atomic energies E_{pa} and E_{pc} were taken from [40]. We have done the calculation for both sets of interatomic matrix elements, which can be determined from the parameters $\eta_{ll'm}$ published by Harrison [37, 38] (see Table 5). One parameter set $\eta_{ll'm}$ was obtained by fitting the tight-binding bands, constructed with sp^3 hybrids as minimal basis, to pseudopotential bands (sp^3 model [37]). The other parameter set $\eta_{ll'm}$ is the result of an adjustment of the so-called sp^3s^* model [38]. In the sp^3s^* model one excited state s^* per atom is added to the basis of the sp^3 model. The effect of the excited states s^* is a better agreement of the tight-binding conduction bands with the actual bands at the edges of the Brillouin zone. Moreover the sp^3s^* model predicts more realistic values for the energy

Table 6

Atomic energies E_{sa}, E_{pa}, E_{sc}, E_{pc} (in eV) (from [40]), lattice constants a_0 (in nm), and deformation potentials (in eV) obtained from (4.10) to (4.12) for the parameters of the sp^3 model (Table 5) for group IV, III–V, and II–VI semiconductors

compound	$-E_{sa}$	$-E_{pa}$	$-E_{sc}$	$-E_{pc}$	a_0	d_0	d'	b	$d'^{v,cs}$	$d_0^{v,cp}$	$d'^{v,cp}$	$b^{v,cp}$	$d'^{c,c}$
C	19.4	11.1	19.4	11.1	0.3567	106.2	6.6	−11.5	0	0	0	0	13.5
SiC	19.37	11.07	14.79	7.58	0.436	66.5	4.2	− 7.2	3.3	25.1	1.6	−2.7	8.0
Si	14.79	7.58	14.79	7.58	0.543	45.8	2.9	− 5.0	0	0	0	0	5.9
Ge	15.15	7.33	15.15	7.33	0.565	42.3	2.6	− 4.6	0	0	0	0	5.4
α-Sn	13.04	6.76	13.04	6.76	0.649	32.1	2.0	− 3.5	0	0	0	0	4.2
BN	26.22	13.84	13.46	8.43	0.3615	95.9	6.0	−10.4	6.2	38.6	2.4	−4.2	10.4
BP	19.22	9.54	13.46	8.43	0.4538	65.1	4.1	− 7.0	2.8	8.5	0.5	−0.9	7.4
BAs	18.91	8.98	13.46	8.43	0.477	59.2	3.7	− 6.4	2.4	4.2	0.3	−0.5	6.9
AlP	19.22	9.54	10.70	5.71	0.5467	37.9	2.4	− 4.1	3.3	24.7	1.5	−2.7	3.8
AlAs	18.91	8.98	10.70	5.71	0.566	36.2	2.3	− 3.9	3.1	21.6	1.4	−2.3	3.6
AlSb	16.02	8.14	10.70	5.71	0.6136	31.8	2.0	− 3.4	2.4	16.6	1.0	−1.8	3.3
GaP	19.22	9.54	11.55	5.67	0.545	38.1	2.4	− 4.1	3.3	24.9	1.6	−2.7	3.8
GaAs	18.91	8.98	11.55	5.67	0.565	36.3	2.3	− 3.9	2.9	21.8	1.4	−2.4	3.6
GaSb	16.02	8.14	11.55	5.67	0.610	32.2	2.0	− 3.5	2.3	16.8	1.1	−1.8	3.5
InP	19.22	9.54	10.14	5.37	0.586	30.5	1.9	− 3.3	3.1	24.9	1.6	−2.7	2.9
InAs	18.91	8.98	10.14	5.37	0.605	29.5	1.8	− 3.2	2.8	22.2	1.4	−2.4	2.8
InSb	16.02	8.14	10.14	5.37	0.647	26.9	1.7	− 2.9	2.4	17.6	1.1	−1.9	2.7
ZnS	24.01	11.60	7.96	4.02	0.541	28.7	1.8	− 3.1	4.0	36.2	2.3	−3.9	2.6
ZnSe	22.86	10.68	7.96	4.02	0.565	27.0	1.7	− 2.9	3.6	32.6	2.0	−3.5	2.4
ZnTe	19.12	9.54	7.96	4.02	0.610	23.6	1.5	− 2.6	3.1	27.6	1.7	−3.0	2.3
CdS	24.01	11.60	7.21	3.99	0.580	22.7	1.4	− 2.5	3.5	33.1	2.1	−3.6	2.1
CdSe	22.86	10.68	7.21	3.99	0.610	20.9	1.3	− 2.3	3.1	29.7	1.9	−3.2	1.9
CdTe	19.12	9.54	7.21	3.99	0.648	19.4	1.2	− 2.1	2.8	25.7	1.6	−2.8	1.7

Table 7

Deformation potentials (in eV) obtained from (4.10) to (4.12), (4.16) for the parameters of the sp^3s^* model (Tables 5 to 6) for group IV, III–V, and II–VI semiconductors

compound	d_0	d'	b	$d'^{v, cs}$	$d_0^{v, cp}$	$d'^{v, cp}$	$b^{v, cp}$	$d'^{c, c}$
C	74.7	4.7	−8.1	0	0	0	0	10.6
SiC	42.2	2.6	−4.6	3.1	26.9	1.7	−2.9	5.7
Si	32.3	2.0	−3.5	0	0	0	0	4.8
Ge	29.8	1.9	−3.2	0	0	0	0	4.2
α-Sn	22.6	1.4	−2.4	0	0	0	0	3.1
BN	60.2	3.8	−6.5	5.5	40.9	2.6	−4.4	7.1
BP	45.1	2.8	−4.9	2.4	9.9	0.6	−1.1	5.5
BAs	41.5	2.6	−4.5	2.1	5.0	0.3	−0.5	5.2
AlP	21.4	1.3	−2.3	2.8	23.5	1.5	−2.5	2.5
AlAs	20.9	1.3	−2.3	2.6	21.1	1.3	−2.3	2.3
AlSb	19.0	1.2	−2.1	2.1	16.7	1.0	−1.8	2.2
GaP	21.5	1.3	−2.3	2.8	23.7	1.5	−2.6	2.5
GaAs	20.9	1.3	−2.3	2.6	21.2	1.3	−2.3	2.4
GaSb	19.2	1.2	−2.1	2.1	16.9	1.1	−1.8	2.3
InP	16.3	1.0	−1.8	2.6	22.4	1.4	−2.4	1.8
InAs	16.1	1.0	−1.7	2.4	20.4	1.3	−2.2	1.7
InSb	15.2	0.9	−1.6	2.1	16.9	1.1	−1.8	1.7
ZnS	13.8	0.9	−1.5	3.3	29.4	1.8	−3.2	1.5
ZnSe	13.1	0.8	−1.4	2.9	26.8	1.7	−2.9	1.4
ZnTe	11.5	0.7	−1.3	2.6	22.8	1.4	−2.5	1.3
CdS	10.6	0.7	−1.2	2.8	26.2	1.6	−2.8	1.1
CdSe	9.9	0.6	−1.1	2.6	23.6	1.5	−2.6	1.0
CdTe	9.2	0.6	−1.0	2.3	20.7	1.3	−2.2	1.0

Table 8

Experimental and calculated values for the deformation potentials d_0, b, d (in eV) for Si and GaAs

		Si	GaAs
ζ (exp.)		0.73 [34]	0.76 [33]
d_0	exp., (Table 1)	40	41
	LCAO, sp^3 model	45.8	36.3
	LCAO, sp^3s^* model	32.3	20.9
	pseudopotential	35	36.4
b	exp., (Table 3)	−2.2	−2.0
	LCAO, sp^3 model	−5.0	−3.9
	LCAO, sp^3s^* model	−3.5	−2.3
	pseudopotential	−2.3	−2.2
d	exp., (Table 4)	−5.1	−5.4
	LCAO, sp^3 model	−5.5	−4.6
	LCAO, sp^3s^* model	−3.9	−2.7
	pseudopotential	−8.7	−4.2

difference between the valence band and the corresponding p-like antibonding states (for example Si: the sp^3 model predicts 6 eV, the sp^3s^* model 3.5 eV for the E_0 gap, which is 3.3 eV). The deformation potentials which are representative for the sp^3 model, are listed in Table 6, the values obtained from the sp^3s^* model are given in Table 7. The sp^3s^* model yields, in general, smaller deformation potentials than the sp^3 model. However, the chemical trends are obviously the same in both cases (see Section 5).

In order to compare the calculated deformation potential d' with experiment, we must evaluate d with (2.10) from the calculated d', d_0, and the experimental value of the internal stress parameter ζ. The results for d, and also for d_0 and b, are compared in Table 8 for two typical materials, Si and GaAs, to experiment and to the results of pseudopotential calculations.

In the following we discuss the determination of the interband deformation potentials $d_0^{v,c}$, $d'^{v,c}$, $b^{v,c}$, $d_0^{c,c}$, and $d'^{c,c}$ defined in (2.11). The matrix elements (2.11) can easily be calculated from the matrix elements of the perturbation Hamiltonian of (4.9).

As mentioned in Section 3, for crystals with point group O_h the interband deformation potentials between valence and conduction band vanish by symmetry. In the case of zincblende-type crystals only the interband deformation potential $b^{v,cs}$ remains zero by symmetry. The coupling between Γ_1^c and Γ_{15}^c produced by perturbations of Γ_{15} symmetry is nonzero even in the germanium structure. In order to calculate the interband deformation potentials we write the p-like (singlet after perturbation) bonding state in the form

$$\psi_p^v = (\beta\psi_{p,a} - \alpha\psi_{p,c}) \tag{4.13}$$

($\alpha, \beta > 0$) and the lowest p-like (singlet) antibonding state

$$\psi_p^c = -(\alpha\psi_{p,a} + \beta\psi_{p,c}) \,, \tag{4.14}$$

where $\psi_{p,a}$ is a normalized linear combination of anion p functions and $\psi_{p,c}$ a normalized combination of cation p functions. Similarly we write for the wave function of the lowest s-type antibonding state ($\mu > 0$, $\nu < 0$)

$$\psi_s^c = -(\mu\psi_{s,a} + \nu\psi_{s,c}) \,. \tag{4.15}$$

With these definitions we obtain from (4.13) to (4.15) and (4.9)

$$\begin{aligned}
d_0^{v,cs} &= d_0^{c,c} = 0 \,, \\
d'^{v,cs} &= \frac{4}{\sqrt{3}} V_{sp\sigma}(\beta\nu + \alpha\mu) \,, \\
d'^{c,c} &= \frac{4}{\sqrt{3}} V_{sp\sigma}(\alpha\nu - \beta\mu) \,, \\
d_0^{v,cp} &= \frac{128}{9\sqrt{3}} [V_{pp\sigma} - V_{pp\pi}] (\beta^2 - \alpha^2) \,, \\
d'^{v,cp} &= \frac{1}{16} d_0^{v,cp} \,, \\
b^{v,cp} &= -\frac{\sqrt{3}}{16} d_0^{v,cp} \,.
\end{aligned} \tag{4.16}$$

The admixture coefficients α, β, μ, ν which appear in (4.16) have to be calculated for the unstrained crystal. They fulfil the relations

$$\alpha^2 + \beta^2 = 1; \quad \frac{\alpha}{\beta} = -\frac{H_{xx}(\delta = 0)}{\frac{E_{pa} - E_{pc}}{2} - \left[\left(\frac{E_{pa} - E_{pc}}{2}\right)^2 + H_{xx}^2(\delta = 0)\right]^{1/2}},$$
$$\mu^2 + \nu^2 = 1; \quad \frac{\mu}{\nu} = -\frac{V_{ss\sigma}}{\frac{E_{sa} - E_{sc}}{2} - \left[\left(\frac{E_{sa} - E_{sc}}{2}\right)^2 + V_{ss\sigma}^2\right]^{1/2}}. \tag{4.17}$$

E_{sa} and E_{sc} are the energies of the free atomic s states of the anion and cation, respectively [40].

The nonzero interband deformation potentials are listed in Tables 6 and 7. Note that the LCAO calculation predicts $d_0^{v,cs} = d_0^{c,c} = 0$ for all zincblende-type compounds. This is not due to symmetry but a consequence of having used the d^{-2} scaling law for the p–s interatomic matrix elements. If a $d^{-\xi}$ rule with $\xi \neq 2$ is assumed the formalism yields $d_0^{v,cs} \neq 0$, $d_0^{c,c} \neq 0$.

4.3 Copper halides

The deformation potentials of the copper halides show anomalies which cannot be explained by extrapolation of the chemical trends of the sp^3-bonded semiconductors shown in Tables 6 and 7.

In Table 9 we list representative experimental values for the deformation potentials. The optical deformation potentials d_0 are taken from [19], the acoustic deformation potentials d, b from [41] and [26]. The values for d, b given in Table 9 include the spin–orbit-dependent part of the deformation potentials which is not completely negligible for these materials [26]. For comparison with our calculations we should extract from them the orbital part of the deformation potentials. However, the separate orbital contributions have only been obtained for CuCl and CuBr [26]. We infer from these few data that the spin–orbit interaction leads to corrections in the deformation potentials of the order of 20%.

Following the chemical trends observed in Tables 2 to 7 for the influence of the ionicity on the deformation potentials one would expect all deformation potentials of the copper halides to be about one order of magnitude larger than experimentally

Table 9

Experimental values for the deformation potentials d_0, b, d of cuprous halides. d_0 has been found to be negative for CuCl and CuBr [65]

	$\lvert d_0\rvert$ [19, 63, 64]	b [41]	d	d_1 [26]
CuCl	2.7, 7, 33	−0.4	0.34 [41]	0.55
		−0.7	0.425 [26]	
CuBr	3.8	−0.25	−0.65 [26]	−0.4
		−0.4	−0.7 [41]	
CuI	1.1	−0.64	−1.4 [26]	—
			−1.8 [41]	

The data include orbital and spin–orbit parts of the deformation potentials. The pure orbital part is known only for d for CuCl and CuBr [26]. The orbital part of d is called d_1.

observed. Also, the negative sign of d_0 for the copper halides and the positive sign of d for CuCl cannot simply be attributed to the high ionicity of these compounds. In this section we want to show that these anomalies result from the admixture of Cu-3d states and halogen-p states in the Γ_{15} valence band (see [43 to 47]).

It is known that the wave function of the top of the valence band of the Cu halides can be written as a linear combination of a halogen-p function ψ_{pa} and copper-4p and 3d functions ψ_{pc} and ψ_{dc} [43, 44, 47],

$$\psi = \alpha\psi_{pa} + \beta\psi_{dc} + \gamma\psi_{pc} . \quad (4.18)$$

Experiments [44] and calculations [43 to 47] give for α^2 the average values 0.23, 0.38, 0.49 for CuCl, CuBr, CuI, respectively, while γ^2 is about 0.02 for all copper halides. As the admixture coefficient γ for the Cu-4p state is small one often neglects it [43]. We shall see, however, that the interaction of Cu and halogen-p states does not play a negligible role for the calculation of intra-valence-band deformation potentials. The contribution of the Cu-4p–halogen-p interaction to the deformation potentials can explain why d_0, d, b are relatively small and why d can be positive for CuCl, but negative for CuBr and CuI. As the admixture of Cu-4p to the valence-band wave function is small we can treat it by perturbation theory. In this way we obtain a simple interpretation for the anomalies described above.

In the following we calculate the intra-valence-band deformation potentials d_0, b, d and compare them with the corresponding experimental values.

According to (4.18) for the calculation of deformation potentials of the upper valence bands we have to take into account three halogen-p orbitals $|p_x\rangle_a$, $|p_y\rangle_a$, $|p_z\rangle_a$, three copper-p orbitals $|p_x\rangle_c$, $|p_y\rangle_c$, $|p_z\rangle_c$, and three d states $|d_{xy}\rangle_c$, $|d_{yz}\rangle_c$, $|d_{zx}\rangle_c$. The remaining d states $|d_{x^2-y^2}\rangle_c$, $|d_{3z^2-r^2}\rangle_c$ correspond to the Γ_{12} valence band, which is about 2 eV lower than the top of the Γ_{15} valence band. Therefore the states $|d_{x^2-y^2}\rangle_c$, $|d_{3z^2-r^2}\rangle_c$ give rise only to second-order interband coupling and are of no interest here.

For the calculation of the intra-valence-band deformation potentials we must obtain α, β, and γ of (4.18). In order to do this, we must diagonalize a 9×9 Hamiltonian matrix. It is also possible to include in this matrix the effect of the perturbation and, in this manner, to determine directly the deformation potential. The diagonal elements of this matrix E_{pa}, E_{pc}, E_{dc} can be determined by fitting tight-binding bands to experimental data or to valence bands calculated with more accurate methods. In the case of CuCl we take the atomic energies from [47]. For CuBr and CuI we have adjusted the tight-binding parameters to LMTO-ASA calculations published in [42]. The atomic energies used as diagonal elements of the tight-binding Hamiltonian are listed in Table 10. For the determination of matrix elements of the Hamiltonian

Table 10

Atomic energies E_{pa}, E_{pc}, E_{dc} (in eV), lattice constant a_0 (in nm), and two-center integrals $V_{pp\sigma}$, $V_{pp\pi}$, $V_{pd\sigma}$, $V_{pd\pi}$ (in eV) for CuCl, CuBr, CuI

compound	E_{pa}	E_{pc}	E_{dc}	a_0	$V_{pp\sigma}$	$V_{pp\pi}$	$V_{pd\sigma}$	$V_{pd\pi}$
CuCl	0	12.75	2.5	0.5406	3.087	−0.876	−1.271	0.305
CuBr	0	10.1	0.5	0.5691	2.786	−0.791	−1.062	0.255
CuI	0	8.6	−0.2	0.6043	2.471	−0.701	−0.861	0.207

The energy E_{pa} of halogen-p is defined to be zero. The determination of all parameters is described in Section 4.3.

between copper-p and halogen-p states we apply the formalism defined in Section 4.2. For the parameters $\eta_{pp\sigma}$, $\eta_{pp\pi}$ we use the values of the sp^{3}s* model (Table 5). The matrix elements connecting halogen-p and copper-d states [37, 39] can be written as a function of the two-center integrals $V^{jj'}_{pd\sigma}$, $V^{jj'}_{pd\pi}$ (4.4) as

$$H_{x,xy} = \sum_{j'=1}^{4} [\sqrt{3} \cos^2(\theta_{j'x}) \cos(\theta_{j'y}) V^{jj'}_{pd\sigma} + \cos(\theta_{j'y}) \left(1 - 2\cos^2(\theta_{j'x})\right) V^{jj'}_{pd\pi}] , \qquad (4.19)$$

$$H_{x,yz} = \sum_{j'=1}^{4} \{\cos(\theta_{j'x}) \cos(\theta_{j'y}) \cos(\theta_{j'z}) [\sqrt{3}\, V^{jj'}_{pd\sigma} - 2V^{jj'}_{pd\pi}]\} .$$

The cosines in (4.19) are the direction cosines of vector $\boldsymbol{r}$ pointing from atom j to the neighboring atom j'. All matrix elements between halogen-p and copper-d which are not explicitly defined in (4.19) can be generated from $H_{x,xy}$ and $H_{x,yz}$ by cyclic permutation of x, y, z. For the off-diagonal matrix elements of (4.19) we take Harrison's $d^{-7/2}$ rule [37], i.e., we assume that the two-center integrals $V^{jj'}_{pd\sigma}$, $V^{jj'}_{pd\pi}$ scale with the distance $d_{jj'}$ between two neighboring atoms j and j' according to (4.4) with $\xi = 7/2$. Values for η_{pdm} can be determined for CuCl from interatomic matrix elements given in [47]. We assume that the interatomic matrix elements for CuBr and CuI can be obtained from the corresponding values for CuCl using the $d^{-7/2}$ scaling rule. In the following we use the symbols $V_{pp\sigma}$, $V_{pp\pi}$, $V_{pd\sigma}$, $V_{pd\pi}$ for the two-center integrals of the undeformed crystals. The values for $V_{pp\sigma}$, $V_{pp\pi}$, $V_{pd\sigma}$, $V_{pd\pi}$ which are the input parameters of our tight-binding calculation can be found in Table 10. With the data given in Table 10 we have calculated the 9×9 Hamiltonian matrix taking into account all states which contribute to the top of the Γ_{15} valence band. We determine the intra-valence-band deformation potentials d_0, d', b from the derivatives of the energy eigenvalues with respect to the corresponding deformations. The results are listed in Table 11.

For comparison with experimental data we must calculate the deformation potential d. In order to do this, the internal strain parameter ζ is needed (2.10). As already mentioned, experimental values for ζ exist only for Ge, Si, and GaAs. In order to estimate the internal strain parameter for the copper halides, we use Martin's calculations [36] which show that ζ increases linearly with ionicity f_i by 0.3 for an ionicity change $\Delta f_i = 1$. If we take $\zeta = 0.73$ and $f_i = 0$ for Si, the internal strain parameter for CuCl should be $\zeta = 0.95$ ($f_i = 0.746$ for CuCl [36]). We assume that this gives a representative value for all copper halides. The theoretical values for d calculated for $\zeta = 0.95$ are listed in Table 11.

Table 11

Deformation potentials d_0, b, d', d (in eV) of copper halides calculated from parameters in Table 9 by diagonalization of the 9×9 Hamiltonian matrix taking into account three halogen-p, three copper-4p, and three copper-3d states. d is determined for the internal strain parameter $\zeta = 0.95$

compound	d_0	b	d'	d
CuCl	−5.45	−0.3	−1.25	0.04
CuBr	−3.5	−0.45	−1.05	−0.22
CuI	−1.85	−0.55	−0.85	−0.41

Tables 9 and 11 show that the experimental and theoretical values of the deformation potentials agree semiquantitatively with each other. The LCAO calculation yields the right order of magnitude and the correct sign in all cases. Especially the theory reproduces the positive sign of d for CuCl and its negative sign for CuBr and CuI. Note that the calculated values of d_0 and d' are negative. In contrast, d_0 and d' are positive for all sp^3-bonded semiconductors.

In order to give a more detailed interpretation of these effects, we estimate the intra-valence-band deformation potentials d_0, d', d, b using perturbation theory. As the contribution of the copper-4p states to the valence band wave function is small we assume the zero-order wave function ψ_0 of the top of the valence band for the unstrained lattice to be solely a linear combination of halogen-p (ψ_{pa}) and copper-3d states (ψ_{dc}),

$$\psi_0 = \alpha_0 \psi_{pa} + \beta_0 \psi_{dc} \,. \tag{4.20}$$

The influence of the copper-4p orbitals can be calculated in perturbation theory. α_0, β_0 in (4.20) ($\alpha_0^2 + \beta_0^2 = 1$) are given by the eigenvectors of a 6×6 Hamiltonian matrix taking into account only halogen-p and copper-3d contributions. The energy eigenvalue which corresponds to ψ_0 for the unstrained crystal will be called E_{pd} in the following. From the 6×6 Hamiltonian matrix which defines ψ_0 we determine deformation potentials $d_0(\mathrm{p, d})$, $d(\mathrm{p, d})$, $d'(\mathrm{p, d})$, $b(\mathrm{p, d})$ according to (2.11). These quantities are the contribution of the coupling between halogen-p and copper-3d states to the deformation potentials d_0, d, d', b, respectively.

On the other hand, if the copper-d states would not be present but only halogen-p and copper-4p would contribute to the wave function of the top of the valence band, the intra-valence band deformation potentials would be given by the expressions (4.16) defined in Section 4.2. For these deformation potentials we introduce the symbols $d_0(\mathrm{p, p})$, $d(\mathrm{p, p})$, $d'(\mathrm{p, p})$, $b(\mathrm{p, p})$.

Taking into account the admixture of copper-4p to the valence band wave function in first-order perturbation theory we obtain for the deformation potentials of the top of tho valcncc band

$$\begin{aligned} d_0 &= d_0(\mathrm{p, d}) + \alpha_0^2 K d_0(\mathrm{p, p}) \,, \\ d' &= d'(\mathrm{p, d}) + \alpha_0^2 K d'(\mathrm{p, p}) \,, \\ b &= b(\mathrm{p, d}) + \alpha_0^2 K b(\mathrm{p, p}) \,, \end{aligned} \tag{4.21}$$

where

$$\begin{aligned} K &= -\frac{[((E_{pa} - E_{pc})/2)^2 + H_{xx}^2(\delta = 0)]^{1/2}}{E_{pd} - E_{pc}} \,, \\ \alpha_0 &= -\frac{V_{x,yz}}{(E_{pa} - E_{dc})/2 - [((E_{pa} - E_{dc})/2)^2 + V_{x,yz}^2]^{1/2}} \,, \\ E_{pd} &= \frac{E_{pa} + E_{dc}}{2} + \left[\left(\frac{E_{pa} - E_{dc}}{2}\right)^2 + V_{x,yz}^2\right]^{1/2} , \end{aligned} \tag{4.22}$$

$$\begin{aligned} d_0(\mathrm{p, d}) &= \frac{16}{9\sqrt{3}} \frac{V_{x,yz}[7V_{pd\sigma} - \frac{5}{3}\sqrt{3}\, V_{pd\pi}]}{[((E_{pa} - E_{dc})/2)^2 + V_{x,yz}^2]^{1/2}} \,, \\ d'(\mathrm{p, d}) &= \frac{4}{9\sqrt{3}} \frac{V_{x,yz}[7V_{pd\sigma} + \frac{13}{3}\sqrt{3}\, V_{pd\pi}]}{[((E_{pa} - E_{dc})/2)^2 + V_{x,yz}^2]^{1/2}} \,, \\ b(\mathrm{p, d}) &= 0 \,. \end{aligned} \tag{4.23}$$

Table 12

Deformation potentials (in eV) and admixture coefficient α_0^2 of copper halides calculated from (4.21), (4.23) for parameters in Table 10

compound	α_0^2	d_0	d_0 (p, p)	d_0(p, d)	d'	d'(p, p)	d'(p, d)	b	b(p, p)
CuCl	0.25	−7.1	8.8	−8.7	−1.37	0.55	−1.47	−0.17	−0.95
CuBr	0.43	−5.8	8.9	−8.3	−1.24	0.56	−1.40	−0.27	−0.96
CuI	0.53	−4.0	8.2	−6.8	−0.97	0.51	−1.14	−0.30	−0.89

$V_{x,yz}$ is equal to the matrix element $H_{x,yz}$ (4.19) for the undeformed crystal,

$$V_{x,yz} = -\frac{4}{3} V_{\mathrm{pd}\sigma} + \frac{8}{3\sqrt{3}} V_{\mathrm{pd}\pi} . \tag{4.24}$$

Table 12 shows the values for α_0^2, d_0, d_0(p, p), d_0(p, d), d', d'(p, p), d'(p, d), b, b(p, p), calculated from (4.22) and (4.23) and parameters from Table 10. Notice that the deformation potentials d_0(p, d), d'(p, d) are negative, whereas the quantities d_0(p, p), d'(p, p) are positive. Consequently according to (4.22) ($K > 0$) the contribution of the interaction between halogen-p and copper-3d states and the contribution of the interaction between halogen-p and copper-4p states tend to compensate each other. This explains why the deformation potentials are relatively small in comparison to most of the zincblende-type semiconductors. The magnitude and the sign of the deformation potentials (see (4.21)) depend strongly on the hybridization of halogen-p and copper-d states. For example, the optical deformation potential d_0 is negative for CuCl, CuBr, CuI although the positive term d_0(p, p) comes out to be slightly larger than the absolute value of d_0(p, d). This is due to the fact that the admixture coefficient α_0^2 of the halogen-p states is small ($\leqq 0.5$).

5. Discussion

We note that the sp^{3}s* LCAO model (Table 7) yields values of d_0 in better agreement with the EPM ones and with experiment than the simpler sp^3 model (Table 8). For the first row of the periodic table the EPM method is not expected to be very good. Nevertheless we obtain for diamond values of d_0 (90 eV) in reasonable agreement with the LCAO ones (75 eV) and with experiment (90 eV). For elements below the first period d_0 shows a trend to decrease with increasing ionicity both in the EPM and the LCAO calculations. This trend is also confirmed by experiment although the value $d_0 = 4$ eV reported for ZnS [32] is probably too low. It was obtained from a fit to *preresonant* Raman cross sections which is not very sensitive to d_0. A slight trend to decrease with increasing ionicity is also shown by $|a|$, $|a(\Gamma_1^c)|$ (Table 2), and by $|b|$ (Tables 3 and 7). This trend appears also for the LCAO calculations of d' (Table 7). It is not so clear for the EPM calculations (Table 3). The agreement with experiment can be considered satisfactory in most cases, the most conspicuous deviation being for the d's of Table 4: the theoretical values are $\approx 60\%$ higher than the experimental ones. This systematic discrepancy between theory and experiment has been noted earlier [34]. Recent theoretical calculations [48 to 50][3]) yield for Ge and Si values of

[3]) In [50] $\zeta = 0.61$ is given. The equilibrium lattice constant, however, is smaller than the experimental one, a fact which should lead to an increase in ζ. We are grateful to O. H. Nielsen for pointing this out to us.

ζ around 0.5, while earlier experiments [51] gave $\zeta = 0.62$. All these facts seem to cast doubts on the recent experimental values of ζ and on the extrapolated values of Table 4. We have also given in this table the values of d calculated with the ζ's of [36], also listed in Table 4. They agree better with experiment than the first set. The LCAO calculations, however, yield values of d in agreement with experiment for the larger values of ζ (Table 8). This is due to the fact that d' is positive for the LCAO calculations and either negative or very small positive for EPM. The difference in the values and signs of d' is not surprising since d' is rather small and, in this case, neglected residual interactions can induce large fluctuations. On the whole the pseudopotential calculations are more reliable than the highly simplified LCAO ones. More experimental determinations of ζ are required to clarify the discrepancies just presented.

We note that $d_0^{v,cs}$ and $d_0^{c,c}$ are exactly zero for our LCAO calculations (4.16). Correspondingly they are also quite small in Table 1 (EPM). The deformation potential $d_0^{v,cs}$ determines the possibility of a ferroelectric instability in these materials [11, 52]. Because of the small value of $d_0^{v,cs}$ this possibility can be excluded for the materials under consideration.

The most reliable experimental information on interband deformation potentials has been obtained for the b and d of InSb from measurements of cyclotron resonance under stress. From [20] we obtain $|d^{v,cs}| = 1.6$ eV. From Tables 1 and 3 we obtain with (2.10) for $\zeta = 0.85$ (Table 4) $d^{v,cs} = 1.2$ eV in excellent agreement with the experimental value. We note that another measurement (Shubnikov-de Haas, [53]) gives $|d^{v,cs}| = 1$ eV. From cyclotron resonance under stress we also find for InSb $|b^{v,cp}| = 2.9$ eV which again compares reasonably well with the calculated value (Table 3) $b^{v,cp} = 1.2$ eV. We should point out that the pseudopotential value of $d^{c,c}$ has been successfully used to interpret the stress dependence of the conduction mass of GaAs [12].

The calculated deformation potentials of the copper halides given in Table 11 (more accurate than those of Table 12) should be compared with the experimental ones (Table 9). The sign and the small magnitude of d_0 is well explained by the calculations. They arise from the partial cancellation between the d_0(p, p) and d_0(p, d) (Table 12). In the case of b extremely good agreement between theoretical and experimental values is obtained. The experimental sign of d is also well explained by the theory although quantitative agreement is poor for CuCl.

Acknowledgement

We would like to thank P. Vogl for making his pseudopotential program available to us and for a number of discussions.

References

[1] J. Bardeen and W. Shockley, Phys. Rev. **80**, 72 (1950).
[2] See, for instance, W. Pötz and P. Vogl, Phys. Rev. B **24**, 2025 (1981).
[3] G. L. Bir and G. E. Pikus, Symmetry and Strain-Induced Effects in Semiconductors, John Wiley & Sons, New York/Toronto 1974.
[4] J. D. Wiley, in: Semiconductors and Semimetals, Vol. 10, Ed. R. K. Willardson and A. C. Beer, Academic Press, 1975 (p. 91).
D. J. Rode, ibid. (p. 1).
[5] M. Cardona, in: Light Scattering in Solids II, Ed. M. Cardona and G. Güntherodt, Springer-Verlag, Heidelberg 1982.
[6] C. Hamaguchi, Phys. Rev. B **11**, 3876 (1975).
[7] H. Presting and H. Vogt, to be published.

[8] L. Pintschovius, J. A. Vergés, and M. Cardona, Phys. Rev. B **26**, 5658 (1982).
[9] M. Chandrasekhar and F. H. Pollak, Phys. Rev. B **15**, 2127 (1977).
[10] P. B. Allen and M. Cardona, Phys. Rev. B **27**, 4760 (1983).
[11] N. Kristoffel and P. Konsin, Soviet Phys. – Solid State **13**, 2113 (1972).
[12] D. E. Aspnes and M. Cardona, Phys. Rev. B **17**, 741 (1978); **16**, 726 (1978).
[13] Landolt-Börnstein, Vol. 17a, Springer-Verlag, Heidelberg 1982.
[14] I. Gorczyca, phys. stat. sol. (b) **108**, 205 (1981).
[15] F. Cerdeira, J. S. DeWitt, U. Rössler, and M. Cardona, phys. stat. sol. **41**, 735 (1970).
[16] P. Melz, J. Phys. Chem. Solids **32**, 209 (1971).
[17] M. Cardona, in: Atomic Structure and Properties of Solids, Ed. L. Burstein, Academic Press, New York 1972 (p. 514).
[18] J. A. Vergés, D. Glötzel, M. Cardona, and O. K. Andersen, phys. stat. sol. (b) **113**, 519 (1982).
[19] F. Meseguer, J. C. Merle, and M. Cardona, Solid State Commun. **43**, 511 (1982).
[20] H.-R. Trebin, M. Cardona, R. Ranvaud, and U. Rössler, in: Physics of Narrow Gap Semiconductors, Ed. J. Raułuskiewicz, M. Gorska, and E. Kaczmarek, PWN, Warszawa 1978 (p. 227).
[21] E. O. Kane, Phys. Rev. **178**, 1368 (1969).
[22] C. Haensel and K. Suzuki, Phys. Rev. B **9**, 4219 (1974).
[22a] N. E. Christensen, Solid State Commun. **50**, 177 (1984).
[23] W. Richter, Springer Tracts mod. Phys. **78**, 121 (1976).
[24] P. Lawaetz, Thesis, Technical University of Denmark, 1978, unpublished.
[25] A. Pinczuk, G. Abstreiter, and M. Cardona, in: Light Scattering in Solids IV, Ed. M. Cardona and G. Güntherodt, Springer-Verlag, Heidelberg 1984.
[26] A. Blacha, S. Ves, and M. Cardona, Phys. Rev. B **27**, 6346 (1983).
[27] J. L. Ivey, Phys. Rev. B **9**, 4231 (1974).
[28] I. Goroff and L. Kleinman, Phys. Rev. **132**, 1080 (1963).
[29] E. Anastassakis and M. Cardona, phys. stat. sol. (b) **104**, 589 (1981).
[30] M. L. Cohen and V. Heine, Solid State Phys. **24**, 37 (1970).
[31] J. A. Sanjurjo, E. López-Cruz, P. Vogl, and M. Cardona, Phys. Rev. B **28**, 4579 (1983).
[32] J. M. Calleja, H. Vogt, and M. Cardona, Phil. Mag. A **45**, 239 (1982).
[33] C. S. G. Cousins, B. J. Sheldon, L. Gerward, B. Selmark, and J. S. Olsen, J. Phys. C **15**, L651 (1982).
[34] H. d'Amour, W. Denner, H. Schulz, and M. Cardona, J. appl. Cryst. **15**, 148 (1982).
[35] C. N. K. Kournelis, G. E. Zardas, C. A. Londos, and D. K. Levenburi, Acta cryst. **A31**, 84 (1975); **A32**, 306 (1976).
[36] R. M. Martin, Phys. Rev. B **1**, 4005 (1970).
[37] W. Harrison, Electronic Structure and the Properties of Solids, Freeman, San Francisco 1980.
[38] W. Harrison, Phys. Rev. B **24**, 5835 (1981).
[39] J. Slater and G. F. Koster, Phys. Rev. **94**, 1498 (1954).
[40] C. F. Fisher, Atomic Data **4**, 301 (1970).
[41] J. L. Deiss, A. Daunois, and O. V. Gogolin, EPS-Meeting, Lausanne 1983, unpublished.
[42] A. Blacha, M. Cardona, N. E. Christensen, and S. Ves, Solid State Commun. **43**, 183 (1982).
[43] M. Cardona, Phys. Rev. **129**, 69 (1963).
[44] A. Goldmann, phys. stat. sol. (b) **81**, 9 (1977).
[45] A. Zunger and M. Cohen, Phys. Rev. B **20**, 1189 (1979).
[46] L. Kleinmann and K. Mednick, Phys. Rev. B **20**, 2487 (1979).
[47] S. Y. Ren, R. E. Allen, J. D. Dow, and I. Lefkowitz, Phys. Rev. B **25**, 1205 (1982).
[48] M. Cardona, K. Kunc, and R. M. Martin, Solid State Commun. **44**, 1205 (1983).
[49] O. H. Nielsen and R. M. Martin, Phys. Rev. Letters **50**, 697 (1983).
[50] B. N. Harmon, W. Weber, and D. R. Hamann, Phys. Rev. B **25**, 5960 (1982).
[51] A. Segmüller and H. R. Neyer, Phys. kondens. Mater. **4**, 63 (1965).
[52] M. Cardona and P. Vogl, Ferroelectrics **53**, 49 (1984).
[53] D. G. Seiler, B. D. Bajaj, and A. E. Stephens, Phys. Rev. B **16**, 2822 (1977).

[54] J. H. Yee and G. Myers, phys. stat. sol. (b) **77**, K81 (1976).
D. L. Rode, Phys. Rev. B **2**, 4036 (1970).
[55] B. A. Weinstein, R. Zallen, M. L. Slade, and A. de Lozanne, Phys. Rev. B **24**, 4652 (1981).
[56] A. Gavini and M. Cardona, Phys. Rev. B **1**, 672 (1970).
[57] A. A. Kaplianskii and L. G. Suslina, Soviet Phys. — Solid State **7**, 1881 (1966).
[58] D. W. Langer, R. N. Euwema, K. Era, and T. Koda, Phys. Rev. B **2**, 4005 (1970).
[59] G. L. Bir, G. E. Pikus, L. O. Suslina, and D. L. Federov, Soviet Phys. — Solid State **12**, 926 (1970).
[60] L. Reggiani, D. Waechter, and S. Zykotynski, Phys. Rev. B **28**, 3550 (1983).
[61] D. G. Thomas, J. appl. Phys. **32**, 2298 (1961).
[62] E. Käräjamäki, R. Laiho, T. Levola, B. Bairamov, A. V. Goltsev, and T. Toporov, Phys. Rev. B **29**, 4515 (1984).
[63] F. Meseguer, J. C. Merle, and M. Cardona, Solid State Commun. **50**, 709 (1984).
[64] J. C. Merle, F. Meseguer, and M. Cardona, Proc. Internat. Conf. Phys. Semicond., San Francisco, 1984, Springer-Verlag, to be published.
[65] F. Meseguer, J. C. Merle, and M. Cardona, Physica (Utrecht) **117/118 B**, 374 (1983).

(Received February 6, 1984)

Addendum in proof:

Pfeffer et al. [66] have recently published a pseudopotential calculation of the deformation potentials a, b, and d of the direct gap of GaAs and a review of existing experimental data. They show that these results ($a = -8.8$, $b = -1.7$, $d = -5.3$ eV) are consistent with an analysis of the measured free-carrier infrared absorption in n-GaAs. The value of a (Γ_V^{15}) has been obtained by Glembocki et al. [67] for GaAs (-7.3 eV) and Si (-6.5 eV) by means of a self-consistent APW calculation.

First principles pseudopotential calculations of the hydrostatic pressure coefficients of several states in Si, Ge, GaAs, InP, CdS, and CdSe (cubic) have been reported by Chang et al. [68]. A comparison to experimental data and to the results of other calculations and numerous references to this work are given.

Christensen [69] has presented LMTO calculations of the various deformation potentials of the highest valence state (Γ_{15}) for the Cu halides. These calculations automatically include the stress dependence of the spin–orbit splitting, i.e. the deformation potentials d_{01}, d_1, b_1 (orbital) and d_{02}, d_2, b_2 (spin–orbit) are calculated. The deformation potentials for the splitting of the Γ_8 component then become $d_{01} + 2d_{02}$, $b_1 + 2b_2$, $d_1 + 2d_2$, while those for the $\Gamma_8 - \Gamma_7$ coupling are $d_{01} - d_{02}$, $b_1 - b_2$, and $d_1 - d_2$. We list in Table 13, page 36, the deformation potentials calculated by Christensen and compare them with LCAO and experimental data. The LCAO treatment, in spite of its simplicity, yields extremely good results.

Cardona et al. [70] have recently analyzed stress-induced spin depolarization experiments by using the interband deformation potentials reported above. The LCAO data of Table 2 of this paper should be multiplied by $\sqrt{3}$ because of a computational error. The coefficient of the stress-induced linear-in-k term V_2 falls for GaAs and GaSb somewhat below the experimental one when calculated by the pseudopotential method and somewhat above when calculated with the LCAO method. Data on stress-induced linear-in-k terms for the copper halides are given in [71, 72]. We have estimated these terms with the LCAO and found them to be an order of magnitude smaller than suggested in [71, 72].

Khan and Allen have discussed in a recent paper [73] the relationship between the deformation potentials discussed here and the coupling constants for the interaction of electrons with long-wavelength acoustic phonons.

[66] P. Pfeffer, I. Gorczyca, and W. Zawadzki, Solid State Commun. **51**, 179 (1984).
[67] O. J. Glembocki, B. M. Klein, H. Krakauer, and S. H. Wei, Bull. Amer. Phys. Soc. **29**, 392 (1984); private communication.
[68] K. J. Chang, S. Froyen, and M. L. Cohen, Solid State Commun. **50**, 105 (1984).
[69] N. E. Christensen, phys. stat. sol. (b) **123**, 281 (1984); **125**, K59 (1984).
[70] M. Cardona, V. A. Maruschak, and T. N. Titkov, Solid State Commun. **50**, 701 (1984).
[71] S. Sakoda and Y. Onodera, J. Phys. Chem. Solids **32**, 1365 (1971).
[72] T. Koda, T. Murahashi, T. Mitani, S. Sakoda, and Y. Onodera, Phys. Rev. B **5**, 705 (1972).
[73] F. S. Khan and P. B. Allen, Phys. Rev. B **29**, 3341 (1984).

Table 13

Intraband deformation potentials d_0, d', d, and b of the highest Γ_{15} valence band of the copper halides (in eV)

		$\lvert d_0 \rvert$	d_{01}	d_{02}	d'	d'_1	d'_1	d	d_1	d_2	b	b_1	b_2
CuCl	experiment[a]) [b])	2,7						0.34	0.56	−0.07	−0.4		
		3.3						0.43			−0.7		
		7.0											
	LMTO[c])	5.92	−6.96	0.46	−1.37	−1.36	−0.005	0.04	0.29	0.11	−0.294	−0.297	0.002
	LCAO[d])		−5.45			−1.25			0.044			−0.3	
CuBr	experiment[a])	3.8						−0.65	−0.42	−0.11	−0.25		
								−0.7			−0.4		
	LMTO[c])	4.26	−5.48	0.608	−1.538	−1.529	−0.005	−0.526	−0.227	−0.148	−0.338	−0.328	−0.005
	LCAO[d])		−3.5			−1.05			−0.22			−0.45	
CuI	experiment[a])	1.1						−1.4			−0.64		
	LMTO[c])	1.09	−2.66	0.79	−1.86	−1.74	−0.06	−1.6	−1.11	−0.25	−0.472	−0.404	−0.034
	LCAO[d])		−1.85			−0.85			−0.41			−0.55	

The relations between d_0, d', d, b and the orbital parts d_{01}, d_1, d_1, b_1 and the spin–orbit parts d_{02}, d_2, d_2, b_2 are given by $d_0 = d_{01} + 2d_{02}$, $d = d_1 + 2d_2$, $d = d_1 + 2d_2$, and $b = b_1 + 2b_2$ [17, 23]. d, d_1, d_2 are determined for the internal strain parameter $\zeta = 0.95$ (see (2.10) and Table 11). The theoretical values for d_0 are negative for all copper halides.

a) See Table 9.
b) [64].
c) [69].
d) See Table 11.

Original Papers

phys. stat. sol. (b) **126**, 37 (1984)

Subject classification: 1.5; 12

Department of Physics, Kabardino-Balkarian State University, Nalchik[1])

The Theory of X-Ray Diffraction Analysis of Elastic-Strain States in Epitaxial Films

By

Yu. P. Khapachev, A. A. Dyshekov, and D. S. Kiselev

The possibility to determine five components of the strain tensor of an expitaxial film is demonstrated. The general expressions are given for the values of the discrepancy of periods of the film and the substrate, and for the components of the stress tensor in cubic heterostructures. Special emphasis is given to the definition of stresses in the layers of a superlattice.

Показана возможность определения пяти компонент тензора деформаций в эпитаксиальных плёнках. Получены общие соотношения для вычисления несоответствия периодов решёток плёнки и подложки, а также компонент тензора напряжений для кубических гетероструктур. Особое внимание уделено определению напряжений в слоях сверхрешёток.

1. Introduction

When heteroepitaxial films are grown from solid solutions of semiconductors, homogeneous elastic strains of the films occur due to the discrepancy of lattice parameters in the film–substrate interface [1]. Such elastic stresses must be taken into account in measuring the real discrepancy of lattice parameters of film and substrate $\beta = \Delta a/a$ and in defining the concentration of the solid solution from β.

Elastic strains and concentrations for a three-component solid solution were defined with the use of an X-ray microanalyzer [2]. However, this can be done by the method of X-ray diffractometry by measuring two reflections: symmetric and asymmetric ones [3].

In [4] the problem of strains in epitaxial films was solved with no regard for the anisotropy of elastic properties. The same problem, but with due regard for anisotropy, was solved in [5, 6]. In order to define the value of the discrepancy in the free state, the correlation factor $\beta = \varrho[(hkl), C_{\nu\mu}]$ was obtained which depends on the orientation of the film and on elastic constants $C_{\nu\mu}$. The discrepancy was found from the experimentally measured components of the strain tensor ε_{ZZ} (Z-axis being directed inside the crystal), the correlation factor was calculated by the following formulas:

$$\beta = \varrho[(hkl), C_{\nu\mu}]\, \varepsilon_{ZZ}\,, \tag{1.1}$$

where

$$\varrho(001) = \frac{C_{11}}{C_{11} + 2C_{12}}, \tag{1.2}$$

$$\varrho(110) = \frac{\frac{1}{2}\, C_{11} + \frac{1}{2}\, C_{12} + C_{44}}{C_{11} + 2C_{12}}, \tag{1.3}$$

$$\varrho(111) = \frac{\frac{1}{3}\, C_{11} + \frac{2}{3}\, C_{12} + \frac{4}{3}\, C_{44}}{C_{11} + 2C_{12}}. \tag{1.4}$$

[1]) Tschernyshevskogo 173, 360004 Nalchik, USSR.

However, the possibility to define the other components of the tensor ε_{ij} and, correspondingly, the components of the stress tensor σ_{ij} was not considered in [4 to 6].

In all these works the two-component heterostructures were analyzed. The similar problem for multi-layer systems was solved in [7], though very rigid constraints were put therein. Namely, the anisotropy of the moduli of elastic compliance was not taken into account, the system was taken to be of band shape, i.e. a certain geometrical form was adopted. All this decreases the precision of the theoretical analysis.

In [8] the general theory of diffraction in a multi-layer periodic crystal is given. According to it, the shape of the curve of diffractional reflection (widths of the satellites, their spacing, etc.) is defined by various structural parameters of the crystal.

In this paper we use the approach of [8] for the analysis of elastic strains in heteroepitaxial superlattices. The purpose of this work is to demonstrate the possibility to analyze the elastic-stress states for two- and multi-layer and periodic heterostructures by the method of non-destructive X-ray diffractometry.

2. Definition of Strains and Stresses in Two-Layer Heterostructures

The angular variable and the deformation factor of a crystal are known to be present in the system of Takagi equations [9] in the form of

$$2\,\Delta\psi \sin 2\theta + 2\,\frac{k_H}{k}\,\nabla_{X,Z}(\boldsymbol{HU})\,, \tag{2.1}$$

where $\boldsymbol{H}$ is the vector of the reciprocal lattice, $\boldsymbol{U}$ the shift vector of atomic planes, $\boldsymbol{k}_H$ the wave vector of the diffracted wave, $\boldsymbol{k}$ the vacuum wave vector of the incident wave, θ the Bragg angle, $\Delta\psi$ the angular distance between diffractional maxima of film and substrate.

Differentiating in (2.1), we obtain

$$\begin{aligned} -\Delta\psi \operatorname{ctg}\theta &= \varepsilon_{ZZ}\cos^2\varphi + \varepsilon_{XX}\sin^2\varphi + \\ &+ \operatorname{ctg}\theta\left(\frac{\partial U_X}{\partial Z}\sin^2\varphi - \frac{\partial U_Z}{\partial X}\cos^2\varphi\right) \pm \\ &\pm \operatorname{ctg}\theta\,\frac{\sin 2\varphi}{2}\,(\varepsilon_{ZZ} - \varepsilon_{XX}) \pm \frac{\sin 2\varphi}{2}\left(\frac{\partial U_X}{\partial Z} + \frac{\partial U_Z}{\partial X}\right). \end{aligned} \tag{2.2}$$

The positive sign in (2.2) corresponds to the case when the crystal surface is at the angle $\theta - \varphi$ to the incident radiation, the negative sign is for the angle $\theta + \varphi$, φ being the angle between atomic planes and crystal surface.

To define the components of the strain tensor ε_{XX}, ε_{ZZ}, $\varepsilon_{XZ} = \frac{1}{2}(\partial U_X/\partial Z + \partial U_Z/\partial X)$ four linearly independent equations are necessary. These equations can be obtained by taking patterns of two orders of symmetric reflection and of one asymmetric reflection in two positions: $\theta + \varphi$ and $\theta - \varphi$. The relation (2.2) was obtained for the case of complanar diffraction in the plane $(X0Z)$. Since the direction of the X-axis is taken arbitrarily, (2.2) is also valid for diffraction in the $(Y0Z)$ plane provided that X-indices are substituted in (2.2) by Y-indices. Therefore, the components ε_{YY} and $\varepsilon_{YZ} = \frac{1}{2}(\partial U_Y/\partial Z + \partial U_Z/\partial Y)$ can be determined by solving three more linearly independent equations for the case of diffraction in the $(Y0Z)$ plane. These equations may be obtained by taking similar patterns after the crystal has been rotated by 90° around the Z-axis. Therefore, five components of the strain tensor ε_{ij} can be defined under the conditions of complanar diffraction.

In order to define the components of the stress tensor σ_{ij} and the real discrepancy β of the lattice parameters we use the generalized Hook's law which takes into account the dependence of deformations in the film–substrate system on both the stress tensor and the discrepancy β_{ij},

$$\varepsilon_{ij} = S_{ijkl}\sigma_{kl} + \beta_{ij}\,, \tag{2.3}$$

where S_{ijkl} is the tensor of elastic compliance. Solving (2.3) for the unknown σ_{kl} and β_{ij} we find the stress in the film and the discrepancies.

Using the condition $\sigma_{ZZ} = 0$, which implies that the upper surface of the film is in the free state, and solving (2.3) for σ_{kl} and β_{ij} we find the following expression for heterostructures of the cubic system degenerated in the (001) orientation:

$$\beta = \varepsilon_{ZZ} + \frac{S_{12}}{S_{11} - S_{12}}\left(\varepsilon_{ZZ} - \frac{\varepsilon_{XX} + \varepsilon_{YY}}{2}\right), \tag{2.4}$$

$$\sigma_{XX} = \frac{\varepsilon_{XX} - \varepsilon_{ZZ}}{S_{11} - S_{12}}, \qquad \sigma_{YY} = \frac{\varepsilon_{YY} - \varepsilon_{ZZ}}{S_{11} - S_{12}}, \tag{2.5}$$

$$\sigma_{XZ} = \frac{2\varepsilon_{XZ}}{S_{44}}, \qquad \sigma_{YZ} = \frac{2\varepsilon_{YZ}}{S_{44}} \tag{2.6}$$

When taking $\varepsilon_{XX} = \varepsilon_{YY}$ in (2.4) we obtain the expression found in [2]. Moreover, if $\varepsilon_{XX} = \varepsilon_{YY} = 0$, then (2.4) is reduced to (1.1) with the correlation factor (1.2) [5].

The above theory is also valid for multi-layer aperiodic structures. However, the deformations and, consequently, stresses and discrepancies must be defined with respect to the substrate the film is grown on. Solving the problem for every pair of adjacent films we can define strains in every film, and, besides this, strains and discrepancies of every film with respect to the substrate.

3. Definition of Strains in the Layers of a Superlattice

As shown in [8] for superlattices (SL) in which the interplane distances and structural factors vary harmonically, the widths of the first-order satellites are not equal to each other and depend on the structural parameters of SL, viz. on the value of the discrepancy $\beta = \Delta a/a$ and on the relative variation of the structural factor $\alpha = \Delta F/F$.

Usually in an X-ray diffraction experiment the period of a SL is defined from the distance between the first-order satellites $\Delta\theta(+1, -1)$ in accord with the kinematic formula [10]

$$T_k = \frac{\lambda}{\Delta\theta(+1, -1)\cos\theta}. \tag{3.1}$$

The average values of such parameters as stress $\langle\sigma\rangle$, discrepancy β, and concentration x of the solid solution can be defined from the angular distance between the main diffraction maximum of the SL and that of the substrate.

The magnitude of the stress in the layers of the SL can be directly determined from X-ray diffraction measurements with regard for the anisotropy of elastic properties. When the composition of the solid solution varies harmonically, the moduli of elastic compliance and, in the first approximation, of stresses in the SL vary similarly,

$$S_{ij} = \langle S_{ij}\rangle\left(1 + \frac{\Delta S_{ij}}{\langle S_{ij}\rangle}\cos\frac{2\pi Z}{T}\right); \qquad i, j = 1, 2\,, \tag{3.2}$$

$$\sigma = \langle\sigma\rangle\left(1 + \frac{\Delta\sigma}{\langle\sigma\rangle}\cos\frac{2\pi Z}{T}\right). \tag{3.3}$$

Here $\langle S_{ij} \rangle$, $\langle \sigma \rangle$ are the average values of the tensors of elastic compliance and of stress over all the depth of the SL, ΔS_{ij} and $\Delta\sigma$ are the amplitudes of S_{ij} and σ, Z is the direction inside the crystal.

We assume that the film and substrate are homogeneously strained in the plane parallel to the heteroboundary and the substrate is thicker than the SL.

In the framework of the assumptions done by (3.2), (3.3) and following the theory of [8] we obtain the following expressions for the difference of widths of the first-order satellites in the symmetric Bragg case [11]:

$$\Delta\varphi(\pm 1) = \begin{cases} \dfrac{\alpha - f^2 ab}{2f \sin 2\theta}; & q < \dfrac{1}{4}, \\ \dfrac{a}{2 \sin 2\theta}; & q > \dfrac{1}{4}, \end{cases} \tag{3.4}$$

where

$$q = \frac{\frac{3}{4} f^4 a^2 b^2 + 2f^4 a^4 + \alpha^2/4 - \alpha f^2 ab/2 - \alpha^2 f^2 a^2/2}{4f^2 a^2}, \tag{3.5}$$

$$f = \frac{T}{4\lambda \sin\theta}, \tag{3.6}$$

$$a = -4 \sin^2\theta(\beta + 2\langle S_{12}\rangle \, \Delta\sigma + 2\, \Delta S_{12}\langle\sigma\rangle), \tag{3.7}$$

$$b = -8 \sin^2\theta \, \Delta S_{12} \, \Delta\sigma . \tag{3.8}$$

(We follow the notation of [8].)

Two situations are possible here in dependence on the signs of α, a, b and, in some cases, on the relation $|\alpha| > f^2 |ab|$. It can occur that either the low-angle satellite is wider than the high-angle one or vice versa [12]. In Table 1 the conditions on the values α, a, b are listed defining which of the two satellites is wider.

The distance between the first-order satellites is

$$\Delta\theta(+1, -1) = \frac{\sqrt{1 + 16f^2 |\chi_H|^2 \eta^2 + \alpha^2/2 - f^2(a^2 + b^2/2)}}{2f \sin 2\theta}. \tag{3.9}$$

If the coefficients a and b can be neglected in comparison with the Fourier component of the polarizability χ_H and $f|\chi_H| \gg |\alpha|$, then (3.9) reduces to the formula for the intersatellite distances which were first derived [13] for a SL produced by ultrasonic waves. For epitaxial SL compounds $A^{III}B^{V}$ a and b are greater than χ_H, as a rule, and than $f|\chi_H| \ll |\alpha|$, then neglecting the second term in (3.9) and accounting for (3.1)

Table 1

low-angle satellite is wider than the high-angle one	high-angle satellite is wider than the low-angle one
at $\alpha > 0$	at $\alpha > 0$
1. $a < 0, b > 0$	$a > 0, b < 0$
2. $a < 0, b < 0, \alpha > f^2 |ab|$	$a > 0, b > 0, \alpha > f^2 ab$
3. $a > 0, b > 0, \alpha < f^2 ab$	$a < 0, b < 0, \alpha < f^2 |ab|$
at $\alpha < 0$	at $\alpha < 0$
4. $a > 0, b > 0$	$a < 0, b < 0$
5. $a < 0, b < 0, |\alpha| > f^2 |ab|$	$a < 0, b > 0, |\alpha| > f^2 |ab|$
6. $a < 0, b > 0, |\alpha| < f^2 |ab|$	$a > 0, b < 0, |\alpha| < f^2 |ab|$

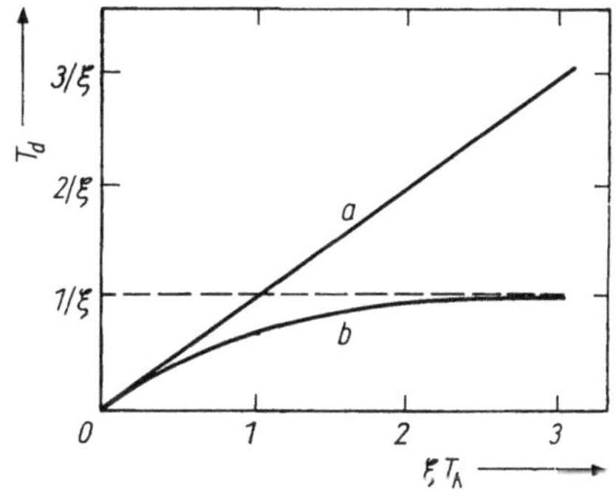

Fig. 1. Dependences of the SL period on ξT_k as calculated by the (a) kinematic and (b) dynamic formulas

and (3.6) we obtain the following dynamic formula defining the SL period:

$$T_d = T_k \sqrt{\frac{1 + \alpha^2/4}{1 + (T\xi)^2}}, \tag{3.10}$$

where

$$\xi = \frac{\sqrt{a^2 + b^2/2 - \alpha ab/2}}{4\lambda \sin \theta}. \tag{3.11}$$

Fig. 1 shows the graphs of the SL periods determined from the kinematic (3.1) and dynamic (3.10) formulas. It is seen that at $T_k\xi > 1$ the results of the two formulas differ essentially.

The analysis of angular positions of satellites with respect to the main diffraction maximum indicates that the middle of the intersatellite distance is shifted from the main maximum at

$$\delta = \frac{f^2ab - \alpha a}{2 \sin 2\theta(1 + 2f^2a^2)}. \tag{3.12}$$

The value of asymmetry increases with the SL period and with $\langle S_{ij} \rangle$, $\Delta\sigma$, and ΔS_{ij} as seen from (3.7) and (3.8).

Formulas (3.4), (3.9), (3.10), (3.12) obtained herein indicate that the parameters of diffraction spectra of a SL, such as the difference of the widths of satellites, their asymmetry with respect to the main maximum, depend on the values of elastic strains and on the anisotropy of elastic constants of this SL.

Thus, solving the proper equations with the use of X-ray diffractometric data about the values $\Delta\varphi(\pm 1)$, $\Delta\theta(+1, -1)$, and δ, we can derive the SL period and the average stress $\langle \sigma \rangle$ as well as the value $\Delta\sigma$, and, correspondingly, the strains in layers of different composition with due regard for the anisotropy of elastic properties.

4. Conclusion

In the present work the possibility to measure five components of the strain tensor for a film–substrate system with the use of the method of X-ray diffractometry is demonstrated. To do this, seven rocking curves must be taken. From the thereby measured five components ε_{ij}: ε_{XX}, ε_{YY}, ε_{ZZ}, ε_{XZ}, ε_{YZ}, using the presented formulas, we can calculate the value of the discrepancy and the components σ_{ij}: σ_{XX}, σ_{YY}, σ_{XZ}, σ_{YZ}. This allows the full analysis of the elastic-strain state of a film–substrate system to be carried out. The amplitudes of the variation of stresses between the layers of a superlattice can be determined with regard for the anisotropy of elastic properties from X-ray diffractometric measurements of the difference of widths of the first-order satellites, intersatellite distance, and of the asymmetry of the satellite positions with respect to the main diffraction maximum.

References

[1] M. G. Milvidskii and V. B. Osvenskii, Kristallografiya **22**, 431 (1977).
[2] I. N. Arsentev, N. A. Bert, S. G. Konnikov, and V. E. Umanskii, Fiz. Tekh. Poluprov. **14**, 96 (1980).
[3] P. V. Petrashen, Fiz. tverd. Tela **16**, 2168 (1974).
[4] K. Ishida, J. Matsui, T. Kamejima, and I. Sakuma, phys. stat. sol. (a) **31**, 255 (1975).
[5] J. Hornstra and W. J. Bartels, J. Crystal Growth **44**, 513 (1978).
[6] W. Bartels and Nijman, J. Crystal Growth **44**, 518 (1978).
[7] C. H. L. Goodman, Ed., Crystal Growth. Theory and Techniques, Vol. 2, Plenum Press, New York/London 1978.
[8] Yu. P. Khapachev, phys. stat. sol. (b) **120**, 155 (1983).
[9] S. Takagi, J. Phys. Soc. Japan **26**, 1239 (1969).
[10] A. Segmuller and A. E. Blakeslee, J. appl. Cryst. **6**, 19 (1973).
[11] Yu. P. Khapachev and A. A. Dyshekov, Zh, tekh. Fiz. **54**, 842 (1984).
[12] Yu. P. Khapachev and G. F. Kuznetsov, Kristallografiya **28**, 27 (1983).
[13] K. P. Assur and I. R. Entin, Fiz. tverd. Tela **24**, 2122 (1982).

(Received July 11, 1984)

phys. stat. sol. (b) 126, 43 (1984)

Subject classification: 1.3 and 1.4; 21

Zentralinstitut für Kernforschung der Akademie der Wissenschaften der DDR Rossendorf[1])

On the Interpretation of Orientation Distributions and Qualitative Ghost Corrections for Hexagonal-Orthorhombic Textures

By

K. HELMING and S. MATTHIES

Specific, symmetry-induced moments that have to be considered interpreting orientation distribution functions as a superposition of texture components are discussed for the case of hexagonal-orthorhombic symmetry. Possibilities of a qualitative ghost correction using geometrical properties (hit numbers) and standard functions are described.

Spezifische, symmetriebedingte Momente, die bei der Interpretation von Orientierungsverteilungsfunktionen als Summe von Texturkomponenten zu beachten sind, werden für den Fall hexagonal-orthorhombischer Symmetrie diskutiert. Möglichkeiten qualitativer Geisterkorrekturen mit Hilfe geometrischer Größen (Trefferzahlen) und Standardfunktionen werden beschrieben.

1. Introduction

The traditional field of texture analysis is the metal research dealing above all with samples of cubic crystal symmetry [1, 2]. However, in the last few years compounds with hexagonal crystal symmetry became also interesting [3 to 5]. Moreover, the texture analysis more and more gains a footing in the geosciences [6] considering samples with lower crystal symmetries yet. Due to the crystal and sample symmetry the G-space of the Eulerian angles describing an orientation g splits into equivalent subregions ("elementary region" ER) [1, 7]. Interpreting the orientation distribution function (ODF) $f(g)$ within such an ER as a superposition of few texture components (model distributions at g_i with a finite width and a certain percentage of intensity) it is necessary to take into consideration specific symmetry relations between the elementary regions in the G-space, especially near their boundary planes. Whereas for cubic crystal symmetry sufficient experience exists, it seems to be useful to describe these relations in more detail for the case of hexagonal symmetry. After that the analysis of the corresponding subgroups of hexagonal symmetry is not problematic.

Commonly the ODF is obtained as a result of a reproduction procedure using pole figure data determined in diffraction experiments [1, 7]. The diffraction experiment leads as a rule to a loss of information on the orientation distribution of the sample [14]. Therefore, the reproduction problem is not an unambiguous one. For the typical case of normal scattering only the reduced ODF $\tilde{f}(g)$ $\left(f(g) \equiv \tilde{f}(g) + \tilde{\tilde{f}}(g)\right)$ can unambiguously be reproduced. This reduced ODF contains as a rule remarkable ghost phenomena falsifying the intensity of the main texture components reflected in $\tilde{f}(g)$ also and creating ghost peaks that do not exist in the true ODF [17, 7].

[1]) Postschließfach 19, DDR-8051 Dresden, GDR.

Practicable quantitative methods of ghost correction, i.e. of a reconstruction of $f(g)$ using certain model assumptions on the ODF or artifical conditions are at the present moment in the phase of methodical development yet and are much more expensive [7 to 9] than the reproduction of $\tilde{f}(g)$ by the harmonic method [1]. On the other hand, statements on orientation distributions concerning more detailed questions than tendencies of the behaviour of the main texture components require elements of a qualitative ghost correction at least. This is especially important for crystal symmetries lower than the cubic one because the falsifying role of the ghost phenomena increases in the former case.

A qualitative ghost correction sufficient for a series of practical questions can be realized in a relatively simple manner using "hit numbers" [10, 11, 17] or $\tilde{f}(g)$-maps calculated for model distributions described by standard functions [12, 13].

Basing on the understanding of the qualitative behaviour and the symmetry properties of the ghost phenomena it is, e.g., possible to interpret an ODF of a sample with hexagonal crystal symmetry [3] widely discussed in literature using up to six components by two components only.

For completeness it is to be added that in the case of normal scattering for certain crystal classes even additional conditions are not able to repair the loss of information on the ODF in the pole figures ("uncorrectable ghosts") because there is a loss of symmetry in the G-space. I.e. $\tilde{f}(g)$ possesses a higher symmetry than $f(g)$ [14]. For the hexagonal crystal symmetry this concerns the crystal classes C_{3h}, C_{6v}, and D_{3h}.

2. Properties of Symmetry of the ODF

Unlike the symmetry properties of direction-dependent physical quantities of a crystal lattice that permit symmetry operations affected by the inversion operation also and being described by one of the 32 possible crystal classes $\mathcal{G}_B$ the properties of crystal symmetry of the ODF are given by one of the eleven point groups G_B containing pure rotational operations only [7, 14]. If there is a conventional sample symmetry [1, 15] it can also be characterized by a pure rotational group G_A. I.e. with $g_{B_j} \in G_B$ ($j = 1, 2, \ldots, N_B$), $g_{A_k} \in G_A$ ($k = 1, 2, \ldots, N_A$)

$$f(g) = f(g_{B_j} g g_{A_k}) \tag{1}$$

is valid.

If the crystal coordinate system K_B is fixed at the crystal lattice and a variant of the Eulerian angles is defined (e.g. $g = \{\alpha, \beta, \gamma\}$ by $0 \leqq \alpha, \gamma < 2\pi$ and $0 \leqq \beta \leqq \pi$, K_A sample coordinate system, $g \triangleq K_A \to K_B$ by the rotations $Z_A(\alpha)$, $Y'_A(\beta)$, $Z''_A(\gamma)$) the $g^{jk} \equiv g_{B_j} g g_{A_k}$ with the resulting Eulerian angles α^{jk}, β^{jk}, and γ^{jk} can explicitly be described by the Eulerian angles of g, g_{B_j}, and g_{A_k}. For $G_A = D_2$ ($N_A = 4$) and $G_B = D_6$ ($N_B = 12$) the corresponding g^{jk} are given in Table 1. Of course, Table 1 contains the g^{jk} for all subgroups of D_6 (C_6, D_3, D_2, C_2) and of $G_A = D_2$(C_2, C_1) also. E.g. for missing sample symmetry ($G_A = C_1$) the column with $k = 1$ contains the corresponding g^{j1}. For $G_B = C_6$ only the cases $j = 1, 2, \ldots, 6$ are to be considered, etc. Therefore, together with the analogous table of the g^{jk} for $G_B = O$ and $C_A = D_2$ given in [12] and containing the subgroups of O (T, D_4, C_4, D_2, C_2, and C_1) also all crystal symmetries for ODF's existing in reality are described. For the (φ_1, Φ, φ_2)-variant of Eulerian angles [1] the relations $\varphi_1 = \alpha + \pi/2$, $\Phi = \beta$, and $\varphi_2 = \gamma - \pi/2$ are valid and the corresponding tables have the same form.

As was mentioned above the G-space can according to (1) be subdivided into $N_A N_B$ equivalent elementary regions. For the hexagonal-orthorhombic case ($G_B = D_6$, $G_A = D_2$, $N_B = 12$, $N_A = 4$) the elementary region can be chosen as

$$\text{ER:} \quad 0 \leqq \alpha, \beta \leqq \pi/2, \qquad 0 \leqq \gamma \leqq \pi/3\,. \tag{2}$$

Table 1

The Eulerian angles of the group elements $g_{B_j} \in G_B = D_6$ ($j = 1, 2, \ldots, N_B = 12$), $g_{A_k} \in G_A = D_2$ ($k = 1, 2, 3, 4 = N_A$), and of the 48 equivalent orientations $g^{jk} = g_{B_j} g g_{A_k} = \{\alpha^{jk}, \beta^{jk}, \gamma^{jk}\}$. At the bottom the twofold rotation axes $B_1, B_2, \ldots, B_6$ with respect to the crystal coordinate system K_B are shown

				$k = 1$ (E)				$k = 2$ (${}^{Z}AC_2$)				$k = 3$ (${}^{Y}AC_2$)				$k = 4$ (${}^{X}AC_2$)			
				α_k	β_k	γ_k		α_k	β_k	γ_k		α_k	β_k	γ_k		α_k	β_k	γ_k	
j (g_B)	α_j	β_j	γ_j	0	0	0		π	0	0		0	π	0		π	π	0	
1 (E)	0	0	0	α	β	γ	1	$\pi + \alpha$	β	γ	13	$\pi - \alpha$	$\pi - \beta$	$\pi + \gamma$	25	$2\pi - \alpha$	$\pi - \beta$	$\pi + \gamma$	37
2 (${}^{Z}BC_6$)	$\pi/3$	0	0	α	β	$\pi/3 + \gamma$	2	$\pi + \alpha$	β	$\pi/3 + \gamma$	14	$\pi - \alpha$	$\pi - \beta$	$4\pi/3 + \gamma$	26	$2\pi - \alpha$	$\pi - \beta$	$4\pi/3 + \gamma$	38
3 (${}^{Z}BC_3$)	$2\pi/3$	0	0	α	β	$2\pi/3 + \gamma$	3	$\pi + \alpha$	β	$2\pi/3 + \gamma$	15	$\pi - \alpha$	$\pi - \beta$	$5\pi/3 + \gamma$	27	$2\pi - \alpha$	$\pi - \beta$	$5\pi/3 + \gamma$	39
4 (${}^{Z}BC_2$)	π	0	0	α	β	$\pi + \gamma$	4	$\pi + \alpha$	β	$\pi + \gamma$	16	$\pi - \alpha$	$\pi - \beta$	γ	28	$2\pi - \alpha$	$\pi - \beta$	γ	40
5 (${}^{Z}BC_3^2$)	$4\pi/3$	0	0	α	β	$4\pi/3 + \gamma$	5	$\pi + \alpha$	β	$4\pi/3 + \gamma$	17	$\pi - \alpha$	$\pi - \beta$	$\pi/3 + \gamma$	29	$2\pi - \alpha$	$\pi - \beta$	$\pi/3 + \gamma$	41
6 (${}^{Z}BC_6^5$)	$5\pi/3$	0	0	α	β	$5\pi/3 + \gamma$	6	$\pi + \alpha$	β	$5\pi/3 + \gamma$	18	$\pi - \alpha$	$\pi - \beta$	$2\pi/3 + \gamma$	30	$2\pi - \alpha$	$\pi - \beta$	$2\pi/3 + \gamma$	42
7 (${}^{B_1}C_2$)	0	π	0	$\pi + \alpha$	$\pi - \beta$	$\pi - \gamma$	7	α	$\pi - \beta$	$\pi - \gamma$	19	$2\pi - \alpha$	β	$2\pi - \gamma$	31	$\pi - \alpha$	β	$2\pi - \gamma$	43
8 (${}^{B_2}C_2$)	$\pi/3$	π	0	$\pi + \alpha$	$\pi - \beta$	$2\pi/3 - \gamma$	8	α	$\pi - \beta$	$2\pi/3 - \gamma$	20	$2\pi - \alpha$	β	$5\pi/3 - \gamma$	32	$\pi - \alpha$	β	$5\pi/3 - \gamma$	44
9 (${}^{B_3}C_2$)	$2\pi/3$	π	0	$\pi + \alpha$	$\pi - \beta$	$\pi/3 - \gamma$	9	α	$\pi - \beta$	$\pi/3 - \gamma$	21	$2\pi - \alpha$	β	$4\pi/3 - \gamma$	33	$\pi - \alpha$	β	$4\pi/3 - \gamma$	45
10 (${}^{B_4}C_2$)	π	π	0	$\pi + \alpha$	$\pi - \beta$	$2\pi - \gamma$	10	α	$\pi - \beta$	$2\pi - \gamma$	22	$2\pi - \alpha$	β	$\pi - \gamma$	34	$\pi - \alpha$	β	$\pi - \gamma$	46
11 (${}^{B_5}C_2$)	$4\pi/3$	π	0	$\pi + \alpha$	$\pi - \beta$	$5\pi/3 - \gamma$	11	α	$\pi - \beta$	$5\pi/3 - \gamma$	23	$2\pi - \alpha$	β	$2\pi/3 - \gamma$	35	$\pi - \alpha$	β	$2\pi/3 - \gamma$	47
12 (${}^{B_6}C_2$)	$5\pi/3$	π	0	$\pi + \alpha$	$\pi - \beta$	$4\pi/3 - \gamma$	12	α	$\pi - \beta$	$4\pi/3 - \gamma$	24	$2\pi - \alpha$	β	$\pi/3 - \gamma$	36	$\pi - \alpha$	β	$\pi/3 - \gamma$	48

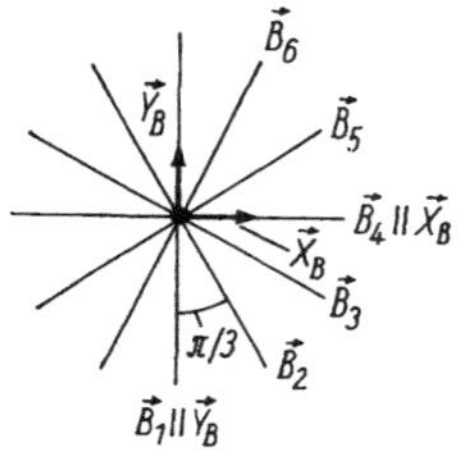

Now, considering the α- and β-boundary planes certain symmetry relations follow from Table 1 and (1) for the ODF values at one and the same plane,

$$\alpha = 0: \qquad f(\{0, \beta, \gamma\}) = f\left(\left\{0, \beta, \frac{\pi}{3} - \gamma\right\}\right), \tag{3}$$

$$\alpha = \frac{\pi}{2}: \qquad f\left(\left\{\frac{\pi}{2}, \beta, \gamma\right\}\right) = f\left(\left\{\frac{\pi}{2}, \beta, \frac{\pi}{3} - \gamma\right\}\right), \tag{4}$$

$$\beta = \frac{\pi}{2}: \qquad f\left(\left\{\alpha, \frac{\pi}{2}, \gamma\right\}\right) = f\left(\left\{\alpha, \frac{\pi}{2}, \frac{\pi}{3} - \gamma\right\}\right), \tag{5}$$

$$\beta = 0: \qquad f(\{\alpha, 0, \gamma\}) = f(\{\alpha + \gamma, 0, 0\}) = f\left(\left\{n\frac{\pi}{3} + \alpha + \gamma, 0, 0\right\}\right)$$

$$= f\left(\left\{n\frac{\pi}{3} - \alpha - \gamma, 0, 0\right\}\right) \tag{6}$$

with $n = 0, 1, 2, 3, 4, 5$ and $0 \leqq n\pi/3 + \alpha + \gamma$ or $n\pi/3 - \alpha - \gamma \leqq 5\pi/6$.

These symmetry relations have to be taken into consideration interpreting the ODF by texture components. E.g. if a component is lying on the $\alpha = 0$ plane ($g_0 = \{0, \beta_0, \gamma_0 \neq \pi/3\}$) then an equivalent peak will, corresponding to (3), be at $g_0^{12,3} = \{0, \beta_0, \pi/3 - \gamma_0\}$ (case Nr. 36 of Table 1). For a sufficiently large full width at half maximum (FWHM) of the component the maximum of the resulting orientation distribution of the two overlapping peaks will be found at $g_{\max} = \{0, \beta_0, \pi/6\}$ and have not to be considered as an independent texture component.

If g_0 is situated near the $\alpha = 0$ plane the tails of the g_0 and $g_0^{12,3}$ peak will overlap at this plane creating for a sufficient great FWHM a cigar-like distribution in the G-space. Analogous situations are found if g_0 is situated in the vicinity of the $\alpha = \pi/2$ or $\beta = \pi/2$ planes.

The symmetry relations (6) for the $\beta = 0$ boundary plane are some more complicated. The reason is that the usually chosen representation of the G-space between orthogonal α-, β-, and γ-coordinate axes does not reflect the property of the G-space that for $\beta = 0$ all α and γ with $\alpha + \gamma =$ const describe one and the same orientation. Therefore, components situated close to this "plane" simulate in such a representation tube-like distributions.

Moreover, for special symmetric points of the boundary planes g_0 and several g_0^{jk} may be identical leading to a multiplicity $M_{g_0} > 1$. The M_{g_0} for the case of hexagonal-orthorhombic symmetry are given in Fig. 1.

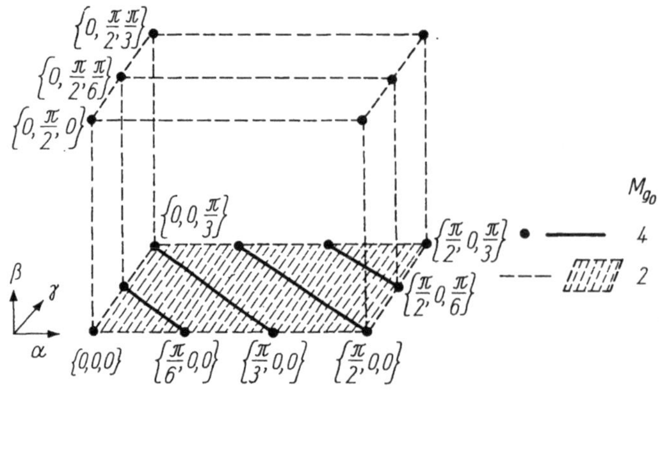

Fig. 1. The multiplicity M_{g_0} for the case of hexagonal-orthorhombic symmetry. Inside the elementary region $M_{g_0} = 1$ is valid

3. "Hit-Numbers" for Qualitative Ghost Correction

As was mentioned above only a reduced ODF $\tilde{f}(g) = f(g)/2 + f_G(g)$ [12] can unambiguously be reproduced from pole figures derived by normal diffraction experiments. The quantitative correction of the "ghost part" $f_G(g)$ of the reduced ODF in order to reconstruct the "true" ODF $f(g)$ is connected with relatively complicated calculations [7 to 9]. However, for a first rough interpretation of the reduced ODF reflecting the most intensive main texture components also, elements of a qualitative ghost correction can at least help to detect ghost peaks connected with certain true components and to estimate the true intensity of these components. The knowledge of the analytical properties of $f_G(g)$ is the basis for such activities [17].

For an orientation peak at g_0 the ghost function $f_G(g_0, g)$ reaches its maximum values at the so-called $\tilde{\omega} = \pi$ surface [16] containing all points of the G-space with the orientation distance $\tilde{\omega} = \tilde{\omega}(g_0, g) = \pi$ (see (7)) from g_0. The shape of this surface is a complicated one and depends on g_0 also. However, there are several characteristic "α- and γ-axes as well as β-tubes" that are in a simple relation to g_0 and facilitating the following analysis [16].

If there exists crystal and sample symmetry each peak at g_0^{jk} will possess its own $\tilde{\omega} = \pi$ surface intersecting with the surfaces of other equivalent positions. Now the "hit number" $H(g_0, g)$ is the number of $\tilde{\omega} = \pi$ surfaces belonging to g_0 and its equivalent orientations g_0^{jk} intersecting in the point g of the G-space. This number gives a good measure for the possible falsification of the true ODF $f(g)$ in the reduced ODF $\tilde{f}(g)$ [10, 11].

For the case of hexagonal-orthorhombic symmetry the $H(g_0, g)$ can be found in a relatively simple manner if g_0 is not situated at a boundary plane of the ER or at the $\gamma = \pi/6$ plane. For g_0 beyond these planes we have (cf. Fig. 1) $M_{g_0} = 1$ and the "effective hit number" $H_e(g_0, g) \equiv H(g_0, g)/M_{g_0}$ equals $H(g_0, g)$. These numbers are given in Fig. 2. The great hit numbers along the γ-axis are to emphasize.

The value of the relation $H_e(g_0, g = \gamma\text{-axis})/H_e(g_0, g_0) = 6/10$ is greater than for analogous relations in the case of cubic-orthorhombic symmetry [16] referring to the increasing role of the ghost phenomena for low crystal symmetries in regions distant from the peak position g_0.

If g_0 is a point of a boundary plane or of the $\gamma = \pi/6$ plane the situation is more complicated. In this case it is better to determine the hit numbers numerically as it will be explained in the next section. For example in Fig. 3 the effective hit numbers belonging to $g_0 = \{0, 0, 0\}$ are given. The mentioned distorted representation

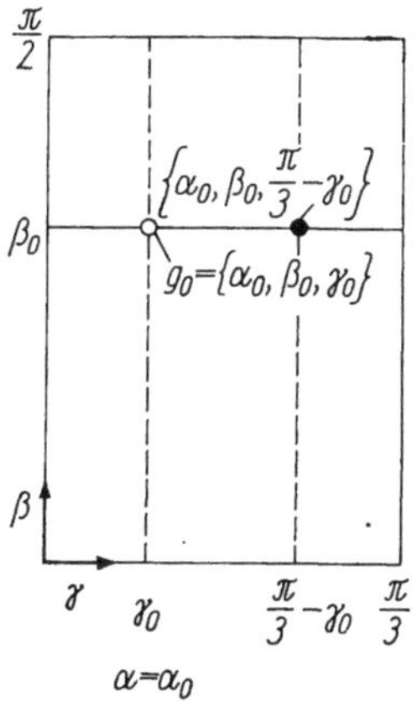

Fig. 2. The effective hit numbers $H_e(g_0, g) \geqq 2$ for g_0 beyond an boundary plane or the $\gamma_0 = \pi/6$ plane. The $\alpha = \alpha_0$ section of the ER is represented. ——— γ-axis, $H_e = 6$; – – – β-tube, $H_e = 2$; ● isolated ghost peak, $H_e = 7$; ○ $g = g_0$, $H_e = 10$. For the α-axis $g = \{\alpha, \beta_0, \gamma_0\}$, $g = \{\alpha, \beta_0, \pi/3 - \gamma_0\}$ and $\alpha \neq \alpha_0$, $H_e = 2$ is valid

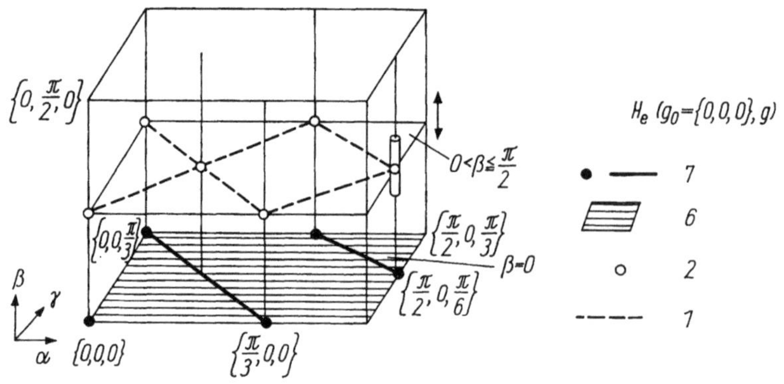

Fig. 3. The effective hit numbers $H_e(g_0, g)$ for $g_0 = \{0, 0, 0\}$, $M_{g_0} = 4$

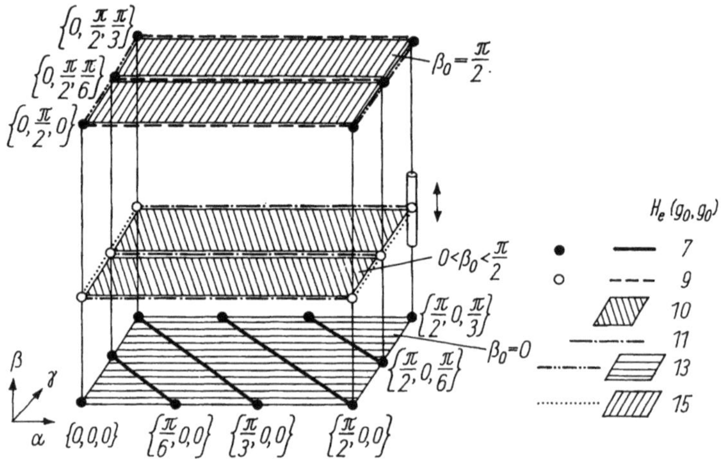

Fig. 4. The effective hit numbers for the peak position $H_e(g_0, g_0)$

of the G-space leads in this case to a "ghost layer" at $\beta = 0$. Such layers will also appear for components with low β_0 and sufficiently great FWHM.

The effective number of "ghost surfaces" intersecting in the peak position $g_0\ (H_e(g_0, g_0))$ is represented in Fig. 4 depending on g_0. For all points of the ER beyond the boundary planes and the $\gamma = \pi/6$ plane we find $H_e(g_0, g_0) = 10$.

4. Numerical Determination of the Hit Numbers

Although the simple geometrical approach of determining the effective hit numbers using the topology of the $\tilde{\omega} = \pi$ surface can be realized for all crystal symmetries (except the cases $G_B = O$ or T leading to ER with a complicated form) the problem needs very careful considerations. Moreover, in reality the texture components possess a finite width (characterized by the FWHM b). Therefore the ghost surfaces will also possess an effective width Δ_0 describing regions with a non-negligible ghost intensity around the geometrical $\tilde{\omega} = \pi$ surface. In order to estimate the intensity of ghost phenomena a hit is now to be understood in the sense that a ghost surface belonging to a peak at g_0 has an orientation distance to the point g lower than Δ_0. The hit numbers discussed above and derived by geometrical considerations correspond to the case $\Delta_0 = 0$.

Explicilty the orientation distance between g and g_0 is given by the relation $(\tilde{\omega} = \tilde{\omega}(g_0, g))$ [12]

$$\begin{aligned}\cos\tilde{\omega} = \tfrac{1}{2}\{&\cos(\alpha-\alpha_0)\cos(\gamma-\gamma_0)[1+\cos\beta\cos\beta_0] - \\ &-\sin(\alpha-\alpha_0)\sin(\gamma-\gamma_0)[\cos\beta+\cos\beta_0] + \\ &+[\cos(\alpha-\alpha_0)+\cos(\gamma-\gamma_0)]\sin\beta\sin\beta_0+\cos\beta\cos\beta_0-1\}\end{aligned} \tag{7}$$

and the hit condition reads

$$\cos\tilde{\omega} \leqq \cos\left(\pi - \frac{\Delta_0}{2}\right) = -\cos\left(\frac{\Delta_0}{2}\right). \tag{8}$$

For a given g all g_0^{jk} (see Table 1 or [12]) have to be examined yielding $H^{\Delta_0}(g_0, g)$.

If there are several (not equivalent) components we must of course investigate the contribution of each component independently.

5. Estimation of Ghost Amplitudes Using the Effective Hit Numbers

The formulae given below estimate only the ghost intensities in the reduced ODF $\tilde{f}(g)$ induced by the texture component at g_0 including all its equivalent positions at g_0^{jk}. If there are more than one nonequivalent texture components the contributions of the various components weighted by their percentage are to sum.

The starting point for the next is the assumption that the intensity distribution $f(b, g_0, g)$ of a texture component at g_0 decreases with increasing orientation distance $\tilde{\omega} = \tilde{\omega}(g_0, g)$ in a bell-shaped manner characterized by the FWHM b and that the ratio ε between the magnitude of f_G at the $\tilde{\omega} = \pi$ surface and of the peak maximum is known,

$$\varepsilon = \frac{f_G(b, g_0, g)|_{\tilde{\omega}=\pi}}{f(b, g_0, g)|_{\tilde{\omega}=0}} \equiv \frac{f_G(\tilde{\omega}=\pi)}{f(\tilde{\omega}=0)}. \tag{9}$$

For the so-called standard functions ε is exactly given by simple analytical expressions [7, 12, 13]. In case of Gauss-like distribution and $b \leqq 30°$ the corresponding asymptotic expressions can be used as a good approximation,

$$\varepsilon \approx \frac{1}{(8\pi S)^{1/2}} \tag{10}$$

with $f(0) \approx (8\pi S^3)^{1/2}$, $f_G(\pi) \approx S$, and $S \equiv \ln 2/(2\sin^2(b/4))$.

Considering the presence of symmetries also the peak intensity in the reduced ODF is (up to the common factor $1/(N_A N_B)$) given by $(\tilde{f} = f/2 + f_G)$

$$M_{g_0}\tilde{f}(b, g_0, g_0) \approx M_{g_0}\frac{f(\tilde{\omega}=0)}{2} + H^{\Delta_0}(g_0, g_0)\, f_G(\tilde{\omega}=\pi)\,. \tag{11}$$

At this the following contributions are neglected:

tails of equivalent components at $g_0^{jk} \neq g_0$,

ghost intensities from equivalent components with an orientation distance $\tilde{\omega} < \pi - \Delta_0/2$ to g_0, and

$f_G(b, g_0, g_0)$ in comparison to $f(b, g_0, g_0)$.

The latter is justifiable for $b \leqq 50°$ in a good approximation. The first neglect may be incorrect if a g_0^{jk} has an orientation distance $\tilde{\omega}$ to g_0 of the order of b.

Now, from (10) and (11) an estimation for the falsification of the peak maximum in the reduced ODF follows,

$$\frac{\tilde{f}(b, g_0, g_0)}{f(b, g_0, g_0)} \approx \frac{1}{2} + H_e^{\Delta_0}(g_0, g_0)\,\varepsilon\,. \tag{12}$$

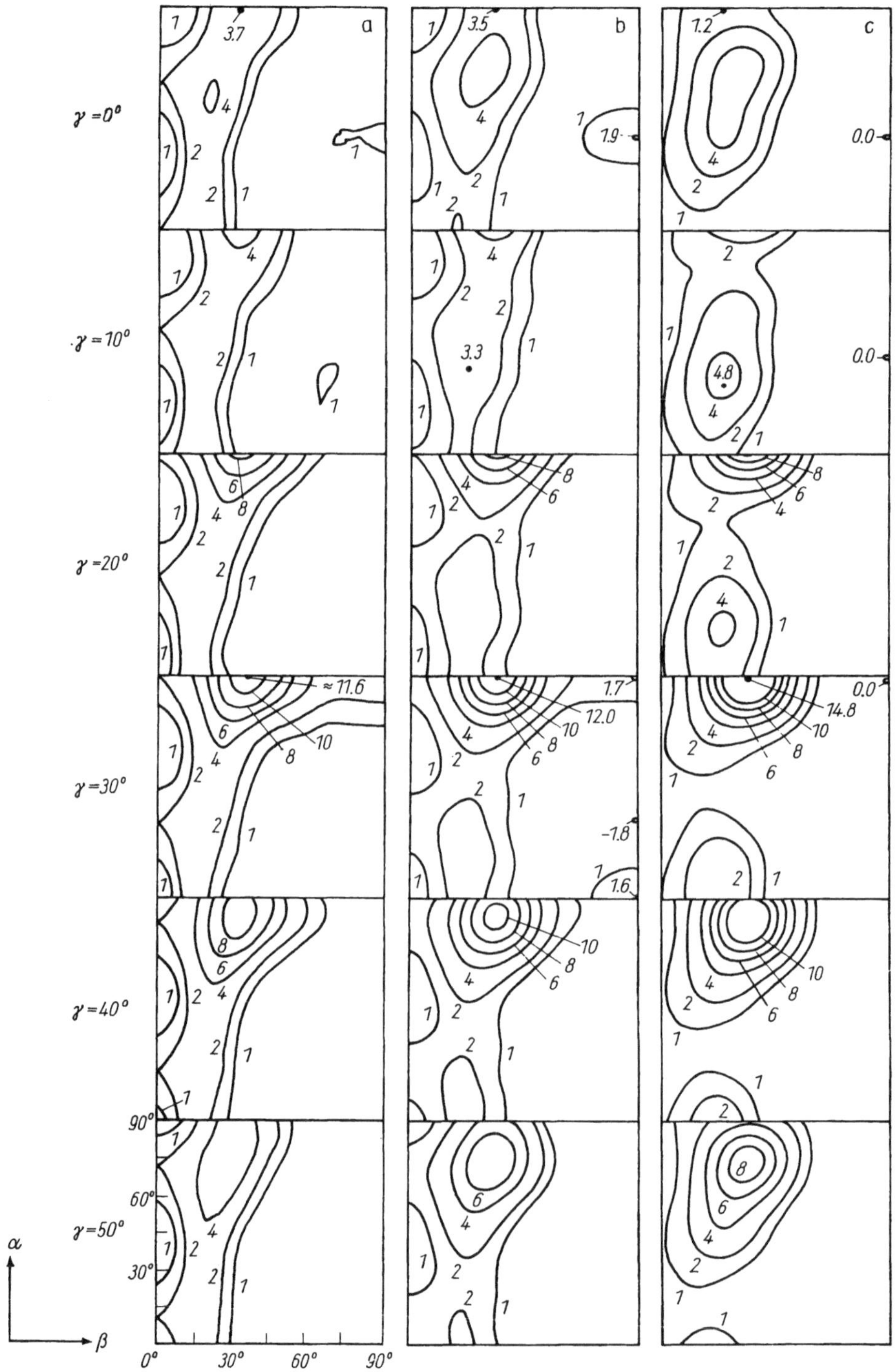

Fig. 5. a) The reduced ODF $\tilde{f}^{\text{exp}}$ for a 90% cold-rolled titanium sheet after Dervin et al. [3]. b) The reduced part of the model distribution represented in c) containing two components (see text)

For the magnitude of an "isolated ghost" at g follows (again up to common factors) approximately

$$M_{g_0}\tilde{f}(b, g_0, g) \approx H^{\Delta_0}(g_0, g)\, f_G(\tilde{\omega} = \pi)\,, \tag{13}$$

neglecting the direct $f(b, g_0^{jk}, g)$ contributions of tails of the equivalent components and the ghost contributions from components at g_0^{jk} with an orientation distance $\tilde{\omega} < \pi - \Delta_0/2$ to g.

Finally with (9), (11), and (13) we get the relation

$$\frac{\tilde{f}(b, g_0, g)}{\tilde{f}(b, g_0, g_0)} \approx \frac{H_e^{\Delta_0}(g_0, g)}{\dfrac{1}{2\varepsilon} + H_e^{\Delta_0}(g_0, g_0)} \tag{14}$$

allowing in a simple way to detect isolated ghosts in the reduced ODF belonging to a component at g_0 basing on the $\tilde{f}$-value for the peak position.

6. Analysis of an Experimental Example

In Fig. 5a the reduced ODF $\tilde{f}^{\exp}(g)$ of a rolled titanium sheet [3] is given. The ODF was reproduced using the harmonic method [1] considering the terms in the series of spherical functions up to $l = L = 16$.

Describing a peak in the G-space the termination errors will be relatively small if its FWHM b obeys the condition $b < 360°/L$ [7]. The order of b can be determined analysing the experimental pole figures. In our case (cf. Fig. 6a) it follows that this condition is obeyed.

For the analysis it is assumed that the main component at $g_0 = \{90°, 35°, 30°\}$ (cf. Fig. 5a) can approximately be described by a Gauss-like distribution with $b = 25°$. For simplicity the "geometrical" hit numbers ($\Delta_0 = 0$) given in Fig. 2 to 4 are used. Now, with $H_e(g_0, g_0) = 9$ (cf. Fig. 4), $\varepsilon \approx 1/27$ (cf. (10)) and $\tilde{f}^{\exp}(g_0, g_0) \approx 11.6$ (cf. Fig. 5a) from (12) follows $f(g_0, g_0) \approx 13.9$ as an estimation of the true ODF value $f(g)$ at $g = g_0$.

Considering the ghost-intensive γ-axis ($g = \{90°, 35°, 0 \leqq \gamma \leqq 60°\}$, $\gamma \neq 30°$) we have $H_e(g_0, g) = 6$. Therefore, from (14) a ghost intensity of the order of $\tilde{f} \approx 6 \times 11.6/(27/2 + 9) = 3.1$ is to be expected. Indeed, along this line $\tilde{f}^{\exp}$ has the minimum value of 3.6 at $g = \{90°, 35°, 0°\}$. For the β-tube ($g = \{90°, 0 \leqq \beta \leqq 90°, 30°\}$, $\beta \neq 35°$) we have $H_e(g_0, g) = 4$ leading to the estimation $\tilde{f} \approx 2.0$. This ghost line can also be seen in Fig. 5a (cf. the upper boundary of the $\gamma = 30°$ section). Analogically for the α-axis $g = \{0 \leqq \alpha < 90°, 35°, 30°\}$ with $H_e(g_0, g) = 2$ it follows $\tilde{f} \approx 1.0$. Because of the small intensity this ghost line is not pronounced in Fig. 5a.

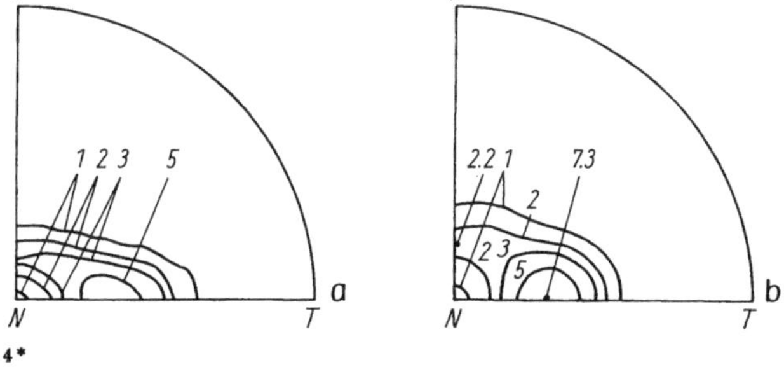

Fig. 6. a) The experimental {0002}-pole figure (Dervin et al. [3]). b) The {0002}-pole figur belonging to the model distribution given in Fig. 5c

The relatively high intensity of $\tilde{f}^{exp}$ along the plane $g = \{0 \leqq \alpha \leqq 90°, \beta \approx 20°, 0 \leqq \gamma \leqq 60.\}$ cannot be explained by ghost phenomena belonging to the g_0-component. For its interpretation it is necessary to take additional components into account. The {0002} pole figure gives us direct information on the α- and β-coordinates of pronounced components (cf. Fig. 6a). Introducing a second component it is already possible to construct a model ODF explaining $\tilde{f}^{exp}$ sufficiently well in the frame of a qualitative analysis, cf. Fig. 5a and 5b or 6a and 6b. This (free from ghost phenomena) model distribution of the two Gauss-like components,

$$g_1 = \{90°, 35°, 30°\}, \quad b_1 = 27.5°, \quad \text{percentage of intensity } 60\%,$$
$$g_2 = \{30°, 25°, 10°\}, \quad b_2 = 26°, \quad \text{percentage of intensity } 30\%,$$

is given in Fig. 5c.

7. Conclusion

As was shown for a first qualitative analysis of ODF's by terms of texture components it is necessary to take carefully into consideration the properties of symmetry in the G-space. Tails of equivalent peaks lying in neighbouring elementary regions may lead to remarkable intensities near the boundaries of the regarded ER and have not to be interpreted as independent components. The same concerns the special effects at $\beta = 0$ or π connected with the distorted representation of the G-space between orthogonal coordinate axes of Eulerian angles.

Analysing reduced ODF's a first qualitative ghost correction can be realized in a relatively simple manner using the hit numbers and avoiding numerically expensive ghost correction procedures. Although in the paper the case of hexagonal crystal symmetry was analysed in more detail, all possible crystal symmetries and the sample symmetries D_2, C_2, C_1 can be considered in the same way. Table 1, the analogic table for $G_B = 0$ in [12] and the working formulae (7) to (14) contain all necessary information for such activities.

References

[1] H. J. Bunge, Texture Analysis in Materials Science, Butterworths, London 1982.
[2] Proc. VI. Internat. Conf. Textures of Materials (ICOTOM-6), Tokyo 1981, Ed. by the Iron and Steel Inst. of Japan 1982.
[3] P. Dervin, J. P. Mardon, M. Pernot, R. Penelle, and P. Lacombe, J. less-common Metals, **55**, 25 (1977).
[4] K. Walther, Dissertation, TU Dresden, 1983.
[5] P. R. Morris and A. J. Heckler, Trans. AIME **245**, 1877 (1969).
[6] H. R. Wenk (Ed.), Preferred Orientation in Deformed Materials, Academic Press, San Diego, in the press.
[7] S. Matthies, Aktuelle Probleme der quantitativen Texturanalyse, ZfK-Rossendorf, Institutspublikation 480 (1982), ISSN 0138-2950.
[8] S. Matthies and G. W. Vinel, phys. stat. sol. (b) **112**, K111, K115 (1982).
[9] P. van Houtte, Textures and Microstructures **6**, 1 (1983).
[10] J. Jura, K. Lücke, and J. Pospiech, Z. Metallk. **71**, 714 (1980).
[11] S. Matthies, Crystal Research and Technol. **16**, 513 (1981).
[12] S. Matthies and F. Wagner, phys. stat. sol. (b) **107**, 591 (1981).
[13] S. Matthies, phys. stat. sol. (b) **112**, 705 (1982).
[14] S. Matthies and K. Helming, phys. stat. sol. (b) **113**, 569 (1982).
[15] H. J. Bunge, C. Esling, and J. Muller, J. appl. Cryst. **13**, 544 (1981).
[16] S. Matthies, Crystal Research and Technol. **15**, 601 (1980).
[17] S. Matthies and J. Pospiech, phys. stat. sol. (b) **97**, 547 (1980).

(Received June 28, 1984)

phys. stat. sol. (b) **126**, 53 (1984)

Subject classification: 1; 20; 21.1

Department of Physics, Nagpur University[1])

XANES and EXAFS Study of the Compound CoAsSe

By

S. CHATTOPADHYAY and C. MANDE

The X-ray absorption near edge fine structure (XANES) and the extended X-ray absorption fine structure (EXAFS) associated with the CoK absorption discontinuity are investigated in the pure metal, in the compound CoAsSe and in two other similar trivalent compounds. The analysis of the results obtained enables to obtain the average bond length (Co-ligand distance) in CoAsSe for which the value is not known from crystallographic work.

Die Röntgenabsorption in der Nähe der Kantenfeinstruktur (XANES) und die ausgebreitete Röntgenabsorptionsfeinstruktur (EXAFS), die mit der CoK-Absorptionsdiskontinuität im reinen Metall verknüpft ist, werden in der Verbindung CoAsSe und in zwei anderen ähnlichen trivalenten Verbindungen untersucht. Die Analyse der erhaltenen Ergebnisse erlaubt die Bestimmung der mittleren Bindungslänge (Co-Ligandenabstand) für CoAsSe, für das der Wert aus kristallographischen Arbeiten nicht bekannt ist.

1. Introduction

It is well known that X-ray absorption discontinuities in solids are usually accompanied by fine structure on their higher-energy side [1 to 5]. The fine structure which extends from about 10 to 40 eV beyond the threshold, has been named XANES (X-ray absorption near edge fine structure) and the fine structure lying beyond it extending up to several hundred and sometimes even up to a few thousand electron volts is named as EXAFS (extended X-ray absorption fine structure). It has been shown that both XANES and EXAFS can give valuable information about the local structure in all types of materials. Applying both these techniques we have studied in the present investigation the average bond length (Co-ligand distance) in the compound CoAsSe for which crystallographic results are not available.

2. Experimental

The X-ray K absorption spectra of cobalt in the pure metal and in the compounds $CoAs_3$, CoAsS, and CoAsSe were photographed using a 0.4 m Cauchois-type bent crystal spectrograph, which was equipped with a well tested mica crystal, oriented to reflect from its (100) planes. A Chirana X-ray tube with tungsten target was used as the source of white radiation. The CuK_{α_1} and NiK_{α_1} lines, which appear as impurity lines and bracket the K discontinuity of cobalt, were used as reference lines. Several microphotometer traces taken at different heights of each spectrum were recorded on a Carl Zeiss MD 100 microdensitometer with magnification ×40. The main K discontinuity was found to be clearly split up into two components K_1 and K_2 in the case of cobalt metal. In the case of the compounds the splitting of the discontinuity was not so good. The wavelength measurements of the maxima and minima in the fine structure were made with reference to the inflection point on the K_1 discontinuity, as per usual practice.

---

[1]) Amravati Road, Nagpur 440010, India.

Cobalt metal of A.R. grade was obtained from British Drug House (BDH) Private Ltd., Bombay. The trivalent cobalt compounds $CoAs_3$, CoAsS, and CoAsSe were prepared in our laboratory using the usual vacuum sealing and sintering technique [6]. The formation of these compounds was checked by taking their X-ray diffraction patterns. Thin films of thickness about 20 μm of cobalt metal were prepared by the vacuum coating technique. The absorbers of the compounds were prepared by spreading uniformly the finely powdered materials between two cellophane tapes.

3. Results and Discussion

The shapes, XANES, and EXAFS of the cobalt K absorption discontinuity in the metal and the compounds, obtained by averaging the microphotometer traces, are shown in Fig. 1. The curves in these figures represent the transmitted intensity and therefore the maxima and minima in these curves correspond to the minima and maxima, respectively, in absorption coefficient. The energies of the fine structure peaks as measured from the K_1 discontinuity are given in Table 1.

According to Levy [7] the absorption minimum at A (maximum in absorption coefficient) corresponds to a transition from the 1s state to a vacant p level and the minimum at B to a kind of ionization or 'escape' energy of the 1s electron from the molecule. Further, the distance ΔE between the minimum at B and the subsequent maximum at β gives a measure of the radius of the first coordination sphere around the central metal atom through the Bragg relation

$$r_1 = \sqrt{\frac{151}{\Delta E}}\,. \tag{1}$$

It can be supposed that r_1 gives approximately the average bond length, i.e. the average distance between the central atom and its nearest ligands.

Mande and Chetal [8] have determined the bond lengths in some cobalt complexes from the fine structure of the K discontinuity, using Levy's method. Kumar et al. [9] and more recently Chattopadhyay and Mande [10] have used this method for the determination of bond lengths in some other cobalt complexes. It has recently been shown [11] that Levy's method gives fairly good results on average bond lengths.

In Table 1 we have given the values of the average bond lengths for the systems studied in this work. We note that the bond lengths obtained by us compare very

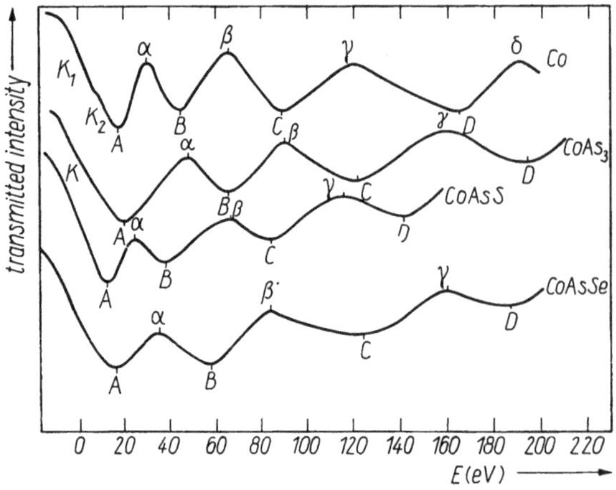

Fig. 1. The XANES and EXAFS associated with the CoK discontinuity for cobalt metal, $CoAs_3$, CoAsS, and CoAsSe

Table 1
Average bond lengths obtained by Levy's method

absorber	$E = (\beta - B)$ (eV)	r_1 (10^{-10} m)	r_1 (10^{-10} m) from X-ray*) diff. data
Co metal	24.52	2.48	2.50
$CoAs_3$	27.67	2.34	2.35
CoAsS	28.20	2.31	2.33
CoAsSe	26.94	2.37	—

*) Wykoff [14].

well with the values obtained from X-ray diffraction data for cobalt metal, $CoAs_3$, and CoAsS.

To interpret the EXAFS Lytle et al. [12] have proposed an elaborate theory which is based on the point scattering model. The physical origin of EXAFS is ascribed to the final state interference between the outgoing photoelectron wave and the back-scattered wave, i.e. the modification of the final states by the presence of the surrounding atoms. From such considerations Lytle et al. have given an expression for the EXAFS function $\chi(k)$, in terms of the coordinating parameters of the absorbing atom, whose periodicity is given by

$$\text{Sin}\,[2kr_j + 2\delta_j(k)]\,, \tag{2}$$

where k is the electron wave vector, r_j the average radial distance from the absorbing atom to the j-th atom, and $\delta_j(k)$ the phase shift. Assuming δ_j to be linear in k, we may write

$$\delta_j = -\alpha_j k + \beta_j\,, \tag{3}$$

where α_j and β_j are constants. Considering the scattering from the first coordination shell only and defining an integer n for the number of maxima and minima we get

$$(\tfrac{1}{2} + n)\,\pi = 2k(r_1 - \alpha_1) + 2\beta_1\,, \tag{4}$$

where $n = 0, 2, 4$ for the maxima and 1, 3, 5 for the minima, and r_1 is the radius of the first coordination shell. The value of k is obtained through the equation

$$k = (0.263E)^{1/2}\,, \tag{5}$$

where E is the energy of the peaks in the EXAFS. By plotting a graph between n and the experimental values of k for the maxima and minima in the EXAFS, the value of $(r_1 - \alpha_1)$ in (4) can be obtained from the slope of the n, k curve. Once the value of α_1 is determined for a standard material of known r_1, one can use this value for determining the bond lengths in chemically similar compounds. We have determined the value of r_1 for CoAsSe by taking the value of α_1 for CoAsS and $CoAs_3$, which are chemically similar to CoAsSe. In Table 2 are given the experimentally determined values of E and k for the different peaks in the EXAFS. Making use of these values the n, k curves were plotted. These curves were found to be straight lines as expected from (4). The bond lengths determined from the slopes of these curves are given in Table 3. The average value of α_1 for the compounds $CoAs_3$ and CoAsS comes out to be 0.34×10^{-10} m which gives $r_1 = 2.38 \times 10^{-10}$ m for the compound CoAsSe. It is

Table 2

The values of E and k for the peaks in EXAFS

max./min.	n	Co metal		$CoAs_3$		CoAsS		CoAsSe	
		E (eV)	k (10^{10} m^{-1})	E (eV)	k (10^{10} m^{-1})	E (eV)	k (10^{10} m^{-1})	E (eV)	k (10^{10} m^{-1})
min. α	1	29.36	2.78	47.92	3.55	28.38	2.48	35.37	3.05
max. B	2	40.39	3.26	60.67	3.99	37.92	3.16	57.05	3.87
min. β	3	64.91	4.13	88.34	4.82	66.12	4.17	83.99	4.70
max. C	4	88.78	4.83	121.38	5.65	80.46	4.60	132.36	5.90
min. γ	5	121.85	5.66	150.91	6.30	115.02	5.50	155.74	6.40
max. D	6	165.86	6.60	194.38	7.15	141.48	6.10	187.80	7.35
min. δ	7	191.11	7.09	—	—	—	—	—	—

Table 3
Average bond lengths obtained from EXAFS study

absorber	slope	$r_1 - \alpha_1$	α_1	r_1 (10^{-10} m)	r_1 (10^{-10} m) from cryst. data
Co metal	1.27	1.99	0.51	—	2.50
$CoAs_3$	1.28	2.01	0.34	—	2.35
CoAsS	1.27	1.99	0.33	—	2.33
CoAsSe	1.30	2.04	—	2.38	—

satisfying to note that our results for the bond length in CoAsSe as determined from XANES and EXAFS studies are very close to each other.

We have calculated the values of the effective charge q in the compounds CoAsS and CoAsSe, using Suchet's method [13]. While the value of q for the former compound comes out to be 0.69 electrons/atom, we find that $q = 0.49$ electrons/atom for CoAsSe. The somewhat larger value of r_1 in this compound as compared to that in CoAsS can be attributed to the decrease in the electrostatic force, which keeps the ligands bound to the central cobalt atom.

References

[1] Y. Cauchois, Les Spectres de Rayon X et a Structure Electronique de la Matière, Gauthier-Villars, Paris 1948.
[2] C. Mande, Presidential Address Indian Sci. Congr. Physics Section, Calcutta 1979.
[3] C. Bonnelle and C. Mande, Advances in X-ray Spectroscopy, Pergamon Press, 1982.
[4] A. Bianconi, L. Incoecia, and S. Stipeich, EXAFS and Near Edge Structure, Springer-Verlag, 1983.
[5] B. K. Agarwal, X-Ray Spectroscopy, Springer-Verlag, 1981.
[6] W. B. Pearson, Handbook of Lattice Spacings and Structures of Metals and Alloys, Pergamon Press, 1958.
[7] R. M. Levy, J. chem. Phys. **43**, 1846 (1965).
[8] C. Mande and A. R. Chetal, Curr. Sci. **35**, 563 (1966).
[9] A. Kumar, A. N. Nigam, and B. D. Srivastava, X-Ray Spectrometry **10**, 25 (1981).
[10] S. Chattopadhyay and C. Mande, X-Ray Spectrometry, in the press, 1984.
[11] M. S. Islam and C. Mande, phys. stat. sol. (a) **81**, 197 (1984).
[12] F. W. Lytle, D. E. Sayers, and E. A. Stern, Phys. Rev. B **11**, 4825 (1975).
[13] J. P. Suchet and F. Baily, Ann. Chim. **10**, 517 (1965).
[14] R. W. G. Wykoff, Crystal Structures, Interscience Publishers, New York 1948.

(Received June 5, 1984)

phys. stat. sol. (b) **126**, 59 (1984)

Subject classification: 2; 8; 22.5.2

Department of Physics (a) and Department of Chemistry (b), Himachal Pradesh University, Simla[1])

A Disorder Model of Melting

II. Alkali Halides[2])

By

B. A. VAID (a), K. C. SHARMA (a), and V. K. SYAL (b)

The disorder model of melting recently proposed, is generalized to include alkali halides. The thermodynamic 'defects' are identified with the (ion-)vacancy pairs existing in molten halides. The agreement of the calculated values of thermodynamic properties viz. thermal expansion coefficient, heat capacity, and compressibility, with the experimental data is found to be quite reasonable.

Das kürzlich vorgeschlagene Fehlordnungsmodell des Schmelzvorgangs wird verallgemeinert und schließt nun Alkalihalogenide ein. Die thermodynamischen Defekte werden als (Ionen)Lehrstellenpaare identifiziert, die in geschmolzenen Halogeniden vorhanden sind. Die Übereinstimmung der berechneten thermodynamischen Eigenschaften, z. B. thermischer Ausdehnungskoeffizient, Wärmekapazität und Kompressibilität, mit experimentellen Werten ist zufriedenstellend.

1. Introduction

In a recent paper [1], referred to as I, the authors have proposed a disorder model of melting for monatomic metallic melts of both types, i.e. those expanding on melting and those expanding on freezing. The model is based on the spontaneous production of thermodynamic 'defects' at the melting point assuming the crystalline phase devoid of 'defects'. The 'defect' in the melt is mainly characterized by the activation energy and activation free volume and must not be interpreted in the sense of geometrically well-defined structures. The volume fraction of 'defects' produced in melts, contrary to what is known from earlier versions of disorder model [2], is not a constant but varies with the structure which is manifested through the radial distribution function or the coordination number. In addition to the basic concept of packing in the crystalline phase, the concept of 'packing' in melts is grossly approximated in terms of the coordination numbers. Further in this model, the lattice and 'defect' dependent shares were assumed to contribute additively to the physical properties of the melt, without any interaction between them. In the framework of the quasi-crystalline picture, the model explains satisfactorily the thermodynamic properties of these systems near the melting point. We intend to apply this model to alkali halides.

There have also been other theories [2] of melting for alkali halides. The dislocation theory of melting [3] has been the most plausible one but there are difficulties in achieving exact results in terms of interatomic forces. A molecular dynamic study [4] of melting of alkali halide crystals has been found to be consistent with the dislocation theory of melting. Recently, an elasticity theory of melting [5] has been proposed in tems of second- and third-order elasticity which includes the fundamental entropy change on melting, arising from the localization of the two sets of shear modes. The

[1]) Summer Hill, Simla 171005, India.

[2]) Part I see phys. stat. sol. (b) **119**, 703 (1983).

cause of melting was the existence of an elastic instability. In this model the liquid state was assumed to be described by a set of local strains which vary randomly both spatially and temporally with well-defined finite mean square values, consequently the contribution to entropy and volume change entered only through third-order elasticity. This model could not yield more information about the thermodynamic properties.

In this paper, we will exploit the disorder model of melting for alkali halides. The basic concepts for the system of interest will be developed in the next section. Basic thermodynamic expressions will be presented in Section 3. Results and a discussion will be given in the last section.

2. Basic Concepts

The main basic concept of the disorder model has been the concept of different packing in different crystalline structures and in the melt. On melting, the thermodynamic 'defects' are spontaneously produced effecting the packing of the system, which is clearly manifested through the change in coordination number. In alkali halides, we make the following observations:

(i) In crystalline phase of LiCl and NaCl, the size of one type of ions is much smaller than the other. The packing of such a system is to be determined by the lattice structure on which bigger ions are located, which turns out to be that with coordination number 12. Further, in molten states, the nearest neighbour coordination number is 5. It implies that on melting, halides with NaCl structure collapse to one, somewhere between the simple cubic and the tetrahedral structure. Having identified this, the volume fraction of 'defects' may be obtained from Table 1 of I. Considering the averages of the two, f comes out to be about 0.8. For the sake of simplicity, we will assume that all halides with NaCl structure behave similarly as discussed above.

(ii) In halides with CsCl structure, the situation is different. Both types of ions are almost of the same size. The nearest-neighbour coordination number in solid phase is 8 corresponding to the b.c.c. structure. In molten state, the coordination number from the experiments comes out to be 6. However, for the sake of consistency with the assumption made for halides with NaCl structure, we assume that the molten state has a structure somewhere between the simple cubic and tetrahedral one. As a consequence, for molten CsCl, f turns out to be 0.65.

(iii) As far as the defects in molten halides are concerned, it is well known [2] from energy considerations keeping electrical neutrality on local scale, that (ion-)vacancy pairs are the most common defects present in these systems. Therefore, we assume the spontaneous production of this particular type of defects, i.e. vacancy pairs, in the molten state.

3. Thermodynamic Properties

Having assumed the presence of (ion) pair defects in molten halides, the concentration of defects in thermal equilibrium, parallel to I, is given by

$$\varkappa = \frac{n_{\mathrm{def}}}{N} = \frac{1}{1 + \exp\,(E/2kT)}, \tag{1}$$

where n_{def} is the number of pair defects and N the total number of ions on the lattice sites. E is the activation energy needed to creat a (ion) pair defect, which may be expressed as [1]

$$E = 2kT_F \ln\left(\frac{f}{\Delta V/V} - 1\right). \tag{2}$$

The defect contribution to the coefficient of thermal expansion $\Delta\alpha$ and heat capacity ΔC_p may be expressed as

$$\Delta\alpha = \frac{f}{T_F} \frac{\frac{E}{2kT_F} \exp\left(\frac{E}{2kT_F}\right)}{\left[1 + \exp\left(\frac{E}{2kT_F}\right)\right]^2}, \tag{3}$$

$$\Delta C_p = \frac{2R\left(\frac{E}{2kT_F}\right)^2 \exp\left(\frac{E}{2kT_F}\right)}{\left[1 + \exp\left(\frac{E}{2kT_F}\right)\right]^2}. \tag{4}$$

The total compressibility may be expressed as in [1].
Assuming f independent of pressure, it yields the defect contribution to compressibility

$$\Delta\chi \equiv \chi^{\mathrm{fl}}_{\mathrm{def}}(T_F) = f \frac{\varkappa^2}{2kT_F} v_{\mathrm{def}} \exp\left[\frac{E}{2kT_F}\right] + \chi^{\mathrm{cr}} f\varkappa\,. \tag{5}$$

In order to have an estimate of v_{def}, we assume, as has been done by Eckstein [6], that the melting process in one and the same substance under variable pressure always starts at the same 'critical' concentration $\varkappa' = \varkappa^{\mathrm{cr}}(T_F)$ of 'defects'. Correspondingly (1) yields

$$T_F \approx -\frac{E^{\mathrm{cr}}(p)}{2k \ln \varkappa'}, \tag{6}$$

which implies

$$\frac{\mathrm{d}T_F}{\mathrm{d}p} = -\frac{v_{\mathrm{def}}}{2k \ln \varkappa'} = -\frac{f v_b}{2k \ln \varkappa'}. \tag{7}$$

We do not have an idea of $\varkappa'$, in contrast to what we have for metallic systems [6]. However, if we recall the Clausius-Clapeyron equation for the change in entropy, i.e.

$$\Delta S = \frac{\mathrm{d}p}{\mathrm{d}T_F} \Delta V = -2 \frac{\Delta V}{V} \frac{R}{f} \ln \varkappa'\,. \tag{8}$$

In this equation $\varkappa'$ appears. At this stage eliminating $\varkappa'$ from (7) and (8),

$$\Delta\chi = \frac{\Delta S f^2 \exp\left(\frac{E}{2kT_F}\right)}{2RT_F \left[1 + \exp\left(\frac{E}{2kT_F}\right)\right]^2 \frac{\Delta V}{V}} \frac{\mathrm{d}T_F}{\mathrm{d}p} + \chi^{\mathrm{cr}} f\varkappa\,. \tag{9}$$

Taking the experimental values of ΔS and $\mathrm{d}T_F/\mathrm{d}p$, the above expression yields the change in compressibility on melting.

4. Results and Discussion

For molten halides, the 'defect' contributions to various thermodynamic properties $\Delta\alpha$, ΔC_p, and $\Delta\chi$ have been calculated and results are displayed in Table 1. Various experimental values have been taken from the Molten Salt Handbook [7]. Some of the

Table 1

	LiCl	NaCl	NaBr	NaI	KCl	KBr	CsCl
$\Delta V/V$ (%)	26.2	25.0	22.4	18.6	17.3	16.6	10.00
T_F (K)	883	1073	1020	933	1043	1007	918
f	0.8	0.8	0.8	0.8	0.8	0.8	0.65
$E/2kT_F$	0.719	0.788	0.944	1.194	1.287	1.340	1.704
$\exp(E/2kT_F)$	2.05	2.20	2.57	3.30	3.62	3.81	5.5
$\Delta\alpha(T_F)$ (10^{-4} K^{-1})	1.43	1.26	1.5	1.82	1.67	1.75	1.57
$\alpha^{cr}(T_F)$ (10^{-4} K^{-1})	2.55	2.49	2.37	2.4	1.83	2.66	2.0
$\alpha^{th}(T_F)$ (10^{-4} K^{-1})	3.98	3.75	3.87	4.22	3.50	4.41	3.57
$\alpha^{exp}(T_F)$ (10^{-4} K^{-1})	2.87	3.49	3.5	3.48	3.84	3.91	3.84
$\Delta C_p(T_F)$ (cal/K)	0.45	0.53	0.71	—	0.70	1.17	1.50
$C_p^{cr}(T_F)$ (cal/K)	14.8	15.2	14.8	—	15.4	15.0	15.6
$C_p^{th}(T_F)$ (cal/K)	15.25	15.73	15.51	—	16.10	16.17	17.1
$C_p^{exp}(T_F)$ (cal/K)	15.0	16.0	15.3	—	17.6	16.4	17.2
dT_F/dp (K/atm)	0.0242	0.0238	0.0287	0.0327	0.0265	0.038	—
ΔS (cal/mol)	5.39	6.3	6.1	6.04	6.0	6.1	—
$\Delta\chi(T_F)$ (10^{-6} cm^2/N)	20.28	19.84	25.22	33.58	24.7	37.22	—
χ^{cr}(NTP) (10^{-7} cm^2/N)	3.35	4.17	5.0	6.55	5.8	6.8	—
$\chi^{th}(T_F)$ (10^{-7} cm^2/N)	23.63	24.01	30.22	40.06	30.50	44.03	—
$\chi^{exp}(T_F)$ (10^{-7} cm^2/N)	27.5	28.7	31.4	37.3	35.6	39.1	—

values, in particular those of compressibility for crystalline phase, are available only at normal temperatures. Within the approximations of this paper, the agreement of the calculated values is quite reasonable with the experimental data in spite of the assumption of the presence of only (ion) pair vacancies in melts. Nevertheless, we do not distinguish between the types of thermodynamic defects, i.e. vacancies or dislocations. Since the derivative of any property is supposed to be one of the better tests of the validity of any theory of melts, the agreement of the results within a range 4 to 16% on compressibility with the experiment gives us more confidence in this model for molten halides.

More recently, Ishii et al. [8] and Shinomoto [9] have proposed different approaches in terms of interatomic forces. We hope to relate the concepts presented in this paper with these more recent approaches in a separate paper.

References

[1] B. A. Vaid, K. C. Sharma, and V. K. Syal, phys. stat. sol. (b) **119**, 703 (1983).
[2] A. R. Ubbelode, The Molten State of Matter, Wiley, New York 1978.
[3] D. Kuhlman-Wilsdorf, Phys. Rev. **140**, A1599 (1965).
[4] M. Amini, D. Finchamo, and R. W. Hockney, J. Phys. C **13**, L221 (1980); J. Phys. C **12**, 4707 (1979).
[5] J. L. Tallon and W. H. Robinson, Phys. Letters A **87**, 365 (1982).
[6] B. Eckstein, phys. stat. sol. **20**, 83 (1967).
[7] J. Janz, Molten Salt Handbook, Academic Press, Inc., London/New York 1967.
[8] T. Ishii et al., Progr. theor. Phys. (Kyoto) **69**, 1074 (1983).
[9] S. Shinomoto, Progr. theor. Phys. (Kyoto) **70**, 687 (1983).

(Received June 19, 1984)

phys. stat. sol. (b) **126**, 63 (1984)

Subject classification: 8 and 11; 6; 22; 22.2.1

Instituto de Física "Gleb Wataghin", Universidade Estadual de Campinas[1])

On Laser-Induced Lattice Heating in a Polar Semiconductor

By

A. R. VASCONCELLOS and R. LUZZI

Nonlinear transport phenomena and relaxation processes in highly photoexcited polar direct-gap semiconductors are considered. The rates of energy transfer between subsystems are calculated and it is shown that for ultra-short laser pulses lattice heating is unimportant. However, for cw laser-light illumination a pronounced enhancement of the lattice temperature is expected when laser power approaches, or surpasses, a critical value, leading the system to melting conditions.

Es werden nichtlineare Transportphänomene und Relaxationsprozesse in optisch hoch-angeregten, polaren, direkten Halbleitern untersucht. Die Geschwindigkeiten des Energietransfer zwischen Subsystemen werden berechnet und es wird gezeigt, daß für ultrakurze Laserimpulse Gitteraufheizung unwesentlich ist. Jedoch wird für Belichtung mit kontinuierlichem Laserlicht eine beträchtliche Erhöhung der Gittertemperatur erwartet, wenn die Laserleistung einen kritischen Wert erreicht oder übersteigt und das System zu Schmelzbedingungen geführt wird.

1. Introduction

Currently a good deal of efforts are devoted to the development of improved quantum generators and other devices based on semiconductors at high level of excitation. There is also a growing interest in the question of laser-induced damage and laser-induced annealing in materials [1]. These facts have made the study of non-equilibrium thermodynamics of highly excited plasmas in semiconductors (HEPS) an area of intense research. Recently we have discussed the kinetics of relaxation processes in HEPS using a non-equilibrium statistical method that allows for the derivation of nonlinear transport equations which describe the evolution of the irreversible phenomena that take place in the system [2]. Ultrafast transient nonlinear transport phenomena and relaxation processes in physical, chemical, and biological systems are presently the object of extensive studies. Advances in ultrafast laser spectroscopy allow for their analysis in the picosecond and even femtosecond time scale [3]. These studies pose new and challenging problems, and these systems become an interesting testing ground for theoretical ideas in non-equilibrium statistical mechanics and thermodynamics of many-body systems.

In this article we complement the studies referred above [2] addressing the question of laser-induced lattice heating of acoustic (A) phonons. We determine the steady-state effective temperature of A phonons, which is dependent on laser intensity and energy relaxation times. We show that there exists a critical condition at a certain laser power threshold at which the effective lattice temperature undergoes a pronounced enhancement. Moreover, the time scales for energy relaxation are also of great relevance in this type of problems, and particularly so too is the question of to what extent and on what time scales the laser-pumped energy is transferred to the lattice. We consider this question through the derivation of the evolution equations for the relevant thermodynamic variables. Our results reinforce a previous conjecture [2] that in polar direct-

[1]) C.P. 6165, 13100 Campinas, Brazil.

gap semiconductors application of short laser pulses of energies of a few J/cm^3 produce weak to moderate lattice heating. To obtain a large increase of the lattice effective temperature it is necessary to use long pulses with energies of the order of tens of J/cm^3.

2. Energy Relaxation Times

Consider a semiconductor illuminated by intense laser radiation with photon energy $\hbar\omega_L$, and kept in thermal contact with a heat reservoir at temperature T_0. Electron–hole pairs are created in processes of absorption of one or two photons ($\hbar\omega_L > E_G$ or $\hbar\omega_L < E_G$ but $2\hbar\omega_L > E_G$ respectively, where E_G is the energy gap). The excess energy optically pumped into the electron system is mainly transferred to the lattice via relaxation processes mediated by deformation potential and Fröhlich interactions in polar semiconductors. The number of pairs varies in recombination processes, and anharmonic effects redistribute the excess energy among the different phonon modes. In intermediate to strong polar semiconductors mutual thermalization of carriers and LO phonons sets in very rapidly, in the picosecond time scale, and it is followed by relaxation to the A phonon system via the anharmonic interaction in, typically, a nanosecond time scale (Aligarte and Luzzi [2]).

Nowadays there exist several theoretical approaches to the question of the dynamics of relaxation phenomena in far from equilibrium many-body systems [4]. The non-equilibrium statistical operator method (NSO) is considered a far reaching generalization of the Chapman-Enskog method in the kinetic theory of gases, and has an appealing structure [5, 6]. We have used extensively Zubarev's method, adapted for the treatment of open systems [7], to discuss ultrafast optical responses and ultrafast transient transport in highly photoexcited semiconductors [2]. The advantages of the method reside in the possibility to describe a large class of experimental situations where the system is arbitrarily away from equilibrium, and where nonlinear, non-local, and memory effects are incorporated from the onset [5]. According to the NSO, for the statistical mechanics description of the highly excited plasma in semiconductors a proper basis set of thermodynamic variables (macrovariables) includes the time-dependent energies of (i) carriers, $E_c(t)$, (ii) longitudinal optical (LO) phonons, $E_{LO}(t)$, (iii) transverse optical (TO) phonons, $E_{TO}(t)$, (iv) acoustic (A) phonons, $E_A(t)$, and (v) the carrier concentration, $n(t)$. The (intensive) non-equilibrium thermodynamic parameters conjugated to these macrovariables, in the sense of the NSO method [5, 6], are the reciprocal effective temperatures of (i') carriers, $\beta_c(t) = 1/k_B T_c^*(t)$, (ii') LO phonons, $\beta_{LO}(t) = 1/k_B T_{LO}^*(t)$, (iii') TO phonons, $\beta_{TO}(t) = 1/k_B T_{TO}^*(t)$, and (iv') A phonons, $\beta_A(t) = 1/k_B T_A^*(t)$, and (v') the quasi-chemical potentials of electrons, $-\beta_c(t)\,\mu_e(t)$, and holes, $-\beta_c(t)\,\mu_h(t)$ [2].

This is an open system in interaction with laser radiation, characterized by photon frequency ω_L, power flux I_L, and pulse duration t_p, and in contact with a thermostat at temperature T_0. It is assumed that the laser and thermostat are ideal reservoirs, i.e. they remain in near steady-state conditions sustaining negligible modifications as a consequence of the coupling with the much smaller system composed by the semiconductor sample.

The non-equilibrium thermodynamic evolution of the HEPS is described by the rate equations for the basis set of macrovariables (i) to (v). Taking into account the different pumping and relaxation channels involved in the experiment, these generalized transport equations (GTE) are

$$\frac{dE_c(t)}{dt} = \left.\frac{dE_c}{dt}\right|_L + \left.\frac{dE_c}{dt}\right|_R + \left.\frac{dE_c}{dt}\right|_{c,A} + \left.\frac{dE_c}{dt}\right|_{c,TO} + \left.\frac{dE_c}{dt}\right|_{c,LO}, \tag{1a}$$

$$\frac{dE_{LO}(t)}{dt} = -\left.\frac{dE_c}{dt}\right|_{c,LO} + \left.\frac{dE_{LO}}{dt}\right|_{LO,A}, \tag{1b}$$

$$\frac{dE_{TO}(t)}{dt} = -\left.\frac{dE_c}{dt}\right|_{c,TO} + \left.\frac{dE_{TO}}{dt}\right|_{TO,A}, \tag{1c}$$

$$\frac{dE_A(t)}{dt} = -\left.\frac{dE_c}{dt}\right|_{c,L} - \left.\frac{dE_{LO}}{dt}\right|_{LO,A} - \left.\frac{dE_{TO}}{dt}\right|_{TO,A} + \left.\frac{dE_A}{dt}\right|_{D}, \tag{1d}$$

$$\frac{dn(t)}{dt} = \left.\frac{dn}{dt}\right|_{L} + \left.\frac{dn}{dt}\right|_{R}, \tag{1e}$$

$dE_c/dt|_L$ and $dE_c/dt|_R$ are the rate of carrier energy variation due to laser pumping and recombination effects, respectivly, $dE_c/dt|_{c,L}$ is due to scattering of carriers with A and TO phonons via deformation potential interaction, $dE_c/dt|_{c,LO}$ is the collision operator resulting from Fröhlich interaction, $dE_{LO}/dt|_{LO,A}$ and $dE_{TO}/dt|_{TO,A}$ are a result of relaxation processes due to anharmonic interactions. $dE_A/dt|_D$ is the collision operator associated to heat diffusion from the hot A phonons. Equation (1 d) contains the contributions of pair creation by laser illumination and pair anihilation in recombination processes. In (4) we have neglected several contributions to the scattering operators like induced recombination, self-absorption, Auger effect, and spatial inhomogeneity.

These equations are (11) of Aligarte and Luzzi [2] augmented by the effects of collisions of carriers with transverse optical phonons[2]), and the equation of evolution for the energy of the acoustic phonons. It was supposed in that paper that the A phonons remained practically at the reservoir temperature T_0; this assumption no longer holds in the cases of illumination by long pulses to be considered here.

Most of the collision operators are given by Aligarte and Luzzi [2] where they were calculated in a low order approximation in Zubarev's NSO method which casts them in the form of Boltzmann-like collision operators in the Born approximation but with distribution functions characterized by the non-equilibrium thermodynamic parameters $T_c^*(t)$, $T_{LO}^*(t)$, $\mu_{e,h}(t)$. The new terms added in (1) are

$$\left.\frac{dE_c}{dt}\right|_{c,TO} = -\gamma n \frac{\nu_{TO} - \nu_{TO}(T_c^*)}{\nu_{TO}(T_c^*)} e^{-\beta\hbar\omega_{TO}/2(\beta\hbar\omega_{TO})^{1/2}} K_1\left(\frac{\beta\hbar\omega_{TO}}{2}\right), \tag{2}$$

where ν_{TO} is the Planck distribution of TO phonons at temperature $T_{TO}^*(t)$, $\nu_{TO}(T_c^*)$ the one in the limit of equilibrium with carriers, K_1 is a Bessel function of the second kind and

$$\gamma = \left(\frac{2}{\pi}\right)^{1/2} \left[\frac{D_e^2 m_e^{3/2} \omega_{TO}^{1/2}}{\pi \hbar^2 \varrho} + \frac{D_h^2 m_h^{3/2} \omega_{TO}^{1/2}}{\pi \hbar^2 \varrho}\right],$$

where $D_{e,h}$ are interaction constants [8], and ϱ is the density of the material. Equation (2) is the result obtained in the limit of a non-degenerate carrier system. Further, we write

$$\left.\frac{dE_{TO}}{dt}\right|_{TO,A} = \frac{E_{TO}(T_{LO}^*) - E_{TO}(T_A^*)}{\tau_{TO,A}}, \tag{3}$$

where $\tau_{TO,A}$ is a phenomenological relaxation time for decay of TO phonons in A pho-

[2]) We thank Prof. M. Cardona (Max-Planck Institute, Stuttgart) for calling our attention to the role of the TO phonons.

nons via anharmonic processes. Finally,

$$\left.\frac{\mathrm{d}E_\mathrm{A}}{\mathrm{d}t}\right|_\mathrm{D} = -\tau_\mathrm{D}^{-1}(T_\mathrm{A}^*)\,[E_\mathrm{A}(T_\mathrm{A}^*) - E_\mathrm{A}(T_0)]\,, \tag{4}$$

which has been derived solving the Fourier heat diffusion equation [12] with a boundary condition based on the assumption that the effective temperature T_A^* is uniform throughout the active volume of the sample with a sharp transition to T_0 at the boundary.

In (4) τ_D is the heat-diffusion relaxation time given by $l^2/D(T_\mathrm{A}^*) = l^2C(T_\mathrm{A}^*)/\varkappa(T_\mathrm{A}^*)$, where D is the diffusion coefficient, C the specific heat, $\varkappa$ the thermal conductivity of the lattice, and l an average linear dimension of the active volume of the sample. This process dissipates away the excess energy which is being transferred to the A phonons via anharmonic processes from the LO phonons overheated through the rapid transfer of carrier excess energy [2]. Since the relaxation time for decay of LO phonons into A phonons decreases while τ_D increases with increasing temperature [12], lattice heating would be favored by growing levels of excitation (in the classical limit $\tau_\mathrm{D}^{-1} \sim 1/T_\mathrm{A}^*$ and $\tau_\mathrm{AN}^{-1} \sim T_\mathrm{A}^{*2}$).

The solution of the set of coupled integro-differential equations (1) depends on each specific problem; i.e. on the physical parameters associated to each semiconductor and the experimental conditions which determine the macroscopic constraints imposed on the sample and fix the initial conditions. Various cases involving GaAs were considered in [2].

In this article we reconsider the rates of carrier energy loss in the augmented macroscopic description of the HEPS we have described, and using a scaling that makes the results independent of the material under consideration. Instead of solving for the explicit time evolution for the macroscopic variables we describe the dependence of the velocity of variation of the carrier energy through the different relaxation mechanisms in terms of the effective carrier temperature and for different values of the phonon effective temperatures. The results are graphically shown in Fig. 1 to 3. θ_0 is the Ein-

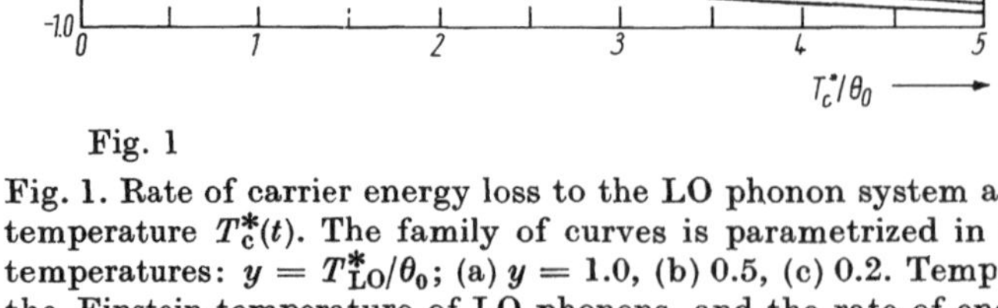

Fig. 1

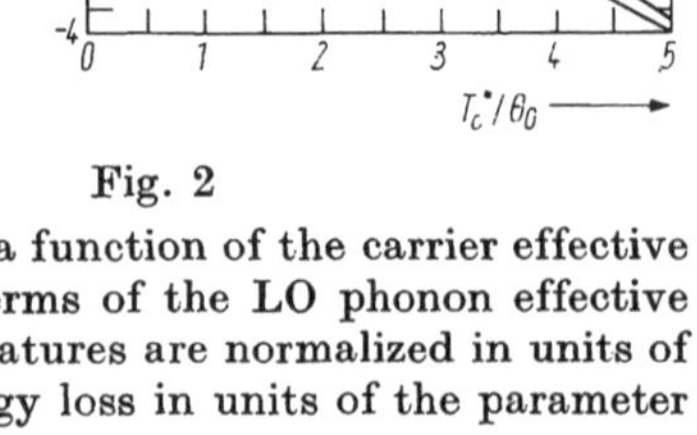

Fig. 2

Fig. 1. Rate of carrier energy loss to the LO phonon system as a function of the carrier effective temperature $T_\mathrm{c}^*(t)$. The family of curves is parametrized in terms of the LO phonon effective temperatures: $y = T_\mathrm{LO}^*/\theta_0$; (a) $y = 1.0$, (b) 0.5, (c) 0.2. Temperatures are normalized in units of the Einstein temperature of LO phonons, and the rate of energy loss in units of the parameter $\gamma_{\alpha,\mathrm{LO}}$ given in the text. (For GaAs, see Table 1)

Fig. 2. Rate of carrier energy loss to the TO phonon system as a function of the carrier effective temperature, and parametrized in terms of the TO phonon temperature: $y = T_\mathrm{TO}^*/\theta_0$, (a) $y = 1.0$, (b) 0.5, (c) 0.2. The rate of energy loss is given in units of $\gamma_{\alpha,\mathrm{TO}}$ as described in the text. (For GaAs, see Table 1)

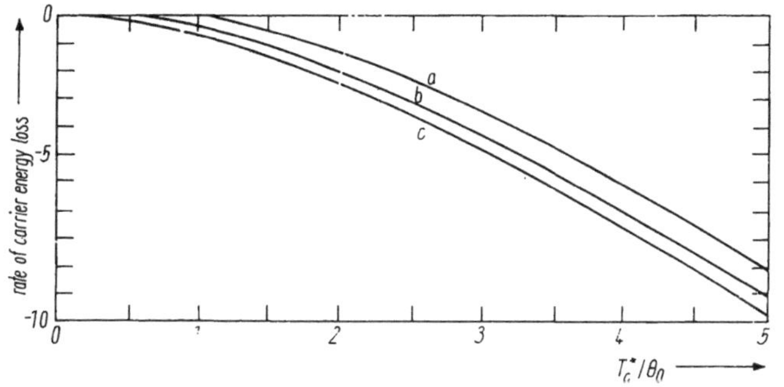

Fig. 3. Rate of carrier energy loss to the A phonon system as a function of the carrier effective temperature, and parametrized in terms of the A phonon effective temperatures: $z = T_A^*/\theta$; (a) $z = 1.0$, (b) 0.5, (c) 0.2. The rate of energy loss is given in units of $\gamma_{\alpha,A}$ as described in the text. (For GaAs, see Table 1)

stein temperature of dispersionless LO phonons, $\theta_0 = \hbar\omega_0/k_B$, and θ_D is the Debye temperature.

Fig. 1 shows the rate of carrier energy loss to the LO phonon system, expressed in units of $\gamma_{\alpha,LO} = (eE_{0\alpha}/2\pi)\,(2\omega_0/\pi m_\alpha)^{1/2}$, where $E_{0\alpha}$ are Fröhlich coupling fields [8], α = e or h for electrons and holes, respectively; (see Table 1 for numerical values in GaAs). These curves are parametrically dependent on the LO phonon effective temperature. As expected the rate of energy loss decreases on carriers cooling, till the value zero is attained at the point of mutual thermalization of carrier and LO phonon subsystems.

Table 1

Scaling parameters for the case of GaAs (in meV/ps)

$\gamma_{e,LO}$	$\gamma_{h,LO}$	$\gamma_{e,TO}$	$\gamma_{h,TO}$	$\gamma_{e,A}$	$\gamma_{h,A}$
14	40	0.4	41	0.01	0.54

LO phonon Einstein temperature, $\theta_0 \approx 430$ K,
A phonon Debye temperature, $\theta_D \approx 276$ K,
TO phonon Einstein temperature, $\theta_{0,TO} \approx 395$ K.

The rate of carrier energy loss to the TO phonons is shown in Fig. 2. The scaling parameter is $\gamma_{\alpha,TO} = D_\alpha^2 m_\alpha^{3/2} \omega_0^{1/2}/\sqrt{2}\,\pi^{3/2}\hbar^2 g$, where D_α is the coupling constant [8] and g is the density of the material (see Table 1 for numerical values in GaAs). A similar behaviour as that of the family of curves of Fig. 1 is observed. Finally Fig. 3 shows $\gamma_{c,A}^{-1}(dE_c/dt)_{c,A}$, where the scaling parameters are $\gamma_{\alpha,A} = \sqrt{2}\,8E_{1\alpha}^2 m_\alpha^{5/2}\omega_0^{3/2}/\pi^{3/2}\hbar^4 g$, with $E_{1\alpha}$ being the coupling strengths [9] (see Table 1 for numerical values in GaAs).

Considering the values for the scaling parameters γ given for the case of GaAs, which may be considered as typical of several technologically relevant polar semiconductors, we note that, as expected, the rate of carrier energy loss via Fröhlich interaction is roughly three orders of magnitude larger than that resulting of collisions with A phonons. As a result thermalization of carriers and LO phonons occurs in the picosecond time scale while final relaxation to the A phonons should follow in a larger than nanosecond time scale. Relaxation to final equilibrium is speeded up through an indirect mechanism consisting of the relaxation of the warmed up LO phonons to A phonons via anharmonic interactions [2]. It should be noted that the interaction of carriers with TO phonons produces, for the case of holes, a rate of energy loss of the same order of magnitude of the one resulting from relaxation processes mediated by Fröhlich interaction.

3. Lattice Heating

We turn next our attention to the change of the A phonon energy in anharmonic processes involving the decay of one LO phonon into two acoustic phonons. This term contributes to the increase of $T_A^*(t)$ at the rate

$$v_c C(T_A^*) \left.\frac{dT_A^*}{dt}\right|_{A,LO} = \tau_{AN}^{-1}(T_A^*)\, \hbar\omega_0 \,[\nu_{LO}(T_{LO}^*) - \nu_{LO}(T_A^*)]\,, \tag{5}$$

where $\nu_{LO}(T) = [\exp(\theta_0/T) - 1]^{-1}$, $C(T_A^*)$ is the A phonon heat capacity per unit volume, and v_c is the unit cell volume. A Debye model is used for the lattice vibrations.

For the case of GaAs if we take $\tau_{AN} \approx 10$ ps [9] it follows that $\theta_0/3\tau_{AN} \approx 14$K/ps. In the experiment reported in [10] the density of energy pumped by the laser ($\hbar\omega_L = 4$ eV) to the carrier system was ≈ 4.2 J/cm³, and the reservoir temperature was $T_0 = 300$ K. A couple of picoseconds after the end of the exciting laser pulse most of the carrier excess energy has been transferred to the LO phonons, with carriers and LO phonons attaining a common effective temperature $T_{LO}^* = T_c^* \approx 340$ K (or $y \approx 0.8$) [2a]. At this delay time the A phonons have no noticeable temperature increase [2a], and then $T_A^* \approx T_0$ ($\approx 0.7\theta_0$). Introducing these numbers into (5) we find $dT_A^*/dt|_{A,LO} \approx 1$ K/ps, what demonstrates the slow heating of A phonons in experiments using ultra-short laser pulses even at the high excitation conditions of the experiment of [10]. To obtain a large increase in the effective temperature of A phonons it becomes necessary to pump energy densities of the order of tens of J/cm³. This implies concentrations of the order of 10^{21} photoinjected carriers per cm³ and higher, what is difficult to obtain in a number of cases, e.g. GaAs, because of optical saturation [11].

Consider now the situation of continuous laser illumination; when a steady-state condition is achieved we have

$$\dot{E}_c + \dot{E}_{LO} + \dot{E}_{TO} + \dot{E}_A = \left.\frac{dE_c}{dt}\right|_L + \left.\frac{dE_c}{dt}\right|_R + \left.\frac{dE_A}{dt}\right|_D = 0\,. \tag{6}$$

We disregard in this equation $dE_c/dt|_R$ [2a], then $dE_A/dt|_D = -dE_c/dt|_L$, and using (4) we obtain

$$E_A(T_A^0) - E_A(T_0) = \tau_D(T_A^0) \left.\frac{dE_c(T_c^0)}{dt}\right|_L\,, \tag{7}$$

where T_A^0 and T_c^0 are the steady-state effective temperatures of A phonons and carriers, respectively. The right hand side of (7) can be expressed in the form $\tau_D(T_A^0)\,\alpha_1(\omega_L)\, I_L$ or $\tau_D(T_A^0)\,\alpha_2(\omega_L)\, I_L^2$ for one- or two-phonon absorption processes respectively, where α_1 and α_2 are the optical absorption coefficients and I_L the laser power flux. We solve (7), for the case of direct one-photon absorption processes, assuming that α_1 is weakly dependent on the carrier effective temperature, using a Debye model to describe the A phonon system and physical parameters typical of GaAs, i.e. α_1 (4 eV) $\approx 10^4$ cm^{-1}, $\theta_D \approx 276$ K, and $v_c \approx 1.8 \times 10^{-22}$ cm³. The results are shown, in the form of the dependence of the steady-state lattice temperature on laser power, in Fig. 4. For τ_D we choose 10^{-9} s, the same for all the three curves corresponding to the three different

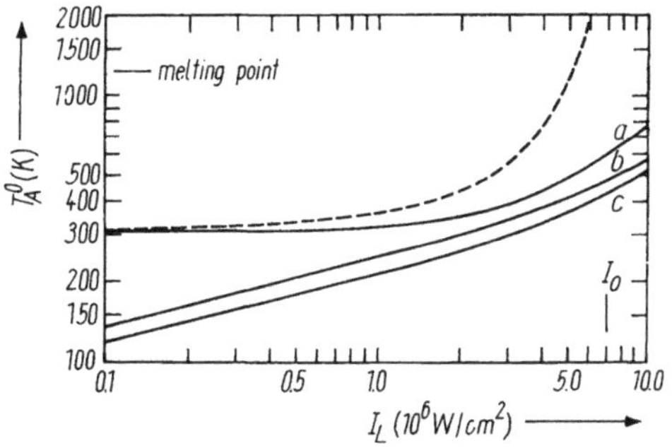

Fig. 4. Dependence of the steady-state effective temperature of A phonons on the power flux of a cw laser. Full lines correspond to the case when the heat diffusion relaxation time is taken as constant, and for three different initial temperatures. (a) $T_0 = 300$, (b) 77, (c) 10 K. The dashed line corresponds to the case of a temperature dependent heat diffusion relaxation time (cf. main text)

values of reservoir temperature T_0, (a) 300 K, (b) 77 K, and (c) 10 K. A constant value for τ_D was used, but, as noted, it should increase with T_A^0, and therefore the values for T_A^0 in Fig. 4 are underestimated. In the non-degenerate limit, typical in laser-annealing processes, an analytic expression for the dependence of T_A^0 on laser power is easily derived from (5) using the fact that $\tau_D(T_A^0) = \tau_D(T_0)\,(T_A^0/T_0)$, and $vC = 3k_B$; this is

$$T_A^0(I_L, I_0) = T^0[1 - (I_L/I_0)]^{-1} \tag{8}$$

with $I_0 = 3k_B T_0/\alpha_1(\omega_L)\,\tau_D(T_0)\,v$. The dependence of T_A^0 on I_L is indicated by the dashed line in Fig. 4, where $I_0 \approx 7 \times 10^6$ W/cm² with $T_0 = 300$ K and $\tau_D(T_0) \approx 1000$ ps.

Equation (8), valid for the model proposed here, indicates a possible trend in laser-illuminated semiconductors consisting of a lattice heating which becomes very effective for laser intensities approaching a critical value I_0. This is simply a consequence of the bottleneck-like effect resulting from the decrease of the rate of heat diffusion out of the active volume of the sample with increasing effective lattice temperature.

4. Concluding Remarks

The results of previous sections lead us to conclude that in intermediate to strong polar semiconductors:

(i) After application of a short laser pulse involving energy transfer of up to a few J/cm³, the photoexcited carriers rapidly relax most of their excess kinetic energy to the LO phonon system, which transfers it to the reservoir via anharmonic interaction with the A phonon system, and the lattice will show moderate to weak heating.

(ii) Application of long pulses producing steady-state conditions may result in appreciable lattice heating, if optical saturation does not exclude the possibility of large energy pumping. Then, as the heat diffusion coefficient decreases while T_A^* increases, there occurs a smooth warming up of the lattice followed by a steep increase in T_A^0 when the laser power flux is of the order of, or larger than, a critical value I_0, and melting should follow.

As a final remark it should be noted that these results do not exclude the possibility that before melting a condensation phenomenon may occur [1]. Such transformation of the excess energy into an ordered structure is possible in far-from-equilibrium conditions governed by nonlinear transport equations like (1) [13]. Theoretical examples of this kind in semiconductor physics are found in the literature involving convection instability (Bonch-Bruevich [14]), formation of superlattice (Bulgadaev and Levinson [14]), and spatial ordering in electron–hole plasmas (Vasconcellos and Luzzi [14]). For a discussion of the question of thermal or athermal laser-induced transitions in semiconductors see [15], and for more recent studies in the femtosecond scale [16].

References

[1] See for example: N. BLOEMBERGER, in: Laser–Solid Interaction and Laser Processing, Ed. S. D. FERRIS, H. J. LEAMY, and J. M. POATE, AIP, New York 1979 (p. 1).
N. BLOEMBERGER, H. KURZ, J. M. LIU, and R. YEN, in: Laser and Electron-Beam Interactions with Solids, Ed. B. R. APPLETON and G.K. CELLER, Elsevier Publ. Co., New York 1982 (p.3).
J. A. VAN VECHTEN, Solid State Commun. **39**, 1825 (1981); in: Laser and Electron-Beam Processing of Materials, Ed. C. W. WHITE and P. S. PERCY, Academic Press, Inc., New York 1980.

[2] A. C. ALGARTE and R. LUZZI, Phys. Rev. B **27**, 7563 (1983).
A. J. SAMPAIO and R. LUZZI, J. Phys. Chem. Solids **44**, 479 (1983).
R. LUZZI and A. R. VASCONCELLOS, in: Semiconductor Processes Probed by Ultra-fast Laser Spectroscopy, Ed. R. R. ALFANO, Academic Press, Inc., New York 1984.
R. LUZZI and A. R. VASCONCELLOS, J. statist. Phys. **23**, 539 (1980).
A. R. VASCONCELLOS and R. LUZZI, Phys. Rev. B **27**, 3874 (1983).

[3] R. R. ALFANO (Ed.), Primary Events in Biology Probed by Ultrafast Laser Spectroscopy, Academic Press, Inc., New York 1983; Semiconductor Processes Probed by Ultrafast Laser Spectroscopy, Academic Press, Inc., New York 1984.

[4] R. ZWANZIG, Kinam **3**, 5 (1981); in: Perspectives in Statistical Physics, Ed. H. J. RAVECHÉ, North Holland Publ. Co., Amsterdam 1981.

[5] D. N. ZUBAREV, Neravnovesnaya staticheskaya termodinamika, Izd. Nauka, Moskwa 1971.

[6] B. ROBERTSON, Phys. Rev. **144**, 151 (1966); **160**, 175 (1968).

[7] J. SEKE, Phys. Rev. A **21**, 2156 (1980).

[8] E. M. CONWELL, Solid State Phys., Suppl. 9, 1 (1967).

[9] R. K. CHANG, J. M. RALSTON, and D. E. KEATING, in: Light Scattering Spectra of Solids, Vol. 1, Ed. G. B. WRIGHT, Springer-Verlag, New York 1969 (p. 369).

[10] C. V. SHANK, D. H. AUSTON, E. P. IPPEN, and O. TESCHKE, Solid State Commun. **26**, 567 (1978).

[11] A. C. ALGARTE, private communication.

[12] e.g.: J. M. ZIMAN, Electrons and Phonons, Clarendon Press, Oxford 1960.

[13] P. GLANSDORFF and I. PRIGOGINE, Thermodynamic Theory of Structure, Stability, and Fluctuations, Wiley, New York 1971.

[14] V. L. BONCH-BRUEVICH, Zh. eksper. teor. Fiz. **67**, 2204 (1974); **71**, 1583 (1976).
S. A. BULGADAEV and J. B. LEVINSON, Zh. eksper. teor. Fiz. **71**, 1987 (1976).
A. R. VASCONCELLOS and R. LUZZI, Phys. Rev. B **29**, 1052 (1984).

[15] J. C. PHILLIPS, Comments Solid State Phys. **10**, 165 (1982).

[16] C. V. SHANK, R. YEN, and C. HIRLINMANN, Phys. Rev. Letters **50**, 454 (1983).

(Received June 14, 1984)

phys. stat. sol. (b) **126**, 71 (1984)

Subject classification: 6; 8; 21

Department of Applied Physics and Mathematics, Mining College, Akita University[1])

Pressure Dependence of the Melting Point of Al

By

T. SOMA, T. ITOH, and H.-MATSUO KAGAYA

The temperature dependence of the mean-square displacement for Al is quantitatively calculated from first principles. The Debye temperature at higher and lower temperatures is also estimated using Debye's model. Then, using the volume dependence of the mean-square displacement, the pressure effect on the melting temperature of Al is studied by Lindeman's melting law, and the melting curve obtained increases as function of the compressed volume.

Die Temperaturabhängigkeit der mittleren quadratischen Verschiebung wird für Aluminium quantitativ aus ersten Prinzipien berechnet. Die Debye-Temperatur wird für höhere und niedrigere Temperaturen mit Hilfe des Debye-Modells abgeschätzt. Unter Benutzung der Volumenabhängigkeit der quadratisch gemittelten Verschiebung wird dann der Einfluß des Druckes auf den Schmelzpunkt von Aluminium nach Lindemans Schmelzgesetz untersucht. Die berechneten Schmelzkurven nehmen mit der Volumenkompression ab.

1. Introduction

The mean-square displacement is important for anharmonic and melting properties. As the temperature is raised, the root-mean-square displacement becomes large, and the contributions to the free energy from the anharmonic term produce the thermal expansion. The increase of the root-mean-square displacement at higher temperatures brings the fusion of the solid lattice by Lindeman's melting law [1]. The pressure effect on the melting point was experimentally reported (see, e.g. [2]), and the qualitative tendency is not unique for all solids. The melting point of covalent crystals decreases under pressure, but that of simple metals increases.

Previously some experimental studies [3 to 5] about the mean-square displacements of Al from the X-ray Debye-Waller factor have been reported, and the effect of pressure on the melting point of Al at lower pressure regions has been also obtained experimentally [6, 7]. On the other hand, theoretical studies (see, e.g. [8, 9]) about the Debye-Waller factor and the melting of Al have been reported, but their obtained results were not conclusive, especially in the phenomenological treatment and in the quantitative stage. In order to obtain the mean-square displacement, the eigenfrequency of the phonon mode is necessary. Previously, we [10] have presented the local Heine-Abarenkov-type model potential with volume-dependent parameters for Al in the perturbational scheme. Then, we have obtained the phonon dispersion curves [11] along the symmetrical directions such as [100], [110], and [111] in good agreement with the observed neutron scattering data, and nearly studied the pressure effect [12] on the phonon frequency of Al. In the present work, we calculate the temperature dependence of the mean-square displacement for Al. Then, using the volume-dependent mean-square displacement, we study the effect of compression on the melting point of Al with f.c.c. phase according to Lindeman's criterion for melting [1].

---

[1]) Akita 010, Japan.

2. Mean-Square Displacements and Lindeman's Criterion for Melting

The mean-square displacement $\langle u^2 \rangle$ is temperature-dependent and can be expressed in terms of the i-th phonon mode $\nu_i(\boldsymbol{q})$ with wave vector $\boldsymbol{q}$ for a monoatomic cubic crystal as

$$\langle u^2 \rangle = \frac{h}{NM} \sum_{i,\boldsymbol{q}} \frac{1}{\nu_i(\boldsymbol{q})} \left\{ n[\nu_i(\boldsymbol{q})] + \frac{1}{2} \right\}, \tag{1}$$

where $n[\nu_i(\boldsymbol{q})]$ is the Bose-Einstein occupation number. Then, the summation is over all the N $\boldsymbol{q}$-points in the Brillouin zone and all three i-branches of the phonon dispersion curves $\nu_i(\boldsymbol{q})$. In order to obtain $\nu_i(\boldsymbol{q})$, we adopt our previous treatment [10 to 12] using the local Heine-Abarenkov-type model potential with volume-dependent parameters in the perturbational scheme.

Table 1

The mean-square displacement $\langle u^2 \rangle$ and $\langle u^2 \rangle^{TA}$ calculated with the total acoustic modes and with only transverse-like acoustic modes at the melting point T_m and absolute-zero temperature $T = 0$ (in 10^{-16} cm^2)

	H	K-L	S-S-T-L	V-S
$\langle u^2 \rangle_{T_m}$	0.0968	0.1007	0.1001	0.1012
$\langle u^2 \rangle_{T_m}^{TA}$	0.0780	0.0808	0.0797	0.0806
$\langle u^2 \rangle_{T=0}$	0.0097	0.0100	0.0099	0.0101
$\langle u^2 \rangle_{T=0}^{TA}$	0.0071	0.0072	0.0072	0.0073

In numerical calculations, we use the following sampling method. Considering the symmetry of the irreducible 1/48 portion of the Brillouin zone, it is sufficient to determine the phonon frequencies in the range $\boldsymbol{q} = 2\pi(q_x, q_y, q_z)/(16a)$, where q_x, q_y, q_z are positive integers and satisfy the inequalities $0 \leqq q_x \leqq q_y \leqq q_z \leqq 8$ and $q_x + q_y + q_z \leqq 12$, and a is the lattice constant. Each point is weighted according to the number of points equivalent to it by symmetry, especially on the surface, edges, and corners of the Brillouin zone. A divergence occurs for the $\boldsymbol{q} \approx \boldsymbol{0}$ acoustic mode phonon in the sum over the wave-vector mesh of (1). Therefore, the contribution to the mean-square displacement owing to the $\boldsymbol{q} \approx \boldsymbol{0}$ mode frequencies is calculated by converting the summation to an integral and assuming a Debye distribution.

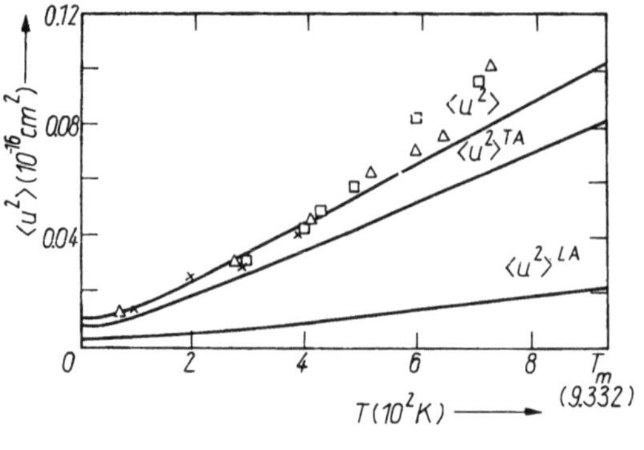

Fig. 1. The mean-square displacement $\langle u^2 \rangle$, $\langle u^2 \rangle^{TA}$, and $\langle u^2 \rangle^{LA}$ of Al obtained using the total, only transverse-like, and only longitudinal-like acoustic modes versus temperature. The points are observed data such as △ [3], □ [4], and × [5]

Table 2

Debye temperature θ_D^{ht} at higher temperatures and θ_D^{lt} at lower temperatures obtained from the mean-square displacement with Debye's model (in K)

	H	K-L	S-S-T-L	V-S	obs.*)
θ_D^{ht}	395.1	387.3	388.5	386.4	—
θ_D^{lt}	417.2	404.7	408.8	400.7	428

*) Kittel [14].

We show the temperature dependence of $\langle u^2 \rangle$ for Al obtained using (1) and with Vashishta-Singwi screening in Fig. 1, where points are observed data [3 to 5], and T_m the melting point at atmospheric pressure. In Fig. 1, the separate contributions $\langle u^2 \rangle^{TA}$ and $\langle u^2 \rangle^{LA}$ to the total mean-square displacement $\langle u^2 \rangle$ from transverse-like and longitudinal-like modes are also given, and we see that the contribution of the transverse-like acoustic modes to the mean-square displacement is predominant for Al. Then, we show in Table 1 the numerical results of $\langle u^2 \rangle$ and $\langle u^2 \rangle^{TA}$ using other screening functions ([10 to 12]) at the melting point T_m and the absolute zero temperature $T = 0$.

In Debye's model, the mean-square displacement $\langle u^2 \rangle$ is expressed as (e.g., see [13])

$$\langle u^2 \rangle = \frac{9\hbar^2 T}{M k_B \theta_D^2} \quad \text{for} \quad T > \theta_D \tag{2}$$

or

$$\langle u^2 \rangle = \frac{9\hbar^2}{4 M k_B \theta_D} \quad \text{for} \quad T \ll \theta_D\,. \tag{3}$$

Using the numerical results of $\langle u^2 \rangle$ in Fig. 1 and Table 1, we obtain the Debye temperature θ_D^{ht} and θ_D^{lt} of Al at higher and lower temperatures. These data are given in Table 2, and comparable to the observed data fit to the specific heat data at low temperatures (e.g., see [14]).

Lindeman [1] proposed that the melting process occurs when the root-mean-square displacement of the lattice vibration $\sqrt{\langle u^2 \rangle}$ reaches a critical fraction of the nearest-neighbour distance. He assumed that this critical fraction was the same for all crystalline solids, but it was later shown (e.g., see [15, 16]) that in various cubic metals and alkali halides this fraction was actually not constant. We define Lindeman's criterion for melting x_m as the ratio of two times the root-mean-square displacement at the

Table 3

Lindeman's criterion for melting x_m and x_m^{TA} calculated with the total acoustic modes and with only transverse-like acoustic modes

	H	K-L	S-S-T-L	V-S
x_m	0.2173	0.2216	0.2210	0.2222
x_m^{TA}	0.1951	0.1985	0.1972	0.1982

melting point to the nearest-neighbour distance R_1 ($= \sqrt{2}a/2$) given by

$$x_m = \frac{2\sqrt{\langle u^2 \rangle}}{R_1}. \tag{4}$$

Because the contribution $\langle u^2 \rangle^{TA}$ of the transverse-like phonon modes to the total mean-square displacement $\langle u^2 \rangle$ was predominant, we investigated the melting process by considering only transverse-like phonon modes. The numerical data of the critical fraction for melting x_m and x_m^{TA} obtained with $\langle u^2 \rangle$ and $\langle u^2 \rangle^{LA}$ in (1) are shown in Table 3. Consequently, we see that the melting process for Al is influenced largely by the transverse-like phonon modes.

3. Volume Effect on the Melting Point

We [12] calculated numerically the phonon frequencies $\nu_i(\boldsymbol{q})$ in dependence on the compressed volume Ω/Ω_0. Using (1), we show the obtained temperature-dependence of the mean-square displacement $\langle u^2 \rangle$ with Vashishta-Singwi screening for the compressed volume in Fig. 2, where full, dash-dotted, and dashed curves represent those at the representative compressed volume $\Omega/\Omega_0 = 0.8$, 0.9, and 1.0, respectively. From Fig. 1, we see that, as the crystal volume is compressed, the decrease of the mean-square displacement is produced due to the increase of the phonon frequencies. Then, we show the relative contribution $\langle u^2 \rangle^{TA}/\langle u^2 \rangle$ and $\langle u^2 \rangle^{LA}/\langle u^2 \rangle$ to the total mean-square displacement from the transverse-like and longitudinal-like modes at the melting point under atmospheric pressure against the compressed volume Ω/Ω_0 in Fig. 3, and see that the contribution of the transverse-like acoustic modes becomes more predominant as the crystal-volume compressed.

Considering the volume-dependence of the mean-square displacement $\langle u^2 \rangle$ and the nearest-neighbour distance R_1 at a constant criterion for melting x_m, we can estimate the volume effect on the melting point by satisfying the following relation:

$$x_m = 2\sqrt{\langle u^2 \rangle_{T_m(\Omega/\Omega_0)}}/R_1(\Omega/\Omega_0)\,, \tag{5}$$

where

$$R_1(\Omega/\Omega_0) = (\Omega/\Omega_0)^{1/3}\, R_1(\Omega/\Omega_0 = 1)\,. \tag{6}$$

The numerical results obtained with Vashishta-Singwi screening for the volume-dependence of the melting point are shown in Fig. 4, where full and dash-dotted curves represent those using the total acoustic modes and only transverse-like acoustic modes, and the points the observed data [6, 7]. The variation width of the obtained results using different forms of exchange and correlation in Fig. 4 are within 5% at the com-

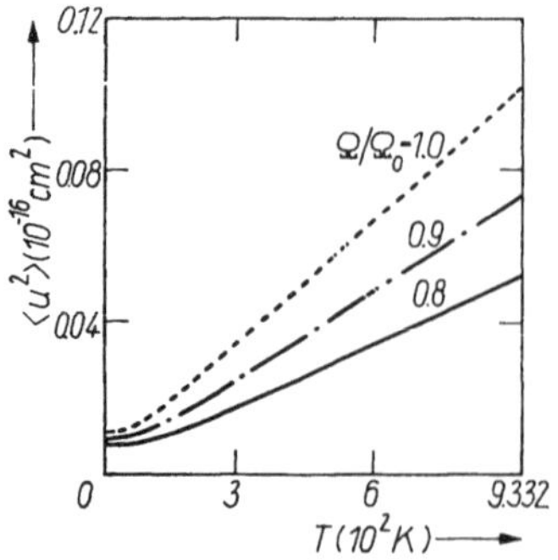

Fig. 2. The mean-square displacement $\langle u^2 \rangle$ under pressure against temperature of Al. Full, dash-dotted, and dashed curves correspond to the compressed volume $\Omega/\Omega_0 = 0.8$, 0.9 and 1.0, respectively

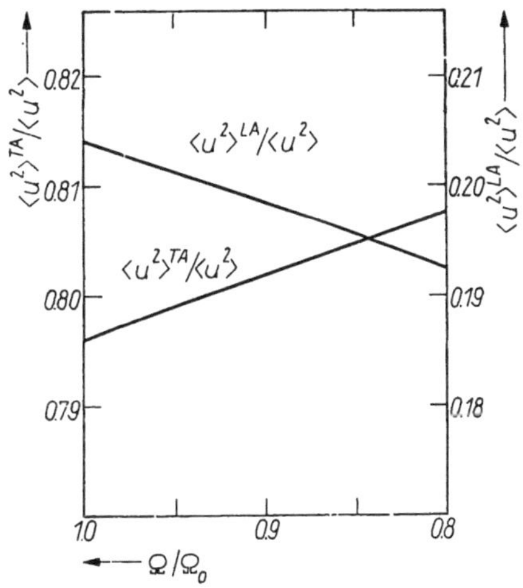

Fig. 3

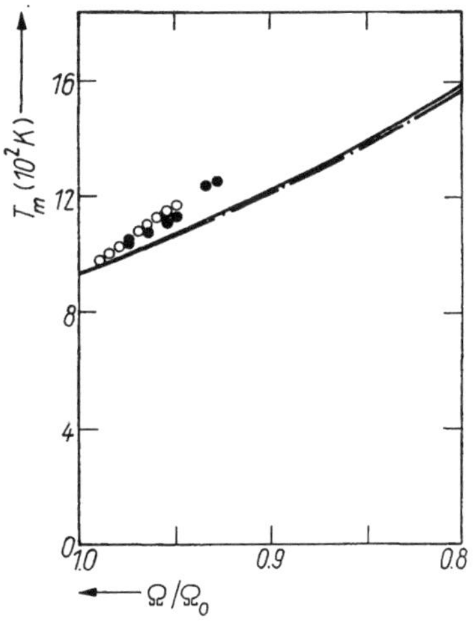

Fig. 4

Fig. 3. The relative ratio $\langle u^2\rangle^{TA}/\langle u^2\rangle$ and $\langle u^2\rangle^{LA}/\langle u^2\rangle$ versus the compressed volume at the melting point T_m under atmospheric pressure

Fig. 4. Melting point T_m versus compressed volume Ω/Ω_0 of Al. Full and dash-dotted curves represent those using the total and transverse-like acoustic modes, and points the observed data; ○ [6] and ● [7]

pressed volume $\Omega/\Omega_0 = 0.8$. From Fig. 4, we see that the melting point of Al increases as the crystal volume is compressed and this qualitative tendency obtained is consistent with the observed data [6, 7]. Then, we see that the melting process of Al is influenced largely by the transverse-like acoustic phonon modes.

References

[1] F. Lindeman, Z. Phys. **11**, 609 (1910).
[2] N. Kawai and Y. Inokuti, Japan. J. appl. Phys. 7, 989 (1968).
[3] E. A. Owen and R. W. Williams, Proc. Roy. Soc. (London) **A188**, 509 (1947).
[4] D. R. Chipman, J. appl. Phys. **31**, 2012 (1960).
[5] P. A. Flinn and C. M. McManus, Phys. Rev. **132**, 2458 (1963).
[6] A. Jayaraman, W. Klement, R. C. Newton, and G. C. Kennedy, J. Phys. Chem. Solids **24**, 7 (1963).
[7] J. F. Cannon, J. Phys. Chem. Ref. Data **3**, 781 (1974).
[8] E. V. Zarochentsev, S. P. Krar-Chuk and T. M. Tarusina, Soviet Phys. — Solid State **18**, 239 (1976).
[9] A. P. G. Kutty and S. N. Vaidya, J. Phys. Chem. Solids **41**, 1163 (1980).
[10] T. Soma, Y. Konno, and H.-Matsuo Kagaya, phys. stat. sol. (b) **117**, 743 (1983).
[11] T. Soma, T. Itoh, and H.-Matsuo Kagaya, phys. stat. sol. (b) **123**, 463 (1984).
[12] T. Soma, T. Itoh, and H.-Matsuo Kagaya, phys. stat. sol. (b) **125**, 107 (1984).
[13] J. M. Ziman, Principles of the Theory of Solids, 2nd edition, Chap. 2, Cambridge University Press, 1976.
[14] C. Kittel, Introduction to Solid State Physics, 5th edition, Chap. 5, John Wiley & Sons, 1976.
[15] A. K. Singh and P. K. Sharma, Canad. J. Phys. **46**, 1677 (1968).
[16] J. S. Reid and T. Smith, J. Phys. Chem. Solids **31**, 2689 (1970).

(Received July 20, 1984)

phys. stat. sol. (b) **126**, 77 (1984)

Subject classification: 12.2; 22.5.2

Department of Physics, Agra College[1])

Temperature Dependences of Third-Order Elastic Constants of CsCl, CsBr, and CsI Crsystals

By

R. K. VARSHNEY[2]) and J. SHANKER

Expressions for third-order elastic constants (TOEC) of CsCl-structure solids are derived taking the thermal contribution into account. Six distinct values of TOEC for CsCl, CsBr, and CsI crystals are calculated at 0 K and at room temperature (300 K). Values of pressure derivatives of second-order elastic constants are also calculated corresponding to both the temperatures. The phonon-pressure contribution is incorporated within the framework of the three-body potential model. The results are discussed and compared with other investigations.

Es werden die Ausdrücke für die elastischen Konstanten dritter Ordnung (TOEC) von Festkörpern mit CsCl-Struktur abgeleitet, wobei der thermische Beitrag berücksichtigt wird. Sechs verschiedene TOEC-Werte für CsCl-, CsBr- und CsJ-Kristalle werden für 0 K und Zimmertemperatur (300 K) berechnet. Werte für die Druckableitungen der elastischen Konstanten zweiter Ordnung werden entsprechend den beiden Temperaturen ebenfalls berechnet. Der Phonondruckbeitrag wird im Rahmen des Dreikörperpotentialmodells einbezogen. Die Ergebnisse werden diskutiert und mit anderen Untersuchungen verglichen.

1. Introduction

There have been several attempts [1, 2] to study the temperature dependence of third-order elastic constants (TOEC) of alkali halide crystals by evaluating various strain derivatives of the vibrational free energy density. They have used two-body potentials [1, 2] which cannot explain the breakdown of Cauchy relations even at 0 K. The expressions used by Ghate [1] had some numerical errors which were later rectified by Mori and Hiki [2]. The correct expressions used by Mori and Hiki are also not capable to explain satisfactorily even the sign for the temperature dependence of second-order elastic constants (SOEC) of alkali halides [3].

These shortcomings have been removed recently [4, 5] by developing an alternative approach based on phonon-pressure theory using the Hildebrand equation of state and taking account of three-body interactions [6]. This approach has been successful in explaining the temperature dependence of SOEC [4, 5] and extended further to evaluate the temperature dependence of TOEC in NaCl-structure alkali halides [7]. In the present work, we extend this formulation to derive TOEC of CsCl-structure solids by taking thermal contributions as well as three-body interactions into account. Numerical values of TOEC are obtained for CsCl, CsBr, and CsI crystals at 0 and 300 K.

---

[1]) Agra 282002, India.

[2]) Department of Physics, M. B. Government (Postgraduate) College, Haldwani 263141, U.P., India.

2. Theory and Method of Calculation

The lattice potential energy Φ, within the framework of the Lundqvist three-body interaction model [6] can be expressed as

$$\Phi = \frac{\alpha_M e^2 \varepsilon(\varepsilon + 16f(r))}{r} + (16V_1(r) + 6V_2(r)) e^2 , \tag{1}$$

where the symbols have their usual meaning [8 to 10].

The Hildebrand equation of state can be written as follows:

$$P = -\frac{d\Phi}{dV} + T\beta B_T , \tag{2}$$

where P is the pressure, T the temperature, β the volume thermal expansion coefficient, and B_T the isothermal bulk modulus.

The expressions for SOEC and the equilibrium condition at any temperature T are obtained from (1) and (2) as mentioned below,

$$C_{11} = 4.1333T\beta B_T + \frac{e^2}{4a^4}\left[\frac{A_1}{6} - 1.7333B_1 + \frac{A_2}{2} - 2.0666B_2 + 3.1336\varepsilon\left(a\frac{\partial f}{\partial r}\right)_0\right], \tag{3}$$

$$C_{12} = -3.0666T\beta B_T + \frac{e^2}{4a^4}\left[\frac{A_1}{6} + 1.3666B_1 + 1.5333B_2 + 3.1336\varepsilon\left(a\frac{\partial f}{\partial r}\right)_0\right], \tag{4}$$

$$C_{44} = -2.0666T\beta B_T + \frac{e^2}{4a^4}\left[\frac{A_1}{6} + 1.3666B_1 + 1.5333B_2\right], \tag{5}$$

$$T\beta B_T = \frac{e^2}{8a^4}\left[0.3392\varepsilon(\varepsilon + 16f(r))_0 + B_1 + B_2\right] . \tag{6}$$

Here suffix at parentheses denotes the terms at equilibrium configuration with $r = a\sqrt{3}$, the equilibrium value of nearest neighbour separation, a being the lattice parameter. When we put $T = 0$, (3) to (6) reduce to the corresponding expressions given by Puri et al. [8]. The short-range force parameters A_1, B_1, C_1, A_2, B_2, C_2 are the same as defined in [8]. Using the method of homogeneous deformation [11], the following expressions for TOEC are obtained:

$$C_{111} = -26.9198T\beta B_T + \frac{e^2}{4a^4}\left[12.1266B_1 + \frac{C_1 - 3A_1}{18} + 13.4599B_2 + \right.$$
$$\left. + \frac{C_2 - 3A_2}{2} + 2.7138\varepsilon\left(a^2\frac{\partial^2 f}{\partial r^2}\right)_0 + 13.1567\varepsilon\left(a\frac{\partial f}{\partial r}\right)_0\right], \tag{7}$$

$$C_{112} = 4.1279T\beta B_T + \frac{e^2}{4a^4}\left[-1.8973B_1 + \frac{C_1 - 3A_1}{18} - 2.0640B_2 + \right.$$
$$\left. + 2.7138\varepsilon\left(a^2\frac{\partial^2 f}{\partial r^2}\right)_0 - 6.2677\varepsilon\left(a\frac{\partial f}{\partial r}\right)_0\right], \tag{8}$$

$$C_{116} = 2.1279T\beta B_T + \frac{e^2}{4a^4}\left[-1.8973B_1 + \frac{C_1 - 3A_1}{18} - 2.0640B_2 + \right.$$
$$\left. + 0.9046\varepsilon\left(a^2\frac{\partial^2 f}{\partial r^2}\right)_0 - 5.3266\varepsilon\left(a\frac{\partial f}{\partial r}\right)_0\right], \tag{9}$$

$$C_{123} = 7.0755T\beta B_T + \frac{e^2}{4a^4}\left[-3.3711B_1 + \frac{C_1 - 3A_1}{18} - 3.5377B_2 + \right.$$
$$\left. + 2.7138\varepsilon\left(a^2\frac{\partial^2 f}{\partial r^2}\right)_0 - 15.9799\varepsilon\left(a\frac{\partial f}{\partial r}\right)_0\right], \quad (10)$$

$$C_{144} = 7.0755T\beta B_T + \frac{e^2}{4a^4}\left[-3.3711B_1 + \frac{C_1 - 3A_1}{18} - 3.5377B_2 + \right.$$
$$\left. + 0.9046\varepsilon\left(a^2\frac{\partial^2 f}{\partial r^2}\right) - 5.3266\varepsilon\left(a\frac{\partial f}{\partial r}\right)_0\right], \quad (11)$$

$$C_{456} = 7.0755T\beta B_T + \frac{e^2}{4a^4}\left[-3.3711B_1 + \frac{C_1 - 3A_1}{18} - 3.5377B_2\right]. \quad (12)$$

The pressure derivatives of SOEC are calculated by using the Birch relations [12].

To evaluate various potential parameters A_1, B_1 etc., we assume that $V_1(r)$ and $V_2(r)$ depend exponentially on the interionic distance r, so that [8]

$$A_1 = -\frac{\sqrt{3}\, aB_1}{\varrho}, \quad (13)$$

$$A_2 = -\frac{2aB_2}{\varrho}, \quad (14)$$

Table 1

Calculated values of potential parameters

crystals	ϱ (10^{-1} nm)	A_1		B_1		C_1	
		(a)*)	(b)**)	(a)	(b)	(a)	(b)
CsCl	0.27318	4.78	4.19	−0.371	−0.322	−61.7	−54.6
CsBr	0.27977	4.89	4.37	−0.373	−0.329	−64.1	−58.1
CsI	0.28426	4.97	4.53	−0.362	−0.326	−68.1	−62.9
crystals	ϱ (10^{-1} nm)	A_2		B_2		C_2	
		(a)	(b)	(a)	(b)	(a)	(b)
CsCl	0.27318	−0.350	−0.402	0.024	0.027	5.21	6.05
CsBr	0.27977	−0.430	−0.505	0.028	0.033	6.50	7.74
CsI	0.28426	−0.354	−0.490	0.022	0.031	5.61	7.86
crystals	ϱ (10^{-1} nm)	$10^{-4}\,(f(r))_0$		$10^{-4}\,a(\partial f/\partial r)_0$		$a^2(\partial^2 f/\partial r^2)_0$	
		(a)	(b)	(a)	(b)	(a)	(b)
CsCl	0.27318	15.3	2.36	94.7	12.6	0.085	0.013
CsBr	0.27977	9.26	11.0	40.1	9.82	0.053	0.065
CsI	0.28426	1.24	15.6	−45.5	1.39	0.008	0.101

*) (a) 0 K values.
**) (b) 300 K values.

Table 2

Calculated values of third-order elastic constants (in 10^{11} Pa)

crystals	C_{111}		C_{112}		C_{166}		C_{123}		C_{144}		C_{456}	
	(a)*)	(b)**)	(a)	(b)	(a)	(b)	(a)	(b)	(a)	(b)	(a)	(b)
CsCl	−1.66	−1.36	−1.15	−0.987	−1.19	−1.01	−1.00	−0.829	−1.02	−0.833	−1.03	−0.834
CsBr	−1.27	−0.912	−1.06	−0.852	−1.03	−0.897	−0.899	−0.716	−0.915	−0.745	−0.922	−0.760
CsI	−1.22	−0.759	−0.881	−0.707	−0.885	−0.757	−0.759	−0.598	−0.773	−0.636	−0.780	−0.656

*) (a) 0 K values.
**) (b) 300 K values.

$$C_1 = \frac{A_1^2}{B_1}, \tag{15}$$

$$C_2 = \frac{A_2^2}{B_2}, \tag{16}$$

where ϱ is the Born-Mayer repulsive hardness parameter. The three-body charge transfer f is also considered to depend exponentially on r [10],

$$f = f_0 \exp\left(-\frac{r}{\varrho}\right). \tag{17}$$

Values of $(f(r))_0$, $(\partial f/\partial r)_0$, $(\partial^2 f/\partial r^2)_0$, and ϱ are calculated using data on SOEC and pressure derivative of bulk modulus given in [8].

3. Results and Discussion

Expressions for SOEC, TOEC, and equilibrium condition obtained in the present paper for CsCl-structure solids are considerably different from the corresponding expressions for NaCl-structure solids. These expressions take into account the effect of three-body interactions as well as the temperature contribution in terms of phonon-pressure theory [4]. The present approach is better than the one adopted by previous workers [1 to 3] as mentioned in the Introduction.

We have performed calculations at $T = 0$ and 300 K. Room temperature values of input data are taken from [8] and 0 K values from Sangster et al. [13]. Volume thermal expansion coefficient and isothermal bulk modulus data are used from Tosi [14]. Calculated values of short-range force constants are given in Table 1. Values of TOEC and pressure derivatives of SOEC are given in Tables 2 and 3. The linear combinations of TOEC for which experimental data [15] are available have been reported in Table 4.

We note from Tables 2 to 4 that the results vary significantly in going from 0 to 300 K. Values of TOEC and pressure derivatives of SOEC are found to present good agreement with experimental data. The agreement could further be improved by including the contribution of the van der Waals interactions. However, the purpose of the present work has been to incorporate explicitly the effect of temperature in the expressions of SOEC and TOEC. As a result of this investigation it is now possible to express the breakdown of Cauchy relations in terms of two separate contributions

Table 3
First-order pressure derivatives of second-order elastic constants (dimensionless)

crystals	$\partial C_{11}/\partial P$			$\partial C_{12}/\partial P$			$\partial C_{44}/\partial P$		
	calculated		experimental	calculated		experimental	calculated		experimental
	(a)*)	(b)**)	[15]	(a)	(b)	[15]	(a)	(b)	[15]
CsCl	4.35	4.89	7.08	5.81	6.39	5.30	4.00	4.47	3.56
CsBr	4.43	4.39	6.62	6.20	6.40	5.26	4.33	4.66	3.68
CsI	5.26	4.42	6.91	6.65	6.47	5.23	4.69	4.85	3.72

*) (a) 0 K values.
**) (b) 300 K values.

Table 4

Linear combinations of third-order elastic constants (in 10^{11} Pa)

crystals	$C_{111} + 2C_{112}$			$C_{123} + 2C_{112}$			$C_{144} + 2C_{166}$		
	calculated		experi-mental	calculated		experi-mental	calculated		experi-mental
	(a)*)	(b)**)	[15]	(a)	(b)	[15]	(a)	(b)	[15]
CsCl	−3.95	−3.34	−4.42	−3.29	−2.80	−2.24	−3.41	−2.85	−2.38
CsBr	−3.33	−2.62	−3.61	−2.97	−2.42	−1.92	−3.03	−2.54	−2.11
CsI	−2.98	−2.17	−3.06	−2.52	−2.01	−1.57	−2.54	−2.15	−1.75

*) (a) 0 K values.
**) (b) 300 K values.

arising from three-body potential and thermal phonon pressure. The main inconsistency of the previous work [8] viz. to use expressions valid at 0 K, along with the room temperature input data, is now removed. The effect of removing this inconsistency on the resulting values of TOEC and pressure derivatives of SOEC has been found significant.

References

[1] P. B. Ghate, Phys. Rev. **139**, A1666 (1965).
[2] S. Mori and Y. Hiki, J. Phys. Soc. Japan **45**, 1449 (1978).
[3] S. Mori, H. Kaburaki, and Y. Hiki, Internal Friction and Ultrasonic Attenuation in Solids, Tokyo University Press, Tokyo 1977 (p. 401).
[4] U. C. Shrivastava, phys. stat. sol. (b) **100**, 641 (1980).
[5] J. Shanker and R. K. Varshney, phys. stat. sol. (b) **114**, K71 (1982).
[6] S. Lundqvist, Ark. Fys. (Sweden) **9**, 435 (1955).
[7] R. K. Varshney and J. Shanker, phys. stat. sol. (b) **122**, 65 (1984).
[8] D. S. Puri, V. K. Garg, and M. P. Verma, phys. stat. sol. (b) **78**, 113 (1976).
[9] J. Shanker, J. P. Singh, and G. D. Jain, phys. stat. sol. (b) **105**, 385 (1981).
[10] R. P. Singh and J. Shanker, phys. stat. sol. (b) **93**, 373 (1979).
[11] D. C. Walkace, Thermodynamics of Crystals, John Wiley & Sons, New York 1972.
[12] F. Birch, Phys. Rev. **71**, 809 (1947).
[13] M. J. L. Sangster, U. Schroder, and R. M. Atwood, J. Phys. C **11**, 1523 (1978).
[14] M. P. Tosi, Solid State Phys. **16**, 1 (1964).
[15] Z. P. Chang, G. R. Barsch, and D. L. Miller, phys. stat. sol. **23**, 577 (1967); Phys. Rev. Letters **19**, 1381 (1967).

(Received July 26, 1984)

phys. stat. sol. (b) **126**, 83 (1984)

Subject classification: 4 and 12.2; 23

Department of Physics, Agra College[1])

Compressibility of Rare-Gas Solids under the Effect of High Pressure

By

J. NARAIN, G. G. AGRAWAL, and J. SHANKER

An interionic potential model is formulated for rare-gas solids by considering the effect of three-body interactions, zero-point energy, and thermal contributions. The model is applied to evaluate compression curves for Ne, Ar, Kr, and Xe, up to a pressure of about 20 kbar at their triple-point temperatures. The calculated and experimental results are found in good agreement with each other.

Ein Interionenpotentialmodell wird für Edelgasfestkörper unter Berücksichtigung von Dreikörperwechselwirkungseffekten, Nullpunktsenergie und thermischen Beiträgen formuliert. Das Modell wird zur Berechnung von Kompressionskurven für Ne, Ar, Kr und Xe bis zu Drücken von etwa 20 kbar bei ihren Tripelpunkttemperaturen angewendet. Die berechneten und experimentellen Ergebnisse befinden sich miteinander in guter Übereinstimmung.

1. Introduction

Rare-gas solids (RGS) are materials of great scientific and technological importance [1]. They are composed of closed-shell atoms and their binding arises mainly from the van der Waals forces [2]. To explain the breakdown of the Cauchy relation between elastic constants in RGS, it is essential to develop an interatomic potential model which takes into account the contribution of many-body interactions. Such attempts have been made independently by Zucker and Paik [3] and by Lehri and Verma [4] using different approaches. The first model [3] takes into account the three-body interactions, viz. the triple dipole, dipole–dipole–quadrupole, dipole–quadrupole–quadrupole, and triple quadrupole interactions [5, 6]. The second model [4] assumes that three-body interactions in RGS arise from the variable induced dipoles. It has been found [7] that both the models yield comparable results for third-order elastic constants and pressure derivatives of second-order elastic constants. However, these models do not take any account of temperature contributions in the elastic constants or equilibrium condition.

In the present paper we revise the model of Lehri and Verma [4] by taking account of the temperature contribution within the framework of the Hildebrand approximation (cf. [8]) so as to make the model capable of studying the temperature dependence of various properties of RGS. The details of the revised model are given in Section 2. We apply this model for the calculation of the equation of state or compression curves for Ne, Ar, Kr, and Xe solids at their triple-point temperatures. Results are discussed in Section 3.

2. Method of Analysis

Within the framework of the Hildebrand approximation we can write [8]

$$\frac{\mathrm{d}\Phi}{\mathrm{d}V} = -P + T\beta B_T \,, \tag{1}$$

[1]) Agra 282002, India.

where Φ is the lattice potential energy, P the cold component of pressure, and the last term represents the thermal phonon pressure. In RGS we take the following expression for the potential energy [4]:

$$\Phi = -\frac{\lambda[\lambda + 12f(r)]}{r^6} + 12W(r) + \frac{9}{8}k_B\theta_D, \tag{2}$$

where the first term on the right-hand side is the van der Waals interaction term including the effect of three-body interaction, the second term is the short-range repulsive potential, and the last term represents the zero-point energy. Using (1) and (2) and the method based on the dynamical matrix in the long-wavelength limit we obtain the following expressions for second-order elastic constants and equilibrium condition:

$$C_{11} = \left[6.7862\frac{\alpha'}{a^6} + 2Aa^2 + \frac{9}{16}k_B\theta_D\gamma + \frac{1}{2}a^3T\beta B_T - \right.$$
$$\left. - 10.2195\frac{\lambda}{a^5}\frac{\partial f(r)}{\partial r}\right]V^{-1}, \tag{3}$$

$$C_{12} = \left[2.930\frac{\alpha'}{a^6} + Aa^2 - \frac{45}{32}k_B\theta_D\gamma - \frac{5}{4}a^3T\beta B_T - \right.$$
$$\left. - 10.2195\frac{\lambda}{a^5}\frac{\partial f(r)}{\partial r}\right]V^{-1}, \tag{4}$$

$$C_{44} = \left[2.9298\frac{\alpha'}{a^6} + Aa^2 + \frac{27}{32}k_B\theta_D\gamma + \frac{3}{4}a^3T\beta_T B\right]V^{-1}, \tag{5}$$

$$B = 0.4516\frac{\alpha'}{\alpha^8} + \frac{9}{32}k_B\theta_D\frac{\gamma}{a^2} + \frac{1}{4}aT\beta B_T, \tag{6}$$

where $\alpha' = \lambda[\lambda + 12f(r)]$, γ is the Grüneisen parameter, $f(r)$ the three-body interaction parameter, k_B the Boltzmann constant, and θ_D the Debye temperature. The short-range force parameters A and B are defined as follows:

$$A = \left[\frac{\partial^2 W'(r)}{\partial r^2}\right]_{r=r_0}, \tag{7}$$

$$B = \left[\frac{1}{r}\frac{\partial W'(r)}{\partial r}\right]_{r=r_0}, \tag{8}$$

where $W'(r)$ is related to repulsive potential $W(r)$ as follows:

$$W'(r) = W(r) + 14.4539\frac{\lambda f(r)}{r_0^6}. \tag{9}$$

Equations (3) to (6) are more general than the corresponding expressions given in [4] in the sense that they are valid at any temperature T. At $T = 0$ (3) to (6) become identical with those equations given by Lehri and Verma [4].

We obtain the compression curves for RGS using the following equation of state derived from (1) and (2):

$$P = \frac{4}{a_0}\left(\frac{V}{V_{0t}}\right)^{-1/3} \left[0.4516\,\frac{\alpha'}{a_0^8}\left(\frac{V}{V_{0t}}\right)^{-8/3} + \frac{9}{32}\,k_B\theta_D\,\frac{\gamma}{a^2}\left(\frac{V}{V_{0t}}\right)^{-2/3} + \right.$$
$$\left. + \frac{1}{4}\,aT\beta B_T + \frac{1}{r\varrho}\exp\left(-\frac{r}{\varrho}\right)\left(b + \frac{\lambda f_0}{r_0^6}\right)\right], \tag{10}$$

where V_{0t} is the volume at the triple-point temperature corresponding to $P = 0$.

In performing calculations we have taken the repulsive energy $W(r)$ and the three-body interaction parameter $f(r)$ to depend exponentially on the interatomic distance r [9], i.e.

$$W(r) = b\exp\left(-\frac{r}{\varrho}\right), \tag{11}$$

$$f(r) = f_0\exp\left(-\frac{r}{\varrho}\right), \tag{12}$$

where b, ϱ, and f_0 are constants.

Table 1

Values of input data for RGS; triple-point temperature T_t in K, volume V_{0t} (at T_t) in 10^{-3} nm³/mol, Debye temperature θ_D in K, Grüneisen parameter γ, and bulk modulus B_0 in kbar, and volume thermal expansion coefficient β in 10^{-4} K^{-1}

RGS	T_t	V_{0t}	θ_D	γ	B_0	β
Ne	24.55	14.06	75	1.72	11.0	20.38
Ar	83.81	24.63	92	2.74	28.6	10.05
Kr	115.70	30.0	72	2.80	34.4	9.24
Xe	161.39	38.6	64	2.90	36.5	6.88

3. Results and Discussion

The model considered in the present work takes account of three-body interactions, zero-point energy, as well as thermal contributions. The expressions for elastic constants and equilibrium condition (equations (3) to (6)) and the equation of state (10) are valid at any temperature T.

Table 2

Calculated values of potential parameters; ϱ in 10^{-1} nm, $\lambda f(r)$ in 10^{-25} Jnm^6, and all other parameters in 10^{-12} Nnm^{-1}

RGS	ϱ	$\lambda f(r)$	α'/a^8	A	B	$(1/a^7)\times$ $\times[\lambda\,\partial f(r)/\partial r]$
Ne	0.214	0.032	157.8	854.9	− 58.1	+ 5.65
Ar	0.231	0.246	364.5	2169.0	−133.3	−11.12
Kr	0.220	0.670	386.4	2537.0	−138.6	−22.30
Xe	0.256	0.690	498.1	3151.0	−186.2	−10.81

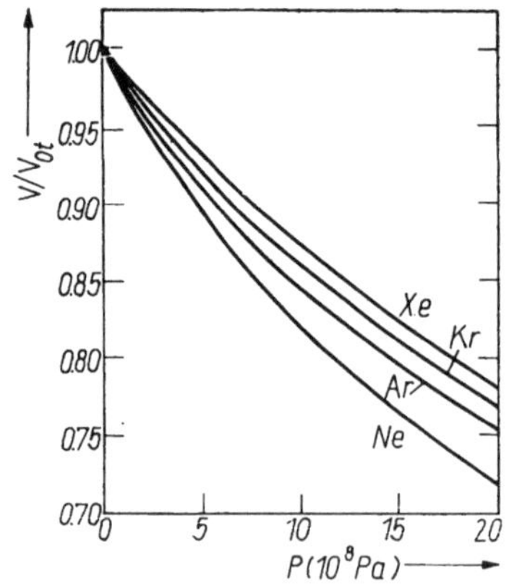

Fig. 1. Compression curves, V/V_{0t}, vs. pressure, P, for rare-gas solids Ne, Ar, Kr, and Xe at their triple-point temperatures given in Table 1

The input data on interatomic distance and elastic constants used in calculations are taken from [4]. Some of the input data taken from [1] are listed in Table 1. Calculated values of potential parameters are given in Table 2.

The compression curves for Ne, Ar, Kr, and Xe solids obtained with the help of (10) are produced in Fig. 1. It is observed from the figure that Ne is least compressible among the solids under study. The results reported in Fig. 1 are in good agreement with the experimental data reported by Anderson and Swenson [10] and Fugate and Swenson [11]. The experimental and calculated values of compression ratio at 20 kbar reported in Table 3 are in good agreement with each other.

Table 3

Compression ratio (V/V_{0t}) at 20 kbar (or 20×10^8 Pa)

RGS	calculated present study	experimental [10, 11]
Ne	0.72	0.68
Ar	0.76	0.77
Kr	0.77	0.78
Xe	0.78	0.79

References

[1] M. L. KLEIN and J. A. VENABLES, Ed., Rare-Gas Solids, Vol. I and II, Academic Press, London/New York 1976/77.
[2] K. S. PITZER, Adv. chem. Phys. **2**, 59 (1959).
[3] I. J. ZUCKER and D. PAIK, J. Phys. C **12**, 771 (1979).
[4] S. LEHRI and M. P. VERMA, phys. stat. sol. (b) **92**, 363 (1979); **98**, 789 (1980).
[5] R. J. BELL, J. Phys. B **3**, 751 (1970).
[6] M. B. DORAN and I. J. ZUCKER, J. Phys. C **4**, 307 (1971).
[7] J. SHANKER and A. K. RAJAURIA, phys. stat. sol. (b) **109**, 711 (1982).
[8] M. P. TOSI, Solid State Phys. **16**, 1 (1964).
[9] R. P. SINGH and J. SHANKER, phys. stat. sol. (b) **93**, 373 (1979).
[10] M. S. ANDERSON and C. A. SWENSON, J. Phys. Chem. Solids **36**, 145 (1975).
[11] R. Q. FUGATE and C. A. SWENSON, J. low-Temp. Phys. **10**, 317 (1973).

(Received July 26, 1984)

phys. stat. sol. (b) **126**, 87 (1984)

Subject classification: 6; 12.1; 21.3

Department of Physics, R.B.S. College, Bichpuri, Agra[1])

Phonons in Face-Centred Cubic Calcium and Strontium

By

S. P. Singh and R. P. S. Rathore

The axially symmetric and unpaired forces are employed to analyse the phonon dispersion and elastic behaviour of face centred cubic calcium and strontium which have so far not been studied adequately. The model with three parameters predicts the results which agree marvellously with the recently measured data.

Die axialsymmetrischen und ungepaarten Kräfte werden benutzt, um die Phononendispersion und das elastische Verhalten des bisher nicht adäquat untersuchten kubisch flächenzentrierten Kalziums und Strontiums zu analysieren. Das Modell mit drei Parametern sagt Ergebnisse voraus, die bemerkenswert gut mit den kürzlich gemessenen Werten übereinstimmen.

1. Introduction

Recently neutron inelastic experiments [1, 2] have yielded the dispersion relations in f.c.c. calcium and strontium. The BvK model extending to eight neighbours has been employed by experimentalists [1] to predict these relations. This study, however, involves too many parameters and adopts an extensive fitting procedure.

A large number of band structure calculations [3 to 6] have revealed that below the Fermi level the electronic bands are relatively simple and almost free-electron like. Because of the proximity of d-bands to the Fermi level, however, the hybridization plays an important role in determining the physical properties of these divalent metals. The generalized pseudopotential [7] theory employed by Moriarty [8] leads to higher results in case of these metals and assigns an f.c.c. structure to barium at normal temperature and pressure. Moriarty has considered empty d-bands in his later first-principle calculations [9]. Animalu [10] has used HAA II [11] and HAA III [12] form factors in his microscopic lattice-dynamical study of these metals. This study, however, loses efficacy because of the dependence of the form factor on a) the nature (local and non-local) of the pseudopotential, b) the correlation factor due to exchange and the correlation effects of the electrons, c) the treatment of the conduction and core electrons, d) the treatment of the orthogonalization, e) the effective valence and mass, and f) the non-linear correlation of the electrons. This study yields results which disagree with the measured data by about 65% at some wave vectors. Taut and Eschrig [13] using a non-local pseudopotential obtained a relatively better agreement with maximum deviation of 16%. This study, however, reproduces T-modes in poor shape.

We have therefore developed a simple model which assumes axially symmetric [14] and unpaired interactions between the first neighbours of the metals. The model employs only three disposable parameters and reproduces the dispersion relations and the elastic behaviour with remarkable success.

---

[1]) Agra 283105, India.

2. Theory

2.1 The axially symmetric forces

Bond stretching and bond bending forces between the nearest neighbours are deployed to express the couplings in terms of A and B. The diagonal and off-diagonal elements thus obtained may be written as

$$\begin{aligned} D^{\mathrm{A}}_{xx}(\boldsymbol{q}) &= 2A[2 - C_xC_y - C_xC_z] + 4B[3 - C_xC_y - C_xC_z - C_yC_z]\,, \\ D^{\mathrm{A}}_{xy}(\boldsymbol{q}) &= 2AS_xS_y\,, \end{aligned} \tag{1}$$

where $C_x = \cos (aq_x/2)$, $S_x = \sin (aq_x/2)$ with q_x as the x-component of the phonon wave vector and a as the lattice constant.

2.2 The unpaired forces

The unpaired forces [15] arising due to the deformation of the most prevalent angle of 60° between the first neighbours are used to determine the following elements of the dynamical matrix:

$$\begin{aligned} D^{\mathrm{U}}_{xx}(\boldsymbol{q}) &= 16K[3 - C_xC_y - C_xC_z - C_yC_z]\,, \\ D^{\mathrm{U}}_{xy}(\boldsymbol{q}) &= 16KS_xS_y\,, \end{aligned} \tag{2}$$

where K is the corresponding force parameter.

The total dynamical matrix may now be written as the sum of these simple matrices, i.e.

$$D(\boldsymbol{q}) = D^{\mathrm{A}}(\boldsymbol{q}) + D^{\mathrm{U}}(\boldsymbol{q})\,. \tag{3}$$

The usual secular determinant is solved to derive the dispersion relations along major symmetry directions.

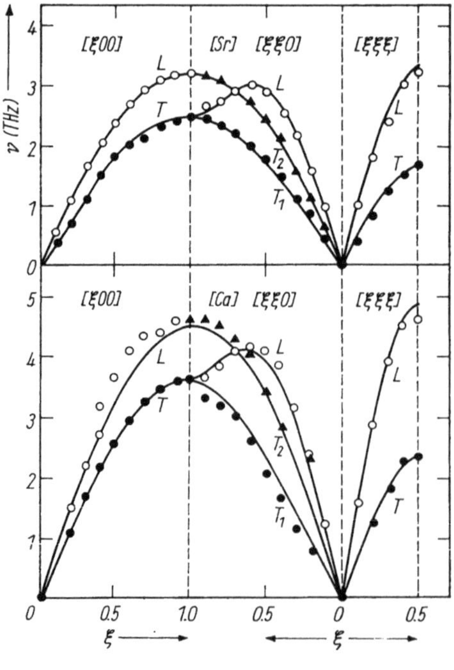

Fig. 1. Dispersion relations in f.c.c. calcium and strontium

Table 1
Input frequencies (in THz) and computed parameters (in N/m)

input data		computed parameters	
calcium	stontium	calcium	strontium
(100) $\nu_L = 4.52$	(100) $\nu_L = 3.2$	$A = 4.7630$	$A = 5.7290$
(110) $\nu_{Ti} = 3.63$	(110) $\nu_{Ti} = 2.5$	$B = -0.6699$	$B = -0.3370$
(111) $\nu_T = 2.36$	(111) $\nu_T = 1.7$	$K = 0.4106$	$K = 0.2870$

3. Calculations and Results

The three disposable parameters of the present model are computed by using three known frequencies, which are given in Table 1 along with the computed values of the parameters. Fig. 1 exhibits the dispersion relations in f.c.c. calcium and strontium.

The relative standard errors for our predictions and also for other theoretical studies [8, 10, 16] are calculated and shown in Table 2. The calculated elastic constants and

Table 2
Relative standard errors for different models

model reference	relative standard errors (%)	
	calcium	strontium
our model	5.3	3.7
Mehrotra et al. [16]	6.7	8.6
Animalu [10]	6.9	9.2
Moriarty [8]	16.4	9.4

their relative percentage deviations from those derived from the slope of the experimental relations are given in Table 3.

Table 3
Elastic constants and their relative percentage deviations

parameter	calcium	strontium
computed values (10^{11} N/m^2)	$C_{11} = 0.24043$	$C_{11} = 0.2418$
	$C_{12} = 0.10689$	$C_{12} = 0.0599$
	$C_{44} = 0.15507$	$C_{44} = 0.1476$
known values*) (10^{11} N/m^2)	$C_{11} = 0.27801$	$C_{11} = 0.2034$
	$C_{12} = 0.18225$	$C_{12} = 0.0791$
	$C_{44} = 0.16304$	$C_{44} = 0.0608$
relative percentage deviations (%)	$R_{11} = 13.5$	$R_{11} = 18.9$
	$R_{12} = 41.4$	$R_{12} = 24.2$
	$R_{44} = 4.9$	$R_{44} = 14.27$

*) Values of [1, 16] are used for Ca and Sr, respectively.

4. Conclusions

The inspection of Table 1 reveals that the bond stretching and unpaired forces account for the repulsive forces whereas bond bending forces stand for the attractive forces between the constituents of these metals. It may be recalled that while using central (radial) forces we require unpaired as well as volume forces to predict the phonon dispersion. Because of the alignment [17] of the electrons to counterbalance the self-repulsion, the bonding forces appropriately account for the forces due to electrons.

References

[1] C. STASSIS, J. HARMON, and R. M. NICKLOW, Phys. Rev. B **27**, 3303 (1983).
[2] L. BUCHENAU, private communication.
[3] A. O. E. ANIMALU and V. HEINE, Phys. Rev. **154**, 535 (1967).
[4] G. JOHENSEN, Solid State Commun. **7**, 731 (1969).
[5] G. JOHENSEN and A. R. MACKINTOSH, Solid State Commun. **8**, 121 (1970).
[6] J. P. JAN and H. L. SKRIVER, J. Phys. F **1**, 805 (1983).
[7] W. A. HARRISON, Phys. Rev. **131**, 1035 (1969).
[8] J. A. MORIARTY, Phys. Rev. B **6**, 4445 (1972).
[9] J. A. MORIARTY, Phys. Rev. B **28**, 4818 (1983).
[10] A. O. E. ANIMALU, Phys. Rev. **161**, 445 (1967).
[11] A. O. E. ANIMALU and V. HEINE, Phil Mag. **12**, 1249 (1965).
[12] A. O. E. ANIMALU, Proc. Roy. Soc. **A294**, 376 (1966).
[13] M. TAUT and H. ESCHRIG, phys. stat. sol. (b) **73**, 151 (1976).
[14] G. W. LEHMAN, T. WOLFRAM, and R. E. DEWAME, Phys. Rev. **128**, 1593 (1962).
[15] M. S. KUSHWAWA and S. S. KUSHWAHA, phys. stat. sol. (b) **87**, 247 (1978); **89**, 583 (1978); **91**, 689 (1979).
[16] K. N. MEHROTRA, G. TAMBAY, and S. K. KAPUR, Acta Ciencia Indica **3**, 89 (1979).
[17] G. H. WANNIER, C. MISNER, and G. SHAY, JR., Phys. Rev. **185**, 963 (1969).

(Received July 26, 1984)

phys. stat. sol. (b) **126**, 91 (1984)

Subject classification: 6; 22.1.2

Institut für Theoretische Physik II, Westfälische Wilhelms-Universität, Münster[1])

Calculation of Grüneisen Constants in Si

By

A. P. MAYER and R. K. WEHNER

Lattice dynamical calculations of the Grüneisen constants for general strain components are performed using the bond charge model of Weber with additional assumptions on anharmonic forces. The model is adjusted to fit data available at the Γ, X, and L points from various experimental techniques. The results along the main symmetry axes indicate that the Grüneisen constants depend sensitively on those model parameters which were originally determined from the phonon dispersion in the harmonic approximation. It is concluded that for a consistent description of phonon dispersion and anharmonic properties the model has to be revised.

Es werden gitterdynamische Berechnungen der Grüneisen-Konstanten für allgemeine Verzerrungen mit dem Bondcharge-Modell von Weber und zusätzlichen Annahmen über anharmonische Kräfte durchgeführt. Das Modell wird an experimentelle Daten aus der Literatur zum Γ-, X- und L-Punkt angepaßt, die mit unterschiedlichen experimentellen Methoden gewonnen wurden. Die Ergebnisse entlang der Hauptsymmetrierichtungen lassen erkennen, daß die Grüneisen-konstanten empfindlich von denjenigen Modellparametern abhängen, die ursprünglich unter Benutzung der harmonischen Näherung durch Anpassen an die Phononendispersion bestimmt wurden. Es wird geschlossen, daß für eine konsistente Beschreibung der Phononendispersion und anharmonischer Eigenschaften das Modell revidiert werden muß.

1. Introduction

Since the model calculations by Dolling and Cowley [1] and Jex [2] various experimental and theoretical studies of silicon have led to increased information about the lattice anharmonicity in this material and effects related to it. On the one hand, more mode Grüneisen constants have been measured in the meantime directly by neutron [3] and light scattering [4 to 7] experiments. Experimental data also became available from the stress dependence of two-phonon bands in infrared absorption [8]. There are, however, discrepancies among some of the data obtained by different experimental techniques, which require further study. On the other hand, there are calculations of mode Grüneisen constants within microscopic theories of phonon dispersion, where the lattice constant enters as a variable parameter [9 to 12]. The results of these calculations are, however, not detailed enough to enable a critical analysis of all available experimental data. Up to now, they yield only limited information about phonon–phonon interactions. To obtain more information about the latter one should consider general strain and not only volume changes under hydrostatic pressure.

The present study is based on the bond charge model of Weber [13] and Nielsen and Weber [14]. Additional assumptions on the anharmonic interactions are made. The two- and three-body potentials of this model have been fitted by Weber within the harmonic approximation to the phonon dispersion. The parameters of these potentials also enter the expressions for the third-order force constants (see, e.g., Leibfried and Ludwig [15] and Maradudin [16]). The Grüneisen constants depend, as we found from our calculations, sensitively on them. This suggests the possibility of testing a harmonic model in calculations of the present type.

[1]) Domagkstr. 75, D-4400 Münster, FRG.

2. Theory

Mode Grüneisen constants describe the strain dependence of the mode frequencies and are defined by

$$\gamma(\boldsymbol{q}j) = -\frac{V}{\omega_{\boldsymbol{q}j}} \frac{\partial}{\partial V} \omega_{\boldsymbol{q}j}(V) \tag{2.1}$$

and

$$\gamma_{\alpha\beta}(\boldsymbol{q}j) = -\frac{1}{\omega_{\boldsymbol{q}j}} \frac{\partial}{\partial \eta_{\alpha\beta}} \omega_{\boldsymbol{q}j}(\eta_{\alpha\beta}) \qquad (\alpha, \beta = x, y, z) . \tag{2.2}$$

Here, V is the crystal volume and the $\eta_{\alpha\beta}$ are the finite strain parameters. The volume Grüneisen constant is related to the strain Grüneisen constants by

$$\gamma(\boldsymbol{q}j) = \frac{1}{3} \left(\gamma_{xx}(\boldsymbol{q}j) + \gamma_{yy}(\boldsymbol{q}j) + \gamma_{zz}(\boldsymbol{q}j)\right) . \tag{2.3}$$

The mode frequencies of the deformed crystal are obtained from the eigenvalues of the strain-dependent dynamical matrix which, to first order in $\eta_{\alpha\beta}$, may be written

$$\omega_{\boldsymbol{q}j}^2 \delta_{jj'} - 2\omega_{\boldsymbol{q}j}\omega_{\boldsymbol{q}j'} \sum_{\alpha\beta} \gamma_{\alpha\beta}(\boldsymbol{q}jj')\, \eta_{\alpha\beta} . \tag{2.4}$$

If an infinitesimal rotation is superimposed on the pure strain, the eigenvectors of (2.4) have to be regarded as the polarization vectors of the modes in the strained crystal refering to the rotated coordinate system. For nondegenerate modes first-order perturbation theory yields

$$\gamma_{\alpha\beta}(\boldsymbol{q}j) = \gamma_{\alpha\beta}(\boldsymbol{q}jj) . \tag{2.5}$$

For degenerate modes the Grüneisen constants are obtained as the eigenvalues of the matrix

$$\Gamma_{\alpha\beta} = \left(\gamma_{\alpha\beta}(\boldsymbol{q}jj')\right) , \tag{2.6}$$

where j and j' are restricted to the degenerate branches ($\omega_{\boldsymbol{q}j} = \omega_{\boldsymbol{q}j'}$). At points in $\boldsymbol{q}$-space, where the degree of degeneracy changes, e.g. at Γ and X, the $\gamma_{\alpha\beta}$ behave to some extent nonanalytically.

The matrix elements in (2.4) are derived from the lattice potential and are given in the notation of Born and Huang [17] by

$$\gamma_{\alpha\beta}(\boldsymbol{q}jj') = -\frac{1}{4\omega_{\boldsymbol{q}j}\omega_{\boldsymbol{q}j'}} \sum_{\varrho\mu\nu} \sum_{\substack{ll' \\ pp'p''}} \frac{w_\mu(p' \mid \boldsymbol{q}j)\, w_\nu^*(p'' \mid \boldsymbol{q}j')}{\sqrt{M_{p'}M_{p''}}} \times$$
$$\times \Phi_{\varrho\mu\nu}\begin{pmatrix} l & l' & 0 \\ p & p' & p'' \end{pmatrix} \mathrm{e}^{i\boldsymbol{q}\cdot \boldsymbol{X}\left(\begin{smallmatrix} l' & 0 \\ p' & p'' \end{smallmatrix}\right)} \left[\delta_{\varrho\alpha} X_\beta \begin{pmatrix} l & 0 \\ p & p'' \end{pmatrix} + \right.$$
$$\left. + \delta_{\varrho\beta} X_\alpha \begin{pmatrix} l & 0 \\ p & p'' \end{pmatrix} - 2K_{\varrho\alpha\beta}(p)\right] \qquad (\mu, \nu, \varrho = x, y, z) . \tag{2.7}$$

In the exponential factor $\boldsymbol{q}$ and $\boldsymbol{X}$ both refer either to the strained or to the unstrained crystal, while all other quantities in this formula are those of the unstrained crystal. The indices p, p', and p'' are running over all particles in the elementary cell. In order to distinguish between the ions and the bond charges we denote them by $\varkappa$ and λ, respectively, and write within the adiabatic approximation

$$\frac{1}{\sqrt{M_p}}\, w_\alpha(p \mid \boldsymbol{q}j) = \begin{cases} \dfrac{1}{\sqrt{M_\varkappa}}\, w_\alpha(\varkappa \mid \boldsymbol{q}j) & (p = \varkappa)\,, \\ -\sum\limits_{\varrho\lambda'} \sum\limits_{\beta\varkappa'} D^{-1}_{\alpha\beta}\begin{pmatrix} \boldsymbol{q} \\ \lambda\ \lambda' \end{pmatrix} D_{\varrho\beta}\begin{pmatrix} \boldsymbol{q} \\ \lambda'\ \varkappa' \end{pmatrix} \dfrac{1}{\sqrt{M_{\varkappa'}}}\, w_\beta(\varkappa' \mid \boldsymbol{q}j) & \\ & (p = \lambda)\,. \end{cases} \tag{2.8}$$

The matrices D are the Fourier-transformed harmonic force constants of the unstrained crystal.

The quantities $K_{\varrho\alpha\beta}(p)$ account for the coupling of the external strain $\eta_{\alpha\beta}$ to the internal strain [18] in linear approximation, where the translation displacements of the sublattices are given by

$$d_\varrho(p) = -\sum_{\alpha\beta} K_{\varrho\alpha\beta}(p)\, \eta_{\alpha\beta}\,. \tag{2.9}$$

Detailed expressions for the coefficients K can be taken from [18]. The connection between $K_{\varrho\alpha\beta}(\varkappa)$ and the internal strain parameter ξ in [13] is given by

$$K_{\varrho\alpha\beta}(\varkappa = 1) = -\frac{a}{8}\, \xi\, |\varepsilon_{\varrho\alpha\beta}| = -K_{\varrho\alpha\beta}(\varkappa = 2)\,, \tag{2.10}$$

with the positions of atoms $\varkappa = 1$ and $\varkappa = 2$ as in [14]. Here a denotes the lattice constant and $\varepsilon_{\varrho\alpha\beta}$ is the Levi-Civita tensor. For the bond charges, which are centres of inversion, the coefficients K vanish.

Model central potentials $\varphi_{pp'}(r)$ have been assumed corresponding to short-range central forces between various pairs of particles. The third-order coupling coefficients depend on them through the parameters

$$A_{pp'} = \frac{1}{r^2}\left\{\varphi''_{pp'}(r) \quad \frac{\varphi'_{pp'}(r)}{r}\right\}, \tag{2.11}$$

$$B_{pp'} = \varphi'''_{pp'}(r)\,. \tag{2.12}$$

In addition to these the long-range Coulomb potentials

$$\varphi^{\mathrm{C}}_{pp'}(r) = \frac{e^2}{4\pi\varepsilon_0}\, \frac{Z_p Z_{p'}}{r} \qquad \left(Z_\lambda = -\frac{1}{2}\, Z_\varkappa\right) \tag{2.13}$$

were included. Furthermore, the third-order terms of the three-body Keating potentials, as employed by Weber up to second order,

$$\varphi_{K_2} = \frac{1}{4}\, C^{(2)} \sum_{l\varkappa\lambda\lambda'}{}' \left[\boldsymbol{x}\begin{pmatrix} l \\ \varkappa \end{pmatrix}\!\lambda\Big) \cdot \boldsymbol{x}\begin{pmatrix} l \\ \varkappa \end{pmatrix}\!\lambda'\Big) - \boldsymbol{X}\begin{pmatrix} l \\ \varkappa \end{pmatrix}\!\lambda\Big) \cdot \boldsymbol{X}\begin{pmatrix} l \\ \varkappa \end{pmatrix}\!\lambda'\Big)\right]^2 \tag{2.14}$$

and the cubic analogue

$$\varphi_{K_3} = \frac{1}{12}\, C^{(3)} \sum_{l\varkappa\lambda\lambda'}{}' \left[\boldsymbol{x}\begin{pmatrix} l \\ \varkappa \end{pmatrix}\!\lambda\Big) \cdot \boldsymbol{x}\begin{pmatrix} l \\ \varkappa \end{pmatrix}\!\lambda'\Big) - \boldsymbol{X}\begin{pmatrix} l \\ \varkappa \end{pmatrix}\!\lambda\Big) \cdot \boldsymbol{X}\begin{pmatrix} l \\ \varkappa \end{pmatrix}\!\lambda'\Big)\right]^3 \tag{2.15}$$

have been taken into account. $\boldsymbol{x}\left(\begin{smallmatrix} l \\ \varkappa \end{smallmatrix}\middle|\lambda\right)$ and $\boldsymbol{X}\left(\begin{smallmatrix} l \\ \varkappa \end{smallmatrix}\middle|\lambda\right)$ denote the actual and the equilibrium difference vector between atom $\begin{pmatrix} l \\ \varkappa \end{pmatrix}$ and its neighbour bond-charge of type λ.

The prime at the summation symbol means that the term with $\lambda = \lambda'$ is to be excluded. The constant factor $C^{(2)}$ is related to the parameter β in [13] by

$$C^{(2)} = \frac{16}{a^2} \beta . \tag{2.16}$$

3. Experimental Data

The measured mode Grüneisen constants considered in this investigation originate from various experimental techniques. They are listed in Tables 1 and 2. A few comments have to be made concerning different sources and on some deviations among the experimental values.

For the $q \approx 0$ acoustic modes the Grüneisen constants have been derived from the ultrasound propagation data of McSkimin and Andreatch [19]. We have determined them either directly from their values of $(\varrho W^2)'$ or, if that quantity was not given for a mode, we used their values of the third-order elastic constants and the relations given here in the Appendix. The quantities listed in Table 1 depend in principle only on the three second-order and six third-order elastic constants. The experimental values, however, represent more than nine different measurements. To fit the model parameters the constants γ, $\gamma_{\alpha\alpha}(\varepsilon, 0, 0)$, and $\gamma_{\alpha\alpha}(\varepsilon, \varepsilon, 0)$ have been used.

For the Grüneisen constant of the Raman mode, $\gamma(\Gamma \mid \mathrm{O})$, the mean value 1.00 of those found by Buchenauer et al. [6] and Weinstein and Piermarini [7] has been adopted (Table 2). This value is slightly higher than the earlier one of 0.9 measured by Anastassakis et al. [4]. We take from [4], however, the measured frequency splitting under uniaxial and shear strain, which can be expressed as

$$\gamma_{xx}(\varepsilon, 0, 0 \mid \mathrm{LO}) = \gamma_{zz}(\varepsilon, 0, 0 \mid \mathrm{TO}_z) = \frac{3p}{p + 2q} \gamma(\Gamma \mid \mathrm{O}) , \tag{3.1}$$

$$\gamma_{xx}(\varepsilon, 0, 0 \mid \mathrm{TO}) = \gamma_{zz}(\varepsilon, 0, 0 \mid \mathrm{LO}) = \gamma_{zz}(\varepsilon, 0, 0 \mid \mathrm{TO}_y) = \frac{3q}{p + 2q} \gamma(\Gamma \mid \mathrm{O}) , \tag{3.2}$$

$$\gamma_{xy}(\varepsilon, \varepsilon, 0 \mid \mathrm{LO}) = \gamma_{yz}(\varepsilon, 0, 0 \mid \mathrm{TO}_{yz}) = \frac{3r}{p + 2q} \gamma(\Gamma \mid \mathrm{O}) . \tag{3.3}$$

The parameters p, q, and r were introduced originally in [18].

To the X-point much attention has been paid in the past because of the remarkably negative Grüneisen constant of the TA mode. Data are available from three different

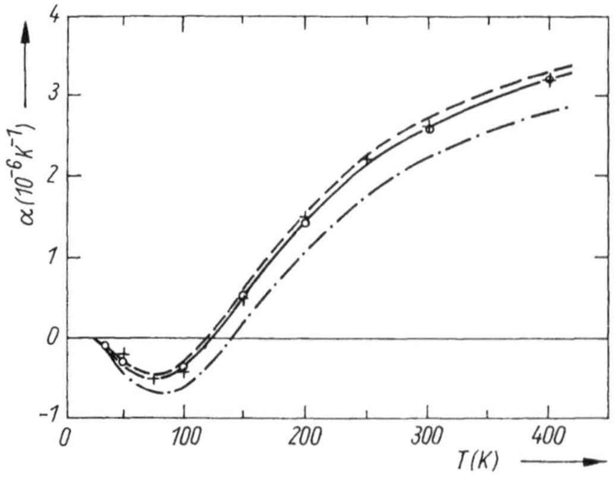

Fig. 1. Thermal expansion of Si for different input values of $\gamma(\mathrm{X} \mid \mathrm{TA})$: -1.40 (— — —), -1.47 (———); -1.74 (— · — · —). (○ [20], + [21])

Table 1

Grüneisen constants of long-wavelength acoustic modes in Si

q	isotropic dilation			uniaxial strain			shear strain		
		exp.	calc.		exp.	calc.		exp.	calc.
$(\varepsilon, 0, 0)$	γ (LA)	1.108	0.983	γ_{xx} (LA)	0.99	0.650			
				γ_{zz} (LA)	1.17	1.149			
	γ (TA)	0.324	0.456	γ_{xx} (TA)	0.91	1.039	γ_{yz} (TA_{yz})	−0.10	−0.145
				γ_{zz} (TA_y)	−0.48	−0.474			
				γ_{zz} (TA_z)	0.55	0.804			
$(\varepsilon, \varepsilon, 0)$	γ (LA)	1.109	1.063	γ_{xx} (LA)	1.40	1.347	γ_{xy} (LA)	0.89	0.979
				γ_{zz} (LA)	0.47	0.493			
	γ ($TA_{x\bar{y}}$)	−0.0487	−0.153	γ_{xx} ($TA_{x\bar{y}}$)	−0.71	−0.754	γ_{xy} ($TA_{x\bar{y}}$)	−0.28	−0.186
				γ_{zz} ($TA_{x\bar{y}}$)	1.27	1.049			
	γ (TA_z)	$\gamma(\varepsilon, 0, 0 \mid TA)$		γ_{xx}(TA_z)	0.21	0.282	γ_{xy}(TA_z)	$\gamma_{yz}(\varepsilon, 0, 0 \mid TA_{yz})$	
				γ_{zz}(TA_z)	$\gamma_{zz}(\varepsilon, 0, 0 \mid TA_z)$				
$(\varepsilon, \varepsilon, \varepsilon)$	γ(LA)	1.081	1.084	$\gamma_{\alpha\alpha}$(LA)	$\gamma(\varepsilon, \varepsilon, \varepsilon \mid LA)$		$\gamma_{\alpha\beta}$(LA)	0.34	0.352
	γ(TA)	0.113	0.117	$\gamma_{\alpha\alpha}$(TA_1)	1.11	1.045	$\gamma_{\alpha\beta}$(TA_1)	−0.24	−0.04
				$\gamma_{\alpha\alpha}$(TA_2)	−0.88	−0.812	$\gamma_{\alpha\beta}$(TA_2)	0.64	0.70

Table 2

Grüneisen constants measured by neutron scattering and optical experiments

q	quantity	exp./ref.		estimated	calc.
≈ 0	$\gamma(\Gamma \mid O)$	0.9	[4]	1.00	1.02
		1.02	[6]		
		0.98	[7]		
	q/p	1.5	[4]		0.72
	$\gamma_{xx}(\varepsilon, 0, 0 \mid LO)$			0.75	0.56
	$\gamma_{xx}(\varepsilon, 0, 0 \mid TO)$			1.125	1.25
	$3r/(p + 2q)$	0.394	[4]		0.40
	$\gamma_{xy}(\varepsilon, \varepsilon, 0 \mid LO)$			0.39	0.41
(1, 0, 0)	$\gamma(X \mid TA)$	−1.4	[5, 7]	−1.47	−1.46
		−1.74	[3]		
	$\gamma(X \mid TO)$	1.5	[7]		1.39
	$\gamma(X \mid TO - TA)$	3.3	[8]		2.87
	$\gamma_{xx}(X \mid TA)$	−2.05	[3]	−1.73	−1.68
	$\gamma_{zz}(X \mid TA)$	−1.59	[3]	−1.34	−1.35
	$\gamma_{yz}(X \mid T)$	±2.33			±0.53
$(\frac{1}{2}, \frac{1}{2}, \frac{1}{2})$.	$\gamma(L \mid TA)$	−1.3	[7]		−1.83
	$\gamma(L \mid TO)$	1.3	[7]		1.17
	$\gamma(L \mid LA + LO)$	0.944	[8]		0.98
	$\gamma(L \mid TA + TO)$	0.553	[8]		0.62

techniques with differences up to 20%. In our calculations the input value of $\gamma(X \mid TA)$ has been varied between the values of −1.4 [5, 7] and −1.74 [3] keeping all other input constant. An estimated value of −1.47 was selected on the basis of two arguments. Firstly, as is shown in Fig. 1, it yielded best agreement with the thermal expansion [20, 21]. Secondly, it was possible to reproduce just this value consistently with the model. The value determined in this way has been used to reduce the experimental values of [3] for $\gamma_{xx}(X \mid TA)$ and $\gamma_{zz}(X \mid TA)$ by a factor 1.47/1.74 to take into account the frequency splitting under uniaxial strain. The value of ±2.33 of [3] for $\gamma_{yz}(X \mid T)$ was considerably outside the typical range of the model. Shear strain data have not been used to fit the model.

Measurements of the stress dependence of two-phonon absorption bands lead directly to Grüneisen constants of combination frequencies [8],

$$\gamma(\boldsymbol{q} \mid j \pm j') = -\frac{V}{\omega_{\boldsymbol{q}j} \pm \omega_{\boldsymbol{q}j'}} \frac{\partial}{\partial V} (\omega_{\boldsymbol{q}j} \pm \omega_{\boldsymbol{q}j'}) = \frac{\omega_{\boldsymbol{q}j}\gamma(\boldsymbol{q}j) \pm \omega_{\boldsymbol{q}j'}\gamma(\boldsymbol{q}j')}{\omega_{\boldsymbol{q}j} \pm \omega_{\boldsymbol{q}j'}}. \tag{3.4}$$

Table 2 includes three quantities of this type.

4. Calculations

Starting with anharmonic ion–ion central forces between first and second neighbours we have obtained results similar to those of Jex [2]. In that work an unsatisfactory feature was the drastic failure of the LA modes. The LA modes improved remarkably when in addition the anharmonic ion–bond charge interaction (parameter $B_{\text{i-bc}}$ of

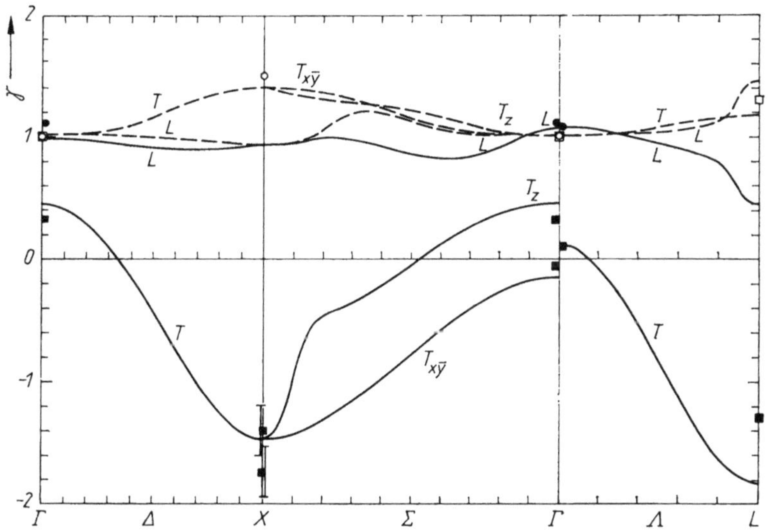

Fig. 2. Grüneisen constants of Si. ———, ●, ■ acoustic modes; – – – ○, □ optical modes; ●, ○ longitudinal; ■, □ transverse. Experimental data as in Table 2

Table 3) was introduced. In this case the second-neighbour ion–ion interaction turned out as unimportant and has been dropped in the subsequent calculations. There remained, however, various discrepancies with the experimental data, particularly of the TA modes. To achieve better agreement we have treated the bond charge–bond charge interaction as anharmonic (parameter B_{bc-bc}) and included the cubic terms of

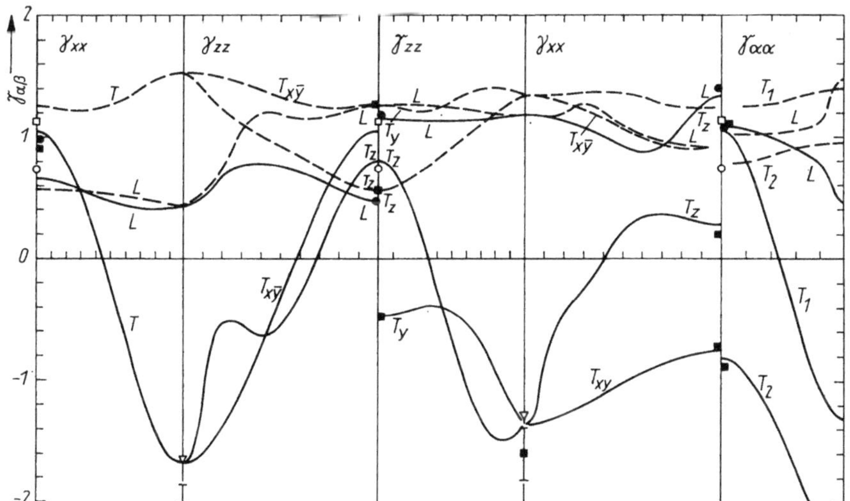

Fig. 3. Grüneisen constants for uniaxial strains (symbols as in Fig. 2). ▽ estimated values (Table 2). Polarization of mode $T_1 \perp \alpha$-axis

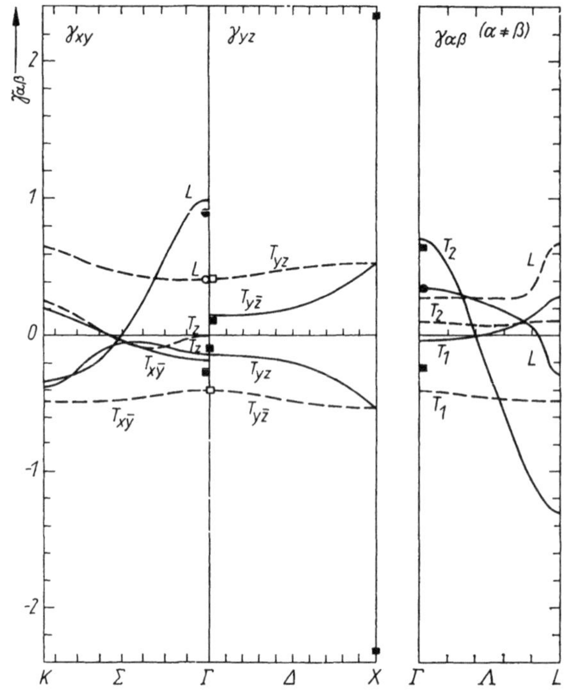

Fig. 4. Grüneisen constants for shear strain (symbols as in Fig. 2). Polarization of mode T_1 in $\alpha\beta$ plane

Table 3

Parameter of two- and three-body potentials

	parameter	a)	b)	c)	units
I	$A_{\text{i-i}}$	1.13	0.215	0.198	$\frac{e^2}{4\pi\varepsilon_0 v_a a^2} \times 10^2$
	$A_{\text{i-bc}}$	−4.14	−4.58	−4.31	
	$A_{\text{bc-bc}}$	—	−0.360	−0.365	
	$C^{(2)}$	1.38	0.645	0.639	
	Z^2_{bc}	0.18	0.005	—	1
II	$B_{\text{i-i}}$		−4.40	−4.41	$\frac{e^2}{4\pi\varepsilon_0 v_a a} \times 10^2$
	$B_{\text{i-bc}}$		0.788	0.867	
	$B_{\text{bc-bc}}$		−0.859	−0.860	
	$C^{(3)}$		77.3	77.4	$\frac{e^2}{4\pi\varepsilon_0 v_a a^3} \times 10^2$

I Entering both second- and third-order force constants.
II Entering only force constants of higher than second order.
a) Parameters fitted to phonon dispersion by Weber [13].
b) Parameters for Grüneisen constants with long-range Coulomb force.
c) Parameters for Grüneisen constants with short-range interactions only (Fig. 2 to 4).

the Keating potential (2.14). We have further taken into account the cubic analogue to the latter interaction (2.15) and incorporated the cubic anharmonicity of the long-range Coulomb forces. All these extensions did not lead to striking improvements. By detailed inspection, it became obvious, however, that the Grüneisen constants depend strongly on the parameters of group I in Table 3. These parameters enter the expressions for the cubic coupling coefficients in (2.7). They contribute considerably, i.e. the so-called Leibfried approximation is not applicable.

Assuming that the parameters of group I, as determined by fitting the phonon dispersion curves [13], may not be adequate for a calculation of anharmonic properties, we let them vary in addition to the purely anharmonic parameters of group II. The parameter values obtained in this way are given in column b of Table 3. It exhibits only a very small value of the bond charge. A special fit with $Z_{bc}^2 = 0.18 =$ const led to different values of the other parameters but only to small changes (1 to 2%) of the Grüneisen constants. It does not seem possible to conclude from the present data the role of the long-range anharmonic forces. The results shown in Fig. 2 to 4 correspond to the parameters of column c in Table 3 ($Z_{bc} = 0$). To obtain a safer estimate of the value of the bond charge it might be helpful to consider the strength of the infrared absorption.

5. Strain-Induced Frequency Splitting of the Degenerate Transverse Modes in Δ and Λ Directions

The Grüneisen constants of Fig. 3 and 4 represent phonon frequency shifts for single strain components $\eta_{\alpha\beta}$. These constants are basically sufficient for calculations of the frequency shifts in case of general strain, i.e. for more than one $\eta_{\alpha\beta}$ being nonzero. For nondegenerate modes this is simply achieved by linear combinations of the $\gamma_{\alpha\beta}$. For the degenerate transverse modes in Δ and Λ directions one has to diagonalize the matrix (2.4) with j and j' referring either to the TA or to the TO modes. From the eigenvalues λ_n of the rank-two matrix

$$\sum_{\alpha\beta} \Gamma_{\alpha\beta}\eta_{\alpha\beta} = \left(\sum_{\alpha\beta} \gamma_{\alpha\beta}(j, j')\,\eta_{\alpha\beta}\right) \tag{5.1}$$

one obtains the relative frequency shifts

$$\frac{\Delta\omega_n}{\omega_j} = -\lambda_n \qquad (n = 1, 2)\,. \tag{5.2}$$

For single strain components $\eta_{\alpha\beta}$, the eigenvalues of the corresponding matrices $\Gamma_{\alpha\beta}$, as shown in Fig. 3 and 4, are denoted by $\gamma_{\alpha\beta}(s)$ $(s = 1, 2)$. In the general case symmetry requirements imply restrictions on the elements of the matrices $\Gamma_{\alpha\beta}$. The conditions following from symmetry can be used to write the eigenvalues λ_n in terms of the quantities $\gamma_{\alpha\beta}(s = 1) \pm \gamma_{\alpha\beta}(s = 2)$. The resulting expressions are:

$\boldsymbol{q} = (\zeta, 0, 0)$

$$\lambda_{1,2} = \gamma_{xx}\eta_{xx} + \tfrac{1}{2}\left(\gamma_{zz}(1) + \gamma_{zz}(2)\right)\{\eta_{yy} + \eta_{zz}\} \pm$$
$$\pm \tfrac{1}{2}\left[\left(\gamma_{zz}(1) - \gamma_{zz}(2)\right)^2 \{\eta_{yy} - \eta_{zz}\}^2 + 4\left(\gamma_{yz}(1) - \gamma_{yz}(2)\right)^2 \eta_{yz}^2\right]^{1/2} \tag{5.3}$$

with

$$\gamma_{xx} = \gamma_{xx}(1) = \gamma_{xx}(2)\,.$$

$\boldsymbol{q} = (\zeta, \zeta, \zeta)$

$$\lambda_{1,2} = \tfrac{1}{2}\left(\gamma_{zz}(1) + \gamma_{zz}(2)\right)\{\eta_{xx} + \eta_{yy} + \eta_{zz}\} +$$
$$+ \left(\gamma_{xy}(1) + \gamma_{xy}(2)\right)\{\eta_{yz} + \eta_{xz} + \eta_{xy}\} \pm$$

$$\pm \tfrac{1}{2} [(\gamma_{zz}(1) - \gamma_{zz}(2))^2 \{\eta_{xx}^2 + \eta_{yy}^2 + \eta_{zz}^2 - \eta_{xx}\eta_{yy} - \eta_{xx}\eta_{zz} - \eta_{yy}\eta_{zz}\} + \\ + 4(\gamma_{xy}(1) - \gamma_{xy}(2))^2 \{\eta_{yz}^2 + \eta_{xz}^2 + \eta_{xy}^2 - \eta_{yz}\eta_{xz} - \eta_{yz}\eta_{xy} - \eta_{xz}\eta_{xy}\} + \\ + 2(\gamma_{zz}(1) - \gamma_{zz}(2)) (\gamma_{xy}(1) - \gamma_{xy}(2)) \{\eta_{xx}(2\eta_{yz} - \eta_{xz} - \eta_{xy}) + \\ + \eta_{yy}(2\eta_{xz} - \eta_{yz} - \eta_{xy}) + \eta_{zz}(2\eta_{xy} - \eta_{yz} - \eta_{xz})\}]^{1/2} . \tag{5.4}$$

Here, $\gamma_{zz}(1)$ and $\gamma_{xy}(1)$ belong to the same eigenvector. To facilitate the interpretation of experimental data, e.g. from neutron scattering measurements, the polarization of the modes under deformation will be described here. The eigenvectors are of the form

$$\boldsymbol{w}'(\varkappa \mid \boldsymbol{q}n) = e_1(n)\, \boldsymbol{w}(\varkappa \mid \boldsymbol{q}j_1) + e_2(n)\, \boldsymbol{w}(\varkappa \mid \boldsymbol{q}j_2) , \tag{5.5}$$

where $\boldsymbol{w}(\varkappa \mid \boldsymbol{q}j_1)$ and $\boldsymbol{w}(\varkappa \mid \boldsymbol{q}j_2)$ are basis vectors [24], given below for the Δ and Λ directions. The coefficients $e_1(n)$ and $e_2(n)$ obey the conditions

$$|e_1(n)|^2 + |e_2(n)|^2 = 1 , \tag{5.6}$$

$$\frac{e_2(n)}{e_1(n)} = \frac{A + \lambda_n - \lambda_m}{B} , \qquad m = \begin{cases} 2, & n = 1 , \\ 1, & n = 2 , \end{cases} \tag{5.7}$$

where A and B depend on $\boldsymbol{q}$ in the following way:

$\boldsymbol{q} = (\zeta, 0, 0)$

Basis

$$\left.\begin{aligned} w_\alpha(\varkappa \mid \boldsymbol{q}j_1) &= \frac{1}{\sqrt{2}} (0, a, -a; 0, b, -b) \\ w_\alpha(\varkappa \mid \boldsymbol{q}j_2) &= \frac{1}{\sqrt{2}} (0, b, b; 0, a, a) \end{aligned}\right. \qquad \begin{cases} |a|^2 + |b|^2 = 1 , \\ |a| = |b| . \end{cases} \tag{5.8}$$

Furthermore

$$A = 2(\gamma_{yz}(j_2) - \gamma_{yz}(j_1))\, \eta_{yz} , \tag{5.9}$$

$$B = (\gamma_{zz}(s = 2) - \gamma_{zz}(s = 1))\, \{\eta_{yy} - \eta_{zz}\} . \tag{5.10}$$

To correlate $s = 1$ and $s = 2$ in the expression for B with the branches in Fig. 3, it should be noticed that the eigenvector belonging to $\gamma_{zz}(1)$ is

$$v_\alpha(\varkappa \mid \boldsymbol{q}1) = \tfrac{1}{2} (0, b + a, b - a; 0, a + b, a - b) . \tag{5.11}$$

In the limit of $\boldsymbol{q} \to 0$ one finds from this

$$\gamma_{zz}(1) = \gamma_{zz}(\mathrm{TA}_y) \qquad (a = b) , \tag{5.12}$$

or

$$\gamma_{zz}(1) = \gamma_{zz}(\mathrm{TO}_z) \qquad (a = -b) , \tag{5.13}$$

respectively.

$\boldsymbol{q} = (\zeta, \zeta, \zeta)$

Basis

$$\left.\begin{aligned} w_\alpha(\varkappa \mid \boldsymbol{q}j_1) &= \frac{1}{\sqrt{2}} (a, -a, 0; b, -b, 0) \\ w_\alpha(\varkappa \mid \boldsymbol{q}j_2) &= \frac{1}{\sqrt{6}} (a, a, -2a; b, b, -2b) \end{aligned}\right. \qquad \begin{cases} |a|^2 + |b|^2 = 1 , \\ |a| = |b| . \end{cases} \tag{5.14}$$

The polarization vectors follow with

$$A = (\gamma_{zz}(j_2) - \gamma_{zz}(j_1))\, \{\eta_{zz} - \tfrac{1}{2} (\eta_{xx} + \eta_{yy})\} + \\ + 2(\gamma_{xy}(j_2) - \gamma_{xy}(j_1))\, \{\eta_{xy} - \tfrac{1}{2} (\eta_{xz} + \eta_{yz})\} , \tag{5.15}$$

$$B = \frac{\sqrt{3}}{2} [(\gamma_{zz}(j_2) - \gamma_{zz}(j_1)) \{\eta_{xx} - \eta_{yy}\} + 2(\gamma_{xy}(j_2) - \gamma_{xy}(j_1)) \{\eta_{yz} - \eta_{xz}\}] . \tag{5.16}$$

If the denominator B in (5.7) vanishes, either the original basis vectors are the appropriate ones or there is no splitting at all.

6. Conclusions

It was found that the Grüneisen constants depend strongly on those parameters of the two- and three-particle potentials which are used to specify the model within the harmonic approximation. It appeared difficult to obtain satisfactory agreement with the experimental data without altering the values of these parameters as given in [13]. For a consistent description of the phonon dispersion and of the Grüneisen constants a revision of the model is suggested. The cubic terms in the lattice potential have been taken into account in so far as they contribute linearly to the dynamical matrix. Changes of the dispersive part of the phonon self-energy have not been worked out. These would be related with changes of the one-phonon and two-phonon spectral densities. To assess their importance one might investigate the pressure dependence of the phonon linewidths similarly to their temperature dependence [25].

Appendix

Relations between third-order elastic constants and Grüneisen constants of long-wavelength acoustic modes

The connection of the dynamical matrix with the stress–strain relations of an elastic medium, i.e. with the elastic constants has been worked out in general form by Thurston and Brugger [23]. It can also be established with the help of the Kun-Huang tensor $\hat{C}$ (Leibfried and Ludwig [22]). For cubic crystals and for $\boldsymbol{q}$ in the main symmetry directions one obtains the following expressions. Here $\boldsymbol{e}(j)$ denotes the polarization vector of the mode j.

$\boldsymbol{q} = (\varepsilon, 0, 0)$

$$j_1 = \text{LA}\,, \qquad \boldsymbol{e}(j_1) = (1, 0, 0)\,, \qquad k = -\frac{1}{2C_{11}},$$

$$\gamma_{xx} = k(3C_{11} + C_{111})\,,$$

$$\gamma_{yy} - \gamma_{zz} = k(C_{12} + C_{112})\,.$$

$$j_2 = \text{TA}_y\,, \qquad \boldsymbol{e}(j_2) = (0, 1, 0)\,,$$

$$j_3 = \text{TA}_z\,, \qquad \boldsymbol{e}(j_3) = (0, 0, 1)\,, \qquad k = -\frac{1}{2C_{44}},$$

$$\gamma_{xx}(j_2) = \gamma_{xx}(j_3) = k(C_{11} + C_{166})\,,$$

$$\gamma_{yy}(j_2) = \gamma_{zz}(j_3) = k(C_{12} + 2C_{44} + C_{166})\,,$$

$$\gamma_{yy}(j_3) = \gamma_{zz}(j_2) = k(C_{12} + C_{144})\,,$$

modes for shear strain η_{yz}:

$$\boldsymbol{e}(\text{TA}_{y\bar{z}}) = \frac{1}{\sqrt{2}} (0, 1, 1):\ \gamma_{yz}(\text{TA}_{yz}) = k(C_{44} + C_{453})\,,$$

$$\boldsymbol{e}(\text{TA}_{y\bar{z}}) = \frac{1}{\sqrt{2}} (0, 1, -1):\ \gamma_{yz}(\text{TA}_{y\bar{z}}) = -\gamma_{yz}(\text{TA}_{yz})\,.$$

$\boldsymbol{q} = (\varepsilon, \varepsilon, 0)$

$$j_1 = \mathrm{LA}\,, \qquad \boldsymbol{e}(j_1) = \frac{1}{\sqrt{2}}\,(1, 1, 0)\,, \qquad k = -\frac{1}{2(C_{11} + C_{12} + 2C_{44})}\,.$$

$$\gamma_{xx} = \gamma_{yy} = \frac{k}{2}\,(4C_{11} + 4C_{12} + 4C_{44} + C_{111} + 3C_{112} + 4C_{166})\,,$$

$$\gamma_{zz} = k(2C_{12} + C_{112} + C_{123} + 2C_{144})\,,$$
$$\gamma_{xy} = k(C_{11} + C_{12} + 4C_{44} + 4C_{166})\,.$$

$$j_2 = \mathrm{TA}_{x\bar{y}}\,, \qquad \boldsymbol{e}(j_2) = \frac{1}{\sqrt{2}}\,(1, -1, 0)\,, \qquad k = -\frac{1}{2(C_{11} - C_{12})}\,,$$

$$\gamma_{xx} = \gamma_{yy} = \frac{k}{2}\,(4C_{11} + C_{111} - C_{112})\,,$$

$$\gamma_{zz} = k(2C_{12} + C_{112} - C_{123})\,,$$
$$\gamma_{xy} = k(-C_{11} + C_{12} + 2C_{44})\,.$$

$$j_3 = \mathrm{TA}_z\,, \qquad \boldsymbol{e}(j_3) = (0, 0, 1)\,, \qquad k = -\frac{1}{2C_{44}}\,,$$

$$\gamma_{xx} = \gamma_{yy} = \frac{k}{2}\,(C_{11} + C_{12} + C_{144} + C_{166})\,,$$

$$\gamma_{zz} = \gamma_{zz}(\varepsilon, 0, 0 \mid \mathrm{TA}_z)\,,$$
$$\gamma_{xy} = \gamma_{yz}(\varepsilon, 0, 0 \mid \mathrm{TA}_{yz})\,.$$

$\boldsymbol{q} = (\varepsilon, \varepsilon, \varepsilon)$

$$j_1 = \mathrm{LA}\,, \qquad \boldsymbol{e}(j_1) = \frac{1}{\sqrt{3}}\,(1, 1, 1)\,, \qquad k = -\frac{1}{3(C_{11} + 2C_{12} + 4C_{44})}\,,$$

$$\gamma_{xx} = \gamma_{yy} = \gamma_{zz} = \gamma\,,$$
$$\gamma = \frac{k}{2}\,(5C_{11} + 10C_{12} + 8C_{44} + C_{111} + 6C_{112} + 2C_{123} + 4C_{144} + 8C_{166})\,,$$

$$\gamma_{xy} = \gamma_{yz} = \gamma_{zx} = k(C_{11} + 2C_{12} + 7C_{44} + 2C_{144} + 4C_{166} + 4C_{456})\,.$$

$$j_2 = \mathrm{TA}_{x\bar{y}}\,, \qquad \boldsymbol{e}(j_2) = \frac{1}{\sqrt{2}}\,(1, -1, 0)\,, \qquad k = -\frac{1}{2(C_{11} - C_{12} + C_{44})}\,,$$

$$j_3\colon\ \boldsymbol{e}(j_3) = \frac{1}{\sqrt{6}}\,(1, 1, -2)\,,$$

$$\gamma_{zz}(j_2) = k(C_{11} + 2C_{12} + C_{112} - C_{123} + C_{166})\,,$$

$$\gamma_{zz}(j_3) = \frac{k}{3}(7C_{11} + 2C_{12} + 4C_{44} + 2C_{111} - 3C_{112} + C_{123} + 2C_{144} + C_{166})\,,$$

$$\gamma_{xy}(j_2) = -k(C_{11} - C_{12} - C_{44} + C_{456})\,,$$

$$\gamma_{xy}(j_3) = \frac{k}{3}\,(C_{11} - C_{12} + 7C_{44} - 4C_{144} + 4C_{166} + C_{456})\,.$$

The above polarization vectors are eigenvectors only for strain η_{zz} and η_{xy}. The eigenvectors appropriate to the other components of the strain tensor are obtained by cyclic permutation of the cartesian indices.

References

[1] G. DOLLING and R. A. COWLEY, Proc. Phys. Soc. (London) **88**, 463 (1966).
[2] H. JEX, phys. stat. sol. (b) **45**, 343 (1971).
[3] J. PRECHTEL, J. KALUS, E. LÜSCHER, L. PINTSCHOVIUS, and R. GHOSH, phys. stat. sol. (b) **93**, 653 (1979).
[4] E. ANASTASSAKIS, A. PINCZUK, E. BURSTEIN, F. H. POLLAK, and M. CARDONA, Solid State Commun. **8**, 133 (1970).
[5] W. RICHTER, J. B. RENUCCI, and M. CARDONA, Solid State Commun. **16**, 131 (1975).
[6] C. J. BUCHENAUER, F. CERDEIRA, and M. CARDONA, in: Proc. 2nd Internat. Conf. Light Scattering in Solids, Paris 1971, Ed. M. BALKANSKI, Flammarion, Paris 1971 (p. 280).
[7] B. A. WEINSTEIN and G. J. PIERMARINI, Phys. Rev. B **12**, 1172 (1975).
[8] J. F. ANGRESS, G. A. GLEDHILL, and A. J. MAIDEN, J. Phys. C **8**, 1136 (1975).
[9] I. ISHIDA, J. Phys. Soc. Japan **39**, 1282 (1975).
[10] T. SOMA, phys. stat. sol. (b) **82**, 319 (1977).
[11] H. WENDEL and R. M. MARTIN, Phys. Rev. B **19**, 5251 (1979).
[12] M. T. YIN and M. L. COHEN, Phys. Rev. Letters **45**, 1004 (1980); Phys. Rev. B **26**, 3259 (1982).
[13] W. WEBER, Phys. Rev. B **15**, 4789 (1977).
[14] O. H. NIELSEN and W. WEBER, Computer Phys. Commun. **18**, 101 (1979).
[15] G. LEIBFRIED and W. LUDWIG, Solid State Phys. **12**, 275 (1961).
[16] A. A. MARADUDIN, phys. stat. sol. **2**, 1493 (1962).
[17] M. BORN and K. HUANG, Dynamical Theory of Crystal Lattices, Clarendon Press, Oxford 1954.
[18] S. GANESAN, A. A. MARADUDIN, and J. OITMAA, Ann. Phys. (USA) **56**, 556 (1970).
[19] H. J. MCSKIMIN and P. ANDREATCH, J. appl. Phys. **35**, 3312 (1964).
[20] H. IBACH, phys. stat. sol. **31**, 625 (1969).
[21] Y. S. TOULOUKIAN, R. K. KIRBY, R. E. TAYLOR, and T. Y. R. LEE, Thermal Expansion, Thermophysical Properties of Matter, Vol. 13, Plenum Press, New York 1977.
[22] G. LEIBFRIED and W. LUDWIG, Z. Phys. **169**, 80 (1960).
[23] R. N. THURSTON and K. BRUGGER, Phys. Rev. A **133**, 1604 (1964).
[24] M. LAX, Symmetry Principles in Solid State and Molecular Physics, Wiley, New York 1974.
[25] J. MENENDEZ and M. CARDONA, Phys. Rev. B **29**, 2051 (1084).

(Received June 14, 1984)

phys. stat. sol. (b) **126**, 105 (1984)

Subject classification: 20.1; 6; 13.5; 22.6

Department of Physics, College of Humanities and Sciences, Nihon University, Tokyo[1])

Radiative Optical Modes of Vibration in NiO Thin Film

By

S. MOCHIZUKI

Radiative optical modes of vibration in NiO thin films on metal substrates are observed at various temperatures from room temperature to 1273 K by reflection and thermal emission methods. A comparison of the experimental spectra and the theoretical ones shows that the correlation between them is good. The upper branch of the polariton dispersion curve of NiO at room temperature is determined experimentally by analysing the reflection spectra due to the virtual mode near the longitudinal optical phonon frequency. The shift in the longitudinal optical phonon frequency with temperature is discussed in terms of thermal expansion and other anharmonic effects.

Die strahlenden optischen Schwingungsmoden von dünnen NiO-Schichten auf Metallsubstraten werden bei verschiedenen Temperaturen von Zimmertemperatur bis 1273 K mit einer Reflexions-Methode und einer Thermoemissions-Methode gemessen. Der Vergleich der experimentellen und theoretischen Spektren bei Zimmertemperatur zeigt gute Übereinstimmung. Die Spektren im Bereich der LO-Phononenfrequenz werden mit einer Virtuell-Moden-Theorie analysiert, und der obere Zweig der Polariton-Dispersions-Kurve bei Zimmertemperatur wird experimentell bestimmt. Die Verschiebung der LO-Phononenfrequenz mit zunehmender Temperatur wird durch thermische Ausdehnung und anharmonische Effekte erklärt.

1. Introduction

The physical properties of nickel monoxide (NiO) have received a great deal of both theoretical and experimental interest. Hitherto, the main interest of physicists was concentrated on the origin of the insulating nature and the antiferromagnetic ordering. To explain the insulating behaviour of compounds with partially filled 3d-bands, various models for the electron states and for the carrier motion have been proposed by many investigators [1, 2]. It is not easy to confirm by measurement of the transport properties only which of these models is most appropriate, because the predominance of a wide 2p-band conduction masks the behaviour of the 3d-electrons with low mobility in the real NiO crystals [3 to 5]. On the other hand, infrared optical spectra provide information on the lattice vibrations and are also useful for estimating the strength of the 3d electron–phonon interactions. As to the infrared spectra of the NiO bulk crystal, only a few studies [6, 7] have been reported and these measurements are limited to room temperature. Since the infrared spectrum is strongly temperature-dependent, it is desirable to measure it over a wide temperature range. More recently, we measured the infrared spectra of bulk crystals [8], together with microcrystals [9], and discussed the temperature dependence of the lattice vibrations in NiO. The purpose of this paper is to report a further study of the infrared optical properties of NiO, using thin-film specimens.

The lattice of NiO has NaCl structure above the Néel temperature (523 K), but below the Néel temperature the crystal transforms to the antiferromagnetic structure which is slightly distorted from the ideal NaCl structure. Since the distortion is quite

[1]) 3-25-40, Sakurajōsui, Setagaya-ku, Tokyo 156, Japan.

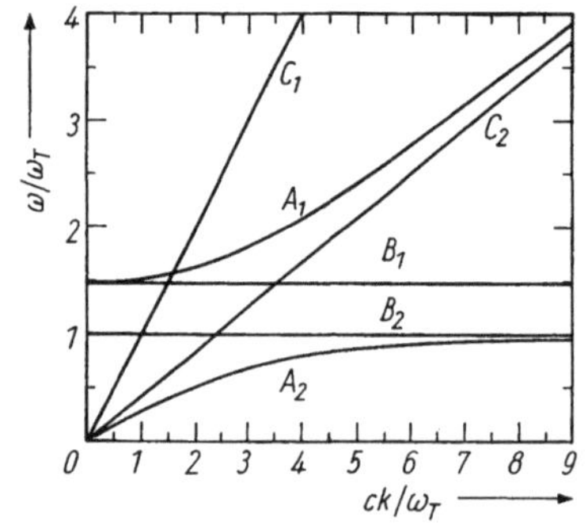

Fig. 1. Polariton dispersion curves of NiO at room temperature. A_1 and A_2 dispersion curves of polaritons, B_1 dispersion line of LO-phonon frequency, B_2 dispersion line of TO-phonon frequency, C_1 dispersion line of light in vacuum, C_2 dispersion line of light in NiO without retardation

small, it is often supposed that the phonon spectrum does not depend on magnetic ordering. The infrared properties of NiO are governed by the vibrational modes in which the sublattice of Ni ions moves opposite to the sublattice of O ions. In the NiO crystal, such a motion is associated with strong electric moments and hence interacts directly with the electric field of the incident radiation. The strong interaction leads to the formation of polariton modes (coupled photon–phonon modes). The polariton modes decay after a certain time through some photon or phonon emission processes. Fig. 1 shows the polariton dispersion curves of NiO at room temperature calculated from the following relation:

$$\left(\frac{ck}{\omega}\right)^2 = \varepsilon(\infty) + \frac{\varepsilon(0) - \varepsilon(\infty)}{1 - \left(\frac{\omega}{\omega_\mathrm{T}}\right)^2}, \tag{1}$$

where $\boldsymbol{k}$ is the wave vector, ω the frequency, c the light velocity in free space, $\varepsilon(0)$ the static dielectric constant, $\varepsilon(\infty)$ the high-frequency dielectric constant, and ω_T the transverse optical phonon frequency at zero wave vector. In the calculation, we used the dispersion parameters as shown below,

$$\varepsilon(0) = 12.4\,, \qquad \varepsilon(\infty) = 5.7\,, \qquad \omega_\mathrm{T} = 400\ \mathrm{cm}^{-1}\,.$$

From the point of Maxwell's equations, the optical property of an ionic crystal thin film is discussed by Fuchs and Kliewer [10]. The theory is valid when the thickness of the film is much smaller than the wavelength of the incident infrared radiation and is larger than the interatomic spacing. The finite thickness of the film causes two types of modes: (i) nonradiative modes with the exponentially damped fields outside the film, (ii) radiative modes with incoming or outgoing oscillatory polarized waves outside the film. The radiative modes are accessible to usual optical measurements and it is argued that the modes are virtual modes [10]. The observation of the radiation due to the virtual modes is made in the region of $\omega > ck$ in the ω–k_x plane, where ω is the virtual-mode frequency and k_x the x-component of the wave vector: the x-axis is parallel to the specimen layer. The virtual-mode equations for p-polarized radiation from a parallel-flat specimen with the thickness $d = 2a$, at oblique angle θ, are

$$-\frac{i\varepsilon\beta_0}{\beta} = \tan \beta a\,, \tag{2}$$

$$\frac{i\varepsilon\beta_0}{\beta} = \cot \beta a \, , \tag{3}$$

$$\beta_0 = \sqrt{\left(\frac{\omega}{c}\right)^2 - k_x^2}\, , \qquad \beta = \sqrt{\left(\frac{\omega}{c}\right)^2 \varepsilon - k_x^2}\, ; \qquad k_x = \frac{\omega \sin \theta}{c}\, , \tag{4}$$

where ε is the frequency-dependent dielectric constant of the film material. For very thin films, $d \ll c/\omega_T$, these virtual-mode equations give the following solutions: (i) OTH-mode centred near the longitudinal optical phonon (LO-phonon) frequency, ω_L, (ii) 2TL-mode centred near the transverse optical phonon (TO-phonon) frequency, ω_T, of the film material. The OTH-mode and the 2TL-mode correspond to the polarization perpendicular and parallel to the specimen layer, respectively. Since k_x is varied with the incident (emission) angle, these virtual mode frequencies are dependent on θ, and then the upper branch of the polariton dispersion curve is experimentally determined by measuring the incident (emission) angle dependence of the reflection (thermal-emission) spectrum due to the OTH-mode. Morever, since the OTH-mode frequency is close to the LO-phonon frequency at zero wave vector for small θ, such measurement at small θ at various temperatures make it possible to plot the temperature dependence of the LO-phonon frequency at zero wave vector.

In the present work, the infrared spectra due to the radiative optical modes of vibration in NiO thin films on metal substrates are observed at various temperatures from room temperature to 1273 K by reflection and thermal-emission methods. The observed spectra are analyzed by the Fuchs-Kliewer theory, and the upper branch of the polariton dispersion curve and the temperature dependence of the LO-phonon frequency of NiO are given with some discussion.

2. Experimental

The NiO thin-film specimens were produced as follows. The nickel films were made by evaporating nickel of 99.98% purity from a heated alumina crucible onto a smooth-surface platinum substrate whose reflectivity coincides with that of the evaporated gold plane mirror within experimental error in the frequency range from 250 to 1000 cm^{-1}. The evaporated nickel on platinum substrates were then oxidized in air at a temperature higher than 1270 K for more than 10 h. Examination using an X-ray diffractometer has indicated the thin films thus obtained to be polycrystalline NiO.

We made infrared measurements using a double-beam grating spectrophotometer (Hitachi-295). Several optical systems in the specimen chamber were chosen. Specifically for the measurements, a variable incident-angle optical system and a 30° emission angle thermal-emission optical system were designed and constructed in our laboratory. A grid-type polarizer (Cambridge Physical Sciences IGP-225) was used for the polarization measurements. No corrections were made for incomplete polarization of the beam caused either by polarizer imperfection or by incomplete collimation of the beam (angular aperture $< 10°$).

3. Results and Discussion

The study of the optical properties of the NiO thin film on the platinum substrate was preceded by that of nickel plates with different thicknesses of the oxidation layer. Typical reflection spectra of the oxidized nickel plates for p- and s-polarized radiation are shown in Fig. 2. The measurements were made at room temperature under 30° incidence condition. The reflectivity spectra (2a, b, c, and d) correspond to the specimen with the weight increment per unit surface area of 8.37×10^{-5}, 2.23×10^{-4},

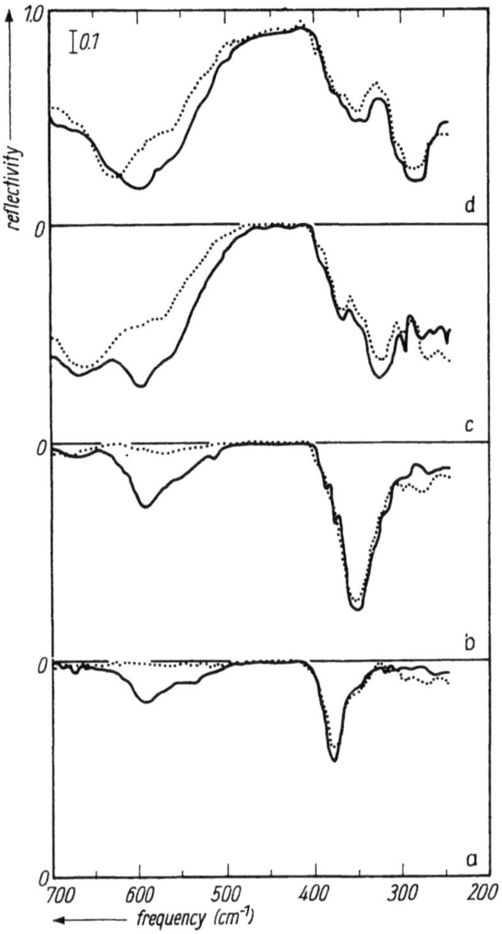

Fig. 2

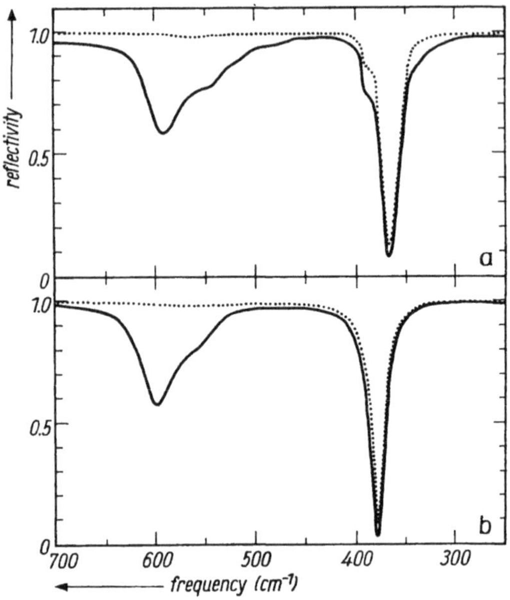

Fig. 3

Fig. 2. Reflectivity spectra of oxidized nickel plates at room temperature under 30° incidence. Solid and dotted curves represent the spectra for p- and s-polarized radiation, respectively. The spectra a), b), c), and d) correspond to the specimens with the weight increment per unit surface area to the oxidation: 8.37×10^{-5}, 2.23×10^{-4}, 8.57×10^{-4}, and 1.17×10^{-3} g/cm^2, respectively

Fig. 3. a) Experimental and b) theoretical reflectivity spectra of NiO thin film of 0.81 μm thickness on a platinum substrate at room temperature under 30° incidence. Solid and dotted curves represent the spectra for p- and s-polarized radiation, respectively

8.57×10^{-4}, and 1.17×10^{-3} g/cm^2, respectively. As shown in Fig. 2a, in the beginning stage of oxidation, two reflection minimum bands appear at 590 and 380 cm^{-1} for p-polarized radiation, while only a reflection minimum band appears at 380 cm^{-1} for s-polarized radiation. Then, with increasing oxidation, the intensity of these reflection minimum bands become stronger and some new bands appear as shown in Fig. 2b to d. At further oxidation, the reflectivity spectra become close to the reststrahlen reflectivity spectrum of the NiO bulk crystal. These transitions in the reflectivity spectra clearly show us the thickness dependence of the optical properties of NiO films on a metal substrate from very thin film to bulk states, though the dependence is not quantitative.

Fig. 3a shows the reflectivity spectra of a NiO film of 0.81 μm thickness on a platinium substrate. The solid curve represents the reflectivity spectrum for p-polarized radiation, while the dotted curve represents that for s-polarized radiation. The reflectivity is measured at room temperature under 30° incidence. We determined the reflectivity by measuring the radiation reflected from the specimen at a given frequency and comparing it with that reflected by a platinum substrate without NiO film, at the same frequency. As seen in this figure, the p-polarized spectrum consists of two reflection minimum bands centred at 592.5 and 370.0 cm^{-1} while the s-polarized spectrum has only one band centred at 370.0 cm^{-1}. These bands are accompanied by some sub-bands. The reflection spectra of the ionic thin film on a metal substrate can be discussed by the virtual-mode theory. In the theory, to relate the optical properties to the virtual-mode equations, the following quantities, L_1 and L_2, are introduced:

$$L_1 = 1 - \frac{i\beta}{\beta_0 \varepsilon(\omega)} \tan \beta d \,, \tag{5}$$

$$L_2 = 1 + \frac{i\beta}{\beta_0} \cot \beta d \,. \tag{6}$$

Writing

$$P_1 = \frac{2 - L_1}{L_1} \,, \tag{7}$$

$$P_2 = \frac{2 - L_2}{L_2} \,, \tag{8}$$

the reflectivity, R, and the emissivity, E, are then

$$R_p = 1 - E_p = |P_1|^2 \,, \tag{9}$$

$$R_s = 1 - E_s = |P_2|^2 \,, \tag{10}$$

for p- and s-polarization, respectively. We calculated the reflectivity spectra by using the following dielectric function $\varepsilon(\omega)$ of the NiO bulk crystal reported previously [8]:

$$\varepsilon(\omega) = \varepsilon(\infty) + \sum_{j=1}^{2} \left[\frac{S_j \omega_j^2}{\omega_j^2 - \omega^2 + i\omega \Gamma_j}\right];$$

$$\omega_1 = 400\ \text{cm}^{-1}\,, \quad \frac{\Gamma_1}{\omega_1} = 0.0276\,, \quad S_1 = 6.40\,, \tag{11}$$

$$\omega_2 = 563\ \text{cm}^{-1}\,, \quad \frac{\Gamma_2}{\omega_2} = 0.139\,, \quad S_2 = 0.126, \quad \text{and} \quad \varepsilon(\infty) = 5.7\,.$$

The result is shown in Fig. 3b. It can be seen in these figures that the theoretical spectra are close to the experimental ones except for the sub-bands around the 370 cm^{-1} band. Since the dielectric function $\varepsilon(\omega)$ used in the calculation was determined from the classical dispersion analysis of the reststrahlen reflectivity spectra of the NiO bulk crystal, the $\varepsilon(\omega)$ expressed as (11) cannot reflect the fine structure. A more detailed treatment is made by considering the phonon–phonon interactions due to the higher-order electric moments or due to the anharmonicity of the lattice potential. According to the well-known theory [11] of anharmonic crystals, if the anharmonicity is weak, the dielectric function is given as

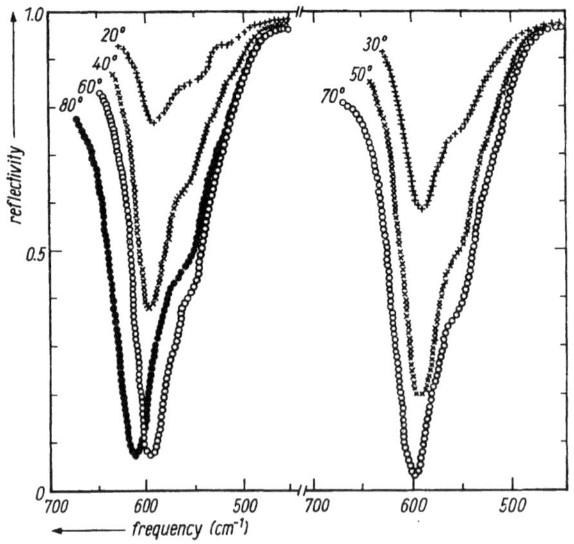

Fig. 4. Reflectivity spectra due to the OTH virtual mode in a NiO thin film of 0.81 μm thickness on a platinum substrate at room temperature as a function of the frequency and the incident angle. The parameters at the curves are values of the incident angle

$$\varepsilon(\omega) = \varepsilon(\infty) + \tilde{\omega}_T(0)^2 \frac{\varepsilon(0) - \varepsilon(\infty)}{\tilde{\omega}_T(\omega)^2 - \omega^2 - 2i\omega_T\Gamma\begin{pmatrix}0\\ j\end{pmatrix}; \omega)}, \tag{12}$$

$$\tilde{\omega}_T(\omega)^2 = \omega_T^2 + 2\omega_T\Delta\left(\begin{matrix}0\\ j\end{matrix}; \omega\right), \tag{13}$$

$$\pi\left(\begin{matrix}0\\ j\end{matrix}; \omega\right) = \Delta\left(\begin{matrix}0\\ j\end{matrix}; \omega\right) - i\Gamma\left(\begin{matrix}0\\ j\end{matrix}; \omega\right), \tag{14}$$

where ω_T is the harmonic dispersion frequency, $\tilde{\omega}_T(\omega)$ the anharmonic dispersion frequency, $\tilde{\omega}_T(0)$ the dispersion frequency in the limiting case $\omega = 0$, and $\pi\left(\begin{matrix}0\\ j\end{matrix}; \omega\right)$ the phonon self-energy of the dispersion oscillator of the mode $\begin{pmatrix}0\\ j\end{pmatrix}$ at frequency ω. According to the theoretical study of ionic crystals, it is found that the self-energy of the TO-mode is strongly frequency-dependent. Such a frequency dependence of the self-energy produces a complicated structure in the infrared optical spectra. The observed sub-bands are thus ascribed to an optical process involving many phonons through the intermediate virtual excitation of the TO-phonon near the centre of the Brillouin zone. At the present stage, the agreement between the observed spectra and those calculated by using the anharmonic dielectric function (12) is not good.

Using the same specimen, we studied the incident-angle dependence of the centre frequency of the reflection minimum band due to the OTH-mode at room temperature. The result is shown in Fig. 4. It is seen in this figure that, as the incident angle increases, the centre frequency shifts to higher frequency and the width of the band increases. The centre frequency of the OTH-mode is theoretically given as one of the solutions in the tangent-type equation (2). The behaviour of the OTH-mode is determined by expanding (2) for small value of βa and by using (4) as follows [10]:

$$W\Omega \frac{\varepsilon - \sin^2\theta}{2} = -i\varepsilon\cos\theta, \tag{15}$$

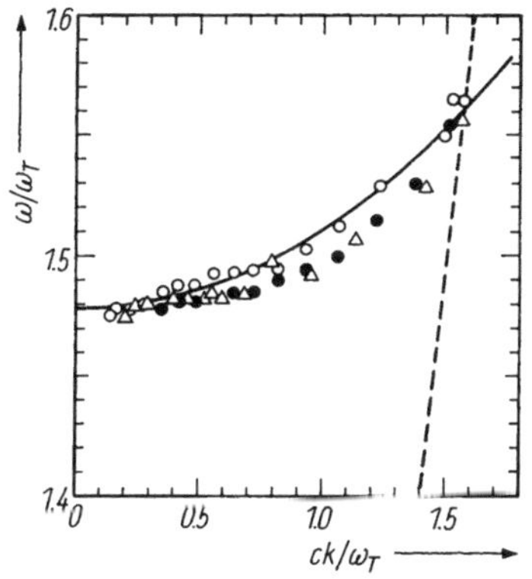

Fig. 5. Upper branch of the polariton dispersion curve of NiO at room temperature. Open circles were determined from the spectrum of 0.81 μm thickness on a platinum substrate. The solid circles were determined from the spectrum of 0.8 μm average thickness with ±0.2 μm thickness fluctuation on a platinum substrate, and the triangles were determined from the spectrum of 0.46 μm thickness on a rough-surface platinum substrate

where W is the reduced thickness, $2\omega_T a/c$, and Ω is the reduced complex virtual-mode frequency given by

$$\Omega = \Omega' + i\Omega'' = \frac{\omega'}{\omega_T} + i\,\frac{\omega''}{\omega_T} = \Omega_p\,e^{i\varphi}\,, \tag{16}$$

where ω' and ω'' are the centre frequency and the half-width of the OTH-mode, respectively. Using the quantity, $A = 2\cos\theta/W\Omega_p$, the dielectric function ε of the film material is given by

$$\varepsilon = \sin^2\theta\,(1 + A\sin\varphi)\,\frac{1 - \dfrac{iA\cos\varphi}{1 + A\sin\varphi}}{1 + 2A\sin\varphi + A^2}\,. \tag{17}$$

We get then the following equation of the upper branch of the polariton dispersion curve:

$$\left(\frac{kc}{\omega_T\Omega_p}\right)^2 = \sin^2\theta\,\frac{1 + A\sin\varphi + \sqrt{1 + 2A\sin\varphi + A^2}}{2(1 + 2A\sin\varphi + A^2)}\,. \tag{18}$$

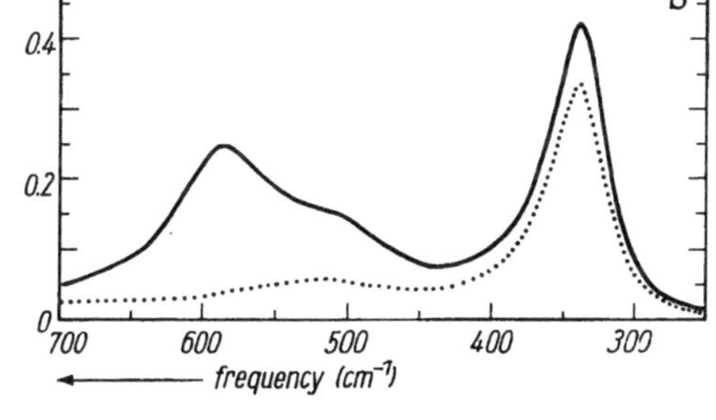

Fig. 6. a) Experimental and b) theoretical emissivity spectra of NiO thin film of 0.81 μm thickness on a platinum substrate at 1273 K under 30° emission. Solid and dotted curves represent the spectra for p- and s-polarized radiation, respectively

This relation enables us to determine experimentally the upper branch of the polariton dispersion curve by using the measured values at various incident angles for p-polarized radiation. Fig. 5 shows the upper branch of the polariton dispersion curve of NiO at room temperature thus obtained. The open circles were obtained from the spectra of a specimen of 0.81 μm thickness. The dashed line is the photon dispersion line in vacuum. The solid curve is the theoretical curve calculated from (1) with the following dispersion parameters:

$$\varepsilon(0) = 12.45, \qquad \varepsilon(\infty) = 5.7, \qquad \omega_T = 400 \text{ cm}^{-1} .$$

In the calculation, we adjusted the value of $\varepsilon(0)$ so that the theoretical curve fits the experimentally determined values. $\varepsilon(0) = 12.45$ is very close to the values 12.4, determined by reststrahlen analysis on NiO bulk crystals. The experimental results agree fairly well with theoretical dispersion curves, though we ignored the effect of the phonon damping due to the anharmonic interactions. It is also found in our experiments that the experimental polariton dispersion curve is somewhat affected by the inhomogeneity of the film thickness and by the surface roughness of the substrate. Particularly, the surface roughness will reduce the polarization perpendicular to the specimen layer and then will lower the centre frequency of the OTH-mode. Typical results are shown in Fig. 5. The solid circles correspond to the specimen of 0.8 μm average thickness with ±0.2 μm fluctuation and the triangles correspond to the specimen of 0.46 μm thickness on a rough-surface platinum substrate. On further increasing the rougness, the spectrum becomes similar to that of NiO microcrystals [9].

Using the same specimen of 0.81 μm thickness, we measured the emissivity spectra at temperatures in the range from 373 to 1273 K under 30° emission. By the way of illustration, the emissivity spectra at 1273 K are shown, together with the theoretical ones at the corresponding temperature, in Fig. 6. The theoretical emissivity spectra were calculated from (9) and (10) by using the $\varepsilon(\omega)$ of the NiO bulk crystal at 1273 K [8]. The observed spectra show considerable broadening and frequency shift. The correlation between the theoretical spectra and the experimental ones is good especially

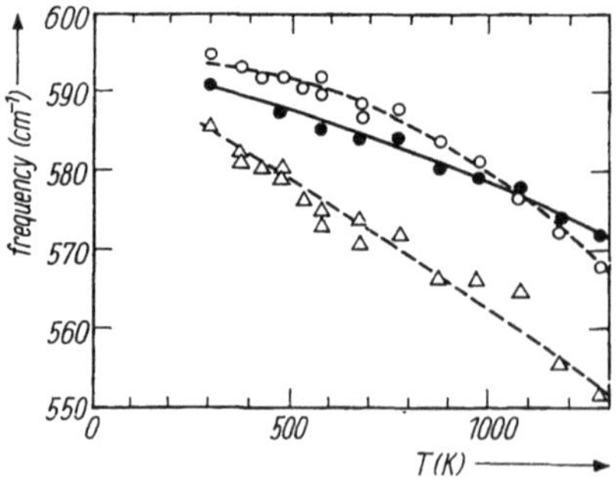

Fig. 7

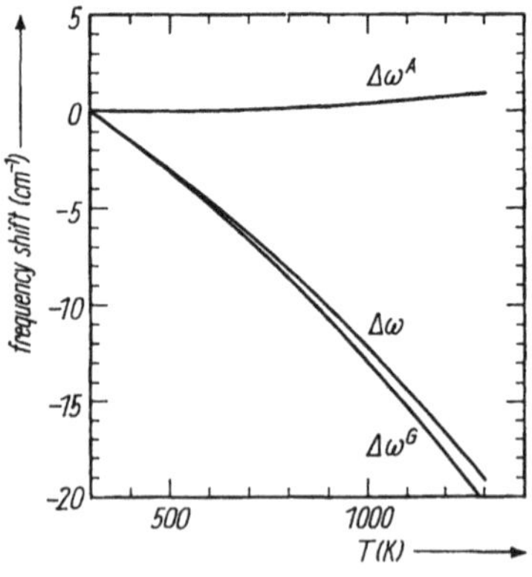

Fig. 8

Fig. 7. Temperature dependence of the LO-phonon frequencies of NiO. Solid circles are the centre frequency of the OTH-mode under 30° incidence (emission), while open circles and triangles are the LO-phonon frequencies defined as the frequency where ε_1 vanishes and that obtained from the Lyddane-Sachs-Teller relation of NiO bulk crystal [8], respectively

Fig. 8. Anharmonic and Grüneisen contributions to the shift of LO-phonon frequency of NiO thin film as a function of temperature

for the peak frequencies. The greater discrepancy for the spectral shapes is probably due to some uncertainty of the fine structure in the dielectric function or some thin-film nature. The temperature dependence of the centre frequency of the OTH-mode is shown, together with those of the LO-phonon frequencies determined from the study of the NiO bulk crystal, in Fig. 7. Solid circles indicate the centre frequencies, ω', of the OTH-mode, while open circles and triangles indicate the LO-phonon frequencies, $\omega_L(\varepsilon_1 = 0)$, of the NiO bulk crystal defined as the frequencies where the real part, ε_1, of the dielectric function vanishes and the LO-phonon frequencies, ω_L (LST), determined by the Lyddane-Sachs-Teller relation, respectively. As for the LO-phonon frequency, it is known that the observed frequency shift $\Delta\omega_L(T)$ in an anharmonic crystal is given as follows:

$$\Delta\omega_L(T) = \omega_L(T) - \omega_L(0) = \Delta\omega_L^G(T) + \Delta\omega_L^A(T) , \tag{19}$$

where $\omega_L(0)$ is the frequency at 0 K, $\Delta\omega_L^G(T)$ is the quasi-harmonic shift due to thermal expansion, and $\Delta\omega_L^A(T)$ is the self-energy shift at a given temperature, T. Since the present measurements are limited to above room temperature, RT, we redefine the observed shift as follows:

$$\Delta\omega_L = \omega_L(T) - \omega_L(\mathrm{RT}) = \Delta\omega_L^G(T) + \Delta\omega_L^A(T) . \tag{20}$$

Using the thermal expansion coefficient, α, we determine $\Delta\omega_L^G(T)$ as follows:

$$\Delta\omega_L^G(T) = \omega_L(\mathrm{RT}) \left[1 - \exp\left(-3\gamma_L \int_{\mathrm{RT}}^{T} \alpha \, \mathrm{d}T\right)\right] , \tag{21}$$

in which the mode Grüneisen constant, γ_L, is calculated from the following equation:

$$\gamma_L = \frac{\left(\dfrac{\partial\omega_L}{\partial T}\right)_p}{3\alpha\omega_L} , \tag{22}$$

by the $\omega_L(T)$-data at low temperatures. In our calculation, we used our empirical formula for the centre frequency, $\omega'(= \omega_L)$,

$$\omega' = 594.6 - 1.159 \times 10^{-4}T - 4.694 \times 10^{-6}T^2 \ (\mathrm{cm}^{-1}) , \tag{23}$$

and the temperature dependence of the lattice parameter, a, as given in the literature [12],

$$a(T) = 0.41644 + 0.4444 \times 10^{-5}T + 0.1132 \times 10^{-8}T^2 \ (\mathrm{nm}) . \tag{24}$$

The result is shown in Fig. 8. It can be seen in this figure that the Grüneisen contribution increases with temperature, but the increase is in the direction opposite to the anharmonic contribution. Maradudin and Fein [13] have considered the effect of the cubic anharmonicity to second order in the perturbation theory and that of quartic anharmonicity to the first order on the self-energy shift. They obtained the expression

$$\Delta\omega_L^A = -\frac{18}{\hbar^2} \sum_{q_1q_2j_1j_2} \left| V\begin{pmatrix} 0 & q_1 & q_2 \\ L & j_1 & j_2 \end{pmatrix} \right|^2 \left[\frac{n_1 + n_2 + 1}{\omega_1 + \omega_2 + \omega_L} + \frac{n_1 + n_2 + 1}{\omega_1 + \omega_2 - \omega_L} + \right.$$
$$\left. + \frac{n_2 - n_1}{\omega_1 - \omega_2 + \omega_L} + \frac{n_2 - n_1}{\omega_1 - \omega_2 - \omega_L} \right] +$$
$$+ \frac{12}{\hbar} \sum_{q_1j_1} V\begin{pmatrix} 0 & 0 & q_1 & -q_1 \\ L & L & j_1 & j_1 \end{pmatrix} (2n_1 + 1) , \tag{25}$$

where $n_1 = n(q_1j_1)$ is the population factor for the phonon mode and $\omega_1 = \omega(q_1j_1)$ is the frequency, and $\boldsymbol{q}$ is the phonon wave vector. The coefficients, V, derive from the lattice potential energy of deformation at constant volume expanded in terms of the phonon normal coordinates. This equation shows that the cubic contributions to the self-energy are necessarily negative, while the quartic terms may be either negative or positive. The small positive values for $\Delta\omega_L^A(T)$ obtained here suggest that the quartic contributions are positive and the cubic contributions are nearly compensated with the quartic ones.

From the discussion above, the infrared optical properties of NiO thin films in the radiative region and the temperature dependence of the LO-phonon frequency are elucidated.

References

[1] J. Yamashita and T. Kurosawa, J. Phys. Soc. Japan **15**, 802 (1960).
[2] R. A. Bari, D. Adler, and R. V. Lange, Phys. Rev. B **2**, 2898 (1970).
[3] D. Adler and J. Feinleib, Phys. Rev. B **2**, 3112 (1970).
[4] W. E. Spear and D. S. Tannhauser, Phys. Rev. B **7**, 831 (1973).
[5] S. Mochizuki, phys. stat. sol. (a) **40**, 119 (1977).
[6] R. Newman and R. M. Chrenko, Phys. Rev. **114**, 1507 (1959).
[7] P. J. Gielisse, J. N. Plendl, L. C. Mansur, S. S. Mitra, R. Mykolajewycz, and A. Smakura, J. appl. Phys. **36**, 2446 (1965).
[8] S. Mochizuki and M. Satoh, phys. stat. sol. (b) **106**, 667 (1981).
[9] S. Mochizuki, phys. stat. sol. (b) **110**, 219 (1982).
[10] R. Fuchs and K. L. Kliewer, Phys. Rev. **140**, A2076 (1965); **144**, 495 (1966); **150**, 573, 589 (1966).
[11] F. Abeles, Optical Properties of Solids, North-Holland Publ. Co., Amsterdam 1972. R. A. Cowley, Adv. Phys. **12**, 421 (1963).
[12] S. P. Srivastava, R. C. Srivastava, I. D. Singh, S. D. Pandey, and P. L. Gupta, J. Phys. Soc. Japan **43**, 885 (1977).
[13] A. A. Maradudin and A. E. Fein, Phys. Rev. **128**, 2589 (1962).

(Received July 16, 1984)

phys. stat. sol. (b) **126**, 115 (1984)

Subject classification: 1.2 and 14.4; 6; 22.8.1

Fachbereich Physik der Universität des Saarlandes, Saarbrücken[1])

Dielectric Dispersion and Dynamics of Nitrite Ions in $K_2Ba(NO_2)_4$

By

H.-G. UNRUH

Dielectric measurements reveal Debye-like dispersions of the main permittivities below room temperature at audio-frequencies, the activation energy of the corresponding motions of nitrite ions being 0.5 eV. The role of the three non-equivalent NO_2 groups in the phase transition at about 203 K is investigated using Raman-spectroscopic results on internal NO_2 modes. A model is proposed that includes freezing of certain NO_2 groups with static disorder.

Die Hauptwerte des Dielektrizitätstensors von $K_2Ba(NO_2)_4$ zeigen unterhalb Raumtemperatur Dispersionen vom Debye-Typ im kHz-Bereich. Die Aktivierungsenergie der dafür verantwortlichen Bewegungen der Nitrit-Ionen wird zu 0,5 eV bestimmt. Mit Hilfe Raman-spektroskopischer Messungen der inneren NO_2-Moden wird der Einfluß der drei nicht-äquivalenten NO_2-Gruppen auf die Phasenumwandlung bei 203 K untersucht. Ein Modell, bei dem NO_2-Gruppen ungeordnet einfrieren, wird vorgeschlagen.

1. Introduction

According to optical and dielectric investigations by Ivanov et al. [1] single crystals of $K_2Ba(NO_2)_4$ undergo two reversible phase transitions at about $T_\alpha \approx 419$ and $T_\beta \approx \approx 203$ K. The high-temperature α-phase has hexagonal symmetry (class 6/mmm) and changes to an orthorhombic β-phase (class mmm) at the ferroelastic transition near 419 K. The symmetry of the γ-phase below 203 K is still unknown. The physical properties of this compound have been further elucidated by Raman [2], caloric [3], and Brillouin measurements [4], which confirmed two transitions of second order. Recently, the room-temperature structure has been determined by Harada [5] showing that the space group is Pbam ($Z = 2$) and that the unit cell has dimensions $a = = 0.6604$, $b = 1.1427$, and $c = 0.6161$ nm. Furthermore, it was suggested that the phase transitions are caused by ordering of orientationally disordered NO_2 groups.

So it might be expected that dipolar interactions play a substantial role in these transitions and the dielectric behaviour should characterize the underlying mechanisms to that extent. In this paper, therefore, precise measurements of the permittivities of single crystals are reported and discussed. Additionally polarized Raman spectra of internal lattice modes are investigated with respect to structural implications.

2. Experimental Results

Slightly yellowish crystals of $K_2Ba(NO_2)_4$ were prepared from aqueous solutions, containing potassium nitrite and barium nitrite at a molar ratio of 2:1, by slow evaporation at about 40 °C. As the crystals were always twinned, samples were carefully selected from untwinned portions and inspected by means of a polarizing microscope. The c-axis was taken parallel to the acute bisectrix, i.e. parallel to the pseudo-

[1]) D-66 Saarbrücken, FRG.

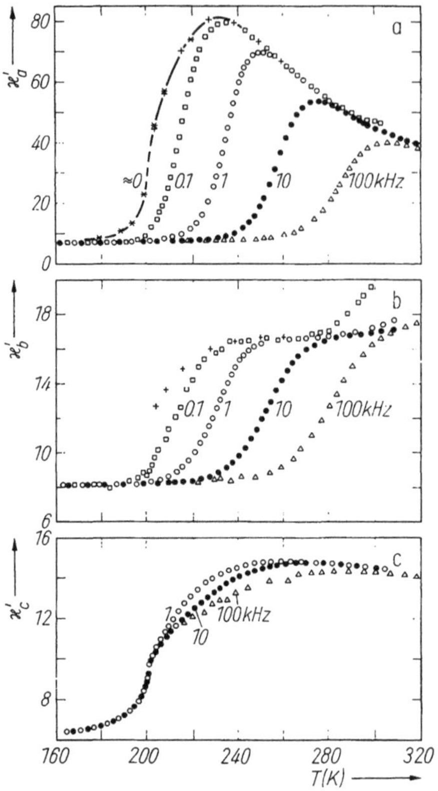

Fig. 1. Temperature dependence of the real part of the permittivities of $K_2Ba(NO_2)_4$ measured along the crystallographic axes a, b, and c, respectively. Crosses and stars refer to quasi-static values of $\varkappa_a'$ and $\varkappa_b'$ (see text)

hexagonal direction, the a-axis orthogonal to the optical plane containing the two optical axes and the b-axis perpendicular to a and c. This choice corresponds to the setting of axes in the space group Pbam.

2.1 Dielectric investigations

Measurements have been done in the temperature range between 12 and 320 K. The real permittivities $\varkappa_i'$ measured along the crystallographic axes at various frequencies are shown in Fig. 1. While below 160 K no anomalies have been found, distinct relaxation phenomena occur at audio-frequencies above T_β. Apart from another relaxation above room temperature, which will not be considered here, by far the largest effect appears along the a-direction. A clear indication of the phase transition, however, is seen in $\varkappa_c'$, only, as $d\varkappa_c'/dT$ seems to change discontinuously at T_β. A similar behaviour could not be detected for $\varkappa_a'$ or $\varkappa_b'$ at finite frequencies. In order to determine the temperature dependence of the static permittivities the dispersion was measured also at constant temperature (Fig. 2) and extrapolated to zero frequency. The values of $\varkappa_a'$ and $\varkappa_b'$ obtained for $\omega \to 0$ are marked by crosses in Fig. 1a and b, respectively. Additionally, the dielectric response along the a-direction has been measured by direct inspection of the dielectric displacement $D_a(E_a)$ at 1 Hz (see stars in Fig. 1a). From the results it has to be inferred that the static dielectric susceptibility along the a-axis shows a broad maximum at about $T_m \approx T_\beta + 25$ K and follows a Curie-Weiss law above T_m as is demonstrated in Fig. 3.

The dispersion of the complex permittivity $\varkappa_a$ is well represented by Cole's formula apart from a small additional relaxation at the high-frequency end. The same holds true for $\varkappa_b$ where the distribution of relaxation times seems to be more pronounced because of the smallness of the main relaxation. Indeed, the permittivities $\varkappa = \varkappa' - j\varkappa''$ could be fitted to a superposition of two relaxations according to

$$\varkappa(\omega) = \varkappa(\infty) + \frac{A}{1 + j\omega\tau_A} + \frac{B}{1 + (j\omega\tau_B)^{1-h}}$$

with $\tau_B < \tau_A$. The most reliable results have been obtained for τ_A from the main dispersion of $\varkappa_a$. The Arrhenius plot of the corresponding relaxation time is shown in

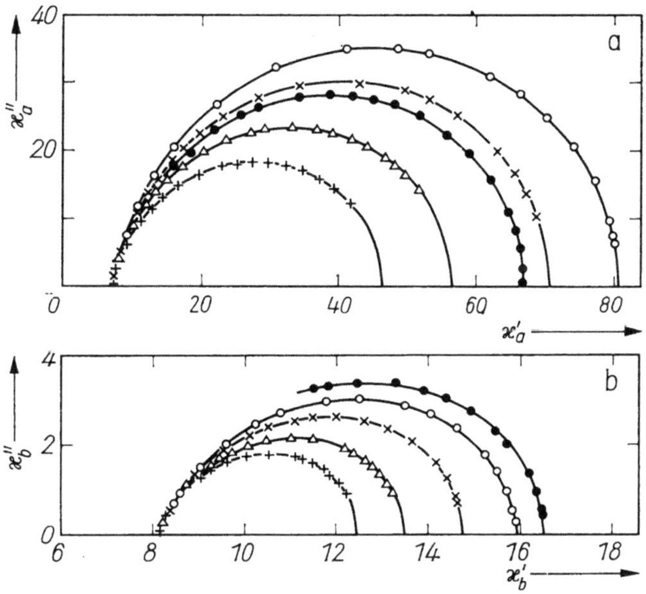

Fig. 2. Dielectric dispersion of $\varkappa_a$ and $\varkappa_b$, respectively, measured in the frequency range from about 0.1 to 20 kHz at various temperatures. ● $T = 260.7$, ○ 228.7, × 217.0, △ 209.7, + 205.5 K

Fig. 4 and the activation energy of the responsible thermally activated process was calculated from the plot as 0.5 eV. Strictly speaking the relaxation time τ has been determined at constant external electric field E_a, a constraint at which relaxing dipoles are coupled by their polarization field. Because of the polydispersive behaviour there is no simple way of evaluating the relaxation time with respect to constant local field [6]. So, although the data obey Arrhenius' law fairly well (Fig. 4), the true activation energy may differ from the value given above by about 10% as has been estimated from the experiments.

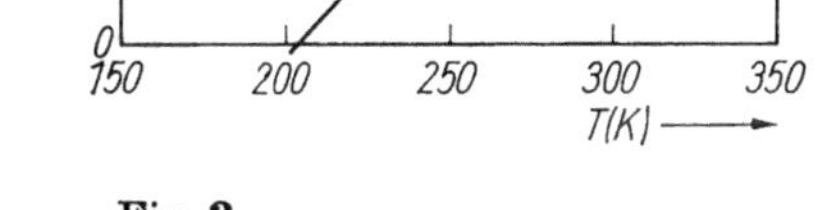

Fig. 3

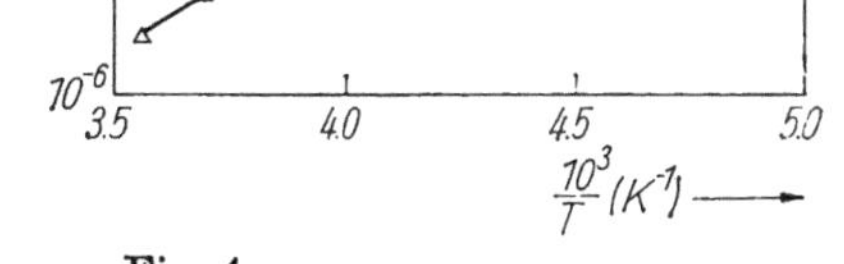

Fig. 4

Fig. 3. Curie-Weiss plot of the permittivity measured along the a-axis. The values of $\varkappa_a(0)$ and $\varkappa_a(\infty)$ were taken from Fig. 1a; for $\varkappa_a(\infty)$ the low-temperature values of $\varkappa_a(10\text{ kHz}, T \lesssim 200\text{ K})$ were extrapolated to higher temperatures

Fig. 4. Arrhenius plot of the relaxation time τ of the main relaxation of $\varkappa_a$. Triangles and circles refer to specimens from different single crystals

It should be mentioned that the relaxations described here are somewhat sample dependent. By comparison of results from various single crystals one has to conjecture that specific defects influence the dispersion. Crystals which are assumed to have less defects show a quantitatively larger dispersion of $\varkappa_a$ and a smaller one of $\varkappa_b$, while no qualitative changes of the main relaxation of $\varkappa_a$ occur. These observations may also explain the distinct discrepancies between the dielectric behaviour reported here and in [1].

2.2 Raman-spectroscopic results

It has been pointed out by Dmitriev et al. [2] that the internal Raman-active modes of the nitrite ion undergo marked changes when T_β is approached. Especially, the deformation mode at about 820 cm^{-1} becomes a doublet at lower temperatures. According to the recent structure determination [5], however, three crystallographically different NO_2 groups exist already at room temperature. The group labelled NO_2 (1) is ordered at ambient temperature, while NO_2 (2) and NO_2 (3) are still disordered with respect to the direction of their dipoles pointing up or down the c- and a-axis, respectively.

As spectroscopic methods should reveal the different species of nitrite groups, polarized Raman spectra have been taken at various temperatures. The results clearly confirm the structural findings as may be seen from Fig. 5. At room temperature (Fig. 5a) the nondegenerate, symmetric NO_2 deformation mode, which is best suited for this investigation because of its small damping, shows up as a triplet in the totally symmetric A_g spectra. In order to assign the three peaks to the NO_2 groups, relative values of the Raman-tensor components for the normal modes of the nitrite ion are necessary. These are unknown for $K_2Ba(NO_2)_4$, but have been determined in case of $NaNO_2$ [7]. Using the values referring to the deformation mode from [7] and noting the orientations of the nitrite groups of the compounds in question [5, 7], the relative intensities of the peaks in Fig. 5a are reasonably well described by setting $\nu_2(NO_2\ (1)) = 817.7$, $\nu_2(NO_2\ (2)) = 814.8$, and $\nu_2(NO_2\ (3)) = 821.5$ cm^{-1}. The frequencies have been

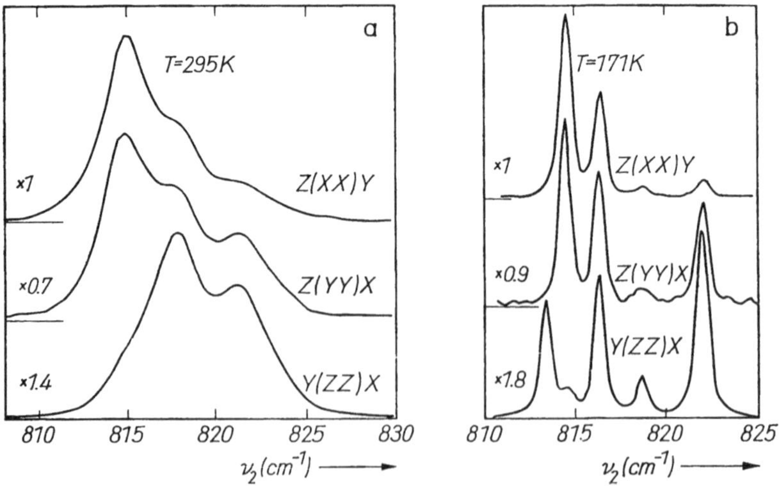

Fig. 5. A_g Raman spectra of the deformation mode of NO_2 groups in $K_2Ba(NO_2)_4$ at a) room temperature and b) 171 K, i.e. below T_β. Intensities are scaled within experimental errors when spectra are multiplied by the numerical factors given. Instrumental widths: 1 cm^{-1} at 295 K and 0.5 cm^{-1} at 171 K

evaluated by a least-squares fit to the spectra. The triplet is split up further when the temperature is lowered to below T_β, where at least five peaks can be distinguished (Fig. 5b). As no soft mode was detected in our spectra around T_β, one may conclude that the motions of further NO_2 groups freeze in within this region.

3. Discussion

Combining the results of the dielectric measurements with structural details (see Fig. 6) the main low-frequency relaxation of the permittivity $\varkappa_a$ can be traced back to the hindered rotation or flipping of NO_2 (3) groups about the c-axis. As the plane of the NO_2 (3) group coincides with the (010) plane only approximately [5], a small contribution of the same relaxation to $\varkappa_b$ has to be expected in agreement with Fig. 1b. The Curie-Weiss-like behaviour of $\varkappa_a'$ above 240 K points to cooperative interactions between NO_2 (3) groups which, however, are suppressed by some other mechanism before an instability occurs. In this connection it is interesting to note that the relative intensities of the deformation mode in (XX), (YY), and (ZZ) spectra do not change essentially for NO_2 (3) on lowering the temperature (see peaks at about 822 cm^{-1} in Fig. 5). So one may infer that these groups freeze without changing their momentary orientation much, but being on a special site with symmetry 2/m at room temperature, the diad axis parallel to the pseudohexagonal c-axis must vanish on freezing, at least locally. Now, NO_2 (1) and NO_2 (3) are arranged in sheets perpendicular to c and it was assumed [5] that these groups are disordered and equivalent in the high-temperature phase above T_α, where NO_2 (1) groups become ordered. At T_β a freezing of NO_2 (3) incorporates an element of frustration in this nearly triangular sublattice of NO_2 groups so that a static disorder of NO_2 (3) groups seems possible. The dielectric results support this interpretation as $\varkappa_a(\omega \to 0)$ decreases drastically below $T_m \approx 230$ K, where a pronounced cusp is observed. At the extrapolated Curie-Weiss temperature, $T_C \approx 205$ K from Fig. 3, most of NO_2 (3) groups can be assumed to be static and to trigger eventually a structural transition involving NO_2 (1) and/or NO_2 (2). The behaviour of $\varkappa_a$ is reminiscent of a dipolar glass although an element of randomness is absent, so that the relaxation stays essentially monodispersive.

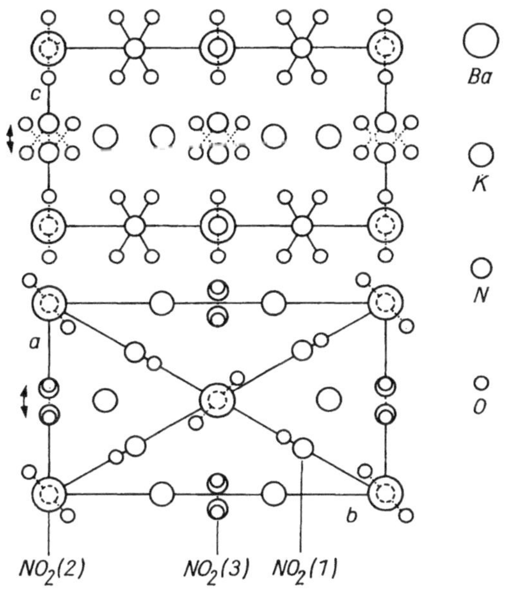

Fig. 6. Projection of the room-temperature structure of $K_2Ba(NO_2)_4$ on (100) and (001) planes according to Harada [5]

The dynamics of NO_2 (2) groups must be assumed to be fast compared to our frequencies of measurement. There is no substantial dispersion of $\varkappa_c$ in the range from 1 to 100 kHz, although NO_2 (2) dipoles are disordered at room temperature, pointing along $\pm c$. On the other hand, an essential decrease of $\varkappa_c'$ takes place with decreasing temperature indicative of the structural transition at about 203 K. In addition the spectral features due to the deformation modes of NO_2 (1) and NO_2 (2) groups become complex below this temperature showing that some splitting into nonequivalent positions occurs. A more detailed analysis of the complete Raman spectra will be given elsewhere.

Acknowledgements

The author is very much indebted to Mrs. A. Engelke for measurements and sample preparation and to Dr. K. H. Ehses for orienting the crystals by precession photographs. This work has been supported by the Deutsche Forschungsgemeinschaft, Sonderforschungsbereich 130 "Ferroelektrika".

References

[1] N. R. IVANOV, L. F. KIRPICHNIKOVA, V. P. KONSTANTINOVA, L. V. SOBOLEVA, and L. A. SHUVALOV, Soviet Phys. — Cryst. **23**, 443 (1978).

[2] V. P. DMITRIEV, L. M. RABKIN, L. V. SOBOLEVA, A. A. CHUBISH, and L. A. SHUVALOV, Soviet Phys. — Cryst. **23**, 435 (1978).

[3] M. CHABIN, F. GILETTA, Y. LUSPIN, and G. HAURET, J. Physique **43**, 337 (1982).

[4] Y. LUSPIN and G. HAURET, Solid State Commun. **46**, 397 (1983).

[5] M. HARADA, J. Phys. Soc. Japan **52**, 3448 (1983).

[6] H. E. MÜSER and J. PETERSSON, Fortschr. Phys. **19**, 559 (1971).

[7] H. H. EYSEL, C. W. VON DER LIETH, G. BERTSCH, and M. H. BROOKER, Mol. Phys. **44**, 395 (1981).

(Received July 26, 1984)

phys. stat. sol. (b) **126**, 121 (1984)

Subject classification: 6; 22.1.2

Physikalisches Institut der Universität Bayreuth[1]) *(a)*
and Institut für Angewandte Kernphysik des Kernforschungszentrums Karlsruhe[2]) *(b)*

Measurement of Phonon Energies in Silicon under Uniaxial Stress

By

J. ESTEL (a), J. KALUS (a), and L. PINTSCHOVIUS (b)

Stress-induced changes in phonon energies of the TA (ξ, ξ, ξ) branch are investigated in silicon by using inelastic neutron scattering. With stress in the [111] direction of 1.01 GPa the energy shift of the phonons changes from 0.5% near the Γ-point to −1.2% at the zone boundary. Comparison with theoretical calculations of Mayer and Wehner [1] shows good agreement.

Mittels inelastischer Neutronenstreuung werden die spannungsinduzierten Änderungen der Phononenenergien des TA-(ξ, ξ, ξ)-Zweiges in Silizium untersucht. Mit einer Spannung in [111]-Richtung von 1,01 GPa ändert sich die Energieverschiebung der Phononen von 0,5% in der Nähe des Γ-Punkts auf −1,2% an der Zonengrenze. Der Vergleich mit theoretischen Berechnungen von Mayer und Wehner [1] zeigt gute Übereinstimmung.

1. Introduction

Most models used so far to describe the lattice dynamics of a solid are based on the harmonic approximation, i.e. in the expansion of the potential energy as a function of atomic displacements all terms higher than second order are neglected. These models often yield accurate values of the vibrational frequencies at a certain temperature and pressure, but cannot describe the temperature and pressure dependence of phonon frequencies nor can they account for the thermal expansion of the lattice. Models aiming at a description of these and related phenomena have to include higher-order terms in the expansion of the crystal potential.

A particularly powerful method for the study of anharmonicity is to apply uniaxial stress to a crystal, for this does not only cause energy shifts, but in addition it can remove the degeneracy of certain phonon branches.

In this paper we report about inelastic neutron scattering experiments which extend the work of Prechtel et al. [2]. We applied uniaxial stress along the [111] direction, and measured stress-induced changes of phonon frequencies for the (ξ, ξ, ξ) transverse acoustic modes. The observed negative mode Grüneisen parameters for phonons with large wave vectors are not unexpected, since it is known that the thermal expansion coefficient in Si is negative for temperatures below 120 K [3]. Large single crystals of Si can be loaded with uniaxial stress of more than $\sigma = 1$ GPa.

In Section 2 we describe the experimental set-up and give a report of the data. In Section 3 we refer to other experiments and make a comparison with the most recent calculation of mode Grüneisen parameters.

---

[1]) Universitätsstr. 30, D-8580 Bayreuth, FRG.

[2]) Postfach 3640, D-7500 Karlsruhe, FRG.

2. Measurements and Results

The energy shifts of the transverse acoustic phonons were investigated at the C3 triple-axis spectrometer at the FR2-reactor in Karlsruhe. The crystal had a length of 175 and a diameter of 18 mm. The crystal was charged with (1.01 ± 0.03) GPa along the [111] axis by a hydraulic press described in [4]. The energy shifts of the Λ(TA) phonons with $\boldsymbol{q} = (\xi, \xi, \xi)$ parallel to the stress direction were determined with constant k_i for the incoming neutrons ($k_i = 2.5127$ Å^{-1}) and with neutron energy gain. Near the zone center two phonons were measured by constant-energy scans because of the high gradient of the dispersion curve in this region.

Table 1

The change of frequency $\Delta\omega/\omega$ for the TA phonon (ξ, ξ, ξ) branch

wave vector in units of $2\pi/a$			$10^3\ \Delta\omega/\omega$
0.1	0.1	0.1	5.43 ± 2.30
0.2	0.2	0.2	− 1.26 ± 0.81
0.3	0.3	0.3	− 8.04 ± 0.42
0.4	0.4	0.4	−11.85 ± 0.52
0.5	0.5	0.5	−12.50 ± 0.60

The uniaxial stress in [111] direction was (1.01 ± 0.03) GPa.

During the experiments the bending stress in the sample was monitored by strain gauges fixed on the crystal surface. The design of the press allowed to keep the bending stresses in all cases below 0.015 GPa. It seems to be justified to assume a homogeneous stress distribution in the neutron-irradiated central part of the sample. (The height of the neutron beam was 5 cm.)

The results are presented in Table 1 and in Fig. 1. The phonon frequencies were deduced from the raw data by means of least-squares fit procedures. For estimating the errors we took into account:

1. Statistical errors and errors resulting from mechanical inaccuracies of the spectrometers.

2. Errors resulting from deadjustment of the crystal after applying the stress. Such deadjustment causes changes $\Delta q_{\parallel}$ and $\Delta q_{\perp}$ of the phonon wave vector parallel and perpendicular to its direction. $\Delta q_{\perp}$ was found to have no noticeable influence, which

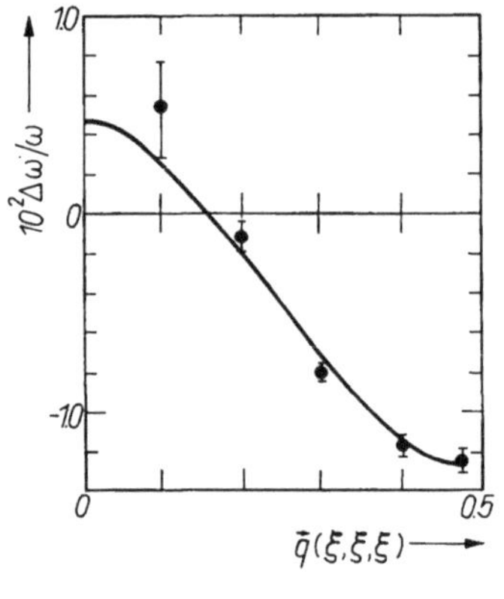

Fig. 1. The measured frequency changes $\Delta\omega/\omega$ for the TA phonon (ξ, ξ, ξ) branch for a uniaxial stress of (1.01 ± 0.03) GPa in [111] direction. The solid curve is calculated according to the theory of Mayer and Wehner [1]

was due both to the small size of $\Delta q_\perp$ and to the fact that we investigated phonons in high-symmetry directions only. The error caused by $\Delta q_{||}$ is dependent on the gradient of the dispersion curve at $\boldsymbol{q}$. These errors turned out to be smaller than the statistical uncertainties. We measured the phonon $\boldsymbol{q} = (0.1, 0.1, 0.1)$ at both sides of the Γ-point. This gave us the value of the zero-point deviation of the spectrometer by which we corrected the raw data.

3. Discussion of the Results

A considerable amount of literature, both experimental and theoretical, has accumulated in recent years on the subject of lattice vibrations in silicon. On the experimental side, we mention as examples work done with the method of inelastic neutron scattering [2, 5, 6], with Raman spectroscopy [7 to 12], and IR absorption [13]. In some experiments [2, 7 to 14] the stress-induced changes in the lattice frequencies were examined. On the theoretical side, many models were tried to explain the experimental facts: shell models [15], Born-von Karman analysis [16], valence force models [17 to 20], bond charge models [21 to 23], and fundamental or first-principles models [24 to 28]. Relatively little work was devoted to the calculation of the equation of state and the Grüneisen parameters [18, 27, 29 to 34].

For an explanation of our result we refer to the most recent calculation of Grüneisen constants in Si of Mayer and Wehner [1]. They calculated the frequency shift $\Delta\omega$ for the phonon $\boldsymbol{q}$ in the branch j for the general strain components $\eta_{\alpha\beta}$ ($\alpha, \beta = x, y, z$),

$$\frac{\Delta\omega}{\omega} = -\sum_{\alpha\beta} \gamma_{\alpha\beta}(\boldsymbol{q}, j)\, \eta_{\alpha\beta} \quad \text{with} \quad \eta_{\alpha\beta} = \sum_{\mu\nu} S_{\alpha\beta,\mu\nu}\sigma_{\mu\nu}\,.$$

$S_{\alpha\beta,\mu\nu}$ are the usual compliance constants, $\gamma_{\alpha\beta}$ is the mode Grüneisen constant related to a strain $\eta_{\alpha\beta}$. The calculation is based on the bond charge model of Weber [23] with additional assumptions on anharmonic forces and has been adjusted to fit the experimental data available for the Γ, X-, and L-points from various experimental techniques. Using this formalism we get for the change of frequencies $\Delta\omega$ of the (ξ, ξ, ξ) phonons with uniaxial stress σ along the [111] direction the relation

$$\frac{\Delta\omega}{\omega\sigma} = -\frac{1}{2}\,\{[\gamma_{xx}(1) + \gamma_{xx}(2)]\,(S_{11} + S_{12}) + [\gamma_{zy}(1) + \gamma_{zy}(2)]\,S_{44}\}\,.$$

The transverse acoustic mode is degenerate in this direction. This degeneracy is not lifted by our special direction of the uniaxial stress. $\gamma_{xx}(1)$, $\gamma_{xx}(2)$, etc. take into consideration the implications caused by this degeneracy. For more details see [1]. For our evaluation shown as solid line in Fig. 1 we took for S_{11} and S_{12} the values of Simons and Wang [35]. We conclude that the model of Mayer and Wehner gives a surprisingly good fit to our data. The absolute values as well as the change in sign are reproduced nearly within the error bars.

Acknowledgements

We are grateful for the financial support of the BMFT.

References

[1] A. P. Mayer and R. K. Wehner, to be published.
[2] J. Prechtel, J. Kalus, E. Lüscher, L. Pintschovius, and R. Ghosh, phys. stat. sol. (b) **93**, 653 (1979).
[3] H. Ibach, phys. stat. sol. **31**, 625 (1969).

[4] J. PRECHTEL, E. LÜSCHER, and J. KALUS, J. Phys. E **10**, 432 (1977).
[5] G. DOLLING, in: Inelastic Neutron Scattering, Vol. 2, IAEA, Vienna 1965 (p. 37).
[6] G. NILSON and G. NELIN, Phys. Rev. B **6**, 3777 (1972).
[7] A. ANASTASSAKIS, A. PINCZUK, E. BURSTEIN, F. H. POLLAK, and M. CARDONA, Solid State Commun. **8**, 133 (1970).
[8] C. J. BUCHENAUER, F. CERDEIRA, and M. CARDONA, Proc. 2nd Internat. Conf. Light Scattering in Solids, Paris 1971, Ed M. BALKANSKI, Flammarion, Paris 1971 (p. 280).
[9] F. CERDEIRA, C. J. BUCHENAUER, F. H. POLLAK, and M. CARDONA, Phys. Rev. B **5**, 580 (1972).
[10] B. A. WEINSTEIN and G. J. PIERMARINI, Phys. Rev. B **12**, 1172 (1975).
[11] W. RICHTER, J. B. RENUCCI, and M. CARDONA, Solid State Commun. **16**, 131 (1975).
[12] M. CHANDRASEKHAR, H. R. CHANDRASEKHAR, M. GRIMSDITCH, and M. CARDONA, Phys. Rev. B **22**, 4825 (1980).
[13] J. F. ANGRESS, G. A. GLEDHILL, and A. J. MAIDEN, J. Phys. C **8**, 1136 (1975).
[14] H. J. MCSKIMIN and P. ANREATCH, J. appl. Phys. **35**, 3312 (1964).
[15] G. DOLLING and R. A. COWLEY, Proc. Phys. Soc. **88**, 463 (1966).
[16] A. D. ZDETSIS and C. S. WANG, Phys. Rev. B **19**, 2999 (1979).
[17] P. N. KEATING, Phys. Rev. **145**, 637 (1966).
[18] P. N. KEATING, Phys. Rev. **149**, 674 (1966).
[19] A. W. SOLBRIG, J. Phys. Chem. Solids **32**, 1761 (1971).
[20] S. L. ALTMANN, A. LAPICCIRELLA, K. W. LODGE, and N. TOMASSINI, J. Phys. C **15**, 5581 (1982).
[21] R. M. MARTIN, Phys. Rev. **186**, 871 (1969).
[22] S. GO, H. BILZ, and M. CARDONA, Phys. Rev. Letters **34**, 580 (1975).
[23] W. WEBER, Phys. Rev. B **15**, 4789 (1977).
[24] P. E. VAN CAMP, V. E. VAN DOREN, and J. T. DEVREESE, Phys. Rev. Letters **42**, 1224 (1979); phys. stat. sol. (b) **93**, 483 (1979).
[25] R. RESTA and A. BALDERESCHI, Phys. Rev. B **24**, 4839 (1981).
[26] M. T. YIN and M. L. COHEN, Phys. Rev. B **26**, 3259 (1982).
[27] K. KUNC, Helv. phys. Acta **56**, 559 (1983).
[28] C. FALTER, M. SELMKE, W. LUDWIG, and W. ZIERAU, J. Phys. C **17**, 21 (1984).
[29] H. JEX, phys. stat. sol. (b) **45**, 343 (1971).
[30] M. I. BELL, phys. stat. sol. (b) **53**, 675 (1972).
[31] T. SOMA, phys. stat. sol. (b) **88**, K69 (1978); **92**, K51 (1979).
[32] H. WENDEL, Solid State Commun. **31**, 423 (1979).
[33] T. SOMA, H. MATSUO, and Y. SAITOH, Solid State Commun. **39**, 913, 1193 (1981).
[34] T. SOMA, Y. KATABIRA, and H. M. KAGAYA, phys. stat. sol. (b) **112**, K99 (1982).
[35] G. SIMONS and H. WANG, Single Crystal Elastic Constants and Calculated Aggregate Properties: A Handbook, MIT-Press, Cambridge (Mass.) 1971.

(Received July 20, 1984)

phys. stat. sol. (b) **126**, 125 (1984)

Subject classification: 2; 4; 20.1; 22.1.1; 22.1.2

Dipartimento di Fisica, Università "La Sapienza", Rome [1])

Study of Bonding Configurations in Amorphous Ge_xSi_{1-x}:H Alloys

By

D. DELLA SALA, C. GIOVANNELLA, and F. EVANGELISTI

The structure of a-Ge_xSi_{1-x}:H alloys produced at two different substrate temperatures by glow-discharge is investigated over the whole range of the gas composition. By comparing IR spectra and chemical analysis, an overall sticking rate three times larger for Ge than for Si is found, an occurrence which has important consequences on the tailoring of the optical gap E_g. The use of a quasi-statistical model accounts for the intensity of the stretching bands of several hydrides present in the structure and permits to differentiate among hydrides differing only in the second neighbor atoms.

La structure des alliages amorphes a-Ge_xSi_{1-x}:H, qui ont été produits à deux différentes températures de déposition avec la méthode du „glow-discharge", a été étudiée pour plusieurs compositions du gaz. La comparaison des spectres d'absorption infrarouge et de l'analyse chimique montre que le contenu rélatif du germanium non-hydrogène est trois fois plus elévé que celui du silicium; un fait très important pour le reglage du gap d'énergie. Un modèle quasi-statistique expliques l'intensité des bandes d'élongation de plusieurs groupes hydrogenès qui sont differents seulement par les atomes seconds voisins.

1. Introduction

While in the recent past the study of amorphous tetrahedral semiconductors was concentrated mainly on a-Si:H, nowadays a large amount of work is also devoted to the production and optimization of photoconductive alloys such as a-Ge_xSi_{1-x}:H [1, 2]. The possibility of varying the optical gap E_g with the atomic composition renders them very attractive for the realization of solar cell structures [3].

By now a-Ge_xSi_{1-x}:H has been the object of several studies based on a variety of experimental techniques [4 to 7]. IR spectroscopy was used to investigate the patterns of hydrogenation, and the IR absorption bands have been assigned to monohydrides and dihydrides [8 to 10]. However, no information is available as yet on the local atomic configuration of the hydrogenated groups, i.e. which atoms are "back-bonded" to Ge–H or Si–H units.

In this paper we report on the influence of the gas composition on the optical and infrared properties of a-Ge_xSi_{1-x}:H. The dependence of the IR band intensity on the atomic composition enables us to distinguish between the cases of one or more Ge or Si atoms back-bonded to the various hydrides. We found that the hydrogen attachment to Ge and Si atoms in the alloy depends linearly on the discharge gas composition r, whereas a more complex behavior determines the atomic composition x.

[1]) Piazzale Aldo Moro 2, 00185 Roma, Italy.

2. Experimental

a-Ge_xSi_{1-x}:H samples were obtained from glow-discharge decomposition of GeH_4 + + SiH_4 mixture, at two substrate temperatures: T_s = 190 and 250 °C. Gas pressure and flow rate were fixed at 0.6 Torr and 8 sccm, respectively; c-Si (for IR spectroscopy) and glass (for visible spectroscopy) were used as substrates. The IR absorption spectra were obtained with a Perkin-Elmer 580B double-beam spectrophotometer (spectral range 4000 to 400 cm^{-1}). The absorption bands in the regions around 2000 and 600 cm^{-1} were deconvoluted by a best-fit procedure, assuming a Gaussian line-shape for each component [11]. The optical absorption edge was measured with a Cary 17 spectrometer, and the optical gap E_g was obtained from usual $(\alpha h\nu)^{1/2}$ versus $h\nu$ plots, according to the theory of Tauc et al. [12]. The atomic composition of the alloys was determined by plasma emission spectroscopy (PES) at CISE Laboratories, Segrate (Milano).

3. Results and Discussion

Typical IR spectra for two samples grown at intermediate gas composition and different T_s are shown in Fig. 1. The arrows mark the various bands the peak frequencies being in agreement with previous assignments [8 to 10]. An exception is the GeH_2 stretching band, observed at 1975 cm^{-1} by some authors [8 to 10] and lying at nearly 1940 cm^{-1} in our spectra. In our spectra this band was generally weak, even for the samples grown at 190 °C which have a larger hydrogen content. Because the much stronger adjacent bands make its deconvolution highly uncertain, it was not studied in detail.

The two wagging bands at 580 and 640 cm^{-1} that receive contributions from the degenerate vibrations of all the hydrides were employed for studying the total hydrogen content bonded to either Ge or Si. For this purpose we used the calibration constants available in literature [13, 14], that relate the intensity of the wagging band to the hydrogen content: $N_H(Ge) = A_{580} \times I_{580}$ with $A_{580} = 1.1 \times 10^{19}$ cm^{-2} and $N_H(Si) = A_{640} \times I_{640}$ with $A_{640} = 1.6 \times 10^{19}$ cm^{-2}, where I_ν is the intensity of each band in cm^{-1}.

For both series of samples grown at different temperatures the relative hydrogen content bonded to germanium, $N_H(Ge)/(N_H(Ge) + N_H(Si))$, corresponds directly to the relative gas composition, $r = [GeH_4]/([GeH_4] + [SiH_4])$, as shown in Fig. 2. Quite on the contrary, the atomic composition of samples with T_s = 250 °C as determined by PES, exhibits a non-linear dependence on r suggesting a deposition rate higher for Ge than for Si (Fig. 3). This behavior can be explained according to a model proposed by Engemann and Fisher [15] for hydrogenated amorphous silicon–carbon alloys.

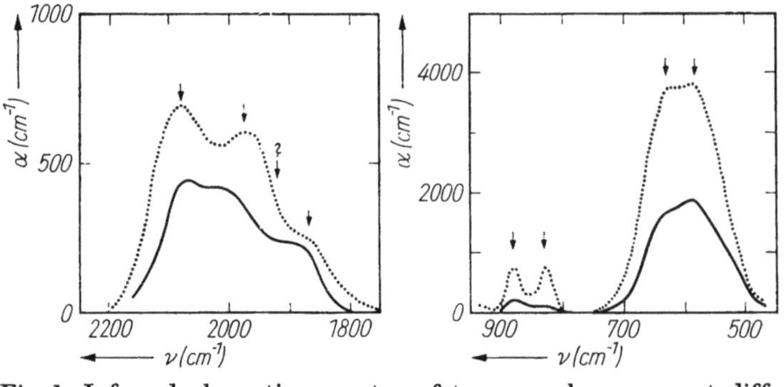

Fig. 1. Infrared absorption spectra of two samples grown at different conditions. Dotted line: T_s = 190 °C, r = 0.65; solid line: T_s = 250 °C, r = 0.50

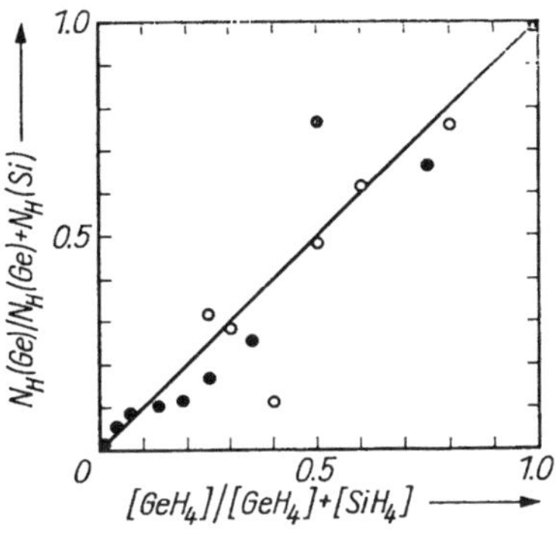

Fig. 2

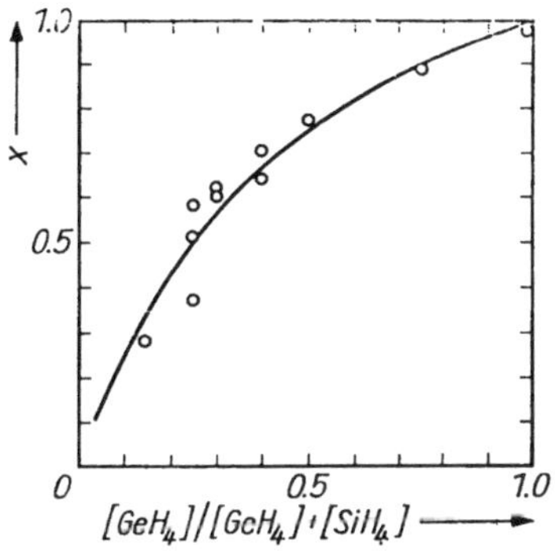

Fig. 3

Fig. 2. Relative hydrogen content bonded to germanium vs. relative gas composition. Open circles $T_s = 250$ °C, closed circles $T_s = 190$ °C

Fig. 3. Atomic composition vs. gas composition for samples grown at $T_s = 250$ °C. The solid line results from the best-fit procedure, as discussed in the text

If we assume that Ge and Si atoms bond to the substrate with independent deposition rates R proportional only to the partial pressure of the corresponding gas we can write $R_{Ge} = f_{Ge} \times p_{GeH_4}$ and $R_{Si} = f_{Si} \times p_{SiH_4}$. Then the atomic composition can be expressed by the equation

$$x = \frac{1}{1 + \alpha(1 - r)/r} \tag{1}$$

with $\alpha = f_{Si}/f_{Ge}$. The best fit of the data of Fig. 3 is obtained with $\alpha = 0.33$. This value supports the preferential attachment of Ge to the substrate and agrees with the ratio of the deposition rates of pure a-Ge:H and a-Si:H films reported by Chevallier et al. [16]. A preferential sticking rate of germanium has been reported by several groups [4, 17, 18] and explained by the weakness of the Ge–H bond, that can be more efficiently dissociated in the discharge.

Equation (1) was derived from data on samples grown at $T_s = 250$ °C, nevertheless from the analysis of the dependence of E_g upon x, we can infer that it is valid also when $T_s = 190$ °C.

The behavior of the optical gap E_g as a function of the discharge gas composition r for our samples grown at $T_s = 190$ °C, is reported in Fig. 4a, together with data by Nozawa et al. [2] for films grown at $T_s = 250$ °C from both a mixture of $GeH_4 + SiH_4$ and of $GeH_4 + Si_2H_6$. The dependence of E_g on r is non-linear with a different

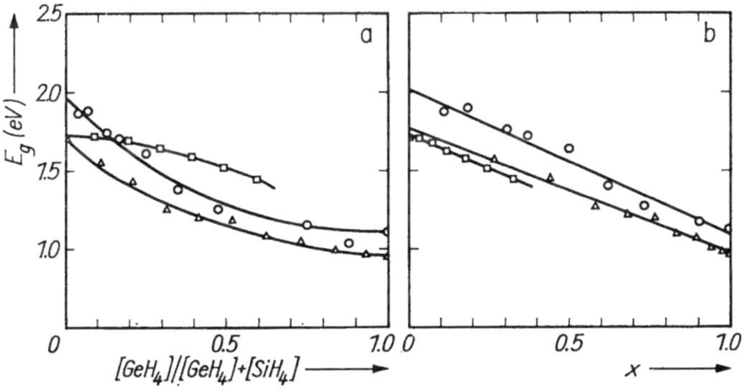

Fig. 4. Behavior of the optical gap as a function of a) gas composition and of b) atomic composition of the alloy. Circles our samples at $T_s = 190$ °C, triangles samples of Nozawa et al. [2] from $GeH_4 + SiH_4$ at $T_s = 250$ °C; squares samples of Nozawa et al. [2] from $GeH_4 + Si_2H_6$ at $T_s = 250$ °C

curvature for the $GeH_4 + Si_2H_6$ case. On the other hand, a linear dependence is obtained (see Fig. 4b) for our samples as well as for the data of [2] concerning the films grown from a mixture of $GeH_4 + SiH_4$, if we plot E_g versus x by using the results of Fig. 3 as a calibration curve. The difference in magnitude of E_g between the two sets of specimens is to be ascribed to the different hydrogen content brought about by the different deposition temperatures. This linear behavior is in agreement with the observations of other authors [1, 16 to 18]. Quite on the contrary, the data of [2] for samples grown by $GeH_4 + Si_2H_6$ mixture require $\alpha = 3$ in order to get the same linearization, suggesting a higher deposition rate for silicon atoms, caused by the use of Si_2H_6 in the discharge.

It was shown by Wieder et al. [19] that it is possible to derive information on the local coordination of the hydrides by combining the knowledge of the atomic percentage x with that of the intensity of the stretching bands. In the following we will apply a similar analysis to the present Ge–Si alloys.

The intensity of the IR stretching bands as a function of the atomic composition x exhibits the dependences shown in Fig. 5a to c. As will become clear in the following, the intensity of each hydrogenated group does not scale simply with the percentage of Ge or Si atoms, and requires a more detailed analysis. First of all, we notice that the frequency and the oscillator strength of the stretching modes of hydrogen can be assumed independent of the configuration for the following reasons: 1. since Ge and Si have the same electronegativity we can exclude that the occurrence of Ge or Si adjacent to a given hydride causes a change of the vibration frequency [20]; 2. the oscillator strength is also unaffected by these changes, because the polarity of the bond remains the same; 3. the variation of the effective mass bonded to hydrogen has little effect, since the stretching modes are strongly localized [21].

If no preferential bonding configuration were present in the alloy, the integrated strength of the stretching bands would vary as a function of x according to the statistical probability of the occurrence of each configuration, that can be easily written by knowing x and r in each sample. Any deviation from the statistical behavior will point to the appearance of preferential grouping. For example, the GeH hydride may exist under several local configurations, $(Ge_3)GeH$, $(Ge_2Si)GeH$, $(GeSi_2)GeH$, $(Si_3)GeH$, where the atoms in parenthesis are the second neighbors of H. With the above assumptions,

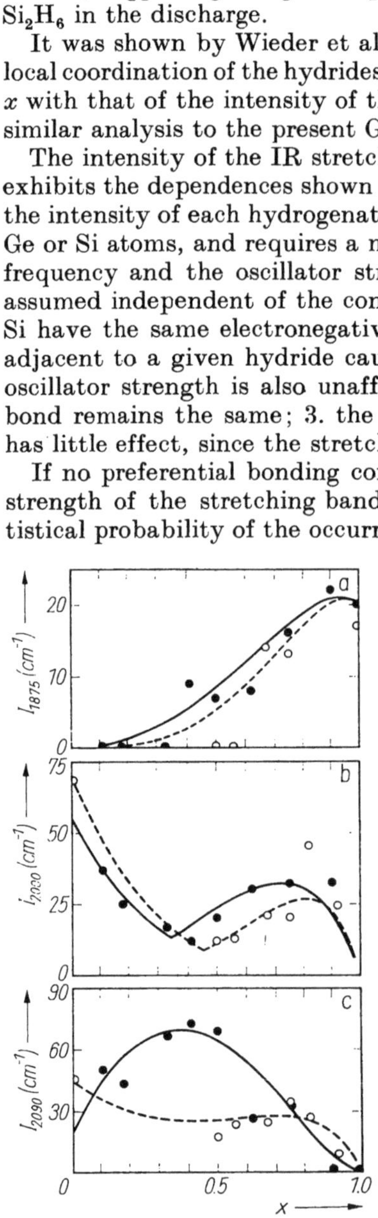

Fig. 5. Dependence of the a) GeH, b) SiH, and c) SiH_2 stretching band intensity on the atomic composition x. Open circles and dashed line refer to samples grown at $T_s = 250$ °C, while closed circles represent samples grown at $T_s = 190$ °C

the dependence on composition can be written as

$$I_{1875}(\mathrm{GeH}) = r \sum_{m=0}^{3} G'_m C_3^m \, x^m (1-x)^{3-m} \,, \tag{2}$$

where each term accounts for the occurrence of one possible configuration. Analogously, for the other hydrides we get

$$I_{1940}(\mathrm{GeH_2}) = r \sum_{m=0}^{2} G''_m C_2^m x^m (1-x)^{2-m} \,, \tag{3}$$

$$I_{2000}(\mathrm{SiH}) = (1-r) \sum_{m=0}^{3} S'_m C_3^m x^m (1-x)^{3-m} \,, \tag{4}$$

$$I_{2090}(\mathrm{SiH_2}) = (1-r) \sum_{m=0}^{2} S''_m C_2^m x^m (1-x)^{2-m} \,. \tag{5}$$

The C's are the statistical weights which account for permutations of identical atoms,

$$C_p^m = \frac{p!}{m!(p-m)!} \,. \tag{6}$$

They normalize the probabilities so that

$$\sum_{m=0}^{p} C_p^m x^m (1-x)^{p-m} = 1 \,. \tag{7}$$

As an example we show in Fig. 6 the component functions $rC_3^m x^m (1-x)^{3-m}$ present in (2). The dotted line represents the sum of the above functions and simulates the behavior of a completely statistical admixture of hydrides. Similar dependences are obtained for (3) to (5).

The G's and S's include the constant oscillator strength as well as the non-statistical effects and account for the possibility that a certain hydride configuration may be favored among the others during the growth process for thermodynamical reasons. A measure of preferential bonding is given by the normalized preferentiality index for each configuration, defined in the case of GeH,

$$P_m(\mathrm{GeH}) = \frac{G'_m}{\sum_{p=0}^{3} G'_p} \,,$$

and similarly for the other hydrides. If no preferential bonding were present, all G's would be identical, and all the P's would have the same value.

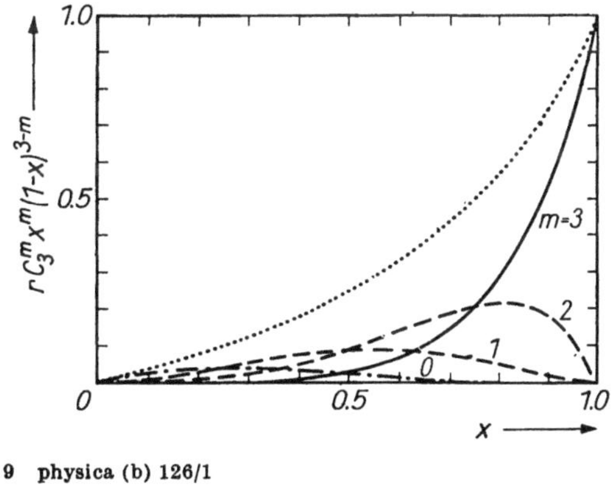

Fig. 6. Basis functions $rC_3^m x^m (1-x)^{3-m}$ used in (2) for the best-fit of the 1875 cm^{-1} band intensity. The dotted line is the sum of the above functions and indicates the behavior of the band for a completely statistical admixture of GeH hydrides

The experimental dependence of the different stretching bands upon x has been fitted with (2) to (5) by using the G's and S's as fitting parameters. The resulting values for the preferentiality index P are reported in Table 1.

Table 1

Values of the preferentiality index P for different local configurations of hydrides. The details of their derivation are described in the text.

hydride local configuration	P ($T_s = 190$ °C)		P ($T_s = 250$ °C)	
$(Si_3)GeH$	0		0	
$(Si_2Ge)GeH$	0.32		0	
$(SiGe_2)GeH$	0.48		0.71	
$(Ge_3)GeH$	0.19		0.29	
	$x < 0.4$	$x > 0.4$	$x < 0.4$	$x > 0.4$
$(Si_3)SiH$	1.00	0	1.00	0
$(Si_2Ge)SiH$	0	0	0	0
$(SiGe_2)SiH$	0	0.33	0	0
$(Ge_3)SiH$	0	0.67	0	1.00
$(Si_2)SiH_2$	0.11		0.20	
$(SiGe)SiH_2$	0.89		0	
$(Ge_2)SiH_2$	0		0.80	

From the data of Fig. 5a we see that the GeH hydrides, corresponding to the 1875 cm^{-1} stretching band, are preferentially bonded to Ge atoms and this trend is enhanced when T_s is raised to 250 °C. In the latter case the deviation from the purely statistical mixture (all P's equal to 0.25) is evident, with the two Si-rich hydride types completely absent.

SiH hydrides follow an unusual dependence on x, as shown in Fig. 5b. A good fit of the data is obtained only by assuming an abrupt change of the weighting factors at $x \approx 0.4$ (see Table 1). The abrupt change in the intensity dependence separates two composition intervals: for $x < 0.4$ there is a strong preferentiality for $(Si_3)SiH$ groups, while for $x > 0.4$ the presence of $(SiGe_2)SiH$ and $(Ge_3)SiH$ groups becomes predominant. The contribution of Ge-rich hydrides is enhanced at $T_s = 250$ °C.

The strongest influence of T_s is observed on the distribution of SiH_2 local configurations (Fig. 5c). The dominant group is $(SiGe)SiH_2$ at $T_s = 190$ °C and is replaced by $(Ge_3)SiH_2$ at $T_s = 250$ °C.

To summarize, the dependence of the stretching bands on x suggests that by raising the deposition temperature all hydrides get an increasing amount of back-bonded germanium, and that this effect is stronger for SiH_2 groups.

4. Conclusions

The detailed analysis of a-Ge_xSi_{1-x}:H alloys by means of IR spectroscopy with the aid of PES and visible spectroscopy, permits to draw the following conclusions.

The Ge atoms as a whole bond to the substrate with a sticking rate three times higher than that of Si atoms, but this preferred attachment is limited only to the unhydrogenated atoms. In fact, the concentration of the respective hydrides is mainly regulated by the gas composition alone.

From the analysis of the preferential bonding configurations it has been found that the deposition temperature influences the local coordination of hydrides. In particular the number of back-bonded Ge increases with temperature and the effect is more pronounced for the SiH_2 hydride.

Finally in close agreement with the behavior in the crystalline case we found a linear dependence of the optical gap on the actual alloy composition.

Acknowledgements

We gratefully acknowledge P. Fiorini for useful discussions, and CISE Laboratories (Milan) for the chemical analysis of the samples.

References

[1] L. Chahed, C. Senemoud, M. L. Theye, J. Bullot, M. Galin, M. Gauthier, B. Bourdon, and M. Toulemonde, Solid State Commun. **45**, 649 (1983).
[2] K. Nozawa, Y. Yamaguchi, J. Hanna, and I. Shimizu, Proc. 10th Internat. Conf. Amorphous and Liquid Semiconductors, Tokyo 1983 (p. 533).
[3] H. Shirahata and Y. Yukimoto, Amorphous Semiconductor Technologies and Devices, Ed. Y. Hamakawa, North-Holland Publ. Co., 1982 (p. 177).
[4] M. Stutzmann, R. J. Nemanich, and J. Stuke, to be published.
[5] A. Morimoto, M. Kumeda, and T. Shimizu, see [2] (p. 537).
[6] S. Hinomura, K. Tsuji, M. Wakagi, T. Ishidate, K. Inoue, and M. Shibuya, see [2] (p. 541).
[7] L. Incoccia, S. Mobilio, M. G. Proietti, P. Fiorini, C. Giovannella, and F. Evangelisti, Phys. Rev., to be published.
[8] D. Bermejo and M. Cardona, J. non-crystall. Solids **32**, 421 (1979).
[9] M. Cardona, phys. stat. sol. (b) **118**, 463 (1983).
[10] W. Paul, D. K. Paul, B. von Roedern, J. Blake, and S. Oguz, Phys. Rev. Letters **46**, 1016 (1981).
[11] P. John, I. M. Odeh, M. J. K. Thomas, M. J. Tricker, F. Riddoch, and J. I. B. Wilson, Phil. Mag. **42**, 671 (1980).
[12] J. Tauc, R. Grigorovici, and A. Vancu, phys. stat. sol. **15**, 627 (1966).
[13] C. J. Fang, K. J. Gruntz, L. Ley, M. Cardona, F. J. Demond, G. Muller, and S. Kalbitzer, Proc. 8th Internat. Conf. Amorphous and Liquid Semiconductors, Cambridge (Mass.) 1979 (p. 255).
[14] H. R. Shanks, F. R. Jeffrey, and M. E. Lowry, Proc. 9th Internat. Conf. Amorphous and Liquid Semiconductors, Grenoble 1981 (p. 733).
[15] D. Engemann and R. Fisher, Appl. Phys. Letters **32**, 567 (1978).
[16] J. Chevallier, H. Wieder, A. Onton, and C. R. Guarnieri, Solid State Commun. **24**, 867 (1977).
[17] J. Bullot, M. Galin, M. Gauthier, and B. Bourdon, J. Physique **44**, 713 (1983).
[18] D. Hauschildt, R. Fischer, and W. Fuhs, phys. stat. sol. (b) **102**, 563 (1980).
[19] H. Wieder, M. Cardona, and C. R. Guarnieri, phys. stat. sol. (b) **92**, 99 (1979).
[20] G. Lucovsky, Solid State Commun. **29**, 571 (1979).
[21] W. B. Pollard and G. Lucovsky, see [14] (p. 353).

(Received July 20, 1984)

phys. stat. sol. (b) 126, 133 (1984)

Subject classification: 11

M. I. Kalinin Leningrad Polytechnical Institute[1])

Group Representation of the Formation Probabilities of Athermal Vacancy Clusters in Atomic Collision Cascades

By

A. I. MELKER and S. N. ROMANOV

Athermal formation of vacancy clusters in displacement cascades is considered in the binary collision approximation. A diagram technique describing the process is suggested. On this basis the group representation of the formation process of vacancy cluster is constructed. The vacancy distribution in the clusters is found in good agreement with the computer simulation results for both f.c.c. and b.c.c. metals.

В приближении бинарных столкновений рассмотрено образование атермических вакансионных кластеров в каскадах смещений. Предложена диаграммная техника, описывающая этот процесс. Построено соответствующее групповое представление процесса образования атермического кластера вакансий. Найдено распределение вакансий по кластерам, которое хорошо согласуется с данными машинных экспериментов в г.ц.к. и о.ц.к. металлах.

1. Introduction

It is well known [1] that the properties of irradiated solids depend not only on the quantity of radiation defects but also on the structure of damage regions created by primary knock-on atoms (PKA). This structure is usually characterized by two types of distributions:

1. the defect (vacancy and interstitial) distribution into clusters,
2. the spacial distribution of defects and their clusters.

The spatial characteristics of displacement cascades in metals and alloys were studied in detail both experimentally by ion-field microscopy [2] and by computer simulation using the Monte-Carlo [3 to 6] and binary-collision [7, 8] methods. The vacancy distribution in athermal clusters in metals was found on the basis of the statistics of computer experiments made by the binary-collision method [9]. There are also works [10, 11] in which the probabilities of the athermal formation of vacancy clusters in displacement cascades are calculated analytically.

In this paper the formation process of the athermal vacancy clusters with the number of vacancies up to ten is examined in the binary-collision approximation, such clusters appearing in the low-energy branches of collision cascades [9], the group representation of this process is constructed and on the basis of this representation the vacancy distribution into clusters is calculated.

2. The Formation Diagrams for Mono-, Di-, Tri-, and Tetra-Vacancies

Consider the formation process of the simplest cluster, a divacancy. According to [10] in the binary-collision approximation an atom moving through a crystal can

[1]) Ul. Polytekhnicheskaya 29, 195251 Leningrad, USSR.

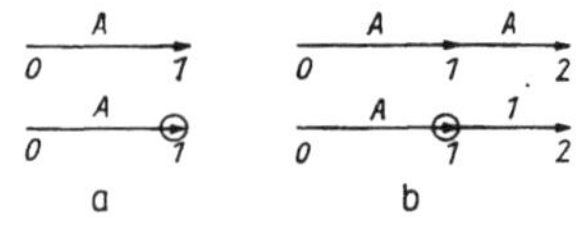

Fig. 1

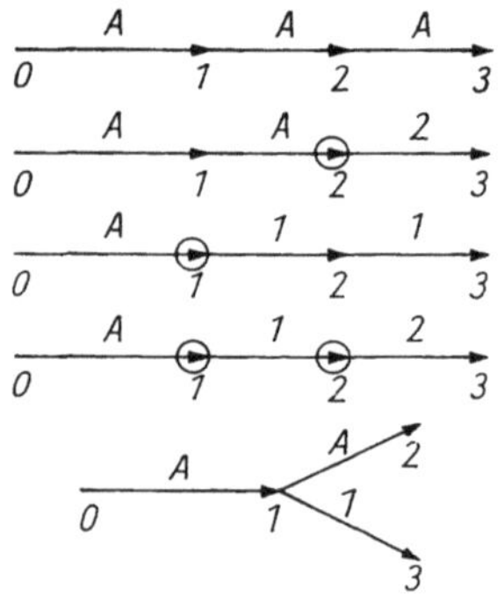

Fig. 2

Fig. 1. a) Mono- and b) divacancy formation diagrams

Fig. 2. Trivacancy formation diagrams

produce a divacancy in two ways: (i) by successively knocking two atoms out of neighbouring lattice sites (the solo mechanism); (ii) by knocking one atom out of a lattice site, with this atom then in turn displacing another atom out of the neighbouring lattice site (the relay-race mechanism). By A let us denote the moving atom and by 1 and 2 the knock-out atoms and the corresponding lattice sites. As in [4, 12] we can use a graph to represent a cascade of displacements producing isolated vacancies and vacancy clusters in a solid. In Fig. 1, the graph parts corresponding to the formation of mono- and divacancies are shown. Here, each arc represents a trajectory of a moving atom, each arrow points into the direction of the formation process, each vertex coincides with a site of a crystal lattice, each symbol over an arc denotes a moving atom, the number of a site is marked by a figure under a vertex, the length of the first arc (0, 1) is arbitrary, the length of the second arc (1, 2) is equal to an

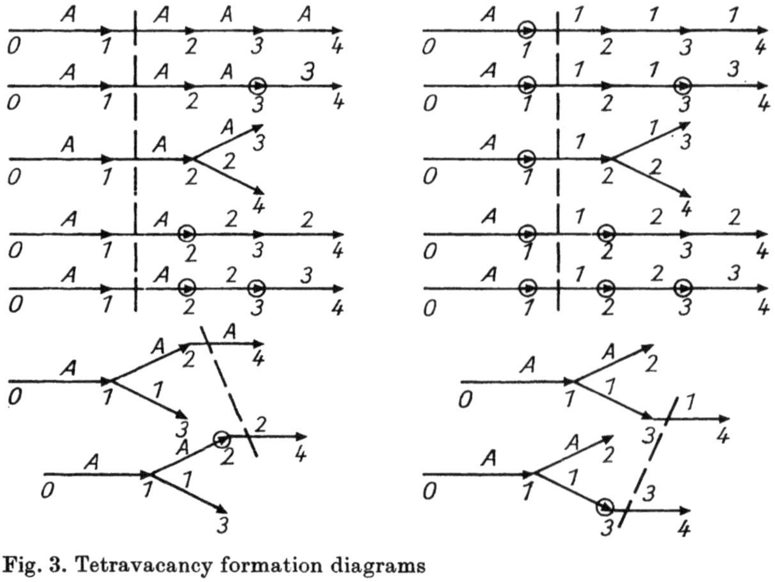

Fig. 3. Tetravacancy formation diagrams

interatomic distance in a crystal lattice. If vertex 1 is encircled it means that the knock-out atom 1 continues the process of defect formation.

Let us turn to a trivacancy. The graph parts corresponding to the formation of an athermal trivacancy are shown in Fig. 2. Here, the lengths of all the arcs except the first ones are equal to an interatomic distance. By examining the diagrams one can see the following. Four diagrams can be got by joining the divacancy diagrams to the monovacancy ones, this process leading to the formation of the trivacancies with a chain structure. The fifth diagram represents the new mechanism of the parallel formation of defects. Here, the incoming atom A produces the vacancy in site 1 and then together with knock-out atom 2 makes two vacancies in neighbouring sites 2 and 3. As a result, the trivacancy with a dendritic structure is formed. By combining the formation diagrams of monovacancies and trivacancies one can construct the formation diagrams of tetravacancies.

It is worth to note that, as it follows from the diagrams shown in Fig. 1 to 3, the formation process of athermal vacancy clusters has the same structure as linear and branched polymers [13]. As in the case of polymers, the diagrams have linear chains, points of branching, main and side chains.

3. Group Representation

The structure of the formation process of athermal vacancy clusters can be represented not only with the help of the diagrams but also by means of tables. By using the language of the set theory, we may say that the incoming and knock-out atoms compile the initial set of producing elements A, 1, 2, 3, ... By double symbols A1, A2, A3, ... , 12, 13, ... , 23, ... let us denote the formation process of monovacancies. Here the first symbol represents an incoming atom, the second one indicates a site which an atom of a lattice is knocked out from. It is obvious that for monovacancies the set of all the formation processes V_1 contains only the processes (elements) of type A1. For divacancies, the elements of set V_2 are processes A1, A2, 12, their number being equal to the quantity of combinations from three producing elements of the initial set in two ones, $C_2^3 = 3$. Similarly for trivacancies $C_4^6 = 6$ and correspondingly V_3(A1, A2, A3, 12, 13, 23), for tetravacancies $C_5^2 = 10$ and V_4(A1, A2, A3, A4, 12, 13, 14, 23, 24, 34).

For example, let us construct the table of the formation process for divacancies (Table 1a). We put all the elements of set V_2 in the upper bordering row, and the elements characterizing the solo and relay-race mechanisms, we write in the second and third rows, respectively. The table structure is such that to the first element A1 of set V_1 one of the uncoincided elements A2 or 12 of set V_2 is joined. Using the same principle, we can construct the table of formation process for tri- and tetravacancies (Tables 1 b and c). Therefore, the set of all the formation processes of a vacancy cluster may be separated into the uncoupled subsets whose elements are the formation processes of clusters of lesser dimensions. In the tables the subsets are disjoined with double lines.

Let us consider each subset as a group of correlated processes independent of processes of another subsets. Then we get the group representation of the athermal formation process for di-, tri-, and tetravacancies in the following form:

$$V_2(A1) \equiv V_2(A1, A2, 12) = V_1(A1)\,\{V_1(A2) + V_1(12)\}\,, \tag{1}$$

$$V_3(A1) \equiv V_3(A1, A2, A3, 12, 13, 23) = V_1(A1)\,\{V_2(A2) + V_2(12)\} + L_3(A1)\,, \tag{2}$$

$$V_4(A1) = V_1(A1)\,\{V_3(A2) + V_3(12)\} + L_3(A1)\,\{V_1(A4) + V_1(14) + V_1(24) + V_1(34)\}\,. \tag{3}$$

Table 1

a) Formation processes of divacancies, b) trivacancies, and c) tetravacancies

a)

A1	A2	A3
A1	A2	
A1		13

b)

A1	A2	A3	12	13	14
A1	A2	13	—	—	—
A1	A2	—	—	—	23
A1	A2	—	—	13	—
A1	—	—	12	13	—
A1	—	—	12	—	23

c)

A1	A2	A3	A4	12	13	14	23	24	34
A1	A2	A3	A4	—	—	—	—	—	—
A1	A2	A3	—	—	—	—	—	—	34
A1	A2	A3	—	—	—	—	—	24	—
A1	A2	—	—	—	—	—	23	24	—
A1	A2	—	—	—	—	—	23	—	34
A1	A2	—	A4	—	13	—	—	—	—
A1	A2	—	—	—	13	14	—	—	—
A1	A2	—	—	—	13	—	—	24	—
A1	A2	—	—	—	13	—	—	—	34
A1	—	—	—	12	13	14	—	—	—
A1	—	—	—	12	13	—	—	24	—
A1	—	—	—	12	13	—	—	—	34
A1	—	—	—	12	—	—	23	24	—
A1	—	—	—	12	—	—	23	—	34

Here subset

$$L_3(A1) = V_1(A1)\, V_1(A2)\, V_1(13) \tag{4}$$

is a branching.

4. Clusters of Greater Dimensions

Having two ways to represent the formation process of athermal clusters, viz. di-, tri-, and tetravacancies, it is not difficult to construct the group representation for clusters of greater dimensions. We give them without deducing. The group representations for penta- and hexavacancies can be written in the form

$$\begin{aligned} V_5(A1) = {} & V_1(A1)\,\{V_4(A2) + V_4(12)\} + \\ & + L_3(A1)\,\{[V_2(A4) + V_2(14) + V_2(24) + V_2(34)] + \\ & + [V_1(A4) + V_1(24)]\,[V_1(15) + V_1(35)]\} + \\ & + L_5(A1, A2) + L_5(A1, 13)\,, \end{aligned} \tag{5}$$

$$\begin{aligned} V_6(A1) = {} & V_1(A1)\,\{V_5(A2) + V_5(12)\} + \\ & + L_3(A1)\,\{[V_3(A4) + V_3(14) + V_3(24) + V_3(34)] + \\ & + [V_1(A4) + V_1(24)]\,[V_2(15) + V_2(35)] + \\ & + [V_2(A4) + V_2(24)]\,[V_1(16) + V_1(36)]\} + \\ & + \{L_5(A1, A2) + L_5(A1, 13)\} \times \\ & \times \{V_1(A6) + V_1(16) + V_1(26) + V_1(36) + V_1(46) + V_1(56)\}\,. \end{aligned} \tag{6}$$

Here $L_5(A1, A2)$ and $L_5(A1, 13)$ are the double branchings in sites 1, 2 and 1, 3 (Fig. 4).

In order to write the group representations not so unwieldy let us introduce the compact system of marking,

$$V_2 = V_1 \cdot 2V_1 , \tag{7}$$

$$V_3 = V_1 \cdot 2V_2 + L_3 , \tag{8}$$

$$V_4 = V_1 \cdot 2V_3 + L_3 \cdot 4V_1 , \tag{9}$$

$$V_5 = V_1 \cdot 2V_4 + L_3(4V_2 + 2V_1 \cdot 2V_1) + 2L_5 , \tag{10}$$

$$V_6 = V_1 \cdot 2V_5 + L_3(4V_3 + 2V_1 \cdot 2V_2 + 2V_2 \cdot 2V_1) + 2L_5 \cdot 6V_1 . \tag{11}$$

It can be shown that the group representations for hepta-, octa-, nana-, and deca-vacancies have the form

$$\begin{aligned} V_7 = {} & V_1 \cdot 2V_6 + L_3(4V_4 + 2V_1 \cdot 2V_3 + 2V_3 \cdot 2V_1) + \\ & + 2L_5(6V_2 + 3V_1 \cdot 3V_1) + 5L_7 , \end{aligned} \tag{12}$$

$$\begin{aligned} V_8 = {} & V_1 \cdot 2V_7 + L_3(4V_5 + 2V_1 \cdot 2V_4 + 2V_2 \cdot 2V_3 + 2V_3 \cdot 2V_2 + 2V_4 \cdot 2V_1) + \\ & + 2L_5(6V_3 + 3V_1 \cdot 3V_2 + 3V_2 \cdot 3V_1) + 5L_7 \cdot 8V_1 , \end{aligned} \tag{13}$$

$$\begin{aligned} V_9 = {} & V_1 \cdot 2V_8 + L_3(4V_6 + 2V_1 \cdot 2V_5 + 2V_2 \cdot 2V_4 + 2V_3 \cdot 2V_3 + \\ & + 2V_4 \cdot 2V_2 + 2V_5 \cdot 2V_1) + L_5(6V_4 + 3V_1 \cdot 3V_3 + 3V_2 \cdot 3V_2 + \\ & + 3V_3 \cdot 3V_1) + 5L_7(8V_2 + 4V_1 \cdot 4V_1) + 20L_9 , \end{aligned} \tag{14}$$

$$\begin{aligned} V_{10} = {} & V_1 \cdot 2V_9 + L_3(4V_7 + 2V_1 \cdot 2V_6 + 2V_2 \cdot 2V_5 + 2V_3 \cdot 2V_4 + \\ & + 2V_4 \cdot 2V_3 + 2V_5 \cdot 2V_2 + 2V_6 \cdot 2V_1) + 2L_5(6V_5 + 3V_1 \cdot 3V_4 + \\ & + 3V_2 \cdot 3V_3 + 3V_3 \cdot 3V_2 + 3V_4 \cdot 3V_1) + 5L_7(8V_3 + 4V_1 \cdot 4V_2 + \\ & + 4V_2 \cdot 4V_1) + 20L_9 \cdot 10V_1 . \end{aligned} \tag{15}$$

Here L_7, L_9 are the triple (Fig. 5) and fourfold (Fig. 6) branchings.

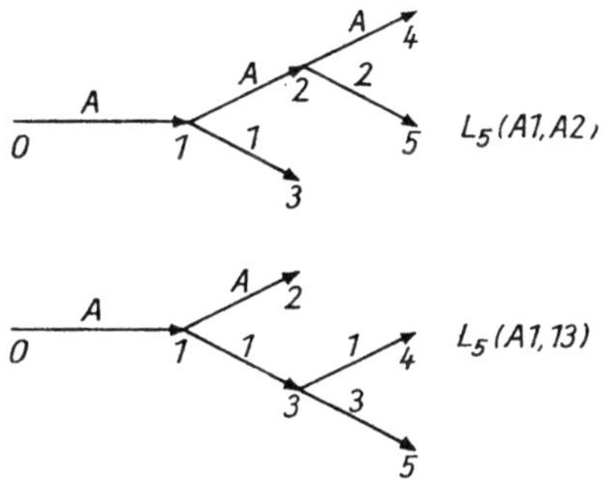

Fig. 4. Double branchings

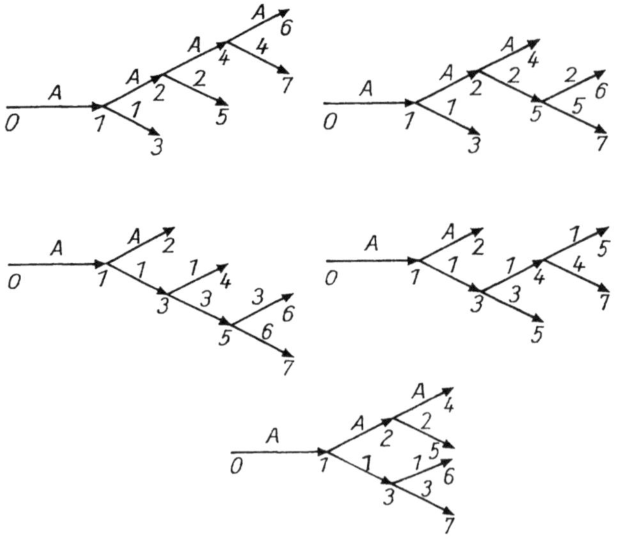

Fig. 5. Triple branchings

We have limited ourselves to decavacancies because the formation probability for n-vacancy clusters in the cascades initiated by a primarily knock-on atom with an energy $E = 0.5$ to 20 keV for $n \geqq 10$ does not exceed 5%, and statistics for such clusters are absent.

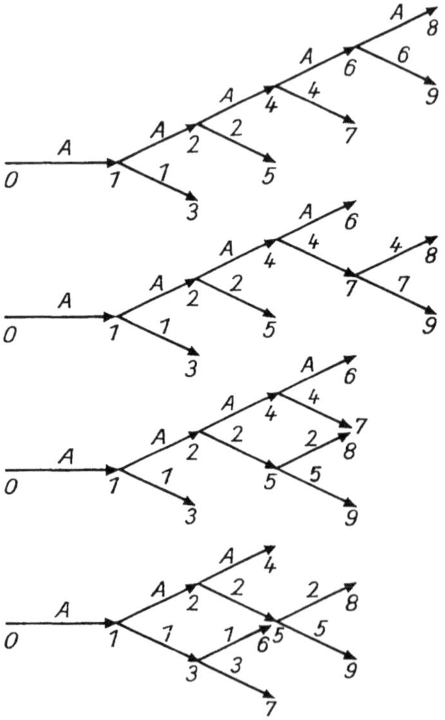

Fig. 6. Fourfould branchings

5. Vacancy Distribution into Clusters

It is known [14] that the cascade function is expressed by

$$\nu(E) = \frac{kE}{E_d}, \tag{16}$$

where $\nu(E)$ is the mean number of Frenkel pairs created by an atom moving with an energy E, E_d is the displacement energy, k is usually $\approx \frac{1}{2}$. Therefore, the energy of a moving atom can be measured in ν-units. If $\nu = 1$, a moving atom produces only one vacancy, but if $\nu = 2$, it can make either two monovacancies or one divacancy. In the latter case the distribution of produced vacancies into clusters can be found from the normalizing condition

$$W_1 + W_2 = 1, \tag{17}$$

where W_i is the probability that a vacancy will be placed in the subset of i-clusters. If $\nu = n$, then analogous reasoning leads to the conclusion that

$$W_1 + W_2 + \dots + W_n = 1. \tag{18}$$

By p let us denote the probability of a vacancy belonging to the monovacancy subset, i.e. $W_1 = p$. Suppose that the probability of vacancy placing in the subset of one or another cluster is determined by the group representation of the athermal

Table 2
Formation probability distributions of vacancy clusters

	i									
ν	1	2	3	4	5	6	7	8	9	10
	p	$2p^2$	$5p^3$	$14p^4$	$42p^5$	$132p^6$	$423p^7$	$1338p^8$	$4464p^9$	$14996p^{10}$
1	1									
2	0.5	0.5								
3	0.392	0.307	0.301							
4	0.347	0.241	0.209	0.203						
5	0.323	0.209	0.168	0.152	0.147					
6	0.308	0.190	0.146	0.126	0.117	0.113				
7	0.298	0.178	0.133	0.111	0.099	0.093	0.088			
8	0.291	0.170	0.124	0.101	0.088	0.081	0.075	0.070		
9	0.286	0.164	0.117	0.094	0.081	0.073	0.067	0.060	0.058	
10	0.282	0.160	0.113	0.089	0.075	0.067	0.061	0.054	0.051	0.048
$\bar{p}_i$	0.403	0.212	0.131	0.088	0.061	0.042	0.029	0.018	0.011	0.005
	$\Sigma = 0.746$			$\Sigma = 0.191$			$\Sigma = 0.058$			
Cu [9]	0.377	0.200	0.160							0.020
	$\Sigma = 0.737$			$\Sigma = 0.213$			$\Sigma = 0.030$			
α-Fe [9]	0.352	0.208	0.129							0
	$\Sigma = 0.689$			$\Sigma = 0.254$			$\Sigma = 0.058$			

formation process of this cluster. Then we get the following equations for the probability p:

$$\begin{aligned} \nu = 1:&\quad p = 1\,, \\ \nu = 2:&\quad p + 2p^2 = 1\,, \\ \nu = 3:&\quad p + 2p^2 + 5p^3 = 1\,, \\ &\ldots\ldots\ldots\ldots\ldots\ldots \\ \nu = 10:&\quad p + 2p^2 + 5p^3 + 14p^4 + 42p^5 + 132p^6 + 423p^7 + \\ &\quad + 1338p^8 + 4464p^9 + 14996p^{10} = 1\,. \end{aligned} \tag{19}$$

The calculated quantities $p_i(\nu)$ are given in Table 2. By averaging $p_i(\nu)$ over the energy spectrum of knock-on atoms in the interval $\nu = 1$ to 10 we get the vacancy distribution into clusters for primary knock-on atoms with the energy $\nu = 10$ (Table 2). Provided [14] $E_d = 25$ eV and $k = \frac{1}{2}$, the values of $\bar{p}_i$ given in Table 2 correspond to the PKA energy which equals 0.5 keV. This distribution is in good agreement with the computer simulation results (two latter rows of the table) both f.c.c. copper and b.c.c. α-iron, although the crystal structure was not plainly taken in consideration in deducing the distribution $\bar{p}_i$.

References

[1] Radiatsionnye defekty v metallakh, Ed. A. T. Luk'yanov, Izd. Nauka, Alma-Ata 1981 (p. 296).
[2] A. L. Suvorov, ibid. (p. 23).
[3] A. I. Melker, ibid. (p. 15).
[4] A. I. Melker and S. N. Romanov, Fiz. i Khim. obrabotki Mater., No. 6, 12 (1979).
[5] A. I. Melker and S. N. Romanov, in: Radiatsionnaya fizika i tekhnologiya, Tula 1979 (p. 128).
[6] A. I. Melker and A. A. Vasilev, Zh. tekh. Fiz. **51**, 634 (1981).
[7] R. Benedek, J. appl. Phys. **52**, 5557 (1981).
[8] A. M. Weissfeld, V. V. Gann, and V. A. Yamnitskii, in: EVM i modelirovanie defektov v kristallakh, FTI, Akad Nauk USSR, Leningrad 1982 (p. 178).
[9] V. M. Agranovich and V. V. Kirsanov, Uspekhi fiz. Nauk **118**, 3 (1976).
[10] S. N. Romanov and A. I. Melker, Zh. tekh. Fiz. **51**, 1252 (1981).
[11] A. I. Melker and S. N. Romanov, phys. stat. sol. (b) **122**, K17 (1984).
[12] A. I. Melker and S. N. Romanov, Zh. tekh. Fiz. **52**, 1362 (1982).
[13] B. M. Yavorskii and A. A. Detlaf, Spravochnik po fizike, Izd. Nauka Moskwa, 1965 (p. 848).
[14] M. W. Thompson, Defects and Radiation Damage in Metals, Cambridge, University Press, 1969.

(Received June 29, 1984)

phys. stat. sol. (b) **126**, 141 (1984)

Subject classification: 10

Department of Mathematics, The University of Southampton[1]) *(a) and Department of Chemistry, Winona State University*[2]) *(b)*

The Grand Partition Function in Defect Statistics

By

P. T. LANDSBERG (a) and S. G. CANAGARATNA (b)

It is explained how the grand partition function method can be used to discuss a) simple defects statistics in which one transition takes place; b) defect statistics in which two transitions take place (Frenkel or Schottky defects); c) more complex statistics. The latter is illustrated by a treatment of a mixture of Schottky and Frenkel defects, not previously discussed in generality.

Es wird erklärt, wie die Methode der großen Verteilungsfunktion benutzt werden kann, um a) simple Defektstatistik, in der ein Übergang stattfindet, b) Defektstatistik, in der zwei Übergänge (Frenkel- oder Schottky-Defekte) stattfinden und c) komplexe Statistiken, zu diskutieren. Die letztere wird durch eine Behandlung einer Mischung von Schottky- und Frenkel-Defekten, die bisher noch nicht allgemein diskutiert wurde, erläutert.

1. Introduction

A knowledge of the concentrations of defects of various types is important for a complete understanding of transport and other rate processes in solids. Their importance in the theory of semiconductors has been recognised for some time now. The theoretical method used in the literature to study this problem has been largely that of the canonical ensemble. It was recently pointed out by one of us [1] that the use of the grand-canonical ensemble in this context has the merit of expeditiousness and absence of mathematical approximations over the free-energy arguments normally employed. Thus, Stirling's approximation can be avoided. It requires the assumption that the number of lattice sites N_L, the number of vacancies n_v and their difference $N_L - n_v$, are all sufficiently large. If Frenkel and Schottky defects are discussed either separately or as occurring jointly in the same system, additional assumptions of this type need to be made. While such assumptions are acceptable in most cases, it is desirable, if possible, to have a treatment which does not require them, and we give one here.

A defence of the use of the grand-canonical ensemble (g.c.e.) in lattice statistics may be unnecessary in view of its wide use for other systems, which would appear to have a constant number of particles. The heat capacity of an electron gas is readily discussed in this framework, for example, by deriving first the Fermi-Dirac distribution by the g.c.e. and then integrating, using a density of states to obtain the total energy. The mean total number of particles in the g.c.e. is then interpreted as the fixed total number of particles one would have in a canonical ensemble treatment. Also, since particles have always a certain sticking probability at surfaces of the container, they have a probability of being removed from the system under study, and in any experiment the total number is therefore known only as a time average. It is this average which is again interpreted as the mean number of the g.c.e. In fact, as long as one is not interested in fluctuations or phase transitions, the same results should be found in the thermodynamic limit by the g.c.e. and by the canonical ensemble.

---

[1]) Southampton S095NH, Great Britain.

[2]) Winona, Minnesota 55987, USA.

Just as the density of a gas treated by the g.c.e. is fixed only on average, so in our case the lattice parameter of the system is fixed only on average. This is due to the interchange of particles in a fixed volume with a reservoir of particles. Assuming a definite lattice structure, it is this average lattice parameter that represents the lattice structure of the system under study. Though the grand partition function (g.p.f.) method has been widely used in the study of adsorption (see, e.g., [2] and [3]), it does not appear to have been used in connection with lattice defects.

The present article illustrates the systematic use of the g.p.f. method in the study of lattice defects. The treatment of such systems has been, up to now, primarily from the statistical thermodynamic point of view (see, e.g., [4]); the purely thermodynamic approach to such systems has been neglected. We shall show elsewhere that the treatment of Born can be extended to systems with lattice imperfections in a natural manner without the need for any new assumptions. An important advantage of such a treatment is that it furnishes a clear definition of the chemical potential of divacancies, defect pairs, etc. In Section 2 the simple (single) defect transition is used as the basis for later discussion. It is shown that further physical conditions (e.g. conservation of mass or charge) lead to a coupling of two or more simple defect transitions and that this leads to a reduction in the number of independent variables. Frenkel and Schottky defects are discussed as examples of such a coupling. This is followed by a treatment of a more general case of the coupling of simple defects. Finally we illustrate the potentialities of the method for treating interactions among defects through the use of a two-site partition function.

2. The Model

In the method of the grand partition function the system is defined by fixed values of the variables V, T, and μ [2, 3]. The system is thus open and the number of particles is subject to fluctuation. Adopting a lattice model of the solid, the volume is replaced by the number of lattice sites as a variable. It is necessary to distinguish between various types of sites, e.g. lattice sites L, interstitial sites I, and surface sites σ. They will be denoted by α, with $\alpha = \mathrm{L}, \mathrm{I}$, or σ. The type of species occupying a given site will be denoted by a superfix i, with $i = +$ for a cation, $-$ for an anion; one may also use an additional symbol (e.g. h for host, D for donor, D^+ for ionised donors). The number of sites of the various types will be denoted by N_α, fixed for each g.c.e., and it will be assumed that these bear a definite relation to one another. The number of α-sites occupied by species i will be denoted by $n_\alpha^{(i)}$. Then N_α will be regarded as independent of $n_\alpha^{(i)}$. This represents a somewhat restrictive assumption only in the case of the surface sites N_σ which could be expected to depend on the total number of occupied surface sites. In this case our assumption means that multilayer surface-site occupation is excluded. But the treatment is readily extended to cover it. For this purpose one would have to introduce separate species of particles associated with a surface site: pairs of particles for two-layer adsorption, triplets of particles for three-layer adsorption, etc. This possible extension is, however, not pursued here, and the number of surface sites $N_\sigma^{(i)}$ occupied by species i is assumed independent of $n_\sigma^{(i)}$ for all i. This is in any case reasonable when $n_\sigma^{(i)} \ll N_\sigma^{(i)}$.

A complete theory using the method of the g.p.f. requires a knowledge of the canonical ensemble partition function of the total crystal. This is a prohibitively difficult task. To simplify matters the following assumptions are made: (i) The degree of freedom relating to positions and motions of the centres of mass of the molecules (the so-called translational degrees of freedom) are separable from all other degrees of freedom;

the latter will be referred to as the internal degrees of freedom. By virtue of this assumption, the system partition function is

$$Q(\text{system}) = Q_{\text{int}}(\text{system})\, Q_{\text{tr}}(\text{system}), \tag{2.1}$$

where Q_{int} refers to internal degrees of freedom and Q_{tr} to the translational degrees of freedom of the system as a whole. (ii) The assumed separation (2.1) suggests that interactions between the occupants of the various sites leave the energy levels pertaining to the internal degrees of freedom unaltered. We therefore assume further that Q_{int} (system) is a product of the internal partition functions $q_{\text{int}}(i)$ of the occupants of the sites, where $q_{\text{int}}(i)$ is independent of the site, it being assumed that it is unaffected by interactions. (iii) Under Q_{tr} we consider the vibrations of the occupants of the lattice sites. Instead of considering the vibration of the crystal as a whole we shall assume that the occupants of the lattice sites vibrate independently of their neighbours. Interactions are assumed to change the position but not the nature of the potential well. We shall further assume that the nature of the potential well in which a species i vibrates depends on i and not on the nature of the site it occupies. The consequence of this assumption is that we may write

$$Q_{\text{tr}} = \Omega Q_{\text{ac}}\,, \tag{2.2}$$

where Ω is the configurational partition function and Q_{ac}, the acoustic partition function, as a product of the vibrational partition function q_{vib} for each occupant of a lattice site. Thus, for Q_{ac} we write

$$Q_{\text{ac}} = \prod_i q_{\text{vib}}(i)\,, \tag{2.3}$$

where $q_{\text{vib}}(i)$ depends on the nature of the potential well. Since the latter is assumed to be left unaltered by interaction, $q_{\text{vib}}(i)$ will depend only on the species (i).

The configurational partition function Ω takes the form

$$\Omega = \sum_j g_j\, \mathrm{e}^{-E_j/kT}\,, \tag{2.4}$$

where g_j is the degeneracy of a configuration with configurational energy E_j. We assume that the most important contribution to Ω in (2.4) comes from the ground state. The index j will then have only one value. Consequently, the configurational partition function for one or more sites may be written thus

$$\Omega = g\, \mathrm{e}^{-E^{(i)}/kT}\,, \tag{2.5}$$

where g is the number of different ways of arranging the species i on the sites. The energy $E^{(i)}$, which changes with the position of the potential well, is the energy of transfer of the species from the lattice site to infinity.

The total partition function for one or more sites is then

$$Q = g q_{\text{int}}^{(i)} q_{\text{vib}}^{(i)}\, \mathrm{e}^{-E^{(i)}/kT} = g Z^{(i)}\, \mathrm{e}^{-E^{(i)}/kT}\,, \tag{2.6}$$

where $Z^{(i)} = q_{\text{int}}^{(i)} q_{\text{vib}}^{(i)}$ depends on i, but is independent of the nature of the site occupied by i.

The absolute activities λ of the various species that are required for the calculation of the $n_\alpha^{(i)}$ are the variables of the theory and are assumed known. At equilibrium the absolute activity of a species i must be independent of the site α occupied, so that $\lambda_\alpha^{(i)} = \lambda^{(i)}$ (say) for all α.

3. The Simple Defect

In a perfect crystal all $\alpha = \mathrm{L}$ sites are occupied and all $\alpha = \mathrm{I}, \sigma$ are vacant. In an actual crystal some L sites will be vacant and some I, σ sites will be occupied. These L vacancies and I, σ occupations are referred to as *defects* since they represent a deviation from the completely ordered arrangement of the perfect crystal. These defects arise from the transfer of a species from a crystal site to bath of this species with a definite chemical potential,

$$i(\alpha\text{-site}) \rightleftharpoons i(\text{bath})\,. \tag{3.1}$$

When the equilibria (3.1) for each of the crystal sites are independent of one another the defects may be termed *simple defects*. When these equilibria are no longer independent owing to various restrictions we may speak of *complex defects*. In either case, under the assumption that there is no interaction between species at different sites, the quantities of interest to us, viz. $n_\alpha^{(i)}$, may be obtained by the method of the single-site partition function. The probability $p_\alpha^{(i)}$ of the occupation of α-sites by i-species is $n_\alpha^{(i)}/N_\alpha^{(i)}$. Here $N_\alpha^{(i)}$ is the number of α-sites available to species i, where of course

$$N_\mathrm{L}^{(\mathrm{D})} = N_\mathrm{L}^{(\mathrm{D}^+)} = N_\mathrm{L}^{(\mathrm{h})} = N_\mathrm{L};$$

$N_\mathrm{L}^{(+)}$ and $N_\mathrm{L}^{(-)}$ are not normally equal to N_L. Also

$$N_\alpha^{(\mathrm{D})} = N_\alpha^{(\mathrm{D}^+)} = N_\alpha^{(\mathrm{h})} = N_\alpha^{(+)} = N_\alpha^{(-)} \quad \text{for} \quad \alpha = \mathrm{I} \text{ or } \sigma\,.$$

The g.p.f. $\Xi_\alpha^{(i)}$ for a single α-site with i-occupation is given by

$$\Xi_\alpha^{(i)} = 1 + Z^{(i)}\lambda^{(i)}\,\mathrm{e}^{-\eta_\alpha^{(i)}}\,, \tag{3.2}$$

where

$$\eta_\alpha^{(i)} \equiv \frac{E_\alpha^{(i)}}{kT}\,. \tag{3.3}$$

We have from the theory of the g.p.f.

$$p_\alpha^{(i)} = \frac{n_\alpha^{(i)}}{N_\alpha^{(i)}} = \frac{\lambda^{(i)}Z^{(i)}\,\mathrm{e}^{-\eta_\alpha^{(i)}}}{1 + \lambda^{(i)}Z^{(i)}\,\mathrm{e}^{-\eta_\alpha^{(i)}}} \tag{3.4}$$

whence also

$$\frac{N_\alpha^{(i)} - n_\alpha^{(i)}}{N_\alpha^{(i)}} = \frac{1}{1 + \lambda^{(i)}Z^{(i)}\,\mathrm{e}^{-\eta_\alpha^{(i)}}}\,. \tag{3.5}$$

According to (3.4) and (3.5) once $\lambda^{(i)}$ and $N_\alpha^{(i)}$ are fixed the $n_\alpha^{(i)}$ are determined. Conversely, the $\lambda^{(i)}$ can be expressed in terms of $n_\alpha^{(i)}/N_\alpha^{(i)}$,

$$\lambda^{(i)}Z^{(i)}\,\mathrm{e}^{-\eta_\alpha^{(i)}} = \frac{n_\alpha^{(i)}}{N_\alpha^{(i)} - n_\alpha^{(i)}}\,, \tag{3.6}$$

which may be written

$$\mu^{(i)} = \mu^{(i)\circledast} + kT \ln \frac{n_\alpha^{(i)}}{N_\alpha^{(i)} - n_\alpha^{(i)}}\,, \tag{3.7}$$

where $\mu^{(i)\circledast} \equiv E_\alpha^{(i)} - kT \ln Z^{(i)}$.

As examples of the applications of these equations, we consider a monatomic crystal. Putting $\alpha = \mathrm{I}, \sigma$ in (3.4) enables us to calculate the number of interstitial and surface atoms. If $\alpha = \mathrm{L}$, (3.5) gives us the number of lattice vacancies $n_{\mathrm{Lv}} = N_\mathrm{L} - n_\mathrm{L}$. We thus have for a monatomic crystal (one species only), writing n_σ, n_I for the total number of occupied surface and interstitial sites,

$$n_\sigma = N_\sigma \frac{\lambda Z\, e^{-\eta_\sigma}}{1 + \lambda Z\, e^{-\eta_\sigma}}, \tag{3.8}$$

$$n_I = N_I \frac{\lambda Z\, e^{-\eta_I}}{1 + \lambda Z\, e^{-\eta_I}}, \tag{3.9}$$

$$n_{Lv} = N_L - n_L = N_L \frac{1}{1 + \lambda Z\, e^{-\eta_L}}. \tag{3.10}$$

Equations (3.8) to (3.10) enable us to calculate the intensive quantities n_σ/N_σ, n_I/N_I, and n_{Lv}/N_L. For this we need a knowledge of the interaction parameters and the energy levels which enter $Z^{(i)}$. These formulae have a structure which is well known in "trapping statistics" [1, 3] which applies to absorption as well as to electron traps in semiconductors; they do not seem to occur in the literature on lattice defects [4].

4. Simplifying Assumptions and Coupling of Transitions

For a pure substance any intensive quantity is a function of T, p. Thus, $\mu = \mu(T, p)$. In the canonical ensemble we select T, N, V as variables, i.e. the intensive variables are T and $c = N/V$. The fact that λ is one of the independent variables in the g.c.e. makes it difficult to use (3.8) to (3.10) to calculate the number of defects since we need to know λZ. It is easier to postulate physically meaningful values for the N_α and $E_\alpha^{(i)}$. An additional assumption will enable us to avoid this difficulty.

The same type of difficulty appears in an ionic solid of type M^+A^-. In the g.c.e. we have to introduce *two* λ's, viz. λ^+ and λ^-, so that it might appear that we have now *three* independent intensive variables T, λ^+, λ^- instead of the two independent variables necessary for a pure substance. However, the condition of electroneutrality introduces *one* condition which brings back the number of independent intensive variables to two,

$$n_L^+ + n_I^+ + n_\sigma^+ = n_L^- + n_I^- + n_\sigma^- . \tag{4.1}$$

Since the n's depend on the λ's, the above equation determines λ^+ in terms of λ^-.

We still have the problem of getting round the difficulty of postulating values for λ, so as to leave only one independent intensive variable, viz. T. The usual way of introducing the extra assumptions is to require that at *every temperature* the number of lattice vacancies is equal to the number of surface and interstitial sites occupied,

$$N_L^{(i)} - n_L^{(i)} = n_I^{(i)} + n_\sigma^{(i)} . \tag{4.2}$$

Equation (4.2) is obviously true at the absolute zero, as can be seen from (3.8) to (3.10) and is likely to be approximately valid at temperatures which are not too high. The above equation amounts to a coupling of the transitions for any species i,

$$i(L) \rightleftharpoons i(\text{bath}) , \quad i(\sigma) \rightleftharpoons i(\text{bath}) , \quad \text{and} \quad i(I) \rightleftharpoons i(\text{bath}) ,$$

so that

$$i(L) \rightleftharpoons i(I) , \quad i(L) \rightleftharpoons i(\sigma) .$$

Since the $n_\alpha^{(i)}$ depend on λ, (4.2) uniquely determines λ as a function of T.

5. Special Cases of Complex Defects

In this section special cases of coupled simple defects will be considered.

One may assume for example that $E_\sigma \gg E_I$ so that we may neglect $n_\sigma^{(i)}$. Equation (4.2) becomes

$$N_L^{(i)} - n_L^{(i)} = n_I^{(i)} . \tag{5.1}$$

Two widely used models in which it is assumed that some types of defects may be neglected from energy considerations are the Frenkel and the Schottky model.

5.1 Frenkel defects

In the Frenkel model for defects in ionic crystals it is assumed that (i) there are only *either* cation *or* anion defects; (ii) surface-site occupation is negligible; (iii) a coupling exists between the basic transitions (3.1). Thus,

$$M^+(L) \rightleftharpoons M^+(\text{bath}); \qquad M^+(I) \rightleftharpoons M^+(\text{bath})$$

are coupled to give

$$M^+(L) \rightleftharpoons M^+(I) .$$

The relevant equations are obtained from (3.9) and (3.10). By virtue of the coupling we have

$$N_L^{(+)} - n_L^{(+)} = n_I^{(+)}$$

whence

$$N_L^{(+)} \frac{1}{1 + \lambda^{(+)} Z^{(+)} e^{-\eta_L^{(+)}}} = N_I^{(+)} \frac{\lambda^{(+)} Z^{(+)} e^{-\eta_I^{(+)}}}{1 + \lambda^{(+)} Z^{(+)} e^{-\eta_I^{(+)}}} . \tag{5.2}$$

This gives us an equation determining $\lambda^{(+)} Z^{(+)}$ in terms of T, $E_L^{(+)}$. The "number of Frenkel defects" $n_F^{(+)}$ may be identified with $n_I^{(+)}$. We can directly eliminate $\lambda^{(+)} Z^{(+)}$ from (3.9) and (3.10) to obtain

$$\frac{[n_F^{(+)}]^2}{\left(N_I^{(+)} - n_F^{(+)}\right)\left(N_L^{(+)} - n_F^{(+)}\right)} = e^{-(E_I^{(+)} - E_L^{(+)})/kT} . \tag{5.3}$$

It will be seen that (5.3) has the form of the expression of the equilibrium constant for the reaction $M^+(L) = M^+(I)$ when the latter is written more fully thus

$$L^{\cdot} + I^{\circ} = L^{\circ} + I^{\cdot} ,$$

where the superscripts "·" and "∘" denote occupancy and vacancy, respectively.

When $n_F^{(+)}$ is very small in comparison with $N_L^{(+)}$ and $N_I^{(+)}$, we have the approximate equation [5]

$$n_F^{(+)} = \left(N_L^{(+)} N_I^{(+)}\right)^{1/2} e^{-(E_I^{(+)} - E_L^{(+)})/2kT} . \tag{5.4}$$

The activation energy of the Frenkel defects in low concentration can thus be obtained from an Arrhenius-type equation. Another way of looking at (5.4) is

$$\frac{n_F^{(+)}}{N_L^{(+)}} \frac{n_F^{(+)}}{N_I^{(+)}} = e^{-(E_I^{(+)} - E_L^{(+)})/kT} \tag{5.5}$$

which is the equilibrium constant of the creation–annihilation reaction

$$V + I = 0 .$$

An explicit expression connecting $\mu^{(+)}$ and $n_F^{(+)}$ may be easily obtained. Putting $N_I^{(+)} = \beta N_L^{(+)}$ in (5.2) and assuming that $n_F^{(+)}$ is small, we have $e^{\eta_L^{(+)}} \ll \lambda^{(+)} Z^{(+)} \ll \ll e^{\eta_I^{(+)}}$, whence

$$\lambda^{(+)} Z^{(+)} = \frac{1}{\sqrt{\beta}} e^{(E_I^{(+)} + E_L^{(+)})/2kT} .$$

Combining this with (5.4) we have

$$\mu^{(+)} = \mu^{\circledast(+)} + kT \ln \frac{n_F^{(+)}}{N_I^{(+)}}, \tag{5.6}$$

where $\mu^{\circledast(+)} = E_I^{(+)} - kT \ln Z^{(+)}$ (see (3.7)).

In the above treatment we have not specified the location of the interstitial defect in relation to a vacancy. If they are bound together so as to remain nearest neighbours, they can be treated by the method of the two-site partition function (see Section 8).

5.2 Schottky defects

In the Schottky model for defects in an ionic solid the following assumptions are made: (i) There are equal cation and anion vacancies; (ii) interstitial site occupation is negligible. These assumptions amount to replacing the scheme

$$M^+(L) \rightleftharpoons M^+(\text{bath}); \quad M^+(\sigma) \rightleftharpoons M^+(\text{bath}),$$

$$A^-(L) \rightleftharpoons A^-(\text{bath}); \quad A^-(\sigma) \rightleftharpoons A^-(\text{bath})$$

with the scheme

$$M^+(L) + A^-(L) \rightleftharpoons M^+(\sigma) + A^-(\sigma) .$$

The following equation is easily derived for the number of Schottky defects $n_S = n_\sigma^{(+)} - n_\sigma^{(-)}$ using (3.6) with $\alpha = L, \sigma$,

$$\frac{n_S^4}{(N_L^{(+)} - n_S)(N_L^{(-)} - n_S)(N_\sigma^{(+)} - n_S)(N_\sigma^{(-)} - n_S)} = e^{-[(E_\sigma^{(+)} - E_L^{(+)}) + (E_\sigma^{(-)} - E_L^{(-)})]/kT} . \tag{5.7}$$

This is the equilibrium constant for the sum of the reactions

$$L^{+\cdot} + \sigma^{+\circ} \rightleftharpoons L^{+\circ} + \sigma^{+\cdot}, \qquad L^{-\cdot} + \sigma^{-\circ} \rightleftharpoons L^{+\circ} + \sigma^{-\cdot} .$$

As for the Frenkel case, we may derive the relation between the chemical potential and the number of Schottky defects for $N_\sigma^+ = N_\sigma^- \equiv N_\sigma$,

$$\mu_S = \mu_S^\circledast + kT \ln \frac{n_S}{N_\sigma}, \tag{5.8}$$

where

$$\mu_S \equiv \mu^{(+)} + \mu^{(-)} \quad \text{and} \quad \mu_S^\circledast \equiv -kT \ln (Z^{(+)} Z^{(-)})^{1/2} + \frac{E_\sigma^{(+)} + E_\sigma^{(-)}}{2}. \tag{5.9}$$

When $n_S \ll N_L$, $n_S \ll N_\sigma$ an appropriate equation analogous to (5.5) may be obtained,

$$\frac{n_S}{N_L} \frac{n_S}{N_\sigma} = \exp \left[- \frac{(E_\sigma^{(+)} - E_L^{(+)}) + E_\sigma^{(-)} - E_L^{(-)}}{2kT} \right].$$

The physics behind the assumption that there are equal cation and anion vacancies is that the cation–anion pair is electrically neutral and would therefore require less energy for displacement from the lattice sites than either an anion or a cation alone. It would be more in keeping with the assumptions of the model to postulate that the cation and anion vacancies are on adjacent sites and that the cation and anion species on the surface are also in pairs. This point does not appear to have been stressed in the literature on defects (see, e.g., [4, 5]). The model would then be indistinguishable from that for a molecular solid. The relevant equations would be formally similar to the ones already derived but the interpretation of the terms would be different. In particular, there would be a difference in the estimates for the entropy, the method taking into account pair formation leading to a lower estimate. It is worth noting that in the traditional combinatorial treatment found in the literature the entropy arising from the Schottky pairs on the surface sites is completely neglected. One of the few places where this neglect is pointed out explicitly is in Kittel's book [6]. The result (5.7) will now be confirmed by using standard theory.

If the number of ways of putting an atom on the surface be $M(n_\sigma)$, the usual theory for the equilibrium number of Schottky defects at temperature T can be derived from the amended free energy

$$F = E - TS = n_\sigma(E_\sigma - E_L) - kT \ln \frac{N_L!\, M(n_\sigma)}{(N_L - n_\sigma)!\, n_\sigma!}$$

$$= n_\sigma(E_\sigma - E_L) - kT[N_L \ln N_L - (N_L - n_\sigma) \ln (N_L - n_\sigma) - - n_\sigma \ln n_\sigma + \ln M]\,,$$

where $N_L^{(+)} = N_L^{(-)} \equiv N_L$, $N_\sigma^{(+)} = N_\sigma^{(-)} \equiv N_\sigma$, and $N_L, n_\sigma, N_\sigma - n_\sigma \gg 1$ have been assumed. This gives

$$\frac{n_S}{N_L - n_S} = e^{-(E_\sigma - E_L)/kT}\, e^{\partial \ln M/\partial n_\sigma} \qquad (n_S \equiv n_\sigma)\,. \tag{5.10}$$

If M is a constant, one finds the usual result. Our result (5.7), however, is equivalent to the choice

$$M = \frac{N_\sigma!}{(N_\sigma - n_\sigma)!\, n_\sigma!}$$

which converts (5.10) to

$$\frac{n_S^2}{(N_L - n_S)\,(N_\sigma - n_S)} = \exp\left(-\frac{E_\sigma - E_L}{kT}\right). \tag{5.11}$$

Now apply this result first to a cation with E_σ replaced by $E_\sigma^{(+)}$ and then to an anion with E_σ replaced by $E_\sigma^{(-)}$, corresponding to the two atomic transfers involved in producing a Schottky defect. One then finds (5.7) as a product of two equations (5.11).

6. Complex Defects. General Case

In this section we shall treat the most general case of complex defects for an ionic solid of the type M^+A^-. This is equivalent to considering a crystal with a mixture of Frenkel and Schottky defects, which does not seem to have been considered before (but see Fowler and Guggenheim [7]). The assumption embodied in (4.2) amounts to a coupling represented by an equilibrium between lattice site occupation and interstitial or surface state occupation by a particle of charge i,

$$i(L) \rightleftharpoons i(I)\,; \qquad i(L) \rightleftharpoons i(\sigma) \qquad (i \text{ denotes positive or negative charge}).$$

Our problem will be to obtain expressions for $\lambda^{(i)}Z^{(i)}$ $(i = \pm)$ and thence to obtain the values of $n_\alpha^{(i)}$ $(\alpha = \mathrm{L, I}, \sigma)$, given a knowledge of the $N_\alpha^{(i)}$ and $\eta_\alpha^{(i)} = E_\alpha^{(i)}/kT$. We have the following equations, (cf. (3.6) and (4.2)):

$$\lambda^{(i)}Z^{(i)} = \frac{n_\alpha^{(i)}}{N_\alpha^{(i)} - n_\alpha^{(i)}}\, \mathrm{e}^{\eta_\alpha^{(i)}} \qquad \text{(six equations)} \tag{6.1}$$

and

$$N_\mathrm{L}^{(i)} - n_\mathrm{L}^{(i)} = n_\mathrm{I}^{(i)} + n_\sigma^{(i)} \qquad \text{(two equations)}. \tag{6.2}$$

We have thus eight equations in eight unknowns which can all be determined. One finds

$$n_\mathrm{I}^{(i)} + n_\sigma^{(i)} = \frac{n_\mathrm{I}^{(i)} N_\mathrm{I}^{(i)}\, \mathrm{e}^{\eta_\sigma^{(i)}} + [n_\mathrm{I}^{(i)}]^2 \left(\mathrm{e}^{\eta_\mathrm{I}^{(i)}} - \mathrm{e}^{\eta_\sigma^{(i)}}\right) + n_\mathrm{I}^{(i)} N_\sigma^{(i)}\, \mathrm{e}^{\eta_\mathrm{I}^{(i)}}}{N_\mathrm{I}^{(i)} \mathrm{e}^{\eta_\sigma^{(i)}} + n_\mathrm{I}^{(i)} \left(\mathrm{e}^{\eta_\mathrm{I}^{(i)}} - \mathrm{e}^{\eta_\sigma^{(i)}}\right)}, \tag{6.3}$$

and by elimination of the $\lambda^{(i)}Z^{(i)}$ we obtain a cubic equation for $n_\mathrm{I}^{(i)}$,

$$\sum_{j=0}^{3} h_j \left(\frac{n_\mathrm{I}^{(i)}}{N_\mathrm{L}^{(i)}}\right)^j = 0\,, \tag{6.4}$$

where

$$h_3 = \left(\mathrm{e}^{\eta_\sigma'^{(i)}} - \mathrm{e}^{\eta_\mathrm{I}'^{(i)}}\right)\left(1 - \mathrm{e}^{\eta_\mathrm{I}'^{(i)}}\right) \mathrm{e}^{2\eta_\mathrm{L}^{(i)}}\,,$$

$$h_2 = \mathrm{e}^{2\eta_\mathrm{L}^{(i)}} \left[\mathrm{e}^{\eta_\mathrm{I}'^{(i)}} - \mathrm{e}^{\eta_\sigma'^{(i)}} + \frac{N_\sigma^{(i)}}{N_\mathrm{L}^{(i)}}\, \mathrm{e}^{2\eta_\mathrm{I}'^{(i)}} + \frac{N_\mathrm{I}^{(i)}}{N_\mathrm{L}^{(i)}}\, \mathrm{e}^{(\eta_\sigma'^{(i)} + \eta_\mathrm{I}'^{(i)})} - \right.$$

$$\left. - 2\, \frac{N_\mathrm{I}^{(i)}}{N_\mathrm{L}^{(i)}}\, \mathrm{e}^{\eta_\sigma'^{(i)}} + \left(\frac{N_\mathrm{I}^{(i)}}{N_\mathrm{L}^{(i)}} - \frac{N_\sigma^{(i)}}{N_\mathrm{L}^{(i)}}\right) \mathrm{e}^{\eta_\mathrm{I}'^{(i)}}\right],$$

$$h_1 = \mathrm{e}^{2\eta_\mathrm{L}^{(i)}}\, \frac{N_\mathrm{I}^{(i)}}{N_\mathrm{L}^{(i)}} \left[\left(2 + \frac{N_\mathrm{I}^{(i)}}{N_\mathrm{L}^{(i)}}\right) \mathrm{e}^{\eta_\sigma'^{(i)}} + \left(\frac{N_\sigma^{(i)}}{N_\mathrm{L}^{(i)}} - 1\right) \mathrm{e}^{\eta_\mathrm{I}'^{(i)}}\right],$$

$$h_0 = -\mathrm{e}^{2\eta_\mathrm{L}^{(i)}} \left(\frac{N_\mathrm{I}^{(i)}}{N_\mathrm{L}^{(i)}}\right)^2 \mathrm{e}^{\eta_\sigma'^{(i)}}.$$

Here $\eta_\alpha' \equiv (E_\alpha - E_\mathrm{L})/kT$ $(\alpha = \mathrm{I}, \sigma)$. Once $n_\mathrm{I}^{(i)}$ is determined, $n_\sigma^{(i)}$ can be found from (6.3) and hence $n_\mathrm{L}^{(i)}$ can be obtained from (6.2).

The fact that $n_\mathrm{I}^{(i)}$ is obtained from a cubic equation raises the interesting possibility that a phase-transition-type behaviour could be obtained. It is hoped to return to this question.

On ignoring the surface sites one should regain our theory for Frenkel defects. However, as N_σ bears a definite ratio to N_L, this limit should not be attempted by letting $N_\sigma \to 0$. Instead it can be reached by the indefinite raising of E_σ, i.e. by letting $\eta_\sigma' \to \infty$. One then finds the relation

$$-(A - B)^2 + B(A - B)^2 + B^2(A - B) \exp \eta_\mathrm{I}'^{(i)} = 0$$

from (6.4), where

$$A \equiv \frac{N_\mathrm{I}^{(i)}}{N_\mathrm{L}^{(i)}}, \qquad B \equiv \frac{n_\mathrm{I}^{(i)}}{N_\mathrm{L}^{(i)}}.$$

This is for $i = +$ easily shown to be equivalent to (5.3), and so checks that the present section represents a generalisation to a mixture of Frenkel and Schottky defects.

7. Application of a Two-Site p.f.

One of the chief advantages of the g.p.f. method is that interactions can be treated by a consideration of the two-site partition function. The latter should give much better results than the single-site partition function but without the increased mathematical complexity of higher-site p.f. or combinatorial methods.

The simplest case where the two-site partition function is useful is in the treatment of divacancies in a monatomic crystal. Here the two-site partition function may be written as

$$\Xi_2 = 1 + 2\lambda_h Z_h \, e^{-\eta_{hv}} + Z_h^2 \lambda_h^2 \, e^{-\eta_{hh}}$$

whence the probability p_{vv} of divacancies is given by

$$p_{vv} = (\Xi_2)^{-1} = e^{-p\tilde{v}/kT} ,$$

where $\tilde{v}$ is the volume for two neighbouring sites. Since $p_{vv} = n_{vv}/(zN_L/2)$, the number of divacancies in a crystal of N_L sites is given by

$$n_{vv} = \frac{zN_L}{2} \, e^{-p\tilde{v}/kT} ,$$

where z is the coordination number of the lattice.

The case where the interstitial ions in Frenkel defects are bound to vacancies may be treated similarly. Consider a lattice site and a neighbouring interstitial site. Then the partition function for these two sites is

$$\Xi_2 = \lambda_h Z_h \, e^{-\eta_L} + \lambda_h Z_h \, e^{-\eta_{Iv}} .$$

The probability p_{Iv} of bound pairs is given by

$$p_{Iv} = \frac{\lambda_h Z_h \, e^{-\eta_{Iv}}}{\lambda_h Z_h \, e^{-\eta_L} + \lambda_h Z_h \, e^{-\eta_{Iv}}} = \frac{e^{-\eta_{Iv}}}{e^{-\eta_L} + e^{-\eta_{Iv}}} .$$

If the concentration of Frenkel defects is small, $p_{Iv} = e^{-(\eta_{Iv} - \eta_L)}$. For a crystal of N_L lattice sites let the number of possible Iv pairs be βN_L where β is the coordination number of an interstitial site. The number of bound Frenkel defects is given by

$$n_{Iv} = \beta N_L \, e^{-(\eta_{Iv} - \eta_L)} .$$

A slightly more complicated case is where the host crystal contains a solute (D) as well as vacancies. If we rule out divacancies and D–D pairs as energetically unfavourable we then have for the g.p.f.

$$\Xi_2 = 2Z_h\lambda_h \, e^{-\eta_{hv}} + 2Z_D\lambda_D \, e^{-\eta_{Dv}} + Z_h^2\lambda_h^2 \, e^{-\varphi(hh)} + 2Z_hZ_D\lambda_h\lambda_D \, e^{-\varphi(hD)} ,$$

where we have written $Z_{AB} = Z_A Z_B \, e^{-\varphi(AB)}$.

Useful relations between the atomic fraction c_v, c_D, and c_{Dv} of vacancies, solute and solute–vacancy pairs may be obtained as follows: We use the following definitions:

$$c_v \equiv \frac{N_v}{N_L}, \qquad c_D \equiv \frac{N_D}{N_L}. \qquad c_{Dv} \equiv \frac{n_{Dv}}{zN_L}, \qquad c_h \equiv \frac{N_h}{N_L}.$$

We have

$$zN_v = n_{hv} + n_{Dv}; \qquad zN_D = n_{Dv} + n_{Dh}; \qquad zN_h = n_{hv} + n_{Dh} + 2n_{hh} .$$

Consequently,

$$c_{\mathrm{v}} = \frac{Z_{\mathrm{h}}\lambda_{\mathrm{h}}\,\mathrm{e}^{-\eta_{\mathrm{hv}}}}{\Xi_2} + \frac{Z_{\mathrm{D}}\lambda_{\mathrm{D}}\,\mathrm{e}^{-\eta_{\mathrm{Dv}}}}{\Xi_2}. \qquad c_{\mathrm{D}} = \frac{Z_{\mathrm{D}}\lambda_{\mathrm{D}}\,\mathrm{e}^{-\eta_{\mathrm{Dv}}}}{\Xi_2} + \frac{Z_{\mathrm{h}}Z_{\mathrm{D}}\lambda_{\mathrm{h}}\lambda_{\mathrm{D}}\,\mathrm{e}^{-\varphi(\mathrm{Dh})}}{\Xi_2},$$

$$c_{\mathrm{Dv}} = \frac{Z_{\mathrm{D}}\lambda_{\mathrm{D}}\,\mathrm{e}^{-\eta_{\mathrm{Dv}}}}{\Xi_2}, \qquad c_{\mathrm{h}} = \frac{Z_{\mathrm{h}}\lambda_{\mathrm{h}}\,\mathrm{e}^{-\eta_{\mathrm{hv}}}}{\Xi_2} + \frac{Z_{\mathrm{D}}Z_{\mathrm{h}}\lambda_{\mathrm{D}}\lambda_{\mathrm{h}}\,\mathrm{e}^{-\eta_{\mathrm{Dv}}}}{\Xi_2} + \frac{(Z_{\mathrm{h}}\lambda_{\mathrm{h}})^2\,\mathrm{e}^{-\varphi(\mathrm{hh})}}{\Xi_2}.$$

Therefore,

$$c_{\mathrm{h}} - (c_{\mathrm{D}} - c_{\mathrm{Dv}}) - (c_{\mathrm{v}} - c_{\mathrm{Dv}}) = \frac{(Z_{\mathrm{h}}\lambda_{\mathrm{h}})^2\,\mathrm{e}^{-\varphi(\mathrm{hh})}}{\Xi_2}$$

and

$$\frac{(c_{\mathrm{v}} - c_{\mathrm{Dv}})\,(c_{\mathrm{D}} - c_{\mathrm{Dv}})}{c_{\mathrm{Dv}}} = \frac{(Z_{\mathrm{h}}\lambda_{\mathrm{h}})^2}{\Xi_2}\,\frac{\mathrm{e}^{\varphi(\mathrm{Dh})}\,\mathrm{e}^{-\eta_{\mathrm{hv}}}}{\mathrm{e}^{-\eta_{\mathrm{Dv}}}} =$$

$$= c_{\mathrm{h}}\left(1 - \frac{c_{\mathrm{D}} - c_{\mathrm{Dv}}}{c_{\mathrm{h}}} - \frac{c_{\mathrm{v}} - c_{\mathrm{Dv}}}{c_{\mathrm{h}}}\right)\mathrm{e}^{\varphi(\mathrm{hh}) + \eta_{\mathrm{Dv}} - (\varphi(\mathrm{Dh}) + \eta_{\mathrm{hv}})}.$$

Since $c_{\mathrm{h}} \gg c_{\mathrm{D}}, c_{\mathrm{v}}$, etc.

$$\frac{(c_{\mathrm{v}} - c_{\mathrm{Dv}})\,(c_{\mathrm{D}} - c_{\mathrm{Dv}})}{c_{\mathrm{Dv}}c_{\mathrm{h}}} = \mathrm{e}^{-[\varphi(\mathrm{Dh}) + \eta_{\mathrm{hv}} - (\varphi(\mathrm{hh}) + \eta_{\mathrm{Dv}})]}.$$

This may be looked upon as an equilibrium constant expression for the equation

$$\mathrm{hh} + \mathrm{Dv} \rightleftharpoons \mathrm{hv} + \mathrm{hD},$$

because we have $c_{\mathrm{v}} = c_{\mathrm{hv}} + c_{\mathrm{Dv}}$, $c_{\mathrm{D}} = c_{\mathrm{hD}} + c_{\mathrm{vD}}$. It corresponds to equation (5.22) of [7] which was given in the form

$$\frac{(c_{\mathrm{v}} - c_{\mathrm{p}})\,(c_{\mathrm{b}} - c_{\mathrm{p}})}{c_{\mathrm{p}}} = \frac{1}{z}\,\mathrm{e}^{g/kT}.$$

This differs from our expression only as regards the constants on the right-hand side and reflects a difference in assumptions made. We allow nearest-neighbour interactions but neglect all others; in the derivation of (5.22) of [8], though interactions are formally taken into account, the entropy of mixing was assumed to be ideal. However, the broad agreement between the two equations is satisfactory since our expression is based on the new results for the concentrations of vacancies, donors, and donor–vacancy pairs. Such expressions are of potential use for the understanding of (i) the part played by impurities in diffusion; (ii) the annealing of quenched solids; (iii) the annealing of irradiated solids, etc.

As a further example we treat the problem of determining the number of pairs formed between the host (h) and a donor (D) in a doped semiconductor. If we assume that the donor may be in the neutral or ionised state (D^+), the problem will be to determine, for a given ratio of host to donor impurities $N_{\mathrm{h}}/N_{\mathrm{D}}^0$, the numbers $N_{\mathrm{D-D}}$, $N_{\mathrm{D-D^+}}$, $N_{\mathrm{D^+-D^+}}$, of D–D, D–D^+, and D^+–D^+ pairs. This problem can be treated adequately by the two-site g.p.f. As a first approximation we neglect the interactions of the occupants of these two-sites with the rest of the lattice. The probability of the pair numbers obtained in this first approximation can be used in the next iteration to calculate the effect of the interaction of the occupants of the two-site with its neighbours: this will give an improved estimate for the pair numbers, but this is not attempted here.

The assumptions of the model are the same as the assumptions (i) to (iii) in Section 2. We postulate only one type of defect site and, as in Section 4, to get round the difficulty of having to specify the values for the λZ's, we shall suppose that at every temperature the number of vacancies equals the number of occupied defect sites.

We shall consider a complex K for any species which occupies one or more sites and denote by g_K the number of distinguishable ways of arranging the complex on the sites. Thus,

$$g_A = 1 \text{ if K} = \text{A occupies one site; A} = \text{h, v, D, or D}^+ , \tag{7.1}$$

$$g_{AA} = 1 . \tag{7.2}$$

Let $\eta(K)$ be the ionisation energy of K divided by kT, and let $\varphi(AB)$ be the interaction energy of A and B when A and B occupy adjacent sites. We then have for any complex

$$Z_K = \frac{g_K}{g_{K^+}} Z_{K^+} e^{-\eta(K)} , \tag{7.3}$$

$$Z_{AB} = g_{AB} Z_A Z_B e^{-\varphi(AB)} . \tag{7.4}$$

Thus, Z_{AB}/Z_{AB^+} can be evaluated by using (7.4) and then (7.3) for K = B, or by just using (7.3) with K = AB. Hence

$$\frac{Z_{AB}}{Z_{AB^+}} = \frac{g_{AB}}{g_{AB^+}} \frac{g_B}{g_{B^+}} e^{-\eta(B) + \varphi(AB^+) - \varphi(AB)} = \frac{g_{AB}}{g_{AB^+}} e^{-\eta(AB)} . \tag{7.5}$$

Using (7.1) in (7.5)

$$\eta(AB) = \eta(B) + \varphi(AB) - \varphi(AB^+) . \tag{7.6}$$

The two-site partition function is

$$\begin{aligned} \Xi = \lambda_v^2 Z_{vv} + \lambda_v(\lambda_h Z_h + \lambda_D Z_D + \lambda_{D^+} Z_{D^+}) + \lambda_h(\lambda_D Z_{hD} + \lambda_{D^+} Z_{hD^+} + \lambda_h Z_{hh}) + \\ + \lambda_D(\lambda_D Z_{DD} + \lambda_{D^+} Z_{DD^+}) + \lambda_{D^+}^2 Z_{D^+D^+} . \end{aligned} \tag{7.7}$$

Our convention about the energy zero shall be such that

$$\lambda_v = Z_v = Z_{vv} = 1 , \quad \varphi(kv) = 0 , \quad k = \text{h, v, D, or D}^+ . \tag{7.8}$$

If Π be the number of adjacent pairs of sites, then by (7.7) and (7.4)

$$n_{KK} = \Pi g_{KK} \lambda_K^2 Z_{KK}/\Xi = \Pi \lambda_K^2 Z_K^2 e^{-\varphi(KK)}/\Xi .$$

For K = D$^+$ and using (7.6) with AB = D$^+$D,

$$n_{D^+D^+} = \Pi \lambda_{D^+}^2 Z_{D^+}^2 e^{\eta(D^+D) - \eta(D) - \varphi(D^+D)}/\Xi .$$

Hence and by similar arguments one finds

$$\begin{aligned} n_{vv} &= \Pi/\Xi , \\ n_{hh} &= \Pi \lambda_h^2 Z_h^2 e^{-\varphi(hh)}/\Xi , \\ n_{hD} &= \Pi 2\lambda_h \lambda_D Z_h Z_D e^{-\varphi(hD)}/\Xi , \\ n_{hD^+} &= \Pi 2\lambda_h \lambda_{D^+} Z_h Z_D e^{\eta(hD) - \varphi(hD)}/\Xi , \\ n_{hv} &= \Pi 2\lambda_h Z_h/\Xi , \\ n_{DD} &= \Pi \lambda_D^2 Z_D^2 e^{-\varphi(DD)}/\Xi , \\ n_{DD^+} &= \Pi 2\lambda_D \lambda_{D^+} Z_D^2 e^{\eta(DD) - \varphi(DD)}/\Xi , \\ n_{D^+D^+} &= \Pi \lambda_{D^+}^2 Z_D^2 e^{\eta(DD^+) + \eta(DD) - \varphi(DD)}/\Xi , \\ n_{Dv} &= \Pi 2\lambda_D Z_D/\Xi , \\ n_{D^+v} &= \Pi 2\lambda_{D^+} Z_D e^{\eta(vD)}/\Xi . \end{aligned} \tag{7.9}$$

Note also that for the equilibrium

$$K^+ + e = K$$

we have

$$\lambda_K = \lambda_{K^+}\lambda_e . \tag{7.10}$$

The absolute activity λ_e of the electrons is related to the effective number of electron states N_c in the conduction band by

$$n_c = N_c \lambda_e \, e^{-\eta_c} , \tag{7.11}$$

where E_c is the energy at the bottom of the conduction band.

For a given number N of lattice sites, Π may be taken as $zN/2$ where z is the co-ordination number of the lattice. In the complete theory the values of the λZ's will have to be provided and (7.9) will permit the calculation of the various pair numbers in terms of the ionisation and interaction energies. As mentioned earlier, the specification of the λZ's is difficult and we get round this by making the assumption $N_v = n_I + n_\sigma$, i. e.,

$$\text{number of lattice vacancies} = \text{number of occupied defect sites} . \tag{7.12}$$

As the unknowns in our problem we have the ten pair numbers in (7.9) and $\lambda_h Z_h$, $\lambda_D Z_D$, $\lambda_{D^+} Z_D$, λ_e, n_c, N, N_h, N_D, and N_v giving a total of 19 unknowns. Equation (7.9) gives ten relations of which only nine are independent since they have been derived from probabilities; (7.10) and (7.11) furnish two more equations. The law of conservation of charge gives us the relation

$$n_c = 2n_{D^+D^+} + n_{DD^+} + n_{D^+h} + n_{D^+v} , \tag{7.13}$$

while the law of conservation of mass gives us the four relations

$$zN_h = 2n_{hh} + n_{hD^+} + n_{hv} + n_{hD} , \tag{7.14}$$

$$zN_D = 2n_{DD} + 2n_{DD^+} + 2n_{D^+D^+} + n_{D^+h} + n_{Dh} + n_{Dv} + n_{D^+v} , \tag{7.15}$$

$$zN_v = 2n_{vv} + n_{hv} + n_{Dv} + n_{D^+v} , \tag{7.16}$$

$$N = N_h + N_D + N_v . \tag{7.17}$$

The values of N_h and N_D (the number of host atoms and donor atoms) gives us two more equations and with (7.12) we have the 19 equations required to evaluate the 19 unknowns. Because we include vacancies we are unable to specify a priori the value of N associated with given values of N_h and N_D; its value has to be determined by the above equations. In practice one is interested only in ratios, e.g. n_{D^+h}/N_h etc.

7.1 Evaluation of the λZ

Because of the large number of unknowns an analytical solution is not possible. We sketch below a procedure that permits an iterative solution for the unknowns. We first evaluate the λZ by the following method. From (7.9), (7.14), and (7.15) we have

$$\frac{N_D}{N_h} = \frac{n_{Dh} + 2n_{DD} + 2n_{DD^+} + 2n_{D^+D^+} + n_{Dv}}{n_{Dh} + n_{hv} + n_{D^+h} + 2n_{hh}} . \tag{7.18}$$

Since N_D and N_h are assumed known and (7.9) gives the pair number occurring in (7.18), this gives us one equation for $\lambda_h Z_h$, $\lambda_D Z_D$, and $\lambda_{D^+} Z_D$.

The second equation comes from (7.13) which may be written

$$n_c = N_D \left(\frac{2n_{D^+D^+}}{N_D} + \frac{n_{DD^+}}{N_D} + \frac{n_{D^+h}}{N_D} + \frac{n_{D^+v}}{N_D} \right) . \tag{7.19}$$

From (7.10) and (7.11)

$$n_c = N_c\, e^{-\eta_c} \frac{\lambda_D Z_D}{\lambda_{D^+} Z_D} \tag{7.20}$$

and since the terms in the r.h.s. of (7.19) are given by (7.9), this gives us another equation for the three λZ's.

We get the third equation by virtue of (7.12) from the total number of vacancies

$$zN_v = 2n_{vv} + n_{vh} + n_{vD} + n_{vD^+}$$

$$= \frac{zN}{2\Xi} (2 + 2\lambda_h Z_h + 2\lambda_D Z_D + 2\lambda_{D^+} Z_D\, e^{\eta(vD)}) . \tag{7.21}$$

Let αN be the number of defect sites associated with N_L lattice sites. Furthermore, assume that for every species

$$Z(\text{defect site}) = \exp(-\chi)\, Z(\text{normal lattice site}) .$$

The number of occupied defect sites is then

$$\eta_I + \eta_\sigma = \frac{\alpha N\, e^{-\chi} (\lambda_h Z_h + \lambda_D Z_D + \lambda_{D^+} Z_D\, e^{\eta(vD)})}{1 + e^{-\chi} (\lambda_h Z_h + \lambda_D Z_D + \lambda_{D^+} Z_D\, e^{\eta(vD)})} . \tag{7.22}$$

The third equation for the three λZ's is obtained by equating N_v from (7.21) to (7.22).

These three equations have to be solved by iteration. We may begin the iteration by using the approximate values of λZ's obtained from the analogues of (7.19) and (7.22) got by a similar argument from the use of the single-site g.p.f. Once the λZ's have been calculated the use of (7.7) and (7.9) gives us the various pair numbers.

8. Comparison with Previous Work

The statistics of defects of semiconductors with a donor impurity has been treated in some detail recently [9]. The method adopted was that of the canonical ensemble, and vacancies were not considered. In contrast we have used the method of the grand-canonical ensemble and also included vacancies in the treatment. The final equations in both methods cannot be solved analytically. A direct comparison of the results of the two methods can, however, be achieved by reference to the predicted ratios of the various pair numbers. For example, we obtain

$$\frac{n_{DD}}{n_{DD^+}} = \frac{1}{2} \frac{n_c}{N_c}\, e^{-\eta(DD)}\, e^{\eta_c} , \tag{8.1}$$

$$\frac{n_{DD^+}}{n_{D^+D^+}} = 2 \frac{n_c}{N_c}\, e^{\eta_c - \eta(DD^+)} . \tag{8.2}$$

The corresponding expressions (11) and (10) of [9] are

$$\frac{n_c}{N_c}\, e^{\varepsilon_1/kT} \quad \text{and} \quad \frac{n_c}{N_c}\, e^{\varepsilon_2/kT} . \tag{8.3}$$

They therefore overestimate the first ratio, underestimate the second ratio, and agree with the present work in so far as the product of (8.1) and (8.2) is concerned. Note that there is agreement as regards the ionisation energies

$$\varepsilon_1 = kT[\eta_c - \eta(DD)] , \qquad \varepsilon_2 = kT[\eta_c - \eta(DD^+)] .$$

The form of the results (8.1) and (8.2) can be obtained at once by applying the mass action law to

$$DD \rightleftharpoons DD^+ + e \quad \text{and} \quad DD^+ \rightleftharpoons D^+D^+ + e .$$

The first reaction then yields, if K is the mass action constant,

$$\frac{n_{\mathrm{DD}}/N_{\mathrm{L}}}{n_{\mathrm{DD^+}}n_{\mathrm{c}}/N_{\mathrm{L}}^2} = K\,,$$

where N_{L} is again the total number of lattice sites so that we have ratios of concentrations on the left-hand side.

Turning now to the ratio $n_{\mathrm{D^+D^+}}/n_{\mathrm{D^+h}}^2$ in the absence of vacancies and interstitials, our equations (7.9), (7.3), and (7.4) give

$$\frac{n_{\mathrm{D^+D^+}}}{n_{\mathrm{D^+h}}^2} = \frac{1}{2zN_{\mathrm{L}}}\,\mathrm{e}^{-\varphi(\mathrm{D^+D^+})-\varphi(\mathrm{hh})-2\varphi(\mathrm{hD^+})}\,, \tag{8.4}$$

where z is the coordination number of the lattice. We have also assumed that Ξ_2 can be approximated by $\lambda_{\mathrm{h}}^2 Z_{\mathrm{hh}}$; this will be a good approximation at low doping. The corresponding equation (9) of [9] gives the ratio $(4/NL)\,\mathrm{e}^{-Q^2/4\pi\varepsilon r_0 kT}$ for $z = 4$. If we make the correspondence

$$\frac{Q^2}{4\pi\varepsilon r_0 kT} \leftrightarrow \varphi(\mathrm{D^+D^+}) + \varphi(\mathrm{hh}) - 2\varphi(\mathrm{hD^+})\,,$$

our pre-factor $1/2z = 1/8$ for $z = 4$ is less than the pre-factor four in [9] which would appear to lead to too large a number of doubly ionised nearest-neighbour donor pairs.

As a separate check on the pre-factors, consider the equilibrium

$$2\mathrm{D^+h} \rightarrow \mathrm{D^+D^+} + \mathrm{hh}\,. \tag{8.5}$$

Paper [9] does not consider the number of hh pairs and so no equilibrium constant can be obtained for (8.5) for their work. Using (7.9), our method, without now using an approximation for Ξ_2, yields

$$\frac{n_{\mathrm{D^+D^+}}n_{\mathrm{hh}}}{(n_{\mathrm{D^+h}})^2} = \frac{1}{4}\exp\left[\eta(\mathrm{DD^+}) + \eta(\mathrm{DD}) \quad 2\eta(\mathrm{hD}) + 2\varphi(\mathrm{hD}) - \varphi(\mathrm{DD})\right]. \tag{8.6}$$

One can compare this result with a simple quasi-chemical treatment which yields, assuming a random arrangement,

$$(N_{\mathrm{D^+D^+}}, N_{\mathrm{D^+h}}, N_{\mathrm{hh}}) = \frac{zN_{\mathrm{L}}}{2}\left[\left(\frac{N_{\mathrm{D}}}{N_{\mathrm{L}}}\right),\quad 2\left(\frac{N_{\mathrm{h}}}{N_{\mathrm{L}}}\right)\left(\frac{N_{\mathrm{h}}}{N_{\mathrm{L}}}\right),\quad \left(\frac{N_{\mathrm{h}}}{N_{\mathrm{L}}}\right)^2\right]$$

This leads to the same pre-factor as in (8.6), in agreement with our more general method.

An advantage of the g.p.f. treatment is that the first approximation outlined here can be used to calculate the pair numbers taking into account the interaction of the species occupying the site with the rest of the lattice. Since the first approximation has given us the probabilities of the different occupations of the site, these probabilities may be used to calculate the average interaction between the pairs and the rest of the lattice; the two-site g.p.f. can be recalculated using these improved energy levels. By contrast, the extension of Chatterjea and Hauser's treatment to higher-order pairs would not be easy either conceptually or mathematically.

We thus see that the g.p.f. method is capable of giving us a more detailed view of the structure of a doped semiconductor while employing fairly elementary mathematics. Thus, we are able to calculate the probabilities of the ten pair numbers. Semiconductors doped with both acceptors and donors can be treated in a similar way without increased mathematical complexity.

References

[1] P. T. Landsberg, in: Handbook of Semiconductors, Vol.1, Ed. T. S. Moss and W. Paul, North-Holland Publ. Co., Amsterdam 1982 (Chap. 8).
[2] C. V. Heer, Statistical Mechanics, Kinetic Theory, and Stochastic Processes, Academic Press, New York 1979.
[3] P. T. Landsberg, in: Statistical Problems in Semiconductors and Phosphors, Ed. M. Désirant, Vieweg, Braunsschweig 1958 (p. 45).
[4] A. B. Lidiard, in: Defects and Their Structure in Nonmetallic Solids, Ed. B. Henderson and A. E. Hughes, Plenum Press New York 1976 (p. 1).
L. A. Girifalco, Statistical Physics of Materials, Wiley, New York 1973.
[5] B. Henderson, Defects in Crystalline Solids, Arnold, London 1972 (Chap. 1.2).
[6] C. Kittel, Introduction to Solid State Physics, 3rd ed., Wiley, New York 1960 (p. 562).
[7] R. A. Fowler and E. A. Guggenheim, Statistical Thermodynamics, Cambridge 1939 (p. 552).
[8] A. B. Lidiard, in: Theory of Imperfect Crystalline Solids, Trieste Lectures 1970, Internat. Atomic Energy Agency, Vienna 1971 (p. 394)
[9] A. Chatterjea and J. R. Hauser, J. Phys. Chem. Solids 17, 1031 (1976).

(Received July 31, 1984)

phys. stat. sol. (b) 126, 157 (1984)

Subject classification: 10.2 and 13.4; 22.1.2

Institute for the Study of Defects in Solids,
Department of Physics, State University of New York at Albany[1])

Comparison of an Extended Interstitial with a Split Interstitial in Silicon

By

K. WESSER[2]), S. N. SAHU, J. P. KARINS[3]), and J. W. CORBETT

Molecular mechanics (MM) is a computer program used in organic chemistry to model the interactions of atoms in a molecule and thereby determine its configuration. The program models the interactions between atoms using two-body potentials, bond-bending and bond-twisting force constants, and van der Waals interactions. In this paper MM is applied to silicon for the first time and used to study defect structures in crystalline silicon. In particular, the Jackson model and the ⟨100⟩ split interstitial are treated and configurations obtained. The resultant structures are then used for molecular orbital calculations using extended Hückel theory (EHT), to give one-electron energy levels in the band gap. It is found that the Jackson interstitial has a high formation energy which does not favor this defect as a high temperature defect in silicon.

Molekül-Mechanik (MM) ist ein Computer-Programm, das in der organischen Chemie benutzt wird, um die Wechselwirkung zwischen Atomen in einem Molekül zu beschreiben und dadurch die Konfiguration zu bestimmen. Das Programm beschreibt die Wechselwirkung zwischen Atomen mit Zwei-Körper-Potentialen, Bindungs-Biegungs- und Bindungs-Drehungs-Kraftkonstanten und der van der Waals-Kraft. In dieser Arbeit wird MM zum ersten Mal auf Silizium angewandt und zur Untersuchung von Defektstrukturen in Siliziumkristallen benutzt. Speziell wird das Jackson-Modell und das ⟨100⟩-aufgespaltene Zwischengitteratom behandelt und Konfigurationen dafür bestimmt. Die Strukturen, die sich ergeben, werden danach für Molekül-Bahn-Berechnungen mit der erweiterten Hückel-Theorie (EHT) benutzt um Einelektron-Energie-Zustände in der Bandlücke zu bestimmen. Es wird gefunden, daß das Jackson-Zwischengitteratom eine hohe Bildungsenergie besitzt und darum nicht als Hochtemperatur-Defekt in Silizium begünstigt.

1. Introduction

The prediction by Seeger and Chik [1] that the high temperature self-diffusion process in silicon is dominated by self-interstitials, considerably enhanced studying the role of these defects in both self- and impurity-diffusion mechanisms. The observed high entropy values (large pre-exponential term in the diffusion coefficient) in the high temperature self-diffusion experiments indicated the possibility of motion of an extended interstitial. In addition, transmission electron microscopy (TEM) results have indicated that swirl defects in silicon are of interstitial nature, suggesting the interstitial as a high temperature defect in silicon. Based on some of these experimentally observed facts, Jackson [2] proposed and Kimerling [3] reported a model for the extended interstitial which has come to be called the Jackson interstitial; it consists of a central five-membered ring (puckered pentagon) surrounded by ten neighboring atoms (two connected to each of the five core atoms distorted from their tetrahedral coordinations) which in turn form part of five- and seven-membered ring

[1]) Albany, New York 12222, USA.

[2]) Present address: Peace Corps, P.O. 707, Monrovia, Liberia.

[3]) Present address: G. E. Electronics Park, Bldg. 3, Syracuse, New York 13221, USA.

chains around the pentagon. Since the number of dangling bonds of such a defect is less than those of a single isolated interstitial atom the enthalpy of formation could decrease at the cost of an entropy increase giving rise to a stable structure, as pointed out by Kimerling [3].

There have been a number of studies related to the electronic [4, 5], structural [6, 7], and thermal [8, 9] properties of some of the simple interstitials such as the ⟨100⟩ split interstitial and their different charge state configurations. In this paper we report and compare the calculated structure and energy levels of two silicon self-interstitials; the results on Jackson interstitial are presented in Section 3 and the following Section 4 deals with the split ⟨100⟩ interstitial. Since the molecular mechanics program is not very familiar to the scientists in the area of "defects in semiconductors", we will sketch some of the salient features of this technique in Section 2. Finally, we conclude this paper with Section 5, with a discussion of the high temperature intrinsic defects in silicon.

2. Molecular Mechanics

The molecular mechanics (MM) [10] program is used by chemists to determine the structure of organic molecules. This method is based on the classical approach of bond stretching, bending, and wagging parameters. For a given atomic configuration, the program optimizes the geometry by moving the atoms around and calculating the total energy (E_t) of the system (with respect to the strain free molecule) given by the following equation:

$$E_t = E_s + E_b + E_\omega + E_{vdW}$$

with E_s the stretching energy, E_b the bending energy, E_ω the torsional energy, and E_{vdW} the van der Waals interaction between pairs of non-bonded atoms; and where these are given by

$$E_s = \sum \frac{K_s}{2} (r - r_0)^2 ,$$

$$E_b = \sum \frac{K_b}{2} (\theta - \theta_0)^2 ,$$

$$E_\omega = \frac{V_1}{2} (1 + \cos \omega) + \frac{V_2}{2} (1 - \cos 2\omega) + \frac{V_3}{2} (1 + \cos 3\omega) ,$$

and

$$E_{vdW} = \varepsilon^* \left[- c_1 \left(\frac{r^*}{r}\right)^6 + c_2 \exp\left(-\frac{c_3 r}{r^*}\right)\right].$$

The various parameters have already been evaluated in the MM program by fitting experimental data on molecules. Cross terms such as stretch–bend and torsion–bend interactions are also included in the program. The stretching energy part of the program is normally supplemented by a cubic term to take into account moderately small deformations, but for deformations such as those encountered in the solid state problems (especially a defect like Jackson interstitial) higher order terms would be needed. We have replaced the stretch energy interactions by a Morse potential [11] given by the expression

$$V_M = \sum_{\text{all bonds}} D \{1 - \exp [-\alpha(r - r_0)]\}^2 ,$$

where D is the energy of dissociation of a covalent bond obtainable from the heat of sublimation, α is given by the formula

$$\alpha = \left(\frac{K}{2D}\right)^{1/2},$$

and K is the force constant, often derived from the compressibility data. The α and D values used for silicon are 0.156 nm and 1.83 eV, respectively. Since the energy changes are most sensitive to bond lengths we have only changed the stretching term. The other terms are left in their original forms.

3. Jackson Interstitial

To model the Jackson interstitial we chose a cluster size of 48 atoms with the extra interstitial connected to form the inner five-membered ring surrounded by ten neighboring atoms. The inner fifteen atoms are relaxed keeping the outer ones fixed. The program then relaxes to the optimum coordinates for the atoms as tabulated in Table 1, along with their regular lattice site positions shown in paranthesis. These coordinates changed very little (less than 1%) when the two outer shell atoms were relaxed in succession in addition to those of the defect. To study the effect of cluster size we reduced the number of atoms in the cluster to 36 and relaxed the inner 15 atoms. Here again the coordinates changed insignificantly. The coordinates in Table 1 are in close agreement with those proposed by Tan [12].

Next we carried out extended Hückel calculations (EHT) using the optimized extended defect coordinates in a 48 atom cluster terminated by 60 hydrogen surface orbitals. The outer 33 silicon atoms were fixed at their regular lattice site positions. The one-electron energy levels of this cluster along with those of a perfect cluster (Si47H60) are shown in Fig. 1. The calculated gap is found much larger than the experimental one, as is typical of cluster calculations. The plethora of energy levels in the gap for the Jackson interstitial is reminiscent of the gap states in amorphous silicon.

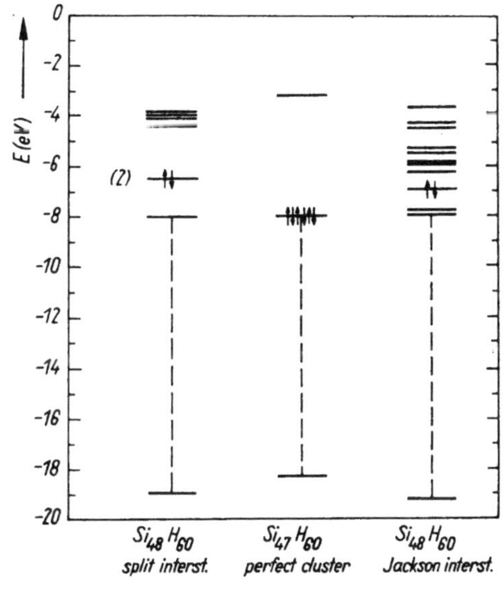

Fig. 1. One-electron energy levels of both Jackson and split ⟨100⟩ interstitials in silicon (Si48H60 clusters used). The perfect cluster (Si47H60) energy levels are also shown for comparison

Table 1

Optimized final cartesian coordinates of the Jackson interstitial (15 atoms)

X (nm)	Y (nm)	Z (nm)	
0.064	0.038	−0.093	(0 0 0)
0.098	0.177	0.088	(1 1 1)
0.083	−0.189	−0.083	(1 −1 −1)
−0.072	−0.201	0.092	(−1 −1 1)
−0.003	−0.004	0.179	(0 0 1)*)
−0.139	0.137	−0.145	(−1 1 −1)
0.007	−0.297	0.289	(0 −2 2)
−0.235	0.014	0.246	(−2 0 2)
0.234	−0.016	0.247	(2 0 2)
−0.001	0.281	0.271	(0 2 2)
−0.271	−0.271	0.001	(−2 −2 0)
0.278	−0.280	0.002	(2 −2 0)
0.282	0.281	0.001	(2 2 0)
−0.004	−0.280	−0.277	(0 −2 −2)
0.246	0.023	−0.246	(2 0 −2)

The first five atoms belong to the puckered pentagon. The equivalent lattice site positions are shown in the paranthesis.

*) The extra interstitial atom.

4. Split <100> Interstitial

In order to draw a comparison between the Jackson and split ⟨100⟩ interstitial, a 48 atom cluster was again chosen having a split ⟨100⟩ interstitial embedded in it. The number of atoms involved in the successive relaxation steps are 2, 6, 18, and 48. When the two innermost atoms of the defect were relaxed the bond length between the split atoms was optimized to 0.217 nm, a value which changed to 0.218 nm and thereafter remained constant in the next three relaxation steps. This can be compared to the distance between the two silicon atoms in the molecular system Si_2Cl_2 (0.230 nm) [13]. The optimum geometry of the inner 18 atoms remained undisturbed when the whole cluster was relaxed. Of the 18 atoms, the outer twelve remained close to their respective lattice sites. The final cartesian coordinates of the six innermost atoms are tabulated in Table 2. The difference in steric energy between the Jackson and the split interstitial is found to be 10.2 eV, the latter being the more stable structure. Once again the final split interstitial coordinates are used to perform extended Hückel calculations in a similar cluster as mentioned above (Si48H60). These one-electron energy levels are also shown in Fig. 1. The split ⟨100⟩ interstitial introduces an energy

Table 2

Optimized final cartesian coordinates of the split ⟨100⟩ interstitial

X (nm)	Y (nm)	Z (nm)
0.109	0.00	0.00
−0.109	0.00	0.00
0.182	0.151	0.151
0.182	−0.151	−0.151
−0.182	0.151	−0.151
−0.182	−0.151	0.151

level in the gap. For the same defect the tight binding calculations of Blount [14] predict an energy level in the upper half (about 0.7 eV from the top of the valence band edge) of the band gap. Because of the finiteness of our cluster size, the band gap is too large and the band edges are less well defined. The gap state thus can be qualitatively identified with other calculated or experimental results. The difference in the total energies of the Jackson and the split interstitial clusters as calculated by the extended Hückel theory is 12.6 eV as compared to their molecular mechanics counterpart value 10.2 eV.

5. Conclusion

We have used the molecular mechanics program to determine the geometrical structure of both an extended (Jackson) and the simple split ⟨100⟩ form of self-interstitials in silicon. Since the ⟨100⟩ split structure is believed to be the most stable of all the self-interstitials [15] we have chosen it to be the suitable candidate for comparison with Jackson interstitial. The electronic structure of both the defects using the optimum coordinates are then obtained through EHT calculations. Split ⟨100⟩ is found to be more stable than Jackson interstitial by 10 eV (molecular mechanics) to 12 eV (EHT).

The idea of an extended interstitial being the mechanism for self-diffusion is not generally accepted (see, for example Car et al. [16]), but to review that topic is beyond this paper. We note simply here that, even though the high entropy factor of the Jackson interstitial may favor it in a high temperature self-diffusion mechanism, the large energy difference in the two structures (as we have shown here) most probably rules out this form of extended interstitial. We view MM and EHT as useful starting points for the study of defects, but better calculations are warranted, such as being done by Masri et al. [15], using a more sophisticated quantum chemical molecular orbital technique.

References

[1] A. Seeger and K. P. Chik, phys. stat. sol. **29**, 455 (1968).
[2] K. A. Jackson, unpublished (1976).
[3] L. C. Kimerling, in: Defects and Radiation Effects in Semiconductors, Ed. J. H. Albany, Inst. of Physics, Bristol 1978 (p. 56).
[4] A. Seeger and W. Frank, in: Radiation Damage and Defects in Semiconductors, Ed. J. E. Whitehouse, Inst. of Physics, London 1973 (p. 262 to 268).
[5] A. Seeger, H. Foll, and W. Frank, Inst. Phys. Conf. Ser. No. 31, 12 (1977).
[6] S. I. Tan, B. S. Berry, and W. Frank, in: Ion Implantation in Semiconductors and Other Materials, Ed. B. L. Crowder, Plenum Press, New York/London 1973 (p. 19).
[7] W. Frank, Inst. Phys. Conf. Ser. No. 23, 23 (1975).
[8] H. J. Mayer, H. Mehrer, and K. Maier, Inst. Phys. Conf. Ser. No. 31, 186 (1977).
[9] F. J. Demond, S. Kabitzer, H. Mannsperger, and H. Damjantschitsch, Phys. Letters A **93**, 503 (1983).
[10] N. L. Allinger, Adv. Phys. org. Chem. **13**, 1 (1976).
[11] P. M. Morse, Phys. Rev. **34**, 57 (1929).
[12] T. Y. Tan, private communication.
[13] R. C. Weast (Ed.), Hdb. of Chemistry and Physics, 59th ed., CRC Press, Cleveland (Ohio) 1978 (p. F217).
[14] E. I. Blount, J. appl. Phys. **30**, 1218 (1959).
[15] P. M. Masri, A. H. Harker, and A. M. Stoneham, to be published.
[16] R. Car, P. J. Kelly, A. Oshiyama, and S. Pantelides, Phys. Rev. Letters **52**, 1814 (1984).

(Received July 12, 1984)

phys. stat. sol. (b) **126**, 163 (1984)

Subject classification: 10.1; 21

Departamento de Materiales, Comisión Nacional de Energía Atómica, Buenos Aires[1])

Divacancy in H.C.P. Metals

By

A. M. MONTI

Static properties of divacancies at first, second, third, and fourth neighbours are investigated in a h.c.p. lattice by using computer simulation techniques. The calculations are performed with short range pair interatomic potentials fitted to some Mg properties. The formation, relaxation, and binding energies as well as the strength dipole tensor of the above defects are computed and the results compared with those of earlier authors. The corresponding formation and binding entropies are also calculated and the defect stability analysed. From the results, the instability of the third neighbour divacancy can be predicted.

Die statischen Eigenschaften von Divakanzen an erster, zweiter, dritter und vierter Nachbarposition werden im h. d. p.-Gitter mittels Computersimulationstechnik untersucht. Die Berechnungen werden mit kurzreichweitigen interatomaren Paar-Potentialen durchgeführt, die an einige Mg-Eigenschaften angepaßt werden, sowohl die Bildungs-, Relaxations- und Bindungsenergien als auch der Tensor der Dipolstärke der obigen Defekte werden berechnet und die Ergebnisse mit denen anderer Autoren verglichen. Die entsprechenden Bildungs- und Bindungsenergien werden ebenfalls berechnet und die Defektstabilität analysiert. Aus den Ergebnissen läßt sich die Instabilität der dritten Nachbardivalenz voraussagen.

1. Introduction

Divacancies may give an important contribution to high-temperature diffusion in metallic lattices [1 to 3]. Also they serve as the necessary nucleation step for loop clustering in a vacancy supersaturated environment [4] and their concentration may be large compared to the single defect concentration after a slow quench. At present, modern diffusion measurement techniques provide accurate results over a large temperature range and the divacancy contribution to matter transport can be separated from the one due to the monovacancy. Measurements of the isotopic effect on the diffusivity, given by the product of the correlation factor f and the isotope factor ΔK, are currently performed to separate the contributions to diffusion from the different mechanisms [2, 3]. Therefore, a theoretical understanding of the static and dynamic divacancy properties is important for studying irradiation behaviour, quenching, and diffusion.

A calculation of the lattice distortion and dynamic of the divacancy based on first principles seems to be impossible at the present time. A computer simulation of the defect lattice static and dynamic properties is then performed. This simulation assumes the validity of an interatomic potential which is based on a fit to a set of perfect or defect lattice physical parameters (such as elastic constants, defect formation energy, etc.). That potential must hold the perfect lattice in equilibrium. For the relatively large distortions occurring at the defect core the validity of the approach is doubtful. However, this simulation procedure, when carefully used, is the only way of studying those regions and some lattice geometry determined properties are revealed.

[1]) Avda. Libertador 8250, 1429 Buenos Aires, Argentina.

By using computer simulation self defect static [6] and dynamic [7] parameters have been calculated in cubic metals. For h.c.p. metals only recently defect computer simulations have been performed by ourselves (see references [5] and [8 to 11]) and other authors [12 to 18]. In [17, 18] lattice static properties of vacancy small clusters are studied by computer simulation. Some of those results for the divacancy are calculated again here by using different interatomic potentials. We shall try throughout the work to emphasize those defect characteristics which depend on the lattice symmetry and differentiate the divacancy in the h.c.p. structure from the one generally studied in f.c.c. structure. By contrast numerical results must be taken only as indicative of general trends because they depend on the interatomic potential used for obtaining them.

Some dynamic parameters, such as the migration energy and the ΔK factor associated with different divacancy jumps, were calculated and discussed by us in a previous paper [5]. The static lattice distortion, volume expansion, and formation, relaxation, and binding energies for divacancies at first, second, third, and fourth neighbours are calculated in the present work and reported in Section 2; whereas the results concerning the vibrational entropies calculated here in three different approximations are discussed in Section 3. The stability of each divacancy type is analysed in Section 4 and finally, the main results are summarized and discussed in Section 5.

2. Lattice Defect Distortion

2.1 Calculation procedure

The program developed by Norgett et al. [19] for defect simulation in metals is used. This program is based on the conjugate gradient method [20] and constitutes the basis of the program DEVIL at Harwell. In the calculation the atomic coordinates of an hexagonal close-packed lattice with the Mg lattice parameter are held in equilibrium by two different interatomic potentials. These potentials are named P5C and P4C (see [5] and [8]) and they are respectively constructed by a set of 5(P5C) and 4(P4C) spline fit cubic functions with continuous first and second derivatives at the matching points. These potentials are very similar and converge to zero with zero first derivative at a cut-off distance slightly smaller than the fourth neighbour distance, but only P5C does it with zero second derivative, too, this being the main difference between them. Both are adjusted to a rigid sphere hexagonal close packed ratio, approximately reproduce the Mg elastic constants, and are fitted to an unrelaxed vacancy formation energy of 0.81 eV [21]. A third potential named SP [8] (for "pseudopotential") based on the ion interaction part of the optimized pseudopotential deduced by Appapillai and Heine [22], is also used. For the SP the c/a ratio appropriate to Mg is adopted, however, the vacancy-formation energy calculated through this potential is 0.24 eV [8], the experimental value [21] being reached in principle by an "ad-hoc" electron-gas volume-dependent contribution. In most of the calculations reported below P5C and P4C are used because of its shorter range which allows for quicker convergency, and its appropriate adjustment to the vacancy-formation energy. As done in [8 to 10] the SP is only used for making comparisons at constant volume.

In order to simulate the defect a perfect h.c.p. lattice region of approximately 3000 atoms (region I) is generated in the computer by the atomic coordinates. The defect core is located at the centre of region I, where two nearest neighbour atoms are removed. There, the remaining host atoms are free to move in order to minimize the crystal energy under an interatomic potential. Region I is surrounded by a perfect lattice shell, region II, where the host atoms remain fixed at their perfect lattice positions. Once a relaxed lattice configuration is achieved the corresponding energy and dipole tensor can be calculated.

The defect formation energy is obtained by subtracting the energy of the perfect lattice from the energy of the relaxed lattice with the defect in I and two extra atoms at the crystal surface. The relaxation part of this energy is calculated by subtracting the energy stored in region I with a defect inside, but holding the atoms at the unrelaxed perfect lattice positions, from the final energy stored in the relaxed region I.

The defect dipole tensor P_{ij} can be calculated by the first moment of the Kanzaki forces K [23], that is

$$P_{ij} = \sum_{l} r_i(l)\, K_j(l)\,, \tag{1}$$

where l runs over all the atoms of region I with atomic perfect lattice coordinates $\boldsymbol{r}(l)$. For a region I sufficiently large, which includes all the host lattice anharmonicity and the full range of the defect–host atom interaction, the defect displacement field $\boldsymbol{u}$ obtained from the simulation can be used to determine the Kanzaki forces [23],

$$K_i(l) = \sum_{l'} \Phi_{ij}(l, l')\, u_j(l')\,, \tag{2}$$

where Φ is the force constant matrix. The forces $\boldsymbol{K}$ are zero when the displacements are harmonic. Using the dipole tensor the crystal change of volume due to the defect can be obtained as [24, 25]

$$V = S_{ii,jk} P_{jk} \tag{3}$$

which depends on the perfect crystal properties only through the compliances $\mathbf{S}$. In lattices with arbitrary symmetry and spherically symmetric defects (isotropic dipole tensor), or for lattices where $C_{33} - C_{13} = C_{11} + C_{12} - 2C_{13}$, see (4.50a) in [25], (3) reduces to the expression generally used,

$$\Delta V = \frac{K}{3} \operatorname{Tr} P\,, \tag{4}$$

where K is the crystal compressibility. As the above-mentioned condition for the elastic constants is verified by the crystal when simulated with P4C [8] and P5C [5], (4) will be used in Section 2.2 to calculate ΔV.

2.2 Static configuration

Divacancies at first (d1), second (d2), third (d3), and fourth (d4) neighbours are schematically shown in Fig. 1 for an ideal h.c.p. structure. For the rigid sphere c/a ratio first and second neighbour vacancies are separated by the same distance a. Then, the nearest neighbour divacancies can adopt two-non-equivalent orientations depending on whether both vacancies are on the same basal plane (d2) or on different ones (d1). This difference in the symmetry of nearest neighbour divacancies does not appear in cubic materials.

At Table 1 we report the formation, relaxation, and binding energies of the divacancies calculated by using the empirical potentials P4C, P5C, and the pseudopotential SP. The formation and relaxation energies calculated for the vacancy are also included. It can be seen that, for simulations carried out with a given potential, d2 has the smallest value of formation energy and d3 the highest one, approximately 10% bigger than the former. The values for d1 and d2 are very similar. As shown in Table 1 these conclusions are independent of the potential, including the results obtained by using SP, where the absolute values are not reliable because this potential does not reproduce the vacancy formation energy [8]. The static divacancy configuration has been previously studied by Sahu et al. [17, 18] by using other interatomic potentials and a numerical

Table 1

Point defect calculations

	empirical potential P4C			empirical potential P5C			pseudopotential SP		
defect	E_F	E_R	E_B	E_F	E_R	E_B	E_F	E_R	E_B*)
vacancy	0.800[8]	−0.010[8]	—	0.799	−0.011	—	0.240[8]	−0.006[8]	—
d1 divacancy	1.468	−0.019	0.132	1.456	−0.019	0.142	0.455	−0.013	0.025
d2 divacancy	1.466	−0.021	0.134	1.454	−0.021	0.144	0.454	−0.014	0.026
d3 divacancy	1.601	−0.016	−0.001	1.622	−0.017	−0.024	0.485	−0.013	−0.005
d4 divacancy	1.597	−0.023	0.003	1.596	−0.024	0.002	0.473	−0.015	0.007

*) E_F formation energy; E_R relaxation energy; E_B binding energy. All the values are given in eV.

Table 2
Dipole tensor and volume change of the lattice defect calculated with empirical potentials

defect	potential	dipole tensor eigenvalues (eV) P_1	P_2	P_3	dipole tensor eigenvectors $\boldsymbol{e}_1$	$\boldsymbol{e}_2$	$\boldsymbol{e}_3$	$\Delta V/\Omega$*)
vacancy	P4C	−0.492	−0.492	−0.499	$\begin{pmatrix}1\\0\\0\end{pmatrix}$	$\begin{pmatrix}0\\1\\0\end{pmatrix}$	$\begin{pmatrix}0\\0\\1\end{pmatrix}$	−0.093
	P5C	−0.515	−0.515	−0.516				−0.097
d1 divacancy	P4C	−0.895	−1.298	−1.044	$\begin{pmatrix}-1/\sqrt{2}\\1/\sqrt{6}\\-1/\sqrt{3}\end{pmatrix}$	$\begin{pmatrix}-1/2\\-\sqrt{3}/2\\0\end{pmatrix}$	$\begin{pmatrix}-1/2\\1/2\sqrt{3}\\\sqrt{2}/\sqrt{3}\end{pmatrix}$**)	−0.202
	P5C	−0.934	−1.335	−1.106				−0.211
d2 divacancy	P4C	−1.059	−1.036	−1.181	$\begin{pmatrix}1\\0\\0\end{pmatrix}$**)	$\begin{pmatrix}0\\1\\0\end{pmatrix}$	$\begin{pmatrix}0\\0\\1\end{pmatrix}$	−0.205
	P5C	−1.105	−1.080	−1.215				−0.212
d3 divacancy	P4C	−0.566	−0.977	−1.039	$\begin{pmatrix}1/\sqrt{2}\\-1/\sqrt{6}\\-1/\sqrt{3}\end{pmatrix}$	$\begin{pmatrix}1/2\\\sqrt{3}/2\\0\end{pmatrix}$	$\begin{pmatrix}1/2\\-1/2\sqrt{3}\\\sqrt{2}/\sqrt{3}\end{pmatrix}$**)	−0.161
	P5C	−0.596	−1.018	−1.077				−0.168
d4 divacancy	P4C	−0.991	−0.991	−1.024	$\begin{pmatrix}1\\0\\0\end{pmatrix}$	$\begin{pmatrix}0\\1\\0\end{pmatrix}$	$\begin{pmatrix}0\\0\\1\end{pmatrix}$**)	−0.188
	P5C	−1.055	−1.055	−1.047				−0.197

*) Ω atomic volume.
**) Eigenvector along the axis of the divacancy.

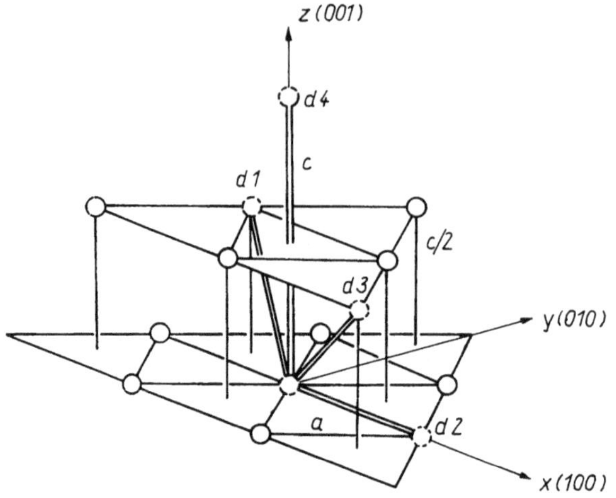

Fig. 1. Divacancies at first (d1), second (d2), third (d3), and fourth (d4) neighbours for an ideal h.c.p. structure

procedure for the relaxation different from the one used by us. They studied only the d1, d2, and d4 configurations and obtained for d2 a formation energy value slightly lower than for d1 in agreement with our results. The divacancy relaxation energies calculated at the present paper are practically twice that of the vacancy. These results are independent of the potential used by us. In addition, they are one order of magnitude higher than the ones obtained by Sahu et al. [18] with a Morse potential. Finally and defining the divacancy binding energy as

$$E_{\mathrm{B}} = 2E_{\mathrm{F}}^{(\mathrm{v})} - E_{\mathrm{F}}^{(\mathrm{d}i)}, \tag{5}$$

it can be seen from Table 1 that the vacancies in orientation d2 and d1 have the larger binding energy values, being in d2 slightly more bound than in d1 in accordance with [18]. The values obtained for the vacancies in orientation d4 are very small and positive and the ones for d3 are also small but negative. The results reported at Table 1 will be used in Section 4 in order to analyse the divacancy stability.

The eigenvalues and eigenvectors of the defect dipole tensor calculated by using (1) as well as the volume change calculated by using (4) are reported in Table 2. Only the empirical potentials were used at the calculations. For each defect, both potentials predict dipole tensor eigenvalues and volume changes rather similar (within a 4%) and the same set of eigenvectors. The largest relaxation volume corresponds to the nearest neighbour divacancies (d1 and d2) and the smallest one to the d3 divacancy. Sahu et al. [18] have also calculated the dipole tensor and the volume change for the d1, d2, and d4 divacancies with the interatomic potential generated by Doneghan and Heald [14]; their values for the volume change are nearly one-half of ours. However, the reason for this disagreement is not clear to us. Taking the value of the dipole tensors reported by those authors in [18], the experimental values of the crystal compressibility ($K = 4.381 \times 10^{-3}$ nm^3/eV) and 23.375×10^{-3} nm^3 for the atomic volume, we obtain values for the volume change nearly in agreement with our results at Table 2. The main difference with those authors remain in the shape of the dipole tensor. They obtain for the d1 and d2 divacancies practically the same set of eigenvectors ((100, (010), (001)); by the defect symmetry this result does not seem appropriate for d1. Table 2 shows that for all the defects studied one of the dipole tensor eigenvectors agrees with the axis of the divacancy.

3. Divacancy Formation Entropy

3.1 General remarks

The vibrational contribution to the formation entropy for a defect in a lattice composed by N atoms is given in the classical limit of high temperatures by

$$S = \frac{k}{2} \ln \left(\prod_{\alpha=1}^{3N} (w_\alpha^0)^2 / \prod_{\alpha=1}^{3N} (w_\alpha)^2 \right) , \tag{6}$$

$M(w_\alpha)^2$ and $M(w_\alpha^0)^2$ being the respective eigenvalues of the defect and perfect lattice force constant matrices and M the atomic masses.

As discussed in [7] and [10] the entropy can be calculated in different approximations; in the present work we have used the following ones:

(i) The Einstein approximation for an unrelaxed lattice; the defect eigenfrequencies are calculated at a lattice with two missing atoms and the remaining ones located at their perfect lattice site.

(ii) The Einstein approximation for a relaxed lattice; the eigenfrequencies are calculated for a relaxed defect lattice where each atom is assumed to vibrate uncoupled with the movement of its neighbours.

(iii) The "cluster" approximation; where a cluster of atoms around the defect at the relaxed lattice is allowed to vibrate in a rigid environment. In this case, the vibrational formation entropy is given by (see [7] and [10])

$$S = \frac{k}{2} \ln (\det \Phi^\circ / \det \Phi) , \tag{7}$$

Φ° being the $3N \times 3N$ force constant matrix of the perfect (Φ°) and defect (Φ) lattice.

3.2 Calculation methods

In order to calculate the entropy in the Einstein approximation the contributions of the defect lattice atoms with eigenfrequencies different from the ones of the perfect lattice must be considered in (6). For an unrelaxed lattice with a few vacant sites and held together by a short-range-pair interaction, the number of atoms to be considered is small. They are those located within the interaction range of the missing atoms. The range of the potentials used here, P4C and P5C, is limited to a distance smaller than the fourth-neighbour atom separation. The three eigenfrequencies of each atom at the perfect lattice are equal and direction-independent, and we shall indicate them as $(Mw^2)^\circ$. The expression of $(Mw^2)^\circ$ as a function of the interaction force constants is given in the Appendix. When a vacancy is created at the unrelaxed lattice, twelve nearest-neighbour atoms change their frequencies to $(Mw_1^2)_1^\mathrm{v}$ and $(Mw_2^2)_1^\mathrm{v}$ and six third neighbours do the same to $(Mw_1^2)_3^\mathrm{v}$ and $(Mw_2^2)_3^\mathrm{v}$. The expressions for those frequencies are also given in the Appendix. Within a nearest-neighbour interaction model those expressions are the same for f.c.c. and h.c.p. lattices, as was previously pointed out in [10].

When a divacancy is created a large number of host atoms change their eigenfrequencies. For the d1 divacancy, five first nearest-neighbour atoms and four third neighbours of each vacancy of the pair contribute to the vibrational formation entropy, together with four nearest neighbours of both vacancies (with eigenfrequencies $(Mw_i^2)_{11}^{\mathrm{d1}}$, $i = 1, 2, 3$, see Appendix), and four atoms that are first neighbours of one vacancy and third neighbours of the other vacancy of the pair (each one with eigenfrequencies $(Mw_i^2)_{13}^{\mathrm{d1}}$, $i = 1, 2, 3$, see Appendix). For an unrelaxed lattice with the ideal c/a ratio considered here, the above description is the same for the d1 and d2 divacancy.

For an d3 divacancy, eight and five atoms are respectively nearest and third neighbours of each vacancy of the pair; and four atoms are nearest neighbours of both vacancies with eigenvalues: $(Mw_i^2)_{11}^{d3}$, $i = 1, 2$, see Appendix. Finally, for the d4 divacancy, nine and three atoms, respectively, first and third neighbours of each vacancy of the pair contribute to the entropy, plus three first and three third neighbours of both vacancies of the pair with eigenfrequencies: $(Mw_i^2)_{11}^{d4}$ and $(Mw_i^2)_{33}^{d4}$, respectively (with $i = 1, 2, 3$).

In order to evaluate the effect due to the lattice relaxation in the formation entropy calculation, the defect equilibrium configurations obtained by computer simulation techniques are also used. In those configurations the eigenfrequencies of a number N of atoms belonging to a spherical zone around the mass centre of the defect are calculated, while the remaining atoms are held fixed at their relaxed sites. The entropy is calculated for different N numbers and then a value for $N \to \infty$ extrapolated (as done in [7] and [10]). Therefore, within the validity of that extrapolation, long-range effects are explicitly included in the calculations. Two approximations are then used to obtain the eigenfrequencies, the Einstein and the "cluster" one. In the Einstein approximation the eigenfrequencies for each one of the N atoms are obtained by diagonalizing the corresponding coupling matrix in the defect relaxed lattice. In the "cluster" approximation the eigenfrequencies of the set of N atoms are calculated by evaluating expression (7), where perfect and defect lattice differ in two vacant sites at the centre of the last one, plus two extra atoms at the surface and the displacement of the remaining host atoms. In this case, although both lattices have an equal number of vibration modes, for equal-size finite regions which include the origin the defect lattice lacks six vibration modes with respect to the perfect one. As (det Φ) depends on the size and shape of the region, equivalent regions for either lattice have to be used for calculating S and the difference in the number of normal modes has to be properly included. For doing that we adopt the approach proposed by Hatcher et al. [7], and so we multiply the perfect-lattice value ln (det Φ°) by $(N - 2)/N$. The S values so obtained are then extrapolated for $N \to \infty$. This approximation allows to estimate the coupling effects in the frequency calculations. The main difficulty with this method is computer storage and time. For example, our present computer facility does not allow us to handle more than approximately 250 atoms. The described calculation procedure was carried out for a crystallite of approximately 8000 atoms (included in the relaxed and unrelaxed regions for the computer simulation) held together by P4C or P5C with a displacement-free surface. Real crystals are generally larger and bounded by a force-free surface. In order to correct the above-calculated constant-volume entropy $S|_v$ with respect to the "true" constant-pressure formation entropy $S|_p$, the following thermodynamic relation due to Eshelby [26] is adopted:

$$S|_p = S|_v + k\alpha\, \Delta V\,, \tag{8}$$

where k stands for the bulk modulus, α for thermal expansion coefficient while ΔV is the calculated defect relaxation volume.

The formation and binding entropy values calculated at the present work are shown in Table 3. In the first two rows of Table 3 the entropy values at constant volume obtained in the Einstein approximation are reported. There, only the contributions of the atoms described at the beginning of this section, that is those atoms within the defect interaction range, are included. These calculations are carried out both for an unrelaxed and a relaxed defect lattice in order to evaluate the effect of the relaxation on the entropy. For comparison the vacancy formation entropy values were also included in the table. For the unrelaxed lattice and considering only a given type of defect both empirical potentials, P4C and P5C, predict the same formation

Table 3

Entropy calculations

approximation	potential	S_F^v	S_F^{d1}	S_B^{d1}	S_F^{d2}	S_B^{d2}	S_F^{d3}	S_B^{d3}	S_F^{d4}	S_B^{d4}*)
Einstein**) unrelaxed lattice	P4C and P5C	1.71	3.20	0.22	3.20	0.22	3.41	0.01	3.44	−0.02
Einstein**)	P4C	1.50	2.90	0.10	2.80	0.20	3.10	−0.10	3.06	−0.06
relaxed lattice	P5C	1.74	3.30	0.18	3.23	0.25	3.46	0.02	3.49	−0.01
Einstein***)	P4C	1.39	2.72	0.06	2.65	0.13	2.98	−0.20	3.21	−0.43
	P5C	1.69	3.15	0.23	3.10	0.28	3.39	−0.01	3.57	−0.19
cluster***)	P4C	1.81	3.22	0.40	3.15	0.47	3.58	0.04	3.71	−0.09
	P5C	1.91	3.57	0.25	3.51	0.31	3.84	−0.02	4.09	−0.27

*) S_F vibrational formation entropy; S_B binding divacancy entropy. Superscripts v, d*i*, refer to vacancy and divacancy type d*i*, respectively. All the values are given in units of the Boltzmann's constant.

**) Values calculated at constant volume including only contributions of the atoms within the defect interaction range.

***) Values calculated at constant pressure and obtained after extrapolation for $N \to \infty$.

entropy value. This is due to the fact that both potentials are adjusted to the same set of force constant values [5, 8] at the first, second, and third neighbour equilibrium distances at the perfect lattice. For the present case, where the c/a ideal ratio and a central force model interaction are considered, there is no difference between the entropy values calculated for the d1 and d2 divacancies at the unrelaxed lattice. In addition, for a nearest-neighbour force model and assuming the transverse force constant $f_t^{(1)}$ to be zero, the entropy has a value of $1.73\,k$ for the vacancy and $3.22\,k$ for the divacancy at nearest-neighbour sites. These numbers exactly agree with the ones for those defects in the f.c.c. structure calculated by Hatcher et al. [7]. A comparison with the values reported in Table 3 for the unrelaxed lattice shows that the main contribution to the defect entropy comes from the nearest-neighbour host atoms.

On the other hand, depending on the structural characteristics of the potentials they predict different entropy values for a given defect when the calculations are carried out at the relaxed lattice. Although the displacement fields around each defect predicted by either potential are very similar, their third derivatives at the relevant displaced atoms are different as it can be seen in [5]. The values at constant volume reported at the first and second rows of Table 3 show that the relaxed lattice under P4C predicts entropy values systematically smaller than the ones calculated at the unrelaxed lattice. By contrast, the relaxed lattice under P5C predicts values systematically higher than the ones at the unrelaxed lattice. In addition, each of the potentials predicts different entropies for the d1 and d2 divacancies, the latter defect having the smaller value. That difference is a consequence of the symmetry of the divacancies. As pointed out in Section 2.2, those configurations are not equivalent having the host atoms near to the defect very different relaxations, and then different Einstein eigenfrequencies.

In the third and fourth rows of Table 3, the formation and binding entropies obtained after extrapolation for $N \to \infty$ when using the Einstein and the cluster approximations are reported. These values are corrected at constant pressure. For each of the approximations and for each of the potentials it is found that

$$S_F^{d2} < S_F^{d1} < S_F^{d3} < S_F^{d4}\,,$$

the value of the d4 divacancy being about 20% higher than the one for d2. The binding entropies are positive for the nearest-neighbour divacancies being slightly larger for d2 than for d1 as also occurs with the binding energies. On the contrary, the binding entropies of the d3 and d4 divacancies are in general negative.

A comparison of the results obtained by the described approximations allows us to assess the relative influence of the coupling components among vibration modes. The calculations indicate that a contribution of over 15% of the total entropy value is due to the cooperative modes which are taken into account when the cluster approximation is used.

4. Stability and Equilibrium Concentration of Divacancies

In order to study the divacancy stability against two free vacancies, the binding free energy: $G_B^{di} = 2G_F^v - G_F^{di}$ was calculated and its sign analysed. Above, $G_F(T, P) = E_F - S_F T$ is the defect Gibb's free energy of formation and, as defined in previous sections, S_F is the formation entropy and E_F the formation energy. This last quantity was used instead of the formation enthalpy because their difference: $H_F - E_F = P\,\Delta V$ (being ΔV the defect formation volume) is small for normal pressures. The calculated values of the formation and binding free energies at the melting temperature ($T_M \approx 913$ K) are reported in Table 4. For those calculations the entropy values obtained in the cluster approximation were used. The results obtained for the divacancies either with P4C or P5C potentials verify the following order:

$$G_F^{d1} < G_F^{d2} < G_F^{d4} < G_F^{d3} \quad (\text{at } T = T_M)\,,$$

and the values obtained for the binding energies G_B are all positive except for the d3 divacancy. The sign of G_B corresponding to each type of defect was analysed for the whole temperature range below the melting point, but no changes with respect to the signs obtained at high temperature were found. On the other hand, in a crystal at thermal equilibrium, the concentration C^{di} of divacancies of type di at temperature T and pressure P is given by

$$C^{di}(P, T) = \tfrac{1}{2} Z\, e^{-G_F^{di}/kT}\,, \tag{9}$$

where the factor $\frac{1}{2}Z$ gives the number of divacancies of type di per lattice site. Expression (9), derived by following [27], assumes N^{di} defects of type di and also N^v vacancies in thermal equilibrium in a crystal with N atoms, being $N^{di} \ll N$ and $N^v \ll N$. Calculations carried out with the values of Table 4 corresponding to P5C, taking $Z = 6$ for d1 and d2 and $Z = 2$ for d4 as deduced by considerations of lattice geometry, give at the melting temperature

$$C^{d1}/C^v \approx 37 \times 10^{-4}\,, \qquad C^{d2}/C^v \approx 36 \times 10^{-4}\,, \qquad C^{d4}/C^v \approx 3.5 \times 10^{-4}\,,$$

these ratios being slightly smaller for P4C.

Table 4

Gibb's free energy calculations at $T = 913$ K

potential	G_F^v	G_F^{d1}	G_B^{d1}	G_F^{d2}	G_B^{d2}	G_F^{d3}	G_B^{d3}	G_F^{d4}	G_B^{d4}*)
P4C	0.657	1.214	0.100	1.218	0.096	1.318	−0.004	1.303	0.011
P5C	0.649	1.175	0.123	1.178	0.120	1.333	−0.035	1.270	0.028

*) G_F Gibb's free energy; G_B binding free energy. Superscripts v and di refer to vacancy and divacancy type di, respectively. All the energy values are given in eV.

5. Results and Discussion

Static properties of four types of divacancies were calculated in the present work by using computer simulation techniques and different interatomic potentials. The main results are summarized and discussed below.

The calculated formation energies of the divacancies favour the d2 divacancy at the basal plane (see Table 1). It seems that this result is a consequence of the h.c.p. lattice geometry because this trend is also observed by other authors [17, 18] with a relaxation procedure and interatomic potentials different from the ones used by us. A negative value of the binding energy for the d3 divacancy is found in the present work for all the potentials used. This result is consistent with the one obtained for the formation energy value, the largest of the calculated set of divacancies, and with the d3 relaxation volume, the smallest of the set.

The formation and binding entropies were calculated here in: (i) the Einstein approximation for an unrelaxed lattice, (ii) the Einstein approximation for a relaxed lattice, and finally (iii) including in the relaxed lattice collective vibration modes, this last being the most reliable approximation [7, 10]. Although the calculated values for the d1 and d2 divacancies in an unrelaxed lattice with the c/a ratio are similar (see Table 3), the relaxation determines significant differences ($\approx 3\%$) in their formation entropies, being $S_F^{d2} < S_F^{d1}$ for all the approximations used for the calculations. Tables 1 and 3 show that, in general, the larger entropies correspond to defects with larger energies (case of d3 and d4); this fact resembles the relations between energy and entropy for the activated state proposed by Zener [28].

Two empirical potentials, P4C and P5C, were used for the above calculations. These potentials are very similar, they have the same interaction range and are fitted to the same parameters, but differ in the fitting to zero of the second derivative at the cut-off distance. This fact determines marked differences in the values calculated with either potential for some physical quantities which depend on the force constant matrix. For instance, Table 3 shows that, when relaxation is allowed, for a given defect the calculations with different potentials predict formation entropy values which differ approximately by 15%, while Table 1 shows that the corresponding difference in formation energies is smaller than 0.7%.

Combining the energies with the entropies corrected at constant pressure calculated here, the Gibb's formation free energy for each one defect was evaluated at the melting temperature. For d1 the smallest value was found and for d3 the largest one. In addition, a negative binding free energy was obtained only for the d3 divacancy. So, the stability of d1, d2, and d4 divacancies against two free vacancies can be predicted, d4 being the least stable of the set. This was previously pointed out by Sahu et al. [17, 18] but they analyzed only the defect formation energy. Comparing these energies for the d1 and d2 configurations, this latter would be found "slightly more stable" than the other. However, the Gibb's free energy through the values of the formation entropy turns round that prediction, the d1 divacancy being then slightly more stable than the other as reported at Table 4. Finally, the negative value of the binding free energy for the d3 divacancy indicates the instability of that type of defect against two free vacancies for the whole temperature range. So, the d3 divacancy would not exist as a defect in thermal equilibrium in a lattice as the one described here.

The divacancy population at high temperatures was found to be significant; approximately eight divacancies at nearest-neighbour sites for 1000 vacancy sites with a contribution one order of magnitude lower due to axial divacancies. The vacancy equilibrium concentration at high temperature as predicted by the empirical potentials P4C and P5C are in good agreement with experimental values obtained by Tzanetakis et al. [29]. So, the equilibrium concentration values here proposed have some degree of confidence.

Appendix

Eigenvalues of the defect force constant matrices

The eigenvalues are indicated as $(Mw_i^2)_m^x$, where

a) the superscript x refers to the type of defect: v stands for vacancy, d1 divacancy at first neighbours, etc.,

b) the subscript m denotes which neighbour of the defect is allowed to vibrate; i.e.: $(Mw_i^2)_{13}^{\mathrm{d1}}$ corresponds to a host atom as first neighbour of one vacancy of the d1-pair and as third neighbour of the other vacancy,

c) the subscript i denotes different eigenfrequencies for given values of x and m.

$$(Mw^2)^0 = 8f_t^{(1)} + 4f_l^{(1)} + 4f_t^{(3)} + 2f_l^{(3)}\,,$$
$$(Mw_1^2)_1^{\mathrm{v}} = 8f_t^{(1)} + 3f_l^{(1)} + 4f_t^{(3)} + 2f_l^{(3)}\,,$$
$$(*)\ (Mw_2^2)_1^{\mathrm{v}} = 7f_t^{(1)} + 4f_l^{(1)} + 4f_t^{(3)} + 2f_l^{(3)}\,,$$
$$(Mw_1^2)_3^{\mathrm{v}} = 8f_t^{(1)} + 4f_l^{(1)} + 4f_t^{(3)} + f_l^{(3)}\,,$$
$$(*)\ (Mw_2^2)_3^{\mathrm{v}} = 8f_t^{(1)} + 4f_l^{(1)} + 3f_t^{(3)} + 2f_l^{(3)}\,,$$
$$(Mw_1^2)_{11}^{\mathrm{d1}} = \tfrac{15}{2} f_t^{(1)} + \tfrac{5}{2} f_l^{(1)} + 4f_t^{(3)} + 2f_l^{(3)}\,,$$
$$(Mw_2^2)_{11}^{\mathrm{d1}} = 6f_t^{(1)} + 4f_l^{(1)} + 4f_t^{(3)} + 2f_l^{(3)}\,,$$
$$(Mw_3^2)_{11}^{\mathrm{d1}} = \tfrac{13}{2} f_t^{(1)} + \tfrac{7}{2} f_l^{(1)} + 4f_t^{(3)} + 2f_l^{(3)}\,,$$
$$(Mw_1^2)_{13}^{\mathrm{d1}} = 7f_t^{(1)} + 4f_l^{(1)} + 3f_t^{(3)} + 2f_l^{(3)}\,,$$
$$(Mw_2^2)_{13}^{\mathrm{d1}} = \tfrac{15}{2} f_t^{(1)} + \tfrac{7}{2} f_l^{(1)} + \tfrac{7}{2} f_t^{(3)} + \tfrac{3}{2} f_l^{(3)} + \tfrac{1}{2}\sqrt{(f_t^{(1)} - f_l^{(1)})^2 + (f_t^{(3)} + f_l^{(3)})^2}\,,$$
$$(Mw_3^2)_{13}^{\mathrm{d1}} = \tfrac{15}{2} f_t^{(1)} + \tfrac{7}{2} f_l^{(1)} + \tfrac{7}{2} f_t^{(3)} + \tfrac{3}{2} f_l^{(3)} - \tfrac{1}{2}\sqrt{(f_t^{(1)} - f_l^{(1)})^2 + (f_t^{(3)} + f_l^{(3)})^2}\,,$$
$$(Mw_1^2)_{11}^{\mathrm{d3}} = 6f_t^{(1)} + 4f_l^{(1)} + 4f_t^{(3)} + 2f_l^{(3)}\,,$$
$$(*)\ (Mw_2^2)_{11}^{\mathrm{d3}} = 7f_t^{(1)} + 3f_l^{(1)} + 4f_t^{(3)} + 2f_l^{(3)}\,,$$
$$(Mw_1^2)_{11}^{\mathrm{d4}} = \tfrac{20}{3} f_t^{(1)} + \tfrac{10}{3} f_l^{(1)} + 4f_t^{(3)} + 2f_l^{(3)}\,,$$
$$(Mw_2^2)_{11}^{\mathrm{d4}} = 6f_t^{(1)} + 4f_l^{(1)} + 4f_t^{(3)} + 2f_l^{(3)}\,,$$
$$(Mw_3^2)_{11}^{\mathrm{d4}} = \tfrac{22}{3} f_t^{(1)} + \tfrac{8}{3} f_l^{(1)} + 4f_t^{(3)} + 2f_l^{(3)}\,,$$
$$(Mw_1^2)_{33}^{\mathrm{d4}} = 8f_t^{(1)} + 4f_l^{(1)} + 2f_t^{(3)} + 2f_l^{(3)}\,,$$
$$(Mw_2^2)_{33}^{\mathrm{d4}} = 8f_t^{(1)} + 4f_l^{(1)} + \tfrac{10}{3} f_t^{(3)} + \tfrac{2}{3} f_l^{(3)}\,,$$
$$(Mw_3^2)_{33}^{\mathrm{d4}} = 8f_t^{(1)} + 4f_l^{(1)} + \tfrac{8}{3} f_t^{(3)} + \tfrac{4}{3} f_l^{(3)}\,.$$

The above $f_l^{(1)}$ and $f_t^{(1)}$ are the longitudinal and the transverse force constants at nearest neighbours, and $f_l^{(3)}$ and $f_t^{(3)}$ the corresponding ones at third-neighbour atoms. Double degeneracy is indicated by (*).

Acknowledgements

The author expresses her sincere thanks to Dr. E. J. Savino for carefully correcting the manuscript and for his invaluable discussions. This work was supported partially by the "Comisión de Investigaciones Científicas de la Provincia de Buenos Aires", Argentina.

References

[1] A. Seeger and H. Mehrer in: Vacancies and Interstitials in Metals, Ed. A. Seeger, D. Schumacher, W. Schilling, and J. Diehl, North-Holland Publ. Co., Amsterdam 1970 (p. 1).
[2] N. L. Peterson and W. K. Chen in: Ann. Rev. Materials Sci., Vol. 3, Ed. R. A. Huggins, R. H. Bube, and R. W. Roberts, Ann. Rev. Inc., Palo Alto 1973 (p. 75).
[3] H. Mehrer, J. nuclear Mater. **69/70**, 38 (1978).
[4] M. Kiritani, J. Phys. Soc. Japan **35**, 95 (1973).
[5] E. J. Savino and A. M. Monti, phys. stat. sol. (b) **121**, 513 (1984).
[6] P. H. Dederichs, C. Lehmann, H. R. Schober, A. Scholz, and R. Zeller, J. nuclear. Mater. **69/70**, 176 (1978).
[7] R. D. Hatcher, R. Zeller, and P. H. Dederichs, Phys. Rev. B **19**, 5083 (1979).
[8] C. N. Tomé, A. M. Monti, and E. J. Savino, phys. stat. sol. (b) **92**, 323 (1979).
[9] A. M. Monti and E. J. Savino, phys. stat. sol. (b) **92**, K39 (1979).
[10] A. M. Monti and E. J. Savino, Phys. Rev. B **23**, 6494 (1981).
[11] A. J. Fendrik, A. M. Monti, and E. J. Savino, phys. stat. sol. (b) **113**, 709 (1982).
[12] R. Migoni, C. N. Tomé, N. Smetniansky-de Grande, and E. J. Savino, Phys. Rev. B **22**, 2658 (1980).
[13] Z. D. Popovic and J. P. Carbotte, J. Phys. F **4**, 351 (1974).
[14] M. Doneghan and P. T. Heald, phys. stat. sol. (a) **30**, 403 (1975).
[15] D. Sahoo and H. K. Sahu, Phys. Rev. B **18**, 6727 (1978).
[16] D. Sahoo and H. K. Sahu, Phys. Rev. B **18**, 6738 (1978).
[17] H. K. Sahu, S. Srinivasan, and K. Krishan, Radiat. Eff. Letters **50**, 73 (1980).
[18] H. K. Sahu, S. Srinivasan, and K. Krishan, Pramana **15**, 189 (1980).
[19] M. J. Norgett, R. C. Perrin, and E. J. Savino, J. Phys. F **2**, L73 (1972).
[20] R. Fletcher and C. M. Reeves, Computer J. **7**, 149 (1964).
[21] C. Mairy, J. Hillairet, and D. Schumacher, Acta metall. **15**, 1258 (1967).
[22] M. Appapillai and V. Heine, Tech. Rep. No. 5 SST/7/1972 University Cambridge (1972).
[23] H. R. Schober and K. W. Ingle, J. Phys. F **10**, 575 (1980).
[24] A. M. Stoneham, J. Phys. C **6**, 223 (1973).
[25] G. Leibfried and N. Breuer in: Point Defects in Metals I, Springer Tracts in Modern Physics, Berlin, Springer 1978 (p. 81).
[26] C. P. Flynn in: Point Defects and Diffusion, Clarendon Press, Oxford 1972 (p. 318).
[27] V. I. Zubov, phys. stat. sol. (b) **106**, K145 (1981).
[28] C. Zener in: Imperfections in Nearly Perfect Crystals, Ed. Shockley, Hollomon, Maurer and Seitz, John Wiley & Sons, New York 1952 (p. 295).
[29] P. Tzanetakis, J. Hillairet, and G. Revel, phys. stat. sol. (b) **75**, 433 (1976).

(Received November 23, 1983)

phys. stat. sol. (b) **126**, 177 (1984)

Subject classification: 10.2 and 14.3; 22.2; 22.4

Institute of Solid State Physics, Academy of Sciences of the USSR, Chernogolovka[1])

Dislocations in $A^{II}B^{VI}$ and $A^{III}B^{V}$ Piezoelectric Crystals

By

I. S. SMIRNOVA

The problem is considered of the distribution of fields produced by dislocations in piezocrystals. The behaviour of the fields is analyzed for all possible dislocations available in b.c.c. crystals (in particular, with the sphalerite lattice) and in crystals with wurtzite lattice (existence of the radial and tangential component $\boldsymbol{E}$; initiation of polarization vortexes). A considerable reduction in the symmetry of elastic stress fields is revealed for mixed dislocations in b.c.c. crystals.

Рассмотрена задача о распределении полей, созданных дислокациями в пьезоэлектрических кристаллах. Проанализирован характер полей для всех возможных дислокаций в г.ц.к. кристаллах (в частности, со структурой сфалерита) и в кристаллах со структурой вюрцита (наличие радиальной и тангенциальной составляющей $\boldsymbol{E}$, возникновение вихрей поляризации). Выявлено существенное понижение симметрии полей упругих напряженний для смешанных дислокаций в г.ц.к. кристаллах.

1. Introduction

In piezoelectric crystals in the absence of volume forces and external charges the strain ε_{kl} and the electric field $\boldsymbol{E}$ are described by the system of equations [1]

$$\frac{\partial \sigma_{ij}}{\partial x_i} = 0 , \quad \operatorname{div} \boldsymbol{D} = 0; \quad \operatorname{rot} \boldsymbol{E} = \boldsymbol{0} , \tag{1}$$

where

$$\sigma_{ij} = c_{ijkl}\varepsilon_{kl} - e_{mij}E_m ; \quad D_i = \varkappa_{ij}E_j + N_i ;$$
$$N_i = e_{ikl}\varepsilon_{kl} ,$$

c_{ijkl}, $\varkappa_{ij}$, e_{mij} are the tensors of elastic, dielectric and piezoelectric constants, respectively. N_i is the polarization caused by the elastic stress field.

In order to determine the electric field induced by the dislocation elastic field we have to supplement (1) with the equation determining the dislocation-induced deformation (the Burgers equation) [2]

$$a_{ijl} \frac{\partial w_{lk}}{\partial x_j} = d_{ik} , \tag{2}$$

where w_{lk} is the tensor of distortions $\varepsilon_{kl} = (w_{lk} + w_{kl})/2$, a_{ijl} is the permutation operator, d_{ik} is the dislocation density tensor. In what follows we shall consider the case of a straight dislocation for which

$$d_{ik} = \tau_i b_k \, \delta(\xi) , \tag{3}$$

$\boldsymbol{\tau}$ is the vector tangential to the dislocation line, $\boldsymbol{b}$ the Burgers vector, ξ the two-dimensional vector radius originating at the dislocation line and lying in the plane perpen-

[1]) Chernogolovka 142432, Moscow District, USSR.

dicular to this line. We shall relate the system of coordinates with the dislocation line by directing the axis along $\boldsymbol{\tau}$, then $\xi = (x, y)$.

Merten was the first to consider the problem of the electric field distributions around a dislocation in piezoelectrics [3]. He calculated the electric potential V employing in (1) the tensor of strain for a dislocation in an isotropic medium without the allowance for the inverse piezoeffect. In [3], however, there was an error in the determination of V due to the fact that in the expressions for the derivatives of the tensor w_{lm}, entering div $\boldsymbol{D} = 0$ no account was taken of the terms containing δ-functional singularities. These singularities necessarily appear in $\partial w_{ik}/\partial x_j$ by virtue of (2) and (3) which determine the specific features of the dislocation-induced deformation. Consequently the field $\boldsymbol{E}$ observed in [3] has no radial component stipulated by the δ-functional bound charge (the corresponding part of div$\boldsymbol{N}$).

The complete problem about dislocations in piezoelectrics with allowance for elastic anisotropy and inverse piezoeffect was set in [4, 5]. In [4] a screw dislocation along the c_4-axis in a cubic crystal is treated, in [5] a screw dislocation in tellurium along the a-axis.

The present work deals with the analysis of the distribution of electric field $\boldsymbol{E}$ and polarisation $\boldsymbol{N}$ around perfect dislocations of all possible types in wurtzite and sphalerite, i.e., edge, screw, 30° and 60°, oriented differently with respect to the crystallographic axes. Part of these dislocations exhibit no piezoelectric effect [6], the others demonstrate diverse $\boldsymbol{E}$ and $\boldsymbol{N}$ symmetries.

The solution of the problem is sought by the Fourier method, enabling one to obtain the result in the general form simpler for further treatment, as compared with that given by Eshelby et al. [7]. We employ the two-dimensional Fourier expansion

$$f(x, y) = \frac{1}{4\pi^2} \iint\limits_{-\infty}^{\infty} \tilde{f}(\alpha, \beta)\, \mathrm{e}^{-i\alpha x - i\beta y}\, \mathrm{d}\alpha\, \mathrm{d}\beta\,, \tag{4}$$

$$\tilde{f}(\alpha, \beta) = \iint\limits_{-\infty}^{\infty} f(x, y)\, \mathrm{e}^{i\alpha x + i\beta y}\, \mathrm{d}x\, \mathrm{d}y\,,$$

for w_{ik} and E_i and we obtain from (1) and (2) the system of inhomogeneous algebraic equations for the Fourier images $\tilde{w}_{ik}$ and $\tilde{E}_i$. The determinant of this system is a polynomial homogeneous in α and β of degree eight for a mixed dislocation, of degree six for an edge dislocation, and of degree four for a screw dislocation. Being homogeneous, all the polynomials can be factorized, and the unknown integrals are easily taken by means of residues.

2. Dislocations in Wurtzite Crystals

Two edge dislocations are of special interest in the mentioned crystals, i.e., basal and prismatic with Burgers vector $\boldsymbol{b} = \boldsymbol{c}$. These dislocations evidence that in one and the same crystal (CdS, for example) the same type of dislocation (edge) gives rise to different configurations of the electric fields caused by the piezoeffect.

The dislocations in question lie in the same direction $[10\bar{1}0]$ but their Burgers vectors are different: for a basal dislocation $\boldsymbol{b} = \frac{1}{3}[\bar{1}2\bar{1}0]$, for a prismatic one $\boldsymbol{b} = [0001]$. The system of coordinates for edge and mixed dislocations will be selected in the following way: the z'-axis will be along the dislocation line, the x'-axis will be in the glide plane. For this system of coordinates the tensors of dielectric- and piezoconstants will assume the following form:

basal

$$\begin{pmatrix} \varkappa'_{11} & 0 & 0 \\ 0 & \varkappa'_{22} & 0 \\ 0 & 0 & \varkappa'_{33} \end{pmatrix}, \quad \begin{pmatrix} 0 & 0 & 0 & 0 & 0 & e'_{112} \\ e'_{211} & e'_{222} & e'_{233} & 0 & 0 & 0 \\ 0 & 0 & 0 & e'_{323} & 0 & 0 \end{pmatrix};$$

prismatic (5)

$$\begin{pmatrix} \varkappa'_{11} & 0 & 0 \\ 0 & \varkappa'_{22} & 0 \\ 0 & 0 & \varkappa'_{33} \end{pmatrix}, \quad \begin{pmatrix} e'_{111} & e'_{122} & e'_{133} & 0 & 0 & 0 \\ 0 & 0 & 0 & 0 & 0 & e_{212} \\ 0 & 0 & 0 & 0 & e'_{331} & 0 \end{pmatrix},$$

where

$$\varkappa'_{11} = \varkappa'_{33} = \varkappa_{11}; \quad \varkappa'_{22} = \varkappa_{33}; \qquad \varkappa'_{11} = \varkappa_{33}; \quad \varkappa'_{22} = \varkappa'_{33} = \varkappa_{11};$$

$$e'_{112} = e'_{332} = e_{15}; \quad e'_{211} = e'_{233} = e_{31}; \qquad e'_{111} = e_{33}; \quad e'_{112} = e'_{133} = e_{31};$$

$$e'_{222} = e_{33} \qquad e'_{212} = e'_{331} = e_{15}\,.$$

In the elastic modulus tensors we have:

basal

$$c'_{1111} = c_{11} \qquad c'_{2222} = c_{33} \qquad c'_{1212} = c_{44} \qquad c'_{1122} = c_{13};$$

prismatic

$$c'_{1111} = c_{33} \qquad c'_{2222} = c_{11} \qquad c'_{1212} = c_{44} \qquad c'_{1122} = c_{13}\,.$$

With account taken of all the tensors the system (1) ,(2) for a basal edge dislocation will assume the form

$$c'_{1111}\frac{\partial w_{11}}{\partial x} + c'_{1122}\frac{\partial w_{22}}{\partial x} + c'_{1212}\left(\frac{\partial w_{12}}{\partial y} + \frac{\partial w_{21}}{\partial y}\right) - e'_{112}\frac{\partial E_x}{\partial y} - e'_{211}\frac{\partial E_y}{\partial x} = 0\,,$$

$$c'_{2211}\frac{\partial w_{11}}{\partial y} + c'_{2222}\frac{\partial w_{22}}{\partial y} + c'_{2112}\left(\frac{\partial w_{21}}{\partial x} + \frac{\partial w_{12}}{\partial x}\right) - e'_{121}\frac{\partial E_x}{\partial x} - e'_{222}\frac{\partial E_y}{\partial y} = 0\,,$$

$$e'_{112}\left(\frac{\partial w_{12}}{\partial x} + \frac{\partial w_{21}}{\partial x}\right) + e'_{211}\frac{\partial w_{11}}{\partial y} + e'_{222}\frac{\partial w_{22}}{\partial xy} + \varkappa'_{11}\frac{\partial E_x}{\partial x} + \varkappa'_{22}\frac{\partial E_y}{\partial y} = 0\,, \tag{6}$$

$$-\frac{\partial w_{11}}{\partial y} + \frac{\partial w_{21}}{\partial x} = -b\delta(x, y)\,, \qquad \frac{\partial w_{22}}{\partial x} - \frac{\partial w_{12}}{\partial y} = 0\,,$$

$$\frac{\partial E_y}{\partial x} - \frac{\partial E_x}{\partial y} = 0\,.$$

In the case (6) we obtain the following expression for

$$E_x(x, y) = \frac{b}{4\pi^2}\iint\limits_{-\infty}^{\infty}\frac{m\alpha^3\beta^2 + l\beta^4}{\Delta(\alpha, \beta)}\,e^{-i\alpha x - i\beta y}\,d\alpha\,d\beta\,,$$

$$E_y(x, y) = \frac{b}{4\pi^2}\iint\limits_{-\infty}^{\infty}\frac{m\alpha^2\beta^3 + l\beta^5}{\Delta(\alpha, \beta)}\,e^{-i\alpha x - i\beta y}\,d\alpha\,d\beta\,,$$

where

$$m = c'_{11}c'_{22}c'_{66} + c'_{12}c'_{66}e'_{21} - c'_{11}c'_{66}e'_{22} - (c'_{12})^2 e'_{16} ,$$
$$l = c'_{12}c'_{66}e'_{22} - e'_{21}c'_{22}c'_{66} ,$$
$$\Delta(\alpha, \beta) = \sum_{i=0}^{6} a_i \alpha^i \beta^{6-i}$$

is the determinant of the Fourier transformed system (6). The constants a_i are expressed as $\hat{c}$, $\hat{\varkappa}$, $\hat{e}$, for example, $a_0 = c'_{22}c'_{66}\varkappa'_{11} + e^{12}_{12}c'_{66}$ etc. By integrating we find

$$E_x(x, y) = \frac{b}{2\pi a_0} \sum_{n=1}^{3} \frac{\lambda_n(x + S_{n1}y) - \varepsilon_n S_{n2}y}{(x + S_{n1}y)^2 + S_{n2}^2 y^2} ,$$
$$E_y(x, y) = \frac{b}{2\pi a_0} \sum_{n=1}^{3} \frac{\lambda_n[S_{n1}(x + S_{n1}y) + S_{n2}y] + \varepsilon_n S_{n2}x}{(x + S_{n1}y)^2 + S_{n2}^2 y^2} . \quad (7)$$

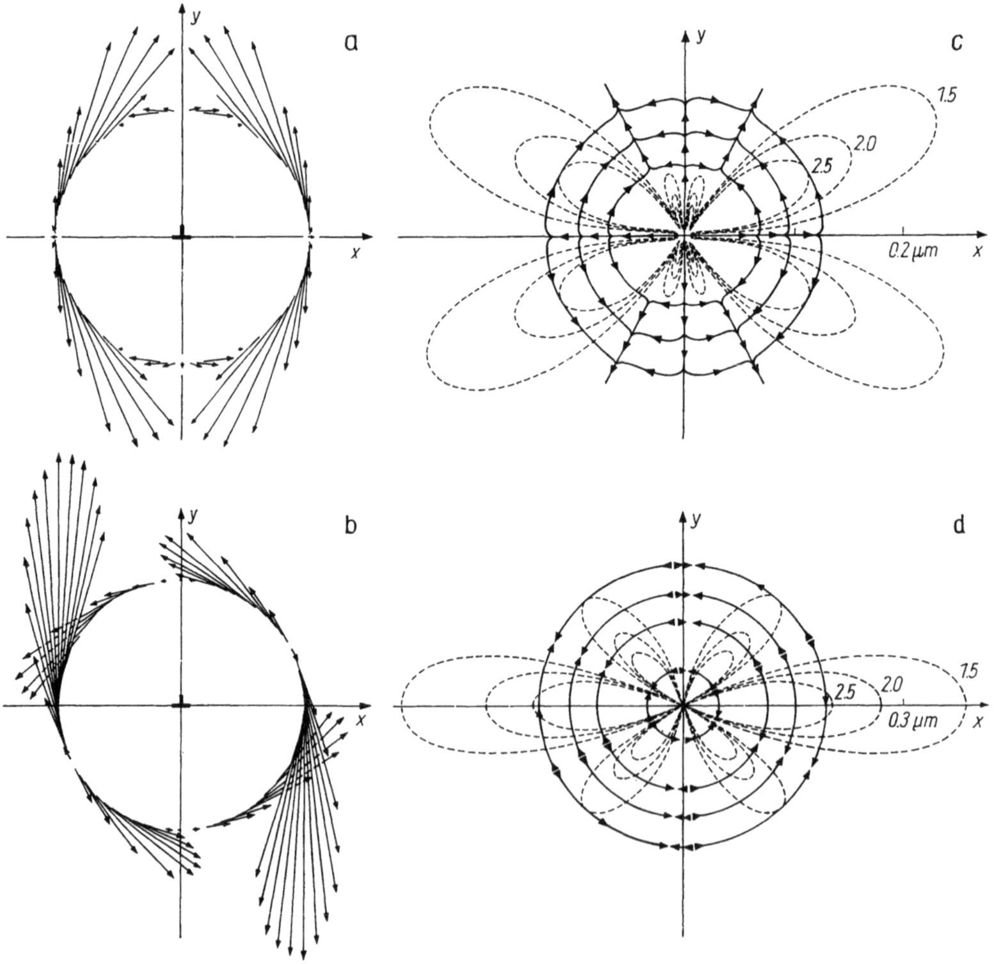

Fig. 1. Electric field of edge dislocations in the wurtzite structure. a) and b) distribution of the electric-intensity vector; c) and d) force lines of this field (solid curves) and isosurfaces (dashed curves) $|\boldsymbol{E}| = (1.5; 2; 2.5) \times 10^6$ V/m; a) and c) basal and b) and d) prismatic ($\boldsymbol{b} = \boldsymbol{c}$) dislocations

Here $S_n = S_{n1} + iS_{n2}$ are the roots of the equation $\Delta(x) = 0$, where $x = \beta/\alpha$, since the coefficients of this equation are real quantities; consequently S_n are complex conjugate. We allow for this in integrating, therefore summation in (7) goes to three rather than to six.

$$\lambda_n = \frac{M_n O_n + N_n P_n}{S_{n2}(O_n^2 + P_n^2)}; \qquad M_n = \mathrm{Re}\,[m(S_{n1} + iS_{n2}) + l(S_{n1} + iS_{n2})^4],$$
$$N_n = \mathrm{Im}\,[m(S_{n1} + iS_{n2}) + l(S_{n1} + iS_{n2})^4], \tag{8}$$
$$\varepsilon_n = \frac{M_n P_n - N_n O_n}{S_{n2}(O_n^2 + P_n^2)}; \qquad O_n = \mathrm{Re} \prod_{j=1}^{6} [S_{n1} + iS_{n2} - S_{nj} - iS_{j2}],$$
$$P_n = \mathrm{Im} \prod_{j=1}^{6} [S_{n1} + iS_{n2} - S_{nj} - iS_{j2}].$$

The solution for a prismatic dislocation is analogous.

To illustrate the obtained results we plot a graph of the electric field distribution of edge dislocations: basal, Fig. 1a, and prismatic, Fig. 1b. The lengths of the arrows correspond to $|\boldsymbol{E}|$ at the distance from the dislocation line $r = 5b$; since $E \sim 1/r$ the distribution at other distances is similar to that presented. Fig. 1 c, d show the configurations of force lines (solid lines) and of the isosurfaces $|\boldsymbol{E}| = \text{const}$ (dashed lines). The maximum value of $|\boldsymbol{E}|$ is about 10^8 V/m for basal and about 10^9 V/m for prismatic dislocations (for $r = 5b$).

The difference in the geometry of the force lines shown in Fig. 1c and 1d, is related to the fact that a basal dislocation alone has a radial component of the electric field equal, according to (7),

$$E_r = \frac{b}{2\pi a_0 r} \sum_{j=1}^{3} \lambda_j ,$$

E_r being about 10^6 V/m with $r = 5b$; for a prismatic dislocation $E_r = 0$. In a basal dislocation the force lines originate in the core region and escape to infinity along the directions where the tangential component of the field is $E_\varphi = 0$. The force lines of a prismatic dislocation originate and terminate on the planes corresponding to a maximum distribution of the bound charge, $\varrho_b = \operatorname{div} \boldsymbol{N}$. Fig. 2 presents the values of $\varrho_b = \text{const}$ for basal (Fig. 2a) and prismatic (Fig. 2b) dislocations. For a basal dislocation ϱ_b is distributed symmetrically about the x- and y-axes, whereas no such symmetry exists for a prismatic dislocation. The electric field configuration (Fig. 1a, b)

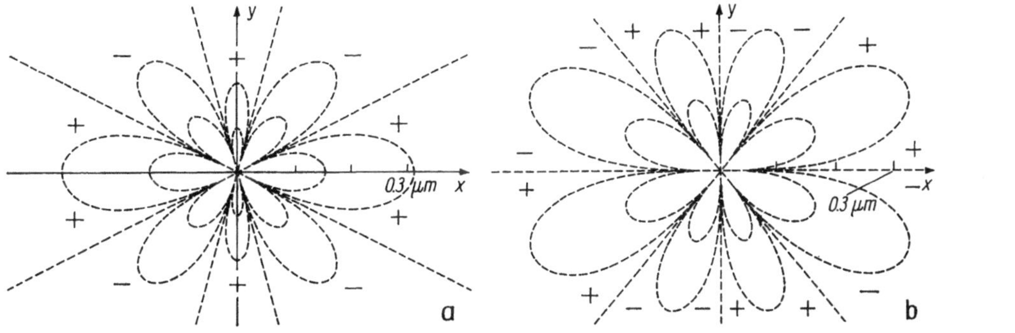

Fig. 2. Contours of constant bound charge density for a) edge basal and b) edge prismatic ($\boldsymbol{b} = \boldsymbol{c}$) dislocations. The inner and outer contours represent 10^5 and 4×10^5 C/m², respectively

reflects the bound charge distribution. The electric field of 60° and 30° dislocations, lying in the basal plane, is different only in magnitude from that considered above for an edge basal dislocation (by a factor of 3/2 and 1/2, respectively).

3. Dislocations in Sphalerite

In cubic piezoelectrics as, normally, in f.c.c. crystals, plastic deformation is caused, principally, by dislocations with the {111} glide plane. It is well known that four types of simple dislocations with such a glide plane can be distinguished, i.e., screw, edge, 60° and 30°. A screw dislocation in sphalerite, just like in wurtzite, has no electric field. The piezoelectric field symmetry of 60° and 30° dislocations is lower than that of the corresponding dislocations in the wurtzite crystal. This is due to the fact that the basal plane, which is the glide plane of the dislocations of interest in hexagonal crystals, is practically isotropic (axis of order six) and the {111} plane in sphalerite has a lower symmetry (axis of order three). The tensors of elastic and piezoelectric constants in the system of coordinates of dislocations in sphalerite have a more complicated form than in wurtzite, therefore the symmetry of electric and even elastic fields of 60° and 30° dislocations in cubic piezoelectrics is lower than that of the corresponding fields in hexagonal piezoelectrics in the wurtzite structure.

3.1 Edge dislocations

An edge dislocation the same as a 30° one lies along the ⟨112⟩ direction. The system of coordinates for the dislocations in question is: z'-axis || $[\bar{1}2\bar{1}]$, x'-axis || $[10\bar{1}]$, y'-axis || [111]. The tensor of the elastic constants for this system (x', y', z') is given elsewhere [9]. The tensor of piezoconstants will assume the form

$$\begin{pmatrix} 0 & 0 & 0 & 0 & e'_{131} & e'_{112} \\ e'_{211} & e'_{222} & e'_{233} & 0 & 0 & 0 \\ 0 & 0 & e'_{333} & e'_{332} & 0 & 0 \end{pmatrix},$$

where

$$e'_{131} = e'_{311} = 2/\sqrt{6}\, e_{14}\,,$$

$$e'_{332} = e'_{112} = e'_{233} = -1/\sqrt{3}\, e_{14}\,,$$

$$e'_{222} = 2/\sqrt{3}\, e_{14}\,, \qquad e'_{333} = -2/\sqrt{3}\, e_{14}\,.$$

The tensor of dielectric constants has the same form in the new system of coordinates as in the (x, y, z) system.

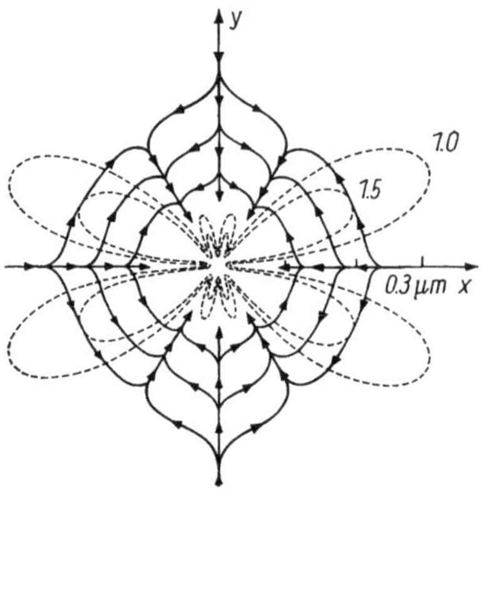

Fig. 3. Electric field of an edge dislocation with the {111} glide plane for the sphalerite structure; force lines (solid), isosurfaces (dashed) $|\boldsymbol{E}| = (1.0; 1.5) \times 10^6$ V/m

It can easily be seen that the system of equations (1), (2), written with account taken of the tensors for edge dislocations in cubic crystals, will have the same form as the system (6). Fig. 3 shows the electric field of an edge dislocation: its force lines and isosurfaces $|\boldsymbol{E}| = \text{const.}$

Inasmuch as the radial component of the electric field E_r of an edge dislocation in cubic piezoelectrics differs in sign from E_r of the analogous dislocation in a hexagonal crystal the force-line patterns of the two dislocations of interest are somewhat different from each other (Fig. 1c and 3). Note that one and the same, for example, Cd dislocation in two crystals (CdTe, cubic, and CdS, hexagonal) has in either case $E_r \approx \approx 10^6$ V/m with $r = 5b$, although $|E_\varphi|$ in CdS is by one order of magnitude (10^8 V/m with $r = 5b$) greater than in CdTe (10^7 V/m with $r = 5b$). The signs of the radial component of a Cd dislocation in the two crystals in question are opposite. If a cubic piezoelectric is assumed to be isotropic in the elastic properties, then E_r in both crystals (in the wurtzite and sphalerite structures) will be of the same sign, i.e., the elastic anisotropy of sphalerite affects the sign of the radial component of the field.

3.2 60° dislocation

We shall estimate the electric field in the system of coordinates: z'-axis $\|$ $[\bar{1}01]$, x'-axis $\|$ $[\bar{1}2\bar{1}]$, y'-axis $\|$ $[111]$. The transition matrix from the standard cubic system to the (x', y', z') system is given elsewhere [10], where also the tensor of the elastic constants for the case under consideration is presented. The tensor of the dielectric constants does not alter its form for the given transformation. In the new system of coordinates the tensor of piezoconstants assumes the following form:

$$\begin{pmatrix} e'_{111} & 0 & 0 & 0 & 0 & e'_{112} \\ e'_{211} & e'_{222} & e'_{233} & 0 & 0 & 0 \\ 0 & 0 & 0 & e'_{332} & e'_{331} & 0 \end{pmatrix},$$

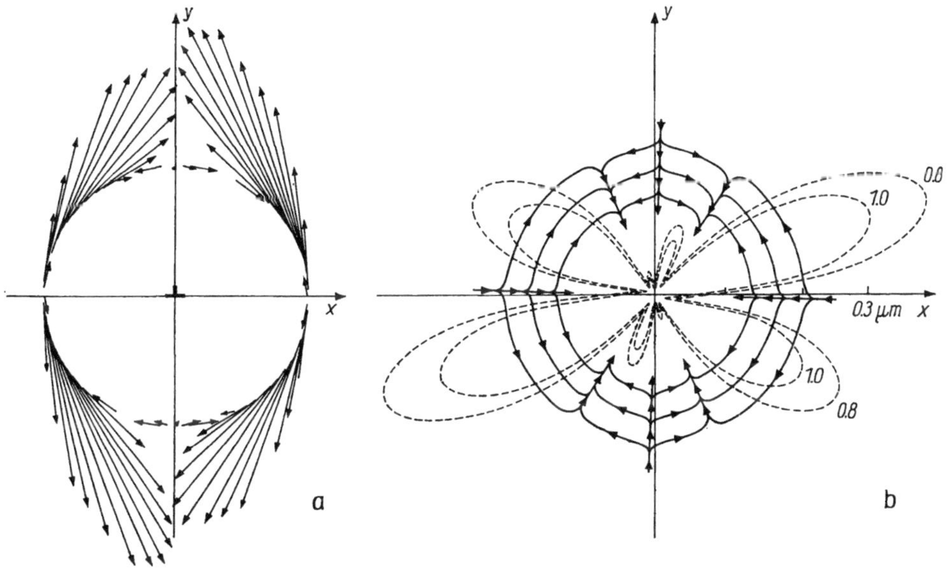

Fig. 4. Electric field of a 60° dislocation in sphalerite. a) Electric intensity, b) force lines (solid), isosurfaces (dashed) $|\boldsymbol{E}| = (0.8; 1.0) \times 10^5$ V/m

where

$$e'_{112} = e'_{211} = e'_{233} = e'_{332} = -1/\sqrt{3}\, e_{14}\,,$$
$$e'_{133} = e'_{331} = -e'_{111} = 2/\sqrt{6}\, e_{14}\,,$$
$$e'_{222} = 2/\sqrt{3}\, e_{14}\,.$$

It can easily be seen that the system (1), (2) for a 60° dislocation can be regarded as two systems: one yields the solution for the screw component of the dislocation, the other for the edge one. The first system does not comprise an electric field, i.e., the screw component of the dislocation will have no electric field. We have to find the electric field of the edge component alone, but the system of equations for it will be more complicated than system (6) due to the presence of the term c'_{16} in the tensor of the elastic constants and of the term e'_{111} in the piezotensor.

Fig. 4 gives the electric vector field of a 60° dislocation in a cubic crystal (Fig. 4a), the force lines of this field and isosurfaces $|\boldsymbol{E}| = \text{const}$ (Fig. 4b). As suggested by the figures, the electric field symmetry of the dislocation in question is lower than for the edge one and $E_r \neq 0$.

3.3 30° dislocation

Fig. 5 shows the electric vector field (Fig. 5a), its force lines and isosurface $|\boldsymbol{E}| = 0.8 \times 10^5$ V/m (inner contour). The radial component of the field is shown by dashed lines (Fig. 5a), its sign is dependent on the crystal anisotropy. The $\boldsymbol{E}$ symmetry of a 30° dislocation is lower than that of an edge one.

3.4 Elastic fields of a dislocation

A decrease of symmetry of the elastic field around 30° and 60° dislocations in anisotropic crystals is of interest from the viewpoint of a number of physical applications. This effect is not related to piezoproperties and is observed in all f.c.c. crystals (in crystals with sphalerite structure, in simple semiconductors of Si, Ge type, and in metals, for example Cu, Pb). Here, the higher the anisotropy factor $A = 2c_{44}/(c_{11} - c_{12})$ is, the more clearly should this effect be manifested.

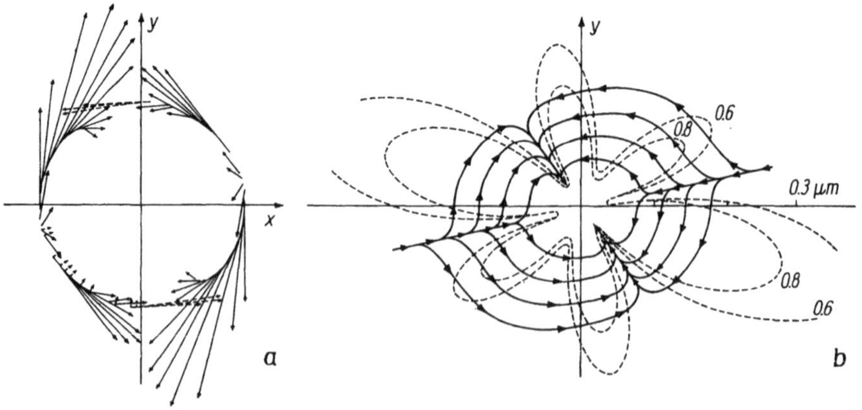

Fig. 5. Electric field of a 30° dislocation in sphalerite. a) Electric intensity, b) force lines (solid), isosurfaces (dashed) $|\boldsymbol{E}| = (0.6; 0.8) \times 10^5$ V/m

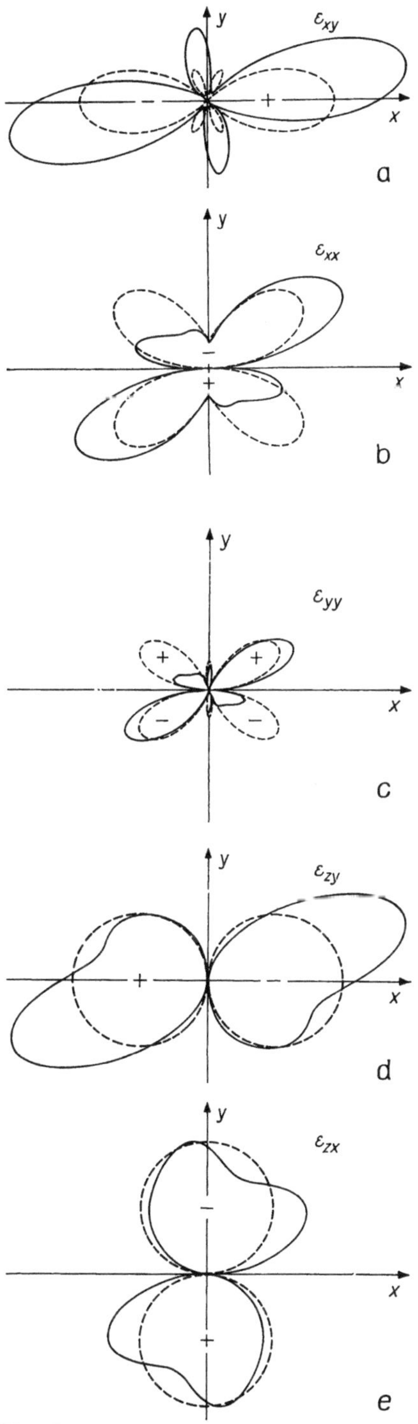

Fig. 6

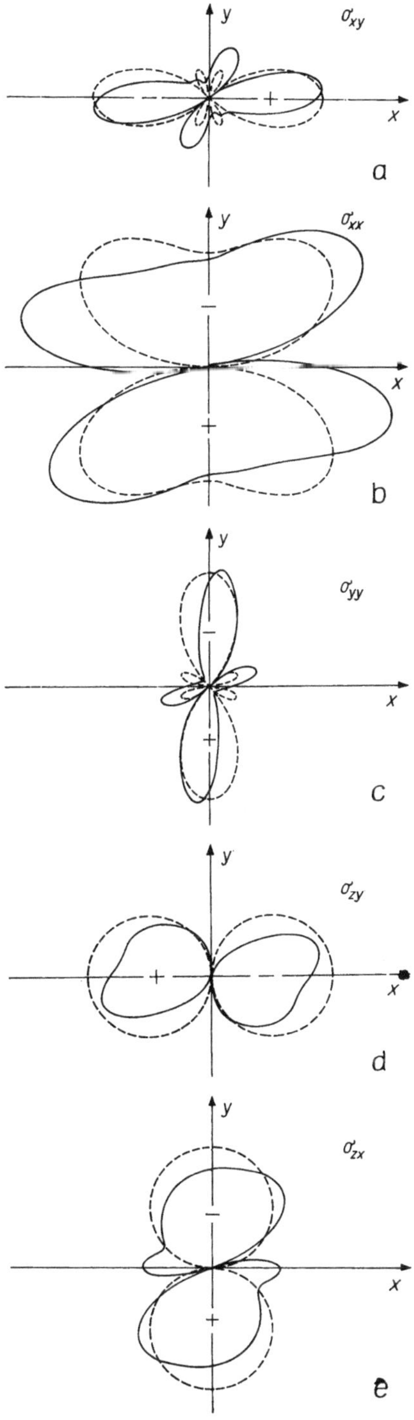

Fig. 7

Fig. 6. Contours of constant deformation around a mixed dislocation; isotropic crystal (dashed), anisotropic crystal (solid)

Fig. 7. Contours of constant stresses around a mixed dislocation; isotropic crystal (dashed), anisotropic crystal (solid)

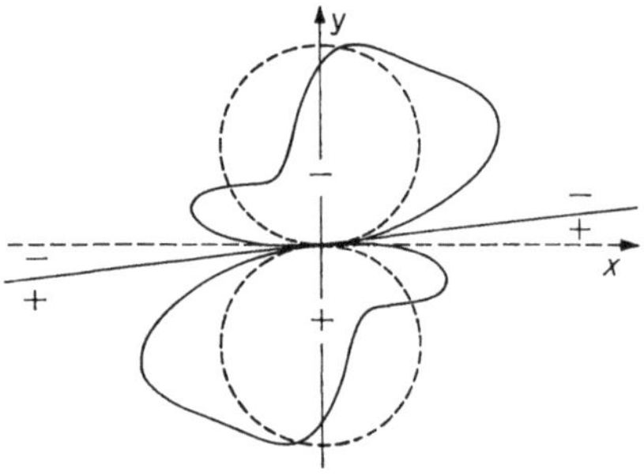

Fig. 8. Contours of constant values of the function div $\boldsymbol{u}$ for a mixed dislocation; isotropic crystal (dashed), anisotropic crystal (solid)

Fig. 6 and 7 present graphs showing constant deformation contours (Fig. 6) and stress σ_{ij} (Fig. 7) of a mixed dislocation in Pb, taken as an example of a strong-anisotropy crystal ($A \approx 4$). For comparison the dislocation elastic fields, obtained for an isotropic continuum, are shown by dashed lines. The figures show that the elastic field symmetry of a mixed dislocation in an anisotropic crystal is lower than that of the same dislocation in an isotropic crystal.

In the isotropic case the dislocation glide plane is free from normal stresses and deformations. In the anisotropic case normal stresses appear on the glide plane of a mixed dislocation: on the one side the material is compressed, on the side other it is extended (Fig. 7 b, c).

Fig. 8 shows the divergence of the displacement div $\boldsymbol{u}$ = const of a mixed dislocation. The dashed line presents the known solution in an isotropic approximation, the solid line presents the isosurface div $\boldsymbol{u}$ = const with account taken of the anisotropy. The straight line is the isosurface div $\boldsymbol{u} = 0$ in an anisotropic crystal. The figure shows that the mixed dislocations in real cubic crystals exhibit asymmetry. A decrease of the div $\boldsymbol{u}$ symmetry might affect the interaction of dislocations with impurities, electrons, the dislocation motion, the Peierls barrier symmetry. A decrease of the elastic field symmetry must be allowed for in the interpretation of the g-factor of the dislocation ESR spectra in Si (Fig. 6 a, b).

4. Polarisation Fields

For a wurtzite-structure crystal the electric field can be calculated from

$$\operatorname{div} \varkappa \boldsymbol{E} = -\operatorname{div} \boldsymbol{N},$$

where $N_i = e_{ikl}\varepsilon_{kl}$ is the polarisation due to the electric field of the dislocation. In this type of crystals there are two screws, each yields $\boldsymbol{E} = \boldsymbol{0}$, but for the screw lying along the a-axis $\boldsymbol{N} \equiv \boldsymbol{0}$, whereas for that lying along the c-axis $\boldsymbol{N} \neq \boldsymbol{0}$ but div $\boldsymbol{N} \equiv 0$. Polarisation of the latter dislocation is presented in Fig. 9a,

$$N_x = -\frac{e_{15}b}{4\pi}\frac{y}{x^2+y^2}, \qquad N_y = \frac{e_{15}b}{4\pi}\frac{x}{x^2+y^2}$$

with

$$\operatorname{div} \boldsymbol{N} \equiv 0, \qquad \operatorname{rot} \boldsymbol{N} = w\delta(x, y), \quad \text{where} \quad w = \frac{e_{15}b}{2}.$$

The dislocation under consideration demonstrates a pure vortex polarisation. For this case $\boldsymbol{E} \equiv \boldsymbol{0}$, $\boldsymbol{D} \neq \boldsymbol{0}$. This situation is characteristic of ferroelectrics. The wall comprising such dislocations (Fig. 9a) can produce a polarisation similar to that produced by the domain boundaries.

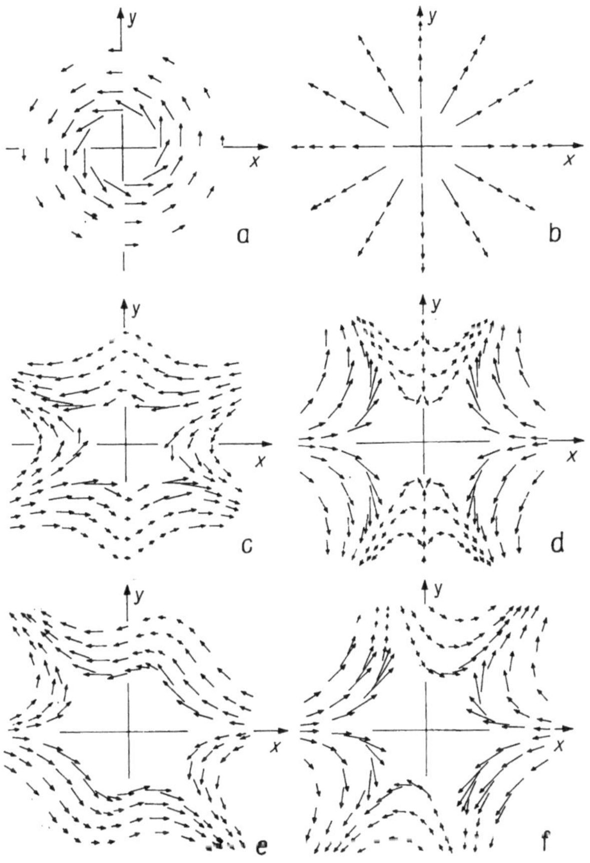

Fig. 9. Polarisation fields. a) Screw dislocation $\boldsymbol{b} = \boldsymbol{c}$ in wurtzite, b) screw dislocation $\boldsymbol{b} = \boldsymbol{a}$ in Te, c) edge basal dislocation, and d) prismatic dislocation $\boldsymbol{b} = \boldsymbol{c}$ in wurtzite, e) 30° dislocation and f) 60° dislocation in sphalerite

A similar screw dislocation (along the c-axis) in a tellurium crystal (different class of symmetry and different tensor of piezoconstant) demonstrates a pure potential polarisation (Fig. 9b). In this case

$$N_x = \frac{e_{14}b}{4\pi} \frac{x}{x^2 + y^2}, \qquad N_y = \frac{e_{14}b}{4\pi} \frac{y}{x^2 + y^2}$$

with

$$\operatorname{div} \boldsymbol{N} = \varrho_b \delta(x, y), \qquad \operatorname{rot} \boldsymbol{N} \equiv \boldsymbol{0}, \quad \text{where} \quad \varrho_b = \frac{e_{14}b}{2}.$$

Here $\boldsymbol{E} \neq \boldsymbol{0}$, $\boldsymbol{D} = \boldsymbol{0}$. Edge and mixed dislocations have a more complicated polarisation $\boldsymbol{N} = \boldsymbol{N}_1 + \boldsymbol{N}_2$, with $\operatorname{rot} \boldsymbol{N}_1 = \boldsymbol{0}$, $\operatorname{div} \boldsymbol{N}_2 = 0$, so that $\boldsymbol{N}_1 = \operatorname{grad} \varphi$, $\boldsymbol{N}_2 = \operatorname{rot} \boldsymbol{s}$.

The potential fraction of polarisation gives rise to an electric field since $\operatorname{div} \boldsymbol{N}_1 = \varrho_b$. One can expect that the solenoidal fraction of polarisation will be manifested in dynamic effects (like the polarisation produced by a transverse wave in ionic crystals [11]).

Fig. 9 c, d show the polarisation of two edge dislocations in wurtzite: Fig. 9d presents the edge dislocation in the basal plane, Fig. 9c the edge dislocation in the prismatic plane. It can easily be seen that the polarisation field flux of the basal dislocation yields a constant, i.e. the basal edge dislocation has a δ-functional bound charge. For this dislocation the polarisation circulation equals zero. For the prismatic dislocation the polarisation field flux equals zero. Note that this dislocation has ions of unlike sign along the extra half-plane and is considered neutral. The polarisation field circulation equals a constant, i.e. the polarisation of an edge prismatic dislocation has the same δ-functional singularity as a pure polarisation vortex (Fig. 9a). The wall, comprising these dislocations, the same as the screw along $\boldsymbol{c}$, produces polarisation similar to that produced by a domain wall.

Fig. 9 e, f present the polarisation vector fields of 30° (Fig. 9e) and 60° dislocations in sphalerite-structure crystals. For both vector fields the circulation and the flux are constant, i.e. the dislocations in question have a δ-functional bound charge, therefore E_r of these dislocations does not equal zero, besides, these dislocations demonstrate a δ-functional polarisation vortex. So, the polarisation vector fields, produced by a dislocation, have, in the general case, potential and solenoidal character, and owing to the specific character of the dislocation-induced deformation these vector fields have δ-functional singularities.

5. Conclusions

The analysis of piezoelectric and polarisation fields of dislocations in hexagonal (wurtzite) and cubic (sphalerite) piezoelectrics suggests that

(i) The polarisation fields of all the dislocations considered have δ-functional singularities: in wurtzite $\oint_S (\boldsymbol{N}, \mathrm{d}\boldsymbol{s}) = \text{const}$ or $\oint_L (\boldsymbol{N}, \mathrm{d}\boldsymbol{r}) = \omega$. In cubic piezoelectrics mixed dislocations exhibit simultaneously both singularities. If $\oint_S (\boldsymbol{N}, \mathrm{d}\boldsymbol{s}) = \text{const}$, then the electric field of this dislocation must have a radial component. The sign of E_r depends on the crystal anisotropy. The dislocation elastic fields obtained for an isotropic continuum are employed for the determination of the piezoelectric effect. This approach simplifies the calculations: it is valid when considering simple dislocations in hexagonal crystals, whereas for cubic crystals a qualitative difference appears in the solutions, e.g. reversal of sign of E_r for all the dislocations, and a decrease of the number of zeros in E_φ for a 30° dislocation.

In piezoelectric crystals besides the long-range elastic field dislocations possess long-range electric and polarisation fields. These fields are very peculiar, that is, in one and the same crystal (CdSe, for example) dislocations of the same type (edge, for example) have marked differences (see Fig. 1 c, d and 9 c, d). It was shown in [12] that in the luminescence spectra dislocations are exhibited differently.

Note that the walls of the dislocation the polarisation fields of which have a δ-functional vortex (Fig. 9 a, c, e), produce a polarisation similar to that produced by a domain boundary.

(ii) In cubic piezoelectrics the elastic field symmetry of mixed dislocations is lower than that suggested by the isotropic approximation. The decrease of the elastic field symmetry of mixed dislocations is characteristic not only of cubic piezoelectrics but, also, of all cubic crystals: f.c.c., b.c.c., NaCl-type crystals. This work presents the calculations performed for f.c.c. crystals.

In certain crystals the stress rosettes (for instance, a shear-stress rosette (Fig. 7a)) can directly be observed in a polarizing microscope. It should be borne in mind that for mixed dislocations in cubic crystals the rosette can have a form different from that suggested by the calculations using the isotropic model.

Note that how large the crystal anisotropy may be, the elastic field symmetry of a pure edge dislocation in cubic crystals does not decrease. In a real crystal a mixed dislocation exhibits asymmetry. This may be manifested in the asymmetric motion, in the dislocation–impurity interaction.

Acknowledgements

The author is grateful to Yu. A. Ossipyan for the constant interest in the work, to V. Ya. Kravchenko for his very useful discussions, and to I. D. Zhukovskaya for the assistance in the calculations.

References

[1] J. E. Nye, Physical Properties of Crystals, Chap. 10, Clarendon Press, Oyford 1967.
[2] R. de Wit, Solid State Phys. **10**, 249 (1960).
[3] L. Merten, Z. Naturf. **19a**, 1161 (1964).
[4] A. M. Kossevich, A. P. Pastour, and E. P. Feldman, Kristallografiya **12**, 916 (1967).
[5] G. Saada, phys. stat. sol. (b) **44**, 717 (1971).
[6] I. S. Smirnova, Fiz. tverd. Tela **15**, 2312 (1973).
[7] I. D. Eshelby, W. T. Read, and W. Shockly, Acta metall. **1**, 251 (1953).
[8] J. Hornstra, J. Phys. Chem. Solids **5**, 129 (1958).
[9] L. J. Teutonico, Acta metall. **11**, 1283 (1963).
[10] J. P. Hirth and J. Lothe, Theory of Dislocations, McGraw-Hill Publ. Co., New York 1968 (p. 406).
[11] M. Born and K. Huang, Dynamical Theory of Crystal Lattice, Clarendon Press, Oxford 1954.
[12] A. V. Bazhenov, Yu. A. Ossipyan, and E. A. Shteinman, Fiz. tverd. Tela **18**, 933 (1976).

(Received July 13, 1984)

phys. stat. sol. (b) **126**, 191 (1984)

Subject classification: 13 and 18; 2; 21.1.1

Department of Solid State Physics, Academy of Mining and Metallurgy, Cracow[1])

Charge Transfer and Concentration Dependence of Magnetization in Amorphous Fe–B Alloys

By

A. Z. MAKSYMOWICZ, T. STOBIECKI, E. JAROCKI, and W. KARAŚ

Measurements are reported of the magnetic moment μ_{Fe} of Fe atoms in amorphous $Fe_{1-x}B_x$ alloys. The observed concentration dependence of μ_{Fe} can be explained qualitatively in terms of a simple rigid-band model assuming a transfer of v electrons per B atom to the Fe d-band. The rigid band is assumed as described by the density of states of densely packed structures. The best agreement is obtained for the charge transfer of about 1.5 electrons per boron atom yielding the observed critical concentration $x = 0.60$ at which magnetism disappears. Details of the density of states and the assumed number of electrons for pure Fe do not change the characteristic features of the μ_{Fe} versus x dependence.

Messungen des magnetischen Moments μ_{Fe} von Eisenatomen in amorphen $Fe_{1-x}B_x$-Legierungen werden durchgeführt. Die beobachtete Konzentrationsabhängigkeit von μ_{Fe} läßt sich qualitativ mit einem einfachen Modell starrer Bänder unter Annahme eines Übergangs von v Elektronen pro B-Atom zum Fe-d-Band erklären. Es wird angenommen, daß das starre Band durch die Zustandsdichten von dichtgepackten Strukturen beschrieben wird. Die beste Übereinstimmung wird für einen Ladungsübergang von etwa 1,5 Elektronen pro Boratom erhalten und ergibt die beobachtete kritische Konzentration von $x = 0.60$ bei der der Magnetismus verschwindet. Einzelheiten der Zustandsdichte und der angenommenen Zahl von Elektronen für reines Eisen ändern den charakteristischen Gang der μ_{Fe}–x-Abhängigkeit nicht.

1. Introduction

The magnetic moment μ_{Fe} per Fe atom in amorphous $Fe_{1-x}B_x$ alloys depends on the concentration x of B atoms. The amorphous transition metal–metalloid systems are usually prepared by rapid quenching from the melt in the form of ribbons. This technique provides stable alloys only in the eutectic range of concentration for x from 0.13 to 0.28. A larger range of concentration from $x = 0.10$ to 0.90 is covered by films produced from the gaseous phase by sputtering [1, 2] or coevaporation [3]. There are no experimental data for the magnetization for alloys with concentration below 10%. For $x > 0.10$ the magnetization is roughly a decreasing function and goes to zero at the critical concentration of about $x = 0.6$. One of the possible mechanisms for the $\mu_{Fe}(x)$ dependence is believed to be a charge transfer of electrons from B atoms to the d-band of Fe atoms [4 to 7]. Due to a strong effect of the hybridization of sp states and transition metal d-states, the Fe d-band is altered and, therefore, one cannot directly apply the b.c.c., iron density of states. Also, the crystallographic structure of amorphous Fe–B alloys resembles the densely packed f.c.c. structure.

No detailed calculations of the electronic structure for amorphous $Fe_{1-x}B_x$ alloys are as yet available. Below we present the simplest rigid-band model calculations of the magnetic moment μ_{Fe} per Fe atom. The d-band is stiff in the sense that the shape of the band remains the same for different concentration from $x = 0.1$ to 0.6. The assumed effective density of states should account both for the hybridization and the closely packed structure of atoms.

[1]) Al. Mickiewicza 30, 30-059 Cracow, Poland.

A simple viewpoint is adopted here to use the f.c.c. density of states as a starting point. Hopefully, the results for magnetization μ_{Fe} versus concentration x are not much sensitive to the density of states.

2. Experimental

We prepared the films by the rf sputtering technique described in detail in [1]. The thickness of the films from 0.4 to 1.5 μm was determined by α-step profiling. The concentrations of the films were found from electron microprobe analysis for $Fe_{1-x}B_x$ ribbons, $0.13 \leqq x \leqq 0.26$, with bulk samples of FeB and pure B samples used as standards. The concentration x was measured with an accuracy of 0.05. The saturation of magnetization was measured at 4.2 K by the SQUID magnetometer in magnetic fields up to 50 kG applied along the normal to the film plane.

3. Numerical

We applied the density of states of the face-centred cubic crystallographic structure of transition metals for the 3d band of iron sites in amorphous densely packed alloys of the Fe–B system [8 to 12] (see also Fig. 1). It is claimed in many papers that the density of states of the amorphous state is rather close to the density of states of f.c.c. metals.

We assume n_0 electrons in the Fe d-band for pure f.c.c. iron. In the calculations we tried $n_0 = 6.8$, 7.2, and 7.4. In the Hartree-Fock approximation, the d-band is split by $U(n_\uparrow - n_\downarrow)$, where U is the Coulomb repulsion integral and the magnetization in Bohr magnetons per Fe atom is $\mu = n_\uparrow - n_\downarrow$. The total number of electrons in the d-band is $n = n_\uparrow + n_\downarrow$. For given U and n one finds self-consistently $n_\uparrow$ and $n_\downarrow$ and so the magnetic moment can be calculated.

For pure f.c.c. iron, we fit U so as to obtain the assumed magnetic moment μ_0, here we used $\mu_0 = 1$, 1.5, 1.8, 2.2, and 2.3 μ_B as trial numbers. When adding B atoms, we believe U does not change much and so U is kept the same for all x in the calculations. We also assume the number v of electrons per B atom supplied to the 3d-band to be in the first approximation x-independent. Then vx electrons from all x atoms of B are transferred to the d-band of all $1 - x$ atoms of Fe. The number of d-electrons at each Fe site is increased by

$$\Delta n = \frac{vx}{1 - x}$$

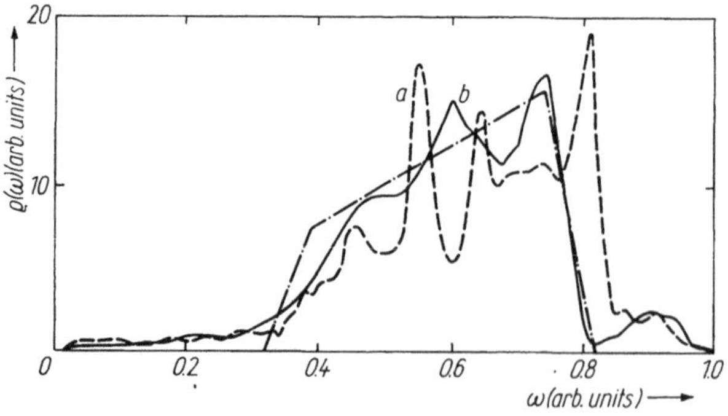

Fig. 1. Density of states of transition metals with face-centred cubic structures; (a) after Wakoh and Yamashita [10], (b) after Snow and Waber [12], (c) simplified density of states used in our calculations

and with increasing concentration the number of d-electrons is

$$n = \frac{n_0 + vx}{1 - x}.$$

Now, for fixed U, v and given n we calculate the magnetic moment μ_{Fe}. There is a critical number of electrons above which no magnetic solution exists; so the value of v determines the critical concentration x_c. In each case we set v to reproduce the observed value $x_c = 0.6$.

4. Results and Discussion

Our experimental results for the magnetic moment of Fe and results of other authors are presented in Fig. 2. Our data are in good agreement with Buschow and van Engen's measurements [3] for coevaporated films and are smaller than the experimental points for ribbons [8, 13] in the range of B concentrations x from 0.10 to 0.28 or for sputtered films with x from 0.15 to 0.40 [7]. All data for amorphous samples show that the crystal values [14, 15] are higher. Generally, the experimental points are scattered and the observed differences are discussed in many papers [16 to 18]. Discrepancies can be caused, especially for films, by experimental errors of magnetization and volume measurements. The estimated maximum error is up to 10%. The accuracy for the concentration is about 5%. Also some strong changes in chemical short-range order observed in Mössbauer experiment for amorphous iron alloys prepared by different methods (such as quenched from gaseous or liquid phases) [16, 18] can be responsible for discrepancies in the magnetic moment of Fe.

In the same Fig. 2 the line represents the best fits obtained for two DOS presented in Fig. 1 and the resulting transfer of electrons from each B atom into the 3d band, $v = 1.5$. Fig. 3 illustrates how different starting parameters influence the $\mu_{Fe}(x)$

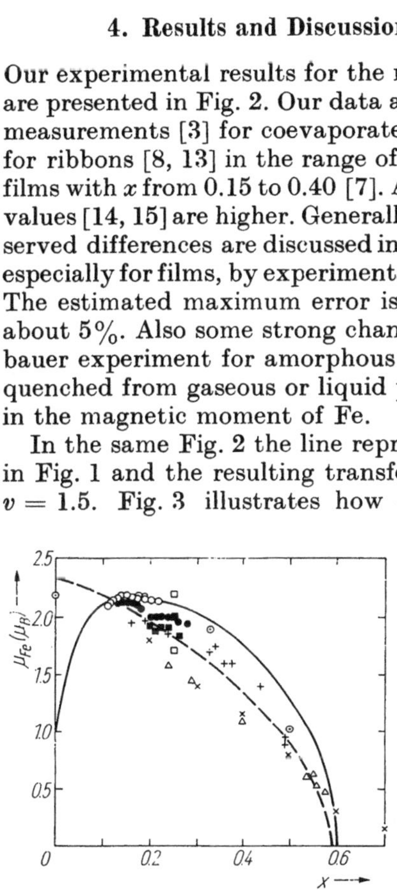

Fig. 2

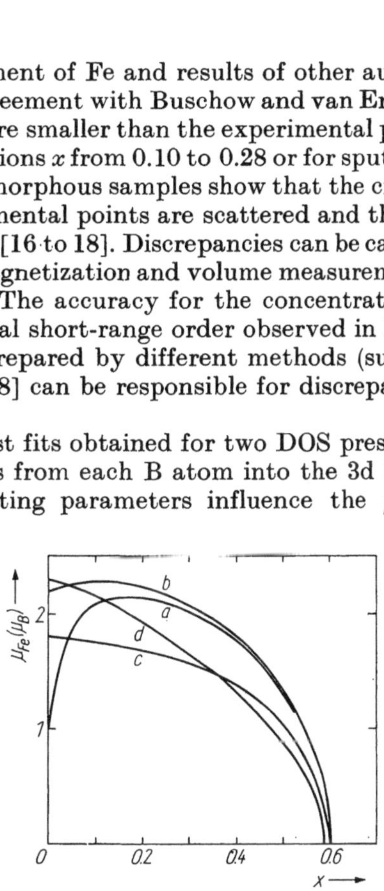

Fig. 3

Fig. 2. Magnetic moment μ_{Fe} per Fe atom vs. concentration x in amorphous $Fe_{1-x}B_x$ alloys. Experimental points are taken from different authors: + [7] films, × [3] films, △ [20] films, ■ [7] ribbons, ○ [13] ribbons, ● [4] ribbons, ⊙ [15] for crystalline Fe, Fe_2B, and FeB, □ [14] for metastable Fe_3B. The solid line is the best fit within the rigid model with simplified density of states in Fig. 1, with $n_0 = 6.8$, $\mu_0 = 1.0\mu_B$, and $v = 1.5$. The dashed line is the best fit using Wakoh's density of states with $n_0 = 7.2$, $\mu_0 = 2.3\mu_B$, and $v = 1.5$

Fig. 3. Some typical calculated $\mu_{Fe}(x)$ curves with the simplified density of states for (a) $n_0 = 6.8$, $\mu_0 = 1.0\mu_B$, $v = 1.5$; (b) $n_0 = 6.8$, $\mu_0 = 2.2\mu_B$, $v = 1.5$; (c) $n_0 = 7.4$, $\mu_0 = 1.8\mu_B$, $v = 1.0$; (d) is for the density of states after Wakoh and Yamashita [10] for $n_0 = 7.2$, $\mu_0 = 2.3\mu_B$, $v = 1.5$

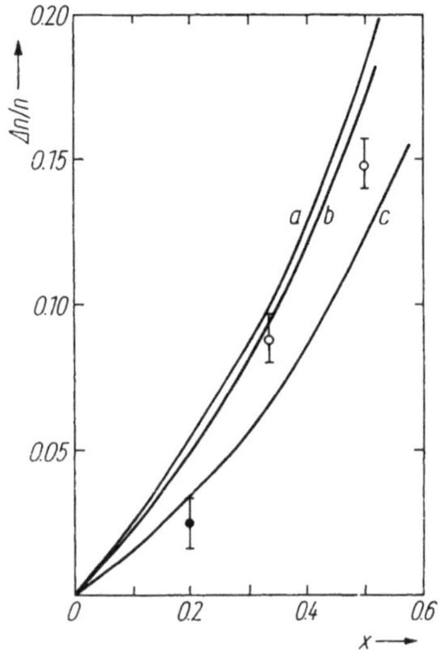

Fig. 4. The relative charge transfer $\Delta n/n$ for (a) $v = 1.5$, $n_0 = 6.8$; (b) $v = 1.5$, $n_0 = 7.2$; (c) $v = 1.0$, $n_0 = 7.2$. The experimental values are taken from Mausli et al. [19]. Full circle amorphous Fe–B, open circles crystalline Fe–B

dependence. The plot of the predicted relative charge transfer $\Delta n/n$ versus concentration x is given in Fig. 4 together with some X-ray data [19].

We conclude that the rigid band model gives a rough, yet qualitatively correct picture of the $\mu_{\text{Fe}}(x)$ dependence. Details of the density of states play a role, but the general features of $\mu_{\text{Fe}}(x)$ remain similar. Different starting parameters n_0 and μ_0 do not change much the $\mu_{\text{Fe}}(x)$ dependence for higher concentrations; major differences occur for small x where no experimental points are available.

It should be mentioned that the characteristic maximum of the $\mu_{\text{Fe}}(x)$ experimental data after Fukamichi et al. [13] also appears in our calculations (see in Fig. 2). The estimated charge transfer lies, at its largest range, between 1.0 and 1.5 electrons per B atom. This sounds reasonable when compared with $v = 1$ evaluated in [7].

Acknowledgements

The work was partly supported by Institute of Physics of the Polish Academy of Sciences. One author (T. S.) expresses his thanks to Prof. H. Hoffmann of Regensburg University for the possibility of performing low-temperature measurements of magnetization and to Dr. F. Stobiecki for the help in sample preparation and SQUID investigations.

References

[1] F. Stobiecki, S. Schwarzl, T. Stobiecki, and H. Hoffmann, J. Physique C 8, 171 (1980).
[2] C. L. Chien and K. M. Unruh, Phys. Rev. B **24**, 1556 (1981).
[3] K. H. J. Buschow and P. G. van Engen, J. appl. Phys. **52**, 3557 (1981).
[4] R. Hasegawa and R. Ray, J. appl. Phys. **49**, 4174 (1978).
[5] F. E. Luborsky, J. L. Watter, and E. P. Wohlfarth, J. Phys. F **10**, 959 (1980).
[6] J. Durand and M. Yung, in: Amorphous Magnetism A, Ed. R. A. Levy and R. Hasegawa, Plenum Press, New York 1977 (p. 275).
[7] G. Beyreuther, G. Enders, H. Hoffmann, U. Korndorfer, W. Oestereicher, K. Röll, and M. Takahashi, J. Magnetism magnetic Mater. **31/34**, 1535 (1983).
[8] E. Colavite, M. De Cresconzi, L. Papagno, L. S. Capuli, G. Chiarello, R. Scarmozzino, and R. Rosei, Solid State Commun. **41**, 545 (1982).
[9] J. H. Wood, Phys. Rev. **126**, 517 (1962).
[10] S. Wakoh and J. Yamashita, J. Phys. Soc. Japan **19**, 1342 (1964).
[11] G. M. Stocks, R. W. Willams, and J. S. Faulkner, Phys. Rev. B **4**, 4390 (1971).
[12] E. C. Snow and J. T. Waber, Acta metall. **17**, 623 (1969).

[13] K. Fukamichi, M. Kikuchi, H. Hiroyoshi, and T. Masumoto, Rapidly Quenched Metals, Vol. 3, Brigton 1978.
[14] I. Vincze, D. S. Boudreaux, and M. Tagze, Phys. Rev. B 19, 4896 (1979).
[15] O. Johnoson, D. J. Joyner, and D. M. Hercules, J. phys. Chem. 84, 542 (1980).
[16] J. Korecki, T. Stobiecki, M. Przybylski, and F. Stobiecki, Nuclear Instrum. and Methods 199, 215 (1982).
[17] N. A. Blum, K. Moorjam, T. O. Pochler, and F. G. Satkiewicz, J. appl. Phys. 52, 1808 (1981).
[18] A. M. van der Kraan and K. H. J. Buschow, Phys. Rev. B 27, 2693 (1982).
[19] P. A. Mausli, G. Burri, S. G. Steinemann, and K. Suzuki, Proc. 4th Internat. Conf. Rapidly Quenched Metals, August 24 to 28, Sendai, Japan.
[20] T. Stobiecki and F. Stobiecki, J. Magnetism magnetic Mater. 35, 217 (1983).

(Received May 14, 1984)

phys. stat. sol. (b) **126**, 197 (1984)

Subject classification: 13.5.4 and 18.4; 22.5

I. V. Kurchatov Institute of Atomic Energy, Moscow[1])

Interactions of Magnons in the Six-Sublattice Antiferromagnet

By

V. L. Safonov

A microscopic model of the six-sublattice antiferromagnet $CsMnF_3$ is considered. Spectra and three-magnon amplitudes of the electronic and nuclear spin waves are calculated. It is shown that in the multisublattice model there are new processes of the low-frequency spin-wave interactions (parallel with other processes that take place in the two-sublattice model too). Contributions of these processes should be taken into account in the theoretical description of nonlinear properties of $CsMnF_3$. Results obtained are compared with experimental data.

Проведен аналитический расчет спектров и амплитуд взаимодействий электронных и ядерных спиновых волн для микроскопической модели шестиподрешеточного антиферромагнетика $CsMnF_3$. Показано, что в этой модели для спиновых волн низкочастотных ветвей спектра наряду с взаимодействиями, присущими двухподрешеточному антиферромагнетику, возникают новые процессы взаимодействия спиновых волн, вклад которых необходимо учитывать при анализе нелинейных свойств $CsMnF_3$. Рассмотрены трехчастичные процессы взаимодействия. Проведено сравнение с экспериментом.

1. Introduction

Theoretically properties of antiferromagnets (AF) in the low-temperature regime ($T \ll T_N$) are usually described in the framework of the two-sublattice model since the effect of the exchange modes in this case is negligible, moreover calculations in this approximation have relatively simple form. However, in the two-sublattice model a priori it is supposed that all magnetic atoms have the same local symmetry as far as in AF, the magnetic cell of which contains more than two magnetic units, the above supposition is incorrect. Taking into account the inequivalence of positions of the magnetic atoms one may find a set of effects (including effects with low-frequency spin waves) which are absent in the two-sublattice antiferromagnet approximation.

In the present paper we calculate spectra and interactions of the electronic spin waves (ESW) and the nuclear spin waves (NSW) in the six-sublattice AF. We assume as a basis the weak-anisotropy AF $CsMnF_3$ which is already the classical object of investigations. This crystal has the spatial group D_{6h}^4 and its magnetic cell contains six Mn^{2+} ions (see Fig. 1) with the following local symmetry: $Mn_{1,4}$: D_{3d} and $Mn_{2,3,5,6}$: C_{3v} [1 to 4].

In the recent work [5] a new process of interaction between NSW and the two quasi-ferromagnetic (f-) ESW in $CsMnF_3$ was concisely communicated. This process is due to the exactly many-sublattice magnetic structure of the AF. The present paper contains the development of the theoretical part of [5].

---

[1]) 123182 Moscow, USSR.

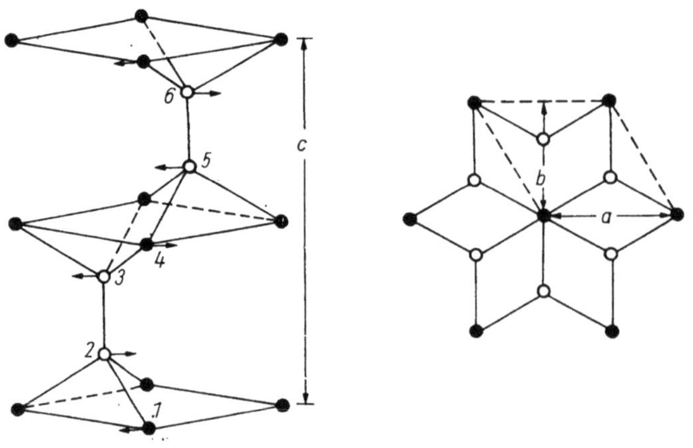

Fig. 1. The magnetic cell of $CsMnF_3$. $a = 0.6213$ nm, $c = 15.074$ nm $b = 5.381$ nm

2. Hamiltonian

The magnetic Hamiltonian of $CsMnF_3$ may be written in the following form:

$$\begin{aligned}\mathcal{H} = \sum [&J_1(\mathbf{S}_1\mathbf{S}_2 + \mathbf{S}_1\mathbf{S}_6 + \mathbf{S}_4\mathbf{S}_3 + \mathbf{S}_4\mathbf{S}_5) + \\ &+ J_2(\mathbf{S}_2\mathbf{S}_3 + \mathbf{S}_5\mathbf{S}_6) - J_3(\mathbf{S}_3\mathbf{S}_5 + \mathbf{S}_2\mathbf{S}_6)] + \\ &+ K_1 \sum_{j=1,4} (S_j^z)^2 + K_2 \sum_{j=2,3,5,6} (S_j^z)^2 - \\ &- A_1 \sum_{j=1,4} \mathbf{S}_j\mathbf{I}_j - A_2 \sum_{j=2,3,5,6} \mathbf{S}_j\mathbf{I}_j - \\ &- \mathbf{H} \sum_{j=1}^{6} (g\mu_B\mathbf{S}_j - g_N\mu_N\mathbf{I}_j)\,,\end{aligned} \tag{1}$$

where the first sum describes intersublattice exchange interactions between nearest neighbours ($J_1/k_B = (6.43 \pm 0.14)$ K, $J_2/k_B = (4.51 \pm 0.51)$ K, $J_3/k_B = (0.66 \pm \pm 0.15)$ K [4]); K_1, K_2 are the phenomenological anisotropy constants; A_1, A_2 are the hyperfine constants ($A_1S/2\pi\hbar = 677$ MHz, $A_2S/2\pi\hbar = 666$ MHz [3]); the last sum in (1) is the Zeeman energy; g and g_N are the electronic and nuclear gyromagnetic ratios for the electron shell (Mn^{2+}) spins $\mathbf{S}_\nu$ and nuclear (^{55}Mn) spins $\mathbf{I}_\nu$, respectively ($\nu = 1, \dots, 6$; $S = I = S/2$).

When the spin operators $\mathbf{S}_\nu$ are expanded in terms of Holstein-Primakoff boson a_ν^+ and a_ν ones from (1) we obtain $\mathcal{H} = E_0 + \mathcal{H}^{(1)} + \mathcal{H}^{(2)} + \mathcal{H}^{(3)} + \dots$, where

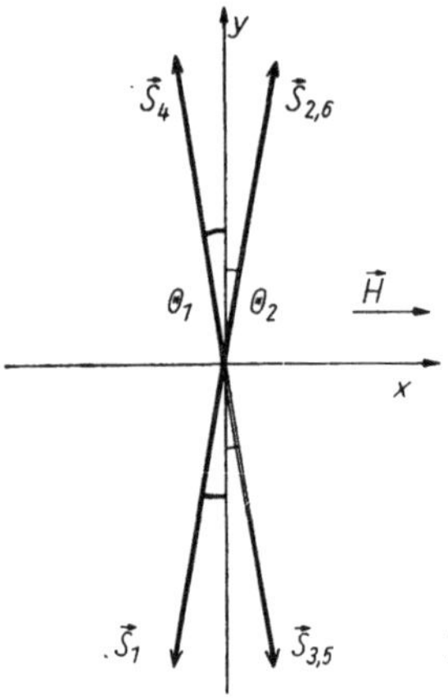

Fig. 2. The equilibrium configuration of the six-sublattice AF ($T \ll T_N$)

each term contains an even number of boson operators. Static orientation of spins (see Fig. 2) one finds from $\mathcal{H}^{(1)} = 0$. This equation may be written as

$$\begin{aligned} &2J_1S\, z_{10} \sin(\theta_1 + \theta_2) = g\mu_B H \cos\theta_1\,, \\ &J_1Sz_{10} \sin(\theta_1 + \theta_2) + J_2Sz_{30} \sin 2\theta_2 = g\mu_B H \cos\theta_2 \end{aligned} \tag{2}$$

($z_{10} = 3$, $z_{30} = 1$ are the numbers of neighbours).
The solution of (2) in moderate magnetic field is

$$\theta_1 \approx \frac{1}{2}\left(1 - \frac{1}{2\lambda}\right)\frac{g\mu_B H}{J_1Sz_{10}}\,, \qquad \theta_2 \approx \frac{1}{4\lambda}\,\frac{g\mu_B H}{J_1Sz_{10}}\,; \tag{3}$$

$$\lambda \equiv \frac{J_2Sz_{30}}{J_1Sz_{10}} \approx 0.234\,.$$

The quadratic part of the Hamiltonian in the Fourier space has the form

$$\begin{aligned} \mathcal{H}^{(2)} = \sum_k \Big\{ &\tilde{A}_{1k}(a_{1k}^+a_{1k} + a_{4k}^+a_{4k}) + \tilde{A}_{2k}(a_{2k}^+a_{2k} + a_{3k}^+a_{3k} + a_{5k}^+a_{5k} + a_{6k}^+a_{6k}) + \\ &+ \Big[\frac{1}{2}K_1S(a_{1k}a_{1-k} + a_{4k}a_{4-k}) + \\ &+ \frac{1}{2}K_2S(a_{2k}a_{2-k} + a_{3k}a_{3-k} + a_{5k}a_{5-k} + a_{6k}a_{6-k}) + \\ &+ J_1S\sin^2\frac{\theta_1+\theta_2}{2}(z_{1k}a_{1k}^+a_{2k} + z_{2k}a_{4k}^+a_{3k} + z_{1k}a_{1k}^+a_{6k} + z_{2k}a_{4k}^+a_{5k}) + \\ &+ J_1S\cos^2\frac{\theta_1+\theta_2}{2}(z_{1k}a_{1k}a_{2-k} + \\ &+ z_{2k}a_{4k}a_{3-k} + z_{1k}a_{1k}a_{6-k} + z_{2k}a_{4k}a_{5-k}) + \\ &+ J_2Sz_{3k}\cos^2\theta_2(a_{2k}a_{3-k} + a_{5k}a_{6-k}) + \\ &+ J_2Sz_{3k}\sin^2\theta_2(a_{2k}^+a_{3k} + a_{5k}^+a_{6k}) - \\ &- J_3Sz_{4k}a_{2k}^+a_{6k} - J_3Sz_{5k}a_{3k}^+a_{5k} + \text{h.c.}\Big]\Big\}, \end{aligned} \tag{4}$$

where

$$\begin{aligned} \tilde{A}_{1k} &= 6J_1S\cos(\theta_1+\theta_2) + K_1S + A_1\langle I_1\rangle + g\mu_B H\sin\theta_1\,, \\ \tilde{A}_{2k} &= 3J_1S\cos(\theta_1+\theta_2) + J_2S\cos 2\theta_2 + 3J_3S + \\ &\quad + K_2S + A_2\langle I_2\rangle + g\mu_B H\sin\theta_2\,; \end{aligned} \tag{5}$$

z_{jk} are the coordination sums,

$$\begin{aligned} z_{\binom{1}{2}k} &= \cos[(\boldsymbol{n}_1 + \boldsymbol{n}_2 \mp \boldsymbol{n}_3)\,\boldsymbol{k}] + \\ &\quad + \cos[(-\boldsymbol{n}_1 + \boldsymbol{n}_2 \mp \boldsymbol{n}_3)\,\boldsymbol{k}] + \cos[(2\boldsymbol{n}_2 \pm \boldsymbol{n}_3)\,\boldsymbol{k}]\,, \end{aligned} \tag{6a}$$

$$z_{3k} = \cos(\xi_c ck)\,; \qquad \xi_c \approx 0.2\,, \tag{6b}$$

$$\begin{aligned} z_{\binom{4}{5}k} &= \cos[(\boldsymbol{n}_1 + \boldsymbol{n}_2 \pm 2\boldsymbol{n}_3)\,\boldsymbol{k}] + \\ &\quad + \cos[(\boldsymbol{n}_1 - \boldsymbol{n}_2 \mp 2\boldsymbol{n}_3)\,\boldsymbol{k}] + \cos[(2\boldsymbol{n}_2 \mp 2\boldsymbol{n}_3)\,\boldsymbol{k}]\,; \end{aligned} \tag{6c}$$

$\boldsymbol{n}_1 = \boldsymbol{a}/2$, $\boldsymbol{n}_2 = \boldsymbol{b}/3$, $\boldsymbol{n}_3 = u\boldsymbol{c}$, $u \approx 0.15$; $\boldsymbol{a}, \boldsymbol{b}, \boldsymbol{c}$ are the basis vectors in Fig. 1. The quantities $\langle I_1 \rangle$ and $\langle I_2 \rangle$ denote mean values of the nuclear spin operators

$$\langle I_\nu \rangle \approx \frac{I(I+1)}{3} \frac{A_\nu S}{k_B T} \qquad (\nu = 1, 2) \,. \tag{7}$$

Interactions of spin waves with each other are defined by $\mathcal{H}^{(n)} (n > 2)$. Later we shall consider the three-magnon processes which are expressed in the spin deviation operators as follows:

$$\begin{aligned} \mathcal{H}^{(3)} = i \left(\frac{S}{2N}\right)^{1/2} \sum_{\boldsymbol{k}_1, \boldsymbol{k}_2, \boldsymbol{k}_3} & (J_1 \sin(\theta_1 + \theta_2) \{z_{11}[(a_{11} - a_{11}^+) \times \\ & \times (a_{22}^+ a_{23} + a_{62}^+ a_{63}) - (a_{21} - a_{21}^+ + a_{61} - a_{61}^+) a_{12}^+ a_{13}] - \\ & - z_{21}[(a_{41} - a_{41}^+) (a_{32}^+ a_{33} + a_{52}^+ a_{53}) - (a_{31} - a_{31}^+ + a_{51} - a_{51}^+) a_{42}^+ a_{43}]\} - \\ & - J_2 \sin 2\theta_2 z_{31} [(a_{21} - a_{21}^+) a_{32}^+ a_{33} - (a_{31} - a_{31}^+) a_{22}^+ a_{23} + \\ & + (a_{61} - a_{61}^+) a_{52}^+ a_{53} - (a_{51} - a_{51}^+) a_{62}^+ a_{63}]) \, \Delta(\boldsymbol{k}_1 + \boldsymbol{k}_2 + \boldsymbol{k}_3) \,. \end{aligned} \tag{8}$$

3. Diagonalization of the Quadratic Part

For simplicity we devide the diagonalization of $\mathcal{H}^{(2)}$ into some stages. After intermediate canonical transformation

$$\begin{aligned} a_1 &= \frac{1}{\sqrt{2}} (\tilde{a}_1 + \tilde{a}_4) \,, \qquad a_4 = \frac{1}{\sqrt{2}} (\tilde{a}_1 - \tilde{a}_4) \,, \\ a_2 &= \frac{1}{2} (\tilde{a}_3 + \tilde{a}_2 + \tilde{a}_5 - \tilde{a}_6) \,, \\ a_3 &= \frac{1}{2} (\tilde{a}_3 - \tilde{a}_2 + \tilde{a}_5 + \tilde{a}_6) \,, \\ a_5 &= \frac{1}{2} (\tilde{a}_3 - \tilde{a}_2 - \tilde{a}_5 - \tilde{a}_6) \,, \\ a_6 &= \frac{1}{2} (\tilde{a}_3 + \tilde{a}_2 - \tilde{a}_5 + \tilde{a}_6) \,, \end{aligned} \tag{9}$$

the quadratic part may be written in the form

$$\mathcal{H}^{(2)} = \mathcal{H}_{12}^{(2)} + \mathcal{H}_{34}^{(2)} + \mathcal{H}_{56}^{(2)} + V^{(2)} \,, \tag{10}$$

where the lower indexes in $\mathcal{H}_{ij}^{(2)}$ show its dependence on the operators $a_{i\boldsymbol{k}}^+$, $a_{i\boldsymbol{k}}$, and $a_{j\boldsymbol{k}}^+$, $a_{j\boldsymbol{k}}$ (the evident forms of $\mathcal{H}_{ij}^{(2)}$ will be written a little later). The interference term $V^{(2)}$ is

$$\begin{aligned} V^{(2)} = \sum_{\boldsymbol{k}} \Bigg\{ & \frac{1}{\sqrt{2}} J_1 S (z_{1\boldsymbol{k}} - z_{2\boldsymbol{k}}) \left[\cos^2 \frac{\theta_1 + \theta_2}{2} (\tilde{a}_{1\boldsymbol{k}} \tilde{a}_{2-\boldsymbol{k}} + \tilde{a}_{3\boldsymbol{k}} \tilde{a}_{4-\boldsymbol{k}}) + \right. \\ & \left. + \sin^2 \frac{\theta_1 + \theta_2}{2} (\tilde{a}_{1\boldsymbol{k}}^+ \tilde{a}_{2\boldsymbol{k}} + \tilde{a}_{3\boldsymbol{k}}^+ \tilde{a}_{4\boldsymbol{k}}) + \text{h.c.} \right] - \\ & - \frac{1}{2} J_3 S (z_{4\boldsymbol{k}} - z_{5\boldsymbol{k}}) (\tilde{a}_{2\boldsymbol{k}}^+ \tilde{a}_{3\boldsymbol{k}} + \text{h.c.}) \Bigg\} . \end{aligned} \tag{11}$$

Note that for $\boldsymbol{k} = \boldsymbol{0}$ we have $V^{(2)} = 0$. Hence in this case six modes are devided in pairs and the diagonalization procedure for the two interacting modes is well known [6]. Thus, the role of $V^{(2)} \sim (ak)^2$ may be defined as a perturbation of normal modes.

Let us rewrite the main formulae for the diagonalization of the quadratic form

$$\mathcal{H}_{12}^{(2)} = \sum_{\boldsymbol{k}} \{\mathcal{A}_{1\boldsymbol{k}} a_{1\boldsymbol{k}}^+ a_{1\boldsymbol{k}} + \mathcal{A}_{2\boldsymbol{k}} a_{2\boldsymbol{k}}^+ a_{2\boldsymbol{k}} + [\tfrac{1}{2} \mathcal{B}_{1\boldsymbol{k}} a_{1\boldsymbol{k}}^+ a_{1-\boldsymbol{k}}^+ + \\ + \tfrac{1}{2} \mathcal{B}_{2\boldsymbol{k}} a_{2\boldsymbol{k}}^+ a_{2-\boldsymbol{k}}^+ + \mathcal{C}_{\boldsymbol{k}} a_{1\boldsymbol{k}}^+ a_{2-\boldsymbol{k}}^+ + \mathcal{D}_{\boldsymbol{k}} a_{1\boldsymbol{k}}^+ a_{2\boldsymbol{k}} + \text{h.c.}]\} \tag{12}$$

by means of the canonical transformation

$$a_{j\boldsymbol{k}} = U_{j1} c_{1\boldsymbol{k}} + V_{j1}^* c_{1-\boldsymbol{k}}^+ + U_{j2} c_{2\boldsymbol{k}} + V_{j2}^* c_{2-\boldsymbol{k}}^+ \tag{13}$$

into thc form

$$\mathcal{H}_{12}^{(2)} = \Delta E_0 + \sum_{\boldsymbol{k}} (\varepsilon_{1\boldsymbol{k}} c_{1\boldsymbol{k}}^+ c_{1\boldsymbol{k}} + \varepsilon_{2\boldsymbol{k}} c_{2\boldsymbol{k}}^+ c_{2\boldsymbol{k}}) \,, \tag{14}$$

where $\varepsilon_{j\boldsymbol{k}}$ $(j = 1, 2)$ is the spin-wave energy,

$$\begin{aligned} \varepsilon_{j\boldsymbol{k}} &= \{\tfrac{1}{2} (\mathcal{A}_{1\boldsymbol{k}}^2 + \mathcal{A}_{2\boldsymbol{k}}^2 - |\mathcal{B}_{1\boldsymbol{k}}|^2 - |\mathcal{B}_{2\boldsymbol{k}}|^2) + |\mathcal{D}_{\boldsymbol{k}}|^2 - |\mathcal{C}_{\boldsymbol{k}}|^2 + (-1)^j P\}^{1/2} \,; \\ P^2 &= [(\mathcal{A}_{1\boldsymbol{k}} + \mathcal{A}_{2\boldsymbol{k}}) \mathcal{D}_{\boldsymbol{k}} - \mathcal{C}_{\boldsymbol{k}}^* (\mathcal{B}_{1\boldsymbol{k}}^* + \mathcal{B}_{2\boldsymbol{k}}^*)]^2 - \\ &\quad - [(\mathcal{A}_{1\boldsymbol{k}} - \mathcal{A}_{2\boldsymbol{k}}) \mathcal{C}_{\boldsymbol{k}}^* - \mathcal{D}_{\boldsymbol{k}} (\mathcal{B}_{1\boldsymbol{k}}^* - \mathcal{B}_{2\boldsymbol{k}}^*)]^2 + \\ &\quad + \tfrac{1}{4} (\mathcal{A}_{1\boldsymbol{k}}^2 - \mathcal{A}_{2\boldsymbol{k}}^2 - |\mathcal{B}_{1\boldsymbol{k}}|^2 + |\mathcal{B}_{2\boldsymbol{k}}|^2)^2 \,. \end{aligned} \tag{15}$$

The coefficients U, V are defined by [6]

$$\begin{aligned} U_{1j} &= -\frac{\Delta_{11}(\varepsilon_{j\boldsymbol{k}})}{\Lambda_j}, \qquad V_{1j} = \frac{\Delta_{12}(\varepsilon_{j\boldsymbol{k}})}{\Lambda_j}, \\ U_{2j} &= -\frac{\Delta_{13}(\varepsilon_{j\boldsymbol{k}})}{\Lambda_j}, \qquad V_{2j} = \frac{\Delta_{14}(\varepsilon_{j\boldsymbol{k}})}{\Lambda_j}, \\ \Lambda_j &= [(-1)^j \, 2\varepsilon_{j\boldsymbol{k}} (\varepsilon_{1\boldsymbol{k}}^2 - \varepsilon_{2\boldsymbol{k}}^2) \, \Delta_{11}(\varepsilon_{j\boldsymbol{k}})]^{1/2} \exp(i\varphi_j) \,, \end{aligned} \tag{16}$$

where $\Delta_{1j}(\varepsilon)$ is the minor of the $(1, j)$ element (of the first line and the j-th column) for the characteristic determinant of the form (12).

The coefficients of $\mathcal{H}_{13}^{(2)}$ $(\varkappa = 1)$ and $\mathcal{H}_{24}^{(2)}$ $(\varkappa = -1)$ are

$$\begin{aligned} \mathcal{A}_{1\boldsymbol{k}} &= \tilde{A}_{1\boldsymbol{k}} \,, \\ \mathcal{A}_{2\boldsymbol{k}} &= \tilde{A}_{2\boldsymbol{k}} + \varkappa J_2 S z_{3\boldsymbol{k}} \sin^2 \theta_2 - \frac{1}{2} J_3 S (z_{4\boldsymbol{k}} + z_{5\boldsymbol{k}}) \,, \\ \mathcal{B}_{1\boldsymbol{k}} &= K_1 S \,, \qquad B_{2\boldsymbol{k}} = K_2 S + \varkappa J_2 S z_{3\boldsymbol{k}} \cos^2 \theta_2 \,, \\ \mathcal{C}_{\boldsymbol{k}} &= \frac{1}{\sqrt{2}} J_1 S (z_{1\boldsymbol{k}} + z_{2\boldsymbol{k}}) \cos^2 \frac{\theta_1 + \theta_2}{2}, \\ \mathcal{D}_{\boldsymbol{k}} &= \frac{1}{\sqrt{2}} J_1 S (z_{1\boldsymbol{k}} + z_{2\boldsymbol{k}}) \sin^2 \frac{\theta_1 + \theta_2}{2}. \end{aligned} \tag{17}$$

For $\mathcal{H}_{56}^{(2)}$ they are $(\nu = 1, 2)$

$$\begin{aligned} \mathcal{A}_{\nu\boldsymbol{k}} &= \tilde{A}_{2\boldsymbol{k}} + J_2 S z_{3\boldsymbol{k}} \sin^2 \theta_2 - (-1)^\nu J_3 S (z_{4\boldsymbol{k}} + z_{5\boldsymbol{k}}) /2 \,, \\ \mathcal{B}_{\nu\boldsymbol{k}} &= K_2 S - (-1)^\nu J_2 S z_{3\boldsymbol{k}} \cos^2 \theta_2 \,, \\ \mathcal{C}_{\boldsymbol{k}} &= 0 \,, \qquad \mathcal{D}_{\boldsymbol{k}} = -J_3 S (z_{4\boldsymbol{k}} - z_{5\boldsymbol{k}}) \,. \end{aligned} \tag{18}$$

Spectra of the ESW are obtained by substituting (17) and (18) into (15),

$$\begin{aligned}
\varepsilon_{1k} &= [(g\mu_B)^2 (H^2 + H_\Delta^2) + f(\boldsymbol{k})]^{1/2} , \\
\varepsilon_{2k} &= [(g\mu_B)^2 (2H_E H_A + H_\Delta^2) + f(\boldsymbol{k})]^{1/2} , \\
\varepsilon_{3k} &\approx \varepsilon_{4k} \approx \varepsilon_{5k} \approx J_1 S z_{10} \sqrt{1 + 2\lambda} , \\
\varepsilon_{6k} &\approx J_1 S z_{10} \sqrt{(1 + 2\lambda + 4\lambda_1)(1 + 4\lambda_1)} ,
\end{aligned} \tag{19}$$

where

$$\begin{aligned}
(g\mu_B H_\Delta)^2 &= \frac{4\lambda}{1 + 2\lambda} J_1 S z_{10} (A_1 \langle I_1 \rangle + 2A_2 \langle I_2 \rangle) , \\
g\mu_B H_E &= \frac{6\lambda}{1 + 2\lambda} J_1 S z_{10} , \\
g\mu_B H_A &= \frac{2}{3} (K_1 + 2K_2) S , \\
f(\boldsymbol{k}) &= \frac{8\lambda}{1 + 2\lambda} (J_1 S z_{10})^2 \frac{1}{z_{10}} [(z_{10} - z_{1k}) + (z_{20} - z_{2k}) + \\
&\quad + \lambda_1 (z_{40} - z_{4k}) + \lambda_1 (z_{50} - z_{5k})] , \\
\lambda_1 &\equiv \frac{J_3 S z_{40}}{2 J_1 S z_{10}} \approx 0.051 .
\end{aligned} \tag{20}$$

Substitution of the experimental data into (20) leads to the following results:

$$\begin{aligned}
H_\Delta^2 &= \frac{6.15 \pm 0.7}{T\,[\mathrm{K}]} \mathrm{kOe}^2 , \\
\sqrt{2H_E H_A} &= (47 \pm 7)\ \mathrm{kOe} , \\
f(\boldsymbol{k})/(g\mu_B)^2 &\approx (\alpha_\perp k_\perp)^2 + (\alpha_\parallel k_\parallel)^2 , \\
\alpha_\parallel &\approx \alpha_\perp = (1.0 \pm 0.13) \times 10^{-5} \mathrm{kOe\ cm} .
\end{aligned} \tag{21}$$

Here $k_\perp$, $k_\parallel$ are the wave number components in the easy plane and in the hard axis; $H_A = (2.7 \pm 0.9)$ kOe [4].

Corresponding experimental values are equal to

$$\begin{aligned}
H_\Delta^2 &= \frac{6.4 \pm 0.2}{T\,[\mathrm{K}]} \mathrm{kOe}^2 \quad [2] , \\
\sqrt{2H_E H_A} &= (41.1 \pm 0.6)\ \mathrm{kOe} \quad [7] , \\
\alpha_\perp &\approx (0.95 \pm 0.1) \times 10^{-5} \mathrm{kOe\ cm} \quad [8] .
\end{aligned} \tag{22}$$

Thus, there is good agreement between the parameters of ESW spectra, calculated from the data of inelastic neutron scattering [4] and the data of NMR experiments [3], with analogous ones obtained by AFMR and parallel pumping investigations.

Now we compute the coefficients U, V using formulae (16) and (17),

$$\begin{aligned}
\tilde{a}_{1k} &= p_{1k}(\tilde{c}_{1k} - \tilde{c}_{1-k}^+) - \xi_1 p_{1k}^{-1}(\tilde{c}_{1k} + \tilde{c}_{1-k}^+) + \dots , \\
\tilde{a}_{3k}/\sqrt{2} &= p_{1k}(\tilde{c}_{1k} - \tilde{c}_{1-k}^+) + \xi_2 p_{1k}^{-1}(\tilde{c}_{1k} + \tilde{c}_{1-k}^+) + \dots , \\
\tilde{a}_{2k} &= p_{2k}(\tilde{c}_{2k} + \tilde{c}_{2-k}^+) - \xi_1 p_{2k}^{-1}(\tilde{c}_{2k} - \tilde{c}_{2-k}^+) + \dots , \\
\tilde{a}_{4k}/\sqrt{2} &= p_{2k}(\tilde{c}_{2k} + \tilde{c}_{2-k}^+) + \xi_2 p_{2k}^{-1}(\tilde{c}_{2k} - \tilde{c}_{2-k}^+) + \dots , \\
p_{\nu k} &\equiv \left[\frac{\lambda J_1 S z_{10}}{(1 + 2\lambda)\,\varepsilon_{\nu k}}\right]^{1/2} \quad (\nu = 1, 2) , \\
\xi_1 &\equiv \frac{1 - 2\lambda}{4(1 + 2\lambda)} , \qquad \xi_2 \equiv \frac{1}{2(1 + 2\lambda)} .
\end{aligned} \tag{23}$$

Here the first and the second low-frequency modes of ESW are taken into account only (because the other modes have exponentially small thermal equilibrium occupancy in the low-temperature regime).

In the final stage of diagonalization we consider the interference term (11) as perturbation. We obtain

$$\tilde{c}_{\nu k} \mp \tilde{c}^+_{\nu-k} = c_{\nu k} \mp c^+_{\nu-k} + u^{(\mp)}_{\nu k}(c_{\mu k} \mp c^+_{\mu-k})$$

$$(\nu = 1\,, \quad \mu = 2) \quad \text{and} \quad (\nu = 2\,, \quad \mu = 1)\,,$$

$$u^{(-)}_{1k} = -u^{(+)}_{2k} = J_1 S \frac{\sqrt{2\varepsilon_{1k}\varepsilon_{2k}}}{\varepsilon^2_{1k} + \varepsilon^2_{2k}} \left[(z_{1k} - z_{2k}) - 2\frac{\lambda_1(1-\lambda)}{1+2\lambda}(z_{4k} - z_{5k})\right], \tag{24}$$

$$u^{(+)}_{1k} = -u^{(-)}_{2k} = \frac{J_1 S}{\sqrt{2\varepsilon_{1k}\varepsilon_{2k}}} \frac{\varepsilon^2_{2k} - \varepsilon^2_{1k}}{\varepsilon^2_{2k} + \varepsilon^2_{1k}} \left[(z_{1k} - z_{2k}) - \right.$$

$$\left. - \frac{2\lambda_1}{1+2\lambda}\left(1 - 2\lambda\frac{\varepsilon^2_{2k}}{\varepsilon^2_{2k} - \varepsilon^2_{1k}}\right)(z_{4k} - z_{5k})\right].$$

Thus, the transformation coefficients of the above diagonalization "ab initio ad finem" are given by (9), (23), and (24). It should be noted that the above relatively visible calculations become possible by taking into consideration the symmetry properties of the magnetic Hamiltonian. In [9] the diagonalization is described with the help of group theory of the more complex magnetic system — the twenty-sublattice ferrimagnet YIG.

4. Three-Magnon Interaction

The three-magnon Hamiltonian (8) in terms of normal spin-wave modes has the form

$$\mathscr{H}^{(3)} = iN^{-1/2} \sum_{k_1,k_2,k_3} \frac{g\mu_B H}{\sqrt{8S}\, z_{10}} \left\{\frac{\sqrt{2}-1}{3}\hat{S}(z_{21} - z_{11})\, P_{11}P_{12}P_{13} - \right.$$

$$- \left[2(\lambda_2 - 1)(z_{11} + z_{21}) + \sqrt{2}\left(\lambda_2 + \frac{1}{2}\right)(z_{13} + z_{23}) + \right.$$

$$\left. + 3(2z_{31} - z_{33})\right] P_{11}P_{12}Q_{23} +$$

$$+ \sqrt{2}\,[(\lambda_2 + 2)(z_{11} + z_{21}) + 3\lambda_2 z_{31}]\, P_{11}Q_{12}P_{23} +$$

$$+ (z_{21} - z_{11})\left(2 - \frac{1}{\sqrt{2}}\right) P_{11}P_{22}P_{23} +$$

$$\left. + \left[\frac{\lambda + 2}{\sqrt{2}}(z_{11} + z_{21}) + \frac{3}{2} z_{31}\right] Q_{21}P_{22}P_{23}\right\} \Delta(\boldsymbol{k}_1 + \boldsymbol{k}_2 + \boldsymbol{k}_3)\,, \tag{25}$$

where

$$P_{\nu k} \equiv p_{\nu k}[c_{\nu k} + (-1)^\nu c^+_{\nu-k}]\,,$$

$$Q_{\nu k} \equiv \xi_1 p^{-1}_{\nu k}[c_{\nu k} - (-1)^\nu c^+_{\nu-k}]\,,$$

$$\lambda_2 \equiv \frac{\sqrt{2}}{1-2\lambda} \approx 2.66 \qquad (\nu = 1, 2)\,,$$

$\hat{S}$ is the symmetrization operator, $\hat{S}z_{\nu_1} = z_{\nu_1} + z_{\nu_2} + z_{\nu_3}$; $\nu = 1, 2$.

Note that the three-magnon Hamiltonian of the two-sublattice AF (see, for example, [10]) contains the interactions {1, 1, 2} and {2, 2, 2} only (the figures in {...} are the numbers of interacting ESW modes), whereas the Hamiltonian (25) contains the {1, 1, 1} and {1, 2, 2} terms, too. The last terms are due to the specificity of the many-sublattice AF, namely to the inequivalence of the exchange interactions in the different positions of the magnetic atoms. If one formally demands $z_{2\boldsymbol{k}} - z_{1\boldsymbol{k}} = 0$ in (25) then the rest of $\mathcal{H}^{(3)}$ will be quite analogous to the three-magnon Hamiltonian of the two sublattice AF.

It should be noted that the {1, 1, 1} three-magnon term in the two-sublattice AF appears by taking into consideration the dipole–dipole interaction [11]. However, its amplitude is exchange weakened. On the contrary, the analogous term in (25) is exchange enhanced.

5. Interaction with the Nuclear Spins

The hyperfine interaction between the electronic spin and the corresponding nuclear spin leads to the coupled electron–nuclear modes. The resulting normal-mode excitations in the NMR frequency region have been called nuclear spin waves [12, 13, 5, 14], we shall consider the lowest mode of NSW which has large frequency pulling. The terms of interest are

$$\mathcal{H}_{\text{en}} = \mathcal{H}_{\text{on}} + \mathcal{H}_{\text{int}}\,,$$

where

$$\mathcal{H}_{\text{on}} = -\hbar\omega_{\text{n}} m_0^z \tag{26}$$

and

$$\mathcal{H}_{\text{int}} = -\frac{\hbar\omega_{\text{n}}}{2\sqrt{N}} \sum_{\boldsymbol{k}} \left(\frac{\lambda}{1+2\lambda} \frac{J_1 z_{10}}{\varepsilon_{1\boldsymbol{k}}}\right)^{1/2} (c_{1\boldsymbol{k}} - c_{1-\boldsymbol{k}}^{+})\,(m_{\boldsymbol{k}}^{-} - m_{-\boldsymbol{k}}^{+})\,;$$

$$\hbar\omega_{\text{n}} \equiv S(A_1 + A_2)/2\,, \tag{27}$$

$$m_{\boldsymbol{k}}^{\alpha} = \sum_{\nu=1}^{6} I_{\nu\boldsymbol{k}}^{\alpha} \qquad (\alpha = \pm, z)\,,$$

$I_{\nu\boldsymbol{k}}^{\alpha}$ are the Fourier components of nuclear spins in their own systems of coordinates.

To divide the electron–nuclear coupling the following unitary transformation is used:[2])

$$\exp(Q_1)\,\mathcal{H}_{\text{en}}\exp(-Q_1) = \mathcal{H}_{\text{on}} + \mathcal{H}_{\text{SN}} + \dots\,;$$
$$Q_1 = \frac{1}{2\sqrt{N}} \sum_{\boldsymbol{k}} \frac{\hbar\omega_{\text{n}}}{\varepsilon_{1\boldsymbol{k}}} \left(\frac{\lambda}{1+2\lambda} \frac{J_1 z_{10}}{\varepsilon_{1\boldsymbol{k}}}\right)^{1/2} (c_{1\boldsymbol{k}} + c_{1-\boldsymbol{k}}^{+})\,(m_{\boldsymbol{k}}^{-} - m_{-\boldsymbol{k}}^{+})\,. \tag{28}$$

As a result one obtains the indirect Suhl-Nakamura Hamiltonian

$$\mathcal{H}_{\text{SN}} = N^{-1} \sum_{\boldsymbol{k}} V_{\boldsymbol{k}}^{(\text{SN})} (m_{\boldsymbol{k}}^{-} - m_{-\boldsymbol{k}}^{+})\,(m_{-\boldsymbol{k}}^{-} - m_{\boldsymbol{k}}^{+})\,;$$
$$V_{\boldsymbol{k}}^{(\text{SN})} = \frac{\lambda}{1+2\lambda} \frac{J_1 z_{10}}{4} \left(\frac{\hbar\omega_{\text{n}}}{\varepsilon_{1\boldsymbol{k}}}\right)^2 \tag{29}$$

---

[2]) The general method for finding the unitary transformations is described in [15].

which in turn is reduced to the diagonal form by means of the unitary transformation [5]

$$\exp(Q)(\mathcal{H}_{\rm on} + \mathcal{H}_{\rm SN})\exp(-Q) = -\hbar\omega_{\rm n} m_0^z + $$
$$+ N^{-1}\sum_{\boldsymbol{k}} \hbar \frac{\omega_{\rm n\boldsymbol{k}} - \omega_{\rm n}}{12\langle I\rangle} m_{\boldsymbol{k}}^- m_{\boldsymbol{k}}^+ + \dots ; \quad (30)$$

$$\langle I\rangle \approx \frac{I(I+1)}{3}\frac{\hbar\omega_{\rm n}}{k_{\rm B}T},$$

$$\omega_{\rm n\boldsymbol{k}} = \omega_{\rm n}\left[1 - \left(\frac{g\mu_{\rm B}H_\Delta}{\varepsilon_{1\boldsymbol{k}}}\right)^2\right]^{1/2},$$

$$Q = \frac{1}{48\langle I\rangle N}\sum_{\boldsymbol{k}} \ln\left(\frac{\omega_{\rm n}}{\omega_{\rm n\boldsymbol{k}}}\right)(m_{\boldsymbol{k}}^- m_{-\boldsymbol{k}}^- - m_{\boldsymbol{k}}^+ m_{-\boldsymbol{k}}^+) .$$

Note that the transformations (28) and (30) which diagonalize the coupled electron–nuclear Hamiltonian (27) mix the unperturbed electron mode with the nuclear components,

$$\exp(Q)\exp(Q_1)(c_{1\boldsymbol{k}} - c_{1-\boldsymbol{k}}^+)\exp(-Q_1)\exp(-Q) =$$
$$= c_{1\boldsymbol{k}} - c_{1-\boldsymbol{k}}^+ - \left(\frac{1}{N}\frac{\lambda}{1+2\lambda}\frac{J_1 z_{10}}{\varepsilon_{1\boldsymbol{k}}}\frac{\omega_{\rm n}}{\omega_{\rm n\boldsymbol{k}}}\right)^{1/2}\frac{\hbar\omega_{\rm n}}{\varepsilon_{1\boldsymbol{k}}}(m_{\boldsymbol{k}}^- - m_{-\boldsymbol{k}}^+) + \dots . \quad (31)$$

This circumstance leads to the origin of new interactions in (25). For example, one gets the interaction between one nuclear and two f-electronic magnons,

$$\mathcal{H}_{\rm n2f}^{(3)} = iN^{-1}\sum_{\boldsymbol{k}_1,\boldsymbol{k}_2,\boldsymbol{k}_3} \Psi_{\rm n2f}(\boldsymbol{k}_1,\boldsymbol{k}_2,\boldsymbol{k}_3)(c_{11} - c_{11}^+)(c_{12} - c_{12}^+) \times$$
$$\times (m_3^- - m_3^+)\,\Delta(\boldsymbol{k}_1 + \boldsymbol{k}_2 + \boldsymbol{k}_3) ,$$
$$\Psi_{\rm n2f}(1,2,3) = g\mu_{\rm B}H\left[\frac{(\sqrt{2}-1)S}{2^{3/2}z_{10}}\right]\left(\frac{\lambda}{1+2\lambda}\frac{J_1 z_{10}}{\varepsilon_{13}}\right)^2 \times \quad (32)$$
$$\times (z_{11} - z_{21} + z_{12} - z_{22} + z_{13} - z_{23})\,\omega_{\rm n}^{3/2}(\omega_{\rm n1}\omega_{\rm n2}\omega_{\rm n3})^{-1/2} .$$

The process of merging the nuclear $(\boldsymbol{k}, \omega_{\rm n\boldsymbol{k}})$ and the f-electronic $(\boldsymbol{q}, \omega_{1\boldsymbol{q}})$ magnons into the f-electronic $(\boldsymbol{k}+\boldsymbol{q}, \omega_{1\boldsymbol{k}+\boldsymbol{q}})$ one is allowed for $k > \hbar\omega_{\rm n0}/g\mu_{\rm B}\alpha$. The damping of the nuclear magnons is

$$\gamma_{\rm n2f} = \frac{\pi V_0}{\hbar} 12\langle I\rangle \int \frac{{\rm d}\boldsymbol{q}}{(2\pi)^3} |2\Psi_{\rm n2f}(\boldsymbol{q}, \boldsymbol{k}+\boldsymbol{q}, \boldsymbol{k})|^2 \times$$
$$\times \frac{\sinh(\hbar\omega_{\rm n\boldsymbol{k}}/2k_{\rm B}T)}{2\sinh(\varepsilon_{1\boldsymbol{q}}/2k_{\rm B}T)\sinh(\varepsilon_{1\boldsymbol{k}+\boldsymbol{q}}/2k_{\rm B}T)}\,\delta(\hbar\omega_{\rm n\boldsymbol{k}} + \varepsilon_{1\boldsymbol{q}} - \varepsilon_{1\boldsymbol{k}+\boldsymbol{q}}) . \quad (33)$$

To calculate (33) we shall use the following simplifications:

$$(z_{1\boldsymbol{q}} - z_{2\boldsymbol{q}})^2 \approx (\bar{a}q)^4 , \qquad \varepsilon_{10} \ll k_{\rm B}T ,$$
$$\hbar\omega_{\rm n} \ll \varepsilon_{10} ; \qquad \bar{a} \sim V_0^{1/3} .$$

V_0 is the volume per one magnetic atom. We also omit simple (but cumbrous) intermediate computations. As a result one obtains

$$\gamma_{\rm n2f} = c_{\rm n2f}\left[1 - \left(\frac{\omega_{\rm n\boldsymbol{k}}}{\omega_{\rm n}}\right)^2\right]^2 \frac{g\mu_{\rm B}H}{\hbar}\frac{H}{\alpha k}\left(\frac{T}{\theta_{\rm N}}\right)^5 ;$$
$$c_{\rm n2f} \approx 10^{-3} , \qquad \theta_{\rm N} \equiv \frac{g\mu_{\rm B}\alpha}{k_{\rm B}V_0^{1/3}} . \quad (34)$$

This expression was first published (without calculations) in [5] in which the nuclear spin-wave relaxation was investigated in $CsMnF_3$. The dependences $\gamma_{n2f}(T, k, \omega_{nk})$ (in addition to other relaxation processes of NSW) are in good agreement with the experiment [5]. The experimental value of c_{n2f} is 3.7×10^{-3}.

There is one more new relaxation process in the six-sublattice antiferromagnet $CsMnF_3$. It is the process of nuclear spin–lattice relaxation which is due to the interaction (32). Calculating the time ($T_1^{(2f)}$) of nuclear spin–lattice relaxation with the above simplifications one gets

$$\frac{1}{T_1^{(2f)}} = c_{2f}\omega_n \left(\frac{g\mu_B H}{\varepsilon_{10}}\right)^2 \left(\frac{T}{\theta_N}\right)^5 ; \qquad c_{2f} \approx 10^{-7} . \tag{35}$$

The dependence $T_1^{-1} \sim T^5$ has been observed in $CsMnF_3$ in [3]. A theory of the nuclear spin–lattice relaxation for the two-sublattice antiferromagnet was developed in [10]. The expression (35) has the same magnitude as results of [10] with $T_1^{-1} \sim T^5$, but it has another magnetic field dependence.

Acknowledgements

It is my pleasant duty to express warmest thanks to my colleagues Prof. V. I. Ozhogin, Dr. A. Yu. Yakubovskii, and Dr. A. V. Andrienko for their close interest in this paper and for their valuable comments. I also wish to thank the members of Academician A. S. Borovik-Romanov's seminar for helpful discussions.

References

[1] A. Zalkin, K. Lee, and D. M. Templeton, J. chem. Phys. **37**, 697 (1962).
[2] K. Lee, A. M. Portis, and G. S. Witt, Phys. Rev. **132**, 144 (1963).
[3] L. B. Welsh, Phys. Rev. **156**, 370 (1967).
[4] D. Khatamian and M. F. Collins, Canad. J. Phys. **55**, 773 (1977).
[5] A. V. Andrienko, V. I. Ozhogin, V. L. Safonov, and A. Yu. Yakubovskii, Zh. eksper. teor. Fiz. **84**, 1158 (1983); Soviet Phys. — J. exper. theor. Phys. **57**, 673 (1983).
[6] V. N. Krivoruchko and D. A. Yablonskii, phys. stat. sol. (b) **103**, K41 (1981).
[7] A. S. Borovik-Romanov, B. Ya. Kotyuzhanskii, and L. A. Prozorova, Zh. eksper. teor. Fiz. **58**, 1911 (1970); Soviet Phys. — J. exper. theor. Phys. **31**, 1027 (1971).
[8] B. Ya. Kotyuzhanskii and L. A. Prozorova, Zh. eksper. teor. Fiz. **65**, 2470 (1973); Soviet Phys. — J. exper. theor. Phys. **38**, 1233 (1974).
[9] I. V. Kolokolov, V. S. Lvov, and V. B. Cherepanov, Zh. eksper. teor. Fiz. **84**, 1043 (1983); Soviet Phys. — J. exper. theor. Phys. **57**, 604 (1983).
[10] V. S. Lutovinov and V. L. Safonov, Fiz. tverd. Tela **23**, 2759 (1981); Soviet Phys. — Solid State **23**, 1615 (1981).
[11] V. S. Lutovinov and V. L. Safonov, Fiz. tverd. Tela **22**, 2640 (1980); Soviet Phys. — Solid State **22**, 1541 (1980).
[12] P. G. de Gennes, P. A. Pincus, F. Hartmann-Boutron, and J. M. Winter, Phys. Rev. **129**, 1105 (1963).
[13] V. I. Ozhogin and A. Yu. Yakubovskii, Zh. eksper. teor. Fiz. **67**, 287 (1974); Soviet Phys. — J. exper. theor. Phys. **40**, 144 (1975).
[14] V. A. Tulin, Fiz. nizk. Temp. **5**, 965 (1979); Soviet Phys. — Low-Temp. Phys. **5**, 455 (1979).
[15] V. L. Safonov, Phys. Letters A **97**, 164 (1983).

(Received June 28, 1984)

phys. stat. sol. (b) **126**, 207 (1984)

Subject classification: 10.1 and 18; 19

Institute of Metal Physics, Academy of Sciences of the Ukrainian SSR, Kiev[1])

Magnetic State of Crystals with Dislocations

By

I. N. Karnaukhov and A. F. Zhuravlev

The inhomogeneous elastic fields of linear dislocations can lead to the formation of the noncolinear magnetic structure in the magnetic crystals under consideration. This structure is calculated in the frame work of the phenomenological approach; its localization region is determined and its formation criteria are discussed. The influence of the dislocation dilatation on the line width of the FMR and NMR is calculated also.

Неоднородные упругие поля прямолинейных дислокаций в парамагнитном состоянии могут приводить к возникновению неколинеарной магнитной структуры сложной конфигурации. В рамках феноменологического рассмотрения рассчитывается магнитное состояние кристалла с прямолинейной дислокацией, определяется область локализации, обсуждаются критерии её образования. Вычисляется уширение линий ФМР и ЯМР под влиянием дислокационной дилатации.

1. Introduction

It is well known that the elastic fields of a dislocation induce electric charges near the dislocation tube [1]. One can expect analogous effects with respect to the magnetic state. Near the point of the magnetic phase transition this effect must be more remarkable if we take into account the arising ordered regions in the disordered phase near dislocations [2]. These ordered regions may form a superstructure which penetrates the whole crystal. The authors [2] have used the concept of a local critical temperature of the phase transition due to the magnetostriction which is connected with the dilatation of the dislocation elastic field.

The magnetic phase in dislocated crystals changes from the ferromagnetic to the paramagnetic state through an intermediate state with ordered areas near the dislocations [2]. The type of the ordering in this intermediate state had not been considered in [2].

In the present work the magnetic state configuration of the dislocated crystals in the intermediate state is considered. It is shown that the sizes of the magnetically ordered areas near the dislocations decrease with increasing temperature and in the paramagnetic phase the magnetic state localization radius is equal to the dislocation core radius. Hence the phase transition to the paramagnetic state is not so sharp. The localization region and the criterion of its formation are determined.

The magnetization inhomogeneities near the dislocations influence the NMR and FMR linewidth. In this paper the mechanism of line broadening of NMR and FMR is regarded and calculated.

2. Thermodynamical Potential and Boundary Conditions

We consider the crystal in the paramagnetic phase with magnetic anisotropy β, exchange interaction between the magnetic atoms J, and inhomogeneous exchange α. Magnetic anisotropy allows to introduce the axis of quantization and determines

[1]) Prospekt Vernadskogo 36, 252680 Kiev 142, USSR.

$c_\uparrow(\boldsymbol{r})$ and $c_\downarrow(\boldsymbol{r})$, the concentrations of magnetic atoms with the spins directed along and opposite to this axis, respectively. In this case we get the local magnetization

$$M_e(\boldsymbol{r}) = \mu_e g S_e c(\boldsymbol{r}) \quad \text{with} \quad c(\boldsymbol{r}) = c_\uparrow(\boldsymbol{r}) - c_\downarrow(\boldsymbol{r}) \,, \tag{1}$$

where μ_e is the electron Bohr magneton, g is the Landé factor.

The elastic deformations of the dislocation are proportional to $1/\varrho$, where ϱ is the distance to the dislocation line in cylindrical coordinates $\{\varrho, \varphi, z\}$, and $0z$ coincides with the dislocation axis $\boldsymbol{\zeta}$. We take into account that at large distances the crystal is magnetically homogeneous,

$$\lim_{\varrho \to \infty} c(\boldsymbol{r}) = c_0 \,. \tag{2}$$

Here c_0 is the homogeneous concentration.

It is not possible to consider the region of the dislocation core, ϱ_0, in continuum approximation. This leads to the conditions

$$c(\varrho_0, \varphi, z) = f(\varphi, z) \,, \tag{3}$$

$$\left.\frac{\partial c(\boldsymbol{r})}{\partial \varrho}\right|_{\varrho = \varrho_0} = l \,, \tag{4}$$

where the unknown functions $f(\varphi, z)$ and l possess screw symmetry and characterize the nonideality of crystal lattice on the dislocation tube.

Taking into account (1), one can obtain the exchange interactions of the magnetic atoms from the exchange Hamiltonian of the system under consideration. In this case the thermodynamical potential is

$$\Phi = \frac{1}{2} \int \mathrm{d}\boldsymbol{r} \left\{ a c^2(\boldsymbol{r}) + \frac{1}{2} P c^4(\boldsymbol{r}) + \alpha [\nabla c(\boldsymbol{r})]^2 + \right.$$
$$\left. + \lambda u_{ij}(\boldsymbol{r}) \frac{\partial c(\boldsymbol{r})}{\partial x_i} \frac{\partial c(\boldsymbol{r})}{\partial x_j} + \gamma \omega(\boldsymbol{r})\, c^2(\boldsymbol{r}) \right\}. \tag{5}$$

Here the parameter a includes mixing energy, entropy term, exchange interaction, energy of magnetic anisotropy, and in accordance with the Landau-Lifshits theory, a can be approximated by $\approx (T - T_C)$, where T and T_C are the temperature and Curie temperature, respectively; P determines the exchange spin–spin interaction of the fourth power and includes a corresponding term of the entropy power set of the magnetic atoms concentration; λ is the inhomogeneous magnetoelastic exchange constant; γ the magnetoelastic constant; $\omega(\boldsymbol{r})$ the the dilatation due to the elastic field of the dislocations; $u_{ij}(\boldsymbol{r})$ are the components of the deformation tensor.

3. Screw Dislocation

In the case of the screw dislocation we have

$$u_{z\varphi}(\boldsymbol{r}) = \frac{b}{2\pi\varrho} \,, \tag{6}$$

where b is the modulus of the Burgers vector, and

$$\omega(\boldsymbol{r}) = \xi \frac{b^2}{\varrho^2} \,, \tag{7}$$

where ξ is determined by the elastic properties of the crystal under consideration.

The equilibrium concentration $c(\boldsymbol{r})$ is obtained from the nonlinear equation which corresponds to the minimum of the energy functional (5)

$$\alpha \, \Delta c(\boldsymbol{r}) + \lambda u_{z\varphi}(\varrho) \frac{1}{\varrho} \frac{\partial^2 c(\boldsymbol{r})}{\partial \varphi \, \partial z} - [a + \gamma \omega(\varrho) + P c^2(\boldsymbol{r})] \, c(\boldsymbol{r}) = 0 \tag{8}$$

with the conditions (2) to (4).

The criterion for the existence of a nontrivial solution of (8) is obtained from the solution of the linearized equation for $c(\boldsymbol{r})$ [3]. This solution describes the magnetically ordered state near the dislocation. In the paramagnetic phase $a > 0$ and in the case of small concentrations of magnetic atoms we can neglect the nonlinear term ($P = 0$) in (8). The solution of the problem in this case is given by

$$c(\boldsymbol{r}) = \sum_k g(k) \, \mathcal{K}_\mu(\varkappa\varrho) \cos\left(kz + k \frac{b}{2\pi} \varphi\right); \tag{9}$$

$$\varkappa^2 = k^2 + \frac{a}{\alpha}, \qquad \mu^2 = \left(\frac{kb}{2\pi}\right)^2 \left(1 + \frac{\lambda}{\alpha}\right) + \frac{\gamma\xi b^2}{\alpha}, \tag{10}$$

$\mathcal{K}_\mu(\varkappa\varrho)$ is the McDonald's function and the coefficients $g(k)$ are determined from (3),

$$g(k) = \frac{1}{\sqrt{2\pi}} \, \mathcal{K}_\mu^{-1}(\varkappa\varrho_0) \int \mathrm{d}z \, \mathrm{d}\varphi \, f(\varphi, z) \cos k\left(z + \frac{b}{2\pi} \varphi\right). \tag{11}$$

If $f(\varphi, z) = \text{const}$, $k = 0$.

The thermodynamic potential value is negative if $\mu^2 = -\tau^2 < 0$ only and this condition is satisfied if $\gamma\xi/\alpha < 0$ ($k = 0$) or $\lambda < -\alpha$ ($\gamma = 0$) as it follows from (11) Hence McDonald's functions of imaginary index correspond to the equilibrium solution. The function $\mathcal{K}_{i\tau}(z)$ rapidly oscillates with decreasing period if $z \to 0$ and if $z > z_\mathrm{m}$ (z_m is the largest root of the equation $\mathcal{K}_{i\tau}(z) = 0$, $z_\mathrm{m} < \tau$), $\mathcal{K}_{i\tau}(z)$ exponentially decreases.

The local Curie temperature $T_\mathrm{C}(\boldsymbol{r})$ which determines the region of validity of solution (9) is

$$T_\mathrm{C}(\boldsymbol{r}) = T_\mathrm{C} - \frac{1}{\varrho^2} \left[\gamma\xi b^2 + \lambda \left(\frac{kb}{2\pi}\right)^2\right]. \tag{12}$$

In contradiction to [2] here the influence of inhomogeneous spin–spin exchange is taken into account.

In the case of low dislocation density the difference between the Curie temperature of the whole dislocated crystal and T_C is negligible. We shall consider this case below.

At $T \approx T_\mathrm{C}$ the inhomogeneous magnetic structure caused by the dislocation exist near the dislocation core. The solution (9) corresponds to the energy minimum if $\varrho < \varrho_\mathrm{d}$ and $c(\boldsymbol{r}) = 0$ when $\varrho \geqq \varrho_\mathrm{d}$,

$$\varrho_\mathrm{d} = \left\{\left[\gamma b^2 \xi + \left(\frac{kb}{2\pi}\right)^2 \lambda\right] (T_\mathrm{C} - T)^{-1}\right\}^{1/2}. \tag{13}$$

The substitution of (9) into (5) gives

$$\Phi = \frac{1}{2} \sum_k \alpha z \frac{\mathrm{d}}{\mathrm{d}z} \ln \mathcal{K}_{i\tau}(z) \Big|_{z=\varkappa\varrho_\mathrm{d}}^{z=\varkappa\varrho_0}, \tag{14}$$

where k is found from (4) which in this case gives

$$\frac{\mathrm{d}}{\mathrm{d}z} \ln \mathcal{K}_{i\tau}(z)\Big|_{z=\varkappa\varrho_0} = \frac{l}{\varkappa}. \tag{15}$$

Looking at (12) and (13) we have that $\Phi < 0$ if $l < 0$.

So in the case $\varrho_d > \varrho_0$ the phase transition to the paramagnetic state goes through the intermediate phase which is determined by the above solution. The paramagnetic phase is formed in the dislocated crystal when $T > T_C(\varrho_0)$ or when the area of the existence of (9) is comparable with the dislocation core size. We point out that analogous solution and boundary conditions are discussed in [4]. The region of localization of the magnetically ordered state is $\varrho_c \sim 1/\varkappa$ as it follows from the asymptotic representation of the McDonald's function. The more interesting case is $\varrho_c \gg \varrho_0$. This corresponds to the temperature region near the Curie point T_C as long as a is small when $T \to T_C$. The calculations yield $k^2 \sim (T - T_C)$ hence the critical index of the spin correlation function $\nu = 1/2$ as in Gauss' model.

It is too hard to investigate the solutions of (15) analytically but it is possible to discuss the shape of the solution (9). The magnetic state of the crystal with the dislocation is noncollinear, has the complex space structure, is localized near the dislocation. The value of $c(r)$ oscillates when $\varrho_c > \varrho > \varrho_0$ corresponding to the behaviour of McDonald's function in this region.

4. Edge Dislocation

Analogous localization of the magnetic state is possible near the edge dislocation. In contradiction to the above case we shall take into account for the edge dislocation only the dilatation,

$$\omega(r) = \frac{b}{2\pi\varrho} \sin\varphi \,. \tag{16}$$

The function $c(\boldsymbol{r})$, which minimizes (5) has been found from

$$\alpha\, \Delta c(\boldsymbol{r}) - [a + \gamma\omega(\boldsymbol{r}) + Pc^2(\boldsymbol{r})]\, c(\boldsymbol{r}) = 0 \,. \tag{17}$$

In this case $c(\boldsymbol{r})$ must satisfy the boundary conditions (2) to (4) where the edge dislocation specifies $f(\varphi, z)$, l, and the condition of the absence of a singularity if $\varrho \to 0$. Except the boundary conditions (2) to (4) we must add

$$\gamma \sin\varphi < 0 \tag{18}$$

which is analogous to the condition $\mu^2 < 0$ in the case of a screw dislocation.

Equation (17) was analysed in details in [3], where the formation of a superconducting region near the edge dislocation was considered. It follows from the analyses [3], that $c(\boldsymbol{r})$ can be represented as a power series of the functions of the approximative Hamiltonian with an attractive model potential which is independent of φ. It is shown in [3] that this power series rapidly converges and as the basic functions are localized near the dislocation, (17) describes the localization of the magnetic atoms near the edge dislocation. The numerical calculations indicate that this solution corresponds to the minimum of (17) for the edge dislocation in the frame of our approach.

5. FMR and NMR Line Width

In contradiction to the previous sections in this one we shall not take into account the magnetic interaction of the atoms. The inhomogeneous distribution of atoms is formed by the volume dilatation effect. So we have [1]

$$c_{e,n}(\boldsymbol{r}) = c^0_{e,n} \exp\left(-\beta^{e,n} \frac{\sin\varphi}{\varrho k_B T}\right), \tag{19}$$

where

$$\beta^{\mathrm{e,n}} = \frac{\mu b}{3\pi}\frac{1+\nu}{1-\nu}\,(v_{\mathrm{s}}^{\mathrm{e,n}} - v_{\mathrm{a}}^{\mathrm{e,n}})\,, \tag{20}$$

ν is the Poisson coefficient; $v_{\mathrm{s}}^{\mathrm{e,n}}$, $v_{\mathrm{a}}^{\mathrm{e,n}}$ are the volumes of mixing and matrix atoms with the electron (e) and nuclear (n) spins, respectively; k_{B} is the Boltzmann constant; μ is the shear modulus.

Let us consider the dipole–dipole interaction of the nuclear spin S_{n} (or electron spin S_{e}) with the spins of the neighbouring atoms

$$W_{\mathrm{e,n}}(\boldsymbol{r}) = -\boldsymbol{M}_{\mathrm{e,n}}(\boldsymbol{r}) \int \mathrm{d}\boldsymbol{R}\,\frac{1}{R^3}\left[\boldsymbol{M}_{\mathrm{e}}(\boldsymbol{r}+\boldsymbol{R}) - \frac{3}{R^2}\,\boldsymbol{R}\big(\boldsymbol{M}_{\mathrm{e}}(\boldsymbol{r}+\boldsymbol{R})\cdot\boldsymbol{R}\big)\right], \tag{21}$$

where μ_{n} is the nuclear Bohr magneton. Suppose that

$$M_i(\boldsymbol{r}) = \mu_i g S_i c_i(\boldsymbol{r})\,, \qquad M_i = \mu_i g S_i c_i^0; \qquad i = \mathrm{n}, \mathrm{e}\,, \tag{22}$$

and that the inhomogeneity of the concentration is determined by the elastic field of the edge dislocation (19) and calculate the value of the dipole–dipole interaction.

To estimate the value of $W(\boldsymbol{r})$ one can take into account the atoms which are at a large distance ϱ from the dislocation line: $\varrho > \varrho_1 = \beta^{\mathrm{e,n}}/k_{\mathrm{B}}T$. The value of ϱ_1, if $\nu = 0.3$, $v_{\mathrm{s}} = 1.2 v_{\mathrm{a}}$, $v_{\mathrm{a}} \sim b^3$, $\mu b^3 = 5$ eV, $k_{\mathrm{B}}T = 0.02$ eV is near to $10a$, where a is the lattice constant.

The expansion of $c(\boldsymbol{r})$ into a power series gives

$$c_i(\boldsymbol{r}) = c_i^0 - c_i^0 \beta^i \frac{\sin\varphi}{\varrho k_{\mathrm{B}} T}. \tag{23}$$

Let us describe the region ϱ_1 near $\varrho > \varrho_1$.

If the magnetic momenta are parallel to the $0y$ axis of the cartesian coordinates, we obtain

$$W_i(\boldsymbol{r}) = -\pi M_i(\boldsymbol{r})\, M_{\mathrm{e}} \cos\varphi_i \sin 2\varphi_i \frac{\beta^{\mathrm{e}}}{k_{\mathrm{B}} T \varrho_i}, \tag{24}$$

where φ_i and ϱ_i are polar coordinates of $M_i(\boldsymbol{r})$.

According to (24) $W(\boldsymbol{r})$ is anisotropic with respect to the orientations of the dislocation line vector $\boldsymbol{\zeta}$ and magnetization, so if $\boldsymbol{M}_{\mathrm{e}} \parallel \boldsymbol{\zeta}$, $W(\boldsymbol{r}) = 0$, and if $\boldsymbol{M}_{\mathrm{e}} \parallel \boldsymbol{b}$ the sign of $W(\boldsymbol{r})$ in (24) is changed.

If the nuclear magnetic moment is situated near the dislocation core, $\varrho_0 \lesssim \varrho \ll \varrho_1$, in accordance with (19) we get from (21),

$$W_i(\boldsymbol{r}) = -2\pi M_i(\boldsymbol{r})\, M_{\mathrm{e}} \left[\frac{1}{2} - \varrho_0 \frac{k_{\mathrm{B}}T}{\beta^{\mathrm{e}}}\, \mathrm{I}_1\!\left(\frac{\beta^{\mathrm{e}}}{\varrho_0 k_{\mathrm{B}} T}\right)\right], \tag{25}$$

where $\mathrm{I}_1(z)$ is the modified Bessel function of the first kind.

One can see from (24), (25) that the dipole–dipole interaction $W(r) \sim 1/\varrho$, where ϱ is the distance to the dislocation line. $W(\boldsymbol{r})$ depends on the orientation of the magnetization vector with respect to the dislocation line. Hence taking into account several glide systems of dislocations one can get the splitting of the FMR (or NMR) signal.

In hexagonal crystals the main glide plane of the dislocations is the basic plane in which we can take the orientation of the dislocation in any direction. In the case of the random orientation of the dislocations the averaged shape of the FMR (or NMR) signal is non-Lorentzian and the frequency and the linewidth of the FMR (or NMR) are renormalized.

The shift of the FMR (or NMR) frequency of the basic plane magnetized ferromagnet is as follows:

$$\Delta\omega_i = \frac{3}{4\hbar} S_e \beta^e \beta^i \frac{n}{(k_B T b)^2} \ln\left(\frac{L}{\varrho_1}\right), \tag{26}$$

where $\hbar$ is Planck's constant, $n = Nb^2L/V$ the density of dislocations, N the number of dislocations, L the dislocation length.

The FMR (or NMR) linewidth at the resonant frequency is determined by the initial value $\Gamma^0_{e,n}$ and by the dislocation dipole–dipole interaction,

$$\Gamma_i^2 = \Gamma_i^0 \sqrt{(\Gamma_i^0)^2 + (\Gamma_i^d)^2}\,, \tag{27}$$

where $\Gamma^d_{e,n} = \Delta\omega_{e,n}$.

The linewidth of the FMR (or NMR) depends on the temperature and is proportional to the dislocation density.

The magnetic momenta which are situated near the dislocation core lead to the increase of the FMR (or NMR) linewidth but this contribution is negligible as their placing area is of the order of $(r/\bar{L})^2$ where $\varrho_0 < r < \varrho_1$ and $\bar{L}$ is the average distance between the dislocations.

As it follows from (27) and the above estimations the dislocation density 10^{10} cm^{-2} corresponds to a linewidth $\Gamma^d \approx 10^{-2}M$ (M is the magnetization of the sample) at high temperatures $k_BT \approx 0.02$ eV. If the temperature is decreased, the value of Γ^d is increased sharply. Hence the inhomogeneous distribution of magnetic atoms due to the elastic fields of dislocations may determine the damping on the FMR (or NMR) frequencies in the low-temperature region.

Let us point out that when $T = 30$ K, the value of $\varrho_1 = 10^2a$ so the contribution to the damping oscillations of the magnetic momenta which are localized near the dislocation core increases also.

6. Conclusions

As follows from the above calculations magnetically ordered area of complicated shape arises near straight dislocations in magnetic crystals in the paramagnetic phase. This means that the ferro–paramagnetic phase transition is realized through a non-collinear inhomogeneous magnetic phase which is connected with the dislocation structure. The configuration of the transition phase is determined by the boundary conditions on the dislocation tubes. The magnetically ordered areas near the straight dislocations must influence the electric conductivity and the magnetoresistance.

The inhomogeneous distribution of the magnetic atoms in the elastic fields of dislocations reflects the magnetic dipole–dipole interaction. The value of this interaction is reversely proportional to the distance to the dislocation. So this dependence leads to an increase of the linewidth of FMR and NMR. The calculated temperature dependence of the FMR and NMR linewidth points out that the damping of the magnetic momentum oscillations increases if the temperature decreases.

References

[1] J. P. Hirth and I. Lothe, Theory of Dislocations, McGraw-Hill Book Co., 1968.
[2] I. M. Dubrowskii and M. A. Krivoglas, Zh. eksper. teor. Fiz. 77, 1017 (1979).
[3] N. N. Nabutovskii and B. Ya. Shapiro, Zh. eksper. teor. Fiz. 75, 948 (1978).
[4] I. M. Dubrowskii, Ukr. fiz. Zh. 20, 1535 (1975); Fiz. nizk. Temp. 8, 373 (1982).

(Received June 27, 1984)

phys. stat. sol. (b) **126**, 213 (1984)

Subject classification: 18.4

Departamento de Física, Universidade Federal de Pernambuco, Recife[1])

Soliton Solution for the Excited State in the Spin-Peierls $S = 1/2$ Antiferromagnetic Chain[2])

By

G. L. M. DASSEN[3]) and M. A. DE MOURA

The ground-state and low energy excitation spectrum of the spin-Peierls $S = 1/2$ antiferromagnetic chain is studied in the adiabatic limit. It is shown that this model corresponds to the semiclassical Gross-Neveu model with only one kind of Fermion and that the lowest excited state corresponds to the usual-kink-soliton of the Φ^4-theory. Its energy is approximately one-third of the singlet–triplet excitation gap.

Nous étudions l'état fondamental et le spectre d'excitation de faible énergie d'une chaine de spins de Peierls $S = 1/2$ pour la limite adiabatique. Nous montrons que ce modèle correspond au modèle semi-classique de Gross-Neveu avec une seule espèce de fermion et avec un premierétat excité correspondant à l'habituel kink-soliton de la théorie Φ^4. Son énergie est approximativement égale un tiers du gap d'excitation singulet–triplet.

1. Introduction

In recent years we have been observing a growing interest in nonlinear excitations in quasi-one-dimensional condensed-matter systems. In the charge-transfer doping mechanisms of polyacetylene Su et al. [1], Rice [2] and Takayama et al. [3] pointed out the importance of nonlinear excitations (solitons) in such systems. These authors showed that polyacetylene can be described by a Hamiltonian containing an electronic hopping and lattice distortion term and further inclusion of the electron-lattice coupling makes it a one-dimensional Peierls system. Since a spin-Peierls system ($S = 1/2$) can be mapped, through the Jordan-Wigner transformation, in a Hamiltonian similar to the one of polyacetylene we expect the existence of solitons in such a system. In a recent letter Suzuki [4] pointed out that the $S = 1/2$ ESR line observed in TCNQ salts can be due to solitons in the spin-Peierls state. In this paper we show, making a detailed calculation, that the lowest excited state in a spin-Peierls system is a soliton. In Section 2 we start from a linear Heisenberg antiferromagnetic ($S = 1/2$) elastic chain, following the treatment by Pytte [5]: the exchange coupling between neighbouring spins is expanded linearly in the interatomic distance and the spin operators are transformed in spinless fermions through the Jordan-Wigner transformation. The four-fermion terms, resulting from the S^zS^z interaction, are then treated by a Hartree approximation.

We continue this scheme considering the phonons in the adiabatic limit, i.e. $M \to \infty$. As a consequence only static distortions of the lattice are allowed, permitting a classical approach of the phonon degrees of freedom. Finally we take the continuum and semi-classical limit. In Section 3 and 4 we solve the Bogolyubov-de Gennes (BdG)

---

[1]) Recife 50000 PE, Brazil.

[2]) Work supported by FINEP, CAPES and CNPq (Brazilian Agencies).

[3]) Permanent address: Departamento de Física, Universidade Federal da Paraíba, Campina Grande 58100, Brazil.

equation in a very simple way and we show in Section 3 that the ground state is dimerized and degenerate. The energy gain per spin due to dimerization, not including the elastic energy, has the well-known BCS-gap dependence on the lattice distortion. Cross and Fisher [6] using the procedure of Luther and Peschel [7] show that this energy is proportional to the 4/3-power in the lattice distortion. In fact our result is a consequence of the Hartree treatment for the four-fermion terms which reduces our model to the XY-model. In Section 4 we show that the excited state solution of the BdG equation corresponds to the usual kink-soliton of the Φ^4-theory. This solution can be interpreted physically as an interpolation between the two equivalent ground-state configurations. We show also that the creation energy of the kink-soliton is approximately one third of the energy gap of the singlet–triplet excitation.

2. The Spin-Peierls Hamiltonian

The usual spin-Peierls Hamiltonian is given as

$$H = \sum_{l=1}^{N} J(l, l+1) \{\mathbf{S}_l \mathbf{S}_{l+1} - \tfrac{1}{4}\} + \sum_q \omega_0(q)\, b_q^+ b_q \,, \tag{1}$$

written for simplicity in one-dimensional notation. Here $\mathbf{S}_l$ is the spin-1/2 operator on site l of the chain, which contains N magnetic atoms; $J(l, l+1)$ is the Heisenberg exchange constant depending on the distance between neighbouring atoms; $b_q^+(b_q)$ is the phonon creation (destruction) operator and $\omega_0(q)$ the unrenormalized phonon frequency.

Following the lines described in the introduction we arrive after the Hartree-approximation at

$$H = \sum_k \varepsilon_k C_k^+ C_k + \frac{1}{\sqrt{N}} \sum_{k,q} i g(k, q)\, Q(q)\, C_k^+ C_{k-q} + \frac{1}{2} M \sum_q \omega_0^2(q)\, Q(q)\, Q(-q) \,, \tag{2}$$

where

$$\varepsilon_k = pJ \cos ka \qquad \left(|k| \leqq \frac{\pi}{a}\right) \qquad (a \text{ is the lattice constant}) \tag{3}$$

and

$$g(k, q) = p\{\operatorname{sen}(ka) - \operatorname{sen}(k-q)\, a\}\, g_0 \qquad \left(|q| \leqq \frac{\pi}{a}\right). \tag{4}$$

$C_k^+(C_k)$ are creation (destruction) operators for spinless fermions, corresponding to the original spin-degrees of freedom, $Q(q)$ is the phonon coordinate of wave vector q, g_0 $(= \nabla_l J(l, l+1))$ a constant appearing in the expansion of $J(l, l+1)$ and $p \approx 1.64$ is a renormalization factor resulting from the Hartree-approximation and approximately constant at low temperatures [5]. Now for weak interaction and low temperatures this Hamiltonian can be subjected to the linearization procedure of Luther and Peschel [7]. It is not difficult to show that this leads to

$$H = \sum_k v_F k (C_{1k}^+ C_{1k} - C_{2k}^+ C_{2k}) + \frac{2 p g_0}{\sqrt{N}} \sum_{\substack{k \\ q \approx 0}} Q\left(q + \frac{\pi}{a}\right) (C_{2k}^+ C_{1k-q} + C_{1k}^+ C_{2k-q}) + \\ + \omega_Q^2 \sum_{q \approx 0} Q\left(q + \frac{\pi}{a}\right) Q\left(-q - \frac{\pi}{a}\right), \tag{5}$$

where we defined the Fermi velocity $v_F = pJa$, $\omega_Q^2 = \omega_0^2(\pi/a) = 4K/M$ (K is the elastic constant), and k-space is limited to $|k| \leq k_F = \pi/2a$. Observe that besides the special $q = 2k_F = \pi/a$ component, only long-wavelength components are retained ($q \approx 0$). The C_{1k} and C_{2k} represent two new independent fields corresponding to right and left going spinless fermions.

Transforming back to real space through

$$\begin{pmatrix} C_{1k} \\ C_{2k} \end{pmatrix} = \frac{1}{\sqrt{L}} \int \mathrm{d}x \, \mathrm{e}^{-ikx} \begin{pmatrix} u(x) \\ v(x) \end{pmatrix}, \tag{6}$$

$$Q(q) = \frac{1}{\sqrt{N}} \sum_q \mathrm{e}^{-iqx} u_l \qquad (x = la) , \tag{7}$$

(u_l is the deviation from equilibrium of the magnetic atom on site l) and taking the continuum limit, keeping only relevant operators, we have

$$H = -iv_F \int \mathrm{d}x \left\{ u^*(x) \frac{\mathrm{d}u(x)}{\mathrm{d}x} - v^*(x) \frac{\mathrm{d}v(x)}{\mathrm{d}x} \right\} + \\ + \int \Delta(x) \{u^*(x) v(x) + v^*(x) u(x)\} \, \mathrm{d}x + \frac{\omega_Q^2}{g^2} \int \Delta^2(x) \, \mathrm{d}x . \tag{8}$$

Here, in the semiclassical limit, $u(x)$ and $v(x)$ are C-number functions and are normalized as

$$\int \{|u(x)|^2 + |v(x)|^2\} \, \mathrm{d}x = 1 . \tag{9}$$

Further, we defined the phonon-order parameter

$$\Delta(x) \equiv 2pg_0(-1)^l u_l \tag{10}$$

which is a smoothly varying function of x and

$$g = 2pg_0 \left(\frac{2a}{M}\right)^{1/2} . \tag{11}$$

3. Dimerization in the Ground State

Varying the Hamiltonian (8) with respect to $u(x)$, $v(x)$, and $\Delta(x)$, after introducing a Lagrange multiplier in virtue of the normalization condition (9), we obtain the equations of motion

$$-iv_F \frac{\mathrm{d}u_\mu(x)}{\mathrm{d}x} + \Delta(x) \, v_\mu(x) = E_\mu u_\mu(x) , \tag{12a}$$

$$-iv_F \frac{\mathrm{d}v_\mu(x)}{\mathrm{d}x} + \Delta(x) \, u_\mu(x) = E_\mu v_\mu(x) , \tag{12b}$$

and the consistency equation

$$S_\mu \{u_\mu^*(x) \, v_\mu(x) + v_\mu^*(x) \, u_\mu(x)\} = -2 \frac{\omega_Q^2}{g^2} \Delta(x) . \tag{13}$$

In these equations $u_\mu(x)$ and $v_\mu(x)$ are single-particle state functions, S_μ means integration for scattering states (in the thermodynamic limit) or summation for bound states and the Lagrange multiplier was identified as the single particle energy E_μ. The system above is known as the equations of Bogolyubov and De Gennes [8].

In order to solve these equations, we introduce the functions $\Phi_1(x)$ and $\Phi_2(x)$ defined through

$$u(x) = \frac{1}{\sqrt{2}} \left(\Phi_1(x) + \Phi_2(x)\right) , \tag{14a}$$

$$v(x) = \frac{1}{i\sqrt{2}} \left(\Phi_1(x) - \Phi_2(x)\right) . \tag{14b}$$

Substitution in (12 a, b) leads to

$$v_{\mathrm{F}} \frac{\mathrm{d}\Phi_1(x)}{\mathrm{d}x} + \Delta(x)\, \Phi_1(x) = iE\Phi_2(x) , \tag{15a}$$

$$v_{\mathrm{F}} \frac{\mathrm{d}\Phi_2(x)}{\mathrm{d}x} - \Delta(x)\, \Phi_2(x) = iE\Phi_1(x) , \tag{15b}$$

from which we obtain the Schrödinger-type equation

$$\frac{\mathrm{d}^2\Phi_2}{\mathrm{d}x^2} + \frac{E^2 - \Delta^2(x) - v_{\mathrm{F}}\Delta'(x)}{v_{\mathrm{F}}^2} \Phi_2 = 0 \tag{16}$$

with

$$\Delta'(x) = \frac{\mathrm{d}\Delta(x)}{\mathrm{d}x} .$$

We can find an immediate solution of (16), if the factor in front of Φ_2 is independent of x, i.e. if

$$\frac{E^2 - \Delta^2(x) - v_{\mathrm{F}}\Delta'(x)}{v_{\mathrm{F}}^2} = \text{constant} = k^2 , \tag{17}$$

then a solution is

$$\Phi_2(x) = A_k\, \mathrm{e}^{ikx} \tag{18}$$

for scattering states.

We now analyse (17),

$$E_k^2 - v_{\mathrm{F}}^2 k^2 - \Delta^2(x) - v_{\mathrm{F}}\Delta'(x) = 0 . \tag{19}$$

A first solution is trivial,

$$\Delta(x) = \Delta(0) = \text{constant}, \tag{20}$$

leading to

$$E_k^2 = v_{\mathrm{F}}^2 k^2 + \Delta_0^2 , \tag{21}$$

the dispersion relation for scattering states.

Clearly this solution corresponds to the ground state of the system because a potential constant in all space is unobservable and cannot generate excitations.

From the definition of $\Delta(x)$, (10), we can now draw the important conclusion that this ground state is dimerized because u_l takes alternatingly the values $\pm\Delta_0$ and is degenerate because l can be even or odd in an infinite chain.

Substituting (18) and (20) into (15 a, b) in order to determine $\Phi_1(x)$, we obtain the scattering wave functions in the ground state through (14 a, b) and the normalization condition (9),

$$u_k(x) = \{2Na|E_k|\}^{-1/2} \frac{\Delta_0}{(|E_k| - v_F k)^{1/2}} e^{ikx} , \tag{22a}$$

$$v_k(x) = -\{2Na\,|E_k|\}^{-1/2} (|E_k| - v_F k)^{1/2} e^{ikx} , \tag{22b}$$

where

$$E_k = -(v_F^2 k^2 + \Delta_0^2)^{1/2} \tag{23}$$

and we choose the minus signal in (21) because occupied states have negative energy in the ground state (occupied Dirac sea). The consistency equation (13) now allows the determination of Δ_0. Substituting the results (22 a, b) into (13), we arrive at

$$\sum_k \frac{1}{|E_k|} = \sum_k (v_F^2 k^2 + \Delta_0^2)^{-1/2} = 2Na \frac{\omega_Q^2}{g^2} . \tag{24}$$

In the limit $N \to \infty$

$$\frac{2\omega_Q^2}{g^2} = \frac{1}{2\pi} \int_{-k_F}^{+k_F} \frac{dk}{\sqrt{v_F^2 k^2 + \Delta_0^2}} = \frac{1}{\pi v_F} \ln \left\{ \frac{v_F k_F + (v_F^2 k_F^2 + \Delta_0^2)^{1/2}}{\Delta_0} \right\}. \tag{25}$$

We now use the "weak-coupling" condition, which is reasonably well satisfied in real spin-Peierls systems,

$$\frac{\Delta_0}{2v_F k_F} \ll 1 , \tag{26}$$

which simplifies (25) to

$$\ln \frac{2v_F k_F}{\Delta_0} = \frac{2\pi v_F \omega_0^2}{g^2} \tag{27}$$

or

$$\Delta_0 = W e^{-1/\lambda} . \tag{28}$$

From (23) we see that Δ_c corresponds to half of the fermion gap. Further we defined the band width $W = 2v_F k_F$ and the coupling constant $\lambda \equiv g^2 2\pi v_F \omega_Q^2 = p g_0^2/\pi J K$. The ground state energy — as compared to the normal, undimerized state (absence of spin–phonon coupling) — is easily calculated as follows:

$$\Delta E = \sum_k E_k - \sum_k E_{0k} + \Delta_{\text{elast}} , \tag{29}$$

where E_k is given by (23), $E_{0k} = -v_F |k|$ is the dispersion in the normal state and Δ_{elast} is the change in elastic energy. Taking the thermodynamic limit leads to

$$\frac{\Delta E}{L} = -\frac{1}{4\pi v_F} \left\{ \Delta_0^2 + 2\Delta_0^2 \ln \frac{2v_F k_F}{\Delta_0} \right\} + \frac{\omega_0^2}{g^2} \Delta_0^2 , \tag{30}$$

where after integration we used the weak-coupling condition (26). Now the second term is canceled exactly by the third in virtue of the gap equation (27).

The final result for the energy gain per spin due to dimerization is thus

$$\frac{\Delta E}{N} = -\frac{\Delta_0^2}{4\pi p J} , \tag{31}$$

where we used $v_F = pJa$, $k_F = \pi/2a$, and $L = Na$.

4. Soliton in the Excited State

In order to determine an excited state of the system we return to (19).

We write it as

$$\int \frac{\mathrm{d}\Delta(x)}{C_k^2 - \Delta^2(x)} = \int \frac{\mathrm{d}x}{v_\mathrm{F}}\,, \tag{32}$$

where

$$C_k^2 \equiv E_k^2 - v_\mathrm{F}^2 k^2\,.$$

Integration yields

$$\Delta(x) = C_k \tanh \frac{C_k(x - x_0)}{v_\mathrm{F}}\,.$$

We now impose $\Delta(x) \to |\Delta_0|$ as $|x| \to \infty$, i.e. at infinity we must recover the ground-state configuration, otherwise the energy of this excited state would be infinite.

In this way we obtain the results

$$\Delta(x) = \pm \Delta_0 \tanh \frac{\Delta_0(x - x_0)}{v_\mathrm{F}} \tag{33}$$

and

$$E_k^2 = v_\mathrm{F}^2 k^2 + \Delta_0^2\,. \tag{34}$$

Equation (33) is the well-known kink-soliton (anti-kink) solution of Φ^4-theory, and represents a type of interpolation between the two equivalent ground-state configurations.

From (34) we see that the dispersion for the fermion scattering states is not modified by the presence of the soliton. For occupied fermion states we thus choose again

$$E_k = -(v_\mathrm{F}^2 k^2 + \Delta_0^2)^{1/2}\,. \tag{35}$$

The eigenfunctions for these scattering states can now be obtained in the same way as in the case of the ground state, using (33). The result is

$$u_k(x) = (2\sqrt{Na}\,|E_k|)^{-1/2} \left\{\Delta_0 \tanh \frac{\Delta_0 x}{v_\mathrm{F}} + i(|E_k| - v_\mathrm{F} k)\right\} \mathrm{e}^{ikx}\,, \tag{36a}$$

$$v_k(x) = -i(2\sqrt{Na}\,|E_k|)^{-1/2} \left\{\Delta_0 \tanh \frac{\Delta_0 x}{v_\mathrm{F}} - i(|E_k| + v_\mathrm{F} k)\right\} \mathrm{e}^{ikx}\,. \tag{36b}$$

From general considerations [9, 10] we know that besides these scattering states there must be a bound state at $k = 0$ with zero energy — the so-called zero-frequency mode —, due to the broken translational symmetry caused by the presence of the soliton.

Using (15b) with $E = 0$, we have

$$v_\mathrm{F} \frac{\mathrm{d}\Phi_2^0}{\mathrm{d}x} - \Delta(x)\, \Phi_2^0 = 0$$

which can be integrated to give

$$\Phi_2^0(x) = B \cosh \frac{\Delta_0 x}{v_\mathrm{F}}\,,$$

where B is a constant of integration. We see however that Φ_2^0 diverges as $|x| \to \infty$, which is forbidden for a physical state, so we must choose

$$\Phi_2^0 = B = 0\,. \tag{37}$$

Equation (15a) gives, with $E = 0$,

$$v_F \frac{d\Phi_1^0}{dx} + \Delta(x)\, \Phi_1^0 = 0$$

or

$$\Phi_1^0(x) = \frac{B'}{\cosh \dfrac{\Delta_0 x}{v_F}} \tag{38}$$

which satisfies the requirements of a bound state. Normalizing and using (14 a, b) we have

$$u^0(x) = \left(\frac{\Delta_0}{4v_F}\right)^{1/2} \operatorname{sech} \frac{\Delta_0 x}{v_F}, \tag{39a}$$

$$v^0(x) = -i \left(\frac{\Delta_0}{4v_F}\right)^{1/2} \operatorname{sech} \frac{\Delta_0 x}{v_F}. \tag{39b}$$

We are now in condition to verify the consistency equation (13). Substituting (36 a, b) and (39 a, b) in (13) we arrive after some straightforward algebra at

$$-2 \frac{\omega_Q^2}{g^2} \Delta(x) = \frac{\Delta_0^2}{4Na} \sum_k \frac{1}{|E_k|^2} \left\{-\frac{4\,|E_k|}{\Delta_0}\right\} \tanh \frac{\Delta_0 x}{v_F}$$

or, with (33),

$$2 \frac{\omega_Q^2}{g^2} = \frac{1}{Na} \sum_k \frac{1}{|E_k|}$$

which is exactly (24) and thus satisfied with the value (28) for Δ_0.

We therefore conclude that the kink (anti-kink)-soliton solution is an exact solution of the system (12 a, b), (13).

The last important quantity we shall calculate is the soliton creation energy, defined as the energy difference between the excited state and the ground state ("vacuum").

We have thus

$$E_s = E_{\text{exc}} - E_v = \left(\sum_k E_k + E_0 + E_{\text{elast}}\right)_{\text{exc}} - \left(\sum_k E_k + E_{\text{elast}}\right)_{\text{vac}}\,. \tag{40}$$

Here $\sum_k E_k$ means the sum over the energy eigenvalues of the scattering states ("valence band"), $E_0 = 0$ is the bound state energy and E_{elast} is the elastic energy.

However, to do the integral over the eigenvalues in the thermodynamic limit, we have to calculate the phase shift of the eigenfunctions in the excited state, because these are not pure plane waves. This is done in the Appendix. The result is

$$\delta(k) = \operatorname{arctg} \frac{\Delta_0}{v_F k} \qquad \left(|k| < \frac{\pi}{2a}\right). \tag{41}$$

We can now write the modified periodic boundary condition

$$k'_n L + \delta(k'_n) = 2\pi n = k_n L\,, \tag{42}$$

where k'_n is the wave vector in the excited state and k_n in the ground state.

We rewrite (40) as

$$E_s = \sum_{\substack{k'_n \\ k_n \neq 0}} E(k'_n) - \sum_{k_n} E(k_n) + \Delta E_{\text{elast}} = \sum_{k'_n} E(k'_n) - \sum_{k_n} E(k_n) + \frac{\Delta_0}{2} + \Delta E_{\text{elast}} . \tag{43}$$

The third term is the correction due to the bound state at $k = 0$ which counts only for one half in the valence band (the other half belongs to the conduction band due to "charge conjugation" symmetry).

Using (42) we have

$$E_s \approx \sum_{k_n} \left\{ E\left(k_n - \frac{\delta(k_n)}{L}\right) - E(k_n) \right\} + \Delta E_{\text{elast}}$$

$$= - \sum_{k_n} \frac{\delta(k_n)}{L} \frac{\mathrm{d}E(k_n)}{\mathrm{d}k_n} + \Delta E_{\text{elast}}$$

$$= \frac{1}{2\pi} \int_{-k_F}^{+k_F} \delta(k) \frac{\mathrm{d}E(k)}{\mathrm{d}k} + \Delta E_{\text{elast}} . \tag{44}$$

The second term is easily calculated as

$$\Delta E_{\text{elast}} = \lim_{L\to\infty} \frac{\omega_Q^2}{g^2} \int_{-L/2}^{+L/2} (\Delta^2(x) - \Delta_0^2)\, \mathrm{d}x = - \frac{2v_F \Delta_0 \omega_0^2}{g^2} .$$

Using this result and (35), (41) in (44) and integrating by parts, we obtain

$$E_s = \frac{-2v_F\Delta_0\omega_Q^2}{g^2} + \frac{\Delta_0}{2} + \frac{\Delta_0 v_F}{\pi} \int_0^{k_F} \frac{\mathrm{d}k}{\sqrt{v_F^2k^2 + \Delta_0^2}} + \frac{1}{\pi} \left[\sqrt{v_F^2k^2 + \Delta_0^2} \operatorname{arctg} \frac{\Delta_0}{v_F k} \right]_0^{k_F} .$$

The first and third term cancel each other exactly in virtue of (25), while the fourth reduces to $(\Delta_0/\pi) - (\Delta_0/2)$ in the weak-coupling limit $\Delta_0/2v_F k_F \ll 1$. We thus have finally for the soliton creation energy

$$E_s = \frac{\Delta_0}{\pi} . \tag{45}$$

As we see, this result is in agreement with the general result of Dashen et al. [11] for the Gross-Neveu model in the semiclassical limit,

$$E_s = \frac{N}{\pi} \Delta_0 .$$

In our case we have $N = 1$, and (8) corresponds to the Gross-Neveu model with only one kind of fermion.

Also the existence of polarons [12, 13] in the spin-Peierls system is not expected, for these become stable only for $N \geqq 2$ (in the adiabatic limit; $M \to \infty$), in contrast to the usual Peierls system where one has $N = 2$, due to the electron spin.

Appendix

The phase shift $\delta(k)$ is most easily calculated using the functions $\Phi_2(x) = A_k\, e^{ikx}$ $(\delta_2(x, k) = 0)$ from (18) and $\Phi_1(x)$ from (15),

$$iE_k\Phi_1(x) = v_F \frac{d\Phi_2}{dx} - \Delta(x)\, \Phi_2(x)$$

or, substituting $\Phi_2(x)$,

$$\Phi_1(x) = -iA_k \frac{iv_F k - \Delta(x)}{E_k} e^{ikx} .$$

Writing

$$\Phi_1(x) = A_k\, e^{i\,(kx + \delta_1(x,k))} ,$$

we identify $\delta_1(x, k)$ as

$$\delta_1(x, k) = -i \ln \left\{\frac{i\Delta(x) - v_F k}{E_k}\right\} .$$

Now

$$\delta_1(k) = \delta_1(x, k)_{x\to\infty} - \delta_1(x, k)_{x\to-\infty} = -i \ln \frac{\Delta_0 + iv_F k}{-\Delta_0 + iv_F k} ,$$

from which

$$\delta_{1k} = 2 \operatorname{arctg} \frac{v_F k}{\Delta_0} .$$

Therefore,

$$\delta(k) = \frac{\delta_1(k) + \delta_2(k)}{2} = \operatorname{arctg} \frac{v_F k}{\Delta_0}$$

for $|k| \leqq \pi/2a$.

References

[1] W. P. Su, J. R. Schrieffer, and A. J. Heeger, Phys. Rev. B **22**, 2099 (1979).
[2] M. J. Rice, Phys. Letters A **71**, 152 (1979).
[3] H. Takayama, Y. R. Lin-Liu, and K. Maki, Phys. Rev. B **21**, 2388 (1980).
[4] N. Suzuki, Phys. Letters A **89**, 208 (1982).
[5] E. Pytte, Phys. Rev. B **10**, 4637 (1974).
[6] M. C. Cross and D. S. Fisher, Phys. Rev. B **19**, 402 (1979).
[7] A. Luther and I. Peschel, Phys. Rev. B **12**, 3908 (1975).
[8] P. G. de Gennes, Superconductivity of Metals and Alloys, W. A. Benjamin, Inc., New York 1966.
[9] R. F. Dashen, B. Hasslacher, and A. Neveu, Phys. Rev. D **10**, 4130 (1974).
[10] R. Jackiw, Rev. mod. Phys. **49**, 681 (1977).
[11] R. F. Dashen, B. Hasslacher, and A. Neveu, Phys. Rev. D **12**, 2443 (1975).
[12] D. K. Campbell and A. R. Bishop, Nuclear Phys. B **200**, 297 (1982).
[13] S. Brazovskii and N. Kirova, Zh. eksper. teor. Fiz., Pisma **33**, 6 (1981).

(Received July 4, 1984)

phys. stat. sol. (b) **126**, 223 (1984)

Subject classification: 13.5.4 and 20.1; 18.3

Physikalisches Institut der Westfälischen Wilhelms-Universität, Münster[1])

Dispersion Relations for Magnon-Polaritons in Uniaxial Ferrimagnetic Crystals

By

L. MERTEN, R. KOTZOTT, and A. LEHMEYER

General dispersion relations are derived for the components of the permeability tensor and the refractive indices of uniaxial ferrimagnetic crystals making use of symmetry properties. The zeros of the polynomials in the dispersion relations can be interpreted as frequencies of special magnons and magnon-polaritons. Some of these frequencies correspond to longitudinal magnons in the limit of large wave vectors.

Es werden allgemeine Dispersionbeziehungen für die Elemente des Permeabilitätstensors und die Brechungsindizes einachsiger ferrimagnetischer Kristalle unter Ausnutzung von Symmetrieeigenschaften hergeleitet. Die Nullstellen der Polynome in den Dispersionsbeziehungen können als Frequenzen spezieller Magnonen und Magnon-Polaritonen gedeutet werden. Einige dieser Frequenzen sind longitudinalen Magnonen im Grenzfall großer Wellenvektoren zuzuordnen.

1. Introduction

Ferrimagnetic crystals in general are of very complex structure and have a large number of magnon and magnon-polariton branches. Therefore, at present it is impossible to get quantitative results for the magnon and magnon-polariton branches by microscopic model theories. In the contrary it is well-known that e.g. for phonon-polaritons a phenomenological theory was very successful to describe quantitatively the dispersion of phonon-polaritons also in very complicated crystals (see e.g. [1]).

In the following it is shown that from some basic assumptions also a complete phenomenological treatment of the dispersion of magnon-polaritons in uniaxial ferrimagnetic crystals is possible. It is proved that even in ferrimagnetic crystals longitudinal magnons exist in the limit of large wave vectors.

2. Basic Equations

In uniaxial ferrimagnetic crystals the magnetic permeability is given by the Polder tensor [2]

$$\boldsymbol{\mu}(\omega) = \begin{pmatrix} \mu_\perp(\omega) & -i\varkappa(\omega) & 0 \\ i\varkappa(\omega) & \mu_\perp(\omega) & 0 \\ 0 & 0 & \mu_\parallel(\omega) \end{pmatrix}. \tag{1}$$

The structure of this tensor results directly from the symmetry of the crystals. From the principle of causality follows that the Polder tensor must obey the symmetry relation [3, 4],

$$\boldsymbol{\mu}(-\omega) = \boldsymbol{\mu}^*(\omega) , \tag{2}$$

where ω is assumed to be real.

[1]) Domagkstr. 75, D-4400 Münster, FRG.

Because magnetism dies out for high frequencies the limit $\omega \to \infty$ of the Polder tensor is the unit tensor **I**,

$$\lim_{\omega\to\infty} \boldsymbol{\mu}(\omega) = \begin{pmatrix} 1 & 0 & 0 \\ 0 & 1 & 0 \\ 0 & 0 & 1 \end{pmatrix} = \mathbf{I} . \tag{3}$$

From (1) and (2) follows that the components of the tensor $\boldsymbol{\mu}(\omega)$ must fulfil the relations

$$\mu_\perp(-\omega) = \mu_\perp(\omega) \tag{4a}$$

and

$$\varkappa(-\omega) = -\varkappa(\omega) , \tag{4b}$$

i.e. $\mu_\perp(\omega)$ must be an even function and $\varkappa(\omega)$ an odd function of ω.

First we treat a wave propagating parallel to the optical axis. For this direction the refractive indices n_1 and n_2 of the two waves are given by [2, 5]

$$n_1^2(\omega) = \frac{c^2k^2}{\omega^2} = \varepsilon_\perp(\omega)\,(\mu_\perp(\omega) + \varkappa(\omega)) = \varepsilon_\perp(\omega)\, f_1(\omega) \tag{5a}$$

and

$$n_2^2(\omega) = \frac{c^2k^2}{\omega^2} = \varepsilon_\perp(\omega)\,(\mu_\perp(\omega) - \varkappa(\omega)) = \varepsilon_\perp(\omega)\, f_2(\omega) , \tag{5b}$$

where

$$f_1(\omega) \equiv \mu_\perp(\omega) + \varkappa(\omega) \tag{5c}$$

and

$$f_2(\omega) \equiv \mu_\perp(\omega) - \varkappa(\omega) . \tag{5d}$$

Regarding (4) both functions are connected by

$$f_2(\omega) = f_1(-\omega) . \tag{6}$$

Due to (3) $f_1(\omega)$ and $f_2(\omega)$ must fulfil the condition

$$\lim_{\omega\to\infty} f_{1,2}(\omega) = 1. \tag{7}$$

3. Zeros and Poles of $f_1(\omega)$, $f_2(\omega)$, $\mu_\perp(\omega)$, and $\varkappa(\omega)$

Now let us assume that $f_1(\omega)$, $\mu_\perp(\omega)$, and $\varkappa(\omega)$ are represented by broken rational functions

$$f_1(\omega) \equiv \frac{N_1(\omega)}{D(\omega)} , \tag{8a}$$

$$\mu_\perp(\omega) \equiv \frac{N_\mu(\omega)}{D(\omega)} , \tag{8b}$$

and

$$\varkappa(\omega) \equiv \frac{N_\varkappa(\omega)}{D(\omega)} , \tag{8c}$$

where $N_1(\omega)$, $N_\mu(\omega)$, $N_\varkappa(\omega)$, and $D(\omega)$ are polynomials of ω.

The denominator $D(\omega)$ must be an even function of ω, because an odd function must be excluded for physical reasons. $D(\omega)$ therefore can be written in the factorized form

$$D(\omega) = \prod_{i=1}^{n} (\omega^2 - \omega_i^2) . \tag{9}$$

Because of (7) the only condition for the polynomial $N_1(\omega)$ is that its degree must coincide with the degree of $D(\omega)$. After factorizing $N_1(\omega)$ therefore must be of the form

$$N_1(\omega) = \prod_{j=1}^{2n} (\omega - \omega_j^0) . \tag{10}$$

The frequencies ω_j^0 can be positive or negative real quantities.

Obviously $N_1(\omega)$ unambiguously can be split into an even part $N_\mu(\omega)$ and an odd part $N_\varkappa(\omega)$,

$$N_1(\omega) = N_\mu(\omega) + N_\varkappa(\omega) , \tag{11a}$$

where $N_\mu(\omega)$ is given by

$$N_\mu(\omega) = \omega^{2n} + a_{2n-2}\omega^{2n-2} + a_{2n-4}\omega^{2n-4} + \dots + a_2\omega^2 + a_0$$
$$= \prod_{k=1}^{n} (\omega^2 - \bar{\omega}_k^2) \tag{11b}$$

and $N_\varkappa(\omega)$ by

$$N_\varkappa(\omega) = a_{2n-1}\omega^{2n-1} + a_{2n-3}\omega^{2n-3} + \dots + a_3\omega^3 + a_1\omega$$
$$= a_{2n-1}\omega \prod_{l=1}^{n-1} (\omega^2 - \tilde{\omega}_l^2) \qquad (n > 1) \tag{11c}$$

with

$$a_{2n-1} = -\sum_{j=1}^{2n} \omega_j^0 . \tag{11d}$$

Inserting (9), (10), (11b), and (11c) into (8a), (8b), and (8c) we finally get

$$f_1(\omega) = \frac{\prod_{j=1}^{2n} (\omega - \omega_j^0)}{\prod_{i=1}^{n} (\omega^2 - \omega_i^2)} , \tag{12a}$$

$$\mu_\perp(\omega) = \frac{\prod_{k=1}^{n} (\omega^2 - \bar{\omega}_k^2)}{\prod_{i=1}^{n} (\omega - \omega_i^2)} , \tag{12b}$$

$$\varkappa(\omega) = a_{2n-1}\omega \frac{\prod_{l=1}^{n-1} (\omega^2 - \tilde{\omega}_l^2)}{\prod_{i=1}^{n} (\omega^2 - \omega_i^2)} \qquad (n > 1) \tag{12c}$$

and combining (6) and (12a),

$$f_2(\omega) = \frac{\prod_{j=1}^{2n} (\omega - [-\omega_j^0])}{\prod_{i=1}^{n} (\omega^2 - \omega_i^2)} . \tag{12d}$$

In agreement with (12b) and (12c) $\mu_\perp(\omega)$ obviously is an even and $\varkappa(\omega)$ an odd function of ω.

4. Interpretation of the Zeros of the Polynomials as Frequencies of Special Magnons and Magnon-Polaritons

In the preceding chapter the frequencies ω_i, ω_j^0, $\bar{\omega}_k$, and $\tilde{\omega}_l$ were introduced from an algebraical point of view as zeros of polynomials, without any physical interpretation. However, in the following we will show that all these frequencies have a definite physical meaning: They are eigenfrequencies of special magnons and magnon-polaritons.

We begin with the frequencies ω_i. If we insert (12a) and (12d) into (5a) and (5b), respectively, the refractive indices for the two waves propagating along the optical axis are given by the formulae

$$n_1^2(\omega) = \frac{c^2k^2}{\omega^2} = \varepsilon_\perp(\omega) \frac{\prod_{j=1}^{2n} (\omega - \omega_j^0)}{\prod_{i=1}^{n} (\omega^2 - \omega_i^2)} \tag{13a}$$

and

$$n_2^2(\omega) = \frac{c^2k^2}{\omega^2} = \varepsilon_\perp(\omega) \frac{\prod_{j=1}^{2n} (\omega - [-\omega_j^0])}{\prod_{i=1}^{n} (\omega^2 - \omega_i^2)} . \tag{13b}$$

The formulae show that the frequencies ω_i are poles of the refractive indices n_1 and n_2. Therefore, the ω_i's must be interpreted as resonance frequencies of the transverse optical magnons propagating along the optical axis. If the eigenfrequencies of these magnons are denoted as $\omega_{\mathrm{TO},i}$, we therefore have

$$\omega_i = \omega_{\mathrm{TO},i} . \tag{14}$$

According to (13a) the zeros ω_j^0 of n_1 are obviously the frequencies of magnon-polaritons for $\boldsymbol{k} = \boldsymbol{0}$,

$$\omega_j^0 = \omega_j(\boldsymbol{k} = \boldsymbol{0}) . \tag{15}$$

Due to (13b) these frequencies always appear in pairs $[\omega_j(\boldsymbol{k} = \boldsymbol{0}), -\omega_j(\boldsymbol{k} = \boldsymbol{0})]$.

The zeros of $\varkappa(\omega)$ in (12c) can be interpreted as follows: For $\omega = \tilde{\omega}_l$ the permeability tensor becomes diagonal, i.e. the two magnon-polaritons are linearly polarized and degenerate, and the properties agree with those in antiferromagnetic crystals.

At last, for the physical interpretation of the zeros $\bar{\omega}_k$ in the numerator of $\mu_\perp$ (see (12b)) we must consider the waves propagating perpendicular to the optical axis. The refractive indices of this two waves are given by [2, 6]

$$n_{\mathrm{I}}^2 = \frac{c^2k^2}{\omega^2} = \varepsilon_{\parallel}(\omega) \frac{\mu_\perp^2(\omega) - \varkappa^2(\omega)}{\mu_\perp(\omega)} , \tag{16a}$$

$$n_{\mathrm{II}}^2 = \frac{c^2k^2}{\omega^2} = \varepsilon_\perp(\omega)\, \mu_{\parallel}(\omega) . \tag{16b}$$

The second wave with n_{II} can be omitted in the following because within linear spin wave theory this wave is a pure electromagnetic wave without any coupling with magnons.

The electric vector $\boldsymbol{E}$ of the first wave is directed parallel to the optical axis. If we fix the direction of the x-axis parallel to the wave vector $\boldsymbol{k}$, the vectors $\boldsymbol{k}$, $\boldsymbol{E}$, and $\boldsymbol{H}$ of this wave are represented by

$$\boldsymbol{k} = (k_x, 0, 0)\,, \tag{17a}$$

$$\boldsymbol{H} = (H_x, H_y, 0)\,, \tag{17b}$$

$$\boldsymbol{E} = (0, 0, E_z)\,. \tag{17c}$$

The components of the magnetic vector $\boldsymbol{H}$ and the electric vector $\boldsymbol{E}$ fulfil the relations [6]

$$H_x = -i\,\frac{n_{\mathrm{I}}\varkappa}{\mu_\perp^2 - \varkappa^2}\,E_z\,, \tag{18a}$$

$$H_y = -n_{\mathrm{I}}\,\frac{\mu_\perp}{\mu_\perp^2 - \varkappa^2}\,E_z \tag{18b}$$

or, after eliminating E_z,

$$\frac{H_y}{H_x} = -i\,\frac{\mu_\perp(\omega)}{\varkappa(\omega)}\,. \tag{18c}$$

Finally, if we substitute $\mu_\perp$ and $\varkappa$ in (16a) and (18c) by expressions (12b) nad (12c), we get

$$n_{\mathrm{I}}^2 = \frac{c^2k^2}{\omega^2} = \varepsilon_{||}(\omega)\,\frac{\prod\limits_{j=1}^{2n}(\omega^2 - [\omega_j^0]^2)}{\prod\limits_{i=1}^{n}(\omega^2 - \omega_{\mathrm{TO},i}^2)\prod\limits_{k=1}^{n}(\omega^2 - \bar{\omega}_k^2)} \tag{19a}$$

and

$$\frac{H_y}{H_x} = -i\,\frac{\prod\limits_{k=1}^{n}(\omega^2 - \bar{\omega}_k^2)}{a_{2n-1}\omega\prod\limits_{l=1}^{n-1}(\omega^2 - \tilde{\omega}_l^2)}\,. \tag{19b}$$

From (19a) follows that for the wave in consideration now both the frequencies $\omega_{\mathrm{TO},i}$ and $\bar{\omega}_k$ are resonance frequencies, corresponding to the limit $k \to \infty$. If in particular the resonance frequency is $\bar{\omega}_k$, we can conclude from (19b) that for $k \to \infty$

$$\frac{H_y}{H_x} \to 0\,. \tag{20}$$

The magnetic vector $\boldsymbol{H}$ in this limit therefore has the form

$$\boldsymbol{H} = (H_x, 0, 0)\,,$$

i.e. is parallel to $\boldsymbol{k}$. The electric vector vanishes in this limit as can be proved directly from (18a).

In words: In the limit $k \to \infty$ the magnon-polariton is longitudinal with the frequency

$$\bar{\omega}_k = \omega_{\mathrm{LO},k}\,. \tag{21}$$

With this interpretation $\mu_\perp$ (see 12b)) can be written as

$$\mu_\perp(\omega) = \frac{\prod\limits_{i=1}^{n}(\omega^2 - \omega_{\mathrm{LO},i}^2)}{\prod\limits_{i=1}^{n}(\omega^2 - \omega_{\mathrm{TO},i}^2)}\,. \tag{22}$$

Equation (22) can be regarded as a generalized Kurosawa relation for uniaxial ferrimagnetic crystals.

Applications of the dispersion relations are given elsewhere [6].

Acknowledgements

We thank Prof. Dr. G. Borstel (Universität Osnabrück) and Dipl.-Phys. R. Voss (Universität Münster) for stimulating discussions. We are indebted to the Deutsche Forschungsgemeinschaft for financial support.

References

[1] R. CLAUS, L. MERTEN, and J. BRANDMÜLLER, Springer Tracts mod. Phys. **75** (1975).
[2] D. POLDER, Phil. Mag. **40**, 99 (1949).
[3] J. M. ZIMAN, Principles of the Theory of Solids, Cambridge University Press 1972 (p. 258).
[4] R. BECKER and F. SAUTER, Theorie der Elektrizität, Vol. 3, Teubner-Verlag, Stuttgart 1969 (p. 117).
[5] D. L. MILLS and E. BURSTEIN, Rep. Progr. Phys. **37**, 817 (1974) (p. 839).
[6] R. VOSS, R. KOTZOTT, and L. MERTEN, to be published.

(Received August 16, 1984)

phys. stat. sol. (b) **126**, 229 (1984)

Subject classification: 18.2; 22.8

Department of Physics, University of Sofia[1])

Some Critical Properties of the s–d Model

By

Julia M. Wesselinowa

A Green's function technique is developed to study the critical behaviour of the s–d model. The temperature and the magnetic field dependence of the spin magnetization are determined. The critical exponents β and δ are numerically calculated. Applying the scaling laws γ and γ_2 are obtained. The critical exponents are in very good agreement with the experimental data and with other theoretical results.

Die Methode der Greenschen Funktionen wird angewendet, um das kritische Verhalten des s–d-Modells zu untersuchen. Die Abhängigkeit der Spinmagnetisierung von der Temperatur und dem magnetischen Feld wird bestimmt. Die kritischen Exponenten β und δ werden numerisch berechnet. Mit Hilfe der Scaling-Gesetze werden γ und γ_2 erhalten. Die kritischen Exponenten sind in sehr guter Übereinstimmung mit experimentellen und theoretischen Ergebnissen.

1. Introduction

A model proposed to describe the properties of ferromagnetic semiconductors (FS) is the s–f (or s–d) model [1 to 3]. This model which has been studied extensively, includes both localized spin Heisenberg ferromagnet and conduction electrons interacting via an s–f (or s–d) contact term. The critical properties of the FS are not so intensively studied. Experimental results of the nonlinear paramagnetic effect in $CdCr_2Se_4$ and in $HgCr_2Se_4$ near the Curie temperature are obtained by Fugiel et al. [4, 5]. The critical exponent $\tilde{\gamma}_2$ determined from these measurements was compared with the critical exponent γ_2 of the expression $\Delta\chi/B^2$ against $T - T_C$ found from the scaling laws, i.e. $\gamma_2 = 2\beta + 3\gamma$ [6]. The critical behaviour of the FS selenospinels $CdCr_2Se_4$, $HgCr_2Se_4$, and $CuCr_2Se_4$ near their Curie points was experimental studied by Zarek [7], who obtained the critical exponents β, γ, and δ for these materials.

The aim of the present paper is to calculate from the s–d model the critical exponents β and δ for the localized-spin magnetization of the ferromagnetic chalcogenide spinels ACr_2X_4 (A = Cd, Zn, Hg and X = S, Se). Applying the scaling laws we obtain γ and other critical exponents. Our results are compared with other theoretical and experimental results.

2. Model and Method

The Hamiltonian of the system is given by

$$H = H_M + H_E + H_I\,. \tag{1}$$

H_M is the Heisenberg Hamiltonian for the ferromagnetically ordered d electrons,

$$H_M = -\frac{1}{2}\sum_{i,j} J_{ij}\mathbf{S}_i\mathbf{S}_j - g\mu_B H \sum_i S_i^z\,, \tag{2}$$

H_E represents the usual Hamiltonian of the conduction band electrons,

$$H_E = \sum_{\mathbf{q},\sigma} (\varepsilon_{\mathbf{q}\sigma} - \mu)\, a^+_{\mathbf{q}\sigma} a_{\mathbf{q}\sigma}\,, \qquad \sigma = \pm 1\,, \tag{3}$$

---

[1]) 5, Boul. A. Ivanov, 1126 Sofia, Bulgaria.

whereas H_I couples the two subsystems by an intra-atomic exchange interaction,

$$H_I = -\frac{I}{2N}\sum_{q,p}[S^+_{q-p}a^+_{p-}a_{q+} + S^-_{q-p}a^+_{p+}a_{q-} + S^z_{q-p}(a^+_{p+}a_{q+} - a^+_{p-}a_{q-})]\,. \tag{4}$$

The retarded Green function to be calculated is defined in matrix form as

$$\tilde{G}_k(t) = -i\theta(t)\,\langle[B_k(t), B_k^+]\rangle = \\ = \begin{pmatrix} \langle\langle S_k^+(t); S_k^-\rangle\rangle & \langle\langle S_k^+(t);\ a^+_{k-}a_{k+}\rangle\rangle \\ \langle\langle a^+_{k+}(t)\,a_{k-}(t); S_k^-\rangle\rangle & \langle\langle a^+_{k+}(t)\,a_{k-}(t);\, a^+_{k-}a_{k+}\rangle\rangle \end{pmatrix}. \tag{5}$$

The operator B_k stands symbolically for the set S_k^+, $a^+_{k+}a_{k-}$. For the approximate calculation of (5) we use a method proposed by Mori and Kawasaki [8] and by Tserkovnikov [9] which is appropriate for spin problems. We have used the same method successfully to study the dynamic properties of the s-d model [2].

3. Critical Exponents

The relative localized-spin magnetization in the direction of the mean field $\langle S^z\rangle(T)$ is equal to

$$\langle S^z\rangle = 2\coth(2\beta E_k) - 0.5\coth(0.5\beta E_k)\,, \tag{6}$$

where $\beta = 1/k_BT$ and E_k is the spin-wave energy which is given by [9]

$$E_k = \frac{\langle[[B_k, H], B_k^+]\rangle}{\langle[B_k, B_k^+]\rangle}\,. \tag{7}$$

We obtain for E_k,

$$E_k = 0.5\,(\varepsilon_{11} + \varepsilon_{22} + \sqrt{(\varepsilon_{11} - \varepsilon_{22})^2 + 4\varepsilon_{12}\varepsilon_{21}}\,) \tag{8}$$

with

$$\varepsilon_{11} = g\mu_BH + \frac{1}{2\langle S^z\rangle}\left\{\frac{1}{N}\sum_q (J_q - J_{k-q})\,(2\langle S^z_qS^z_{-q}\rangle - \langle S^-_{k-q}S^+_{k-q}\rangle) + \right. \\ \left. + \frac{I}{N^2}\sum_{q,p}(\langle S^-_{p-q}a^+_{p+}a_{q-}\rangle + \langle S^z_{q-p}a^+_{p+}a_{q+}\rangle - \langle S^z_{q-p}a^+_{p-}a_{q-}\rangle)\right\},$$

$$\varepsilon_{12} = \frac{1}{2\varrho}\left\{-\frac{I}{N}\sum_q(\langle S^z_{q-k}a^+_{k+}a_{q+k+}\rangle - \langle S^z_{-q}a^+_{q-}a_{k-}\rangle) - \right. \\ \left. - \frac{I}{2N}\sum_q(\langle S^-_qa^+_{q+}a_{k-}\rangle + \langle S^-_{k+q}a^+_{k+}a_{k-q-}\rangle)\right\},$$

$$\varepsilon_{21} = \frac{1}{2\langle S^z\rangle}\left\{-\frac{I}{N}\sum_q(\langle S^z_qa^+_{q+}a_{k+}\rangle - \langle S^z_{k+q}a^+_{k-}a_{k-q-}\rangle) + \right. \\ \left. + \frac{I}{2N}\sum_q(\langle S^+_qa^+_{k-}a_{q+}\rangle + (S^+_{k+q}a^+_{k-q-}a_{k+}\rangle)\right\},$$

$$\varepsilon_{22} = 2\mu_BH - \frac{1}{2\varrho}\left\{\frac{I}{2N}\sum_q(\langle S^+_{q-k}a^+_{k-}a_{q+}\rangle - \langle S^+_{k-q}a^+_{q-}a_{k+}\rangle) - \right. \\ - \frac{I}{2N}\sum_q(\langle S^z_{k-q}a^+_{q+}a_{k+}\rangle - \langle S^z_{q-k}a^+_{k-}a_{q-}\rangle) + \\ \left. + \frac{I}{2}\langle S^z_0a^+_{k-}a_{k-}\rangle - \frac{I}{2}\langle S^z_0a^+_{k+}a_{k+}\rangle\right\}.$$

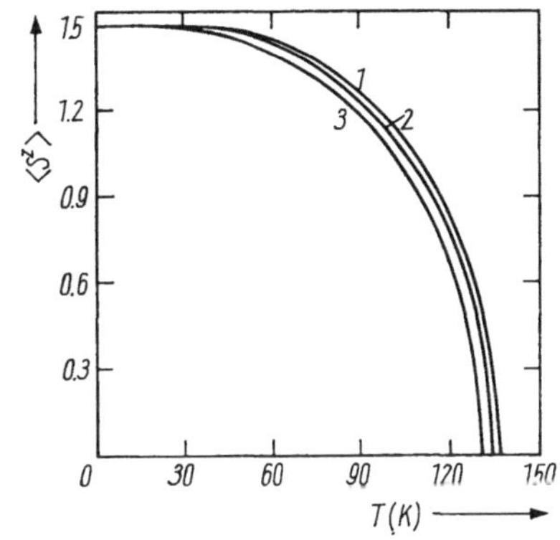

Fig. 1

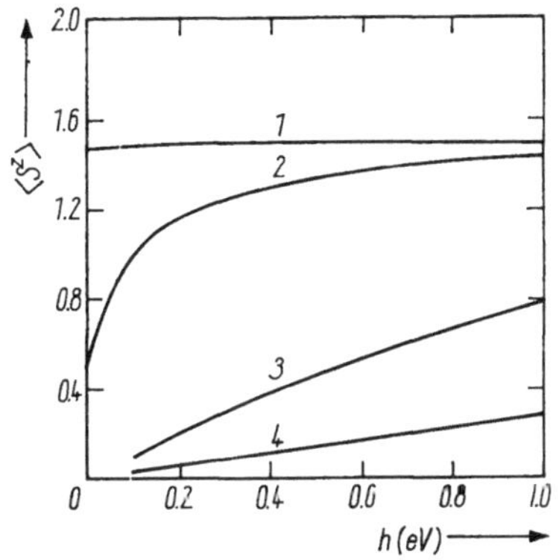

Fig. 2

Fig. 1. Relative localized-spin magnetization $\langle S^z \rangle$ for $CdCr_2Se_4$ as a function of temperature T for different conduction band-width values W ($J_0 = 0.0001$ eV, $I = 0.5$ eV, $T_C = 132.5$ K). (1) $W = 0.1$, (2) 0.5, (3) 1.0 eV

Fig. 2. Relative localized-spin magnetization $\langle S^z \rangle$ for $CdCr_2Se_4$ (parameters as in Fig. 1) as a function of magnetic field $h \equiv g\mu_B H$ for $W = 0.1$ eV and for different temperatures T. (1) $T = 50$, (2) 130, (3) 400, (4) 1000 K

In the generalized Hartree-Fock approximation (GHFA) (8) leads to an optical and to an acoustic spin-wave branch. For $\boldsymbol{k} = \boldsymbol{0}$ we obtain the two solutions,

$$E_{\text{opt}} = 2\mu_B H + I(\varrho + \langle S^z \rangle) \,, \qquad E_{\text{ac}} = g\mu_B H \,, \tag{9}$$

where ϱ is the conduction-electron magnetization [2].

The temperature dependence of $\langle S^z \rangle$ was calculated numerically using parameters appropriate to the ferromagnetic chalcogenide spinels [10, 11] for $H = 0$ (Fig. 1). $\langle S^z \rangle$ shows a steeper slope near T_C in comparison with the MFA. It is valid $T_C(\text{GHFA}) < T_C(\text{MFA})$. Fig. 2 shows a plot of $\langle S^z \rangle$ as a function of h ($h \equiv g\mu_B H$) for $W = 0.1$ eV and for several values of T.

It should be noted here that we have not made an expansion with respect to powers of $1/Z$ (Z is the number of nearest neighbours), which would not be valid in the vicinity of T_C, and so we have taken into account all summation terms.

The critical exponent β describing the order parameter $\langle S^z \rangle$ is defined as [12]

$$\langle S^z \rangle = A \left(\frac{T_C - T}{T_C} \right)^\beta \quad \text{for} \quad T \to T_C^- \,, \tag{10}$$

whereas changes of magnetization with the effective magnetic field intensity h satisfy the dependence

$$\langle S^z \rangle = B h^{1/\delta} \,, \qquad T \approx T_C \,. \tag{11}$$

β is independent of the magnitude of the spin, but dependent on the dimensionality of the lattice. We have calculated numerically β and δ for a three-dimensional cubic lattice in the temperature range $T_C - 40\text{ K} < T < T_C - 5\text{ K}$ from

$$\ln \langle S^z \rangle = \ln A + \beta \ln (1 - T/T_C) \,, \tag{12}$$

$$\ln \langle S^z \rangle = \ln B + (1/\delta) \ln h \,, \tag{13}$$

and obtained the following values,

$$\beta = 0.361 \pm 0.006\,, \qquad A = 1.819 \pm 0.006; \tag{14}$$

$$\delta = 4.471 \pm 0.006\,, \qquad B = 1.506 \pm 0.006\,. \tag{15}$$

Applying the scaling laws [12] the exponent γ for the longitudinal susceptibility may be expressed in terms of β and δ,

$$\gamma = \beta(\delta - 1) = 1.253\,. \tag{16}$$

For the critical exponent γ_2 [6] we obtain

$$\gamma_2 = 2\beta + 3\gamma = 4.482\,. \tag{17}$$

In the critical region, i.e. $5 \times 10^{-4} < \varepsilon = (T_C - T)/T_C < 5 \times 10^{-3}$ Zarek [7] has found experimentally by magnetic balance for $CdCr_2Se_4$ $\beta = 0.34 \pm 0.02$, $\delta = 4.2 \pm 0.3$, and $\gamma = 1.29 \pm 0.02$; for $HgCr_2Se_4$ $\beta = 0.34 \pm 0.02$, $\delta = 4.1 \pm 0.3$, and $\gamma = 1.30 \pm 0.02$; and for $CuCr_2Se_4$ $\beta = 0.37 \pm 0.02$, $\delta = 4.2 \pm 0.3$, and $\gamma = 1.32 \pm 0.02$. For the critical exponent $\gamma_2 = 2\beta + 3\gamma$ he obtained $\gamma_2 = 4.58$ for $HgCr_2Se_4$ and $\gamma_2 = 4.55$ for $CdCr_2Se_4$.

In the MFA $\beta = 0.5$, $\delta = 3$, $\gamma = 1$, and $\gamma_2 = 4$. This failure of the mean field approximation in the critical region is well known in the theory of phase transition.

Ferer et al. [13] determined β and γ theoretically from the three-dimensional Heisenberg model using the high-temperature expansion (HTSE) to $\beta = 0.385 \pm 0.025$ and $\gamma = 1.43 \pm 0.01$. For γ_2 they have $\gamma_2 = 5.06$.

Using the HTSE technique, too, Lee and Stanley [14, 15] and Baltzer et al. [11] obtained for the three-dimensional Heisenberg model $\beta = 0.365 \pm 0.035$ and $\gamma = 1.36 \pm 0.04$.

Le Guillou and Zinn-Justin [16] calculated the critical exponents of the n-vector model through field-theoretical methods and presented new and more precise values. In the case $n = 3$ (isotropic Heisenberg model) they are

$$\beta = 0.3647 \pm 0.0012 \quad \text{and} \quad \gamma = 1.3866 \pm 0.0012\,.$$

Menyuk et al. [17] concluded from their measurements (using a vibrating-coil magnetometer) on the FS EuO that $\gamma = 1.29$. This value disagrees both with the estimates of γ from high-temperature series expansion and with the work of Als-Nielsen et al. [18] who studied EuO and also EuS using neutron scattering and obtained for both EuO and EuS, $\gamma = 1.39$, but is in agreement with the experimental data of Zarek [7] for γ for the ferromagnetic selenospinels.

Basing on (14) to (17) it may be concluded that the s–d model describes very well the critical properties of the FS and that the method and the approximation applied here give values for the critical exponents, which are in very good agreement with the experimental data for the ferromagnetic chalcogenide spinels and with other theoretical results based on the Heisenberg model. It will be of interest to calculate the critical exponent γ from the longitudinal susceptibility and to compare it with that obtained from the scaling laws in this work, which will be reported elsewhere.

References

[1] W. NOLTING, phys. stat. sol. (b) **96**, 11 (1979).
[2] JULIA M. WESSELINOWA, phys. stat. sol. (b) **120**, 585 (1983).
[3] A. MAUGER and D. L. MILLS, Phys. Rev. B **29**, 3815 (1984).
[4] B. FUGIEL, J. ZIOŁO, and A. CHELKOWSKI, Solid State Commun. **40**, 733 (1981).
[5] B. FUGIEL, J. ZIOŁO, and M. DRZAZGA, Phys. Rev. B **28**, 6470 (1983).

[6] B. Fugiel and J. Ziolo, Phys. Letters **A100**, 259 (1984).
[7] W. Zarek, Acta Phys. Polon. **A52**, 657 (1977).
[8] H. Mori and K. Kawasaki, Progr. theor. Phys. (Kyoto) **27**, 529 (1962).
[9] Yu. A. Tserkovnikov, Teor. mat. Fiz. **7**, 250 (1971).
[10] C. Haas, Phys. Rev. **168**, 531 (1968).
[11] P. K. Baltzer, P. J. Wojtowicz, M. Robbins, and E. Lopatin, Phys. Rev. **151**, 367 (1966).
[12] H. E. Stanley, Introduction to Phase Transitions and Critical Phenomena, Clarendon Press, Oxford 1971.
[13] M. Ferer, M. A. Moore, and M. Wortis, Phys. Rev. B **4**, 3954 (1971).
[14] M. H. Lee and H. E. Stanley, Proc. Internat. Conf. Magnetism, Grenoble 1970.
[15] M. H. Lee and H. E. Stanley, Phys. Rev. B **4**, 1613 (1971).
[16] J. C. Le Guillou and J. Zinn-Justin, Phys. Rev. Letters **39**, 95 (1977).
[17] N. Menyuk, K. Dwight, and T. B. Reed, Phys. Rev. B **3**, 1689 (1971).
[18] J. Als-Nielsen, O. W. Dietrich, W. Kunnmann, and L. Passell, Phys. Rev. Letters **27**, 741 (1971).

(Received August 1, 1984)

phys. stat. sol. (b) 126, 235 (1984)

Subject classification: 13 and 18; 8

Sektion Physik der Karl-Marx-Universität Leipzig[1])

Critical Study of the Static Functional Integral Method in the Hubbard Model

By

B. Lorenz and H. Fehske

The effect of the static approximation in the functional integral representation of the paramagnetic Hubbard model is critically investigated. The exactly solvable two-site model is treated within the one- and two-field static functional integral method and some thermodynamic properties (local magnetic moment, specific heat c_v) are compared with the exact solution for various interaction strengths and electron densities. It is shown that the static functional integral theory yields unphysical results for c_v in the strong interacting system at small densities. The two-field and one-field methods are preferable in the low- and high-temperature regions, respectively.

Der Einfluß der statischen Näherung in der Funktionalintegraldarstellung des paramagnetischen Hubbardmodells wird kritisch untersucht. Das exakt lösbare Zweiplatzmodell wird mit der Ein- und Zweifeldmethode behandelt und einige thermodynamische Eigenschaften (lokales magnetisches Moment, spezifische Wärme c_v) werden für verschiedene Werte der Wechselwirkung und der Elektronendichte mit der exakten Lösung verglichen. Im Bereich großer Wechselwirkung und kleiner Teilchendichten liefert die statische Näherung falsche und unphysikalische Ergebnisse für c_v. Die Zweifeldmethode ist bei tiefen und die Einfeldmethode bei hohen Temperaturen geeigneter.

1. Introduction

In recent years the magnetic properties of transition metals and alloys have attracted increasing attention. Despite much experimental and theoretical work a current understanding of itinerant magnetism has not been achieved. Neither the conventional Stoner theory nor the localized spin models yield a satisfactory description of the magnetic behaviour of these materials. The non-integral value (in units of μ_B) of the ground state magnetization, the Curie-Weiss behaviour of the paramagnetic spin susceptibility, and the temperature dependence of the magnetization observed in the transition metals are characteristic features of both localized and band magnetism.

As the theoretical basis for an interpolating description the Hubbard model [1] in its functional integral (FI) representation is frequently used [2 to 6]. These spin fluctuation theories have provided a fairly accurate description of itinerant magnetic phenomena. Methodically, the Hubbard-Stratonovich transformation is used to linearize the interaction and to express the partition function of the Hubbard model by a functional integral over time-dependent auxiliary fields. The FI is evaluated within various approximations. In the static approximation it is reduced to an ordinary integral which is treated by further approximations, e.g. the saddle point approximation [2], the self-consistent single-site approximation [3], etc.

Unfortunately, the FI representation of the Hubbard Hamiltonian depends on the number and kinds of auxiliary fields introduced by the Hubbard-Stratonovich transformation. Usually, local charge and spin fields connected with the charge density

---

[1]) Karl-Marx-Platz, DDR-7010 Leipzig, GDR.

operator $n_{i\uparrow} + n_{i\downarrow}$ and the spin density operator $n_{i\uparrow} - n_{i\downarrow}$, respectively, are used in the two-field method. However, with the identity for Fermi operators $n_{i\sigma}^2 = n_{i\sigma}$ the FI can be expressed by only one field per site (charge or spin field). This one-field method yields a simpler expression for the partition function [7]. In an exact FI representation both methods are equivalent but evaluating the FI approximately different results can be obtained. Until now the effects of the various approximations, especially that of the static one, have not been investigated very thoroughly. For the Anderson model of a magnetic impurity it has been shown that there may arise considerable errors due to the violation of the Pauli principle in the static approximation [8]. Furthermore, in this model the two-field static FI scheme is superior to the one-field methods since it represents the first corrections to some exactly solvable limits rigorously [8]. A similar statement is obtained for the Anderson model treated in the static and saddle point approximations [9] where the Hartree-Fock result is reproduced by the two-field method only.

In order to investigate the effect of the static approach in the FI theory of the Hubbard model the evaluation of exactly solvable problems is of particular interest. Therefore, we consider the two-site Hubbard model for various electron densities. In the next section the two-site problem is treated rigorously and in the static FI theory using the one- and two-field methods. The results for some thermodynamic functions are given in Section 3 and compared with the exact solution. The main conclusions are presented in Section 4.

2. Two-Site Hubbard Model

The two-site Hamiltonian is written as

$$\begin{aligned} H &= H^{(0)} + U(n_{1\uparrow}n_{1\downarrow} + n_{2\uparrow}n_{2\downarrow}) , \\ H^{(0)} &= -t \sum_\sigma (c_{1\sigma}^+ c_{2\sigma} + c_{2\sigma}^+ c_{1\sigma}) - \left(\mu + \frac{U}{2}\right) \sum_\sigma (n_{1\sigma} + n_{2\sigma}) . \end{aligned} \tag{1}$$

$n_{i\sigma}$ is the occupation number operator for an electron at the site i with spin projection σ, U is the intraatomic Coulomb integral, and $H^{(0)}$ describes the hopping of electrons between the two sites with the overlap parameter t. Since we are working in the grand-canonical ensemble the chemical potential μ is introduced in order to fix the mean electron density per site n. It is convenient to shift the chemical potential by the constant $U/2$. Hence, for $n = 1$ we have $\mu = 0$.

The exact solution and the thermodynamic properties of (1) are discussed in [10, 11]. In the FI formalism the two-site model has the advantage that it can be treated in the static approach without applying additional approximations. Hence, in contrast to the infinite model we are able to investigate the effect of the static approximation separately. If the results of this paper should be relevant also for the infinite Hubbard model, we have to consider those physical quantities which are qualitatively insensitive to the size of the system. Since there is no magnetic long-range order in a finite lattice we restrict our discussion to the behaviour in the paramagnetic phase. The approximate treatment of finite and infinite Hubbard chains for $n = 1$ has shown that some thermodynamic functions, e.g. the specific heat, the entropy, the local magnetic moment, etc. of the model (1) are very similar to those of the infinite system [10 to 13]. The local magnetic moment $L_0 = \frac{3}{4} \langle (n_{i\uparrow} - n_{i\downarrow})^2 \rangle$ of (1) shows the same qualitative temperature dependence as that of a large lattice. Hirsch [14] has found only a week size dependence of L_0 in two dimensions. Furthermore, some available limits ($t = 0$ or $U = 0$) for L_0 in the infinite model are reproduced by the model (1) [11]. The specific heat of (1) shows the temperature dependence which is characteristic of the infinite lattice. Especially, at larger U-values there appears a typical low-temperature peak due to the destruction of magnetic short-range order which is known for the infinite model to be an effect of the hopping-induced effective Heisenberg interaction. Therefore, we investigate the local magnetic moment and the specific heat of (1) in the static FI approach and compare the results with the exact solution.

2.1 Exact solution

The grand partition function of (1) is expressed as

$$Z = 1 + (x + x^3)\, S_1 + x^2 S_2 + x^4\,, \tag{2}$$

with

$$S_1 = 4e^{\beta U/2} \cosh\{\beta t\}, \qquad x = e^{\beta\mu}\,,$$

$$S_2 = 1 + 3e^{\beta U} + 2e^{\beta U/2} \cosh\{\tfrac{1}{2}\beta \sqrt{U^2 + 16t^2}\}\,.$$

$\beta = 1/k_B T$ is the inverse temperature.

The chemical potential μ is determined by the condition

$$\frac{\partial}{\partial\mu}\left(-\frac{1}{\beta}\ln Z\right) = 2n\,, \tag{3}$$

from which we get the equation of fourth degree

$$x^4(2 - n) + x^3(1.5 - n)\, S_1 + x^2(1 - n)\, S_2 + x(0.5 - n)\, S_1 - n = 0\,. \tag{4}$$

The local magnetic moment can be expressed by the local correlation function

$$L_0 = \tfrac{3}{4}\langle (n_{i\uparrow} - n_{i\downarrow})^2\rangle = \tfrac{3}{4}(n - 2\langle n_{i\uparrow} n_{i\downarrow}\rangle) \tag{5}$$

and is calculated from (2) according to

$$L_0 = \frac{3}{4}\frac{1}{\beta}\frac{\partial \ln Z}{\partial U}\,. \tag{6}$$

Let us note that in (5) we have used the identity $n_{i\sigma}^2 = n_{i\sigma}$. An alternative formula for L_0 is given by

$$L_0 = \frac{3}{4}\frac{1}{\beta^2 Z}\frac{\partial^2 Z}{\partial h_i^2}\bigg|_{h_i=0}\,. \tag{7}$$

h_i denotes an external field that couples to the local spin. Formulas (6) and (7) are equivalent definitions of L_0. However, in an approximate treatment of Z this equivalence can be destroyed. In a previous paper [15] we have shown that in the static FI theory better results are obtained from (6) rather than from (7).

The specific heat per particle in the grand-canonical ensemble is given by

$$\frac{c_v}{k_B} = \frac{\beta^2}{2n}\left\{\frac{\partial^2 \ln Z}{\partial\beta^2} - \frac{\left(2n - \dfrac{\partial^2 \ln Z}{\partial\beta\,\partial\mu}\right)^2}{\dfrac{\partial^2 \ln Z}{\partial\mu^2}}\right\}. \tag{8}$$

2.2 Static one-field theory

Rewriting the interaction term of (1) as

$$n_{i\uparrow} n_{i\downarrow} = \tfrac{1}{2}(n_{i\uparrow} + n_{i\downarrow}) - \tfrac{1}{2}(n_{i\uparrow} - n_{i\downarrow})^2 \tag{9}$$

and applying the Hubbard-Stratonovich transformation the partition function of (1) in the static approximation takes the form

$$Z^{\mathrm{st}} = \iint\limits_{-\infty}^{\infty} \mathrm{d}\xi_1\, \mathrm{d}\xi_2 \frac{\beta U}{2\pi}\, \mathrm{e}^{-\frac{\beta U}{2}(\xi_1^2+\xi_2^2)}\, \mathrm{Tr}\, \mathrm{e}^{-\beta H_{\mathrm{L}}}\,. \tag{10}$$

H_{L} is a one-particle Hamiltonian describing the coupling of the fields ξ_i to the local spin operator and is given by

$$H_{\mathrm{L}} = H^{(0)} - U \sum_{j=1}^{2} \sum_{\sigma=\pm} (\sigma\xi_j - \tfrac{1}{2})\, n_{j\sigma}\,. \tag{11}$$

Note that in the infinite Hubbard model further approximations have to be applied in order to evaluate the trace and the integrals for the calculation of Z^{st} [7]. For the model (1) the trace in (10) can be performed and we get

$$\mathrm{Tr}\, \mathrm{e}^{-\beta H_{\mathrm{L}}} = 4\left[\cosh\left\{\frac{\beta U}{2}(\xi_1+\xi_2)\right\} + \cosh\left\{\frac{\beta U}{2}\sqrt{(\xi_1-\xi_2)^2 + \left(\frac{2t}{U}\right)^2}\right\}\right]^2. \tag{12}$$

Inserting (12) into (10) and calculating one integral after a proper substitution analytically the partition function (10) is represented in the form (2) with

$$\begin{aligned} S_1 &= 4\mathrm{e}^{\beta U/4}\sqrt{\frac{\beta}{\pi U}}\int\limits_{-\infty}^{\infty} \mathrm{d}\alpha\, \mathrm{e}^{-\beta\alpha^2/U} \cosh\{\beta\sqrt{\alpha^2+t^2}\}\,, \\ S_2 &= 2\left[1 + \mathrm{e}^{\beta U} + \sqrt{\frac{\beta}{\pi U}}\int\limits_{-\infty}^{\infty} \mathrm{d}\alpha\, \mathrm{e}^{-\beta\alpha^2/U} \cosh\{2\beta\sqrt{\alpha^2+t^2}\}\right]. \end{aligned} \tag{13}$$

The integrals in (13) are easily calculated numerically. The chemical potential, the local moment, and the specific heat are determined according to (4), (6), and (8), respectively.

It is worth noting that besides (9) there is an alternative one-field representation given by

$$n_{i\uparrow} n_{i\downarrow} = -\tfrac{1}{2}(n_{i\uparrow} + n_{i\downarrow}) + \tfrac{1}{2}(n_{i\uparrow} + n_{i\downarrow})^2\,. \tag{14}$$

Using (14) two auxiliary fields are introduced which couple to the charge density operators at sites 1 and 2. The partition function in the static approximation is expressed in an analogous way as above by (2) with

$$\begin{aligned} S_1 &= 4\mathrm{e}^{3\beta U/4}\sqrt{\frac{\beta}{\pi U}}\int\limits_{-\infty}^{\infty} \mathrm{d}\alpha\, \mathrm{e}^{-\beta\alpha^2/U} \cosh\{\beta\sqrt{t^2-\alpha^2}\}\,, \\ S_2 &= 2\mathrm{e}^{\beta U}\left[2 + \sqrt{\frac{\beta}{\pi U}}\int\limits_{-\infty}^{\infty} \mathrm{d}\alpha\, \mathrm{e}^{-\beta\alpha^2/U} \cosh\{2\beta\sqrt{t^2-\alpha^2}\}\right]. \end{aligned} \tag{15}$$

2.3 *Static two-field theory*

In the representations (9) and (14) the identity $n_{i\sigma}^2 = n_{i\sigma}$ has been used. As shown for the Anderson model [8, 9] in this scheme considerable errors arise in the static approximation. The relation

$$n_{i\uparrow}n_{i\downarrow} = \frac{1}{4}\,(n_{i\uparrow} + n_{i\downarrow})^2 - \frac{1}{4}\,(n_{i\uparrow} - n_{i\downarrow})^2 \tag{16}$$

does not employ the Fermi character of the operators and was preferred by many authors [2 to 6]. With (16) two kinds of fields are introduced by the transformation to the FI. One field is connected with the local spin operator and the other one couples to the charge density operator. Then the static partition function becomes

$$Z^{\text{st}} = \int\limits_{-\infty}^{\infty} \mathrm{d}\xi_1\,\mathrm{d}\xi_2\,\mathrm{d}\eta_1\,\mathrm{d}\eta_2 \left(\frac{\beta U}{4\pi}\right)^2 \mathrm{e}^{-\frac{\beta U}{4}(\xi_1^2+\xi_2^2+\eta_1^2+\eta_2^2)}\,\mathrm{Tr}\,\mathrm{e}^{-\beta H_{\mathrm{L}}}\,, \tag{17}$$

$$H_{\mathrm{L}} = H^{(0)} - \frac{U}{2}\sum_{j=1}^{2}\sum_{\sigma=\pm}(\sigma\xi_j + i\eta_j)\,n_{j\sigma}\,. \tag{18}$$

i denotes the imaginary unit.

The Hamiltonian (18) is non-hermitian. The complex eigenvalues of H_{L} are given by

$$E_{1/2} = -z_1 \pm \sqrt{z_2^2 + t^2} - \left(\mu + \frac{U}{2}\right), \qquad E_{3/4} = -E_{1/2}^* - (2\mu + U)\,, \tag{19}$$

with

$$z_{1/2} = \frac{U}{4}\,[\xi_1 + i\eta_1 \pm (\xi_2 + i\eta_2)]\,,$$

from which we get

$$\mathrm{Tr}\,\mathrm{e}^{-\beta H_{\mathrm{L}}} = \sum_{j=1}^{4} \mathrm{e}^{-\beta E_j}\,. \tag{20}$$

Inserting (20) into (17) two of the four integrals are solved analytically and by a suitable transformation of the integration variables the partition function (17) is expressed by (2) with

$$S_1 = 32\,\mathrm{e}^{\beta U/2}\,\frac{\beta}{\pi U}\int\limits_0^{\infty}\mathrm{d}r\int\limits_0^{\pi/2}\mathrm{d}\varphi\; r\,\mathrm{e}^{-2\beta r^2/U}\cosh\{\beta a\}\cos\{\beta b\}\,,$$

$$S_2 = 2\,\mathrm{e}^{\beta U} + 16\,\mathrm{e}^{\beta U/2}\,\frac{\beta}{\pi U}\int\limits_0^{\infty}\mathrm{d}r\int\limits_0^{\pi/2}\mathrm{d}\varphi\; r\,\mathrm{e}^{-2\beta r^2/U}\,(\cosh\{2\beta a\} + \cos\{2\beta b\})\,, \tag{21}$$

with

$$a = \sqrt{(R_1 + R_2)/2}\,, \qquad b = \sqrt{(R_1 - R_2)/2}\,,$$

and

$$R_1 = \sqrt{r^4 + 2t^2r^2\cos(2\varphi) + t^4}\,, \qquad R_2 = |t^2 + r^2\cos(2\varphi)|\,.$$

The twofold integrals (21) can be solved numerically with sufficient accuracy and the thermodynamic quantities of interest are calculated from (4), (6), and (8).

3. Results and Discussion

In order to investigate a broad spectrum of the model parameters we calculate the temperature dependences of the local moment and the specific heat for different interaction strengths ($U/t = 0.5;\, 4;\, 8$) and electron densities ($n = 1;\, 0.6;\, 0.2$). The parameters $U/t = 0.5$, 4, and 8 are representative for the small, intermediate, and large interaction regions, respectively. Because of the symmetry properties of the Hubbard model we can restrict our calculations to $n < 1$, where $n = 1$ corresponds to the half-filled band case in the infinite system. The particle densities $n = 1$ and $n < 1$ will be discussed separately.

3.1 $n = 1$

In Fig. 1 the temperature dependence of L_0 as obtained in the FI approaches described in Section 2 is compared with the exact solution. Obviously, in the low-temperature region the two-field static approximation yields the best result, whereas the one-field methods are even qualitatively incorrect for small U. For instance, the zero-temperature limit of L_0 calculated from (13) is zero in contrast to the exact result (Fig. 1a). This behaviour is confirmed by a ground-state analysis. The U-dependence of $L_0(T = 0)$ and the ground-state energy E_0 are shown in Fig. 2 and 3, respectively. The correct value of L_0 in the band limit ($U = 0$) is only reproduced by the two-field approach which also yields the best fit to the exact ground-state energy. Especially, the first-order corrections to E_0 near the band limit are given correctly by the two-field method only. A similar statement has been obtained for the Anderson model [8]. The results of the one-field approach according to (13) are discussed more extensively in [16].

At larger temperatures the local moments in the one-field approximations tend rapidly to the exact one. However, in the two-field approach L_0 deviates appreciably and the exact high-temperature limit $L_0(T = \infty) = 3/8$ is not obtained. We have calculated this limiting value analytically from (21) and get $L_0(T = \infty) = 3(1 + 1/\pi)/8$. That is, in the high-temperature region the static two-field scheme is not superior to the one-field schemes as one should from cliam the ground-state discussion. Note that the alternative calculation of L_0 in the two-field scheme according to (7) yields the high-temperature limit $L_0 = 3(1/2 + 1/\pi)/8$ which is smaller than the exact value. A similar behaviour is also observed in the infinite lattice [17]. In the one-field method (13) the formulas (6) and (7) yield identical results and the correct limit $L_0(T = \infty) = 3/8$.

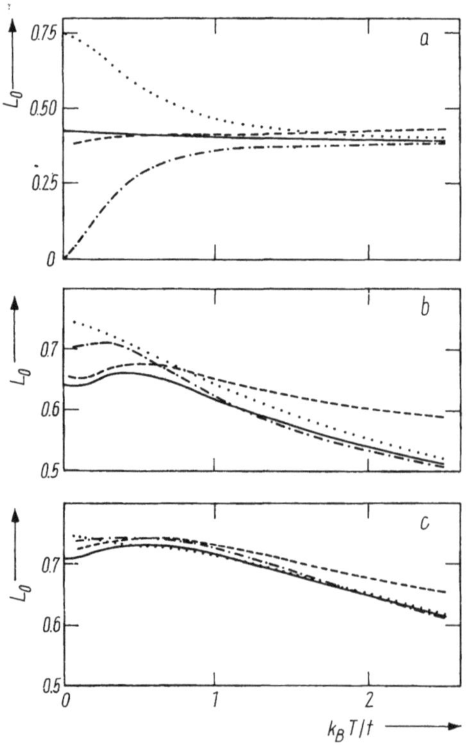

Fig. 1. Local magnetic moment L_0 of the two-site Hubbard model vs. temperature for $n = 1$ and a) $U = 0.5t$, b) $4t$, and c) $8t$. ——— exact result, — — — two-field static approximation according to (16), ·—·—· spin-field static approximation according to (9), ·········· charge-field static approximation according to (14)

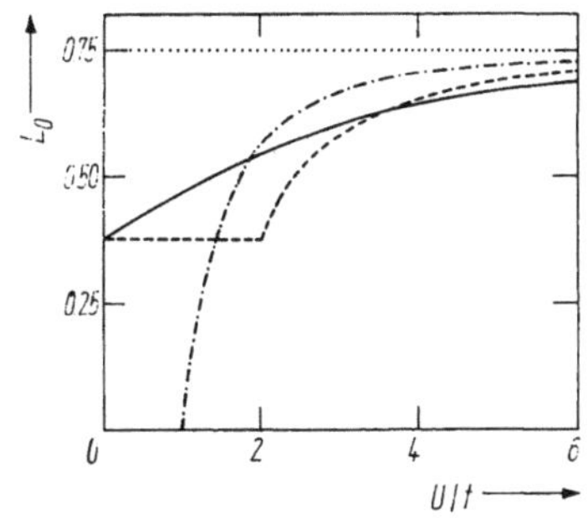

Fig. 2. Ground-state local moment as function of interaction strength U. Notations as in Fig. 1

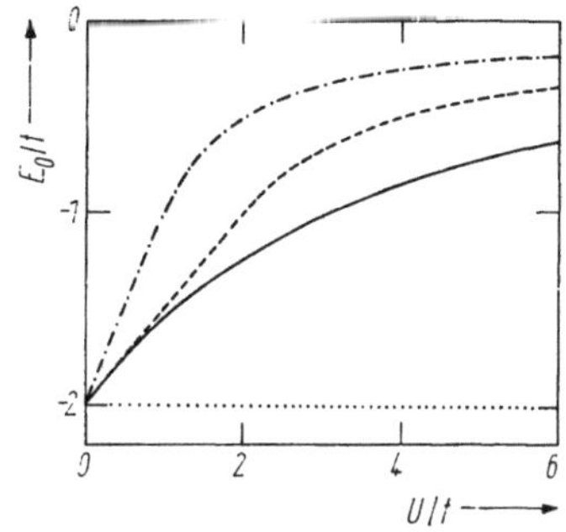

Fig. 3. Ground-state energy of the two-site model. Notations as in Fig. 1

The specific heat is shown in Fig. 4. For small values of U there is a reasonable agreement of all approximate methods with the exact result (Fig. 4a). In the case of intermediate and large U (Fig. 4b, c) there is a considerable overestimation of c_v in

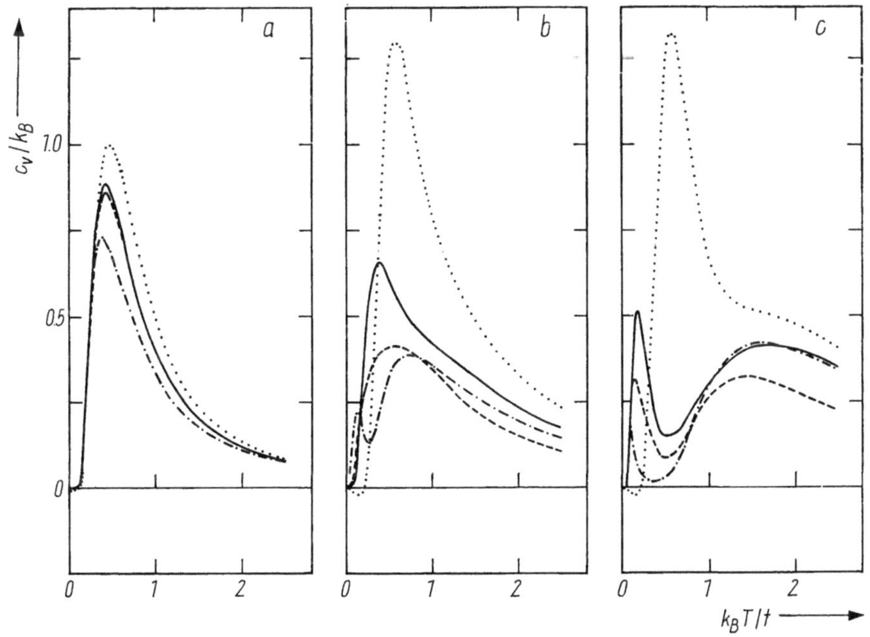

Fig. 4. Temperature dependence of the specific heat per particle c_v for $n = 1$ and a) $U = 0.5t$, b) $4t$, c) $8t$. Notations as in Fig. 1

the one-field approach using the charge fields (15). This yields a negative low-temperature entropy and a negative c_v at very low T. Furthermore, the characteristic two-peak structure of c_v at $U = 8t$ is not well resolved. In the one-field approximation (13) and in the two-field scheme two peaks are obtained (Fig. 4c). The latter method results in a better low-temperature description and the former one resolves the low-temperature peak of c_v already at $U = 4t$ (Fig. 4b). Obviously, this behaviour is strongly connected with the excitation spectrum. In the one-field approach (13) the excitation energy due to the destruction of magnetic short-range order is too small compared with the exact one resulting in a shift of the narrow peak of c_v to lower temperatures (Fig. 4b, c). In the high-temperature region the specific heat in the static two-field method is considerable smaller than the exact one confirming the conclusion that the high-temperature description in this scheme is not very accurate.

3.2 $n < 1$

For $n = 0.6$ and $n = 0.2$ the temperature dependence of L_0 is shown in Fig. 5 and 6, respectively. The results of the comparative study of the different FI schemes are

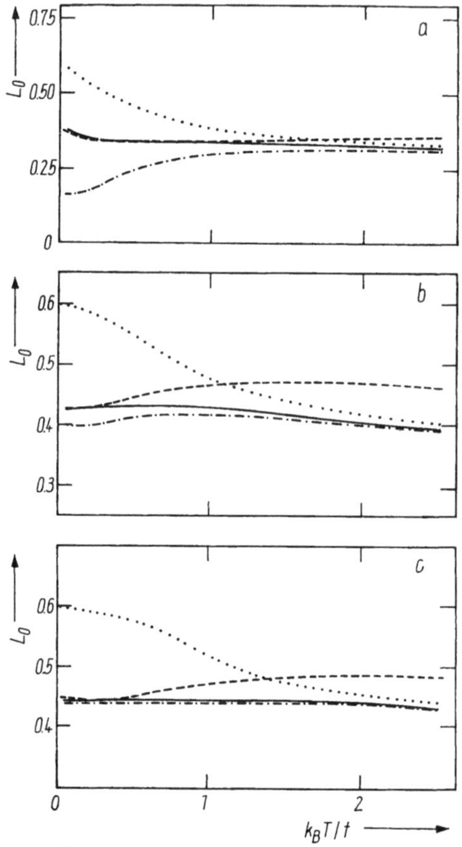

Fig. 5

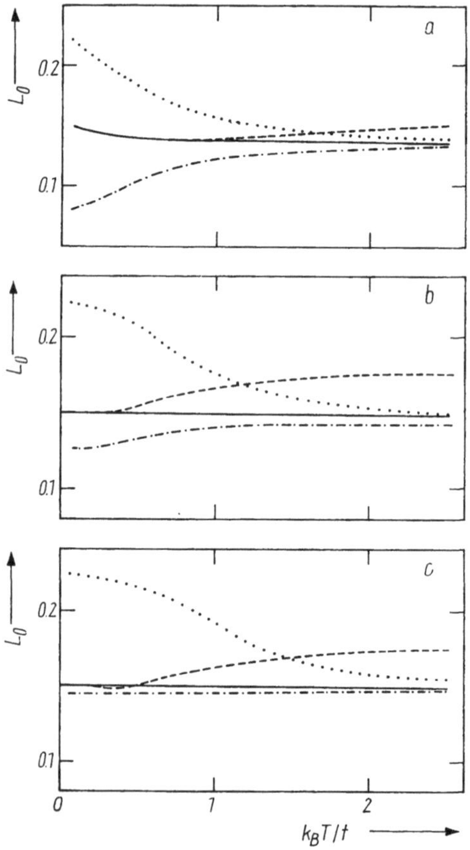

Fig. 6

Fig. 5. Local magnetic moment vs. temperature for $n = 0.6$ and a) $U = 0.5t$, b) $4t$, c) $8t$. Notations as in Fig. 1

Fig. 6. Local magnetic moment vs. temperature for $n = 0.2$ and a) $U = 0.5t$, b) $4t$, c) $8t$. Notations as in Fig. 1

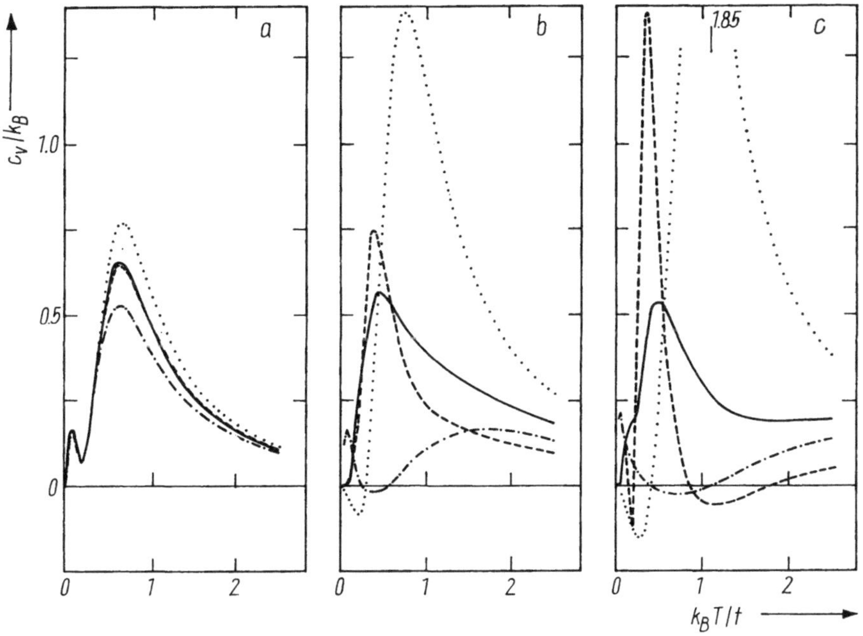

Fig. 7. Specific heat c_v vs. temperature for $n = 0.6$ and a) $U = 0.5t$, b) $4t$, c) $8t$. Notations as in Fig. 1

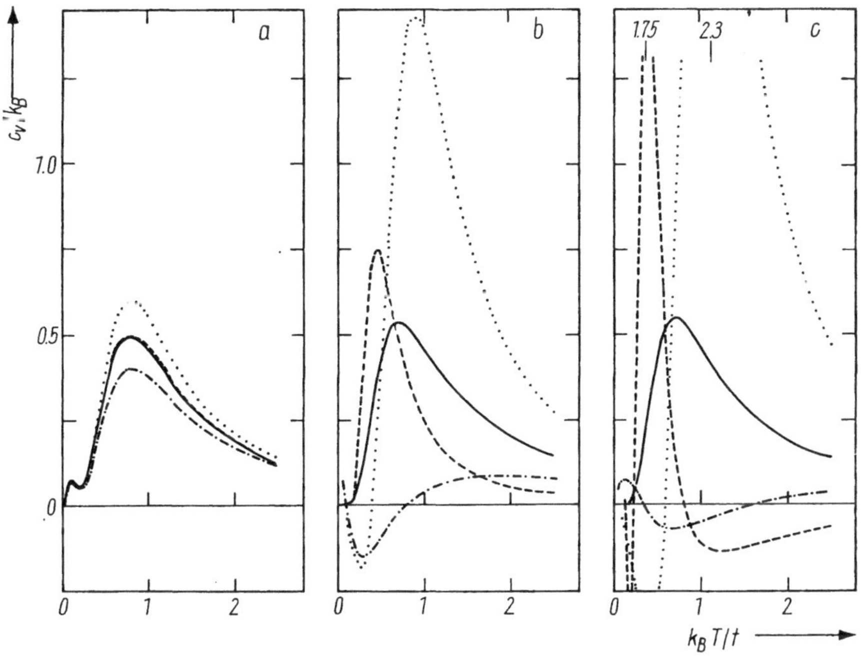

Fig. 8. Specific heat c_v vs. temperature for $n = 0.2$ and a) $U = 0.5t$, b) $4t$, c) $8t$. Notations as in Fig. 1

very similar to the case $n = 1$. The two-field method describes L_0 rather accurate in the low-temperature region. However, considerable deviations and a false limiting value are obtained for intermediate and high temperatures. The results of the one-field approximations are bad for low T and agree quite well at high T. Especially, the high-temperature limit is given correctly.

As to the specific heat the situation is not so clear. The specific heat versus temperature curves are represented in Fig. 7 and 8. In the small U-region (Fig. 7a, 8a) c_v is qualitatively well described. The curve of the two-field approach nearly coincides with the exact one. Note that the small peak at very low T is an effect of the smallness of the system (1) and is a consequence of a discrete low-lying excitation due to the noninteger value of n. In an infinite lattice these excitations are gapless and cause a faster increase of c_v at low T.

For $U = 4t$ the specific heat is only qualitatively well described by the two-field scheme. The peak of c_v is overestimated and shifted to lower T (Fig. 7b, 8b). For intermediate and larger $T c_v$ is appreciably underestimated. In the one-field approximations c_v becomes negative at smaller T and the overall behaviour of the exact specific heat is not reproduced. Hence, the one-field method is not able to describe the thermodynamics of (1). The same statement is true also for larger interaction (Fig. 7c, 8c). Moreover, the two-field scheme yields for $U = 8t$ negative c_v values in the low- and intermediate temperature regions (Fig. 7c, 8c) indicating the break-down of the static FI theory as applied to the model (1).

4. Conclusions

On the basis of the results discussed in the previous section we can estimate the range of validity of the various static FI approximations. Since we have considered physical quantities which do not depend strongly on the size of the lattice (cf. Section 1) and some results for the two-site model in the static approximation are also obtained in the paramagnetic phase of the infinite Hubbard model (e.g. negative low-temperature entropy [6]), we expect the main statements to be valid also in the latter case. Our conclusions are summarized as follows:

(i) The static two-field approach describes the thermodynamic behaviour of the model (1) quite well for small interaction strengths. In the intermediate and large interaction regions the results are qualitatively correct for $n = 1$ although considerable deviations from the exact result appear at higher temperatures. For $n < 1$, however, the approach is bad and thermodynamic stability conditions are violated at large U (negative specific heat).

(ii) The static one-field approximations for small U yield qualitatively good results for c_v but not for the low-temperature local moment. For intermediate and large interactions there are appreciable quantitative errors and, for $n < 1$, the specific heat becomes negative in some temperature regions. However, in contrast to the two-field method, the results for c_v and L_0 in the high-temperature region are fairly accurate and the correct limits for L_0 are reproduced.

(iii) As a general tendency all the static FI approaches give unphysical results for c_v in the strong interacting system with small electron densities. The two-field and one-field methods yield the best results in the high- and low-temperature regions, respectively.

It is worth noting that besides the representations of the Hubbard interaction (9), (14), (16) considered in Section 2 there are other possibilities which introduce the square of the vector spin operator $(\boldsymbol{S}_i)^2$ [18, 19]. However, the latter method yields incorrect results even in the atomic limit ($t = 0$) whereas with (9), (14), or (16) the static approximation is an exact procedure for $t = 0$. The static approximation neglects the anticommutativity of the Fermi operators. In the atomic limit of (1) with (9), (14), or (16) there appear only occupation number operators which are commutative. Using the vector spin decomposition, however, the components of the spin operator $\boldsymbol{S}_l$ do not commute and the static approach does not reproduce the limit $t = 0$.

Hubbard [4] has proposed a representation which combines the vector character of the spin with the commutative properties of the atomic-limit operators and which is able to give the correct atomic-limit result in the static approximation. In this method the spin vector is projected onto an arbitrary unit vector $\boldsymbol{e}$ and the partition function is averaged over all directions of $\boldsymbol{e}$ in order to preserve the spin rotational invariance of the Hubbard Hamiltonian. For this method we can show that the ground-state results are identical to those of the corresponding scalar representations (9) or (16). Within the single-site approximation for the infinite system it has been pointed out [20] that the scalar and projected vector field methods are equivalent in the paramagnetic state. Differences appear in the description of long-range magnetic ordering which is not considered here.

References

[1] J. Hubbard, Proc. Roy. Soc. A **276**, 238 (1963).
[2] H. Hasegawa, J. Phys. Soc. Japan **46**, 1504 (1979).
[3] H. Hasegawa, J. Phys. Soc. Japan **49**, 178, 963 (1980).
[4] J. Hubbard, Phys. Rev. B **19**, 2626 (1979); B **20**, 4584 (1979).
[5] S. N. Evangelou, H. Hasegawa, and D. M. Edwards, J. Phys. F **12**, 2035 (1982).
[6] S. N. Evangelou and D. M. Edwards, J. Phys. C **16**, 2121 (1983).
[7] M. Sakoh and M. Shimizu, J. Phys. Soc. Japan **40**, 974 (1976).
[8] D. J. Amit and H. Keiter, J. low-Temp. Phys. **11**, 603 (1973).
[9] C. A. Macedo, M. D. Coutinho-Filho, and M. A. De Moura, Phys. Rev. B **25**, 5965 (1982).
[10] U. Bernstein and P. Pincus, Phys. Rev. B **10**, 3626 (1974).
[11] B. Lorenz, phys. stat. sol. (b) **119**, 555 (1983).
[12] H. Shiba, Progr. theor. Phys. (Kyoto) **48**, 2171 (1972).
[13] H. Takano and M. Suzuki, Physica (Utrecht) **109**, A618 (1981).
[14] J. E. Hirsch, Phys. Rev. Letters **20**, 1900 (1983).
[15] B. Lorenz and H. Fehske, Proc. 20th Winter School Theoretical Physics, Karpacz 1984, to be published.
[16] H. Fehske and B. Lorenz, J. Phys. C, in the press.
[17] H. Fehske, phys. stat. sol. (b) **120**, 611 (1983).
[18] M. Cyrot, J. Physique **33**, 125 (1972).
[19] T. Moriya and Y. Takahashi, J. Phys. Soc. Japan **45**, 397 (1978).
[20] H. Hasegawa, Solid State Commun. **39**, 1229 (1981).

(Received August 2, 1984)

phys. stat. sol. (b) 126, 247 (1984)

Subject classification: 13.5

Institute of Physics, Academy of Sciences of Vietnam, Hanoi[1])

On the Damping of the Excitonic Polaritons

By

HOANG NGOC CAM, NGUYEN VAN HIEU, and NGUYEN AI VIET

The general properties of the damping constants of polaritons in the two-branch and three-branch models are considered in the framework of the Green's function method. In particular, the expressions for the damping constants of the polaritons in terms of those of the excitons are established.

Рассмотрены общие свойства констант затухания поляритонов в двух и трехветленной моделях в рамаках метода функции Грина. В частности установлены их выражения через константы затухания экситонов.

1. Introduction

There are three different approaches in the theory of the excitonic polaritons in the direct band gap semiconductors:

1. The dielectric formalism in which the polaritons appear as the poles of the dielectrical constant of the solid [1].
2. The Hamiltonian diagonalization method based on the use of the Bogolyubov transformation to rewrite the Hamiltonian of the exciton–photon system as that of a system of free quasi-particles [2 to 4].
3. The Green function approach in which the polaritons appear as the poles of the corresponding Green functions [5].

In studying the polariton interaction processes both the second and third methods might be applied. For excitons without damping they are equivalent [5, 6]. When the damping of the excitons must be taken into account the Green function method is more convenient. In this work we apply the Green function approach of [5] to establish the relations between the damping constants of the polaritons in the two-branch and three-branch models. We assume that the damping constants of the excitons are known. They have been calculated, for example, in [7, 8]. Our results for the two-branch model agrees with those of [1].

2. Two-Branch Model

First we consider the two-branch model in which there is a transition of the photon into only one bound state of the electron–hole pair, i.e. into only one exciton state. Denote by $E(k)$ the energy of the exciton with the momentum k, by ε_∞ the high frequency dielectric constant, and by $f(k)$ the k-dependent effective constant of the exciton–photon transition. The latter has been calculated in [5]. We assume that the photon is absorbed by the semiconductor only through its transition into the exciton which undergoes different recombination processes. In other words, except for the transition of the photon into the exciton we neglect the direct interaction of the photon with other quasi-particles. Therefore, the photon damping is negligible in the absence

[1]) Nghia Do, Tu Liem, Hanoi, Vietnam.

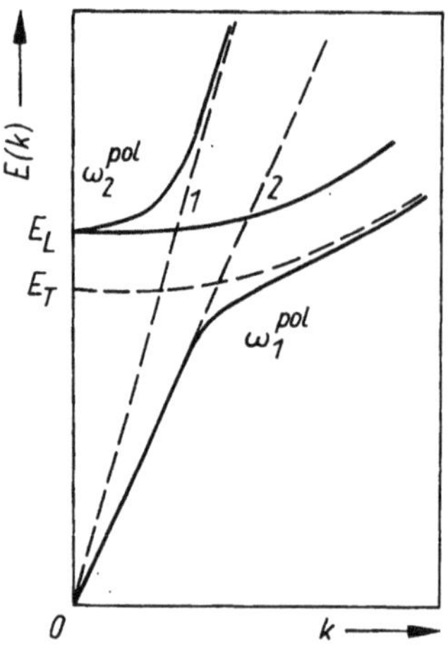

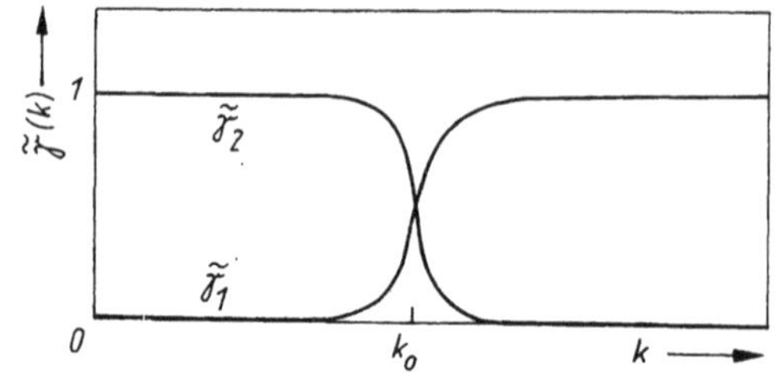

Fig. 1 Fig. 2

Fig. 1. Dispersion curves of polaritons in the two-branch model. (1) $E = k\varepsilon_\infty^{-1/2}$, (2) $k\varepsilon_\infty^{-1/2}(1 + f^2)\, E^{2-1/2}$

Fig. 2. Damping constants of polaritons in the two-branch model. $k_0 \approx E_T\varepsilon_\infty^{1/2}\,(1 + f^2/E_T^2)^{1/2}$

of the exciton–photon transition. Denote by $\Gamma(k)$ the halfwidth of the exciton. It may depend on the exciton momentum k. The Fourier transforms of the Green functions of the bare exciton with the damping and of the bare photon, i.e. of the exciton (with the damping) and the photon without their mutual transition are equal to

$$G_0(k, \omega) = \frac{1}{\omega^2 - E(k)^2 - i\omega\Gamma(k)}, \tag{1}$$

$$D_0(k, \omega) = \frac{1}{\varepsilon_\infty\omega^2 - k^2}.$$

The Fourier transform $G(k, \omega)$ of the exact Green function of the (dressed) exciton is determined by the inhomogeneous equation

$$G = G_0 + f^2\omega^2 G_0 D_0 G\,. \tag{2}$$

The solution of this equation can be written in the form

$$G(k, \omega) = \sum_{i=1}^{2} \frac{F_i(k)}{\omega^2 - \omega_i^{\text{pol}}(k)^2 - i\omega\gamma_i(k)}, \tag{3}$$

where the index $i = 1, 2$ labels the polariton branches,

$$\omega_{1,2}^{\text{pol}}(k)^2 = \frac{1}{2}\left\{E(k)^2 + \frac{k^2 + f(k)^2}{\varepsilon_\infty} \mp \right.$$
$$\left. \mp \left[\left(E(k)^2 + \frac{k^2 + f(k)^2}{\varepsilon_\infty}\right)^2 - \frac{4k^2E(k)^2}{\varepsilon_\infty}\right]^{1/2}\right\} \tag{4}$$

are the squared energies of the polaritons [2 to 5], and

$$\gamma_1(k) = [E(k)^2 - \omega_1^{\text{pol}}(k)^2]\,[\omega_2^{\text{pol}}(k)^2 - \omega_1^{\text{pol}}(k)^2]^{-1}\,\Gamma(k)\,,$$
$$\gamma_2(k) = [\omega_2^{\text{pol}}(k)^2 - E(k)^2]\,[\omega_2^{\text{pol}}(k)^2 - \omega_1^{\text{pol}}(k)^2]^{-1}\,\Gamma(k) \tag{5}$$

are their halfwidths [1]. The k-dependence of $\omega_{1,2}^{\text{pol}}$ and of the ratios

$$\tilde{\gamma}_{1,2}(k) = \frac{\gamma_{1,2}(k)}{\gamma_1(k) + \gamma_2(k)}$$

are plotted in Fig. 1 and 2.

From (5) it follows that the damping constants $\gamma_1(k)$ and $\gamma_2(k)$ of the polaritons must satisfy the relation

$$\frac{\gamma_1(k) - \gamma_2(k)}{\gamma_1(k) + \gamma_2(k)} = \frac{2E(k)^2 - \omega_1^{\text{pol}}(k)^2 - \omega_2^{\text{pol}}(k)^2}{\omega_1^{\text{pol}}(k)^2 - \omega_2^{\text{pol}}(k)^2} \tag{6}$$

which could be checked experimentally. The functions $F_i(k)$ are expressed in terms of $E(k)$, k, ε_∞, and $f(k)$. The longitudinal–transverse splitting

$$E_{\text{LT}} = E_{\text{L}} - E_{\text{T}}\,;$$
$$E_{\text{L}} = \omega_2^{\text{pol}}(0)\,, \qquad E_{\text{T}} = E(0)\,,$$

equals

$$E_{\text{LT}} = [E(0)^2 + f(0)^2]^{1/2} - E(0)\,. \tag{7}$$

3. Three-Branch Model

It is straightforward to generalize the above results to the case of the three-band model in which there exist the transitions with the effective constants $f_i(k)$, $i = 1, 2$, of the photon into two exciton states with the energies $E_i(k)$, $i = 1, 2$, and the halfwidths $\Gamma_i(k)$, $i = 1, 2$. The energies of three polariton branches and their halfwidths are denoted by $\omega_l^{\text{pol}}(k)$ and $\gamma_l(k)$, $l = 1, 2, 3$. We have

$$\begin{aligned}
\omega_1^{\text{pol}}(k)^2 &= -d(k)\cos\left(\frac{\varphi(k)}{3} - \frac{\pi}{3}\right) + \frac{a(k)}{3}\,,\\
\omega_2^{\text{pol}}(k)^2 &= d(k)\cos\varphi(k) + \frac{a(k)}{3}\,,\\
\omega_3^{\text{pol}}(k)^2 &= d(k)\cos\left(\frac{\varphi(k)}{3} - \frac{\pi}{3}\right) + \frac{a(k)}{3}\,,
\end{aligned} \tag{8}$$

where

$$\varphi(k) = \arccos\left\{\frac{2a(k)^3 - 9a(k)\,b(k) + 27c(k)}{2[a(k)^2 - 3b(k)]^{3/2}}\right\}, \tag{9}$$

$$\begin{aligned}
d(k) &= \tfrac{2}{3}\,[a(k)^2 - 3b(k)]^{1/2}\,,\\
a(k) &= \frac{k^2}{\varepsilon_\infty} + E_1(k)^2 + E_2(k)^2 + f_1(k)^2 + f_2(k)^2\,,\\
b(k) &= \frac{k^2}{\varepsilon_\infty}\,E_1(k)^2 + \frac{k^2}{\varepsilon_\infty}\,E_2(k)^2 + E_1(k)^2\,E_2(k)^2 +\\
&\quad + f_1(k)^2\,E_2(k)^2 + f_2(k)^2\,E_1(k)^2\,,\\
c(k) &= \frac{k^2}{\varepsilon_\infty}\,E_1(k)^2\,E_2(k)^2
\end{aligned} \tag{10}$$

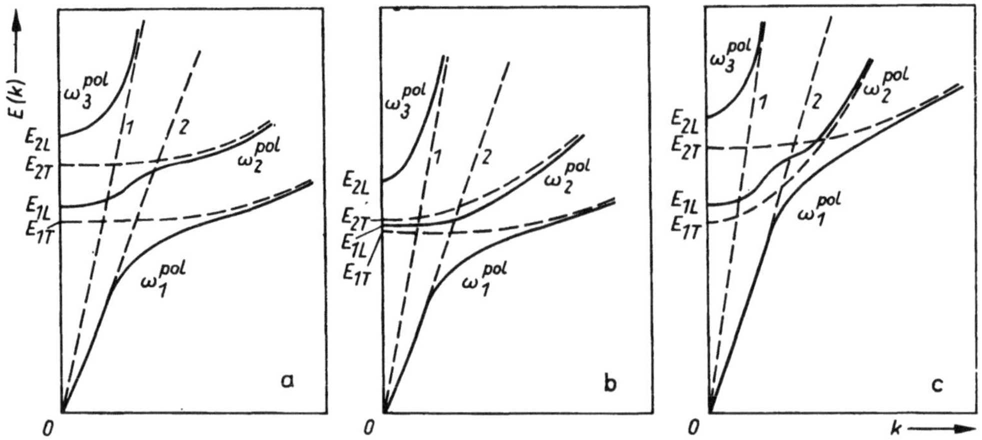

Fg. 3. Dispersion curves of polaritons in the three-branch model. a) $E_{2T} - E_{1T} \gg f_i(0)$, $M_2 < M_1$, (1) $E = k/\sqrt{\varepsilon_\infty}$, (2) $k\varepsilon_\infty^{-1/2}(1 + f_1^2/E_1^2)^{-1/2}(1 + f_2^2/E_2^2)^{-1/2}$. b) $E_{2T} - E_{1T} \ll f_i(0)$, $M_2 < M_1$; (1) $E = k\varepsilon_\infty^{-1/2}$, (2) as in a). c) $E_{2T} - E_{1T} \gg f_i(0)$, $M_2 > M_1$, $(i = 1, 2)$; (1) and (2) as in b)

and

$$\gamma_l(k) = R_l(k)\,\Gamma_1(k) + S_l(k)\,\Gamma_2(k)\,, \tag{11}$$

$$\begin{aligned} R_1(k) &= [\omega_2(k)^2 - \omega_3(k)^2]\, Q_2(\omega_1(k)^2)\,, \\ R_2(k) &= [\omega_3(k)^2 - \omega_1(k)^2]\, Q_2(\omega_2(k)^2)\,, \\ R_3(k) &= [\omega_1(k)^2 - \omega_2(k)^2]\, Q_2(\omega_3(k)^2)\,, \end{aligned} \tag{12}$$

$$\begin{aligned} S_1(k) &= [\omega_2(k)^2 - \omega_3(k)^2]\, Q_1(\omega_1(k)^2)\,, \\ S_2(k) &= [\omega_3(k)^2 - \omega_1(k)^2]\, Q_1(\omega_2(k)^2)\,, \\ S_3(k) &= [\omega_1(k)^2 - \omega_2(k)^2]\, Q_1(\omega_3(k)^2)\,, \end{aligned} \tag{13}$$

$$Q_i(\omega_l(k)^2) = \frac{1}{\Delta(k)}\left\{\omega_l(k)^4 - \omega_l(k)^2\left[\frac{k^2}{\varepsilon_\infty} + E_i(k)^2 + f_i(k)^2\right] + \frac{k^2}{\varepsilon_\infty}\,E_i(k)^2\right\}, \tag{14}$$

$$\begin{aligned} \Delta(k) = {}& \omega_1(k)^4\,[\omega_2(k)^2 - \omega_3(k)^2] + \\ & + \omega_2(k)^4\,[\omega_3(k)^2 - \omega_1(k)^2] + \omega_3(k)^4\,[\omega_1(k)^2 - \omega_2(k)^2]\,; \end{aligned} \tag{15}$$

$$i = 1, 2\,, \qquad l = 1, 2, 3\,.$$

It is easy to verify that the polariton damping constants $\gamma_l(k)$ satisfy following relations

$$\begin{aligned} &\frac{[S_n(k) - R_n(k)]\,\gamma_l(k) - [S_l(k) - R_l(k)]\,\gamma_n(k)}{\gamma_1(k) + \gamma_2(k) + \gamma_3(k)} = \\ &= R_l(k)\,S_n(k) - S_l(k)\,R_n(k)\,; \qquad l, n = 1, 2, 3\,. \end{aligned} \tag{16}$$

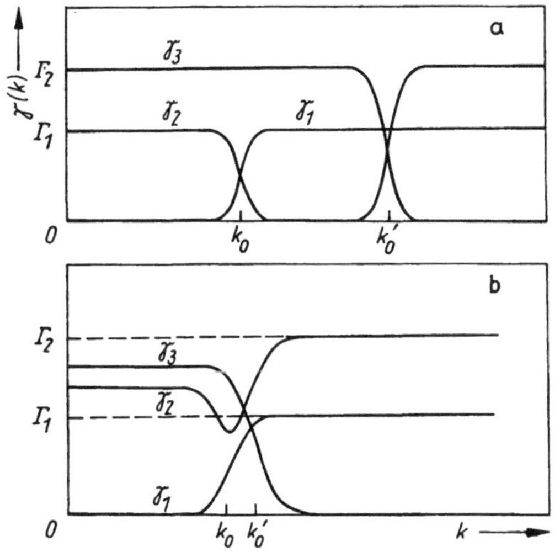

Fig. 4. Damping constants of polaritons in the three-branch model a) $E_{2T} - E_{1T} \gg f_i(0)$, $\tilde{E}_{iT} \approx E_{iT}(1 + f_i^2/E_i^2)^{1/2}$; b) $E_{2T} - E_{1T} \ll f_i(0)$ $(i = 1, 2)$. $k_0 \approx \tilde{E}_{1T}\varepsilon_\infty^{1/2}$, $k_0' \approx \tilde{E}_{2T}\varepsilon_\infty^{1/2}$

From the above results we can calculate the values of the longitudinal–transverse splittings

$$E_{iLT} = E_{iL} - E_{iT}\,, \qquad E_{iL} = \omega_{i+1}^{\text{pol}}(0)\,, \qquad E_{iT} = E_i(0)\,; \qquad i = 1, 2$$

and obtain

$$E_{iLT} = \frac{1}{\sqrt{2}}\,\{E_1(0)^2 + f_1(0)^2 + E_2(0)^2 + f_2(0)^2 + {} + (-1)^i\,[(E_1(0)^2 + f_1(0)^2 - E_2(0)^2 - f_2(0)^2)^2 + 4f_1(0)^2\,f_2(0)^2]^{1/2}\}^{1/2} - E_i(0)\,; \quad i = 1, 2\,. \tag{17}$$

At small values of the momentum k, $k \approx 0$, expressions (11) to (15) are reduced into the approximate formulae

$$\gamma_1(k) \approx 0\,, \tag{18}$$

$$\gamma_2(k) \approx \frac{E_2(0)^2 + f_2(0)^2 - \omega_2^{\text{pol}}(0)^2}{\omega_3^{\text{pol}}(0)^2 - \omega_2^{\text{pol}}(0)^2}\,\Gamma_1(k) + \frac{E_1(0)^2 + f_1(0)^2 - \omega_2^{\text{pol}}(0)^2}{\omega_3^{\text{pol}}(0)^2 - \omega_2^{\text{pol}}(0)^2}\,,$$

$$\gamma_3(k) \approx -\,\frac{E_2(0)^2 + f_2(0)^2 - \omega_3^{\text{pol}}(0)^2}{\omega_3^{\text{pol}}(0)^2 - \omega_2^{\text{pol}}(0)^2}\,\Gamma_1(k) - \frac{E_1(0)^2 + f_1(0)^2 - \omega_3^{\text{pol}}(0)^2}{\omega_3^{\text{pol}}(0)^2 - \omega_2^{\text{pol}}(0)^2}\,.$$

The momentum dependence of the photon energies and of their halfwidths, i.e. the form of the corresponding dispersion curves, for some relative values of $f_i(0)$, E_{iT}, E_{iL} and

$$M_i = \left[2\,\frac{\mathrm{d}^2E_i(k)}{\mathrm{d}k^2}\right]\Bigg|_{k=0}\,; \qquad i = 1, 2\,,$$

and for constant Γ_i given in Fig. 3 and 4. The curves plotted in Fig. 3b can be applied to the direct band gap semiconductors with the degenerate valence band and the allowed electromagnetic transition between the valence and conduction bands.

The role of the polaritons in several optical processes has been investigated by many authors [9 to 16]. It is very interesting to measure the halfwidths of the polaritons at different momentum values and to check the above results experimentally.

References

[1] Y. OSAKA, S. IMAI, and Y. TAKEUTI, J. Phys. Soc. Japan **24**, 236 (1968).
[2] S. I. PEKAR, Zh. eksper. teor. Fiz. **33**, 1022 (1957); **34**, 1176 (1958).
[3] V. M. AGRANOVICH, Zh. eksper. teor. Fiz. **37**, 430 (1959).
[4] J. J. HOPFIELD, Phys. Rev. **112**, 1555 (1958).
[5] NGUYEN VAN HIEU, Ann. Phys. (USA) **127**, 179 (1980).
[6] HA VINH TAN and NGUYEN TOAN THANG, Czech. J. Phys., in print.
[7] W. C. TAIT and R. L. WEIHER, Phys. Rev. **178**, 1404 (1969).
[8] V. V. HIZHNYAKOV, phys. stat. sol., **34**, 421 (1969).
[9] J. VOIGT, phys. stat. sol. (b) **64**, 549 (1974).
[10] F. HENNEBERGER and J. VOIGT, phys. stat. sol. (b) **76**, 313 (1976).
[11] V. MAY, K. HENNEBERGER, and F. HENNEBERGER, phys. stat. sol. (b) **94**, 611 (1979).
[12] V. MAY and J. RÖSELER, phys. stat. sol. (b) **102**, 533 (1980).
[13] G. FISHMAN, J. Lum. **18/19**, 289 (1979).
[14] R. BONNEVILLE and G. FISHMAN, Phys. Rev. B **22**, 2008 (1980).
[15] B. SERMEGE and G. FISHMAN, Phys. Rev. B **23**, 5107 (1981).
[16] Y. NOZUE, M. ITOH, and K. CHO, J. Phys. Soc. Japan. **50**, 889 (1981).

(Received May 2, 1984)

phys. stat. sol. (b) **126**, 253 (1984)

Subject classification: 2; 20.1; 22.1.2

Centre for Surface and Vacuum Investigation, Moscow[1])

Calculation of Some Characteristics of an Intrinsic Random Field in Hydrogenated Amorphous Silicon

By

N. M. KHAMIDULLINA

A calculation is presented of some characteristics of an intrinsic random field in a-Si:H needed to interpret statistical and transport properties of this disordered heteropolar material. Numerical results are obtained for the characteristic energies related to the density of states and optical tails, etc. The results are in a satisfactory agreement with the experimental data available. On the basis of the estimated atomic radial distribution function parameters the problem of the medium-range order in a-Si:H is discussed.

Настоящая работа посвящена расчёту некоторых характеристик собственного случайного поля в a-Si:H, связанных со статистическими и кинетическими свойствами этого неупорядоченного гетерополярного материала. С помощью проведённых расчётов получены значения величин "хвоста" плотности состояний в запрещённой зоне и "хвоста" коэффицента междузонного поглощения света и др. Полученные результаты хорошо согласуются с экспериментальными данными. На основе оценки параметров радиальных функций распределения атомов обсуждается вопрос о промежуточном порядке в a-Si:H.

1. Introduction

This paper presents the calculation of some statistic and kinetic characteristics of a random field in a-Si:H caused by random elements of the atomic structure of the substance as such and thus called intrinsic [1].

a-Si:H is known to be a partly heteropolar material (especially when the hydrogen concentration is high enough). This influences various material characteristics as compared to those of a "pure" a-Si. This paper is devoted to the calculation of the binary correlation function of a-Si:H. The results are used to evaluate the characteristic energy entering the Urbach rule.

2. Statement of the Problem. Basic Equations

One of the most important characteristics of a random field is its binary correlation function, $\psi(\boldsymbol{r}, \boldsymbol{r}')$. For heteropolar materials, the hydrogenated amorphous silicon being one of them, this function may be defined as follows [1]:

$$\psi(\boldsymbol{r}) = \frac{1}{(2\pi)^3} \sum_{\mathrm{a,a'}} (n_\mathrm{a} n_{\mathrm{a'}})^{-1/2} \int \mathrm{d}\boldsymbol{q}\; V_\mathrm{a}(\boldsymbol{q})\; V_{\mathrm{a'}}(-\boldsymbol{q})\; S_{\mathrm{aa'}}(\boldsymbol{q})\; \mathrm{e}^{\mathrm{i}\boldsymbol{qr}} \,. \tag{1}$$

Here index a labels the type of atoms (ions), n_a is the average concentration of the atoms of the type a, and $n_\mathrm{a} = \Omega_\mathrm{a}^{-1}$. $V_\mathrm{a}(\boldsymbol{q})$ is the Fourier transform of the atomic

---

[1]) 1 Ezdakov pereulok, 117 334 Moscow, USSR.

(ion) local pseudopotential defined by

$$V_a(\boldsymbol{q}) = \frac{1}{\Omega_a} \int d\boldsymbol{r}\, e^{-i\boldsymbol{q}\boldsymbol{r}}\, V_a(\boldsymbol{r})\,. \tag{2}$$

Furtheron, $S_{aa'}(\boldsymbol{q})$ is the well-known interference function [2]. It may be presented in the form

$$S_{aa'}(\boldsymbol{q}) = \delta_{aa'} + (n_a n_{a'})^{1/2} \int d\boldsymbol{r}\, [g_{aa'}(\boldsymbol{r}) - 1]\, e^{i\boldsymbol{q}\boldsymbol{r}}\,, \tag{3}$$

where $\delta_{aa'}$ is the Kronecker symbol and $g_{aa'}(\boldsymbol{r})$ the atomic binary correlation function; according to [3],

$$\langle n_a(\boldsymbol{r}_1)\, n_{a'}(\boldsymbol{r}_2)\rangle = \frac{1}{\Omega_a \Omega_{a'}}\, g_{aa'}(\boldsymbol{r}_1 - \boldsymbol{r}_2) + \frac{1}{(\Omega_a \Omega_{a'})^{1/2}}\, \delta(\boldsymbol{r}_1 - \boldsymbol{r}_2)\,, \tag{4}$$

$n_a(\boldsymbol{r})$ being the local density of the a-type particles.

As we consider a system with no long-range order the correlation between distant atoms should disappear: $[g_{aa'}(r) - 1] \to 0$ when $r \to \infty$. Hence the function $S(q) \to 1$ when $q \to \infty$ [1].

For what follows the $S(q)$ behaviour at $q \to 0$ is of particular importance. It is known [3] that at thermodynamic equilibrium

$$\lim_{q \to 0} S_{aa'}(q) = (n_a n_{a'})^{1/2}\, T\varkappa\,, \tag{5}$$

where T is the absolute temperature in energy units and $\varkappa$ is the isothermal compressibility.

However, in case of amorphous substances, having some statistic fluctuations of non-thermodynamic character, (5) does not generally hold and $S(0)$ should be considered as an empirical parameter (it may as well be in excess of the right part of (5)).

In a homogeneous and isotropic system considered here expressions (1) and (3) take the form

$$\psi(R) = \frac{1}{2\pi^2} \sum_{a,a'} (n_a n_{a'})^{-1/2} \int_0^\infty \frac{\sin qR}{qR}\, q^2 S_{aa'}(q)\, V_a(q)\, V_{a'}(q)\, dq\,, \tag{6}$$

$$S_{aa'}(q) = \delta_{aa'} + \frac{4\pi}{(\Omega_a \Omega_{a'})^{1/2}} \int_0^\infty \frac{\sin qr}{qr}\, r^2 [g_{aa'}(r) - 1]\, dr\,. \tag{7}$$

$g_{aa'}(r)$ will further be called a radial distribution function.

3. Characteristics of Radial Distribution Functions in a-Si:H

Possible effects of the short-range order on $\psi(R)$ are characterized by the functions $S_{aa'}(q)$. The latter are to be defined either directly from the experiments [2] or by using (3), (7) with the help of some model or experimentally obtained functions $g_{aa'}(r)$ (see, for example, [4]). A useful analytic approximation may be obtained by presenting $g_{aa'}(r)$ in the form of the sum of Gaussians,

$$g_{aa'}(r) = \begin{cases} \sum_{i=1}^{K} A_i \exp\left[-\left(\frac{r - R_i}{\Delta_i}\right)^2\right] & \text{when} \quad r < R_{aa'}\,, \\ 1 & \text{when} \quad r \geqq R_{aa'}\,. \end{cases} \tag{8}$$

The quantities $R_{aa'}$ in (8) will be called the correlation radii of the radial distribution functions, K is the number of the coordination peaks considered; R_i, Δ_i are the coordinate and the width of the i-th peak, respectively.

The values of A_i are defined by the normalization conditions

$$\frac{4\pi}{\Omega_{a'}} \int_i g_{aa'}(r)\, r^2\, dr = N_i\,. \tag{9}$$

Here integration is carried out with respect to each i-th coordination maximum and N_i is the relevant coordination number.

To interpret some properties of amorphous Si and Ge, including the construction of radial distribution functions (RDF), several structural models were offered [5]. The authors of [6, 7] offered models of the closely packed Wigner-Seitz cells either leading to a diamond-type lattice or to a pentagonal, atomic ring structure called an amorphon. Another structural model was based upon the idea of a random network of atoms [8, 9]. RDFs based on these models were in good agreement with some experiments.

Electron diffraction on a-Si films was studied experimentally by Moss and Graczyk [10, 11]. Their experiments allowed to define the coordinates, widths, and coordination numbers of the first two maxima of the $g_{Si-Si}(r)$ function (see Table 1) as well as showed the total disappearance of the third coordination maximum present on the corresponding curve for crystalline Si.

Some parameters of the function $g_{H-H}(r)$ were obtained in [12] on the basis of a computer modelling of the a-Si:H structure. According to this data the coordinate of the first and only well-distinguishable peak of $g_{H-H}(r)$ was ≈ 0.235 nm while the peak width was ≈ 0.030 nm. The RDF peak describing the second coordinating sphere was rather smeared and could hardly be traced.

As to the functions $g_{H-Si}(r)$ and $g_{Si-H}(r)$ the centre coordinate of the first maximum was found to be equal to the length of the Si–H bond. According to [13]

$$R_{SiH} \approx 0.148 \text{ nm}\,. \tag{10}$$

To calculate A_i entering the functions g_{H-Si} and g_{Si-H} the values of Δ_i are needed. Unfortunately, no relevant experimental data is known to us. Therefore, we assumed Δ_1 to be of the same order of magnitude as the relevant quantities entering the functions g_{Si-Si} and g_{H-H} and to make an estimate assumed $\Delta_1 = 0.020$ nm.

To define the positions of the second and the following coordinate maxima (if any) one may use some known parameters of $g_{H-H}(r)$ and $g_{Si-Si}(r)$. Indeed, the distance between the first and second coordinate peaks of the function $g_{H-Si}(r)$, for example, is roughly equal to the coordinate of the centre of the first coordination sphere of $g_{Si-Si}(r)$. According to [10, 11] this coordinate is 0.235 nm. Similar assumptions lead to an approximate evaluation of the position of the second coordination maximum of $g_{Si-H}(r)$: the distance between it and the first peak is of the same order of magnitude as the centre of the first maximum of $g_{H-H}(r)$.

It is obvious from the experimental data mentioned above that there is no sense in taking into account more than two coordination maxima if one wants to describe the RDF of these samples.

For convenience all the known parameters of the functions $g_{aa'}(r)$ of a-Si:H films are given in Table 1.

The values of the correlation radii $R_{aa'}$ may be estimated using the following expression:

$$S_{aa'}(0) = \delta_{aa'} + \frac{4\pi}{(\Omega_a \Omega_{a'})^{1/2}} \int_0^{R_{aa'}} [g_{aa'}(r) - 1]\, dr\,. \tag{11}$$

Table 1

radial distribution functions	R_1 (nm)	R_2 (nm)	Δ_1 (nm)	Δ_2 (nm)	N_1	N_2	A_1	A_2	correlation radii $R_{aa'}$, (nm)
$g_{Si-Si}(r)$	0.235	0.386	0.012	0.023	4	12	9.61	5.58	0.43
$g_{H-H}(r)$	0.235	—	0.030	—	1	—	$\frac{0.96}{C_H}$	—	$0.169 \sqrt[3]{\frac{2}{C_H}}$
$g_{Si-H}(r)$	0.148	0.383	0.020	?	1*)	?	$\frac{3.63}{C_H}$	?	$0.169 \sqrt[3]{\frac{N_1 + N_2}{C_H}}$
$g_{H-Si}(r)$	0.148	0.383	0.020	?	1	?	3.63	?	$0.169 \sqrt[3]{1 + N_2}$

*) The value of N_1 is given for the SiH bond configuration; for SiH_2, SiH_3, and SiH_4 groups the values of N_1 are equal to 2, 3, and 4, respectively.

Taking account of (9) one obtains

$$S_{aa'}(0) = \delta_{aa'} + \sqrt{\frac{\Omega_{a'}}{\Omega_a}}\,(N_1 + N_2) - \frac{4\pi R_{aa'}^3/3}{(\Omega_a \Omega_{a'})^{1/2}} \tag{12}$$

or

$$\frac{4\pi R_{aa'}^3}{3} = (\Omega_a \Omega_{a'})^{1/2} \left\{\delta_{aa'} + \sqrt{\frac{\Omega_{a'}}{\Omega_a}}\,(N_1 + N_2) - S_{aa'}(0)\right\}. \tag{13}$$

According to [4] the quantity of $S_{Si-Si}(0) \ll 1$ in a-Si. Assuming that all the components $S_{aa'}(0)$ are of the same order of magnitude we put $S_{aa'}(0) \ll 1$ in a-Si:H as well. Such a conclusion follows also from an approximate underestimation of the results of the calculation based on (5), then $S(0) \approx 6 \times 10^{-4}$.

Thus (12) may be simplified as follows:

$$\frac{4\pi R_{aa'}^3}{3} = (\Omega_a \Omega_{a'})^{1/2} \left\{\delta_{aa'} + \sqrt{\frac{\Omega_{a'}}{\Omega_a}}\,(N_1 + N_2)\right\}. \tag{14}$$

Hence we obtain

$$\left.\begin{aligned} R_{Si-Si} &= R_a \sqrt[3]{1 + N_1 + N_2} \approx 0.43\ (\text{nm})\,, \\ R_{H-H} &= R_a \sqrt[3]{\frac{1 + N_1}{C_H}} \approx 0.169 \sqrt[3]{\frac{2}{C_H}}\ (\text{nm})\,, \\ R_{Si-H} &= R_a \sqrt[3]{\frac{N_1 + N_2}{C_H}} \approx 0.169 \sqrt[3]{\frac{N_1 + N_2}{C_H}}\ (\text{nm})\,, \\ R_{H-Si} &= R_a \sqrt[3]{N_1 + N_2} \approx 0.169 \sqrt[3]{1 + N_2}\ (\text{nm})\,. \end{aligned}\right\} \tag{15}$$

Here $R_a = 0.169$ nm is the Si atomic radius [17]; the symbol C_H is defined by $C_H = n_H/n_{Si}$, n_H and n_{Si} being the average concentrations of H and Si atoms. The value of C_H depends on the film preparation conditions; in most cases known to us C_H varies from 3 to 30% [14 to 16]. Note that by definition

$$\Omega_{Si} = \frac{4\pi R_a^3}{3}.$$

This allows us to estimate n_{Si} and Ω_{Si} entering the expression for $\psi(R)$,

$$\Omega_{Si} = 0.2 \times 10^{-22}\ \text{cm}^3 \quad \text{and} \quad n_{Si} = 5 \times 10^{22}\ \text{cm}^{-3}\,. \tag{16}$$

It is obvious from (15) that all $R_{aa'}$ correlation radii are of the same order of magnitude and do not exceed ≈ 10 nm. Taking into account that in the materials considered the screening radius r_0 ranges usually from 10^2 to 10^3 nm we see that

$$\frac{R_{aa'}}{r_0} \ll 1\,. \tag{17}$$

It seems to follow from here that in the a-Si:H samples considered there is no medium-range order defined in [18] by the condition $r_c/r_0 \gtrsim 1$; $r_c = \max\{R_{aa'}\}$.

In a later paper [19] the same author introduces the medium-range order parameter otherwise, using the parameter $y = r_c/\Omega_{Si}^{1/3}$; in this case the presence of the medium-range order in the system is defined in terms of the inequality $y \gg 1$. Under the conditions considered in this paper the value of y may well be up to 30.

We find it reasonable, following for example [20], to relate the violation of the medium-range order to the technological defects occurring in amorphous materials in the process of preparation. The a-Si:H films can possess either internal cavities or/and pebble-shaped inhomogeneities of about 100 nm, or transverse "columns" of about 10 μm in diameter piercing through the films and containing an excessive quantity of hydrogen compared to the surrounding "tissue" [15, 21, 22].

So some additional investigations may be necessary both to clarify the nature and the cause of the medium-range order and to give a final analytical definition (if possible at all).

Nevertheless, condition (17) simplifies significantly the calculation of a binary correlation function of a random field [1, 18] expressing it in terms of the same known quantities.

4. Effective Charge of the Si–H Bond

When calculating $\psi(R)$ we shall use to the simplest expression for the Fourier transform of local pseudopotentials for the Si and H atoms, that is the screened Coulomb potential

$$V_a(q) = -\frac{4\pi Z_a e}{\varepsilon\Omega_a(q^2 + r_0^{-2})}, \tag{18}$$

where ε is the static dielectric permittivity ($\varepsilon = 12$ for Si), taking into account the screening by valence electrons; Z_a is the effective charge of an a-type atom (in units of electron charge). The screening radius r_0 provides the phenomenological account of the screening mechanism known in the physics of semiconductors.

It should be noted that the chemical bond in the material considered is not totally polar, so the values of Z_a are generally no integers. To estimate them one may use the electronegativities of Si and H.

There are several empirical ways to estimate the electronegativities. Most frequently the following ones are used [23, 24]:

$$\text{(i)}\quad \Delta E_{AB} = 23(\chi_A - \chi_B)^2 \quad \text{(kcal)}\,, \tag{19}$$

where χ_A, χ_B are the electronegativities of A and B elements, respectively, ΔE_{AB} is an additional bond energy resulting from the polarity of the A–B chemical bond.

$$\text{(ii)}\quad R_{AB} = r_A + r_B - 0.009\,|\chi_A - \chi_B| \quad \text{(nm)}\,, \tag{20}$$

Table 2

bond configuration	effective charges
SiH	$\lvert Z_H\rvert = \lvert Z_{Si}\rvert \approx 0.1$
SiH_2	$2\lvert Z_H\rvert = \lvert Z_{Si}\rvert \approx 0.14$
SiH_3	$3\lvert Z_H\rvert = \lvert Z_{Si}\rvert \approx 0.17$
SiH_4	$4\lvert Z_H\rvert = \lvert Z_{Si}\rvert \approx 0.19$

where R_{AB} is the A–B polar bond length; r_A, r_B are the covalent radii of A and B atoms, respectively.

To be specific we consider the Si–H bond, where

$$|Z_{Si}| = |Z_H| = Z\,.$$

By definition of ΔE_{AB} and R_{AB}

$$\Delta E_{AB} = \frac{Z^2e^2}{\varepsilon R_{AB}}\,. \tag{21}$$

Using now the chemical data [23, 25] $\chi_{Si} = 1.8$; $\chi_H = 2.2$; $r_{Si} = 0.117$ nm; $r_H = 0.037$ nm, we obtain

$$\Delta E_{SiH} \approx 0.09\ \text{eV}\,, \qquad R_{SiH} \approx 0.15\ \text{nm}\,.$$

Then for the effective charge Z one has

$$Z = 0.097 \approx 0.1\,. \tag{22}$$

It should be noted that besides the SiH configurations other hydrogen and silicon bond configurations are possible — SiH_2, SiH_3, and SiH_4 (silane). Taking account of the electric neutrality condition and using (21) one obtains the data of Table 2.

The value obtained for Z_H in case of silane ($Z_H \approx 0.05$) agrees with that given in [26].

5. Estimate of Some Random Field Parameters in a-Si:H

Under the conditions described above $\psi(R)$ is given by

$$\psi(R) = \frac{8e^4}{\varepsilon^2}\int_0^\infty q^2\,\frac{\sin qR}{qR}\,\frac{lq}{(q^2 + r_0^{-2})^2}\sum_{a,a'}\frac{Z_aZ_{a'}}{(\Omega_a\Omega_{a'})^{1/2}}\,S_{aa'}(q)\,. \tag{23}$$

At $R/r_0 \ll 1$ this takes the form

$$\psi(R) = \frac{2\pi e^4 r_0}{\varepsilon^2}\sum_{a,a'}\frac{Z_aZ_{a'}}{\sqrt{\Omega_a\Omega_{a'}}}\left\{S_{aa'}(0) + \frac{3}{2}\,\frac{\Omega_{aa'}}{\sqrt{\Omega_a\Omega_{a'}}}\times\left(\frac{R_{aa'}}{r_0}\right)^2 Y_{aa'} - \frac{1}{2}\left(\frac{R}{r_0}\right)^2\left[S_{aa'}(0) + \frac{5}{6}\,\frac{\Omega_{aa'}}{\sqrt{\Omega_a\Omega_{a'}}}\left(\frac{R_{aa'}}{r_0}\right)^2 Y_{aa'}\right]\right\}. \tag{24}$$

Here

$$\Omega_{aa'} = \frac{4\pi R_{aa'}^3}{3}\,,$$

$$Y_{aa'} = \int_0^1 x^4[g_{aa'}(x) - 1]\,dx\,,$$

where $x = r/R_{aa'}$.

The asymptotic behaviour of $\psi(R)$ at $R/r_0 \gg 1$ is described by

$$\psi(R) = \frac{2\pi e^4 r_0}{\varepsilon^2} e^{-\frac{R}{r_0}} \sum_{a,a'} \frac{Z_a Z_{a'}}{\sqrt{\Omega_a \Omega_{a'}}} \left\{S_{aa'}(0) + \frac{1}{2} \frac{\Omega_{aa'}}{\sqrt{\Omega_a \Omega_{a'}}} \left(\frac{R_{aa'}}{r_0}\right)^2 \left(1 - \frac{2}{R/r_0}\right) Y_{aa'}\right\}. \quad (25)$$

Equations (24) and (25) may be simplified if account is taken of (17) and of the fact that the integrals $Y_{aa'}$ are of the order of one. We obtain

$$\psi(R) \approx \frac{2\pi e^4 r_0}{\varepsilon^2} N_{\text{eff}} \left(1 - \frac{1}{2}\left(\frac{R}{r_0}\right)^2\right) \quad \text{at} \quad \frac{R}{r_0} \ll 1\,, \quad (26)$$

$$\psi(R) \approx \frac{2\pi e^4 r_0}{\varepsilon^2} N_{\text{eff}}\, e^{-\frac{R}{r_0}} \quad \text{at} \quad \frac{R}{r_0} \gg 1\,, \quad (27)$$

where

$$N_{\text{eff}} = \sum_{a,a'} \frac{Z_a Z_{a'}}{\sqrt{\Omega_a \Omega_{a'}}} S_{aa'}(0)\,. \quad (28)$$

Assuming, for example, $|Z_H| = |Z_{Si}| = Z = 0.1$, $n_H \approx 0.1 n_{Si}$, and $S_{aa'}(0) = S(0) = 10^{-2}$ one obtains the estimate

$$N_{\text{eff}} = Z^2 S(0)\, n_{Si} (1 + \sqrt{C_H})^2 \approx 9 \times 10^{18}\ \text{cm}^{-3}\,. \quad (29)$$

This may be used to evaluate the quantity $\psi^{1/2}(0) \equiv \psi_1^{1/2}$ [1] characterizing the mean square fluctuation of the electron potential energy in a random field (the latter determines in particular the range of the density of states tail),

$$\psi_1^{1/2} = \left(\frac{2\pi e^4 r_0 N_{\text{eff}}}{\varepsilon^2}\right)^{1/2} \approx 0.27\ \text{eV}\,. \quad (30)$$

Here and further we assume $r_0 = 10^2$ nm.

This estimate is not too far from the experimentally obtained value for a-Si:H: $\psi_1^{1/2} \approx (0.20$ to $0.26)$ eV [16].

6. Tail of the Interband Light Absorption Coefficient in a-Si:H

The frequency dependence of the interband light absorption coefficient in hydrogenated amorphous silicon in the frequency range of $\hbar\omega < E_g - 2\bar{W}$ is described by the well-known Urbach rule [27],

$$\alpha(\omega) \sim \exp\left(-\frac{E_g - \hbar\omega}{\bar{W}}\right). \quad (31)$$

The characteristic energy $\bar{W}$ is independent both of the frequency of light and temperature; and on the other hand, it depends somewhat on the film preparation conditions ($\bar{W} \approx (0.05$ to $0.09)$ eV, the most frequently met value is $\bar{W} = 0.06$ eV [28, 29]).

The results obtained make it possible to estimate the value of $\bar{W}$ in a-Si:H. To this end one may use the general expression for $\psi_1^{1/2}/\bar{W}$ [27] in a substance with random Coulomb field,

$$\frac{\psi_1^{1/2}}{\bar{W}} \sim \frac{\sqrt{2\pi}}{1.1} \left(\frac{r_0}{a_B}\right)^{1/2} (N_{\text{eff}} a_B^3)^{1/10}\,, \quad (32)$$

where

$$a_B = \frac{\hbar^2 \varepsilon}{m_n e^2}.$$

Putting $m_n = 0.3m_0$ and using the value of N_{eff} obtained above we find

$$\frac{\psi_1^{1/2}}{\overline{W}} \approx 5.8 ,$$

hence (cf. (30))

$$\overline{W} \approx 0.05 \text{ eV} . \tag{33}$$

Acknowledgement

The author is grateful to Prof. V. L. Bonch-Bruevich for his continuous interest in this work and useful discussions.

References

[1] V. L. Bonch-Bruevich, I. P. Zvyagin, R. Keiper, A. G. Mironov, R. Enderlein, and B. Esser, Electronnaya teoriya neuporyadochennykh poluprovodnikov. Izd. Nauka, Moscow 1981.
[2] M. A. Krivoglaz, Teoriya rasseyaniya rentgenovskikh luchey i teplovykh neitronov realnimy kristallamy, Izd. Nauka, Moscow 1967.
[3] L. D. Landau and I. M. Lifshits, Statisticheskaya fizika, Part I, Izd. Nauka, Moscow 1970.
[4] N. C. Halder and R. L. Wourms, Z. Naturf. 30a, 55 (1975).
[5] N. F. Mott and E. A. Davies, Electronic Processes in Non-Crystalline Materials, Clarendon Press, Oxford 1971.
[6] R. Grigorovici and R. Manaila, Thin Solids Films **1**, 343 (1968).
[7] M. V. Coleman and D. J. D. Thomas, phys. stat. sol. **24**, K111 (1967).
[8] R. Zallen, Phys. Rev. **173**, 824 (1968).
[9] D. E. Polk, J. non-crystall. Solids **5**, 365 (1971).
[10] S. C. Moss and J. E. Graczyk, Phys. Rev. Letters **23**, 1167 (1969).
[11] S. C. Moss and J. E. Graczyk, Proc. of the 10th Internat. Conf. Physics of Semiconductors, Cambridge (Massachusetts), Ed. S. P. Keller, J. C. Hensel, and F. Stern, 1970 (p. 658).
[12] A. MacKinnon and B. Kramer, Physikalisch-Technische Bundesanstalt, 1980 (preprint).
[13] J. C. Knights, G. Lucovsky, and R. J. Nemanich, Phil. Mag. B **37**, 467 (1978).
[14] E. C. Freeman and W. Paul, Phys. Rev. B **18**, 4288 (1978).
[15] J. A. Reimer, R. W. Vaughan, and J. C. Knights, Phys. Rev. Letters **44**, 193 (1980).
[16] N. F. Mott, J. Phys. C **13**, 5433 (1980).
[17] V. Heine, M. L. Cohen, and D. Weaire, Teoriya pseudopotenziale, Izd. Mir, Moscow 1973.
[18] V. D. Iskra, Zh. eksper. teor. Fiz. **82**, 646 (1982).
[19] V. D. Iskra, phys. stat. sol. (b) **119**, 515 (1983).
[20] H. Fritsche, Fundamental Physics of Amorphous Semiconductors, Proc. Kyoto Summer Institute, Kyoto (Japan), September 8 to 11, 1980, Ed. F. Yonezawa, Springer-Verlag, Berlin/Heidelberg/New York 1981 (p. 1).
[21] W. Paul, ibid. (p. 72).
[22] J. Knights, J. non-crystall. Solids **35/36**, 159 (1980).
[23] L. Pauling, Obschaya khimiya, Izd. Mir, Moscow 1974.
[24] U. Kosman, Vvedeniye v kvantovuyu khimiyu, Izd. inostrannoy Literatury, Moscow 1960.
[25] Svoystva elementov, Spravochnik, Ed. G. W. Samsonov, Izd. Metallurgiya, Moscow 1976.
[26] G. Lucovsky, see [20] (p. 87).
[27] V. L. Bonch-Bruevich, Uspekhi fiz. Nauk **140**, 583 (1983).
[28] C. D. Cody, B. Abeles, C. R. Wronsky, and B. Brooks, J. non-crystall. Solids **35/36**, 463 (1980).
[29] T. I. Pankove, F. H. Pollak, and C. Schnabolk, J. non-cryst. Solids **35/36**, 459 (1980).

(Received June 7, 1984)

phys. stat. sol. (b) **126**, 261 (1984)

Subject classification: 20.1; 22.4.1

Pedagogical High School, Schumen[1]) *(a)*
and Central Institute of Electron Physics,
Academy of Sciences of the GDR, Berlin[2]) *(b)*

Peculiarities of Polarization Properties of CdS Single Crystals around the Isotropic Point

By

T. N. Dimov (a), I. A. Iliev (a), and H. Lange (b)

In dependence on sample position between crossed polarizer and analyzer two effects can be distinguished in CdS crystals: the conventional birefringence and the mode coupling between ordinary and extraordinary waves. Both effects are clearly identified by comparing the ellipticity parameters. The mode coupling shows up as a spectral structure consisting of two peaks separated by a pronounced minimum at the isotropic point for as-grown CdS platelets. The ellipticity parameters in this case exhibit principally the same features as the optical activity. The occurrence of this effect is due to non-diagonal components of the dielectric susceptibility tensor. These components can arise as a result of higher-order terms of the dielectric susceptibility tensor and/or internal stress. Expressions are derived which very well describe the spectral shape as well as the influence of an external magnetic field on the spectral characteristics due to mode coupling.

В зависимости от положения кристаллической пластинки сульфида кадмия, помещенной между скрещенными поляризатором и анализатором, наблюдается два эффекта: обыкновенное двойное лучепреломление и переброс энергии между модами необыкновенной и обыкновенной волны (связь модов) в изотропной точке (и.т.). Эти эффекты однозначно идентифицированы на основе их эллиптических характеристик. Эффект связи модов в и.т. проявляется в спектрах CdS-кристаллов, выращенных по методу Фрерихса, как структура, составленная из двух максимумов, разделенных глубоким минимумом в и.т. Свойства эллипсометрических параметров в этом случае принципально подобны тем, как у оптической активности. Проявление этого эффекта связано с недиагональными компонентами тензора диэлектрической восприимчивости, которые являются результатом рассмотрения членов второго порядка в разложении тензора диэлектрической восприимчивости в ряд по волновом векторе и/или обусловлены присутствием внутренных механических напрежений в кристалле. Полученные теоретические выражения хорошо описывают спектральную форму, а также и влияние внешнего магнитного поля на спектралные характеристики эффекта связи модов в и.т.

1. Introduction

The II–VI compounds with hexagonal structure exhibit pronounced polarization anisotropies at the absorption edge. Selection rules allow the lowest electronic transitions to occur at smaller energy for $\boldsymbol{E} \perp \boldsymbol{c}$ (Γ_{9v}–Γ_{7c} transitions) than for $\boldsymbol{E} \parallel \boldsymbol{c}$ (Γ_{7v}–Γ_{7v} transitions). A special consequence of the polarization anisotropies of wurtzite-type crystals is a crossing of the ordinary ($\boldsymbol{E} \perp \boldsymbol{c}$) and extraordinary ($\boldsymbol{E} \parallel \boldsymbol{c}$) refractive indices ($n_o$ and n_e, respectively) below the absorption edge at a definite energy (for CdS 2.37 eV at 300 K), called isotropic point (i.p.); this is because n_o grows faster than n_e towards the absorption edge. The spectral dependences of the

[1]) 9700 Schumen, Bulgaria.

[2]) Rudower Chaussee 5, DDR-1199 Berlin, GDR.

refractive indices of hexagonal CdS and their variation with temperature and pressure are known from numerous studies [1 to 5]. The temperature and pressure dependence of the position of the i.p. has been studied as well [6].

The existence of the i.p. provides the possibility of mode coupling between ordinary and extraordinary electromagnetic waves due to some external or internal symmetry-reducing perturbation or by higher-order terms in the $\boldsymbol{k}$-expansion of the tensor of dielectric susceptibility $\boldsymbol{\chi}(\omega, \boldsymbol{k})$ giving rise to non-diagonal components of $\boldsymbol{\chi}(\omega, \boldsymbol{k})$. The consequences of mode coupling around the i.p. by non-diagonal components of the susceptibility tensor have been studied by various authors, both experimentally and theoretically. Hopfield and Thomas [7] observed absorption lines at the i.p. of CdS which they attributed to mode coupling. Henry [8] studied mode coupling at the i.p. of CdS experimentally and theoretically under the influence of external stress and magnetic field and could provide an explanation of the results of Hopfield and Thomas. Detailed experimental results on the stress-induced mode coupling are reported by Laurenti and Rouzeyre [9] where uniaxial stress was applied at 45° with respect to the c-axis and the sample was placed between crossed polarizers with the c-axis parallel to the polarization plane of the polarizer or analyzer. The results were in good agreement with the theoretical predictions of Henry [8]. In [10, 11] Laurenti et al. observed a mode coupling already without external perturbation on CdS, CdSe, and CdS_xSe_{1-x} mixed crystals which showed up as a narrow line of transmitted intensity at the i.p. for the sample in the above measuring geometry, and which was interpreted as being due to internal stress. Similar results are also reported in [12] on CdS at 77 K.

The present paper implements the existing experimental information on mode coupling in CdS single crystals and reports spectral features in polarized light around the i.p. at 300 K which can hardly be explained with the theoretical relationships given in literature. Moreover, we found that the influence of the external magnetic field on the mode coupling cannot be described on the same footing as external stress, a supposition put forward by Henry [8]. We therefore reconsider the theoretical basis and derive expressions which give a better quantitative description of the presently available experimental data of mode coupling around the i.p. of CdS.

2. Experimental

The experimental studies were carried out at room temperature in the spectral range from 515 to 560 nm on plate-shaped CdS single crystal grown from the vapour phase by a modified Frerichs method [13], the $(1\bar{2}10)$ face was perpendicular to the x-axis of the crystal coordinate system. The sample size was 1.5×3 mm², and the thickness varied from 65 to 95 μm. The crystals were carefully mounted on the crystal holder by avoiding bending and distortions.

A parallel beam of monochromatic light (SPM-2 monochromator) was incident on crossed polaroid polarizers with the sample in between. Goniometer readings of angular positions of polarizers and crystal holder were possible with an accuracy of 0.1°. Signals were recorded with a type P12 FQS52A photomultiplier. Measurements in magnetic field were carried out up to 16 kG.

The thickness homogeneity of the crystals was checked by recording the transmitted light intensity as a function of the sample rotation angle at an interference maximum of birefringence. The symmetry of the polar diagram was taken as indication of plane parallelism of the crystal faces.

Interference spectra in polarized light were measured for diagonal and parallel position of the sample with respect to the polarization plane of polarizer and analyzer.

The former position is characterized by the angle $\pi/4$ and the latter one by the angle 0 or $\pi/2$ between c-axis and polarization plane of the polarizer. The intensity of the light transmitted through the polarizer–crystal–analyzer (P–C–A) system is described by

$$I = I_0 \left[\cos^2 (\beta - \alpha) - \sin 2\alpha \sin 2\beta \sin^2 \frac{\delta}{2}\right], \tag{1}$$

where α and β are the angles between c-axis and the polarization plane of analyzer and polarizer, respectively; $\beta - \alpha = \pi/2$ in case of crossed polaroids. $\delta = (\pi d/\lambda)(n_e - n_o)$ is the phase difference, λ the wavelength, d the sample thickness, and I_0 the intensity of incident light. The influence of analyzer absorption on the spectra was eleminated by evaluating the data according to [17]

$$\Gamma = \frac{(I_\perp/I_\parallel)_b - (I_\perp/I_\parallel)_a}{1 + (I_\perp/I_\parallel)_b - (I_\perp/I_\parallel)_a}. \tag{2}$$

$I_\perp$ and $I_\parallel$ are the intensities for crossed and parallel polaroids, respectively, index a denotes the intensity ratio $I_\perp/I_\parallel$ without crystal and index b that with crystal. In our case $(I_\perp/I_\parallel)_a = 0$.

3. Experimental Results

3.1 Diagonal sample position

In this case the interference pattern is due to the birefringence of the sample. Fig. 1a shows the birefringence spectra determined by (1) and (2) for a CdS crystal of thickness $d = (65 \pm 5)$ μm. Number and energetic position of interference maxima depend on sample thickness. The decrease of intensity towards shorter wavelengths is caused by the absorption edge tail. The minimum connected with the i.p. occurs at (523 ± ± 0.5) nm for all crystals. Ellipticity and azimuth for the sample are shown in Fig. 1 b and c. For wavelengths longer than λ_A (see Fig. 1) azimuth and ellipticity exhibit

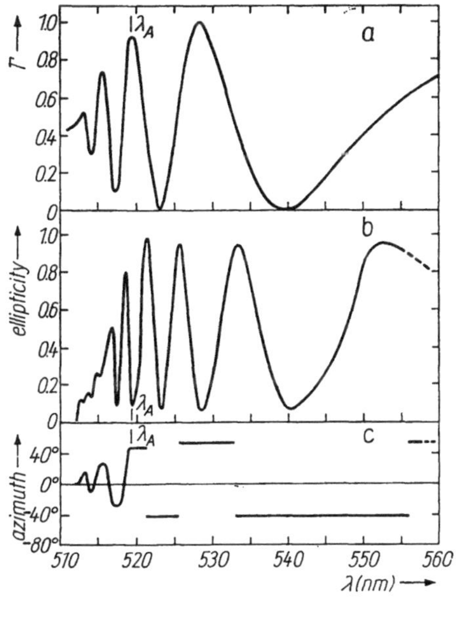

Fig. 1. a) Interference spectrum due to birefringence of a CdS single crystal of thickness $d = (65 \pm 5)$ μm in diagonal position between crossed polarizer, $T = 300$ K; b) and c) parameters of the polarization ellipse: b) ellipticity, c) azimuth

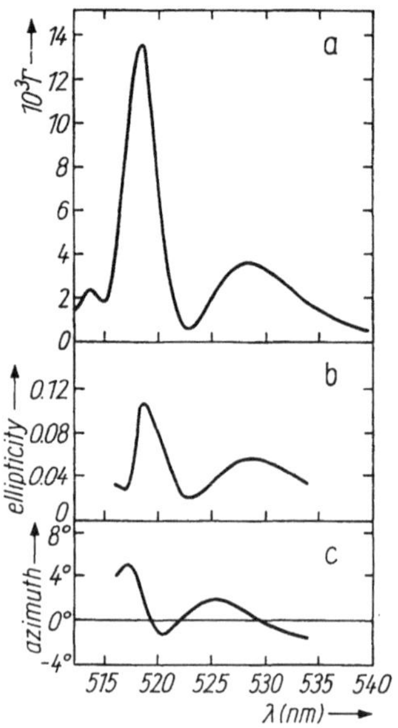

Fig. 2

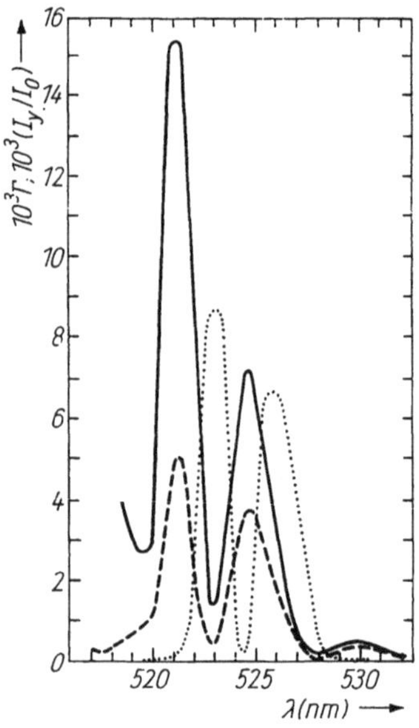

Fig. 3

Fig. 2. a) Interference spectrum due to mode coupling around the i.p. of a CdS single crystal of thickness $d = (65 \pm 5)$ μm in parallel position between crossed polarizers; b) and c) parameters of the polarization ellipse: b) ellipticity, c) azimuth

Fig. 3. Interference spectrum around the i.p. for different polarizations of the indicent light for a CdS ($T = 300$ K) single crystal of thickness $d = (75 \pm 5)$ μm in parallel position between crossed polarizers. The dotted line I_y/I_0 is the result of theoretical calculations according to formula (18). Parameters used in theoretical calculations are: $d = 7.5 \times 10^{-3}$ cm, $\varrho = 10^7$ cm^{-2}, the refractive index data are taken from [2] (——— $\boldsymbol{E} \perp \boldsymbol{c}$, – – – $\boldsymbol{E} \parallel \boldsymbol{c}$)

features typical of non-absorbing birefringent materials [14]. Deviations from this behaviour at shorter wavelengths are due to the influence of the absorption tail. At the positions of azimuth sign reversal the transmitted light is circularly polarized. Minima in ellipticity (Fig. 1b) are connected with interference minima and maxima (Fig. 1a), at these wavelengths the transmitted light is linearly polarized. Measurements in magnetic field (up to 16 kG) revealed no change of the interference pattern.

3.2 Parallel sample position

In this experimental arrangement a signal was only recorded around the i.p. The spectral dependence of light intensity transmitted through the (P–C–A) system in this case is characterized by two well pronounced maxima separated by a deep minimum exactly at the i.p. for the as-grown CdS platelets (Fig. 2). The position of maxima depended on the sample thickness, the maximum on the shorter-wavelength side of the i.p. was always higher than the other one. For thicker crystals additional weak maxima were observed on both sides of the main ones. The occurrence of this spectral pattern was independent of light polarization. The differences seen in the spectral shape for $\boldsymbol{E} \perp \boldsymbol{c}$ and $\boldsymbol{E} \parallel \boldsymbol{c}$ (Fig. 3) are due to differences in the absorption edge position. The maximum intensity of transmitted light is by two to three orders of magnitude lower than that at diagonal position. The amplitudes varied from sample to sample. The spectral features around the i.p. were very sensitive to mechanical

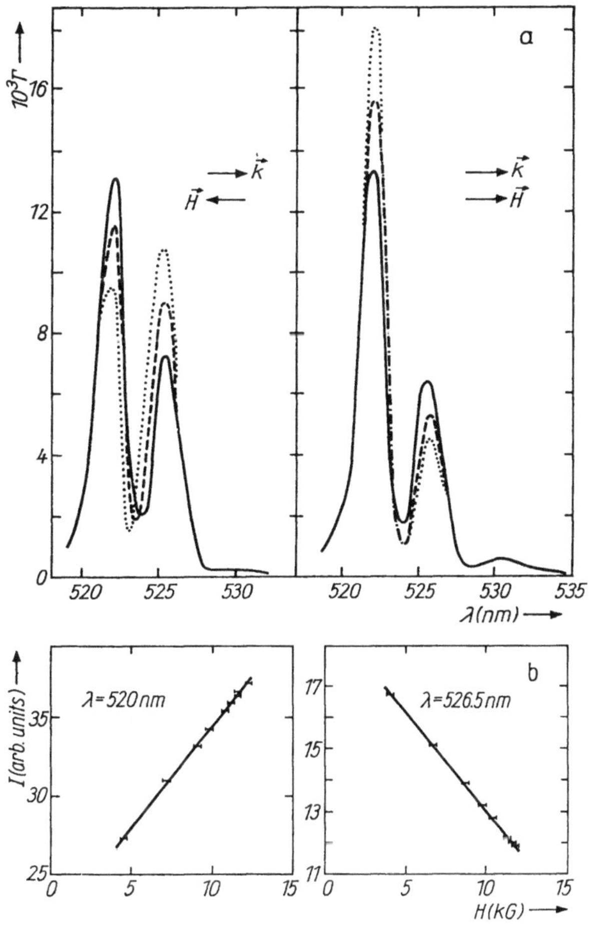

Fig. 4. a) Influence of a magnetic field on the interference spectrum around the i.p. for a CdS single crystal of thickness $d = (90 \pm 5)$ μm in parallel position between crossed polarizers for two fields directions; ——— $H = 1.6$, – – – 6.6, ······ 10 kG; b) dependence of the height of high and low-energy peak on magnetic field strength

distortions which could smear out the double-peak structure changing it into a single-peak one of lower intensity with maximum at the i.p. The spectral dependences of azimuth and ellipticity in case of parallel position show a behaviour typical for gyrotropic crystals [17, 16].

Applying a magnetic field perpendicular to the $(1\bar{2}10)$ face leads to an intensity decrease of one maximum and simultaneously to an increase of the other one. With magnetic field reversal the direction of intensity change also reverses (Fig. 4). The intensity changes are linearly related to the magnetic field strength. This behaviour is independent of light polarization.

4. Discussion

As evidenced by the spectral characteristics of amplitude ratios and ellipticity parameters presented in the last section, the polarization spectra observed in parallel and diagonal sample positions have different physical origin. In case of diagonal position the observed spectral dependences are in correspondence with the well understood birefringence of uniaxial crystals [14]. This phenomenon is associated with the diagonal components of the tensor of dielectric susceptibility. In case of parallel sample position the spectral characteristics of ellipticity parameters show them to be of gyrotropic nature; they have their origin in the non-diagonal components of the dielectric susceptibility tensor. As crystals of 6mm symmetry do not exhibit natural activity [15], in perturbation-free crystals non-diagonal components of the dielectric susceptibility tensor are due to k^2-terms in the expansion of $\boldsymbol{\chi}(\omega, \boldsymbol{k})$:

$$\chi_{ij}(\omega, \boldsymbol{k}) = \chi_{ij}(\omega) + i\gamma_{ijl}(\omega)\, k_l + \alpha_{ijlm} k_l k_m \,. \tag{3}$$

For 6mm symmetry $\chi_{ij}(\omega) = 0$ for $i \neq j$, $\gamma_{ijl} = 0$ (no natural activity); thus non-diagonal components in χ_{ij} are caused by the tensor α_{ijlm} [18]. Another reason for the occurrence of non-diagonal components of the dielectric susceptibility tensor are symmetry-lowering external or internal perturbations (e.g. stress, electric or magnetic field).

Henry [8] has derived an expression for the light intensity transmitted through crossed analyzer and polarizer with the sample in between in parallel position for the case when an external stress or a magnetic field is applied, and has assumed that the magnetic field produced the same effect as the external stress. He has shown that due to the mode coupling around i.p. the light intensity transmitted through the (P–C–A) system is given by the following expression:

$$I_\mathrm{T} = I_0 \frac{|\beta|^2}{|\beta|^2 + \left(\dfrac{\Delta k}{2}\right)^2} \sin^2 \left\{\left[|\beta|^2 + \left(\frac{\Delta k}{2}\right)^2\right]^{1/2} d\right\}, \tag{4}$$

where $\beta = 2\pi\chi\omega^2/c^2k_z = 2\pi\chi\omega^2/c^2k_y$ is the coupling parameter, χ is the non-diagonal component of the susceptibility tensor, k_z and k_y are given by (8), c is the light velocity, and $\Delta k = k_y - k_z$ is the difference between the wave numbers of the y- and z-modes. A two-peak structure as observed in the present experiments follows from this expression for $|\beta|\, d \approx n\pi$ where $n = 1, 2, \ldots$ We anticipate that the coupling parameter is a very small quantity in unperturbed, as-grown CdS single crystals as (i) the higher-order terms in the expansion of $\boldsymbol{\chi}$ in powers of k are by orders of magnitude smaller than the diagonal ones and (ii) the built-in mechanical stress in the as-grown crystals is small. However, for $|\beta|\, d \leqq 1$ a single-peak structure of the spectral shape of transmitted light intensity follows from (4) with a maximum at the i.p. which is contrary to our observation. Moreover, we found that the magnetic field has an influence on the two-peak structure which could not be described by (4) if simply assuming that the magnetic field enters into $|\beta|$. We therefore recalculated the intensity transmitted through the (P–C–A) system in case of parallel sample position using another approach than Henry did.

4.1 Mode coupling without magnetic field

We adopt the same starting relations as used by Henry:

$$\frac{\partial^2}{\partial x^2} E_y + \frac{\omega^2}{c^2} E_y = -4\pi \frac{\omega^2}{c^2} P_y\,, \qquad \frac{\partial^2}{\partial x^2} E_z + \frac{\omega^2}{c^2} E_z = -4\pi \frac{\omega^2}{c^2} P_z\,, \tag{5}$$

where

$$\begin{pmatrix} P_y \\ P_z \end{pmatrix} = \begin{pmatrix} \chi_{yy} & \chi_{yz} \\ \chi_{zy} & \chi_{zz} \end{pmatrix} \begin{pmatrix} E_y \\ E_z \end{pmatrix}. \tag{6}$$

Then equation (5) can be rewritten as

$$\frac{\partial^2}{\partial x^2} E_y + k_y^2 E_y - \varrho E_z = 0\,, \qquad \frac{\partial^2}{\partial x^2} E_z + k_z^2 E_z - \varrho E_y = 0 \tag{7}$$

with the components of wave vector

$$k_y^2 = \frac{\omega^2}{c^2} (1 + 4\pi\chi_{yy}) , \qquad k_z^2 = \frac{\omega^2}{c^2} (1 + 4\pi\chi_{zz}) , \tag{8}$$

and the coupling parameter

$$\varrho = -4\pi \frac{\omega^2}{c^2} \chi_{yz} = -4\pi \frac{\omega^2}{c^2} \chi_{zy} . \tag{9}$$

From (7) the following equation is obtained for E_y:

$$\frac{\partial^4}{\partial x^4} E_y + (k_y^2 + k_z^2) \frac{\partial^2}{\partial x^2} E_y + (k_y^2 k_z^2 - \varrho^2) E_y = 0 . \tag{10}$$

The characteristic equation of the above differential equation has the following roots:

$$\begin{aligned} r_1 &= \frac{1}{\sqrt{2}} \sqrt{- (k_y^2 + k_z^2) + \sqrt{(k_y^2 - k_z^2)^2 + 4\varrho^2}} , \\ r_2 &= -\frac{1}{\sqrt{2}} \sqrt{- (k_y^2 + k_z^2) + \sqrt{(k_y^2 - k_z^2)^2 + 4\varrho^2}} , \\ r_3 &= \frac{1}{\sqrt{2}} \sqrt{- (k_y^2 + k_z^2) - \sqrt{(k_y^2 - k_z^2)^2 + 4\varrho^2}} , \\ r_4 &= -\frac{1}{\sqrt{2}} \sqrt{- (k_y^2 + k_z^2) - \sqrt{(k_y^2 - k_z^2)^2 + 4\varrho^2}} , \end{aligned} \tag{11}$$

which, under the condition $\chi_{yz}, \chi_{zy} \ll \chi_{yy}, \chi_{zz}$ are complex conjugate

$$r_1 = i\gamma_2 , \qquad r_2 = - i\gamma_2 , \qquad r_3 = i\gamma_1 , \qquad r_4 = -i\gamma_1 , \tag{12}$$

γ_1 and γ_2 are defined by

$$\begin{aligned} \gamma_1 &= \frac{1}{\sqrt{2}} \sqrt{k_y^2 + k_z^2 + \sqrt{(k_y^2 - k_z^2)^2 + 4\varrho^2}} , \\ \gamma_2 &= \frac{1}{\sqrt{2}} \sqrt{k_y^2 + k_z^2 - \sqrt{(k_y^2 - k_z^2)^2 + 4\varrho^2}} . \end{aligned} \tag{13}$$

The solution of (10) is

$$E_y = iC_1 \sin \gamma_1 d + C_2 \cos \gamma_1 d + iC_3 \sin \gamma_2 d + C_4 \cos \gamma_2 d . \tag{14}$$

The coefficients C_1 to C_4 are derived from the boundary conditions [19]

$$E_y(0) = 0 , \qquad \left.\frac{\partial E_y}{\partial x}\right|_{x=0} = 0 , \qquad E_z(0) = E_0 , \qquad \left.\frac{\partial E_z}{\partial x}\right|_{x=0} = 0 \tag{15}$$

and are given by

$$C_1 = C_3 = 0 , \qquad C_2 + C_4 = 0 , \qquad C_2\gamma_1^2 + C_4\gamma_2^2 + E_0\varrho = 0 \tag{16}$$

from which

$$C_2 = -C_4 = \frac{E_0\varrho}{\gamma_2^2 - \gamma_1^2} . \tag{17}$$

Thus the following equation is obtained for the transmitted light intensity in

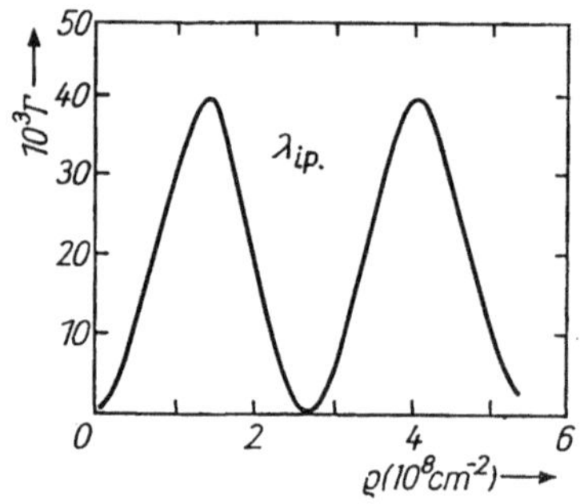

Fig. 5. Variation of the transmitted light intensity as a function of the coupling parameter according to formula (18) at the i.p.

parallel geometry:

$$\frac{I_y}{I_0} \approx \left(\frac{E_y}{E_0}\right)^2 = \frac{\varrho^2}{(k_z^2 - k_y^2)^2 + 4\varrho^2} (\cos \gamma_1 d - \cos \gamma_2 d)^2 . \tag{18}$$

I_y is the intensity of light polarized perpendicular to the optical c-axis transmitted through the (P–C–A) system, I_0 is the intensity of incident light polarized parallel to the c-axis. Formula (18) has a shape similar to (4) derived by Henry [8]. It consists of two factors, the first plays the role of a filter function with a maximum at the i.p. and a width depending on the coupling parameter. The second has oscillatory character in both expressions (4) and (18). The differences between filter function and oscillatory term in (4) and (18) are due to (i) the use of Maxwell boundary conditions (15) in our treatment and (ii) the introduction of simplifying assumptions at a later stage than Henry did. Expression (18) reproduced an analogous oscillatory dependence of the transmitted light intensity on the coupling parameter as in [8]. This is exemplified by Fig. 5 where the light intensity I_y/I_0 at the i.p. is plotted as a function of ϱ for a fixed sample thickness of $d = 0.0075$ cm. Formula (18) well reproduces the observed two-peak spectrum around the i.p. for very small coupling parameter values to be expected for unstressed, as-grown plate-shape CdS single crystals which are distinguished by a high crystallographic perfection. This could not be accomplished with relation (4). The dotted line in Fig. 3 presents the I_y/I_0 spectrum calculated according to formula (18) for a crystal with the following parameter values: thickness 7.5×10^{-3} cm, $\varrho = 1 \times 10^7$ cm^{-2}. The refractive index data have been taken from [2]. The calculated spectrum agrees with the experimental one as regards spectral shape, amplitudes, and amplitude ratios. The difference in minimum position is connected with the difference in i.p. position in [2] (524.5 nm) and in our studies ((523 $\pm$ 0.5) nm). The calculations have been also carried out for the situations considered by Henry [8]. It could be shown that all spectral shapes obtained by Henry can be well reproduced by formula (18) for appropriate coupling parameter values.

4.2 Influence of an external magnetic field

The influence of magnetic field on the spectral structure around the i.p. at parallel sample position is calculated on the basis of Faraday rotation. We assume a linear superposition of mode coupling and Faraday rotation. Around the i.p. the crystal is regarded as being composed of "two crystals". The "first" being in magnetic field is considered isotropic with a refractive index $n = (n_e + n_o)/2$. This is justified in so far as around the i.p. the crystal can be represented as quasi-isotropic because of the small birefringence. The "second crystal" is anisotropic and outside the magnetic

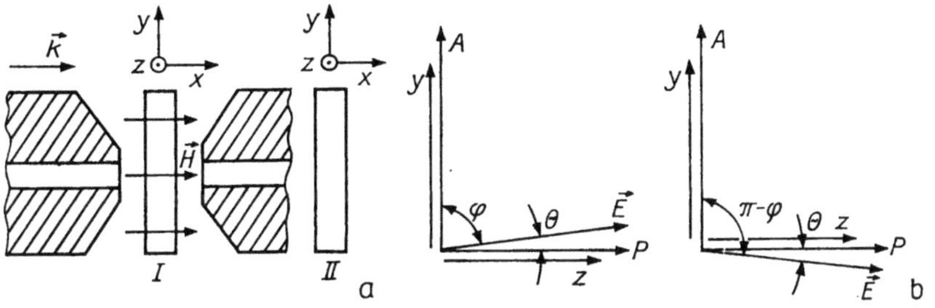

Fig. 6. a) Scheme illustrating the formulation of the boundary conditions in case of superposition of Faraday rotation on mode coupling around in i.p. b) Dependence of the Faraday rotation on the direction of the magnetic field along the x-axis

field and exhibits the usual refractive indices n_e and n_o (see Fig. 6a). The electric vector of the linearly polarized electromagnetic wave is rotated by the angle $\theta = Vd$ by the "first crystal" (V Verdet constant). The components of E can be written as (see Fig. 6b)

$$E_y = E_0 \sin\theta = E_0 \cos\varphi\,, \qquad E_z = E_0 \cos\theta = E_0 \sin\varphi\,, \tag{19}$$

where

$$\varphi = \varphi_1 = 90^\circ - \theta\,.$$

For the "second crystal" (without magnetic field) the same considerations as in Section 4.1 apply; the boundary conditions are now

$$E_y(0) = E_0 \cos\varphi\,, \quad \left.\frac{\partial E_y}{\partial x}\right|_{x=0} = 0\,, \quad E_z(0) = E_0 \sin\varphi\,, \quad \left.\frac{\partial E_z}{\partial x}\right|_{x=0} = 0\,. \tag{20}$$

From (7)

$$E_z(x) = \frac{1}{\varrho}\frac{\partial^2 E_y}{\partial x^2} + \frac{k_y^2}{\varrho} E_y\,.$$

The constants C_1 to C_4 in (14) derived from these boundary conditions are

$$\begin{aligned} &E_0 \cos\varphi = C_2 + C_4\,,\\ &E_0 \sin\varphi = \frac{1}{\varrho}(-C_2\gamma_1^2 - C_4\gamma_2^2) + \frac{k_y^2}{\varrho}(C_2 + C_4)\,,\\ &C_1\gamma_1 + C_3\gamma_2 = 0\,,\\ &C_1\gamma_1^3 + C_3\gamma_2^3 = 0 \end{aligned} \tag{21}$$

yielding

$$\begin{aligned} &C_1 = 0\,,\\ &C_2 = E_0\left[\frac{-(k_z^2 - k_y^2)\cos\varphi}{2\sqrt{(k_y^2 - k_z^2)^2 + 4\varrho^2}} - \frac{\varrho \sin\varphi}{\sqrt{(k_y^2 - k_z^2)^2 + 4\varrho^2}} + \frac{1}{2}\cos\varphi\right],\\ &C_3 = 0\,,\\ &C_4 = E_0\left[\frac{(k_z^2 - k_y^2)\cos\varphi}{2\sqrt{(k_y^2 - k_z^2)^2 + 4\varrho^2}} + \frac{\varrho \sin\varphi}{\sqrt{(k_y^2 - k_z^2)^2 + 4\varrho^2}} + \frac{1}{2}\cos\varphi\right]. \end{aligned} \tag{22}$$

Equation (14) in this case gets the form

$$\frac{E_y}{E_0} = \left[\frac{\varrho}{\sqrt{(k_y^2 - k_z^2)^2 + 4\varrho^2}} (\cos\gamma_2 d - \cos\gamma_1 d)\right] \sin\varphi + \\ + \left[\frac{1}{2}(\cos\gamma_1 d + \cos\gamma_2 d) + \frac{(k_z^2 - k_y^2)(\cos\gamma_2 d - \cos\gamma_1 d)}{2\sqrt{(k_y^2 - k_z^2)^2 + 4\varrho^2}}\right] \cos\varphi\,. \quad (23)$$

Reversing the magnetic field direction the angle between E and the analyzer becomes $\varphi_2 = 90° + \theta$ ($\varphi_2 = \pi - \varphi$) (see Fig. 6b). Then the constants C_1 to C_4 follow as

$$\begin{aligned} C_1 &= 0\,, \\ C_2 &= E_0 \left[\frac{-\varrho\sin\varphi}{\sqrt{(k_y^2 - k_z^2)^2 + 4\varrho^2}} + \frac{(k_z^2 - k_y^2)^2 \cos\varphi}{2\sqrt{(k_y^2 - k_z^2)^2 + 4\varrho^2}} - \frac{1}{2}\cos\varphi\right], \\ C_3 &= 0\,, \\ C_4 &= E_0 \left[\frac{\varrho\sin\varphi}{\sqrt{(k_y^2 - k_z^2)^2 + 4\varrho^2}} - \frac{(k_z^2 - k_y^2)\cos\varphi}{2\sqrt{(k_y^2 - k_z^2)^2 + 4\varrho^2}} - \frac{1}{2}\cos\varphi\right]. \end{aligned} \quad (24)$$

and the intensity ratio becomes

$$\frac{E_y}{E_0} = \left[\frac{\varrho}{\sqrt{(k_y^2 - k_z^2)^2 + 4\varrho^2}} (\cos\gamma_2 d - \cos\gamma_1 d)\right] \sin\varphi - \\ - \left[\frac{1}{2}(\cos\gamma_1 d + \cos\gamma_2 d) + \frac{(k_z^2 - k_y^2)(\cos\gamma_2 d - \cos\gamma_1 d)}{2\sqrt{(k_y^2 - k_z^2)^2 + 4\varrho^2}}\right] \cos\varphi\,. \quad (25)$$

Calculations were performed with the same values of d and ϱ as in the former case without magnetic field. Using the known Verdet constant of CdS the Faraday rotation for $H = 10$ kG is about 1°, as was also verified experimentally. The results are plotted in Fig. 7. As seen from this figure and Fig. 4, the theoretical results well agree with

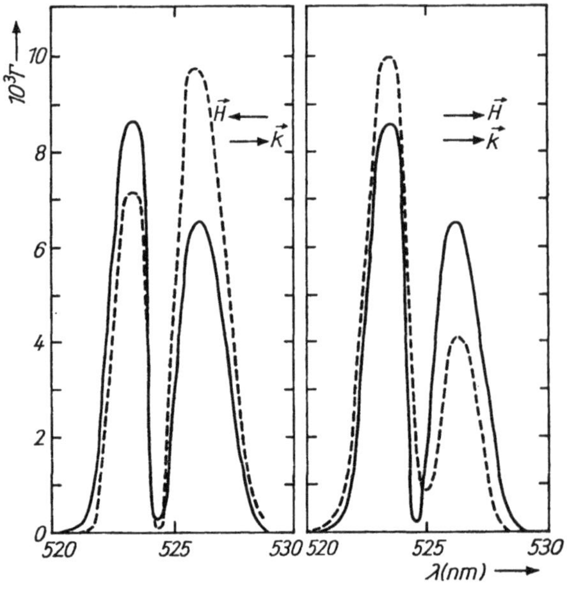

Fig. 7. Theoretically calculated interference spectrum around the i.p. for a CdS single crystal in parallel position between crossed polarizers with (— — —) and without (———) magnetic field for both field directions according to (23) and (25). Parameters used in calculations: $d = 7.5 \times 10^{-3}$ cm, $\varrho = 10^7$ cm^{-2}; the refractive index data are taken from [2]

the experimental ones as regards magnitude of intensity variation and the direction of intensity change as a function of magnetic field direction. The intensity in the maximum depends linearly on magnetic field, in accord with the experiment.

5. Conclusions

CdS platelets grown by the Frerichs method, when placed between crossed polarizers, for both polarization directions of incident light ($\boldsymbol{E} \perp \boldsymbol{c}$ and $\boldsymbol{E} \parallel \boldsymbol{c}$) exhibit an effect around the i.p. at 300 K which is due to mode coupling between ordinary and extraordinary waves at the i.p. This mode coupling originates else from the spatial dispersion of second order in $\boldsymbol{k}$ and/or from symmetry lowering of the crystal by external stress. The spectral structure due to mode coupling has two maxima separated by a deep minimum at the i.p. The properties characterizing this structure are pronounced sensitivity to external mechanical perturbations and ellipticity parameters with features similar to those of optical activity. The magnetic field causes specific changes of both maxima which depend on the magnetic field direction. The line shape of the experimental spectra can be described on the basis of a theoretical treatment of energy transfer at the i.p.

Acknowledgements

The authors are very grateful to Dr. M. Iliev and Dr. St. Stoyanov from the Faculty of Physics of Sofia University for stimulating discussions.

References

[1] S. J. Czyzak, W. M. Baker, R. C. Crane, and J. B. Hove, J. Opt. Soc. Amer. 47, 240 (1957).
[2] H. Gobrecht and A. Bartschat, Z. Phys. 156, 131 (1959).
[3] T. M. Bieniewski and S. J. Czyzak, J. Opt. Soc. Amer. 53, 496 (1963).
[4] A. A. Reza, G. A. Babonas, and A. J. Shileika, Fiz. Tekh. Poluprov. 9, 1494 (1975).
[5] A. A. Reza, G. A. Babonas, and A. J. Shileika, phys. stat. sol. (b) 72, 421 (1975).
[6] D. W. Langer, J. appl. Phys. 37, 3530 (1966).
[7] J. J. Hopfield and D. G. Thomas, Phys. Rev. Letters 15, 22 (1965).
[8] C. H. Henry, Phys. Rev. 143, 627 (1966).
[9] J. P. Laurenti and M. Rouzeyre, J. appl. Phys. 52, 6484 (1981).
[10] J. P. Laurenti, K. C. Rustagi, and M. Rouzeyre, Appl. Phys. Letters 28, 212 (1976).
[11] J. P. Laurenti, K. C. Rustagi, and M. Rouzeyre, J. appl. Phys. 48, 203 (1977).
[12] L. E. Solovev and V. S. Rudakov, Vestnik Gos. Leningradsk. Univ. 4, 23 (1968).
[13] P. Flögel, Kristall und Technik 3, 161 (1968).
[14] I. M. Nagibina, Interferentsia i diffraktsiya sveta, Izd. Mashinostroenie, Leningrad 1974.
[15] B. K. Veinshtein, Sovremennaya kristallografiya, Vol. 4, Fizicheskie svoistva kristallov, Izd. Nauka, Moskva 1981.
[16] A. P. Konstantinova, N. R. Ivanov, and B. N. Grechusnikov, Kristallografiya 14, 283 (1969).
[17] J. Kobajashi, T. Takahashi, and T. Hosokawa, J. appl. Phys. 49, 809 (1978).
[18] V. N. Agranovich and V. L. Ginzburg, Kristallooptika s uchetom prostranstvennoi dispersii i teoriya eksitonov, Izd. Nauka, Moskva 1979.
[19] R. P. Feynman, R. B. Leighton, and W. Sands, Lecture Notes on Physics, Vol. 7, Izd. Mir, Moscow 1977.

(Received August 8, 1984)

phys. stat. sol. (b) **126**, 273 (1984)

Subject classification: 4 and 13.1; 20.1; 22.8

Department of Physics, Institute of Chemical Technology, Pardubice[1])

The Width of the Energy Gap of $Tl_3^IB^VC_4^{VI}$-Type Compounds

By

K. Čermák

The Tl_3BC_4 (B = V, Nb, Ta; C = S, Se) compounds satisfy the $8N$ valence rule of Mooser and Pearson and belong to semiconductors. An increase in the value of energy gap E_g with the substitution of V by Nb or Ta in these compounds can be ascribed to the increasing bond ionicity due to the increasing difference of electronegativity of the anionic and cationic part of the compounds.

Die Tl_3BC_4 (B = V, Nb, Ta; C = S, Se)-Verbindungen folgen der $8N$-Valenzregel von Mooser und Pearson und gehören zu den Halbleitern. Das Anwachsen des Wertes der Energielücke E_g mit der Substitution von V durch Nb oder Ta in diesen Verbindungen kann der wachsenden Bindungsionizität infolge der wachsenden Differenz der Elektronegativität des Anionen- und Kationenanteils dieser Verbindungen zugeschrieben werden.

1. Introduction

The compounds of $Tl_3^IB^VC_4^{VI}$, where B = V, Nb, Ta and C = S, Se, belong to the group of ternary isomorphous compounds with the space group of I$\bar{4}$3m. Preparation and some basic physical properties of these compounds were studied previously [1 to 3]. The compounds of Tl_3BC_4 reveal nonmetallic properties and an "anomalous" behaviour of the energy gap increasing with increasing atomic weight (or atomic number) of the element B(V, Nb, Ta), while usually (e.g. in "classical" $A^{II}B^{VI}$ or $A^{III}B^V$ compounds) the energy gap of semiconducting compounds decreases with increasing atomic weight of atoms in the compounds. In this paper we suggest an explanation of this "anomalous" behaviour of the Tl_3BC_4 compounds.

2. Experimental

The preparation of the Tl_3VS_4, Tl_3NbS_4, and Tl_3TaS_4 compounds and their single crystals as well as the preparation of samples for optical measurements is described elsewhere [1 to 3]. The spectral dependence of reflectivity of the Tl_3BS_4 compounds was measured in the vinicity of the absorption edge (Fig. 1). The spectral dependence of optical transmittance of Tl_3BS_4 compounds was measured also in the vicinity of the absorption edge (Fig. 2). A sharp decrease in the transmittance of Tl_3BS_4 compounds is associated with the fundamental absorption edge of the compounds and therefore from its position we have estimated the value of the optical gap E_g of the compounds shown in Table 1 together with the values of E_g reported by Belyaev [4] and melting points T_m of the compounds.

3. Discussion

The band theory of solids does not afford simple criteria for the prediction of semiconducting properties of solids from their chemical composition only. Another approach to the problem of predicting semiconducting properties of elements and com-

[1]) Leninovo nám., 53210 Pardubice, Czechoslovakia.

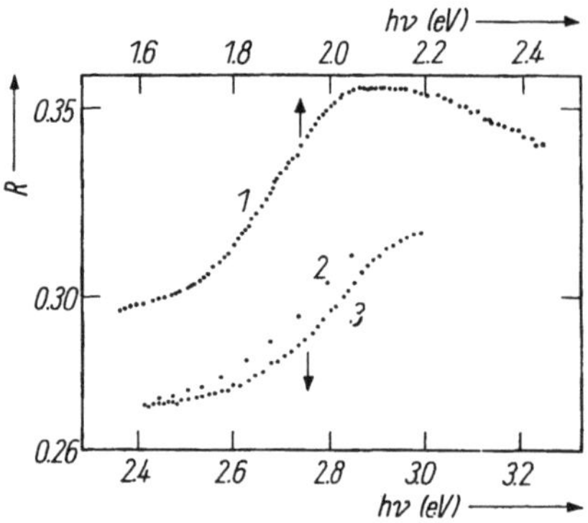

Fig. 1

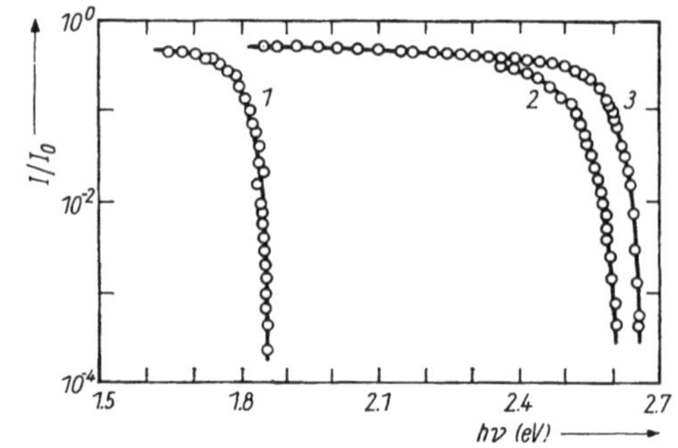

Fig. 2

Fig. 1. Reflectivity spectra of (1) Tl_3VS_4, (2) Tl_3NbS_4, and (3) Tl_3TaS_4
Fig. 2. Optical transmittance of (1) Tl_3VS_4, (2) Tl_3NbS_4, and (3) Tl_3TaS_4

pounds takes into account the chemical forces between the atoms of a compound under study. The model of a semiconducting bond introduced by Mooser and Pearson [5 to 7] makes it possible to predict and discuss semiconducting properties of solids. The focus point of their model is the general $8N$ valence rule [8],

$$\frac{n_e + b_a - b_c}{n_a} = 8\,,$$

where n_e is the total number of valence electrons in the compound, n_a the number of anions, b_a the number of electrons in anion–anion bonds, b_c the number of electrons in cation–cation bonds and nonbonding electrons altogether. All the values are calculated per formula unit. The compounds obeying this rule cannot reveal metallic behaviour.

Table 1
The width of the energy gap E_g of Tl_3BS_4 compounds

compound	E_g (eV)	E_g [4] (eV)	T_m [10] (°C)
Tl_3VS_4	1.8	1.8	505 ± 5
Tl_3NbS_4	2.5	2.58	540 ± 5
Tl_3TaS_4	2.6	2.60	550 ± 5

Application of the valence rule requires the knowledge of the atomic configuration in the structure of the compound. The structural arrangement in Tl_3BC_4 (Fig. 3) can be considered to be that of the substituted CsCl structure with contraction of four chalcogen atoms around the transition metal. The B atoms are tetrahedrally coordinated by chalcogen atoms and from the values of interatomic distances in the compound we can assume a prevailing covalent character of the B–C bonds. Thus, for the following discussion we can neglect bonding forces between C–C, B–B, Tl–B, and Tl–Tl atoms.

The measurement of magnetic susceptibility of the compounds [9] revealed their diamagnetic character; the contribution of the paramagnetic term to the total magnetic susceptibility decreases with increasing atomic number of the transition metal B

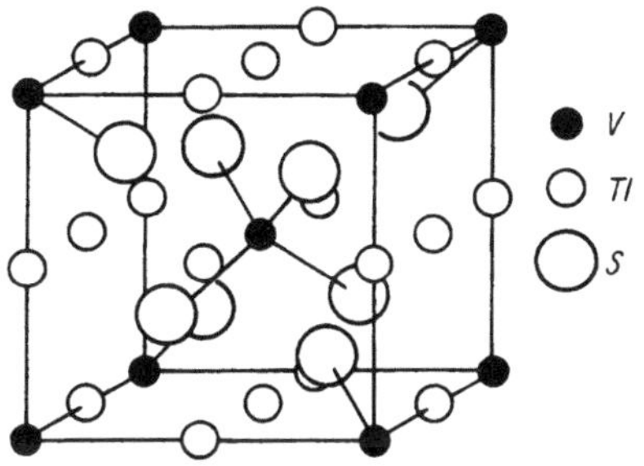

Fig. 3. Structure of Tl_3VS_4

($\chi_{para} = 29 \times 10^{-6}$ for Tl_3VS_4 and 11×10^{-6} for Tl_3TaS_4). As the transition metal is in the +V valence state and the Tl atom in +I state we assume loosing of electrons from d-orbitals of the transition metal and the nonbonding state of 6s electrons of Tl atoms. Thus the transition atoms and Tl atoms have vacant bonding orbitals and all their electrons are paired, which is confirmed by a negative value of the magnetic susceptibility. If the bonding between neighbouring vacant bonding orbitals would take place, the bonding paths would penetrate the whole structure resulting in the metallic conductivity of the compounds.

When we take into account the pair of nonbonding electrons of the Tl atoms, the $8N$ rule is fulfilled for the Tl_3BC_4 compounds giving thus semiconducting behaviour of the compounds. This prediction was confirmed experimentally by the measurement of the temperature dependence of electrical conductivity of the compounds and by the positive sign of charge carriers in these compounds [10].

At present there is no method for an exact calculation of the energetic spectrum and the energy gap of ternary compounds. Moreover, in our case the calculation is complicated by the presence of d-orbitals which can form narrow bands in the forbidden band. Several criteria were suggested for a semiquantitative calculation of the energy gap of solids. The Batsanov method [11], applicable for a relatively large group of compounds, is based on a simple relation for the calculation of the energy gap of the compound BA,

$$E_{BA} = E_A + E_B + aX_{AB} - b\bar{n} ,$$

where E_A and E_B are energy gaps of the elements A and B, respectively, $X_{BA} = X_A - \bar{X}_B$ is the difference in electronegativity of anions X_A and the mean value of cations $\bar{X}_B$, and $\bar{n}$ the mean principal quantum number of the compound. The value of $\bar{n}$ is calculated from the relation derived by Pearson [8], $\bar{n} = \sum c_i n_i / \sum c_i$, where n_i is the principal quantum number of the i-th atom and c_i the number of atoms per formula unit of the compound. The mean value of the electronegativity of the cationic part is defined similarly, $\bar{X}_B = \sum c_i X_i / \sum c_i$. The constants a and b for sulphide and selenide compounds have the values of 5.3 and 1.3 (S) and 6.3 and 1.0 (Se), respectively [11].

For the estimation of the energy gap of the Tl_3BC_4 compounds we used for V, Nb, and Ta the values of electronegativity obtained by Haissinski [12] and for the others

Table 2

Electronegativity values X of the elements

element	Cu	Tl	V	Nb	Ta	S	Se
X	1.8	1.5	1.8	1.6	1.4	2.6	2.4

the values of Batsanov [11]. Using the relation of Batsanov [11] and the above values of electronegativity (Table 2) we obtained the values of the energy gap of Tl_3BC_4 compounds given in Table 3.

Table 3
The values of the energy gap E_g (calculated by the Batsanov method) and mean atomization energy $\overline{H}_s$ of the Tl_3BC_4 compounds

compound	E_g (eV)	$\overline{H}_s$ (kJ mol^{-1})	compound	E_g (eV)	$\overline{H}_s$ (kJ mol^{-1})
Tl_3VS_4	2.5	270.0	Tl_3VSe_4	2.0	234.9
Tl_3NbS_4	2.6	295.5	Tl_3NbSe_4	2.2	261.2
Tl_3TaS_4	2.7	303.0	Tl_3TaSe_4	2.4	268.4

The calculation of the energy gap using the values of electronegativity is approximative due to the approximative values of the electronegativity which has been calculated by various methods and various authors and therefore there are various systems of electronegativity values. In spite of the approximative character of the method of Batsanov, the estimated values of E_g of the series of the Tl_3BC_4 compounds make it possible to discuss the relation of bonding character and energy gap of the compounds.

When we compare the values of E_g of the Tl_3BC_4 compounds calculated by the Batsanov method (Table 3) with those obtained from optical data (Table 1), we can see the same general trend in the increase of E_g when going from V to Nb and Ta. The substitution of sulphur by selenium in Tl_3BC_4 compounds results in a decrease of E_g of the ternary selenides against the sulphides. As we used for the calculation the difference of electronegativities as a criterion of bond ionicity, we can conclude that the increase in E_g from Tl_3VS_4 to Tl_3TaS_4 is in agreement with the increase in the bond ionicity. An increase in the metallic character of the bonds (represented according to Mooser and Pearson by increasing principal quantum number) is not so high as to compensate the increasing bond ionicity of the compounds.

The above conception of an increased bond ionicity in Tl_3TaS_4 is supported by a lower value of reflectivity and higher melting point of Tl_3TaS_4 in comparison with Tl_3NbS_4 and Tl_3VS_4.

The increase in E_g with increasing atomic number of the transition metal is in agreement with the observed decrease in the paramagnetic contribution to the sus-

Table 4
The values of the energy gap E_g of Cu_3BC_4 compounds (B = V, Nb, Ta; C = S, Se)

compound	colour [14]	E_g [15] (eV)	E_g [16] (eV)	E_g (eV)
Cu_3VS_4	black	1	1.3	2.2
Cu_3NbS_4	green	2.5 to 3.0		2.4
Cu_3TaS_4	yellow	3.0		2.5
Cu_3VSe_4	black			1.4
Cu_3NbSe_4	red			1.6
Cu_3TaSe_4	green			1.8

ceptibility value and with increasing value of the mean atomization energy H_s according to Sadagopan and Gatos [13] (Table 3).

A similar behaviour was observed for the sulvanite-type compounds Cu_3BC_4 (B = V, Nb, Ta; C = S, Se), which can be considered as the analogues of the ternary Tl chalcogenides with Tl substituted by Cu. In Table 4 the colour [14], the estimated values of E_g by Hulliger [15], the values of E_g obtained from optical data [16], and the E_g-values calculated by the Batsanov method of Cu_3BC_4 compounds are given. The data in Table 4 show that also in Cu_3BC_4 compounds an increase in the atomic number of the transition metal B results in an increase of the E_g-values.

As we mentioned above, an increase in the values of E_g in sulvanite-type compounds and ternary Tl chalcogenides and a few other compounds with increasing atomic number of constituent atoms was taken as "anomalous" in comparison with other semiconducting compounds like $A^{III}B^V$, $A^{II}B^{IV}$, etc., or their mixed ternary compounds [15]. But in the $A^{III}B^V$ or $A^{II}B^{VI}$ compounds electronegativity of the cationic part increases with increasing atomic number (e.g. $X_{Zn} < X_{Cd} < X_{Hg}$; $X_{Al} < X_{Ga} < X_{In}$) and thus the energy gap E_g decreases with decreasing difference in the electronegativity of cationic and anionic parts of the compound, that means with decreasing ionicity of the compound. In the ternary Tl chalcogenides or sulvanite-type compounds and several other compounds of transition metals due to a decrease in electronegativity of the transition metal with increasing atomic number ($X_V > X_{Nb} > X_{Ta}$; $X_{Ti} > X_{Zr}$), the electronegativity difference $X_A - X_B$ increases. Thus, the ionicity of these compounds also increases and if this increase in ionicity is not sufficiently compensated by an increasing metallic character of the compound (due to an increase in the mean principal quantum number), the energy gap increases. Therefore, we assume that from this point of view we need not consider the studied Tl chalcogenides or sulvanite-type compounds in contrast to $A^{III}B^V$ or $A^{II}B^{VI}$ compounds, because both the groups show the same tendency in the dependence of E_g on the bond ionicity.

References

[1] K. Čermák, Čes. Čas. Fys. **A18**, 436 (1968).
[2] K. Čermák, Čes. Čas. Fys. **A19**, 149 (1969).
[3] K. Čermák, Czech. J. Phys. **B34**, 88 (1984).
[4] B. V. Belyaev, Fiz. Tekh. Poluprov. **9**, 133 (1975).
[5] E. Mooser and W. B. Pearson, Phys. Rev. **101**, 1608 (1956).
[6] E. Mooser and W. B. Pearson, J. Electronics **1**, 629 (1956).
[7] E. Mooser and W. B. Pearson, Progr. Semicond. 5, 103 (1960).
[8] W. B. Pearson, Acta cryst. **17**, 1 (1964).
[9] M. Matyáš and K. Čermák, Czech. J. Phys. **B19**, 1305 (1969).
[10] K. Čermák, Collect. Czech. chem. Commun. **34**, 3605 (1969).
[11] S. S. Batsanov, Zh. strukt. Khim. **5**, 927 (1964).
[12] M. Haissinski, J. Phys. Radium **7**, 7 (1946).
[13] V. Sadagopan and H. C. Gatos, Solid State Electronics **8**, 529 (1965).
[14] A. E. van Arkel and C. Crevecoeue, J. less-common Metals **5**, 177 (1963).
[15] F. Hulliger, Helv. phys. Acta **34**, 379 (1961).
[16] D. Petritis, G. Martinez, C. Levy-Clement, and O. Gorochov, Phys. Rev. B **23**, 6773 (1981).

(Received June 20, 1984)

phys. stat. sol. (b) **126**, 279 (1984)

Subject classification: 20; 21.4

Department of Physics, East Texas State University, Commerce[1]) *(a)*
and Department of Physics, Indian Institute of Technology, Bombay[2]) *(b)*

Extended X-Ray Absorption Fine Structure in RAl_2Ga_2 Intermetallics[3])

By

T. K. HATWAR (a), B. D. PADALIA (b), M. N. GHATIKAR (b) and D. R. CHOPRA (a)

The near-edge structure (NES) and the extended X-ray absorption fine structure (EXAFS) associated with the K-absorption edge of Ga and L_3-edges of the rare earths (R) (where R = La, Ce, Pr, Nd, Eu, and Yb) are studied in RAl_2Ga_2 intermetallics using a 600 mm curved crystal spectrograph. The results obtained are discussed in the light of the recent point scattering theory of EXAFS proposed by Lytle, Sayers, and Stern. The EXAFS data are used to obtain the near-neighbour distances in these newly synthesized compounds.

Die kantennahe Struktur (NES) und die ausgedehnte Röntgenabsorptionsfeinstruktur (EXAFS), die mit der K-Absorptionskante von Ga und den L_3-Kanten von Seltenen Erden (R) verknüpft sind (wobei R = La, Ce, Pr, Nd, Eu und Yb ist), werden für RAl_2Ga_2-Intermetalle unter Benutzung eines gekrümmten 600 mm-Kristallspektrographen untersucht. Die erhaltenen Ergebnisse werden im Licht der neueren von Lytle, Sayers und Stern vorgeschlagenen Punktstreutheorie der EXAFS diskutiert. Die EXAFS-Werte werden benutzt, um die Abstände nächster Nachbarn in diesen neu synthetisierten Verbindungen zu bestimmen.

1. Introduction

We have synthesized RAl_2Ga_2 compounds (where R is the rare earth, La, Pr, Nd, Eu, and Yb) in addition to the already reported $CeAl_2Ga_2$ [1]. These compounds crystallize in $ThCr_2Si_2$ structure [2]. NMR and X-ray measurements were reported earlier [3, 4]. This paper reports the investigation on the extended X-ray absorption fine structure (EXAFS) of these new intermetallics. The EXAFS has some advantages for these compounds because the EXAFS of both the L_3-edges of the rare earths and the K-edge of Ga can be measured. This allows us to determine the local structure about each atom.

The absorption spectra of X-ray edges can be divided into two intervals [5]. The near-edge structure (NES) extends about 40 eV beyond the threshold followed by the EXAFS region which extends typically 1000 eV. In a single-particle approximation, the absorption coefficient $\mu(E)$, where E denotes the energy of incident photons, depends on both the density of the available unoccupied states $N(E)$ and the transition probability $P(E)$, $\mu(E) \sim N(E)\, P(E)$. Within a few tens of eV above the absorption edge, $P(E)$ varies slowly as a function of energy [6]. So the NES directly reflects the density distribution of empty or partially empty states just above the Fermi level. The shape of the EXAFS is governed by the crystal structure that surrounds the absorbing atom. We have used NES in the K-edge of Ga and L_3-edges of R to obtain information

[1]) Commerce TX 75428, USA.

[2]) Bombay 400076, India.

[3]) This article is based on T. K. Hatwar, Ph. D. Thesis, Indian Institute of Technology, Bombay 1981 (unpublished).

regarding the conduction band in these intermetallics and the EXAFS to estimate near-neighbour distances employing the theory of Stern [5] and Lytle et al. [7].

2. Experimental

A 600 mm curved mica crystal ((2̄01) plane) X-ray spectrograph of Cauchois type was used in the present work [8]. The experimental arrangement is adequate [9] for the EXAFS work on the concentrated samples (0.2 mg/mm^2) used in the present work. A careful shielding arrangement minimized the background scattering at the photographic plate enabling us to obtain well-resolved peaks in the EXAFS. The dispersion on the spectrogram was about 8 X units/mm. The energies of the maxima and minima in the fine structures are directly measured from the microphotometer recording taken at 100 × magnification. Several spectra have been recorded for each sample to eliminate the effect of fluctuations due to grain size and thickness effect. The estimated errors in the measurement of energies are less than 0.4 eV.

3. Results and Discussion

Fig. 1 and 2, respectively, show L_3-level spectra of R and K-level spectra of Ga in RAl_2Ga_2 intermetallics. As the wavelengths corresponding to the lighter rare earths, La and Ce, fall beyond 0.21 nm, these could not be recorded using the present spectrograph with 600 mm radius of curvature. The L_2-edge of Yb occurs within 40 eV on the high-energy side of the Ga K-edge, and the fine structure associated with Ga K-edge in $YbAl_2Ga_2$ could not be recorded. The successive absorption maxima in the fine structure are denoted by A, B, C, ... etc., while the minima are marked α, β, γ, ... etc.

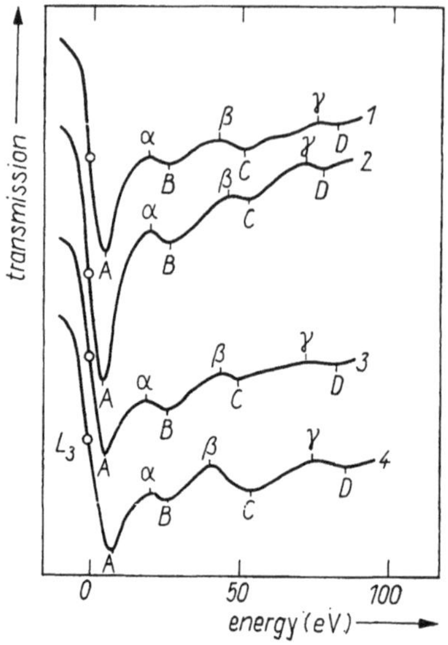

Fig. 1

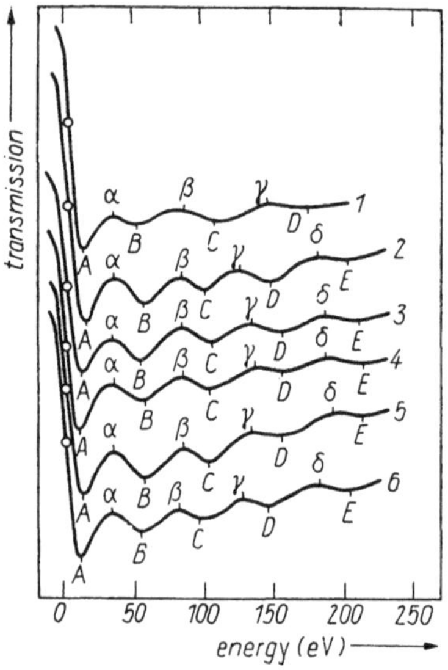

Fig. 2

Fig. 1. Shape of the L_3-edge and the associated EXAFS of rare earths in (1) $PrAl_2Ga_2$, (2) $NdAl_2Ga_2$, (3) $EuAl_2Ga_2$, (4) $YbAl_2Ga_2$

Fig. 2. Shape of the K-edge and the associated EXAFS of Ga in (1) Ga metal, (2) $LaAl_2Ga_2$, (3) $CeAl_2Ga_2$, (4) $PrAl_2Ga_2$, (5) $NdAl_2Ga_2$, (6) $EuAl_2Ga_2$

Table 1

Extended fine structure of K absorption edge of Ga in pure metal and RAl_2Ga_2 intermetallics measured in eV

n	max/min	Ga metal	$LaAl_2Ga_2$	$CeAl_2Ga_2$	$PrAl_2Ga_2$	$NdAl_2Ga_2$	$EuAl_2Ga_2$
0	A	13.0	12.8	12.2	12.5	11.8	12.0
1	α	31.9	32.5	32.3	34.5	33.7	33.0
2	B	49.6	52.5	52.3	54.2	55.8	53.0
3	β	81.4	80.5	80.8	83.2	85.0	81.0
4	C	102.8	98.5	99.6	100.3	100.7	95.0
5	γ	140.0	123.0	129.0	130.0	129.0	122.8
6	D	170.0	146.0	151.6	152.0	152.6	143.9
7	δ	—	177.0	180.0	182.0	187.0	181.0
8	E	—	197.0	204.0	211.0	212.0	—

Fig. 1 and 2 indicate that the L_3-edge of R and the K-edge of Ga are both characterized by the presence of resonance "white lines" (peak A) near the absorption threshold. The dominant cause of the white lines has been shown to be due to the dipole-allowed transitions to the final states of proper symmetry above the Fermi level E_F. In the case of rare earths, the peak A corresponds to an electron transition from the 2p level into conduction states having both s- and d-partial wave characters. However, the s-partial wave contribution in the L_3-spectra can be ignored because of the much smaller $2p \rightarrow \varepsilon s$ matrix elements when compared with $2p \rightarrow \varepsilon d$. Thus the observed structures of the L_3-spectra arise from strong features in the d-projected density of states. The K-edge of Ga is ascribed to $1s \rightarrow np$ transition; it thus probes the density of states with p-symmetry above the Fermi level. The existence of a prominent peak A in both metals [4, 10] and their intermetallics indicates the presence of a high density of states of d- and p-symmetry in the conduction band of R and Ga, respectively. The formation of chemical bonds among the constituents affects the occupancy and/or hybridization of the conduction states, and the changes can be revealed in the white-line intensities and widths. The white lines for compounds are sharper and more pronounced than those observed for metals [10]. The narrowing of the white lines can be ascribed to the increase in the interatomic distances and the hybridization of the constituent bands in RAl_2Ga_2 compounds [11].

Both the K-edge of the Ga and the L_3-edges of the R exhibit the marked EXAFS in RAl_2Ga_2 compounds. The measured energies of the EXAFS peak are given in Tables 1 and 2. Stern [5] and Lytle et al. [7] have proposed a rigorous theory of the EXAFS

Table 2

Extended fine structure of L_3 absorption edge of R in RAl_2Ga_2 intermetallics measured in eV

n	max/min	$PrAl_2Ga_2$	$NdAl_2Ga_2$	$EuAl_2Ga_2$	$YbAl_2Ga_2$
0	A	5.4	4.7	5.0	8.1
1	α	19.2	20.3	20.2	21.3
2	B	27.2	27.7	26.6	24.2
3	β	43.1	45.2	44.1	40.2
4	C	50.6	53.0	50.2	53.4
5	γ	75.0	72.4	73.5	73.9
6	D	81.5	78.1	84.0	85.0

based on the scattering mode. Fourier inversion of the EXAFS data is a general and powerful technique to determine many physical parameters; however, it is possible to determine nearest-neighbour distances and phase shifts from a simple graphical method [7]. This is possible because the first shell scattering usually dominates the EXAFS curve. So if only the major EXAFS peaks are used, the period will be that of the first coordination shell. In the case of $ThCr_2Si_2$ structure [2], Th atom lies in the centre of the coordination sphere surrounded by eight Cr atoms. On the other hand, Si atom is surrounded by four Cr atoms as its nearest neighbours. Since RAl_2Ga_2 belong to the $ThCr_2Si_2$ structure, the EXAFS of the L_3-edges of the R provide R–Ga distances while the EXAFS of the K-edge of the Ga gives Ga–Al distances in these intermetallics. The method of estimating the bond lengths is discussed below.

According to the graphical method of Lytle et al. [7], the EXAFS can be visualized as a kind of electron diffraction with the electron source being the absorbing atom. Considering back scattering from the first coordination shell only, the maxima and minima are governed by the argument of the sine term given by [7]

$$\tfrac{1}{2} n\pi = 2k(R_1 - \alpha_1) + \beta_1 ,$$

where k is the electron wave vector, R_1 the average radial distance (bond length), α_1 the phase shift, and β_1 a constant. In the above equation $n = 0, 2, 4, \ldots$ for maxima and $1, 3, 5 \ldots$ for minima. The slope of the plot between k and n for maxima and minima can be written as $M = (4/\pi)\,(R_1 - \alpha_1)$. Thus k versus n plot can be utilized to obtain the value of R_1 or α_1 if either is known. Using the crystallographic value of 0.244 nm [12] for R_1, we obtain the value of $\alpha_1 = 0.142$ nm for Ga metal. Since phase shift α_1 depends upon the chemical bonding [7], the use of this value for calculating Ga–Al distances in RAl_2Ga_2 compounds would be erroneous. For this purpose we used the following alternative method for estimating the bond lengths, thus avoiding the uncertainty in the values of α_1 in the compounds. The Lytle-Sayers-Stern plots are used to establish the second absorption maximum B and the subsequent minimum β. These values are then used to calculate the near-neighbour distances employing the relation $R_1 = (151/\Delta E)^{1/2}$, where ΔE is the energy difference between the second absorption maximum B and the subsequent minimum β [13]. Such a procedure has yielded a very good estimate of the bond lengths [14]. We obtained a value of 0.242 nm in agreement with the accepted value 0.244 nm for Ga metal. The same procedure was extended to the EXAFS of the L_3-edges of R. The estimated values of Ga–Al and R–Ga distances are given in Table 3. To find the variation of α_1 with the change of chemical

Table 3
Estimated average bond lengths (R_1) and phase shifts α_1 in RAl_2Ga_2 compounds

compound	Ga–Al bond length (nm)	R–Ga bond length (nm)	phase shift α_1 of Ga in RAl_2Ga_2 (nm)
Ga metal	0.242*)	—	0.142
$LaAl_2Ga_2$	0.273	—	0.151
$CeAl_2Ga_2$	0.273	—	0.151
$PrAl_2Ga_2$	0.271	0.349	0.150
$NdAl_2Ga_2$	0.270	0.345	0.148
$EuAl_2Ga_2$	0.274	0.350	0.151
$YbAl_2Ga_2$	0.269	0.350	0.150

*) Ga–Ga bond length.

bonding, the α_1's in RAl_2Ga_2 compounds were calculated from the estimated Ga–Al bond lengths and are included in Table 3. According to the Lytle et al. theory, the variation in α_1-values reflects changes in chemical bonding. Table 3 shows that the α_1-value for pure Ga metal is less than that in the RAl_2Ga_2 compounds. Thus assuming Ga metal to be 100% covalent, the increased values of α_1 in RAl_2Ga_2 compounds show that iono-covalent bonds are formed in the RAl_2Ga_2 compounds. These results are consistent with the previous measurements of the X-ray absorption edge shifts on the same compounds [4]. Similar increase in the value of α while going from metal to compound was also noted by others in the case of Fe [7].

Thus the present study of the fine structures associated with the K-edge of Ga and L_3-edges of R is quite successful in providing the information about the conduction band, the nature of the chemical bond, and various bond lengths in RAl_2Ga_2 series of compounds.

Acknowledgement

The authors express their sincere thanks to Ms. Susan Willbern for typing the manuscript.

References

[1] O. S. Zarechnyuk, P. I. Kripyakevich, and E. I. Gldyshevskii, Soviet Phys.-Cryst. **9**, 706 (1965).
[2] Z. Ban and M. Sikirica, Acta cryst. **18**, 594 (1965).
[3] E. V. Sampathkumaran, L. C. Gupta, R. Vijayaraghavan, T. K. Hatwar, M. N. Ghatikar, and B. D. Padalia, Mater. Res. Bull. **15**, 939 (1980).
[4] M. N. Ghatikar, T. K. Hatwar, B. D. Padalia, E. V. Sampathkumaran, L. C. Gupta, and R. Vijayaraghavan, phys. stat. sol. (b) **106**, K89 (1981).
[5] E. A. Stern, Phys. Rev. B **10**, 3027 (1974).
[6] L. V. Azaroff and D. M. Pease, X-Ray Spectroscopy, Chap. 6, McGraw-Hill Publ. Co., New York 1974.
[7] F. W. Lytle, D. E. Sayers, and E. A. Stern, Phys. Rev. B **11**, 4835 (1975).
[8] T. K. Hatwar, M. N. Ghatikar, B. D. Padalia, S. K. Malik, and R. Vijayaraghavan, phys. stat. sol. (b) **103**, 159 (1981).
[9] P. A. Lee, P. H. Citrin, and B. M. Kincaid, Rev. mod. Phys. **53**, 769 (1981).
[10] S. N. Gupta, V. P. Vijayavargiya, B. D. Padalia, B. C. Tripathi, and M. N. Ghatikar, phys. stat. sol. (b) **82**, 603 (1977).
[11] F. V. Hillebrecht, J. C. Fuggle, P. A. Bennet, Z. Zolnierek, and Ch. Freiburg, Phys. Rev. B **27**, 2179 (1983).
[12] R. C. Weast (Ed.), Hdb. Chem. Phys., 59th ed., CRC Press, Inc., 1978/79 (p. 216).
[13] R. M. Levy, J. chem. Phys. **43**, 1845 (1965).
[14] B. K. Agrawal and V. Balkrishnan, Phys. Rev. B **28**, 2852 (1983).

(Received June 19, 1984)

phys. stat. sol. (b) **126**, 285 (1984)

Subject classification: 13.1; 22.5

Institute of Physics, Polish Academy of Sciences, Warsaw[1])

Electronic Band Structure of CdF_2

Photoemission Experiment and Pseudopotential Calculations

By

B. A. ORŁOWSKI and P. PLENKIEWICZ[2])

Energy distribution curves (EDC) of electrons photoemitted ($h\nu_1 = 21.22$ eV and $h\nu_2 = 40.8$ eV) from the valence band (F^- 2p) and the Cd 4d band of a CdF_2 crystal are measured and the valence and conduction band structures are calculated using the local empirical pseudopotential method (EPM). Comparison of the measured and calculated data show that the three main maxima measured in the valence band correspond well to the maxima in X_5, L_3, and X_5 obtained in the calculated histogram of the valence band density of states. The influence of the final-state structure on the position of peaks in the valence band is not observed on EDCs obtained for both, $h\nu_1$ and $h\nu_2$ photoemission exciting energies. For Cd 4d band the wide splitting of 1.46 eV is obtained only for $h\nu_1 = 21.22$ eV while it is not obtained on EDC for $h\nu_2 = 40.8$ eV. This Cd 4d band splitting may be caused by the influence of the final density-of-states maximum obtained in conduction band in X_1- point. The results obtained are compared with reflectivity data available in the literature.

Die Energieverteilungskurven (EDC) von photoemittierten Elektronen ($h\nu_1 = 21{,}22$ eV und $h\nu_2 = 40{,}8$ eV) aus dem Valenzband (F^- 2p) und dem Cd-4d-Band von CdF_2-Kristallen werden gemessen und die Valenz- und Leitungsbandstruktur wird mit der lokalen empirischen Pseudopotentialmethode (EPM) berechnet. Der Vergleich der gemessenen und berechneten Werte zeigt, daß die drei Hauptmaxima, die im Valenzband gemessen werden, gut mit den Maxima in X_5, L_3 und X_5 übereinstimmen, die im berechneten Histogramm der Valenzbandzustandsdichte erhalten werden. Der Einfluß der Struktur der Endzustände auf die Lage der Maxima im Valenzband wird nicht in den EDC's beobachtet, die sowohl aus mit $h\nu_1$ als auch $h\nu_2$ angeregten Photoemissionsenergien erhalten werden. Für das Cd-4d-Band wird die breite Aufspaltung von 1,46 eV nur für $h\nu_1 = 21{,}22$ eV erhalten, während sie in der EDC für $h\nu_2 = 40{,}8$ eV nicht beobachtet wird. Diese Cd-4d-Bandaufspaltung kann durch den Einfluß des Maximums der Dichte der Endzustände im Leitungsband im X_1-Punkt verursacht werden. Die erhaltenen Ergebnisse werden mit Reflexionswerten aus der Literatur verglichen.

1. Introduction

Cadmium fluoride (CdF_2) has been the subject of recent research due to the simultaneous appearance of two main properties, visible and ultraviolet light transparency ($h\nu < 6$ eV) and semiconductivity after doping and thermal treatment. These properties make CdF_2 an interesting material for different applications, e.g. light-emitting devices (electroluminescence in the visible range [1] or light energy converter [2]).

Cadmium fluoride belongs to the group of the crystals with fluorite structure [3]. The bonding in CdF_2 is highly ionic. Cadmium $5s^2$ electrons are mainly bound to fluorine 2p states and F^- 2p electrons create the valence band. The conduction band is built mainly of empty Cd^{2+} $5s^2$ states. The Cd 4d electrons lie at a distance 1.8 eV

[1]) Al. Lótnikow 32/46, 02-668 Warsaw, Poland.

[2]) Present address: Institute of Physics, Technical University of Wrocław, Wybrzeże Wyspiańskiego 27, Wrocław, Poland.

below the valence band [4]. The light absorption edge is determined by the energy of exciton creation with maximum of absorption at $h\nu = 7.6$ eV [5]. The direct band gap was approximated by 8.4 eV (Γ_{15}–Γ_1 [4]), giving the exciton binding energy of 0.8 eV, consistent with the results for CaF_2 crystal [6] and the large-polaron model. The indirect gap X_2–Γ_1 is approximately equal to 7.8 eV [4]. High electronic conductivity of CdF_2 crystals is obtained by doping with trivalent atoms, e.g. Y, Yb, In, or most of the rare-earth metals and applying a conversion procedure [3, 7]. This property is presumably related to the very high electronic affinity of CdF_2 ($\chi = 4$ eV [8, 9]) as compared with the affinity of almost zero for most of the other fluorites [10]. The conversion procedure results in producing shallow donor levels (0.1 eV below the conduction band edge) and at room temperature they supply the conduction band with electrons. A model proposed to explain these properties [10] was strongly confirmed by the experimental results obtained on alloys such as $Cd_{1-x}Ca_xF_2$:Y and $Cd_{1-x}Pb_xF_2$:Y, where the insulator–semiconductor transition was predicted and observed for $x = 0.07$ and 0.15, respectively [11].

One purpose of the paper is to explain some particular details of the obtained UPS experimental results. The electronic band structure was precisely calculated in a particular region of electron energy and momentum space where the origin of these details is expected. As an example of these details we took the influence of the final states on the Cd 4d band "splitting" which occurs in energy distribution curves (EDCs) measured for $h\nu = 21.22$ eV, but does not occur in EDCs measured for $h\nu = 40.8$ eV. The shares of final states in the UPS optical transitions occur very distincly in three-component crystals like in $Cd_{1-x}Mn_xF_2$, where the increase of x brings about remarkable changes in the distribution of the final states causing big differences in the EDCs of the Cd 4d band and valence band obtained for $h\nu = 21.22$ and 40.8 eV [12]. For electrons excited from the valence band with energy $h\nu = 21.22$ eV the occurrence of a structure in the density of final states is predicted by the band structure calculation, while for $h\nu = 40.8$ eV we can expect this structure to be negligible.

In the present paper we applied the empirical pseudopotential method to calculate the valence and conduction band structure of CdF_2 crystals. The histogram of the density of states obtained on the basis of band structure calculations is compared with the UPS measured data and with reflectivity results available from literature. For this kind of comparison we need to know the positions of the band edge in the absolute energy scale relatively to the vacuum level. The absolute position of the conduction band edge was determined in the photoemission yield experiment for CdF_2:Y [9] and for internal photoemission measurements [8]. The absolute position of the valence band edge was determined in UPS results presented in this paper.

Section 2 of the paper presents the conditions of the experiment and the obtained experimental results. Section 3 describes the method of calculation and the parameters used in the calculation. Section 4 compares the measured and calculated data with the reflectivity data taken from literature.

2. Experimental Conditions

The commercial angle-resolved spectrometer ADES-400 produced by Vacuum Generator Scientific, Ltd., was used to perform the experiment in UHV conditions ($p = 10^{-10}$ Torr). The spectrometer is equipped with UPS, XPS, and LEED facilities together with a movable 150° hemispherical electron energy analyser rotating around the sample in the horizontal plane. The energy resolution of the analyser was kept at 0.1 eV for UPS experiments and 1 eV in XPS regime. The angle resolution was 2°. The analyser measured the electrons photoemitted perpendicular to the crystal surface.

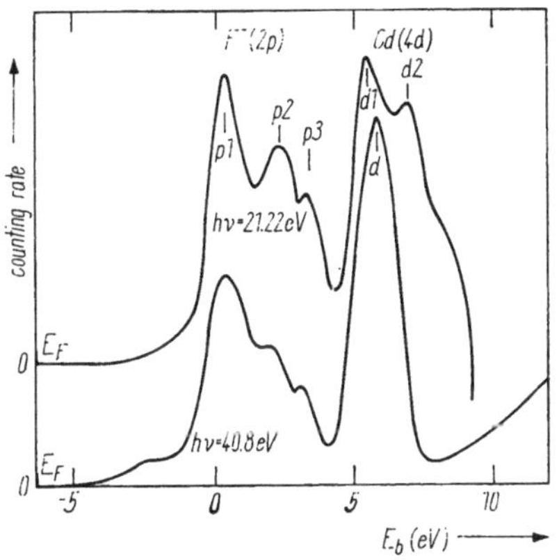

Fig. 1. Energy distribution curves (EDCs) obtained for CdF_2 crystal for $h\nu = 21.22$ and 40.8 eV (E_b mean binding energy)

For UPS measurements, a de-excited resonance source was used with two-stage differential pumping. The HeI ($h\nu = 21.22$ eV) and HeII ($h\nu = 40.8$ eV) excitation photon beams were separated in an ultraviolet grazing incidence monochromator V.G. UVM 100 connected in series with an uv lamp. The angle of light incidence was 45° measured with respect to the surface normal.

Data acquisition and storing have been performed by a V.G. Data System 4025 on-line computer based on a DEC LSJ 11/2 processor. For data handling standard V.G. programs have been used. The samples were clipped to the standard UM 20 manipulator which enables two independent rotations in two perpendicular planes and the heating of the sample up to 600 °C by a resistivity heater. An Au sample was used to calibrate the energy scale and to determine the Fermi level position.

The CdF_2 crystals used in the experiment were obtained by the modified Bridgman method [13].

The crystal possesses a small contamination by Y (0.01%) to obtain sufficient conductivity, to avoid charging effects. Plates of the crystal with diameter 7 and thickness 1 mm were cut on a wire saw and were mechanically and chemically polished. The crystal surface was cleaned in the spectrometer chamber by argon bombardment (2 keV, 20 μA, 20 min) and annealed at 120 °C of 1 h. Before this cleaning procedure the contamination on the surface was detectable in XPS (C1s, S2p, and C12p), while after the cleaning no surface contamination was seen and the UPS ($h\nu = 21.22$ eV) spectra showed a small scattered electron peak. We regard the crystal surface prepared in this way as properly clean to take the ARUPS data.

The ARUPS measurements were used as indicator of the crystalline surface quality when the crystal surface was being prepared. The lack of changes of the AREDCs structure with the change of the angle of measurement was regarded as an indicator of a lack of monocrystallinity of the crystal surface. In the case of CdF_2 crystal surfaces the AREDCs obtained were nearly identical for different angles. They can be regarded as corresponding to a polycrystalline surface and can be treated as approching angle-integrated EDC corresponding to the total density of states.

These EDCs obtained for $h\nu = 21.22$ and 40.8 eV for the CdF_2 surface are presented in Fig. 1.

3. Band Structure Calculations

The band structure of CdF_2 has been previously calculated by Albert et al. [14] by combining the tight-binding and the pseudopotential methods. The empirical pseudopotential method (EPM) is widely applied for band structure calculation of metals and

Table 1

Atomic pseudopotential form factors V_{Cd} and V_F

G $(l_1 l_2 l_3)$	G^2 $(2\pi/a)$	V_{Cd} (eV)	V_F (eV)
111	3	−0.972	—
200	4	−0.9145	−3.911
220	8	1.226	−2.219
311	11	0.03	—
222	12	0.006	−0.486
400	16	0.0	−0.621
331	19	0.0	—
420	20	0.0	0.12

semiconductors. It has been also successful in getting the band structure corresponding well to the experimental data. The EPM method was also successfully applied to the band structure calculations of such wide-gap ionic materials as MgO [15, 16] and NaCl and KCl [17, 18].

In this paper we have used the local EPM method to calculate the band structure of the CdF_2 crystal. The pseudopotential matrix elements have been calculated in the plane wave representations without including spin-orbit effects and neglecting the d-states.

Atomic pseudopotential form factors (see Table 1) for Cd have been determined using data and simple expressions taken from Plenkiewicz and Dowgiałło-Plenkiewicz [19].

Table 2

Comparison of the band position energies (in eV) calculated by Albert et al. [14] and in this work

	state	Albert et al. [14]	this work
valence band	Γ_{15}	0	0
	Γ'_{25}	−1.3872	−2.30*)
	X'_2	0.6936	1.1325
	X_5	0.3944	0.2898
	X'_5	−2.4888	−2.8288
	X_1	−2.8288	−4.2305
	L_3	−0.816	−1.5796
	L'_3	−1.0064	−1.7267
	L_1	−1.6184	−2.7576
	L'_2	−1.6728	−2.4541
conduction band	Γ_1	9.38	8.600
	Γ_{12}	18.82	18.085
	Γ'_{25}	18.71**)	18.563
	X'_4	13.518	13.861
	X_1	13.43	14.82
	X_3	15.518	17.968
	L_1	13.178	15.059
	L'_2	13.274	17.569

*) Extrapolated value.

**) Γ_{15}-state.

The starting values of atomic pseudopotentials of fluorine were estimated from those for oxygen by means of the Lin-Kleinman model [20] and the bare-ion pseudopotential theory. The secular equation was solved using the perturbation technique proposed by Löwdin [21] and modified by Brust [22]. The matrix was 65×65 and about 70 plane waves have been taken into account as a perturbated set. The band structure calculation has been performed for three axes of high symmetry, λ, Δ, and Z. The coordinates of these axes are shown in Fig. 2.

The resulting band structure was fitted to the experimental data by small changes of the atomic pseudopotential form factors both for Cd and F. It was noticed that only the pseudopotential form factors for the smallest G ($G^2 < 20$) play an important role in the calculations. The final values of atomic pseudopotential form factors used in calculations are given in Table 1.

The resulting band structure of CdF_2 is shown in Fig. 2. As is seen the structure of the valence band is similar to that calculated by Albert et al. [14]. The comparison between the energy values obtained by the EPM method and those calculated by Albert et al. is given in Table 2.

In Fig. 2 also the histogram of the density of states is demonstrated.

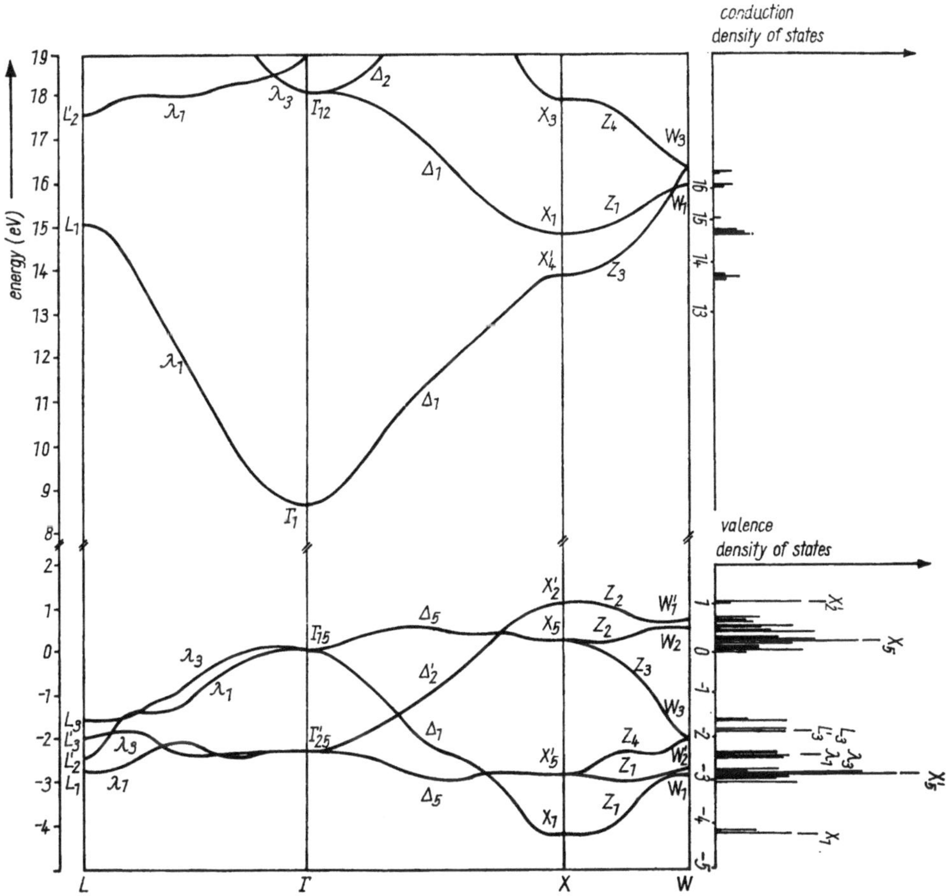

Fig. 2. Electronic band structure calculation by the local EPS method and histogram of the density of states corresponding to it

4. Comparison of Theoretical and Experimental Results

In Fig. 2 the valence and conduction band structures obtained by the local EPM calculation is presented together with the histogram of the density-of-states distribution. The direct band gap was taken as Γ_1–Γ_{15} = 8.6 eV and the indirect band gap was obtained from the calculation as X_2'–Γ_1 = 7.47 eV. The sum of the direct band gap and the electron affinity [9] gives the photoemission threshold energy equal 11.47 eV, which corresponds to the value 11.7 eV measured in UPS.

Table 3

Comparison of the distances between peaks measured in UPS with related distances obtained in the calculation (in eV)

experiment		theory		corresp. distances
Raisin et al. [5]	this work	Albert et al. [14]	EPM (this work)	
	5.2	3.52	5.36	X_2'–X_1
3.0	p1–p3 2.99	2.88	3.12	X_5–X_5'
1.8	p1–p2 2.03		2.02	X_5–L_3'
		2.01		X_5–L_1
1.2	p2–p3 0.96		1.10	L_3'–X_5'
		0.87		L_{21}–X_5'

The Cd 4d electron interaction with F^- 2p valence electrons is neglected in the calculation and it leads to a shifting down of the lower part of the valence band and to an increase of the valence band-width to the value 5.36 eV (see Table 3).

The calculation done by Albert et al. [14] leads to the valence band-width value 3.52 eV. The Cd 4d electrons influence mainly the lower part of the valence band, while for the upper part of the valence band this influence can be neglected. In Fig. 3

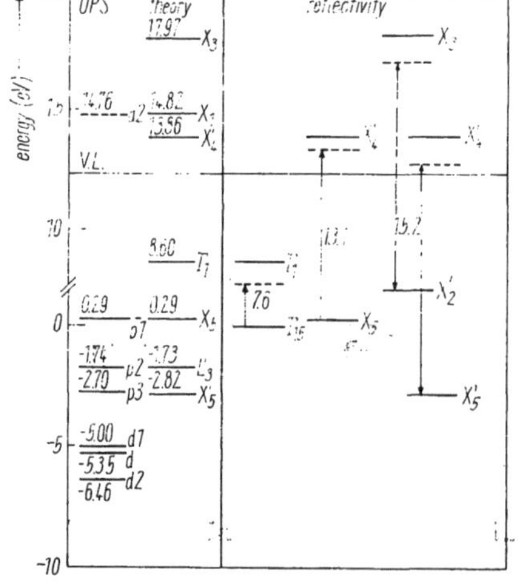

Fig. 3. Comparison of the peak positions in UPS with the corresponding calculated maxima of the histogram of the density of states and with the reflectivity data (V.L. vacuum level)

Table 4

Comparison of the measured energy of the reflectivity peaks with the corresponding distances in the band structure calculation (in eV)

exciton energy	Albert et al. [14]		EPM (this work)	
	ΔE	ΔE bands	ΔE	ΔE bands
7.6	9.4	Γ_{15}–Γ_1	8.6	Γ_{15}–Γ_1
13.1	14.82	X_2'–X_3	13.57	X_5–X_4'
15.2	18.0	X_5'–X_3	16.69	X_5'–X_4'
			16.83	X_2'–X_3

the EDC positions of peaks obtained for $h\nu = 40.8$ eV are located on the energy scale in such a way that the p1 peak position (see Fig. 1) coincides with the maximum of the density of states calculated in X_5 (see Fig. 2). Table 3 presents the comparison of the measured energy distances between peaks p1, p2, and p3 with the calculated density-of-states maxima in the histogram (Fig. 2). To compare the UPS and calculated band structure data on the same energy scale for conduction and valence bands, it was assumed that the photoemitted electron gains only the photon energy $h\nu$ and that the many-body effects which could lead to an additional change of its energy (like e.g. in [23]) can be neglected. In Fig. 3 the position of the main peaks obtained in the UPS experiment (first column) and the corresponding calculated positions of the density-of-states histogram maxima (second column) are presented and compared with the reflectivity data. For this comparison it was assumed that the experimentally observed energy positions of the reflectivity peaks are associated with the creation of an exciton which corresponds to the energy of the expected interband transitions lowered by the exciton binding energy. The excitons are expected to deviate strongly from hydrogenic-like excitons and to have a binding energy of about 1 eV. The measured reflectivity peak positions [5, 24, 25] and the distances between bands corresponding to them are presented in Table 4.

As was mentioned before, the parts of EDCs in Fig. 1 corresponding to the Cd 4d electrons are different for $h\nu = 40.8$ eV (single sharp peak) and for $h\nu = 21.22$ eV (double peak). For $h\nu = 40.8$ eV the electrons are excited much above the vacuum level to the region of the conduction band where a rich structure of the final density of states is not expected. For $h\nu = 21.22$ eV the electrons are excited about 3 eV above the vacuum level where the structure of the density of states was found from the band structure calculation at the X-point. We assume that the real position of the Cd 4d band is -5.35 eV obtained for $h\nu = 40.8$ eV. Two peaks d1 and d2 obtained for $h\nu = 21.22$ eV occur because the electrons excited from d-states over X_1-states partly relax to the X_1-states. This relaxation will lead to the appearance of the additional d2 peak marked in the first column of Fig. 3 at the energy -6.45 eV in the valence band and in 14.76 eV in the conduction band. The experimental value 14.76 eV corresponds to the value 14.82 eV calculated for X_1. The reflectivity peak measured for 15.2 eV could correspond equally to the X_5'–X_4' and X_2'–X_3 transitions.

5. Summary

The experimental results (UPS, XPS, and reflectivity) obtained for CdF_2 crystals are compared with the results of electronic structure calculations for this crystal. The data gave a consistent picture of the electronic structure of CdF_2 crystals.

The influence of the conduction band density-of-states maximum in X_1 on the obtained EDC for $h\nu = 21.22$ eV leads to the appearance of the double Cd 4d peak split by 1.46 eV. The local EPM was found as a proper method for band structure calculation of the ionically bound, wide band gap CdF_2 crystal. The pseudopotential form factors for Cd and F were determined.

Acknowledgements

Authors appreciate helpful discussions with Doc. Dr. hab. J. M. Langer. Authors express their gratitude to Dr. W. Chab for nice cooperation and to T. Langer for supplying us with crystals.

References

[1] J. M. Langer, A. Lemańska-Bajorek, A. Suchocki, W. Walukiewicz, and B. Wiktor, J. Lum. **24/25**, 889 (1981).
[2] A. Stąpor and J. M. Langer, Phys. Rev. B **25**, 3407 (1982).
[3] J. W. Hodby, Crystals with the Fluorite Structure, Ed. W. Hayes, Clarendon Press, Oxford 1974.
[4] B. A. Orłowski and J. M. Langer, phys. stat. sol. (b) **91**, K53 (1979); Acta phys. Polon **A63**, 107 (1983).
[5] C. Raisin, J. M. Berger, S. Robin-Kandare, G. Krill, and A. Amamou, J. Phys. C **13**, 1835 (1980).
[6] T. Tomiki and T. Miyata, J. Phys. Soc. Japan **27**, 658 (1969).
[7] J. M. Langer, T. Langer, G. L. Pearson, B. Krukowska-Fulde, and U. Piekara, phys. stat. sol. (b) **66**, 537 (1974).
[8] J. Garbarczyk, B. Krukowska-Fulde, T. Langer, and J. M. Langer, J. Phys. D **11**, L17 (1978).
[9] B. A. Orłowski, J. M. Langer, and M. Wasiucionek, Proc. 7th Internat. Vacuum Congr. and 3rd Internat. Conf. Solid Surfaces, Vienna 1977, Vol. 1, Ed. R. Dobrozemsky, F. Rudenauer, F. P. Viehböck, and A. Breth, F. Berger und Söhne, Horn 1977 (p. 679).
[10] J. M. Langer, Postępy Fiz. **31**, 435 (1980).
[11] J. E. Dmochowski, I. Kosacki, and J. M. Langer, Radiat. Eff. **72**, 139 (1983).
[12] B. A. Orłowski and W. Chab, Solid State Commun. **44**, 819 (1982).
[13] B. Krukowska-Fulde and T. Niemyski, J. Crystal Growth **1**, 183 (1968).
[14] J. P. Albert, C. Jouanin, and C. Gout, Phys. Rev. B **16**, 4619 (1977).
[15] M. L. Cohen, P. J. Lin, D. M. Roessler, and W. C. Walker, Phys. Rev. **155**, 992 (1967).
[16] D. M. Roessler and W. C. Walker, Phys. Rev. **159**, 733 (1967).
[17] C. Y. Fong and M. L. Cohen, Phys. Rev. Letters **21**, 22 (1968).
[18] C. Y. Fong and M. L. Cohen, Phys. Rev. **185**, 1168 (1969).
[19] P. Plenkiewicz and B. Dowgiałło-Plenkiewicz, phys. stat. sol. (b) **92**, 379 (1979).
[20] P. I. Lin and L. Kleinman, Phys. Rev. **142**, 478 (1966).
[21] P. Löwdin, J. chem. Phys. **19**, 1396 (1951).
[22] D. Brust, Phys. Rev. **134**, A1337 (1964).
[23] R. T. Poole, J. A. Nicholson, J. Liesegang, J. G. Jenkin, and R. C. G. Leckey, Phys. Rev. B **20**, 1733 (1979).
[24] A. J. Bourdillon and J. H. Beaumont, J. Phys. C **9**, L73 (1976).
[25] J. M. Berger, G. Leveque, and J. Robin, C.R. Acad. Sci. (France) **B279**, 509 (1974).

(Received July 13, 1984)

phys. stat. sol. (b) **126**, 293 (1984)

Subject classification: 2 and 13.1; 19; 20.1; 22.6

Department of Physics, Brunel University, Uxbridge[1])

The Spin Density and Optical Properties of Amorphous Thin Arsenic Silicate Films

By

S. K. J. AL-ANI and C. A. HOGARTH

The spin density of amorphous As_2O_5/SiO thin films is found to be lower than that of SiO films deposited under similar conditions. The measurements show a decrease in the spin density of an amorphous As_2O_5/SiO film with increasing temperature which can be explained in terms of some annealing of dangling bonds. The fundamental absorption edge of these films at different temperatures is investigated and analysed assuming optical absorption by indirect transitions. The infrared absorption spectra indicate some chemical bonding between the two oxides. Experimental data on the wavelength dependence of the refractive index are presented. The dispersion of the refractive index follows a single-oscillator model. Some effects of room temperature aging of the films are also presented.

La concentration des spins électroniques mesurée dans les couches minces de As_2O_5/SiO est plus basse que dans les couches minces de SiO préparées sous les mêmes conditions. Quand la température augmente il y a une décroissance du densité des spins dans une couche mince amorphe de SiO, un résultat qui peut être expliqué par conséquence de la réparation de la structure des couches minces après un traitement à température peu élevée. Les seuils d'absorption dcs couches minces à quelques températures sont mesurés et analysés avec la théorie d'absorption optique par les transitions non-directes. Les spectres d'absorption dans la région infra-rouges donnaient l'évidence des jonctions chimiques entre les deux oxydes. Les résultats pour l'indice rélative de réfraction n_f et pour sa variation avec la longueur d'onde, la dispersion de n_f en termes d'un modèle mono-oscillatoire, et les effets de vieillissement, sont présentées.

1. Introduction

The optical properties of amorphous non-metallic solids may be described by two constants, the index of refraction n_f, and the extinction coefficient k which is related to the optical absorption coefficient $\alpha(\omega)$ at an angular frequency of radiation ω by the equation

$$\alpha(\omega) = \frac{4\pi k}{\lambda}, \tag{1}$$

where $\omega = 2\pi\nu$, λ is the wavelength, and ν the frequency of the radiation in vacuum. The values of n_f are in accordance with that derived from the wavelength position of the interference maxima,

$$2n_f d = m\lambda, \tag{2}$$

where d is the thickness of the film and m the order of the interference. It is known from simple dispersion theory that in the region of low absorption, the index of refraction is expected to be given [1] in a single-oscillator model by

$$n_f^2 - 1 = \frac{S_0\lambda_0^2}{1 - (\lambda_0/\lambda)^2}, \tag{3}$$

where S_0 is the average oscillator strength and λ_0 the average oscillator position.

[1]) Uxbridge, Middlesex UB8 3PH, Great Britain.

For many glassy and amorphous non-metallic materials, the absorption edge can be divided into two regions. For absorption coefficients $\alpha(\omega)$ of less than 10^4 cm^{-1}, there is usually an Urbach [2] tail where $\alpha(\omega)$ depends exponentially on photon energy $\hbar\omega$ as

$$\alpha(\omega) = \alpha_0 \exp\left(\frac{\hbar\omega - \hbar\omega_0}{E_e}\right), \tag{4}$$

where α_0 is a constant and E_e is often interpreted as the width of the tails of localised states in the gap region. Further up the edge where $\alpha(\omega) > 10^4$ cm^{-1}, the absorption takes the form [3, 4]

$$\alpha(\omega)\,\hbar\omega = B(\hbar\omega - E_{opt})^r\,, \tag{5}$$

where B is a constant, E_{opt} the optical energy gap of the material, and the exponent r may take values 2, 3, $\frac{1}{2}$, $\frac{3}{2}$ indicating the type of electron transitions in $\boldsymbol{k}$-space. Equation (5) with $r = 2$ has been successfully applied to many amorphous thin oxide films [5 to 7] and chalcogenide glasses [8].

Arsenic pentoxide (As_2O_5) is a well recognized glass-former [9, 10]. Binary glasses can have many interesting physical properties when As_2O_5 is one of the constituent oxides [11 to 15]. As_2O_5 has been suggested [16] as a target material for the cyclotron production of carrier-free ^{77}Br, but the literature [17 to 19] gives only incomplete information. Nevertheless, some research on thin dielectric films was devoted to the study of the system As_2O_3/SiO_2 because it is a good arsenic-diffusion source into silicon for the fabrication of p–n junction [20, 21].

Evaporated As_2O_5 films are unstable when exposed to air and a mixture with silicon monoxide can produce more stable amorphous films. Although no theoretical work has been reported on the band structure of this mixed material, an experimental attempt has been made on its preparation [22] but no detailed measurements have been displayed. In order to gain some insight into its electronic band structure, a systematic study of the optical and electron spin resonance properties of amorphous thin films of arsenic silicate glass, As_2O_5/SiO, is reported. The results are combined and interpreted in the light of current theories of amorphous materials.

2. Experimental Work

Thin evaporated layers of As_2O_5/SiO were deposited on polished silicon discs and on Corning 7059 glass substrates, at a pressure of the order of 0.665 mPa in a Balzers BA510 coating unit using the traditional co-evaporation technique established by Hogarth and Wright [22]. Tantalum boats were used for the evaporation of SiO while As_2O_5 was evaporated from boron nitride crucibles. The arsenic pentoxide necessitated a long outgassing period because of its affinity to water. The substrate temperature during the evaporation was of the order of 375 K. The film thickness was carefully determined by multiple-beam interferometry using Fizeau fringes of equal thickness. Details of absorption measurements in the wavelength range 190 to 850 nm as well as the temperature effect have been described by Al-Ani et al. [23]. The optical data near the infrared up to 3000 nm were obtained at room temperature using a Beckman Acta M-series spectrophotometer MIV. Discs of KBr 2 cm in diameter were prepared by hot pressing in vacuo. Thin As_2O_5/SiO films were then deposited on these discs and their infrared absorption spectra were recorded using a Perkin-Elmer 577 grating infrared double-beam spectrometer. The ESR measurements were made using a Varian E-3 EPR spectrometer working near 3 cm and well-known standard procedures. The absorption was calibrated using an aqueous solution of $CuSO_4 \cdot 5\,H_2O$ (0.203 g/250 cm^3).

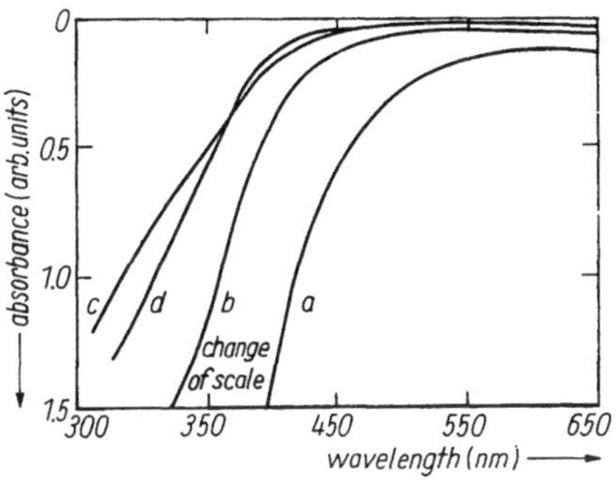

Fig. 1. Absorbance as a function of wavelength for amorphous As_2O_5/SiO films (thickness for curves a, b, and c 740, d 1020 nm). Composition: (a) 100, (b) 95, (c) and (d) 90 mol% SiO

The structure of our films was studied using a JEOL, JEM 7 electron microscope with a magnification of 38000 ×. Film compositions up to 20 mol% As_2O_5/80 mol% SiO were found to be stable and their optical characteristics could be obtained.

3. Results and Discussion

3.1 The fundamental absorption edge

The optical absorption spectra near the fundamental absorption edge of thin films of amorphous SiO and As_2O_5/SiO are shown in Fig. 1. The spectra were not corrected for reflection loss because the absorption coefficient $\alpha(\omega)$ is very high and the reflection correction is not very important. Fig. 2 displays the exponential behaviour in accordance with (4) and the values of E_e are estimated and listed in Table 1. The optical data in Fig. 1 fit (5) with $r = 2$ and indicate that non-direct transitions are involved. Fig. 3

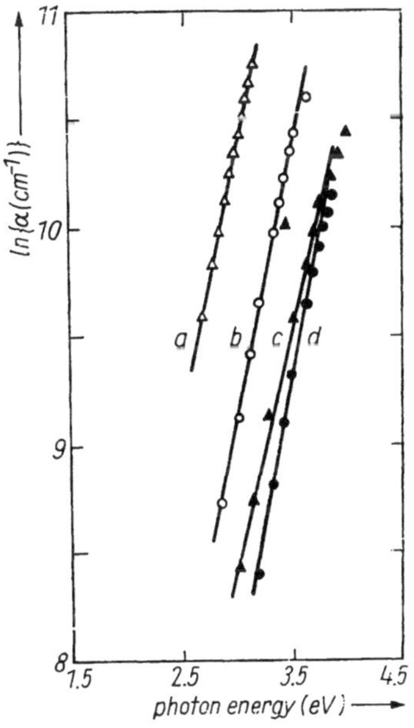

Fig. 2

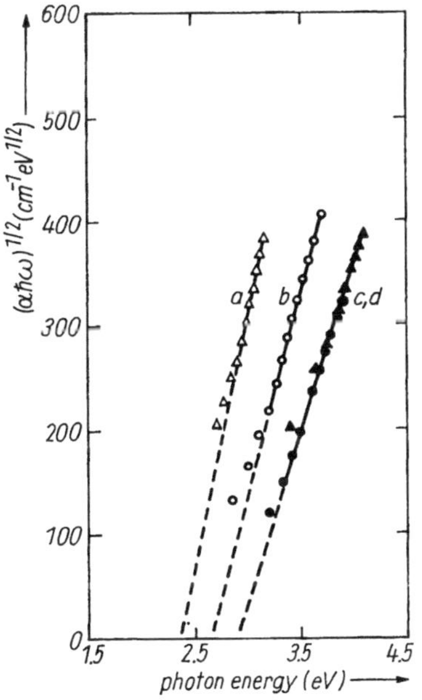

Fig. 3

Fig. 2. Data of Fig. 1 plotted in accordance with the Urbach law

Fig. 3. Data of Fig. 1 plotted in accordance with theory for indirect transitions

Table 1

Room temperature optical parameters of amorphous As_2O_5/SiO thin films of thickness $\approx$ 740 nm

film composition	E_{opt} (eV)	E_e (eV)	B (10^5 cm^{-1} eV^{-1})
100 mol% SiO	2.35	0.40	2.50
95 mol% SiO/5 mol% As_2O_5	2.63	0.40	1.52
90 mol% SiO/10 mol% As_2O_5	2.85	0.45	1.06
90 mol% SiO/10 mol% As_2O_5 of thickness $\approx$ 1020 nm	2.85	0.37	1.06

shows the parameter $(\alpha\hbar\omega)^{1/2}$ versus $\hbar\omega$ and the extrapolation of the well-defined linear part intersects the abscissa at E_{opt}. The values of E_{opt} and B are also recorded in Table 1. For a fixed film composition, the thickness variation has not affected the value of E_{opt} as is evident from Fig. 3 and Table 1. The origin of the exponential tail absorption is still a matter of conjecture. The most acceptable interpretation of the form of the Urbach rule (4) results from the consideration of the broadening of the absorption edge by random electric fields [24, 25]. These fields are caused by disorder, i.e. they are induced by potential fluctuations due to the lack of long-range order by charged defects [26, 27]. The addition of As_2O_5 to the mixture increases the degree of disorder in the system and causes an increase in both the optical energy gap and the widths of the tails of localized states as well as a reduction in the value of the constant B.

A few As_2O_5/SiO specimens were preserved at room temperature for about six months. This aging process shifts the position of the absorption edge towards shorter wavelength and increases the value of E_{opt}. Table 2 summarizes some of these results. Investigations of the aging process in the thin films are of interest because of their applications in opto-electronic systems [28], and indeed many research workers have examined the effect of aging on the structure and physical properties of thin films [29 to 31]. The aging of As_2O_5/SiO films relaxes the results of the carefully defined preparation conditions such as the rapid rate of film deposition and leads to the increase in the value of the E_{opt}. This change may be assumed to be associated with changes at the grain boundaries during the aging or by the absorption of OH groups from absorbed water vapour or by the absorption of oxygen. The processes can influence the light absorption coefficient, causing changes in the optical properties.

Table 2

Aging effect on the optical energy gap of amorphous As_2O_5/SiO thin films

sample thickness (nm)	E_{opt} (eV) (fresh film)	E_{opt} (eV) (aged film)
1000	2.58	2.63
717	2.58	2.68
746	2.63	2.65
941	2.60	2.64
715	2.53	2.58

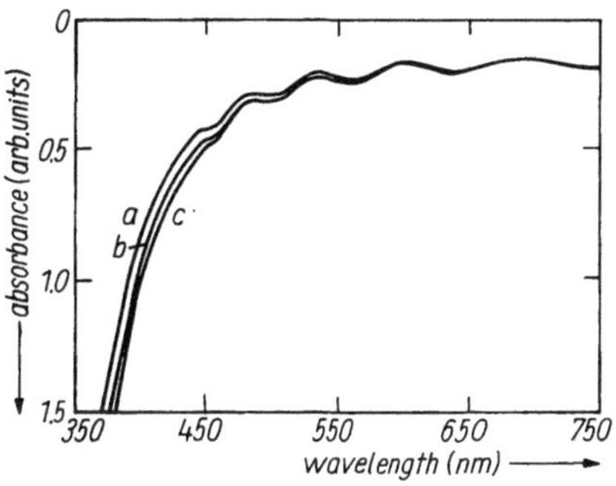

Fig. 4. Absorbance as a function of wavelength for a As_2O_5/SiO thin film (1123 nm) at three temperatures, (a) $T = 95$, (b) 295, and (c) 480 K

Another possibility is that the specimens including SiO_x are partially decomposed with age to form different structures.

Fig. 4 shows the variation of the absorbance with wavelength for an amorphous As_2O_5/SiO film at three temperatures. As the temperature of the specimen is increased, the absorption edge is seen to move towards higher wavelengths. These data are replotted as $\ln \alpha$ versus $\hbar\omega$ in Fig. 5 and the values of E_e at different temperatures are estimated and listed in Table 3. The slope of the Urbach tails is considered to be independent of temperature, up to 200 °C. This fact is known to be common to many amorphous semiconductors [32]. Furthermore, in Fig. 6, the parameter $(\alpha\hbar\omega)^{1/2}$ is plotted versus $\hbar\omega$ and the derived values of E_{opt} as well as the constant B are also

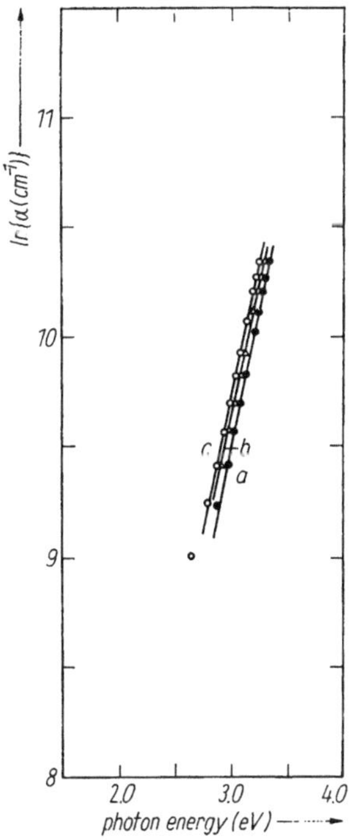

Fig. 5

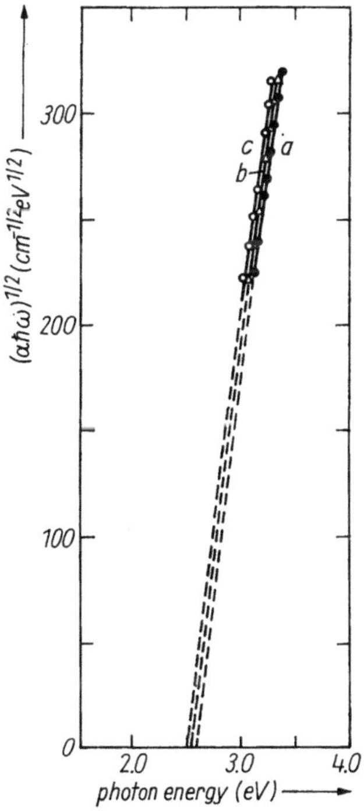

Fig. 6

Fig. 5. Data of Fig. 4 plotted in accordance with the Urbach law

Fig. 6. Data of Fig. 4 plotted in accordance with the theory of indirect transitions

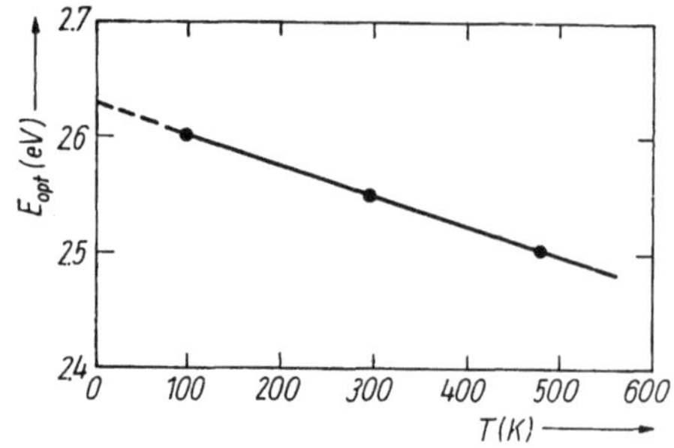

Fig. 7. Data derived from Fig. 6 showing E_{opt} as a function of temperature

included in Table 3. Fig. 7 shows plots of E_{opt} as a function of temperature. From this slope, the temperature coefficient, γ, of E_{opt} is found to be -2.59×10^{-4} eVK^{-1}. Actually, the γ-value obtained here is relatively smaller than those in chalcogenide glasses [33] and almost of the same range obtained in other amorphous oxide films [32]. The constant B obtained from the gradient of Fig. 6 is found to be independent of temperature. These results suggest that as the temperature is increased the disorder in the sample increases and the band tailing starts at lower energies but reaches the same value of energy at E_e. Consequently the effective values of mobility gap and optical energy gap are expected to be reduced as demonstrated above.

Table 3

Temperature effect on the fundamental absorption edge of an amorphous As_2O_5/SiO thin film

sample temp. (K)	E_{opt} (eV)	E_e (eV)	B (10^5 cm^{-1} eV^{-1})
95	2.60	0.38	1.60
295	2.55	0.39	1.60
480	2.50	0.38	1.60

3.2 The refractive index

The refractive index of a solid is an important physical property which gives valuable information on the electronic polarizability and local field inside the material. It has

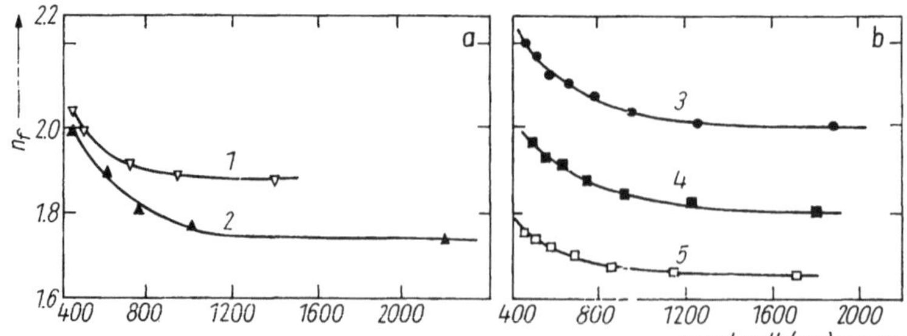

Fig. 8. Refractive index n_f, as a function of wavelength for a) two thin films of composition 96 mol% SiO/4 mol% As_2O_5 and thickness (1) 746 nm and (2) 633 nm; b) three films each 1 μm thick, composition: (3) 96, (4) 93, (5) 90 mol% SiO, remainder As_2O_5

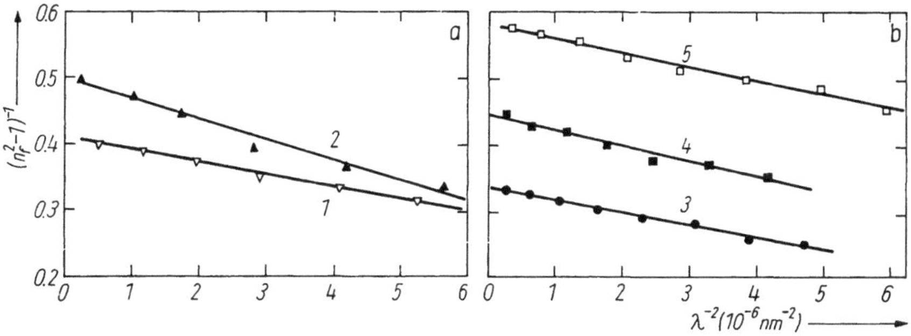

Fig. 9. Data of a) Fig. 8 a and b) Fig. 8b plotted in accordance with theory for a single-oscillator model

also been used for determining the density of colours in thin films [34]. The physical dependence of refractive index on the structure is often related to the density and electronic structure differences among the phases [35]. The density effect can be estimated as $(n_f^2 - 1) \sim 1/a_0^3$ where a_0 is the lattice parameter. The dispersion curves in the visible and near-infrared regions are given in Fig. 8 for a series of As_2O_5/SiO films and follow the dependence predicted by classical theory for a single oscillator [36]. Both the refractive index and extinction coefficient decrease rapidly to approach constant values at long wavelengths. The rapid decrease occurs in the vicinity of the absorption edge. The constancy of the refractive index at long wavelengths may be characteristic of a highly damped system or a system with little free-carrier absorption [37]. The validity of assumption (3) is checked by plotting the quantity $(n_f^2 - 1)^{-1}$ against λ^{-2} as in Fig. 9. The parameter S_0 can then be derived from the slope of the straight line while λ_0 follows from the infinite wavelength intercept. Fig. 9 gave, in all the cases, linear relationships for the mixtures examined and the parameters n_f, S_0, E_0, and the refractive index dispersion parameter E_0/S_0 are calculated and listed in Table 4. The accuracy claimed here is within the acceptable range for the techniques used where the error may be dominated by the thickness measurement. The refractive index of amorphous As_2O_5/SiO films is dependent on its thickness and composition (Fig. 8) as has been found for other semiconductors [38, 39]. Our results are of the same order as for other amorphous thin oxide films [38, 40] and for other materials [1]. Temperature has not affected the refractive index. This is seen from the constancy of the fringe positions in Fig. 4. Aging, on the other hand, has changed the fringes which indicates that the refractive index is different from that of the fresh film.

Table 4

Optical parameters derived from refractive index measurements

sample composition	thickness (nm)	S_0 (10^{13} m^{-2})	E_0 (eV)	E_0/S_0 (10^{-14} eV m^2)	n_f
96 mol% SiO/4 mol% As_2O_5	633	3.31	5.02	15.2	1.74
	746	5.78	6.01	10.4	1.87
	941	5.96	5.55	9.3	1.99
93 mol% SiO/7 mol% As_2O_5	1010	4.21	5.37	12.8	1.80
90 mol% SiO/10 mol% As_2O_5	1010	4.65	6.44	13.9	1.65

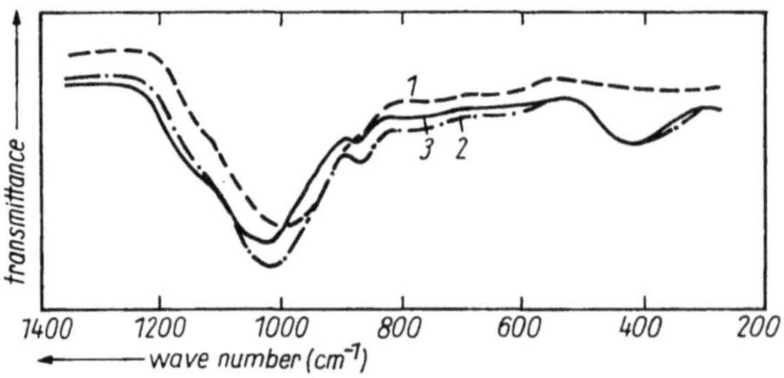

Fig. 10. Infrared absorption spectra of three films. Composition: (1) 100 mol% SiO, (2) 95 mol% SiO/5 mol% As_2O_5, (3) 87 mol% SiO/13 mol% As_2O_5

3.3 *Infrared absorption*

Fig. 10 shows the infrared spectra of arsenic silicate, As_2O_5/SiO, amorphous thin films which differ from that of pure SiO by the appearance of additional absorption bands. Three main absorption peaks appear in these spectra and as the As_2O_5 content is increased, a small shift to higher wave numbers is observed. The small shoulder (absorption peak $\approx 870\ cm^{-1}$) points to some chemical interaction between the two oxides. For a low ratio of R/p (rate of evaporation/pressure in the system), one usually observes absorption peaks near 2.8 and 2.9 μm, characteristic of the assembly Si–O–H and H_2O. This refers to the porosity of highly oxidised films allowing diffusion of water into the structure of the film. These peaks were absent in Fig. 10 indicating that the R/p ratio is high. Table 5 summarises some of these results. In the As_2O_3/SiO_2 system [41], the absorption band characteristic of the vibrations of pure SiO_2 at 1060, 800, and 450 cm^{-1} has not been perturbed, neither in position nor in intensity by the presence of As_2O_3. In Fig. 10, however, it is shown that the bands do not remain invariant in position as well as in intensity after the co-evaporation process. The combined infrared and ultraviolet spectroscopy results simply suggest that As_2O_5 and SiO are co-deposited as a chemical mixture with possible Si–O–As linkages.

3.4 *Electron spin resonance ESR*

It is well known that thin amorphous films of silicon oxide produced by evaporation of silicon monoxide powder vary strongly in composition, depending upon the deposition rate and the oxygen partial pressure. Timson and Hogarth [42] measured spin densities between 10^{18} and $10^{20}\ cm^{-3}$ which correspond to samples with varying oxygen

Table 5
Infrared absorption spectra of amorphous As_2O_5/SiO thin films deposited at pressure ≈ 0.133 mPa

film composition	infrared absorption peaks (cm^{-1})		
100 mol% SiO	1000	—	—
96 mol% SiO/4 mol% As_2O_5	1020	870	420
94 mol% SiO/6 mol% As_2O_5	1030	870	430
90 mol% SiO/10 mol% As_2O_5	1030	875	400
87 mol% SiO/13 mol% As_2O_5	1030	880	430
95 mol% SiO/5 mol% As_2O_5 deposited at $p \approx 5.33$ mPa	1020	—	400

content covering the range of composition from SiO_2 to SiO. They further suggested that the ESR signal originates in unpaired electrons on unsatisfied bonds. Because the unsaturated bonds of oxygen atoms in silicon oxide films may give rise to the ESR signal, Paez et al. [43] have recently measured the spin concentrations of SiO_x films due to an electric field applied simultaneously with the ESR measurements.

We have easily found a strong ESR signal at $g \approx 2.004$ in evaporated films of silicon oxide and the resonance was generally symmetrical. The lineshape of ESR signals in amorphous SiO and As_2O_5/SiO thin films were nearly, but not exactly, Lorentzian in shape.

The spin densities n_d for amorphous SiO and As_2O_5/SiO films were calculated and listed in Table 6. The measured spin concentration for amorphous SiO film was ascribed to dangling bonds and was of the order of 10^{19} cm^{-3} in good agreement with published results [42, 44, 45] for the higher values of R/p. The absence of the hyperfine doublet due to ^{29}Si may suggest [43] that the unpaired electrons are localised on the oxygen rather than on the silicon atoms. This means that the amorphous silicon oxide network produced by the evaporation process contains chain endings of SiO radicals with measured spin densities of the order of 10^{19} cm^{-3}. The spin density in amorphous As_2O_5/SiO films was found to be lower by an order of magnitude than in amorphous SiO films. This is attributed to the effect of the arsenic oxide which was to take up the free bond and thus effectively reduce the defect concentration within the film. This is in support of the findings for the infrared absorption. For an amorphous As_2O_5/SiO sample of nearly the same composition and different thicknesses it was found that the thicker film had the higher value of n_d (Table 6). This confirms earlier findings that the ESR signal is a function of the sample thickness [46]. Aging some of these films is found to have a slight decrease in the dangling bond density which is a characteristic of high R/p films [42, 47].

Table 6

Room temperature ESR measurements of amorphous As_2O_5/SiO thin films

sample thickness (nm)	film composition	n_d (10^{18} cm^{-3})
700	100 mol% SiO	15.5
300	96 mol% SiO/4 mol% As_2O_5	7.19
320	93 mol% SiO/7 mol% As_2O_5	1.02
150	96 mol% SiO/4 mol% As_2O_5	22.0
548 (as evaporated)	87 mol% SiO/13 mol% As_2O_5	1.68
548 (aged film)	87 mol% SiO/13 mol% As_2O_5	1.22
840 (as evaporated)		4.06
840 (aged film)		3.10

The ESR measurements of amorphous As_2O_5/SiO of thickness $\approx$ 802 nm at different temperatures

sample temp. (K)	n_d (10^{18} cm^{-3})
473	4.20
353	5.37
295	5.77
liquid nitrogen	7.45

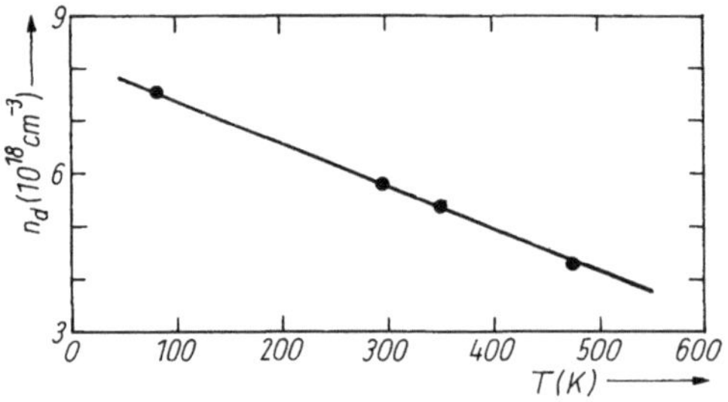

Fig. 11. Variation of electron spin density with temperature for an As_2O_5/SiO film 802 nm thick

The variation of the spin density with annealing temperature for an amorphous As_2O_5/SiO sample is illustrated in Fig. 11 and the data are also listed in Table 6. The spin density of amorphous Si films is found to decrease with increasing temperature [48]. After reaching a minimum between 200 to 400 °C it has started to increase until 550 °C has been reached. The spin density of co-evaporated As_2O_5/SiO film is seen to decrease linearly with increasing annealing temperature. The decrease of n_d with annealing up to 475 K should be due to the annealing out of dangling bonds and/or the saturation of dangling bonds by oxygen present in the sample as a contamination.

3.5 Concluding remarks

The structural characteristics are expected to be important in the study of the physical properties of thin evaporated films. Evaporated SiO films are well-established and have amorphous structures, while the As_2O_5/SiO film presented in Fig. 12 shows some evidence for polycrystallinity.

Finally while arsenic pentoxide thin films do not present radioactivity hazards as do thorium fluoride thin films [49], nevertheless the operator can be in serious danger if the necessary precautions are not taken. Thus special care was required when handling this compound and when cleaning apparatus that had been used for the evaporation of As_2O_5.

A model due to internal mechanical stress has been proposed [31] to explain the aging effect on SiO_x films. However, a model which embodies the changes in the

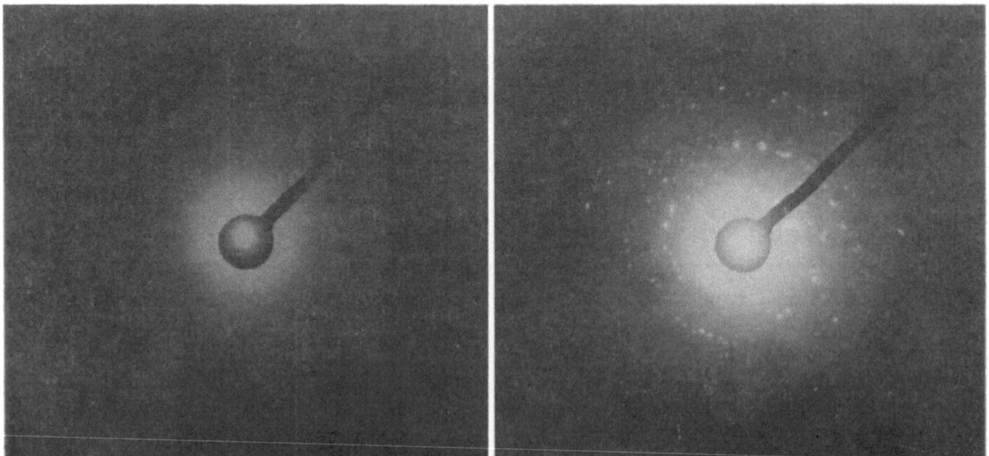

a b

Fig. 12. Electron diffraction pattern for an As_2O_5/SiO thin film showing a) amorphous character and b) some polycrystallinity

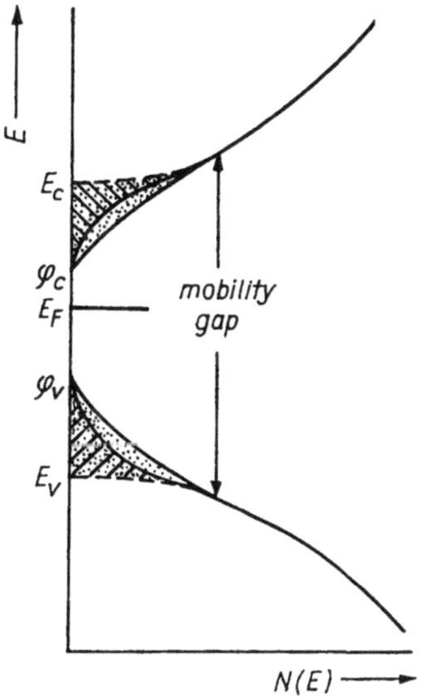

Fig. 13. Density-of-states diagram for an amorphous As_2O_5/SiO thin film. E_c, E_V represent band edges in possible crystalline material, E_F is the Fermi energy, φ_c, φ_V represent the extent of band tailing. The curvature between the mobility edge and φ_V or φ_c depends on the temperature

physical properties due to aging which are observed can only be a somewhat crude approximation without actual measurements of the effect of aging on the structure of amorphous As_2O_5/SiO thin films.

The results may be understood, at least qualitatively, with reference to the type of density-of-states functions for these amorphous materials as shown in Fig. 13. The mobility gap and the optical energy gap are both related to the point of inflection in the curve of $N(E)$ against E, the electron energy, which determines when band tailing into localized states becomes significant. Our data is consistent with the interpretation that both the width of the exponential edge E_e and the optical band gap E_{opt} are controlled by the amount of disorder, structural and thermal, in the network. In terms of Fig. 13 and as the structural disorder increases, the band tailing starts at higher energies but extends further into the forbidden gap. This may explain the increase of the values of E_{opt} and E_e with the addition of As_2O_5. In the case of films of constant composition but with varying temperature leading to increased thermal disorder, the band tailing starts at lower energies but remains invariant into the forbidden gap as shown by the band tails in Fig. 13. Hence the value of E_{opt} is expected to reduce but E_e will remain almost invariant with temperature.

The decrease of spin density n_d which is expected to be related to the density of localized states around the Fermi level E_F suggests the possibility of a shift of E_F. By annealing below 200 °C, E_F shifts downward [50] along with the decrease of n_d. Furthermore, the downward shift of E_F and the decrease of n_d with annealing may suggest that localised states due to dangling bonds in the upper energy region in the gap are probably annealed out more easily than those in the lower one. Further controlled electrical transport measurements would be needed to confirm this possibility.

Acknowledgements

We should like to thank Dr. K. Lott of the Department of Chemistry, Brunel University, for assistance during the ESR measurements and Dr. B. R. Orton, Safety Officer, of the Department of Physics, for guidance in handling the arsenic pentoxide.

References

[1] M. Di Domenico and S. H. Wemple, J. appl. Phys. **40**, 720 (1969).
[2] F. Urbach, Phys. Rev. **92**, 1324 (1953).
[3] J. Tauc, R. Grigorovici, and A. Vancu, phys. stat. sol. **15**, 627 (1966).
[4] E. A. Davis and N. F. Mott, Phil. Mag. **22**, 903 (1970).
[5] C. A. Hogarth and M. Y. Nadeem, phys. stat. sol. (a) **68**, K181 (1981).
[6] S. K. J. Al-Ani, K. I. Arshak, and C. A. Hogarth, J. Mater. Sci. **19**, 1737 (1984).
[7] S. K. J. Al-Ani, C. A. Hogarth, and M. Ilyas, J. Mater. Sci. Letters **3**, 391 (1984).
[8] N. F. Mott and E. A. Davis, Electronic Processes in Non-Crystalline Materials, 2nd ed., Clarendon Press, Oxford 1979 (p. 497).

[9] J. W. Mellor, A Comprehensive Treatise on Inorganic and Theoretical Chemistry, Vol. 9, Longmans, Green & Co., New York 1929.
[10] L. G. Van Uitert and W. H. Grodkiewicz, J. Amer. Ceram. Soc. **64**, 690 (1981).
[11] T. Minami, K. Imazawa, and M. Tanaka, J. non-crystall. Solids **42**, 469 (1980).
[12] W. H. Grodkiewicz, H. M. O'Bryan, G. A. Pasteur, L. Pressman, S. Singh, L. G. Van Uitert, and G. Zydzik, Mater. Res. Bull. **16**, 373 (1981).
[13] W. H. Grodkiewicz, H. M. O'Bryan, L. Pressman, S. Singh, L. G. Van Uitert, and G. Zydzik, J. non-crystall. Solids **44**, 405 (1981).
[14] D. Kumar and D. Chakravorty, J. Phys. D **13**, 1331 (1980).
[15] Z. Mazurak, W. Strek, and B. Jezowska-Trzebiatowska, Acta. phys. Polon. **A54**, 203 (1978).
[16] A. D. Nunn and S. L. Waters, Internat. J. appl. Radiation and Isotopes **26**, 731 (1975).
[17] H. Barten, J. appl. Cryst. **14**, 68 (1981).
[18] Y. Mizokawa, H. Iwasaki, R. Nishitani, and S. Nakamura, J. Electron Spectroscopy related Phenomena **14**, 129 (1978).
[19] L. L. Kazmerski, P. J. Ireland, S. S. Chu, and Y. T. Lee, J. Vacuum Sci. Technol. **17**, 521 (1980).
[20] D. B. Lee, Solid State Electronics **10**, 623 (1967).
[21] J. Wong and C. A. Angell, Glass Structures by Spectroscopy, Marcel Dekker, Inc., New York 1976 (p. 509).
[22] C. A. Hogarth and L. A. Wright, Proc. 9th Internat. Conf. Phys. Semiconductors, Moscow 1968, Izd. Nauka, Leningrad 1968 (p. 1274).
[23] S. K. J. Al-Ani, C. A. Hogarth, and D. N. Waters, phys. stat. sol. (b) **123**, 653 (1984).
[24] J. Tauc, Mater. Res. Bull. **5**, 721 (1970).
[25] J. D. Dow and D. Redfield, Phys. Rev. B **5**, 594 (1972).
[26] K. L. Bhatia and P. H. Bhat, phys. stat. sol. (b) **62**, 301 (1974).
[27] A. A. Andreev, Fiz. tverd. Tela **18**, 53 (1976).
[28] Z. Porada and E. Schabowska, Thin Solid Films **110**, 1 (1983).
[29] P. A. Timson and J. Beynon, Vacuum **22**, 233 (1972).
[30] H. Birey, J. appl. Phys. **48**, 5209 (1977).
[31] K. Hübner, A. V. Shendrik, and A. M. Praulinsh, phys. stat. sol. (b) **118**, 239 (1983).
[32] A. Nakamura and S. Yamada, Appl. Phys. **24**, 55 (1981).
[33] M. Nunoshita, H. Arai, T. Taneki, and Y. Hamakawa, J. non-crystall. Solids **12**, 339 (1973).
[34] S. K. Deb, Phil. Mag. **27**, 801 (1973).
[35] M. Wohlecke, V. Marrelo, and A. Onton, J. appl. Phys. **48**, 1748 (1977).
[36] T. S. Moss, Optical Properties of Semiconductors, Butterworths, London 1959 (p. 15).
[37] S. K. Bahl and K. L. Chopra, J. appl. Phys. **40**, 4940 (1969).
[38] V. Dolocan and F. Iova, phys. stat. sol. (a) **64**, 755 (1981).
[39] B. Broberg and S. Lindgren, J. appl. Phys. **55**, 3376 (1984).
[40] V. Dolocon and E. Dolocan, Ann. Univ. Bucureşti, Ser. Fiz. **30**, 71 (1981).
[41] J. Wong and M. Chezzo, J. Electrochem. Soc. **118**, 1540 (1971).
[42] P. A. Timson and C. A. Hogarth, Thin Solid Films **8**, 237 (1971).
[43] R. Paez, L. C. Hernandez, J. I. Jaramillo, P. Prieto, J. Rodriguez, and G. Fritsch, J. magnetic Resonance **29**, 251 (1978).
[44] R. Paez and G. Fritsch, 2nd Latin-America Symp. Magnetic Resonance, Bogota 1976.
[45] J. Pivot, D. Morell, J. Tardy, J. A. Roger, and C. H. S. Dupuy, Thin Solid Films **28**, 357 (1975).
[46] M. H. Brodsky and R. S. Title, Phys. Rev. Letters **23**, 581 (1969).
[47] L. R. Callard, Proc. I.E.R.E./I.E.E. Conf. Application of Thin Films in Electronic Engineering, London 1966 (paper No. 15).
[48] H. K. Lee and C. Lee, Solid State Commun. **42**, 71 (1982).
[49] W. Heitmann and E. Ritter, Appl. Optics **7**, 307 (1968).
[50] S. Hasegawa, S. Yazaki, and T. Shimizu, Solid State Commun. **26**, 407 (1978).

(Received June 6, 1984)

phys. stat. sol. (b) **126**, 305 (1984)

Subject classification: 13.5.1 and 20.1; 22.6

Institut für Physik, Universität Dortmund[1]) *(a)*
and Physics Department, Second University of Rome (b)

Temperature Dependence of the Exciton Lineshape of GeO_2

By

D. FRÖHLICH (a), H.-J. WEBER (a), R. WILLE (a), and R. FRANCINI (b)

Two-photon measurements are presented of the 1S exciton lineshape in GeO_2 between 4.5 and 150 K. With a frequency-doubled dye laser and a Raman-shifted Nd:YAG laser the exciton-energy around 4.65 eV can be scanned. The temperature dependence of the total line width can be explained by a theoretical model of exciton–phonon coupling. Furthermore the results reveal an unexpected asymmetry, which seems to originate from a non-uniform contribution of optical phonons.

Es werden Zwei-Photonenmessungen der Linienbreite des 1S-Exzitons in GeO_2 im Temperaturbereich von 4,5 K bis 150 K vorgestellt. Mit einem frequenzverdoppelten Dye-Laser und einem mittels stimulierten Ramaneffekt energetisch verschobenen Nd-YAG Laser kann die Exzitonenergie um 4,65 eV durchgestimmt werden. Die Temperaturabhängigkeit der Gesamtlinienbreite kann mit einem theoretischen Modell für Exziton–Phonon-Kopplung erklärt werden. Weiterhin zeigen die Ergebnisse eine unerwartete Asymmetrie, die in einem unsymmetrischen Beitrag der optischen Phononen zu liegen scheint.

1. Introduction

Materials with direct forbidden band gap like Cu_2O, SnO_2, and GeO_2 are very well suited to be studied by two-photon absorption (TPA). As was shown by Uihlein et al. [1] the TPA of the excitonic structure of Cu_2O yields detailed information on the valence band structure. In SnO_2 Fröhlich et al. [2] were able to clarify the symmetry assignment of the valence band by two-photon measurements with polarized light. GeO_2 should be another candidate for a forbidden band gap. Indeed Stapelbroek and Evans [3] found a very weak 1S exciton line which they assigned to a quadrupole transition between even-parity bands. From a preliminary analysis of this quadrupole-allowed transition they suggest Γ_3^+ symmetry for the highest valence band. Our detailed polarization analysis of the TPA spectra confirms this assignment. Though this symmetry assignment is a nice by-product of our investigations we want to stress in this paper another promising application of TPA. We use this nonlinear optical method to study the temperature dependence of the lineshape in order to get information on exciton–phonon interaction. For two reasons TPA is well suited to gain reliable data on the exciton linewidth:

(i) As a nonlinear process TPA yields rather small absorption coefficients as compared to one-photon allowed transitions. The TPA spectra are therefore measured in crystals of at least a few mm thickness. Corrections for reflection at the surface can be neglected.

(ii) Since the two photons coupled into the crystal are in the transparent region, where the refactive index n can be assumed to be constant, the resonances occur at definite k-vectors. The total k-vector can even be preset within certain limitations by a proper choice of the relative orientation of the incoming photon beams [4].

[1]) Otto-Hahn-Str., D-4600 Dortmund 50, FRG.

In one-photon spectra ω and $\boldsymbol{k}$ in the crystal cannot be chosen independently at all since they are strictly coupled by the dielectric function $\varepsilon(\omega, \boldsymbol{k})$. In one-photon processes the absorption line reflects the total polariton resonance spanning a rather large region of $\boldsymbol{k}$-space whereas in TPA the resonance is measured at a definite k-vector on a polariton branch.

In the following we will first give information about our experimental set-up. After presentation of our results we will discuss the analysis of the temperature dependence of the lineshape.

2. Experiments

Tetragonal GeO_2 crystals of 0.5 to 1 cm^3 volume have been grown by the top-seed technique. Usually they show a slightly brownish colour which is probably caused by a small platinum content. The incorporation of this impurity is inevitable because the Na_2O containing melt attacks the platinum crucible. The addition of Na_2O is necessary to reduce the viscosity of the melt and to lower the melting point below the phase transition temperature between the hexagonal quartz and the tetragonal rutile modifications of GeO_2. Of course another impurity is sodium. A chemical analysis of our crystals revealed a concentration of about 70 ppm of Pt and Na [5]. Further imperfections which may influence the optical properties arise from the high anisotropy of the growth process. This manifests itself in the strong preference of [001] as the growth direction. In order to control the influence of the mentioned imperfections on our results we have studied crystals with different degrees of colouring and with

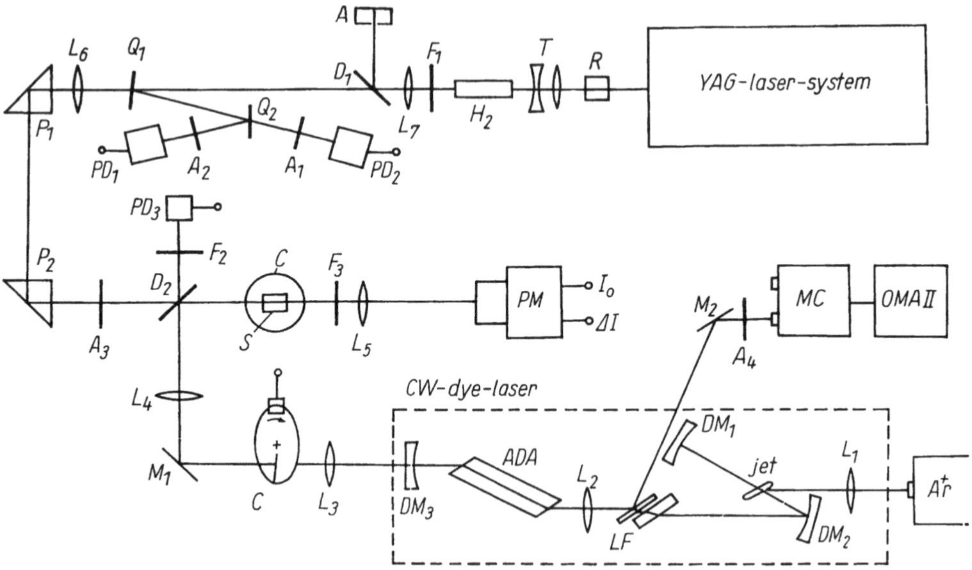

Fig. 1. Diagram of the experimental set-up. R rotator; T telescope; H_2 hydrogen cell; F_1, F_2, F_3 optical filters; L_1, L_2, L_3, L_4, L_5, L_6, L_7 lenses; A absorber; D_1 dielectric mirror reflecting the YAG laser; D_2 dielectric mirror transparent for dye laser and IR radiation reflecting the UV radiation; Q_1, Q_2 quartz plates as beam splitters; P_1, P_2 prisms; A_1, A_2, A_3, A_4 neutral glass filters for attenuation; PD_1, PD_2, PD_3 photodiodes; M_1, M_2 Al mirrors; C chopper; PM S-20 photomultiplier; MC 1m-monochromator; OMA optical multichannel analyser; LF two-plate Lyot filter; DM_1, DM_2, DM_3 dye laser mirrors; ADA SHG crystal; S sample; C cryostat

different orientations. Several samples with optically polished (001), (110), and (100) faces have been prepared with thicknesses between 2 and 4 mm. We did not notice any dependence of the linewidth and of the energetic position of the 1S exciton on the degree of coloring or on the orientation.

The point symmetry of the tetragonal GeO_2 is D_{4h}. Assuming Γ_3^+ and Γ_1^+ symmetry for the upper valence band and the lowest conduction band, respectively, the symmetry of S excitons including spin is given by

$$\Gamma^s = \Gamma_3^+ + \Gamma_4^+ + \Gamma_5^+ ,$$

where Γ_3^+ represents a singlet state and $\Gamma_4^+ + \Gamma_5^+$ triplet states. In one-photon absorption the singlet state is observable only by quadrupole excitation whereas it is dipole-allowed in TPA. The triplet states can be excited in a magnetic field. The selection rules [6] for the two-photon excitation of a Γ_3^+ state are given by the condition

$$(l_1 l_2 - m_1 m_2) \neq 0 ,$$

where l_ν and m_ν are the components of the polarization vector of light source ν and are referred to the crystallographic axes x and y, respectively.

The energy of both photons had to be chosen in such a way that the following two requirements were fulfilled:

a) The sum of both photons must yield about 4.65 eV, the energetical position of the exciton in GeO_2.

b) The wavelength of one laser should be as short as possible to increase the two-photon transition probability due to a nearly resonant denominator.

Furthermore it is necessary to tune the wavelength of one laser and to work with at least one high-power light source.

The experimental set-up is presented schematically in Fig. 1. Concerning the electroncis and data handling it is very similar to the two-photon set-up described earlier [1, 8].

The tunable light source was a dye-jet laser with intracavity frequency doubling. With a DCM-dye we obtained wavelengths between 621.2 and 624.5 nm by rotating a two-plate Lyot filter which was driven by a microprocessor-controlled stepper motor. The wavelength was monitored with an OMA II videcon system which detected a reflected beam from the Lyot filter after it had passed a 1 m-monochromator. The nonlinear crystal ammonium dihydrogen arsenate (ADA) was operated in the non-critical phase-matching mode at a fixed angle of 90° between its optical axis and the incident laser beam. Phase matching at different wavelengths was achieved by varying the temperature of the crystal. Our experience has shown that it is possible to work with ADA up to 130 °C for hours without destroying the ADA crystal. The crystal faces (cut at the Brewster angle) were protected by quartz flats with a drop of index matching oil between crystal and quartz. The noncritical phase matching and the high intracavity power resulted in a high second-harmonic efficiency. Pumping the dye with an argon laser of about 7.5 W we obtained an UV output power up to 2 mW.

As high-power light source we used a Nd:YAG laser. Its wavelength was shifted to 1.908 μm (0.64966 eV) by stimulated Raman emission of H_2 gas at about 40 bar. Thus, together with the tunable light source possessing a center-wavelength of 311 nm (4 eV), we were able to scan the GeO_2 exciton around 4.65 eV. The resolution of 0.3 meV was limited by the spectral width of the dye laser system.

The UV beam was mechanically chopped to produce pulses of about 30 μs length. These pulses were used to synchronize the Nd:YAG laser and were recorded by means of a S-20 photomultiplier. The change of transmission ΔI induced by the short pulse of the high-power laser (10 ns) was electronically separated from the long UV pulse using a clipping network. The samples were kept in a variable temperature cryostat using helium as contact gas. The temperature stability was about 0.5 K.

3. Experimental Results

In Fig. 2 we present a TPA spectrum of GeO_2 at 4.5 K. The narrow line at 4.639 eV is assigned to the 1S exciton whereas the broader structure at 4.665 eV is probably due to a 2S exciton transition. We measured the temperature dependence of the 1S exciton absorption between 4.5 and 150 K. To get a feeling for its qualitative behavior the results taken at three different temperatures are presented in Fig. 3. With rising temperature different effects occur: Besides a decrease of the peak height we notice a shift of its position to lower energies. We assume that this behavior is due to the influence of thermal expansion on the band structure. In Fig. 4 we present its temperature dependence quantitatively. The main purpose of the present work is to study the exciton–phonon coupling as it manifests itselfs in the linewidth Γ. Fig. 3 indicates that Γ increases drastically with temperature. In addition we observed that the lineshape shows a strong asymmetry. Below about 55 K the low-energy part of the resonance line is broader than the high-energy side, at temperatures higher than 55 K this behavior is reversed. In order to analyse quantitatively this asymmetric shape the TPA absorption spectrum at each temperature has been fitted with a modified Lorentzian curve

$$S(E) = \frac{H}{1 + (E - E_0)^2/(B(1 + \alpha\theta(E - E_0)))^2}\,. \tag{1}$$

$S(E)$ is the TPA signal as a function of energy. E, H, E_0, and B describe the usual parameters of a Lorentzian curve and α determines together with the step function $\theta(E - E_0)$ the asymmetry. In this way we obtain two different halfwidths of size B and $B(1 + \alpha)$ and the total halfwidth is $\Gamma = B + B(1 + \alpha)$. The θ-function has been defined in the usual way and α can be positive or negative. Examples of fitted curves are presented in Fig. 3.

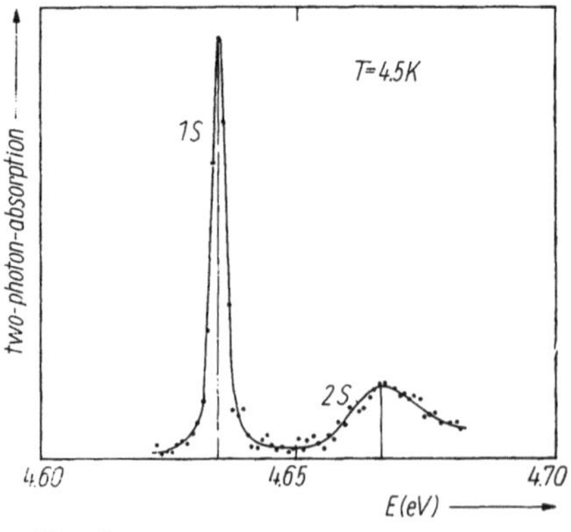

Fig. 2

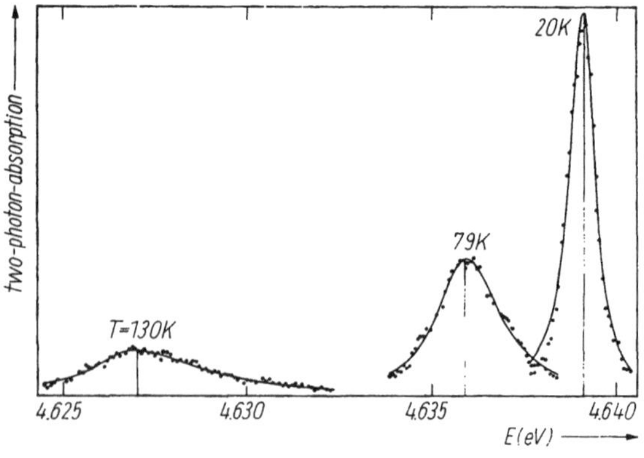

Fig. 3

Fig. 2. TPA spectrum of GeO_2 at 4.5 K

Fig. 3. Temperature dependence of the 1S resonance; the solid lines are fitted asymmetric Lorentz curves

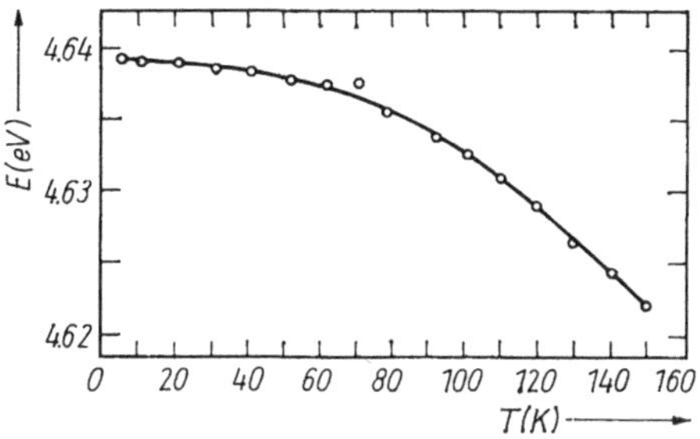

Fig. 4

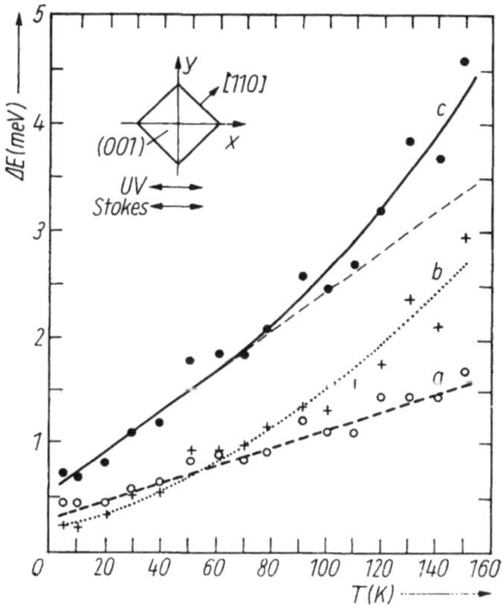

Fig. 5

Fig. 4. Temperature dependence of peak shift

Fig. 5. Temperature dependence of the halfwidths. (a) (o---o---o) halfwidth on the low-energy side, (b) (+···+···+) halfwidth on the high-energy side, (c) (●——●——●) total halfwidth

In Fig. 5 the temperature dependence of the fitted linewidths is shown separately for the low- and high-energy sides (curves a and b). The results have been obtained with the configuration which is illustrated in the inset. The dotted and the dashed lines are fitted polynomials of order two. We stress the fact that the quadratic term of the low-energy curve has turned out to be negligibly small. As mentioned above the low- and high-energy side linewidths intersect at a temperature of approximately 55 K. In spite of the scattering of experimental points this intersection as well as the general trends of curves a to c were verified with measurements on different samples and orientations.

4. Discussion

The theory of exciton lineshape [8, 9] shows that the increase of linewidth with rising temperature is caused by exciton–phonon interaction. Following the theory of Toyozawa [9], the total linewidth Γ is described by

$$\Gamma = \Gamma_0 + aT + \frac{b}{\exp\left(\hbar\omega_{\mathrm{LO}}/kT\right) - 1}. \tag{2}$$

The linear term represents the influence of acoustic phonons, whereas the broadening caused by optical phonons increases with their occupation number $\bar{n} = 1/(\exp(\hbar\omega_{\mathrm{LO}}/kT) - 1)$.

The constant value Γ_0 takes into account the broadening by internal strains and impurities. In order to compare Toyozawa's theory with our experiments we have fitted (2) to the measured values of Γ. To avoid an unreasonably large number of free parameters we have chosen a fixed phonon energy of 42.8 meV. This is the value Alberts and Roessler [10] have determined by infrared reflectance measurements as the largest energy of a longitudinal phonon. The fit yields the following values for the free parameters in (2): $\Gamma_0 = 0.58$ meV, $a = 0.019$ meV/K, $b = 24.6$ meV.

The result is illustrated by curve c in Fig. 5. Identifying a linear increase with a dominant influence of acoustic phonons as suggested by (2) the following interpretation is possible. In the low-temperature region the slope is determined by acoustic phonons and even at 150 K, the highest temperature at which the exciton could be detected with reasonable accuracy, these phonons are the dominant origin of broadening. With respect to the asymmetry of the line, theoretical considerations [9] suggest that broadening is stronger on the high-energy side, due to the asymmetric behavior of the oscillator strength as well as due to the increase of density of states in the exciton bands towards higher energies. From our results we conclude that there exists a further asymmetric contribution which we attribute to optical phonons. On the low-energy side the linear slope shows that the influence of acoustic phonons is dominant whereas the optical phonons show a pronounced influence only on the high-energy side.

These effects do not explain the asymmetry below 55 K. Although this asymmetry is only small, we are sure that it is not an experimental artifact. At the moment we are not able to give a reasonable interpretation for this unusual behavior.

Acknowledgement

We thank Dr. D. Frölich from the University of Hannover for supplying main parts of the frequency doubling system.

References

[1] CH. UIHLEIN, D. FRÖHLICH, and R. KENKLIES, Phys. Rev. B **23**, 2731 (1981).
[2] D. FRÖHLICH, R. KENKLIES, and R. HELBIG, Phys. Rev. Letters **41**, 1750 (1978).
[3] M. STAPELBROEK and B. D. EVANS, Solid State Commun. **25**, 959 (1978).
[4] D. FRÖHLICH, E. MOHLER, and P. WIESNER, Phys. Rev. Letters **26**, 554 (1971).
[5] H. J. WEBER, Mater. Res. Bull. **17**, 1313 (1982).
[6] M. M. DENISOV and V. P. MAKAROV, J. Phys. C **5**, 2651 (1972).
[7] D. FRÖHLICH and M. SONDERGELD, J. Phys. E **10**, 761 (1977).
[8] Y. TOYOZAWA, Progr. theor. Phys. (Kyoto) **20**, 53 (1958).
[9] M. SCHREIBER and Y. TOYOZAWA, J. Phys. Soc. Japan **51**, 1528, 1537 (1982).
[10] D. M. ROESSLER and W. A. ALBERS, JR., J. Phys. Chem. Solids **33**, 293 (1972).

(Received August 3, 1984)

phys. stat. sol. (b) **126**, 311 (1984)

Subject classification: 13.1 and 13.4; 22.1.2

Institute of Nuclear Physics, Academy of Sciences of the Uzbekian SSR, Tashkent[1] *(a) and Institute of General and Inorganic Chemistry, Academy of Sciences of the USSR, Moscow (b)*

Green's Function Method in the Equivalent Orbital Basis

Point Defects and Complexes of Two Point Defects in Silicon

By

A. Sh. Makhmudov (a), Z. M. Khakimov (a), A. A. Levin (b), M. S. Yunusov (a), and P. K. Khabibullaev (a)

A version of the Green's function method on the basis of equivalent orbitals (EO) for the calculation of localized states in diamond-like semiconductors is described. The band structure of silicon is obtained in terms of the tight-binding method including a bond interaction through the fourth-next neighbours. The electronic states of point defects in silicon (vacancy, phosphorus, and sulphur impurities) and of their complexes (divacancy, phosphorus + vacancy) are calculated.

Описан вариант метода функций Грина в базисе эквивалентных орбиталей для расчета локализованных состояний в алмазоподобных полупроводниках. Получена зонная структура кремния в рамках метода сильной связи с учетом взаимодействия связей вплоть до четвертых соседей. Рассчитаны электронные состояния точечных дефектов (вакансия, фосфор, сера) и их комплексов (дивакансия, фосфор + вакансия) в кремнии.

1. Introduction

Various methods have been proposed [1 to 3] to investigate the electronic states of deep impurities and other point defects in semiconductors. Now the most consistent method of the localized state calculation is the Green's function one. This method has been developed mainly in two versions: the Slater-Koster approach [4] and the continuous fraction technique [5]. For the first time the approach [4] has been applied in [6] for the electronic states calculation of vacancy and divacancy in silicon. The further development of this approach has been prolonged for a long time. Only comparatively recently the main ideas of Slater-Koster method [4] have been developed and applied to the study of deep local centres in [7 to 10]. The equivalent orbitals (EO) [7, 10] and the atomic orbitals [8] (and more recently [9]) have been chosen as the basis functions.

In this paper the version [7, 10] of Green's function method proposed earlier for the calculation of localized states (in the energy gap of a host crystal) in diamond-like semiconductors is described in detail and its application to deep localized centres in silicon is considered. This approach is based on the combination of the Slater-Koster approach [4] with the semiempirical method of equivalent orbitals [11, 12].

2. Green's Function Method in the Equivalent Orbital Representation

It is known that first of all the solutions of one-electron Schrödinger equation for a (perfect) crystal have to be found in Green's function method (GFM),

$$H_0 \Psi_{nk} = E_{nk} \Psi_{nk} \,, \tag{2.1}$$

[1]) 702132 Tashkent, USSR.

where Ψ_{nk} and E_{nk} are eigenfunctions and eigenvalues of the effective one-electron Hamiltonian H_0 of the perfect crystal.

To describe a perfect crystal we use the tight-binding method and EO basis. As usually [11] a diamond-like lattice is divided into A- and B-sublattices. Eight EO are associated with each B-atom: four bonding EO $\varphi_j(\boldsymbol{r} - \boldsymbol{R})$ ($j = 1, 2, 3, 4$) and four antibonding EO $\varphi_j(\boldsymbol{r} - \boldsymbol{R})$ with ($j = 5, ..., 8$) constructed of sp^3 hybrid orbitals of the B-atom and of its four A-type neighbours. Sometimes we label bonding EO also by $\varphi_j^+(\boldsymbol{r} - \boldsymbol{R})$ and antibonding EO by $\varphi_j^-(\boldsymbol{r} - \boldsymbol{R})$, then ($j = 1, ..., 4$).

The basis Bloch functions are written as

$$\psi_j(\boldsymbol{k} \mid \boldsymbol{r}) = \frac{1}{\sqrt{N}} \sum_{\boldsymbol{R}} \varphi_j(\boldsymbol{r} - \boldsymbol{R})\, e^{i\boldsymbol{k}\boldsymbol{R}} . \tag{2.2}$$

Then eigenfunctions Ψ_{nk} of Hamiltonian H_0 will be the combinations of functions (2.2),

$$\Psi_{nk} = \sum_{j=1}^{8} C_{jn}(\boldsymbol{k})\, \psi_j(\boldsymbol{k} \mid \boldsymbol{r}) . \tag{2.3}$$

We find the eigenfunction of Hamiltonian $H = H_0 + V$ of a crystal with defect as

$$\Psi = \sum_{j=1}^{8} \sum_{\boldsymbol{R}} C_{j\boldsymbol{R}} \varphi_j(\boldsymbol{r} - \boldsymbol{R}) . \tag{2.4}$$

Then the corresponding eigenvalue problem

$$H\Psi = (H^0 + V)\, \Psi = \varepsilon \Psi \tag{2.5}$$

can be written using the Green's operator $G = (\varepsilon - H_0)^{-1}$,

$$\Psi = GV\Psi , \tag{2.6}$$

or in matrix form

$$(\tilde{G}\tilde{V} - \tilde{I})\, \tilde{C} = 0 , \qquad \tilde{G} = \tilde{C}^0(\varepsilon \tilde{I} - \tilde{E}_0)^{-1}\, \tilde{C}^{0*} . \tag{2.7}$$

Here $\tilde{I}$ is the unit matrix, $\tilde{E}_0$ the matrix of eigenvalues E_{nk} of operator H_0, $\tilde{C}_0$ the matrix with columns being the coefficients of the expansion of eigenfunctions on EO (2.3); $\tilde{C}$ is the column of coefficients $C_{j\boldsymbol{R}}$. The matrix elements G of Green's operator have the form

$$G_{jR,j'R'} = \frac{1}{N} \sum_{nk} \frac{C_{jn}(k)\, C^*_{j'n}(k)}{\varepsilon - E_{nk}}\, e^{ik(\boldsymbol{R}-\boldsymbol{R}')} . \tag{2.8}$$

For the localized defect in the matrix $\tilde{V}$ of the perturbation operator all elements are equal to zero except those corresponding to a finite number of vectors $\boldsymbol{R}$. Using the corresponding numeration these elements form the $M \times M$ submatrix $\tilde{V}_M$ of some dimension M placed in the left upper corner of the $\tilde{V}$-matrix. Then taking into consideration (2.7) the determination of the localized states comes to solving the secular equation

$$\det |\tilde{G}_M \tilde{V}_M - \tilde{I}| = 0 , \tag{2.9}$$

where $\tilde{G}_M$ is the corresponding $M \times M$ submatrix of the $\tilde{G}$-matrix (2.8).

3. Parametrization of Band Structure of Diamond-Like Semiconductors

For a detailed description of the host crystal bands in terms of the tight-binding or EO method it is necessary to account for interactions of distant neighbours (or to

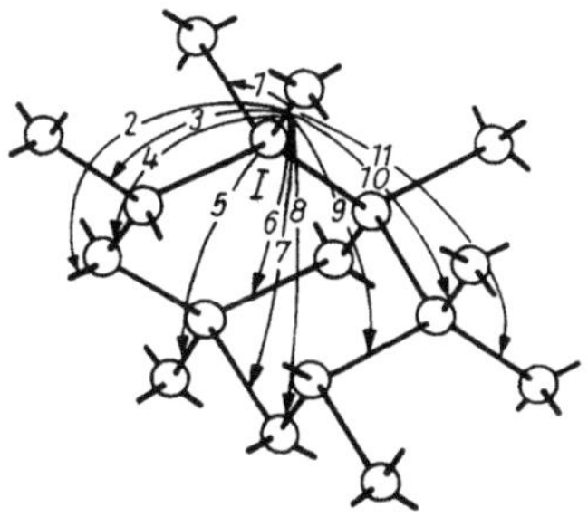

Fig. 1. Diamond-like crystal fragment, illustrating all possible types of interactions between EO in the limits of three atomic coordination spheres: 1 β, 2 β_w, 3 β_g, 4 β_t, 5 β_z, 6 β_q, 7 β_l, 8 β_n, 9 β_k, 10 β_p, 11 β_m, where $\beta = \beta_A$, $\beta_w = \beta_{wA}$ etc., or $\beta = \beta_B$, $\beta_w = \beta_{wB}$ etc., depending on the type (A or B) of atom I

include additional d or s* orbitals [27, 28]). With this in mind we parametrized the band structure in silicon taking into consideration the interatomic interaction up to third neighbours.

As is known, the dispersion law E_{nk} for the crystal with diamond-like lattice accounting for bonding and antibonding EO is determined by solving the secular equation of eighth order [11, 12],

$$\begin{vmatrix} H_{jj'}^{+,+} - E\delta_{jj'} & H_{jj'}^{+-} \\ H_{jj'}^{+-} & H_{jj'}^{--} - E\delta_{jj'} \end{vmatrix} = 0; \qquad j, j' = 1, \dots, 4\,, \tag{3.1}$$

where

$$\begin{aligned} H_{jj'}^{++} &= \langle \psi_j^+(\boldsymbol{k} \mid \boldsymbol{r})| \, H^0 \, |\psi_{j'}^+(\boldsymbol{k} \mid \boldsymbol{r})\rangle \,, \\ H_{jj'}^{--} &= \langle \psi_j^-(\boldsymbol{k} \mid \boldsymbol{r})| \, H^0 \, |\psi_{j'}^-(\boldsymbol{k} \mid \boldsymbol{r})\rangle \,, \\ H_{jj'}^{+-} &= \langle \psi_j^+(\boldsymbol{k} \mid \boldsymbol{r})| \, H^0 \, |\psi_{j'}^-(\boldsymbol{k} \mid \boldsymbol{r})\rangle \,. \end{aligned} \tag{3.2}$$

Here the matrix elements (3.2) are expressed as Hamiltonian matrix elements in EO basis $\langle \varphi_j^\pm(\boldsymbol{r} - \boldsymbol{R})| \, H^0 \, |\varphi_{j'}^\pm(\boldsymbol{r} - \boldsymbol{R})\rangle$. The latter are expressed in terms of Coulomb (α^s, α^p) and resonance ($\beta_{ss}, \beta_{sp}, \beta_{pp}$, and β_π) integrals in the basis of atomic orbitals (AO).

The system of matrix elements (3.2) accounting for the bond interaction through the fourth neighbours (necessary for interatomic interaction through third neighbours) includes the following types of Coulomb and resonance integrals (α, β, β_t, β_g, β_p, β_w, β_k, β_q, β_m, β_l, β_n, β_z), corresponding to the various types of relative position of EO in the range of three atomic coordination spheres (Fig. 1):[2])

$$\begin{aligned} \alpha &= (1 + \nu_j \nu_{j'}) \, (\alpha^s + 3\alpha^p)/8 + (\nu_j + \nu_{j'}) \, (\beta_{ss}^{(1)} + 2\sqrt{3}\,\beta_{sp}^{(1)} + 3\beta_{pp}^{(1)})/8 \,, \\ \beta &= (\nu_B + \nu_A \nu_j \nu_{j'}) \, (\alpha^s - \alpha^p)/8 + (\nu_j + \nu_{j'}) \, (\beta_{ss}^{(1)} + \frac{2}{\sqrt{3}} \beta_{sp}^{(1)} - \beta_{pp}^{(1)})/8 + \\ &\quad + (\nu_A + \nu_B \nu_j \nu_{j'}) \, (\beta_{ss}^{(2)} + 2\sqrt{2}\,\beta_{sp}^{(2)} + 2\beta_{pp}^{(2)} + \beta_\pi^{(2)})/8 \,, \\ \beta_t &= (\nu_B \nu_{j'} + \nu_A \nu_j) \, (\beta_{ss}^{(1)} - \frac{2}{\sqrt{3}} \beta_{sp}^{(1)} + \frac{1}{3} \beta_{pp}^{(1)} - \frac{8}{3} \beta_\pi^{(1)})/8 + \\ &\quad + (1 + \nu_j \nu_{j'}) \, (\beta_{ss}^{(2)} - 2\beta_{pp}^{(2)} + \beta_\pi^{(2)})/8 \,, \\ \beta_g &= (\nu_B \nu_{j'} + \nu_A \nu_j) \, (\beta_{ss}^{(1)} - \frac{2}{\sqrt{3}} \beta_{sp}^{(1)} + \frac{1}{3} \beta_{pp}^{(1)} + \frac{4}{3} \beta_\pi^{(1)})/8 + \\ &\quad + (1 + \nu_j \nu_{j'}) \, (\beta_{ss}^{(2)} + \sqrt{2}\,\beta_{sp}^{(2)} - \beta_\pi^{(2)})/8 + (\nu_B \nu_j + \nu_A \nu_{j'}) \times \\ &\quad \times (\beta_{ss}^{(3)} + \frac{6}{\sqrt{11}} \beta_{sp}^{(3)} + \frac{9}{11} \beta_{pp}^{(3)} + \frac{20}{11} \beta_\pi^{(3)})/8 \,, \end{aligned}$$

[2]) In (3.3) the upper indexes (1), (2), and (3) of resonance integrals in AO basis describe their accessibility to first, second, and third neighbours, respectively.

$$\beta_p = (1 + \nu_j\nu_{j'})\,(\beta_{ss}^{(2)} + 3\beta_\pi^{(2)})/8 + (\nu_j + \nu_{j'})\,(\beta_{ss}^{(3)} + \frac{6}{\sqrt{11}}\beta_{sp}^{(3)} + \frac{9}{11}\beta_{pp}^{(3)} - \frac{24}{11}\beta_\pi^{(3)})/8\,, \tag{3.3}$$

$$\beta_k = (\nu_B + \nu_A\nu_j\nu_{j'})\,(\beta_{ss}^{(2)} - \beta_\pi^{(2)})/8 + (\nu_j + \nu_{j'})\,(\beta_{ss}^{(3)} + \frac{2}{\sqrt{11}}\beta_{sp}^{(3)} - \frac{3}{11}\beta_{pp}^{(3)} + \frac{8}{11}\beta_\pi^{(3)})/8\,,$$

$$\beta_m = (\nu_B + \nu_A\nu_j\nu_{j'})\,(\beta_{ss}^{(2)} - \sqrt{2}\,\beta_{sp}^{(2)} - \beta_\pi^{(2)})/8 + (\nu_B\nu_j + \nu_A\nu_{j'}) \times (\beta_{ss}^{(3)} - \frac{2}{\sqrt{11}}\beta_{sp}^{(3)} - \frac{15}{11}\beta_{pp}^{(3)} + \frac{4}{11}\beta_\pi^{(3)})/8\,,$$

$$\beta_q = (\nu_B + \nu_A\nu_j\nu_{j'})\,(\beta_{ss}^{(2)} - \sqrt{2}\,\beta_{sp}^{(2)} - \beta_\pi^{(2)})/8 + (\nu_B\nu_{j'} + \nu_A\nu_j) \times (\beta_{ss}^{(3)} - \frac{2}{\sqrt{11}}\beta_{sp}^{(3)} - \frac{15}{11}\beta_{pp}^{(3)} + \frac{4}{11}\beta_\pi^{(3)})/8\,,$$

$$\beta_w = (\nu_B + \nu_A\nu_j\nu_{j'})\,(\beta_{ss}^{(2)} - 2\sqrt{2}\,\beta_{sp}^{(2)} + 2\beta_{pp}^{(2)} + \beta_\pi^{(2)})/8\,,$$

$$\beta_l = (\nu_B\nu_{j'} + \nu_A\nu_j)\,(\beta_{ss}^{(3)} - \frac{6}{\sqrt{11}}\beta_{sp}^{(3)} + \frac{5}{11}\beta_{pp}^{(3)} + \frac{16}{11}\beta_\pi^{(3)})/8\,,$$

$$\beta_n = (\nu_B\nu_{j'} + \nu_A\nu_j)\,(\beta_{ss}^{(3)} - \frac{2}{\sqrt{11}}\beta_{sp}^{(3)} + \frac{1}{11}\beta_{pp}^{(3)} - \frac{32}{11}\beta_\pi^{(3)})/8\,,$$

$$\beta_z = (\nu_B\nu_{j'} + \nu_A\nu_j)\,(\beta_{ss}^{(3)} - \frac{10}{\sqrt{11}}\beta_{sp}^{(3)} + \frac{25}{11}\beta_{pp}^{(3)} - \frac{8}{11}\beta_\pi^{(3)})/8\,.$$

Here ν_A, ν_B, ν_j, and $\nu_{j'}$ are equal to 0, +1, −1; namely

$$\nu_j = \begin{cases} +1 & \text{if} \quad \varphi_j = \varphi_j^+\,, \\ -1 & \text{if} \quad \varphi_j = \varphi_j^-\,, \end{cases} \qquad \nu_{j'} = \begin{cases} +1 & \text{if} \quad \varphi_{j'} = \varphi_{j'}^+\,, \\ -1 & \text{if} \quad \varphi_{j'} = \varphi_{j'}^-\,. \end{cases} \tag{3.4}$$

ν_A and ν_B are introduced to discriminate the integrals β_A and β_B, β_{tA} and β_{tB}, and others, i.e.

$$\beta_A = \beta \quad \text{at} \quad \nu_A = 1 \quad \text{and} \quad \nu_B = 0\,, \qquad \beta_B = \beta \quad \text{at} \quad \nu_A = 0 \quad \text{and} \quad \nu_B = 1\,. \tag{3.5}$$

Table 1

q_1, q_2, and q_3 dependence on j (see the text and (3.6))

j	q_1	q_2	q_3
1	q_x	q_y	q_z
2	$-q_x$	$-q_y$	q_z
3	q_x	$-q_y$	$-q_z$
4	$-q_x$	q_y	$-q_z$

Table 2

q_1, q_2, and q_3 dependence on j and j' (see the text and (3.7))

j	j'	q_1	q_2	q_3
1	2	q_x	q_y	q_z
	3	q_y	q_z	q_x
	4	q_x	q_z	q_y
2	3	$-q_x$	q_z	$-q_y$
	4	$-q_y$	q_z	$-q_x$
3	4	q_x	$-q_y$	$-q_z$

Table 3

Parameters for Si energy band calculation (in eV)

parameter	first neighbours	second neighbours	third neighbours
β_{ss}	−2.516	0.110	0.155
β_{sp}	−1.802	−0.119	−0.010
β_{pp}	−2.330	−0.403	−0.045
β_{π}	−0.576	−0.079	−0.022

Using the parameters α, β and others, matrix elements (3.2) have the following form (+ and − indexes have not been disposed for matrix elements, to account them (3.4), (3.5) are used):

$$\begin{aligned} H_{jj} = \alpha &+ \beta_{tA}\,\{e^{i(q_1+q_2)} + e^{i(q_1+q_3)} + e^{i(q_2+q_3)}\} + \\ &+ \beta_{tB}\{e^{-i(q_1+q_2)} + e^{-i(q_1+q_3)} + e^{-i(q_2+q_3)}\} + \\ &+ 2\beta_p\{\cos(q_1 - q_2) + \cos(q_1 - q_3) + \cos(q_2 - q_3)\} + \\ &+ \beta_{nA}\{e^{2iq_1} + e^{2iq_2} + e^{2iq_3}\} + \beta_{nB}\{e^{-2iq_1} + e^{-2iq_2} + e^{-2iq_3}\} + \\ &+ \beta_{zA}\{e^{i(2q_1+q_2+q_3)} + e^{i(q_1+2q_2+q_3)} + e^{i(q_1+q_2+2q_3)}\} + \\ &+ \beta_{zB}\{e^{-i(2q_1+q_2+q_3)} + e^{-i(q_1+2q_2+q_3)} + e^{-i(q_1+q_2+2q_3)}\}\,, \end{aligned} \tag{3.6}$$

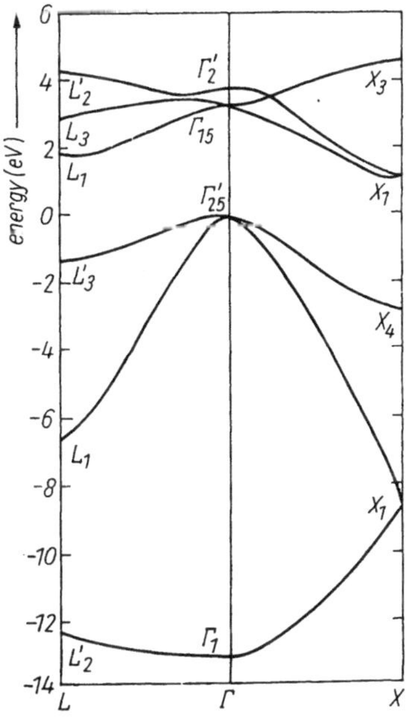

Fig. 2. Energy bands of silicon calculated by the EO LCAO method with parameters given in Table 3

$$\begin{aligned} H_{jj'} = {} & \beta_{\rm B} + \beta_{\rm A}\, {\rm e}^{i(q_1+q_2)} + \beta_{\rm gA}\{{\rm e}^{i(q_1+q_3)} + {\rm e}^{i(q_2+q_3)}\} + \\ & + \beta_{\rm gB}\{{\rm e}^{i(q_1-q_3)} + {\rm e}^{i(q_2-q_3)}\} + \beta_{\rm kA}\{{\rm e}^{2iq_1} + {\rm e}^{2iq_2}\} + \\ & + 2\beta_{\rm kB}\cos(q_1 - q_2) + \beta_{\rm qA}\,\{{\rm e}^{i(2q_1+q_2+q_3)} + {\rm e}^{i(q_1+2q_2+q_3)}\} + \\ & + \beta_{\rm qB}\{{\rm e}^{-i(q_1+q_3)} + {\rm e}^{-i(q_2+q_3)}\} + \beta_{\rm wA}\,{\rm e}^{2i(q_1+q_2)} + \\ & + \beta_{\rm wB}\,{\rm e}^{-i(q_1+q_2)} + \beta_{\rm mA}\,\{{\rm e}^{i(2q_1+q_2-q_3)} + {\rm e}^{i(q_1+2q_2-q_3)}\} + \\ & + \beta_{\rm mB}\,\{{\rm e}^{-i(q_1-q_3)} + {\rm e}^{-i(q_2-q_3)}\} + \beta_{\rm lA}\{{\rm e}^{2iq_3} + {\rm e}^{i(q_1+q_2+2q_3)}\} + \\ & + \beta_{\rm lB}\{{\rm e}^{-2iq_3} + {\rm e}^{i(q_1+q_2-2q_3)}\}\,; \qquad j' \neq j\,. \end{aligned} \tag{3.7}$$

Here q_1, q_2, and q_3 are expressed in terms of q_x, q_y, q_z components of vector $\boldsymbol{q} = (a/2)\,\boldsymbol{k}$ in accordance with Tables 1 and 2 for the diagonal and non-diagonal matrix elements, respectively (a is the lattice constant).

In (3.3) which help to determine matrix elements (2.6), (2.7), Coulomb integrals $\alpha_{\rm Si}^{\rm s}$, $\alpha_{\rm Si}^{\rm p}$ in AO basis have been taken from data of Roetti-Clementi (see [29]). The resonance integrals in AO basis have been chosen so that the resulting dispersion law represented in a best way the band structure of silicon [13]. The results of the described host band parametrization are presented in Table 3 and in Fig. 2.

4. Matrix Elements of the Perturbation Potential

The calculation of matrix elements of the perturbation potential due to the presence of a localized defect is possible either by non-empirical or by semiempirical procedures. A non-empirical calculation of the matrix elements of the perturbation potential needs the explicit form of this potential and of Wannier functions, which causes difficulties. Thus, in [6] prior to [7, 8] as the vacancy perturbation potential the pseudopotential of the silicon ion with reverse sign multiplied by a scale factor has been used. The matrix elements of this potential have been calculated in Wannier function basis. Nevertheless, such a consideration of vacancy electron states was unsuccessful and the results turned out to be in contrast with the experiment. Simultaneously, semiempirical procedures are free of difficulties, connected with the explicit perturbation potential and basis functions as the necessary matrix elements are found from experimental data.

The calculation of the matrix elements of the perturbation potential used here is based on a semiempirical approach. The perturbation potential is presented as the difference of two Hamiltonians: $V = H - H^0$, where H corresponds to the crystal with a defect, H^0 to a perfect crystal. The resulting operator V then assigned by the matrix elements in the basis of the perfect crystal EO, which are calculated in dependence on a concrete type of defect.

5. Isolated Point Defects and Complexes of Two Point Defects in Silicon

Now let us consider the concrete realization of the discussed procedure of the calculation of electron states for some defects. We investigate an isolated vacancy, sulphur, phosphorus substitutional impurity, and phosphorus + vacancy in silicon. For the band structure description of the perfect silicon crystal the parameter values from Table 3 have been used. The dispersion laws and the Bloch functions obtained by means of parameters from Table 3 have been used in the matrix element calculation of the Green's function. For this purpose the special point technique [10, 14] has been used.

5.1 The isolated neutral vacancy and phosphorus and sulphur substitutional impurities in silicon

The isolated vacancy has been modelled by a "pseudo-impurity" which has zero values of the Coulomb integrals α^s, α^p and zero values of some resonance integrals depending on the accepted perturbation potential "radius". The calculation has been carried out in two variants (I) and (II). In variant I the resonance integrals are equal to zero inside the first coordination sphere of the defect, and in variant II in a wider region accounting for Si atoms from the first and second coordination spheres of the defect (note that in contrast to [8], here the "pseudo-impurity" orbitals are preserved in the basis and the unphysical, infinitely large repulsive potential in a vacant site is not introduced).

Using (2.3) and the definition of the perturbation operator $V = H - H^0$ its matrix can be found easily. Thus, in variant I after corresponding numeration of the orbitals it has the form of an 8×8 matrix

$$\tilde{V} = \begin{vmatrix} V_{jj'}^{++} & V_{jj'}^{+-} \\ V_{jj'}^{-+} & V_{jj'}^{--} \end{vmatrix} ; \qquad j, j' = 1, \dots, 4 , \tag{5.1}$$

where

$$\begin{aligned} V_{jj} &= \tfrac{1}{8} (\alpha^s + 3\alpha^p) - \alpha , \\ V_{jj'} &= \tfrac{1}{8} \nu_j \nu_{j'} (\beta_{ss}^{(2)} + 2\sqrt{2}\, \beta_{sp}^{(2)} + 2\beta_{pp}^{(2)} + \beta_{\pi}^{(2)}) - \beta_B ; \qquad j' \neq j , \\ V_{jj}^{+-} &= V_{jj}^{-+} = -\tfrac{1}{8} (\alpha^s + 3\alpha^p) . \end{aligned} \tag{5.2}$$

Here all the parameters in the right-hand part of (5.2) correspond to the atoms of the host crystal (silicon). In analogy, the 32×32 matrix of the V-operator in variant II is written down by calculating the matrix of the H- and H^0-operators in the 32 EO basis that form the defect localization region.

The solution of (2.9) for an isolated vacancy in silicon allowing for perturbation of 8 (variant I) and 32 (variant II) EO gives local levels in the band gap with the energy of $E_v + 0.19$ eV and $E_v + 0.20$ eV, respectively, and T_2 symmetry. (The symmetry is easily discovered after investigating the level splitting at small variations of the corresponding parameters or after making some calculation in the symmetrized basis.) As can be seen, allowing for the perturbation of 32 orbitals in variant II gives small changes in vacancy level position compared to variant I.

Phosphorus and sulphur are close to silicon in the Periodic System. Thus investigating sulphur and phosphorus impurities it can be supposed that the parameters of interatomic interaction between impurity and host atoms (Si) are approximately the same as that between the silicon atoms.

Accordingly the matrix elements of the perturbation potential will depend on the difference of Coulomb integrals of impurities and silicon atoms and the perturbation operator will be an 8×8 matrix of (5.1) type where now

$$\begin{aligned} V_{jj} &= \tfrac{1}{8} (\Delta\alpha^s + 3\Delta\alpha^p) , \\ V_{jj'} &= \tfrac{1}{8} (\Delta\alpha^s - \Delta\alpha^p) ; \qquad j' \neq j , \end{aligned} \tag{5.3}$$

$\Delta\alpha^s$ and $\Delta\alpha^p$ is the difference between the Coulomb integrals of the impurity atom and the host atom (Si).

Except the approximation pointed out the attempt can be made at least qualitatively to allow for the existence of one (for phosphorus) and two (for sulphur) additional valence electrons compared to silicon atoms. From the simplified "chemical" point of view, if we consider only bonds of the impurity with its neighbours in the first coordination sphere $\begin{matrix}\mathrm{Si}\\\mathrm{Si}\end{matrix}\!\!>\!\mathrm{A}\!<\!\!\begin{matrix}\mathrm{Si}\\\mathrm{Si}\end{matrix}$ it is possible to suppose that the "additional" electrons which cannot be positioned on four bonding EO are distributed along the corresponding antibonding EO. The antibonding influence of these electrons is obviously equivalent to the decrease of resonance integrals in the limits of each bonding EO $\varphi^+ = 1/\sqrt{2}$ (χ impurity + χ_{Si}) on $\gamma \approx \frac{1}{8}$ in case of phosphorus impurity and $\approx \frac{1}{4}$ in case of sulphur impurity. Thus allowing for a non-isoelectronic character of the substitution changes the resonance parameters only slightly and as a result the matrix elements become

$$V_{jj}^{++} = \frac{1}{8}(\Delta\alpha^{\mathrm{s}} + 3\Delta\alpha^{\mathrm{p}}) - \frac{\gamma}{4}(\beta_{\mathrm{ss}}^{(1)} + 2\sqrt{3}\,\beta_{\mathrm{sp}}^{(1)} + 3\beta_{\mathrm{pp}}^{(1)}),$$
$$V_{jj'}^{++} = \frac{1}{8}(\Delta\alpha^{\mathrm{s}} - \Delta\alpha^{\mathrm{p}}) - \frac{\gamma}{4}\left(\beta_{\mathrm{ss}}^{(1)} + \frac{2}{\sqrt{3}}\beta_{\mathrm{sp}}^{(1)} - \beta_{\mathrm{pp}}^{(1)}\right); \qquad j' \neq j. \tag{5.4}$$

The calculation of the localized levels for impurities have been made both accounting for the antibonding influence of additional electrons (calculation III for phosphorus and V for sulphur) and not accounting for it (calculation IV for phosphorus and VI for sulphur).

The solution of (1.9) for phosphorus in calculation III in agreement with the experiment gives a shallow localized level in the energy gap of silicon ($E_{\mathrm{c}} - 0.06$ eV). At the same time calculation IV gives a level with ionization energy $E_{\mathrm{c}} - 0.12$ eV. In case of sulphur in calculation VI the level $E_{\mathrm{c}} - 0.42$ eV has been obtained and in calculation V $E_{\mathrm{c}} - 0.65$ eV.

5.2 The isolated [111] *divacancy and* [111] *"phosphorus + vacancy" complex in silicon*

In our calculation the matrix elements of perturbation potential for a [111] divacancy have been estimated in analogy to the case of a vacancy. Two neighbouring Si atoms have been substituted for a "pseudo-impurity" (in [111] direction). Again two variants (calculations VII, VIII) have been considered, where the resonance integrals were supposed to be equal to zero for two coordination spheres of each defect site. According to the calculation in both cases we obtain a deep level, $E_{\mathrm{v}} + 0.31$ eV (VII) and $E_{\mathrm{v}} + 0.33$ eV (VIII).

Now we consider the complex phosphorus + vacancy (E-centre): the complex axis is directed to the [111] direction. In this case one of these sites again is substituted by the pseudo-impurity and for the neighbouring site the Coulomb integrals $\alpha_{\mathrm{Si}}^{\mathrm{s}}, \alpha_{\mathrm{Si}}^{\mathrm{p}}$, are substituted by the corresponding values for phosphorus atoms: $\alpha_{\mathrm{P}}^{\mathrm{s}}, \alpha_{\mathrm{P}}^{\mathrm{p}}$. The resonance integrals for $\mathrm{P}\!<\!\!\begin{matrix}\mathrm{Si}\\\mathrm{Si}\\\mathrm{Si}\end{matrix}$ bonds have not been changed, this corresponds to the presence of an unshared pair of electrons placed on the fourth hybride AO, directed to a vacant site. Two variants have been considered: IX and X corresponding to application of variants I and II for the "vacancy" component of the complex. According to the obtained results in both cases the deep level with energy $E_{\mathrm{v}} + 0.16$ eV appears (IX and X).

6. Discussion of the Results

It is evident from Sections 5.1 and 5.2 that, in spite of the fact that lattice relaxation has not been taken into account, the described calculation reproduces correctly the dependence of the local level energy on the type of the defect. In agreement with experimental data [15, 16] vacancy, divacancy, and PV complex give a single deep level in the energy gap of silicon and the P impurity gives a single shallow level. Again in agreement with experiment the S impurity level lies deeper than the P impurity level, and divacancy level and PV complex level are deeper than the levels of single vacancy and impurity phosphorus atom.

It is noticeable also that the account of the antibonding state of the additional (fifth) phosphorus electron leads to the "improvement" of its level position in the energy gap of silicon (variant IV).

For a more detailed discussion of the obtained data we shall concentrate ourselves mainly on the results of Green's function methods. Because of the well-known disadvantages of the molecular-cluster method we shall not consider the data of this calculation in detail. Analogically to [18 to 22] we have not discussed the problem of the quantitative agreement with experiment for the single-particle levels obtained by the Greeen's function method. It is known that the experimental levels include lattice distortions (and also many-electron effects) and methods of precise qualitative account of some of them still have been developed only for a vacancy [17].

6.1 Vacancy in silicon

A number of works has been devoted to the theoretical and experimental study of a vacancy in silicon, so the electron structure of this defect has been investigated in detail now. In [18] the vacancy levels have been studied by a continuous fraction Green's function technique in [8, 19], in the frame of the non-self-consistent method of Slater-Koster and in [20, 21] in the frame of the (non-empirical) self-consistent Slater-Koster method. In all cases as well as in our study, it has been discovered that in the energy gap of Si for an unrelaxed vacancy there is only the (single-particle) T_2 level. The obtained values of this level are represented in Table 4. It is seen from this table that the value obtained in our work is close to the one obtained in the frame of other non-self-consistent Slater-Koster techniques. Though we should note that the difference between non-self-consistent and self-consistent values of the T_2-level is not still uniquely evident to show the reliability of the self-consistent data from Table 4. As pointed out in [22] this difference is explained not by the self-consistency procedure, but by the variation of the conduction band of the Si host crystal (in the self-consistent calculation [20 to 22] the procedure of the host crystal band calculation is changed compared to non-self-consistent ones [8, 19]). If it were really so, at accurate enough accounting of the band structure of the host crystal the non-self-consistent scheme can provide good results. In this connection we should point out

Table 4
Calculated values of the T_2 level for the unrelaxed neutral vacancy in silicon (in eV)

	present	[8]	[19]	[18]	[20]	[17]	[22]
method	non-self-consistent GFM (Slater-Koster method)	non-self-consistent GFM	non-self-consistent GFM	continuous-fraction method	self-consistent GFM	self-consistent GFM	self-consistent GFM
energy	$E_V + 0.19$	$E_V + 0.27$	$E_V + 0.24$	$E_V + 0.11$	$E_V + 0.62$	$E_V + 0.7$	$E_V + 0.8$

the strict fitting of the host crystal band in this paper. On the other hand, the Si band structure data used in the self-consistent calculation [20 to 22] are absent in literature, that is why it is difficult to judge about the quality of the host crystal band reproduction.

6.2 Phosphorus and sulphur impurities in Si

For neutral phosphorus and sulphur impurities in a silicon lattice site calculations by the Green's function method have been carried out in [9]. Effective mass [23] and molecular cluster [24] calculations have been published, too. The values of the localized level [23] $E_c - 0.042$ eV for P, $E_c - 0.3$ eV for S, and $E_c - 0.66$ eV for S obtained by the effective mass method are in good agreement with those obtained in the present paper and also with the experimental data. The value $E_c - 0.4$ eV for the A_1-level of S obtained in [9] agrees with our result again, but P in this calculation gives no level in the gap.

6.3 Divacancy and E-centre in Si

For a divacancy in Si calculations by the Green's function method are presented in [6, 19, 26]. In [6] for Si:V_2 the level in the Si energy gap has not been obtained. In [19] for the [111] divacancy in the Si energy gap two close levels of E_u- and E_g-types have been obtained by the non-self-consistent Green's function method. Their energies are equal to $E_v + 0.24$ and $E_v + 0.25$ eV, respectively, and this practically coincides with the single-vacancy level calculated in the same work. The level obtained in the present work with energy of $E_v + 0.31$ eV or $E_v + 0.33$ eV for the [111] divacancy has approximately the same energy as that obtained in [19], though it is ≈ 0.1 eV higher than a separate vacancy level. We have not discussed the analogical divacancy Si:V_2 calculations in [25] by the non-self-consistent Green's function method, because it refers not to a [111]-, but to a [110] divacancy.

It is necessary to point out that except the Green's function method calculations by the molecular cluster method [26] have been published recently. In contrast to the experiment this calculation has led to three localized states in the band gap with significantly different energies.

Finally, for the E-centre (Si:P + vacancy) we have no other calculation by the Green's function method, only an X_α scattering wave calculation by the molecular-cluster method has been performed [26]. For [111] E-centre two levels are obtained in the gap: $E_v + 0.5$ eV and $E_v + 0.8$ eV (the experiment gives one level). As we have pointed out earlier, we have obtained only one level $E_v + 0.16$ eV in the present work.

References

[1] V. A. Telezhkin and K. B. Tolpigo, Fiz. Tekh. Poluprov. **16**, 1337 (1982).
[2] S. T. Pantelides, Rev. mod. Phys. **50**, 797 (1978).
[3] M. Jaros, Adv. Phys. **29**, 409 (1980).
[4] G. F. Koster and G. C. Slater, Phys. Rev. **95**, 1167 (1954); **96**, 1208 (1954).
[5] V. F. Masterov and B. E. Samorukov, Fiz. Tekh. Poluprov. **2**, 625 (1978).
[6] G. Callaway and A. J. Hughes, Phys. Rev. **156**, 860 (1967); **164**, 1043 (1967).
[7] A. Sh. Makhmudov, M. S. Yunusov, and A. A. Levin, Proc. Internat. Conf. Rad. Effects and Defects in Semiconductors, Nice (France) 1978; Dokl. Akad. Nauk. UzbSSR **8**, 34 (1978).
[8] J. Bernholc and S. T. Pantelides, Phys. Rev. B **18**, 1780 (1978).
[9] H. P. Hjalmarson, P. Vogl, D. J. Wolford, and J. D. Dow, Phys. Rev. Letters **44**, 810 (1980).
[10] A. Sh. Makhmudov, Avtoreferat kand. dissertatsii, Tashkent 1979.

[11] A. A. LEVIN, Vvedenie v kvantovuyu khimiyu tverdogo tela, Izd. Khimiya, Moscow 1974.
[12] A. A. LEVIN, Solid State Quantum Chemistry, McCoren-Hill, New York 1978.
[13] I. M. TSIDILKOVSKII, Zonnaya struktura poluprovodnikov, Izd. Nauka, Moscow 1978.
[14] D. CHADI and M. L. COHEN, Phys. Rev. B **8**, 5747 (1973).
[15] A. MILNS, Primesi s glubokimi urovnyami v poluprovodnikakh, Izd. Mir, Moscow 1977.
[16] G. D. WATKINS, Physica (Utrecht) **118B**, 9 (1983).
[17] G. A. BARAFF, E. O. KANE, and M. S. SCHLUTER, Phys. Rev. **21**, 5662 (1980).
[18] P. PECHEUR, E. KAUFFER, and M. GERL, in: Defects and Radiation Effects in Semiconductors, Ed. J. H. ALLANY, Inst. Phys., London 1978 (p. 174).
[19] D. N. TALWER and C. S. TING, J. Phys. **15**, 6573 (1982).
[20] N. O. LIPARI, J. BERNHOLS, and S. T. PANTELIDES, Phys. Rev. Letters **43**, 1354 (1979).
[21] G. A. BARAFF and M. SCHLUTER, Phys. Rev. **19**, 4945 (1979).
[22] J. BERNHOLS, N. LIPARI, and S. T. PANTELIDES, Phys. Rev. Letters **41**, 895 (1978).
[23] A. M. OMELIANOVSKII and V. I. FISTUL, Primesi perekhodnikh metallov v poluprovodnikakh, Izd. Metallurgia, Moscow 1983.
[24] F. SHUZUKI and K. MINAMI, phys. stat. sol. (b) **62**, 585 (1974).
[25] S. T. PANTELIDES, I. IVANOV, M. SCHEFFLER, and J. P. VIGNERON, Physica (Utrecht) **116B**, 18 (1983).
[26] A. FARRIZO, J. R. LEITE, and M. J. CALDAS, Physica (Utrecht) **116B**, 90 (1983).
[27] S. G. LOUIE, Phys. Rev. B **22**, 1933 (1980).
[28] P. VOGL, H. P. HJALMARSON, and J. D. DOW, J. Phys. Chem. Solids **44**, 365 (1983).
[29] J. E. LOWTER, Phys. Rev. B **15**, 3928 (1977).

(Received March 15, 1984)

phys. stat. sol. (b) 125, 323 (1984)

Subject classification: 13.4 and 19; 22.8.1

Département de Physique, Université de Hacettepe, Beytepe, Ankara[1] *(a)*
et Laboratoire de Radioélectricité et Théorie du Solide, Aubière (b)

Le spectre de RPE de l'ion Mn^{2+} dans le tartrate de cadmium ($CdC_4H_4O_6 \cdot 5\,H_2O$)

Par

M. KORKMAZ[2] (a), B. AKTAŞ (a) et M. DUPONT (b)

Nous avons étudié, à température ambiante, le spectre de RPE de l'ion Mn^{2+} dans un monocristal de $CdC_4H_4O_6 \cdot 5\,H_2O$ préparé par diffusion. L'ion Mn^{2+} entre dans le réseau cristallin comme substituant de Cd^{2+}. Les sites de l'ion "impureté" ne sont pas équivalents vis à vis de la symétrie, mais sont magnétiquement équivalents.

Mn^{2+} doped single crystals of $CdC_4H_4O_6 \cdot 5\,H_2O$, grown by diffusion technique, are investigated by EPR at room temperature. Mn^{2+} enters the lattice as a substitutional impurity and is trapped at two magnetically equivalent but physically inequivalent sites.

1. Introduction

La RPE permet d'étudier en détail les symétries du champ électrique produit à l'emplacement d'un ion magnétique, par les atomes qui l'entourent. Les ions cuivre et manganèse ont été souvent utilisés comme sondes dans ce genre d'étude [1 à 6].

Les spectres d'absorption optique et de RPE des éléments de transition de la première série, inclus dans des monocristaux de sels de l'acide tartrique ont été beaucoup étudiés.

Le premier travail de RPE a été fait par Wakim et al. [7] sur l'ion Mn^{2+} introduit dans le réseau du tartrate de calcium ($CaC_4H_4O_6 \cdot 4\,H_2O$). Plus tard, Murphy et Bohandy [8], Lau et Lin [9] ont étudié plusieurs ions d'éléments de transition dans le tartrate de calcium et le tartrate de strontium ($SrC_4H_4O_6 \cdot 4\,H_2O$). Enfin, nous avons publié récemment les résultats d'une étude des composés de $SrC_4H_4O_6 \cdot 3\,H_2O$ [10] et de $PbC_4H_4O_6$ [11] dopés au cuivre. Il nous a semblé intéressant d'étendre cette étude à d'autres tartrates.

Dans ce travail, nous présentons les résultats obtenus pour les paramètres de l'Hamiltonien de spin et ceux du champ cristallin; l'ion étudié est Mn^{2+} dans le tartrate de cadmium pentahydraté ($CdC_4H_4O_6 \cdot 5\,H_2O$).

2. Echantillons et méthode

Pour préparer les monocristaux de tartrate de cadmium pentahydraté, on a laissé diffuser les solutions aqueuses de $Cd(NO_3)_2$ (contenant un peu de nitrate de manganèse) et de $Na_2C_4H_4O_6$, à travers un gel préparé avec de l'acide tartrique et du silicate de sodium. Le gel était contenu dans un tube capillaire en verre qui reliait les deux solutions.

Des cristaux de dimensions suffisantes se sont formés en une quinzaine de jours aux deux extrémités du tube capillaire.

[1]) Ankara, Turkey.

[2]) A qui on doit adresser toute correspondance.

L'étude aux rayons X de ces cristaux a montré qu'il s'agit bien de tartrate de cadmium pentahydraté ($a = 0{,}6129$ nm, $b = 1{,}2314$ nm, $c = 0{,}7627$ nm, $\beta = 116°14'$, $Z = 2$) [12].

Le spectre a été relevé avec un spectromètre Varian (E-L 9″) fonctionnant dans la bande X ($\approx 9{,}2$ GHz). Le signal d'un cristal de DPPH placé dans la cavité nous a servi de marqueur pour g ($= 2{,}0036$). La fréquence micro-onde était mesurée par un fréquencemètre Hewlett-Packard, de type 5245L.

La rotation de l'échantillon autour d'un axe perpendiculaire au champ magnétique s'est faite à l'aide d'un goniomètre. On a choisi comme axes expérimentaux a^*, b, c: b et c sont les axes cristallographiques orthogonaux et a^* est perpendiculaire à b et c.

Les mesures ont été faites, pour la plupart, à température ambiante, et les autres à 77 K.

3. Résultats et discussions

Pour une orientation quelconque du cristal, le spectre montre l'existence de deux espèces de sites, identiques, mais orientés différemment. Cela vient de ce que la maille du cristal contient deux molécules.

La variation angulaire des intensités des raies hyperfines dépend du type des transitions électroniques. Le sextet central (M: $\frac{1}{2} \to -\frac{1}{2}$) de chaque complexe présente l'anisotropie la plus marquée.

Quand le champ magnétique a la direction de l'axe principal z d'un complexe, l'intensité des raies hyperfines de ce dernier est maximum. Dès que le champ s'éloigne de cette direction, l'intensité diminue et un certain nombre de raies interdites ($\Delta M = \pm 1$, $\Delta m = \pm 1$), bien que très faibles, commencent à apparaître.

Comme on le fait d'ordinaire, nous avons cherché les écarts maxima entre les sextets M: $-\frac{3}{2} \to -\frac{5}{2}$ et M: $\frac{5}{2} \to \frac{3}{2}$ pour trouver les axes principaux z des complexes. Les cosinus directeurs de z, rapportés aux axes a^*, b, c sont respectivement $-0{,}313$, $0{,}915$ et $0{,}256$ pour l'un des types de site; $0{,}516$, $0{,}836$ et $0{,}188$ pour l'autre. Aux erreurs d'expérience près, ces axes coincident avec ceux que nous avons trouvés pour $CdC_4H_4O_6 \cdot 5\,H_2O$, dopé au cuivre, dans une étude en cours. La Fig. 1 montre le spectre enregistré lorsque $\boldsymbol{H}$ est parallèle à l'axe principal z de l'un des complexes.

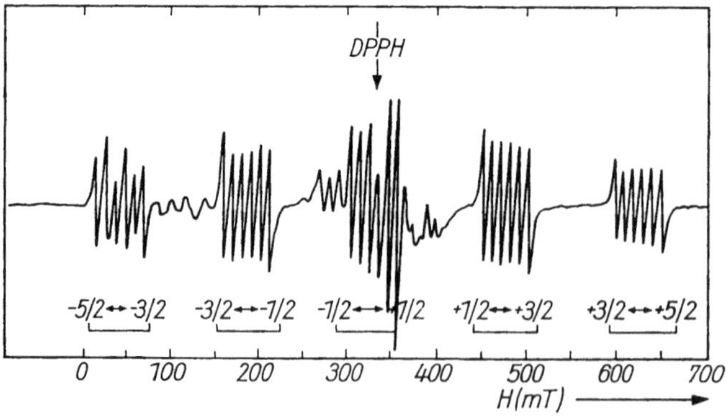

Fig. 1. Spectre de Mn^{2+} dans un monocristal de $CdC_4H_4O_6 \cdot 5\,H_2O$, à température ambiante. Le champ magnétique est parallèle à l'axe z d'un des complexes magnétiques de Mn^{2+}. (Constante de temps 0.3 s, amplitude de modulation 0.5 mT, gain de recepteur 1.25×10^2, puissance de micro-ondes 3.0 mW, temps de balayage 16 min, fréquence de modulation 100 kHz, température 20 °C, frequence de micro-ondes 9.264 GHz)

La comparaison entre les raies hyperfines (Fig. 1) montre que l'élargissement inhomogène dans les sextets M: $\pm\frac{5}{2} \to \pm\frac{3}{2}$ et M: $\pm\frac{3}{2} \to \pm\frac{1}{2}$ est notablement plus grand que celui qui apparaît dans la transition centrale M: $+\frac{1}{2} \to -\frac{1}{2}$. Cela indique une quasi perfection cristalline de l'échantillon.

La largeur de raie ne dépend que très peu de l'orientation du cristal. Elle est à peu près de 0.8 mT pour $\boldsymbol{z} \parallel \boldsymbol{H}$.

La dépendance angulaire du spectre indique que les "complexes" magnétiques ont une symétrie orthorhombique. C'est pourquoi nous avons utilisé l'Hamiltonien de spin suivant qui convient pour cette symétrie [13]:

$$\begin{aligned}\mathcal{H} &= \beta(g_x H_x \hat{S}_x + g_y H_y \hat{S}_y + g_z H_z \hat{S}_z) + \\ &+ (B_2^0 \hat{O}_2^0 + B_2^2 \hat{O}_2^2 + B_4^0 \hat{O}_4^0 + B_4^2 \hat{O}_4^2 + B_4^4 \hat{O}_4^4) + \\ &+ (A_x \hat{S}_x \hat{I}_x + A_y \hat{S}_y \hat{I}_y + A_z \hat{S}_z \hat{I}_z) .\end{aligned}$$

Ici, le premier terme représente l'interaction avec le champ magnétique extérieur, le second, l'interaction avec le champ électrique cristallin et le dernier, l'interaction hyperfine. $\hat{O}_k^m$ et B_k^m sont les "opérateurs de Stevens" et les paramètres du champ cristallin. L'introduction de termes supplémentaires (interaction quadrupolaire, etc.) n'est pas nécessaire ici.

Pour déterminer les paramètres de l'Hamiltonien, nous avons mesuré les valeurs du champ magnétique à la résonance pour chacune des transitions. Les spectres utilisés sont ceux qu'on obtient lorsque $\boldsymbol{H}$ est parallèle aux axes principaux z, y et x de l'un des complexes. Les calculs ont été faits par diagonalisation de $\mathcal{H}$ et les valeurs trouvées sont dans le Tableau 1.

Tableau 1

Paramètre de l'Hamiltonien de spin de Mn^{2+} (300 K) dans $CdC_4H_4O_6 \cdot 5\,H_2O$

$A_x = -(9{,}70 \pm 0{,}10)$ mT	$A_y = -(9{,}65 \pm 0{,}10)$ mT	$A_z = -(9{,}60 \pm 0{,}10)$ mT
$g_x = 1.980 \pm 0{,}001$	$g_y = 1{,}992 \pm 0{,}001$	$g_z = 2{,}002 \pm 0{,}001$
$B_2^0 \left(= \frac{D}{3}\right) = -(23{,}5 \pm 0{,}1)$ mT	$B_2^2 (= E) = (3{,}7 \pm 0{,}5)$ mT	$B_4^0 = (0{,}0035 \pm 0{,}0002)$ mT
$B_4^2 = -(0{,}0200 \pm 0{,}0002)$ mT	$B_4^4 \approx 0$	

Les signes relatifs s'obtiennent à l'aide des déplacements du 2ème ordre. Si, par exemple, $D\ (= 3B_2^0)/A < 0$; l'écart entre les raies hyperfines d'un sextet diminue vers les champs bas et augmente du côté des champs élevés, quand le champ magnétique est parallèle à l'axe principal z [14].

Nous avons calculé les valeurs du champ pour chaque transition au moyen des données du Tableau 1 pour comparer avec les valeurs expérimentales que l'on mesure aux orientations $\boldsymbol{H} \parallel \boldsymbol{z}$, $\boldsymbol{H} \parallel \boldsymbol{y}$, $\boldsymbol{H} \parallel \boldsymbol{x}$ pour les mêmes transitions. Les différences moyennes entre valeurs calculées et expérimentales sont de 0,12 mT($\boldsymbol{H} \parallel \boldsymbol{z}$), 0,90 mT($\boldsymbol{H} \parallel \boldsymbol{y}$) et 0,49 mT($\boldsymbol{H} \parallel \boldsymbol{x}$). L'accord est bon et montre que notre méthode de calcul des paramètres est satisfaisante.

Le paramètre $E (= B_2^2)$ qui représente l'écart à la symétrie tétragonale, est relativement faible: le champ cristallin est encore rhombique, même après l'introduction, de l'impureté dans le cristal. La substitution se fait très probablement sur les sites de Cd^{2+} ($r(Cd^{2+}) = 0{,}097$ nm, $r(Mn^{2+}) = 0{,}080$ nm).

Par conséquent, la géométrie des sites de l'ion Mn^{2+} est déterminée par les atomes du voisinage qui forment des complexes de type $[MnO_m\,(H_2O)_n]^{2+}$ avec l'ion Mn^{2+}.

Tableau 2

Paramètres de l'Hamiltonien de spin de Mn^{2+} dans divers tartrates

composé	g_x	g_y	g_z	g_{moyen}	A_x	A_y	A_z	A_{moyen}	D	E	ionicité (%)	références
$CdC_4H_4O_6 \cdot 5\ H_2O$	1,980 ±0,001	1,992 ±0,001	2,002 ±0,001	1,995	−9,70 ±0,10	−9,65 ±0,10	− 9,60 ± 0,10	−9,65	−70,5 ± 0,3	3,7 ± 0,5	94,5	ce travail
$CaC_4H_4O_6 \cdot 4\ H_2O$	1,992 ±0,003	1,992 0,003	1,992 0,003	1,992	−9,64 ±0,10	−9,64 ±0,10	− 8,96 ± 0,10	−9,41	52,6 ± 0,3	9,3 ± 1	92,8	[7, 9]
$SrC_4H_4O_6 \cdot 4\ H_2O$	2,002 ±0,002	2,002 ±0,002	2,002 ±0,002	2,002	−9,40 ±0,10	−9,40 ±0,10	−10,20 ± 0,10	−9,67	52,8 ± 0,2	8,3 ± 2	94,5	[8]
$SrC_4H_4O_6 \cdot 3\ H_2O$	1,9781 ±0,0005	1,9781 ±0,0005	2,0012 ±0.0005	1,986	−9,49 ±0,10	−9,49 ±0,10	− 9,62 ± 0,10	−9,53	48,3 ± 0,1	10,1 ± 0,3	93,3	[8]
$LiNH_4C_4H_4O_6 \cdot H_2O$	2,010 ±0,004	2,005 ±0,004	2,007 ±0,002	2,007	−9,15 ±0,05	−9,00 ±0,10	− 8,80 ± 0,05	−8,98	60,9 ± 0,2	−10,9 ± 0,2	91,3	[19]

Température voisine de la température ambiante ($\approx$ 300 K); les paramètres sont donnés en mT, sauf les g.

Comme la structure cristalline de tartrate de cadmium n'a pas encore été complètement établie, il n'est pas possible de trouver les nombres n et m à l'aide de la seule RPE.

On voit, d'après les résultats du Tableau 1, que le facteur de Landé possède une petite anisotropie; g a une valeur très proche de celle de l'électron libre mais les paramètres du champ cristallin ne sont pas nuls: l'état fondamental de l'ion dans le réseau n'est donc pas exactement un état $^6S_{5/2}$. La différence entre la valeur $g = 2,0023$ pourrait venir de liaisons covalentes avec les éléments du complexe $[MnO_m \cdot (H_2O)_n]^{2+}$; mais il n'est pas facile d'en tirer des renseignements sur la nature de ces liaisons. D'après Müller [15], les interactions entre l'ion Mn^{2+} et ses voisins immédiats ont beaucoup plus d'influence sur la constante hyperfine que sur $D\ (= 3B_2^0)$ et g pour lesquels les interactions à grande distance peuvent être importantes.

Il semble donc clair que la constante hyperfine A de l'ion Mn^{2+} diminue quand le degré de covalence des liaisons dans le complexe augmente. La relation entre A et les liaisons de Mn^{2+} a été étudiée [16, 17] en traçant la courbe A = fonction du degré de covalence. Nous avons trouvé une valeur de 94,5% pour l'ionicité en utilisant la courbe de Matumura [17] et en prenant pour A la moyenne des valeurs données dans le Tableau 1, c'est à dire $A = 9{,}65$ mT.

Dans le Tableau 2, figurent les valeurs des paramètres de l'Hamiltonien de spin et valeurs d'ionicité des divers sels de l'acide tartrique. L'ionicité de ces sels varie de 91,3% à 94,5%: les liaisons de Mn^{2+} y sont donc plutôt ioniques et ce caractère est encore renforcé dans le $CdC_4H_4O_6 \cdot 5\ H_2O$. Cela pourrait venir de ce que le nombre de molécules d'eau coordonnées à l'ion Mn^{2+} est plus grand.

Quant à $D(= 3B_2^0)$ et $E(= B_2^2)$, ils varient respectivement de 48,2 à 70,5 mT et de 10,9 à 3,7 mT. Le tartrate de cadmium a un D plus grand et un E plus petit; dans ce sel, le champ électrique à l'endroit de l'ion Mn^{2+} est presque axial et il est plus intense que ceux qu'on a trouvés jusqu' à présent dans les autres sels de l'acide tartrique.

Le signe de D, dans le tartrate de cadmium est opposé à celui qu'on trouve dans les autres tartrates. Cela a également été observé pour Mn^{2+} et Cr^{3+} dans des sulfates doubles [5, 18]; on l'attribue à la nature trigonale de la modification du champ cristallin.

Bibliographie

[1] D. Suryanarayana et J. Sobhanadry, J. magnetic Resonance **14**, 1 (1974).
[2] R. S. Eachus et W. G. Mc Dugle, Inorg. Chem. **12**, 1561 (1973).
[3] V. J. Folen, Phys. Rev. **139**, A1961 (1965).
[4] R. E. Donovan et A. A. Vuylsteke, Phys. Rev. **127**, 76 (1962).
[5] V. K. Jain et P. Venkateswarlu, J. Phys. C **12**, 865 (1979).
[6] J. Mankowitz et W. Low, Phys. Rev. B **2**, 28 (1970).
[7] F. G. Wakim, H. K. Henisch et H. A. Atwater, J. chem. Phys. **42**, 2619 (1965).
[8] J. C. Murphy et J. Bohandy, J. chem. Phys. **46**, 1215 (1967); **52**, 3301 (1970).
[9] P. W. Lau et W. C. Lin, J. chem. Phys. **59**, 3981, 3988 (1973).
[10] M. Korkmaz et B. Aktas, J. Phys. Chem. Solids **44**, 651 (1983).
[11] M. Korkmaz, R. Bessège et A. Gavaix, à paraître.
[12] J. S. Hopwood et A. W. Nicol, J. appl. Cryst. **5**, 437 (1972).
[13] A. Abragam et B. Bleaney, Electron Paramagnetic Resonance of Transition Ions, Chap. 7, Clarendon Press, Oxford 1970.
[14] W. Low, Solid State Phys., Suppl. **2**, 72 (1960).
[15] K. A. Müller, Proc. 16th Coll. Ampère, Bucharest 1970, publié par I. Ursu.
[16] E. Simanek et K. A. Müller, J. Phys. Chem. Solids **31**, 1027 (1970).
[17] O. Matumura, J. Phys. Soc. Japan **14**, 108 (1959).
[18] A. Manoogian et A. Leclerc, J. chem. Phys. **63**, 4450 (1975).

(Received June 14, 1984)

phys. stat. sol. (b) **126**, 329 (1984)

Subject classification: 13.4 and 13.5.1; 22.2; 22.4

Laboratoire de Physique des Semiconducteurs, Université de Metz[1]) *(a)*
and Laboratoire de Spectroscopie et d'Optique du Corps Solide,
associé au C.N.R.S. n° 232, Université Louis Pasteur, Strasbourg[2]) *(b)*

Neutral Bound Excitons in a Low Magnetic Field

By

F. DUJARDIN (a), B. STÉBÉ (a), and G. MUNSCHY (b)

The diamagnetic shift of an exciton bound to a neutral donor or acceptor shallow impurity in direct gap semiconductors is studied using the previously obtained Page and Fraser type wave function. The results are compared with the experimental observations in InP for the exciton–neutral donor complex and in GaAs and ZnTe for the exciton–neutral acceptor complex. In all cases qualitative agreement is found while for InP a very good agreement is obtained if an isotropic experimental effective hole mass is used.

Nous étudions le déplacement diamagnétique d'un exciton lié à une impureté neutre peu profonde de type donneur ou accepteur dans les semiconducteurs à transitions directes. A cet effet, nous utilisons la fonction d'onde variationelle de Page et Fraser que nous avons déterminée précédemment. Nous comparons nos résultats avec les observations expérimentales dans InP pour le donneur et dans GaAs et ZnTe pour l'accepteur. Dans l'ensemble, nous trouvons un accord qualitatif tandis que pour InP, nous obtenons un très bon accord à condition de prendre une valeur moyenne de masse de trou déduite de l'expérience.

1. Introduction

Since the pioneering work of Thomas and Hopfield [1], the use of a magnetic field has become a common tool for the identification of excitons bound to impurities. But till now, the lack of accurate enough envelope wave functions has not allowed one to take full advantage of all the information available from the many magneto-optical experiments involving excitons bound to neutral impurities [2]. It is only recently that even in the absence of any external perturbation the stability of these four-particle complexes has been proved by means of strictly variational calculations [3 to 5]. Thus only approximate models could be proposed in the past in order to estimate the diamagnetic shift of neutral bound excitons. In particular, when the interaction between an exciton and a neutral acceptor impurity can be represented by an effective short-range potential, the binding energy does not depend on the magnetic field, so that the transition energy shifts of free and bound excitons are equal [6, 7]. Experimental observations in CdS [1], GaSb [9], and GaAs [6, 7] seem to confirm this assumption. It has also been suggested that for deep impurities, the transition shift of an exciton bound to a neutral acceptor could be identified with the energy shift of a neutral donor [10, 11] as it results from more recent experimental data on GaSb [10] and GaAs [11]. Finally, other measurements in CdTe [12] seem to support the fact that the two kinds of neutral bound excitons give rise to a diamagnetic shift similar to that of a free exciton. All these approximations suppose in fact that the magnetic shifts of the electron–hole or the hole–hole correlation energies may be neglected.

[1]) Ile du Saulcy, 57045 Metz, France.

[2]) 5, rue de l'Université, 67084 Strasbourg Cedex, France.

In a recent paper [4], we have shown that it is essential to take into account these correlations in order to obtain a stable binding. It seems therefore useful to study the influence of a uniform magnetic field on the ground state of a neutral bound exciton by using our previously obtained [4] fully correlated Page and Fraser type wave function. In this preliminary work, we restrict ourselves to the case of low and intermediate fields and to shallow impurities within the effective mass approximation in a simple non-degenerate two-band model. We discuss explicitly the case of the exciton neutral acceptor complex (A^0, X), which is quite analogous to that of the exciton neutral donor complex (D^0, X) by interchanging the electrons and the holes.

2. Theory

The effective envelope Hamiltonian may be simplified by using the usual atomic units $a_A = \varepsilon \hbar^2/m_h^* e^2$ for length, $2\,|E_A| = m_h^* e^4/\varepsilon^2 \hbar^2$ for energy and the dimensionless effective magnetic field parameter $\gamma_A = \hbar\omega_C/2\,|E_A|$, where $\omega_C = eB/m_{hc}^*$ is the effective hole cyclotron frequency related to the magnetic field directed along the z-axis. ε is an appropriate dielectric constant taking into account the polarization effects. By using the Lorentz gauge and neglecting the spins as well as the electron–hole exchange interaction [13], the Hamiltonian of the system is given by

$$H = H_0 + H_1' + H_2', \tag{1}$$

where H_0 is the zero-field Hamiltonian

$$H_0 = T_h + \sigma^{-1} T_e + V \tag{2}$$

with

$$T_h = -\tfrac{1}{2}\left(\nabla_1^2 + \nabla_2^2\right), \qquad T_e = -\tfrac{1}{2}\nabla_e^2, \tag{3}$$

$$V = -\frac{1}{r_1} - \frac{1}{r_2} + \frac{1}{r_e} + \frac{1}{r_{12}} - \frac{1}{r_{1e}} - \frac{1}{r_{2e}}, \tag{4}$$

whereas H_1' and H_2' denote the orbital magnetic terms

$$H_1' = \frac{\gamma_A(-L_{1z} - L_{2z} + \sigma^{-1} L_{ez})}{2}, \tag{5}$$

$$H_2' = \frac{\gamma_A^2(\varrho_1^2 + \varrho_2^2 + \sigma^{-1}\varrho_e^2)}{8} \tag{6}$$

Table 1

Effective magnetic fields γ_A, γ_D, and γ_X for acceptors, donors, and excitons corresponding to a magnetic flux density $B = 10$ T in various semiconductors

substance	ε	m_e^*/m_0	γ_1	γ_D	γ_A	γ_X
AlSb	12.0	0.18	4.15	0.19	0.11	0.58
GaP	10.75	0.17	4.20	0.17	0.09	0.50
GaAs	12.56	0.067	7.65	1.50	0.39	3.43
GaSb	15.7	0.045	11.80	5.18	1.46	12.14
InP	12.4	0.080	6.28	1.02	0.26	2.30
InAs	14.6	0.023	19.67	17.14	3.50	36.15
InSb	17.9	0.014	35.08	69.55	16.77	154.64
Ge	15.36	0.038	13.35	6.95	1.79	15.79
ZnS	8.1	0.28	2.54	0.04	0.02	0.10
ZnSe	9.1	0.14	3.77	0.18	0.05	0.42

The values of ε, m_e^*, and $\gamma_1 = m_0/m_h^*$, m_h^* being the isotropic hole mass, are taken from [14 to 16].

with $\sigma = m_e^*/m_h^*$, $\varrho_i^2 = x_i^2 + y_i^2$, and where L_{iz} are the projections of the angular momentum operators of the three mobile particles. The symmetry properties of the Hamiltonian (1) show that the energies of the two complexes (A^0, X) and (D^0, X) are related by

$$\frac{E_{(A^0,X)}(\sigma, \gamma_A)}{E_A} = \frac{E_{(D^0,X)}(\sigma^{-1}, \gamma_D)}{E_D} . \tag{7}$$

where the effective fields γ_D and γ_A are assumed to have the same numerical value. γ_D is defined in the same manner as γ_A. In particular, for a given magnetic field, $\gamma_D = \gamma_A/\sigma^2$. The linear magnetic term H_1' gives rise to the orbital Zeeman splitting, whereas the two terms H_1' and H_2' are in general responsible for the diamagnetic shift. Their influence depends on the value of the effective magnetic field γ, which for acceptor impurities may often be one order of magnitude smaller than for donor impurities as results from Table 1.

For small enough values of the effective magnetic field γ_A, the magnetic terms H_1' and H_2' may be computed as perturbations with respect to the zero-field Hamiltonian H_0, the eigenvalues of which have been previously determined [4] by using the following 35-term variational wave function,

$$\begin{aligned} \varphi(r_{ij}) &= \psi(kr_{ij}) , \\ \psi(r_{ij}) &= \sum_{mnpq} c_{mnpq}(1 + P_{12}) \, |mnpq\rangle , \\ |mnpq\rangle &= r_1^m r_{1e}^n r_e^p r_{12}^q \exp(-\alpha r_1 - \beta r_2 - r_e) , \end{aligned} \tag{8}$$

c_{mnpq}, α, β, and k being the variational parameters, P_{12} the operator of permutation of the indices 1 and 2, and m, n, p, q positive integers with $m + n + p + q < 3$. Because this function does not depend on the angles, only the quadratic term has to be considered at first order of perturbation, though the two terms H_1' and H_2' may give non-zero contributions at higher orders of perturbation. These latter contributions, which need the knowledge of all the discrete and continuous eigenfunctions of the Hamiltonian H_0 are expected to be negligible. At the present stage, we restrict ourselves to the first-order contribution of the quadratic term H_2' which leads to the following total energy shift:

$$\Delta E_{\text{per}}(\gamma_A, \sigma) = \gamma_A^2 \, |E_A| \, F_{\text{per}}(\sigma) , \tag{9}$$

where the function F_{per} has been determined using the zero-field ground-state wave

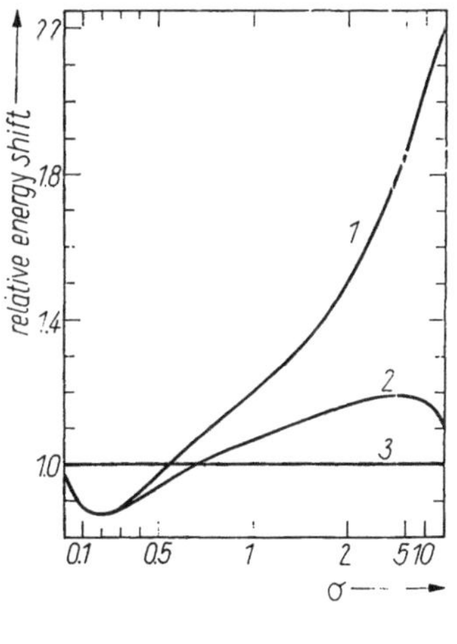

Fig. 1. Energy shift of the (A^0, X) complex obtained by using the perturbation theory, (1) $\Delta E/\Delta E_X$, (2) $\Delta E/(\Delta E_X + \Delta E_A)$; and (3) $\Delta E/(\Delta E_X + \Delta E_A)$, comparison with the theoretical prediction of Willmann et al. [7] ($\sigma = m_e^*/m_h^*$)

function (8). In the same approximation, the acceptor and exciton diamagnetic shifts are given by $\Delta E_A = \gamma_A^2 |E_A|/2$ and $\Delta E_X = \gamma_X |E_X|/2$, with $\gamma_X = \gamma_A(1 + \sigma^{-1})^2$. Our results reported in Fig. 1 show that in the low-field approximation, the energy shift ΔE of the neutral bound excitons, may be rather different from that of free excitons and the binding energy shift $\Delta W = \Delta E - (\Delta E_A + \Delta E_X)$ is in general non-negligible and depends on the mass ratio, unlike a previous assumption [6, 7]. We remark also that the (A^0, X) complex becomes more stable for σ-values less than 0.69. Yet, the above conclusions depend on the validity of the perturbation theory. For example the condition $\Delta E/|E| < 0.1$ is realized if $\gamma_A < 0.1$ for $\sigma = 0.5$ and if $\gamma_A < 0.4$ for $\sigma = 10$.

In order to extend somewhat the validity range of the previous results, we have determined variationally the eigenvalues of the Hamiltonian (1) using the wave function (8) but by varying only the linear parameters c_{mnpq}. The corresponding energy shifts become then

$$\Delta E_{\text{var}}(\gamma_A, \sigma) = \gamma_A^2 |E_A| F_{\text{var}}(\gamma_A, \sigma) . \tag{10}$$

In Fig. 2 we have reported the variational as well as the perturbation energies with respect to the first generalized Landau level $E_L = \gamma_A(2 + \sigma^{-1}) |E_A|$ of the complex. In Fig. 3 we compared the (A^0, X) transition energy shift $\Delta h\nu = \Delta E - \Delta E_A$ to that of a free exciton, $\Delta h\nu_X = \Delta E_X$ and that of the effective mass (A^0, X) pseudo-donor model, $\Delta h\nu = \Delta E_{A^+} + \Delta E_D - \Delta E_A$. To determine the energy shifts of the exciton, the donor, and the acceptor impurities, we have used the highly accurate results of Cabib et al. [17] as well as those of Larsen [18] for that of the positively charged acceptor A^+. It appears that, in the effective mass approximation, the exciton and the bound exciton shifts may be equal only in some particular circumstances: i.e. $\sigma \approx 0.5$. Moreover, the bound-exciton shifts increase with the effective magnetic field and take

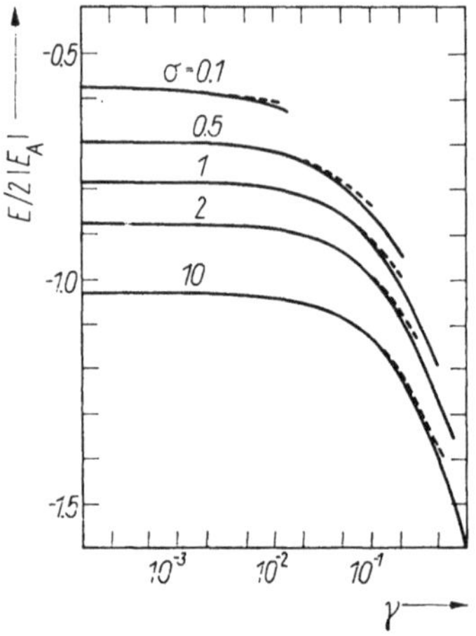

Fig. 2

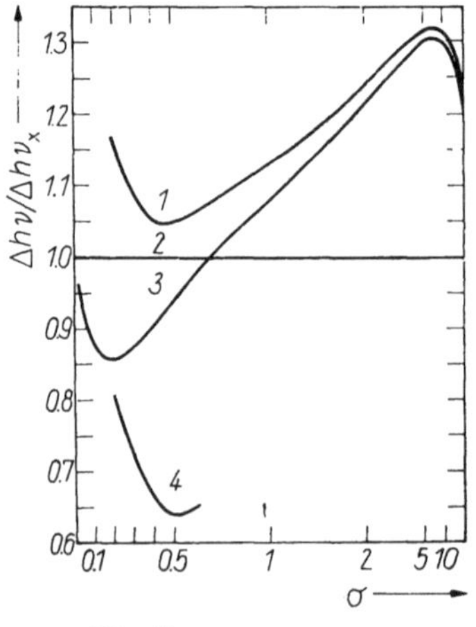

Fig. 3

Fig. 2. Energies of the (A^0, X) complex obtained by the variation method (solid lines) and the perturbation method (dashed lines). The zero of energy is at the first Landau level. γ is the effective magnetic field

Fig. 3. Comparison between the theoretical transition energy shifts of the (A^0, X) complex ($\Delta h\nu = \Delta E(A^0,X) - \Delta E_A$) obtained by different methods ((1) variational, $\gamma = 0.1$; (3) perturbation) and those of the free exciton ($\Delta h\nu_X = \Delta E_X$). Curves 2 and 4 refer, respectively, to the theoretical prediction of Willmann et al. [7] and to the "effective mass" pseudo-donor model ($\Delta h\nu = \Delta E_{A^+} + \Delta E_D - \Delta E_A$; $\gamma = 0.1$)

higher values for the (D^0, X) complex than for the (A^0, X) complex in most of the semiconductors.

3. Comparison with Experiment

It may be questionable to compare the results of our present rather crude model with the experimental observations. Nevertheless, in the past, different attempts [6, 7, 11] have been made to interpret the observed transition shifts in Sn-doped GaAs. The theoretical prediction of Willmann et al. [7] is based upon the deuteron model which does not depend on the mass ratio $\sigma = m_e^*/m_h^*$ and is in agreement with the present theory (Fig. 1) only for $\sigma \approx 0.5$ which corresponds roughly to the case of GaAs. From the results shown in Table 2, it can be seen that the agreement between

Table 2

Comparison between theoretical and experimental diamagnetic shifts for the (A^0, X) complex in GaAs:Sn (1.507 eV) [6] and ZnTe:Cu (2.3750 eV) [19 to 22] and the (D^0, X) complex in InP [23] for an effective magnetic field of $\gamma = 0.1$

complex	substance	m_e^*/m_0	γ_1	m_0/m_h^*	ε	B (T)	ΔE_X (meV)	ΔE_D (meV)	$\Delta h\nu^{th}$ (meV)	$\Delta h\nu^{exp}$ (meV)
(A^0, X)	GaAs	0.0665 [8]	7.65 [8]		12.56 [8]	2.54	1.04	0.38	1.09	0.39 [6, 7]
(A^0, X)	ZnTe	0.16 [20]	3.93 [20]		10.1 [20]	14.92	2.30	0.64	2.45	1.61 [20, 22]
(D^0, X)	InP	0.08 [14]	6.28 [14]		12.4 [14]	0.98	0.114		0.142	0.075 [23]
				2.02 [23]	12.1 [23]	10.3	0.057		0.078	

theory and experiment is only qualitative, mainly because the effective mass theory is not well adapted to deep impurities, although the shifts observed with different impurities seem to be the same [7]. Indeed, in this approximation, the mean radii of the orbitals are overestimated, so that the theoretical diamagnetic shifts are too large. On the other hand, for deep impurities, the pseudo-donor model seems to be a good approximation, because in this case the transition energy shift $\Delta h\nu$ may be identified with the donor energy shift ΔE_D. But this latter shift cannot be identified with the effective mass pseudo-donor transition energy shift because the contribution $\Delta E_{A^*} - \Delta E_A \approx 0.30$ meV is not negligible. In the case of ZnTe:Cu [19 to 22] where the Cu-acceptor impurity is less deep, we get a better agreement with the experimental observations [20 to 22]. The remaining discrepancy may be due to the fact the valence bands of GaAs and ZnTe are degenerate at the centre of the Brillouin zone and cannot be accurately described by a simple isotropic hole mass parameter. This can be seen when we compare the results obtained for the (D^0, X) complex in InP (Table 2). By using $m_h^* = 1/\gamma_1$, the results are not very satisfactory but with an experimental "mean" hole mass [23] we find a good agreement between the theoretical and experimental diamagnetic shifts. This fact shows that the degeneracy effects may be very important. On the other hand, it seems necessary to extend the present low-field theory to higher values of the effective magnetic field.

Acknowledgements

We would like to thank Prof. C. Klingshirn, Dr. W. Maier, and Dr. G. Schmieder for giving us information about their results prior to publication.

References

[1] D. G. THOMAS and J. J. HOPFIELD, Phys. Rev. **128**, 2135 (1962).
[2] A. BALDERESCHI, in: Theoretical Aspects and New Developments in Magneto-Optics, Ed. J. T. DEVREESE, Plenum Press, New York/London 1980 (p. 291).
[3] J. ADAMOWSKI, S. BEDNAREK, and M. SUFFCZYNSKI, J. Phys. C **12**, L325 (1979).
[4] B. STÉBÉ and G. MUNSCHY, Solid State Commun. **35**, 557 (1980).
[5] T. D. CLARK, K. K. BAJAJ, W. M. THEIS, and D. E. PHELPS, phys. stat. sol. (b) **110**, 341 (1982).
[6] W. DREYBRODT, F. WILLMANN, M. BETTINI, and E. BAUSER, Solid State Commun. **12**, 1217 (1973).
[7] F. WILLMANN, W. DREYBRODT, and M. BETTINI, Phys. Rev. B **8**, 2891 (1973).
[8] D. D. SELL, S. E. STOKOWSKI, R. DINGLE, and J. V. DILORENZO, Phys. Rev. B **7**, 4568 (1973).
[9] E. J. JOHNSON and H. Y. FAN, Phys. Rev. **139**, A1991 (1965).
[10] W. RÜHLE and D. BIMBERG, Phys. Rev. B **12**, 2385 (1975).
[11] W. SCHAIRER, D. BIMBERG, W. KOTTLER, K. CHO, and W. SCHMIDT, Phys. Rev. B **13**, 3452 (1976).
[12] A. M. WHITE, P. J. DEAN, B. DAY, and F. MANSFIELD, phys. stat. sol. (b) **74**, K9 (1976).
[13] B. STÉBÉ and G. MUNSCHY, Solid State Commun. **40**, 663 (1981).
[14] N. D. LIPARI and M. ALTARELLI, Solid State Commun. **33**, 47 (1980).
[15] P. LAWAETZ, Phys. Rev. B **4**, 3460 (1971).
[16] A. BALDERESCHI and A. LIPARI, Phys. Rev. B **9**, 1525 (1974).
[17] D. CABIB, E. FABRI, and G. FIORIO, Solid State Commun. **9**, 1517 (1971).
[18] D. LARSEN, Phys. Rev. B **20**, 5217 (1979).
[19] H. VENGHAUS and P. J. DEAN, Phys. Rev. B **21**, 1596 (1980).
[20] W. MAIER, Ph. D. Thesis, Karlsruhe (FRG) 1982.
[21] C. KLINGSHIRN, Lectures Notes in Physics **177**, 214 (1983).
[22] G. SCHMIEDER, W. MAIER, P. L. LIU, and C. KLINGSHIRN, J. Lum. **28**, 357 (1983).
[23] W. RÜHLE and W. KLINGENSTEIN, Phys. Rev. B **18**, 7011 (1978).

(Received February 16, 1984)

phys. stat. sol. (b) 126, 335 (1984)

Subject classification: 2 and 13

Sektion Physik der Technischen Universität Dresden[1])

Non-Exponential Localization in an Inhomogeneously Disordered Chain

By

R. HAYN and W. JOHN

The behaviour of localized states is studied for a semi-infinite chain with decreasing amount of disorder along the chain. It is shown that for slowly decreasing disorder all states are localized. In contrast to the case of homogeneous disorder, the states show non-exponential localization and may be characterized by a generalized Lyapunov-coefficient.

Das Verhalten der lokalisierten Zustände in einer halbunendlichen Kette mit abnehmender Unordnung entlang der Kette wird untersucht. Es wird gezeigt, daß für langsam abnehmende Unordnung alle Zustände lokalisiert sind. Im Gegensatz zu einer Kette mit homogener Unordnung sind die Zustände jedoch nicht exponentialartig lokalisiert. Ihre Charakterisierung erfolgt durch einen verallgemeinerten Lyapunov-Koeffizienten.

1. Introduction

Now it is well known that the spectrum of a wide class of one-dimensional disordered systems is a pure dense-point spectrum and the corresponding eigenstates are exponentially localized [1]. However, this statement is valid only for homogeneous systems. In the following we study the behaviour of states in an inhomogeneously disordered semi-infinite chain where the size of disorder decreases along the chain. The motivation for our consideration is as follows. Any D-dimensional system may be transformed using the recursion method in an equivalent chain model [2]. Starting from a given state $|u_0\rangle$ we obtain from successive operations with the Hamiltonian H,

$$b_{n+1}|u_{n+1}\rangle = H|u_n\rangle - a_n|u_n\rangle - b_n|u_{n-1}\rangle\,, \tag{1}$$

a new normalized and orthogonal basis set $|u_n\rangle$. The coefficients a_n and b_n are given by

$$a_n = \langle u_n|\,H\,|u_n\rangle\,, \qquad b_n = \langle u_{n-1}|\,H\,|u_n\rangle\,, \qquad b_0 = 0\,. \tag{2}$$

Thus, in the new basis set $|u_n\rangle$ the Hamiltonian has a particularly simple form of a tri-diagonal matrix, which corresponds to a semi-infinite tight-binding chain model with site energies a_n and hopping matrix elements b_n between nearest neighbours. The eigenstate of the equivalent chain model

$$|\psi\rangle = \sum_{n=0}^{\infty} \psi_n|u_n\rangle \tag{3}$$

may be obtained from the recursion relation

$$b_{n+1}\psi_{n+1} = (E - a_n)\,\psi_n - b_n\psi_{n-1}\,. \tag{4}$$

Any localized state in the original D-dimensional lattice corresponds to a localized state in the equivalent chain model, because the recursion method realizes a unitary transformation. The essential information needed from the recursion procedure to study the spectrum of the system is only the set of coefficients a_n and b_n.

[1]) Mommsenstr. 13, DDR-8027 Dresden, GDR.

For a disordered system the coefficients a_n and b_n are random quantities and may be calculated only numerically. Nevertheless the recursion method is very effective in calculating the local density of states, because a limited number of coefficients is sufficient to reproduce the density of states fairly well. As opposed to that the asymptotic behaviour of the recursion coefficients a_n and b_n for large n must be known in order to study the localization problem [3 to 7]. For the Bethe lattice Haydock [3] using perturbation theory has shown that the mean value of the recursion coefficients converges with $n \to \infty$ and the mean-square deviations $(\Delta a_n)^2$ and $(\Delta b_n)^2$ behave like

$$(\Delta a_n)^2 \sim (\Delta b_n)^2 \sim Z_n\,, \tag{5}$$

where Z_n is the number of lattice sites in the n-th shell of the Bethe lattice. For real lattices only a very limited number of recursion coefficients can be calculated from numerical reasons. Stein and Krey [5] have calculated the recursion coefficients for the square lattice up to $n = 64$ and for the diamond lattice up to $n = 32$. Their numerical results suggest an asymptotic behaviour of the mean-square deviation as

$$(\Delta a_n)^2 \sim (\Delta b_n)^2 \sim n^{-\alpha} \tag{6}$$

with $\alpha \approx D - 1$ for small disorder. This result may be understood at least qualitatively by simple structural arguments [6, 7]. Using a limited number of recursion coefficients Stein and Krey have studied the localization problem numerically by an exact real space renormalization procedure [5]. However, as far as we know, no analytical results have been obtained till now for the localization problem using the recursion method. As a first step in this direction we consider a simple model with a given probability distribution function for the recursion coefficients. We assume that there are no statistical correlations between the recursion coefficients at different sites in the chain model for large n. Furthermore we consider only site-diagonal disorder with a power-law behaviour for the mean-square deviation like (6). Recently Simon [8] has shown that for slowly decreasing disorder ($\alpha < 1$) the non-homogeneous chain has only a point spectrum. In the following we demonstrate that the states within the band of the corresponding ordered system are not exponentially localized for $\alpha < 1$ and show a continuous transition to power-law localization for $\alpha = 1$.

The organization of the paper is as follows. In Section 2 we describe the models studied in this paper. In Section 3 we introduce a generalized Lyapunov coefficient which describes in a proper way the non-exponential growth of particular solutions for the non-homogeneous chain model. The generalized Lyapunov coefficient is calculated analytically, and the results are compared with numerical calculations. In Section 4 the behaviour of eigenstates is discussed. The Thouless formula for the localization length is generalized to the non-homogeneous chain, too. For $\alpha = 1$ we find power law localization and a localization edge.

2. Model

We consider in the following an inhomogeneously disordered chain and take into account only site-diagonal disorder,

$$H = \sum_{n=0}^{\infty} \{\varepsilon_n |n\rangle\langle n| + |n\rangle\langle n+1| + |n+1\rangle\langle n|\}\,. \tag{7}$$

The site energies ε_n in (7) are random variables with a probability distribution function $P_n(\varepsilon_n)$ which depends explicitly on the site n. Especially, we study two models which are characterized by different probability distribution functions $P(\varepsilon)$.

Model A: Lorentzian distribution function

$$P_{l\Gamma_n}(\varepsilon_n) = \frac{\Gamma_n}{\pi(\varepsilon_n^2 + \Gamma_n^2)}\,; \qquad \Gamma_n = \Gamma n^{-\alpha}\,, \tag{8}$$

with a width Γ_n which falls off according to a power law with increasing n.

Model B: Any probability distribution function $P_n(\varepsilon_n)$ with finite moments and a mean-square deviation σ_n which decreases along the chain,

$$\sigma_n = \langle \varepsilon_n^2 \rangle = \sigma n^{-\alpha}; \qquad \langle \varepsilon_n \rangle = 0 . \tag{9}$$

Finally (see Section 3.3) we generalize the results that we have obtained for these two models to a more general class of models.

As a special case for $\alpha = 0$ the models A and B describe a homogeneous disordered chain which shows only exponentially localized states. If α is sufficiently large, only a finite number of sites is really perturbed at the beginning of the chain and we expect extended states. Therefore, there must be a transition between localized and extended states if α is increased. For the critical value of α where the transition occurs, we expect $\alpha = 1$ because for $\alpha > 1$ the sum of the mean-square deviations (9) $\sum_n^{\infty} \sigma_n$ is finite.

3. Generalized Lyapunov Coefficient

3.1 Model A

Let us start our investigation of the non-homogeneous chain (7) by considering the growth of particular solutions ψ_n of the stationary Schrödinger equation

$$\psi_{n+1} = (E - \varepsilon_n)\, \psi_n - \psi_{n-1}\,, \qquad \psi_0 = 0\,, \qquad \psi_1 = 1 \tag{10}$$

within the band of the corresponding ordered system ($E^2 < 4$). Introducing the state ratio

$$Z_{n+1} = \frac{\psi_{n+1}}{\psi_n}\,, \tag{11}$$

we obtain from(10) the recurrence relation

$$Z_{n+1} = (E - \varepsilon_n) - \frac{1}{Z_n}. \tag{12}$$

In the case of model A where ε_n is a random variable with a Lorentzian distribution function $P_{i\Gamma_n}(\varepsilon_n)$, Z_n also belongs to a Lorentzian distribution $P_{s_n}(Z_n)$ with s_n given by [1]

$$s_{n+1} = E + i\Gamma_n - \frac{1}{s_n}. \tag{13}$$

From the probability distribution $P_{s_n}(Z_n)$ the growth of the wave function may be calculated by

$$\langle \ln |\psi_N| \rangle = \sum_{n=1}^{N} \int \mathrm{d}Z_n \ln |Z_n|\; P_{s_n}(Z_n) \tag{14}$$

$$= \sum_{n=1}^{N} \ln |s_n| \,. \tag{15}$$

Because we are interested in the asymptotic behaviour of the wave function for large n where Γ_n (8) is small, we may assume that Γ_n changes slowly enough, so that s_n can follow the change of Γ_n. With $s_{n+1} \approx s_n$ we obtain from (13) for the asymptotic behaviour for large n

$$s_n = \frac{E + i\Gamma_n}{2} + \sqrt{\left(\frac{E + i\Gamma_n}{2}\right)^2 - 1}\,. \tag{16}$$

In Section 3.3 the validity of the approximation (16) will be discussed in more detail. Taking into account the leading n-dependent terms we get for $E^2 < 4$ and $\alpha \leqq 1$

$$\langle \ln |\psi_N| \rangle = \frac{\Gamma}{\sqrt{4 - E^2}} \sum_{n=1}^{N} n^{-\alpha} \tag{17}$$

$$= \frac{\Gamma}{\sqrt{4 - E^2}} \frac{N^{1-\alpha} - 1}{1 - \alpha}; \qquad \alpha \leqq 1 . \tag{18}$$

From (18) it follows that the amplitude of the wave function increases for $\alpha \leqq 1$. Moreover, the behaviour of the amplitude changes continuously from an exponential growth ($\alpha = 0$) to a power-law growth ($\alpha = 1$). Indeed, using the identity $\ln N = \lim_{x \to 0} (N^x - 1)/x$ from (18) follows for $\alpha = 1$

$$\langle \ln |\psi_N| \rangle = \frac{\Gamma}{\sqrt{4 - E^2}} \ln N . \tag{19}$$

For $\alpha > 1$ we find from (18) that the amplitude of the wave function remains finite, which indicates the existence of extended states.

Until now, we have considered states within the band of the corresponding ordered system ($E^2 < 4$). For $E^2 > 4$ the asymptotic behaviour of the wave function is determined by the corresponding ordered system. From (15) and (16) follows in this case for any value of $\alpha \neq 0$ and $N \to \infty$

$$\langle \ln |\psi_N| \rangle = N \ln \frac{E + \sqrt{E^2 - 4}}{2} . \tag{20}$$

The disorder contributes to the asymptotic behaviour of the wave function by terms which may be neglected in comparison with (20). Therefore, the amplitude increases exponentially for $E^2 > 4$. Exactly at the band edge $|E| = 2$ of the ordered system for the asymptotic behaviour of the wave function from (15) and (16) follows

$$\langle \ln |\psi_N| \rangle = \sqrt{2\Gamma} \frac{N^{1-\alpha/2} - 1}{1 - \alpha/2}; \qquad \alpha \leqq 2 . \tag{21}$$

3.2 Perturbation theory

Qualitatively the same results for the asymptotic behaviour of the wave function as (18) to (20) may be derived for the case of model B. For this purpose we use the perturbation theory for the Green function as in the case of homogeneous disorder [9]. The eigenvalues E_ν of a finite chain of $N - 1$ sites are given by the zeros of $\psi_N(E)$ with the initial condition $\psi_0 = 0$, $\psi_1 = 1$. Therefore, from (10) follows [1]

$$\ln |\psi_N(E)| = \sum_{\nu=1}^{N-1} \ln |E - E_\nu| . \tag{22}$$

Using the Green function G_N of the chain with $N - 1$ sites, we obtain from

$$\operatorname{Tr} G_N(E + i\varepsilon) = \sum_{\nu=1}^{N-1} (E + i\varepsilon - E_\nu)^{-1} \tag{23}$$

and (22) the relation

$$\frac{\mathrm{d}}{\mathrm{d}E} \langle \ln |\psi_N(E)| \rangle = \operatorname{Re} \operatorname{Tr} \langle G_N(E + i\varepsilon) \rangle ; \qquad \varepsilon \to 0 . \tag{24}$$

Since we are interested in the asymptotic behaviour of $\ln |\psi_N|$, we substitute in (24) G_N by the Green function G of the infinite system and obtain finally

$$\frac{\mathrm{d}}{\mathrm{d}E} \langle \ln |\psi_N(E)| \rangle = \mathrm{Re} \sum_{n=1}^{N} \langle G_{nn}(E + i\varepsilon) \rangle ; \qquad \varepsilon \to 0 . \tag{25}$$

Because for $n \to \infty$ the disorder (9) is weak, we evaluate (25) by perturbation theory. Up to second order we obtain from (25) for $E^2 < 4$,

$$\frac{\mathrm{d}}{\mathrm{d}E} \langle \ln |\psi_N(E)| \rangle = \frac{\mathrm{d}}{\mathrm{d}E} \frac{\sigma}{2(4 - E^2)} \sum_{n=1}^{N} n^{-\alpha} . \tag{26}$$

Taking into account the behaviour of $\langle \ln |\psi_N| \rangle$ for $|E| \to \infty$ (20) we get for $\alpha \leqq 1$

$$\langle \ln |\psi_N(E)| \rangle = \frac{\sigma}{2(4 - E^2)} \frac{N^{1-\alpha} - 1}{1 - \alpha} , \tag{27}$$

which may be compared with (18).

However, it must be mentioned that higher-order terms of the perturbation series diverge for special energies. For instance, for $E = 0$ the 4th order of perturbation theory gives a term which is proportional to $\sigma^2(N^{1-\alpha} - 1)^2$. Of course the same situation occurs for the homogeneous chain. Detailed investigations by Kappus and Wegner [10] show that the perturbation theory breaks down at zero energy. Nevertheless second-order perturbation theory gives reasonable results for the growth of the wave function also for $E = 0$.

The growth of the wave function for $\alpha \leqq 1$ and $E^2 < 4$ (18) and (27) may be characterized for both models A and B by a generalized Lyapunov coefficient,

$$\tilde{\gamma}(E) = \lim_{N \to \infty} \frac{1 - \alpha}{N^{1-\alpha} - 1} \langle \ln |\psi_N(E)| \rangle , \tag{28}$$

which is independent of α and corresponds for $\alpha = 0$ to the usual Lyapunov exponent $\gamma(E)$ [1].

Fig. 1 shows as an example the numerical results of $\langle \ln |\psi_N| \rangle$ for $E = 1$ and different values of α. The calculations are performed for a rectangular probability distribution function $P_n(\varepsilon_n)$ with a mean-square deviation (9) $\langle \varepsilon_n^2 \rangle = 25/(12n^\alpha)$. The ensemble average is calculated from 40 equivalent chains. As may be seen, the agreement of the numerical calculations with the result of perturbation theory, given by the solid line, is satisfactory.

As mentioned above, the perturbation theory breaks down at zero energy. Numerically we find for $E = 0$ that the generalized Lyapunov coefficient is about 20% smaller than the corresponding result of perturbation theory (cf. (27)).

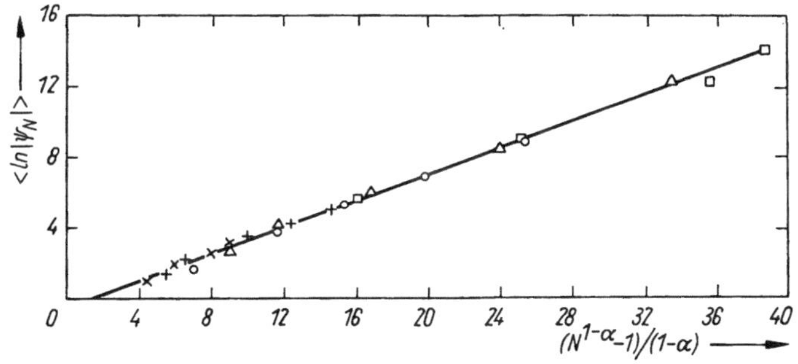

Fig. 1. Growth of the amplitude of the wave function for model B with $\langle \varepsilon_n^2 \rangle = 25/(12n^\alpha)$. × $\alpha = 0.50$, + 0.45, ○ 0.40, △ 0.35, □ 0.30

3.3 *Generalization*

Certainly, our results for the behaviour of the wave function are valid also for more general models of disorder, including non-diagonal disorder. Let us assume that in the case of weak disorder the Lyapunov exponent $\gamma(E)$ of a homogeneous chain is given by

$$\gamma(E, A) = \lim_{N\to\infty} \frac{1}{N} \langle \ln |\psi_N| \rangle = AB(E) + O(A^2) , \tag{29}$$

where A characterizes the amount of disorder ($A = 0$ corresponds to the ordered system). Furthermore we only consider energies within the band of the ordered system. If the parameter of the disorder A decreases along the chain like $A_n = \sigma n^{-\alpha}$ we get an inhomogeneously disordered chain. In analogy to the discussion in Section 3.1 we approximate the inhomogeneous system by an array of homogeneous chains and calculate the growth of the wave function by

$$\langle \ln |\psi_N(E)| \rangle = \sum_{n=1}^{N} A_n B(E) . \tag{30}$$

The approximation (30) for the asymptotic behaviour of the wave function is valid, if the change of $A_nB(E)$ within an interval $\Delta n \gg [A_nB(E)]^{-1}$ is smaller than $A_nB(E)$. Thus, we obtain the inequality

$$n \gg \frac{\alpha n^{\alpha}}{\sigma B(E)} , \tag{31}$$

which is certainly fulfilled for $n \to \infty$ and $\alpha < 1$. Therefore, the generalized Lyapunov coefficient of an inhomogeneous chain with $\alpha < 1$ may be calculated from (30), if the Lyapunov exponent of the corresponding chain with weak homogeneous disorder is known. Comparing (30) and (28) we obtain for the generalized Lyapunov coefficient

$$\tilde{\gamma}(E) = \sigma B(E) . \tag{32}$$

For $\alpha > 1$ inequality (31) is certainly not valid for $n \to \infty$. Our numerical calculations show for $\alpha > 1$ oscillating wave functions as for the corresponding ordered system. In contrast to this result, from (18) and (30) follows that $\langle \ln |\psi_N| \rangle$ reach a constant value for $N \to \infty$.

4. Localization of Eigenstates

From the behaviour of the wave function (18), (27) we conclude that for $\alpha < 1$ all eigenstates are localized. For the envelope of the eigenstate $\psi_{n\nu}$ which is localized at site n_ν we make use of the ansatz

$$|\psi_{n\nu}| \sim \exp\left(- \tilde{\gamma}_\nu \frac{|n^{1-\alpha} - n_\nu^{1-\alpha}|}{1 - \alpha} \right), \tag{33}$$

where the degree of localization $\tilde{\gamma}_\nu$ is given by

$$\tilde{\gamma}_\nu = - \lim_{N\to\infty} \frac{1 - \alpha}{N^{1-\alpha} - 1} \ln |\psi_{1\nu}\psi_{N\nu}| . \tag{34}$$

Intuitively it may be expected that the degree of localization is given by the generalized Lyapunov coefficient $\tilde{\gamma}(E)$.

Following the arguments of Thouless [11] for the homogeneous chain we obtain quite analogically for the inhomogeneously disordered chain

$$\tilde{\gamma}_\nu = \lim_{N\to\infty} \frac{1-\alpha}{N^{1-\alpha}-1} \sum_{\substack{\mu \\ (\mu \neq \nu)}} \ln |E_\nu - E_\mu| \,, \tag{35}$$

which is identical with $\tilde{\gamma}$ (28), (22) if we neglect the restriction on the summation in (35).

Therefore, we conclude that the behaviour of eigenstates for $E^2 < 4$ changes continuously from exponential localization for homogeneous disorder ($\alpha = 0$) to power-law localization for $\alpha = 1$. For the special case $\alpha = 1$, where the asymptotic behaviour of the wave function (33) is given by

$$|\psi_{n\nu}| \sim \left(\frac{n_\nu}{n}\right)^{\pm\tilde{\gamma}_\nu} \quad \text{for} \quad n \gtrless n_\nu \,, \tag{36}$$

the states are normalized only for $2\tilde{\gamma}_\nu \gtrless 1$. Therefore, using (32) we find for $\sigma \leqq 4$ localization edges E_c given by

$$E_c = \pm \sqrt{4 - \sigma} \,. \tag{37}$$

For $E^2 > 4$ the states are exponentially localized for any value of $\alpha \neq 0$ and the corresponding Lyapunov exponent is given by (20).

Summarizing our results we have shown that in a chain model with slowly decreasing disorder $\alpha < 1$ — where the usual Lyapunov exponent is zero — all states are localized. Only in the special case $\alpha = 1$ we find a localization edge and power-law localization. Of course, due to simplifying assumptions on the probability distribution function of the chain parameters, the relevance of our results for the D-dimensional case is an open question.

References

[1] K. Ishii, Progr. theor. Phys. (Tokyo), Suppl. **53**, 77 (1973).
[2] R. Haydock, J. Phys. A 7, 2120 (1974).
[3] R. Haydock, Phil. Mag. **B37**, 97 (1978).
[4] D. Weaire and C. H. Hodges, J. Phys. C **12**, L375 (1979).
[5] J. Stein and U. Krey, Z. Phys. **B34**, 287 (1979).
[6] R. Haydock, Phil. Mag. **B43**, 203 (1981).
[7] T. A. L. Ziman, Phys. Rev. B **26**, 7066 (1982).
[8] B. Simon, Commun. math. Phys. **87**, 253 (1982).
[9] D. J. Thouless, in: Ill-Condensed Matter, Ed. R. Balian, R. Maynard, and G. Toulouse, North-Holland, Amsterdam/New York 1979 (p. 31).
[10] M. Kappus and F. Wegner, Z. Phys. **B45**, 15 (1981).
[11] D. J. Thouless, J. Phys. C **5**, 77 (1972).

(Received August 16, 1984)

phys. stat. sol. (b) **126**, 343 (1984)

Subject classification: 13.4 and 19; 22.4.1

Sektion Physik der Humboldt-Universität zu Berlin[1]), *Applikationsgruppe*

Optically Induced EPR of Conduction and and Donor Band Electrons in Zinc Sulphide

By

D. BACKS and A. HERTRAMPF

Conventional EPR of mobile electrons as well as optical EPR excitation and quenching spectra of variously doped ZnS single crystals are reported. The different kinetic behaviour and the EPR parameters of the resonance lines are discussed involving donor and conduction band electrons. From the optical EPR excitation and quenching spectra and from ac-microwave photoconductivity measurements a simple band model is proposed for the optical transitions.

Es werden konventionelle EPR-Untersuchungen an beweglichen Elektronen sowie optische EPR-Anregungs- und Tilgungsspektren von verschiedenen dotierten ZnS-Einkristallen vorgestellt. Das unterschiedliche kinetische Verhalten und die EPR-Parameter werden für Donator- und Leitungsbandelektronen diskutiert. Ein einfaches Bandmodell für die optischen Übergänge wird unter Berücksichtigung der EPR-Anregungs- und Tilgungsspektren und ac-Mikrowellen-Photoleitfähigkeitsuntersuchungen vorgeschlagen.

1. Introduction

In our investigation of optically excited EPR of lead-doped ZnS we used a model involving the existence of donor and conduction band electrons [1, 2]. In this paper we want to present optically excited EPR of conduction band electrons (CBE) and donor band electrons (DBE) in a lot of variously doped ZnS single crystals to confirm the A-centre model and the optical excitation processes.

Beside the EPR of CBE and DBE in lead-doped ZnS we made investigations of mobile electrons in Zn–Al-, I-, and Ag-doped ZnS. In some cases we are able to compare our measurements with results published elsewhere [3, 4].

The great importance of donor and conduction band electrons is based on their participitation in optical processes (D–A recombination or photoconductivity). The experimental proof of the connection of weakly bound electrons and D–A recombination transitions was given in a series of ODMR investigations (e.g. [5 to 8].)

But there is still another important aspect in the physical behaviour of DBE and CBE in ZnS or ZnSe. With the systematic investigation of the manganese impurity in materials used as active layers in ac driven thin film electroluminescence (ACTFEL) structures, an interaction of manganese and donor impurities was observed, resulting in an influence of the manganese EPR spectra [9] on the ODMR of donor states in ZnSe [10] and on optical transitions in ZnS [11].

In this paper we present our EPR investigations on donor and conduction band electrons in ZnS in connection with a discussion of optical excitation and quenching spectra (Section 4) and some first results on the kinetics of these centres. In a forthcoming paper we will give some detailed results on the kinetics and the temperature dependence of the EPR of mobile electrons in zinc sulphide [12].

[1]) Invalidenstraße 110, DDR-1040 Berlin, GDR.

2. Samples and Experiments

All the samples used and described here were grown by using an iodine transport technique with the doping elements added in the growing process or by subsequent diffusion. Due to the relatively low growing temperatures (850 to 1000 °C) crystals of pure cubic or polytype structures were obtained. This could be shown by measuring the manganese EPR spectra, observed in all the crystals as a natural impurity and therefore used as a very sensitive probe to determine crystal structures [13 to 15].

For the optical excitation of the EPR centres a 500 W high-pressure mercury lamp in connection with a 10 cm water filter and a colour glass UG 5 as well as quartz lenses were used. EPR spectra, with the exception of excitation and quenching curves, were measured by using a Varian E-4 spectrometer in combination with a magnetometer MJ 110R (Radiopan, Poland) for g-factor determination.

For measuring the EPR excitation spectra we used a 150 W xenon high-pressure lamp and a high-intensity Bausch & Lomb monochromator. The experimental procedure was relatively simple, because of the fast relaxation of DBE and CBE (fast compared to other optically excited EPR centres in ZnS, as A-centres, transition metal ions, or $^2S_{1/2}$-centres). So a continuous registration of the excitation spectra was possible, without using a two-beam method, as was necessary for other centres [2] or for the quenching measurements. In the latter case a 200 W high-pressure mercury lamp was used for the excitation of DBE and CBE and a 150 W Varian xenon lamp was taken to record the quenching spectra.

EPR registration of the optical excitation and quenching spectra was done by using a home-made straight-line spectrometer with commercial constituents and a modulation frequency of 63 kHz. A block diagram of this simple experimental arrangement is shown in Fig. 1. The cavity is a JES-UCV-2 with four rods for the field modulation and a quartz window for optical illumination.

In some cases ac-microwave photoconductivity spectra were taken to compare them with the EPR excitation spectra. A detailed description of the technique used here is given in [16].

3. Experimental Results

EPR of donor and conduction band electrons was measured for a series of variously doped ZnS single crystals. Spectra of mobile electrons were found in Al–Zn-, I-, Pb-, Ge-, and Ag-doped crystals. The group-IV elements and silver are important only to distinguish between the different samples — there is no immediate influence of these elements on the DBE and CBE spectra. The EPR spectra measured will be given in the following section, completed by excitation and quenching spectra and photoconductivity as far as possible.

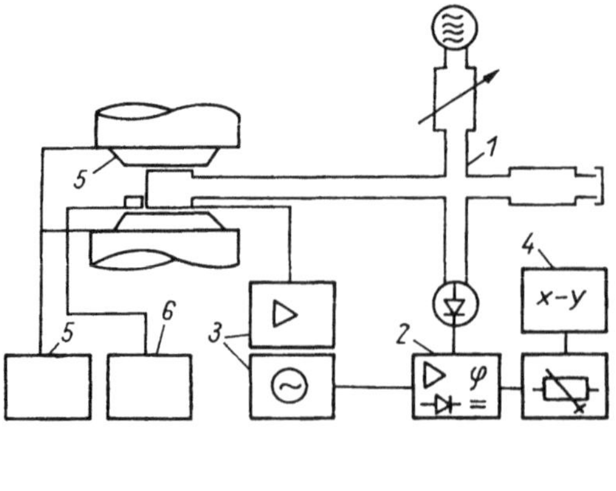

Fig. 1. Block diagram of the simple home-made straight-line spectrometer, consisting of a 1 microwave bridge, 2 lock-in amplifier, 3 radio frequency field modulation, 4 x–y recorder, 5 electromagnet and its current source, and a 6 NMR oscillator

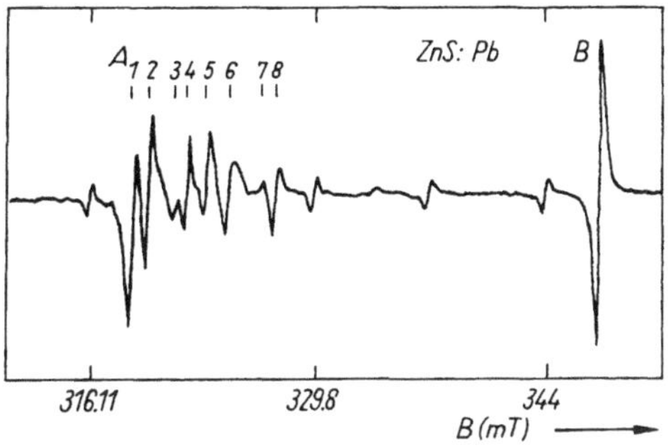

Fig. 2

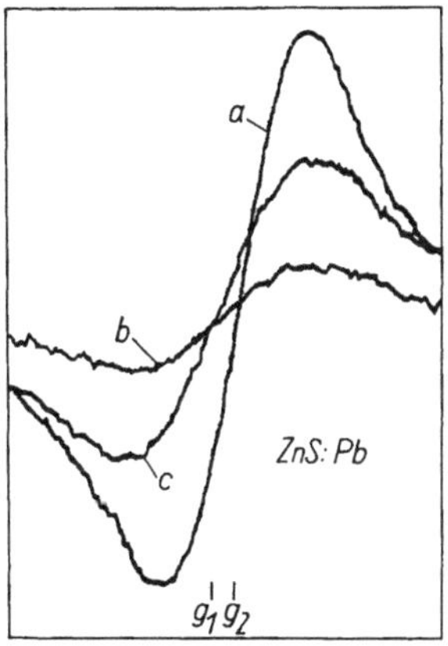

Fig. 3

Fig. 2. EPR spectrum of the lead A-centre (resonance lines A1 to A8) and the resonance line of mobile electrons on the high-field side of the spectrum (line B). The lines which are not labelled are due to Mn^{2+} and Cr^{+}. The spectrum was taken at 77 K with UV excitation and arbitrary orientation of the crystal with respect to the magnetic field

Fig. 3. g-factor deviation and variation in the linewidth of the EPR line of mobile electrons in ZnS:Pb, I. (a) With full UV excitation, (b) 10 s, and (c) 30 s after switching off the UV excitation. Curve c was recorded with a higher gain

3.1 Donor and conduction band electrons in ZnS:Pb, I

The EPR spectra of the lead A-centre and the DBE and CBE are given in Fig. 2. In this configuration there is an arbitrary orientation of the crystal to the magnetic field, with no loss of information due to the isotropic behaviour of the DBE and CBE resonance lines. Besides the optically active centres the hyperfine structure of Mn^{2+} can be seen in Fig. 2.

In accordance with [3] the resonance line on the high-field side of the spectrum proves to be a superposition of two resonances, which can be distinguished by their g-factors and linewidths. An experimental separation of these two lines can be done by varying the excitation light intensity. In Fig. 3 the variation of the line position and halfwidth is shown in dependence on optical excitation and registration time. Curve a was taken with full UV excitation, curve b 10 s, and curve c 30 s after the light had been switched off. The observation of the broad line is already possible in the dark.

Two effects may be seen from Fig. 3 immediately. The first is a g-factor deviation and the second a broadening of the resonance line. As will be discussed in Section 4 the line with the lower g-factor and the smaller linewidth is due to conduction band electrons (g_2) and the other line to donor band electrons (g_1). The measured g-factors and halfwidths are:

$$g_1 = 1.8850\,, \qquad g_2 = 1.8846\,,$$
$$\Delta B_1 = 0.69\ \text{mT}\,, \qquad \Delta B_2 = 0.49\ \text{mT}\,.$$

In other samples (doped with lead, but no observation of A-centre) only the resonance line of CBE was observed. The g-factor was $g_2 = 1.8839$, the halfwidth $\Delta B_2 = 0.49$ mT. The conclusion, that this resonance is caused by conduction band electrons only, can

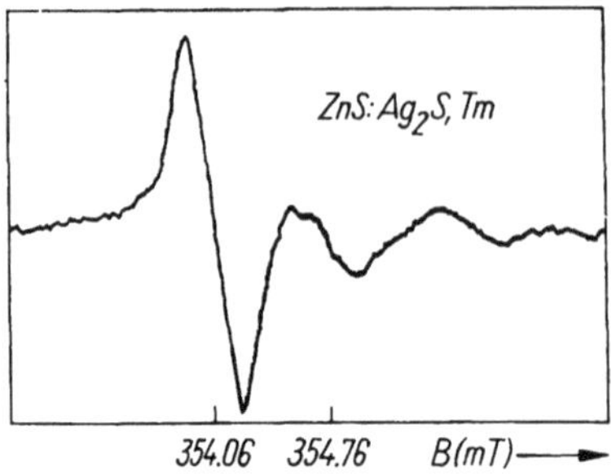

Fig. 4

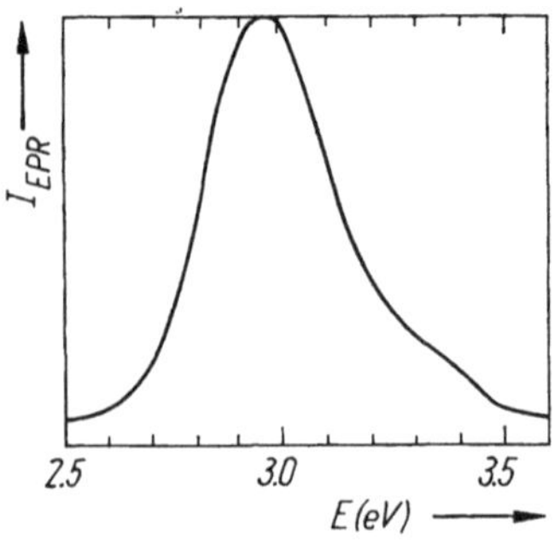

Fig. 5

Fig. 4. The isotropic photosensitive EPR line in Ag-doped ZnS with a g-factor of 1.8494 and two unknown weaker resonances

Fig. 5. The optical excitation spectrum of the EPR of mobile electrons in ZnS ($T = 77$ K)

be drawn from the kinetics of the centres. The resonance line disappears immediately after the excitation light has been switched off.

3.2 EPR spectra of ZnS:Zn, Al, ZnS:Ge, I, and ZnS:Ag_2S

In ZnS:Zn, Al a resonance line with a g-factor $g = 1.8848$ was measured in accordance with [3]. In varying the light intensity we were able to show that the observed line is also a superposition of DBE and CBE resonances, shown by different linewidths. ΔB_1 was 0.72 and $\Delta B_2 = 0.57$ mT. Because of a very small shift of the line no separation of the g-factors was possible.

In germanium-doped ZnS a resonance line was observed similar to that found in lead-doped ZnS. A g-factor of 1.8834 and a linewidth of $\Delta B_2 = 0.46$ mT have been determined together with a rapid decay of the EPR line when the UV excitation is switched off.

In the series of EPR of mobile electrons we want to include a resonance line measured in Ag-doped ZnS. The g-factor of this line is $g = 1.8494$, the linewidth $\Delta B_1 = 0.35$ mT. In Fig. 4 this resonance line, together with two, up to now unknown

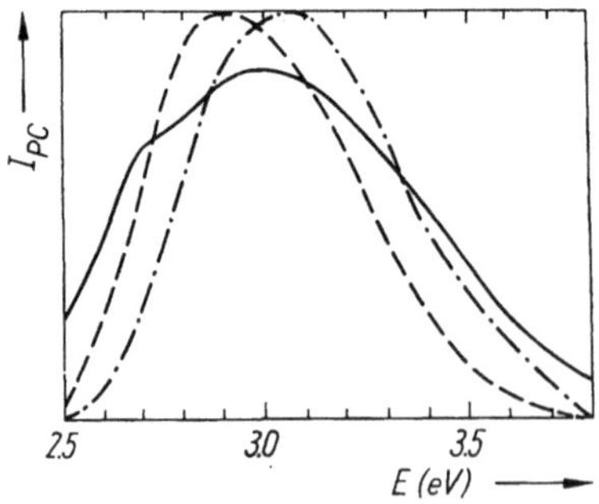

Fig.6

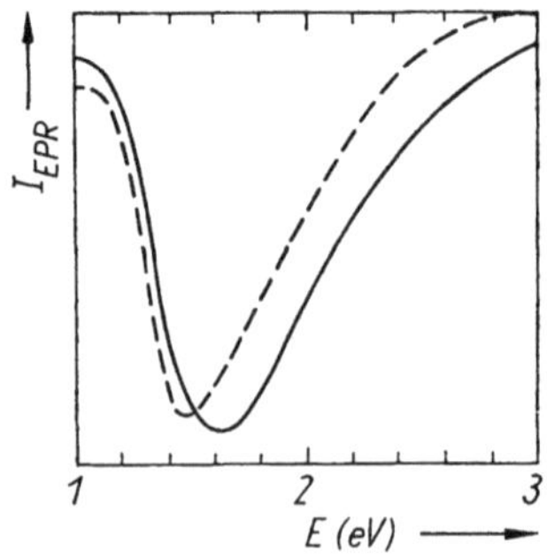

Fig. 7

Fig. 6. ac-microwave photoconductivity spectra of ZnS:Zn, Al (full line) and ZnS:Ge, I (dashed line) at 300 K and ZnS:Ge, I (dashed-dotted line) at 77 K

Fig. 7. Optical EPR quenching spectra of resonances of mobile electrons in ZnS:Ge, I (dashed line) and ZnS:Zn, Al (full line)

weaker resonances, is shown. Besides these EPR lines the spectra of the Ag–V_S-centre, for the first time described by Dieleman et al. [17], of Fe^{3+} [18] and Fe^+ [19], as well as Mn^{2+} have been observed in that sample. In the spectra of Fe^{3+}, already visible at 77 K, and Fe^+, measured at 20 K, a very strong influence of crystal irregularities was shown.

The optical excitation spectra of the EPR of DBE and CBE in ZnS:Zn, Al and ZnS:Ge, I are shown in Fig. 5. From there we took a low-energetic threshold of 2.5 and 2.6 eV, respectively, for the excitation and excitation maxima at about 3.0 eV. In Fig. 6 the ac-microwave photoconductivity spectra of ZnS:Zn, Al and ZnS:Ge, I at 300 K and ZnS:Ge, I at 77 K are shown. Only in some cases it was possible to observe photoconductivity in ZnS at 77 K because of a very strong decrease of the effect with decreasing temperature.

The quenching spectra of the DBE and CBE EPR of ZnS:Ge, I and ZnS:Zn, Al are given in Fig. 7. For both spectra a low-energetic threshold of about 1.1 to 1.2 eV can be determined.

4. Discussion

If there are two resonance lines with nearly equal g-factors, as it is the case for lead-or aluminium-doped ZnS, a simple model for the origin of the resonances can be used. In [3] this has been done for iodine-doped crystals.

The most striking difference in the behaviour of both resonance lines is in the decay time. The narrow line with g_2 has a very rapid decay time ($\tau \lesssim 1$ ms as estimated from the time constant of the E-4 spectrometer), whereas the broad one has a slower decay time ($\tau > 1$ ms). Assigning the narrow one (g_2) to conduction band electrons and the broad one (g_1) to donor band electrons a simple explanation can be given for the experimental results:

(i) With full UV excitation there are more electrons in the conduction band compared to the number with low excitation. From this it already can be concluded that the narrow line with g_2 and ΔB_2 is due to conduction band electrons.

(ii) Following Elliott [18] the spin–lattice relaxation time T_1 for charge carrier EPR and nearly identical g-factors is proportional to the mobility of the charge carriers. In the conduction band the mobility of the electrons is higher than in a donor band, consequently the CBE resonance line is narrower than that for DBE.

An estimation of the g-factors of mobile electrons can be done by using a theory of Roth et al. [19]. From this the g-factor is given by

$$g = 2\left[1 - \left(\frac{m}{m^*} - 1\right)\frac{\lambda}{3E_g + 2\lambda}\right]$$

with E_g the band gap, λ the spin–orbit splitting of the valence band, and m^* the effective mass.

In Roth's theory an interaction of the lowest conduction band Γ_{1c} with the valence band Γ_{15v}, split by spin–orbit interaction, is considered, supposing that the wave function of an electron in a shallow donor band is identical with that of a conduction band electron. A calculation of the g-factor of conduction band electrons of iodine-doped ZnS yielded $g = 1.97$ [4]. In experiment $g = 1.882$ was determined [3]. For the calculation in [4] the following parameters were used: $E_g = 3.84$ eV, $\lambda = 0.065$ eV, $m^*/m = 0.28$.

Considering the interaction of the conduction band Γ_{1c} with the energetically higher conduction band Γ_{15v} in addition to the interaction with Γ_{15v} Cardona [20] derived a formula for the g-factor on the basis of a $\boldsymbol{k} \cdot \boldsymbol{p}$ perturbation theory. Taking the formula given in [20] we found $g = 1.83$ for mobile electrons in ZnS, which gives a better fit to the experimental values. Using the empirically derived relation for the g-factor of conduction electrons — a modified form of the Roth et al. formula — a value of $g = 1.90$ is predicted for ZnS [23].

For all resonances of mobile electrons given here or in the literature (Table 1) a characteristic hyperfine structure due to an interaction with the donor ions is to be expected. In no case a hyperfine interaction was observed. This is thought to be a concentration-dependent effect. If the concentration of donors is high enough there will be an overlapping of the wave functions of the donor centres and no hyperfine structure is observable. The formation of donor bands above a distinct donor concentration C_D will result in a contact of the donor band at a critical concentration C_c [21]. If there is some knowledge about the energetic position of the donor levels in the band gap and on the dielectric constant of the host material, some estimations on C_c can be done [3]. For CdS a value of 2×10^{18} cm^{-3} was determined with $E_D = 0.03$ eV.

Table 1

EPR parameters of mobile electrons in zinc sulphide

crystal structure	doping	g-value g_1	g-value g_2	ΔB (mT)	ref.
cubic	Cl, Br	1.883		0.6	[22]
	I	1.88225		0.23 to 0.54	[3]
	Cl	1.8835		0.24 to 0.54	[3]
	Al–Zn	1.8849		0.44	[3]
polytype	Pb, I	1.8850	1.8846	0.49 to 0.69	this work
		1.8839		0.49	this work
	Zn–Al	1.8848		0.57 to 0.72	this work
	Ge, I	1.8834		0.46	this work
	Ag	1.8494		0.35	this work

For the explanation of the excitation mechanism we will use a model involving the existence of acceptor levels, probably Cu^{2+} or twofold negatively charged zinc vacancies. From these acceptor levels electrons will be excited by illumination with light energies $E_{ex} > 2.6$ eV into the conduction band or donor levels (Fig. 8, transition 1). By emission of visible radiation (E_{D-A}) the electrons will recombine with the acceptor states, following the well-known mechanism of D–A recombination. For this recombination process a different behaviour for variously doped samples was observed. A very

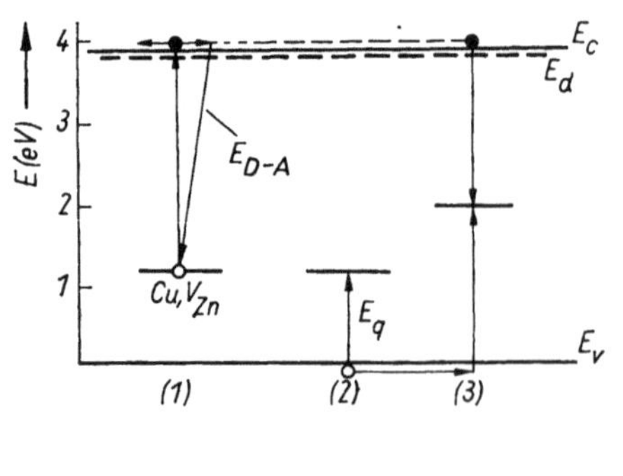

Fig. 8. Band model for the optical transitions in the EPR excitation and quenching process of mobile electrons

intense green emission was observed for ZnS:Ge, I and ZnS:Zn, Al, probably due to Cu acceptors, a blue emission for ZnS:Pb, I crystals. Considering the fact that in the latter crystals the EPR of Pb A-centres was observed, it can be concluded that the zinc vacancies are involved in the optical transitions. A confirmation of transition 1 in Fig. 8 is given by the photoconductivity spectra especially for ZnS:Ge, I at 77 K, because there are identical threshold energies for EPR excitation and photoconductivity.

The quenching process is shown schematically in Fig. 8, transition 2. By absorption of light with energy $E_q > 1.2$ eV an electron is excited from the valence band to the acceptor levels. The hole, created in the valence band will take part in a radiationless recombination via an unknown centre U in the band gap (transition 3).

Acknowledgements

We would like to thank Dr. R. Boyn and Dr. J. Dziesiaty for a critical reading of the manuscript. We are grateful to Dr. R. Stößer from the Department of Chemistry of the Humboldt-University at Berlin for providing the possibility to make measurements at the Varian spectrometer.

References

[1] J. Dziesiaty and D. Backs, phys. stat. sol. (b) **100**, 509 (1980).
[2] J. Dziesiaty, D. Backs, and D. Mantwitz, phys. stat. sol. (b) **107**, 553 (1981).
[3] K. A. Müller and J. Schneider, Phys. Letters (Netherlands) **4**, 288 (1963).
[4] J. Schneider, B. Dischler, and A. Räuber, J. Phys. Chem. Solids **29**, 451 (1968).
[5] J. R. James, B. C. Cavenett, J. E. Nicholls, J. J. Davies, and D. J. Dunstan, J. Lum. **12/13**, 447 (1976).
[6] P. Dawson and B. C. Cavenett, J. Lum. **18/19**, 853 (1979).
[7] J. E. Nicholls, J. J. Davies, B. C. Cavenett, J. R. James, and D. J. Dunstan, J. Phys. C **12**, 361 (1979).
[8] K. M. Lee, K. P. O'Donnel, and G. D. Watkins, Solid State Commun. **41**, 881 (1982).
[9] M. Lieberenz, R. Stösser, and D. Backs, to be published.
[10] J. Kluge and J. Donecker, phys. stat. sol. (a) **81**, 675 (1984).
[11] D. D. Thong and O. Goede, private communication.
[12] R. Stösser, M. Lieberenz, P. Heblik, and D. Backs, to be published.
[13] L. M. Matarrese and Ch. Kikuchi, J. Phys. Chem. Solids **1**, 117 (1956).
[14] J. Schneider, S. R. Sircar, and A. Räuber, Z. Naturf. **18a**, 980 (1963).
[15] B. Lambert, T. Buch, and B. Clerjaud, Solid State Commun. **10**, 25 (1972).
[16] D. Backs, Exper. Tech. Phys. **32**, 347 (1984).
[17] J. Dieleman, S. H. de Bruin, C. Z. van Doorn, and J. H. Haanstra, Philips Res. Rep. **19**, 311 (1964).
[18] R. J. Elliott, Phys. Rev. **96**, 266 (1954).
[19] L. M. Roth, B. Lax, and S. Zwerdling, Phys. Rev. **114**, 90 (1959).
[20] M. Cardona, J. Phys. Chem. Solids **24**, 1543 (1963).
[21] A. R. Hutson, Phys. Rev. **108**, 222 (1957).
[22] P. H. Kasai and Y. Otomo, Phys. Rev. Letters **7**, 17 (1961); J. chem. Phys. **37**, 1263 (1962).
[23] A. N. Chakravarti, Indian J. pure appl. Phys. **5**, 238 (1967).

(Received June 25, 1984)

phys. stat. sol. (b) **126**, 351 (1984)

Subject classification: 13.1 and 20.1; 10.2; 22.2

Laboratoire de Structure et Propriétés de l'Etat Solide, associé au C.N.R.S. nº 234, Université des Sciences et Techniques de Lille, Villeneuve d'Ascq[1])

Free-Carrier Optical Absorption Induced by Dislocation Scattering Mechanisms in III–V Compounds

By

D. VIGNAUD and J. L. FARVACQUE

The influence of dislocations on the free carrier absorption in the case of III–V compounds are determined theoretically taking into account the piezoelectric scattering potential associated with dislocations and the eventual existence of a deep flat band. Then the possibility to observe such effects is discussed and these calculations are illustrated by experimental results obtained on InSb.

Der Einfluß von Versetzungen auf die Absorption freier Träger in III–V-Verbindungen wird theoretisch bestimmt, wobei das mit Versetzungen assoziierte piezoelektrische Streupotential und die eventuelle Existenz eines tiefen Flachbandes berücksichtigt werden. Die Möglichkeit, derartige Effekte zu beobachten, wird diskutiert und diese Rechnungen werden mit experimentellen Ergebnissen an InSb erläutert.

1. Introduction

Free-carrier optical absorption phenomena correspond to electronic transitions within one energy band of a metal or a semiconductor. Because of the extension of the Brillouin zone and the subsequent relatively smooth variations of the dispersion curves $E(\mathbf{k})$, such optical transitions are necessarily assisted, the incident photons being unable to transfer the momentum required for these collisions. The absorbing material itself has then to provide for the momentum exchange either by direct creation or annihilation of phonons or by collisions at scattering centres. Such mechanisms have to be seen as the high-frequency limit of ac conductivity phenomena (i.e.: $\Omega > \Omega_p$, the plasma frequency).

Obviously, introducing scattering centres into a material, keeping constant its free-carrier density, must result in an increase of the absorption coefficient.

It is well-known that dislocations are responsible for a large decrease of the free-carrier mobility and one expects that they can also enhance noticeably the efficiency of the intraband absorption by their scattering potentials, as well as they could decrease the free-carrier mobility. None the less dislocations may also introduce deep bands in the forbidden band gap which, when ionized, strongly decrease the free-carrier density, and thereby the free carrier absorption.

These two kinds of dislocation effects act in opposite senses on the free-carrier absorption and it will not be evident to predict the variations of the absorption coefficient without self-consistent calculation.

Recently, edge-type dislocations in InSb have been shown to be responsible for a large decrease of the intraband optical absorption, whereas no specific absorption peak which would correspond to transitions between some extrinsic dislocation levels and the crystal bands, could be observed [1].

---

[1]) Bat. C6, 59655 Villeneuve d'Ascq Cédex, France.

Owing to the previous remark, the decrease of the free-carrier absorption is, none the less, an indirect evidence of extrinsic states introduced by plastic deformation and one supposes then that the direct transitions could not be observed because of too small optical cross-sections.

These extrinsic states observed indirectly, can be associated with dislocations, but also with further point defects created by the dislocation motion or more probably with both of them. As will be shown in the discussion, it looks difficult to separate qualitatively these two possible origins. None the less, the aim of this paper is to establish the expression of the intraband absorption connected with dislocations only, in order to define its characteristics.

With the help of numerical calculations one then discusses the possibilities to observe experimentally the free-carrier intraband absorption in the case of InSb and GaAs, and to use this kind of experiment to determine the position of the dislocation flat band eventually introduced in the energy band gap.

2. The Free-Carrier Absorption Coefficient

Recently, a calculation based on the use of the dielectric function has been developed [2], generalizing the previous method of Gerlach et al. [3] to determine the ac conductivity of a multi-carrier system (mainly electrons, light and heavy holes) in presence of any type of scattering centres.

When the carrier-carrier scattering mechanism is considered, this calculation may be extended, following a procedure proposed by Gerlach et al. [4]. It leads to a set of linear equations, from which the mobility of each carrier family can be obtained. When this scattering mechanism is neglected, one directly obtains the conductivity $\tilde{\sigma}_l$ of each l-th carrier family by the simple expression

$$\tilde{\sigma}_l(\Omega) = \frac{N_l e^2}{i m_l^* \Omega - \varphi_l(\Omega)}\,, \tag{2.1}$$

where N_l and m_l^* are the free carrier density and the effective mass of the l-th free carrier set, respectively, and where

$$\varphi_l(\Omega) = \frac{iD}{(2\pi)^3\, \varepsilon_0 \varepsilon_L N_l \Omega} \int \frac{(\boldsymbol{k}\cdot\boldsymbol{u})^2}{k^2}\, |\varrho_{sc}(\boldsymbol{k})|^2 \left\{\frac{X_l(\boldsymbol{k},\Omega)}{\boldsymbol{\varepsilon}_T(\boldsymbol{k},\Omega)} - \frac{X_l(\boldsymbol{k},0)}{\boldsymbol{\varepsilon}_T(\boldsymbol{k},0)}\right\} \mathrm{d}^3\boldsymbol{k}\,, \tag{2.2}$$

where D and $\varrho_{sc}(\boldsymbol{k})$ are the density and the Fourier transform of the scattering charge istribution, respectively. $X_l(\boldsymbol{k}, \Omega)$ and $\boldsymbol{\varepsilon}_T(\boldsymbol{k}, \Omega)$ are the susceptibility of the l-th free carrier family and the dielectric function of the form $\boldsymbol{\varepsilon}_T(\boldsymbol{k}, \Omega) = 1 + \Sigma_l X_l(\boldsymbol{k}, \Omega)$, respectively. The static part of the dielectric function $\boldsymbol{\varepsilon}_L$ has been reported in the pre-factor of the integral (2.2). Formula (2.2), when applied to the special case of one free-carrier family only, gives again the classical result obtained by Gerlach et al. [5].

As soon as one is concerned with applied fields, the frequency of which are higher than the plasma frequency of the material (i.e. $\Omega \gg \Omega_p$), it is easy to take off from (2.1) the real part of the conductivity which, owing to 2.2, turns out to be

$$\tilde{\sigma}_l'(\Omega) = \frac{De^2}{m_l^{*2}(2\pi)^3\, \varepsilon_0 \varepsilon_L \Omega^3}\, J_l(\Omega)\,, \tag{2.3}$$

where

$$J_l(\Omega) = -\int \frac{(\boldsymbol{k}\cdot\boldsymbol{u})^2}{k^2}\, |\varrho_{sc}(\boldsymbol{k})|^2\, \mathrm{Im}\left\{\frac{X_l(\boldsymbol{k},\Omega)}{\boldsymbol{\varepsilon}_T(\boldsymbol{k},\Omega)}\right\} \mathrm{d}^3\boldsymbol{k}\,. \tag{2.4}$$

Relations (2.3) and (2.4) allow one to determine the optical absorption coefficient $\alpha(\Omega)$ given by

$$\alpha_l(\Omega) = \frac{\tilde{\sigma}_l'(\Omega)}{\varepsilon_0 nc}, \tag{2.5}$$

where n is the refractive index of material and c the light velocity. This gives

$$\alpha_l(\Omega) = \frac{De^2}{(2\pi)^3\, \varepsilon_0^2 \varepsilon_L nc\Omega^3} \frac{J_l(\Omega)}{m_l^{*2}}. \tag{2.6}$$

For any frequency, the imaginary part of $X_l(\boldsymbol{k}, \Omega)/\varepsilon_T(\boldsymbol{k}, \Omega)$ is, generally, a complicated function which self-consistently takes into account the simultaneous screening effects of the scattering potentials by each carrier family. However, for optical frequencies $\Omega \gg \Omega_p$, the real part of the dielectric function $\varepsilon_T = 1 - \Omega_p^2/\Omega^2$ tends towards unity and one just obtains

$$\mathrm{Im}\left\{\frac{X_l(\boldsymbol{k}, \Omega)}{\varepsilon_T(\boldsymbol{k}, \Omega)}\right\} \approx X_l''(\boldsymbol{k}, \Omega), \tag{2.7}$$

where $X_l''(\boldsymbol{k}, \Omega)$ is the imaginary part of the l-th free-carrier family susceptibility.

Assuming that the free carriers can be represented in the effective-mass approximation, one in the standard way obtains from the Lindhard's dielectric function [6]

$$X_l'' = -\frac{(2\pi)^{1/2} m_l^{*1/2} N_l e^2}{\hbar \varepsilon_0 \varepsilon_L (KT)^{1/2}\, k^3} \mathrm{sh}\left(\frac{\hbar\Omega}{2KT}\right) \exp\left\{-\frac{\hbar^2 k^2}{8 m_l^* KT} - \frac{m_l^* \Omega^2}{2KTk^2}\right\}. \tag{2.8}$$

The simple relation (2.7) obtained for $\Omega \gg \Omega_p$ means that each contribution $\alpha_l(\Omega)$ does not depend, indeed, on the presence of the other free-carrier families, since for high frequencies one does not get any more static screening of the scattering centres by the free carriers. In that sense, the total absorption coefficient $\alpha_T(\Omega)$ will be given by the sum of independent terms $\alpha_l(\Omega)$,

$$\alpha_l(\Omega) = \frac{De^4 N_l}{(2\pi)^{5/2}\, \varepsilon_0^3 \varepsilon_L^2 ncm_l^{*3/2}} \frac{\mathrm{sh}\left(\frac{\hbar\Omega}{2KT}\right)}{(KT)^{1/2}\, \Omega^3 \hbar} J_l'(\Omega), \tag{2.9}$$

where

$$J_l'(\Omega) = \int \frac{(\boldsymbol{k}\cdot\boldsymbol{u})^2}{k^5} |\varrho_{sc}(\boldsymbol{k})|^2 \exp\left\{-\frac{\hbar^2 k^2}{8 m_l^* KT} - \frac{m_l^* \Omega^2}{2KTk^2}\right\} \mathrm{d}^3\boldsymbol{k}. \tag{2.10}$$

These relations (2.9 and 10) depend on the knowledge of the free-carrier density N_l of the l-th free-carrier set and on the type of the scattering centres for which one just has to know the Fourier transform $\varrho_{sc}(\boldsymbol{k})$ of the static bare charge distribution.

3. Scattering Potentials Associated with Dislocations in III–V Compounds

It is now well established that dislocation scattering potentials associated with dislocations can be classified into two categories [7]. The first one is connected with the presence of deep levels associated with the dislocations and are known under the name of "core effects". The second one is connected with the local deformation fields introduced by the dislocations.

a) The core effects are associated with the thermal ionization of the dislocation deep bands, when they exist, and transform the dislocation lines into "charged" wires, the

Coulomb potential of which constitutes an origin of scattering effects. These core effects are the prominent ones in the case of covalent semiconductors. For any semiconductor, core effects will be present as soon as dislocations will be responsible for the existence of deep bands.

b) The existence of the dislocation strain fields leads to two kinds of scattering potentials. The first one is a consequence of the atom displacement in the neighbourhood of the dislocations, which destroy locally the translation invariance of the crystal-potential. This effect, known under the name of "deformation potential", is generally small when compared to the "core effects". It could be evidenced for elemental semiconductors in the vicinity of the temperature at which the dislocations are neutral [8] and also for binary dislocations in the case of tellurium [9 to 11].

The most important effect connected with the dislocation strain fields results, indeed, from the piezoelectric coupling in non-centro-symmetric lattices. This latter scattering potential has been seen as the most effective one for dc conductivity in III–V compounds [12, 13]. In particular, it was shown in [13] that even Coulomb potentials describing the "core effects", could be neglected to calculate the remaining scattering potentials of dislocations. Following these previous calculations, one only takes into account, in the following, the piezoelectric scattering potentials as the only scattering mechanism. Let us just remark, before to go further, that this assumption only applies for "edge"-type dislocations since piezoelectric potentials vanish for screw dislocations [14].

The presence of a deep, flat band associated with the dislocations will none the less be considered in the view to calculate the actual free-carrier density N_l in the presence of dislocation trapping.

It has been shown in [14] that the elastic anisotropy could be forgotten when establishing the expression of the unscreened piezoelectric charge distribution of the dislocation which is given by

$$\varrho_{\text{sc}}(\boldsymbol{r}) = \frac{b_{\text{e}} e_{14}}{2\pi \sqrt{6}\,(1-\nu)}\, F(\Phi)\, \frac{1}{r^2} \tag{3.1}$$

with

$$F(\Phi) = \sin 4\Phi + (1 - 2\nu) \sin 2\Phi + 2^{1/2}(1 + \nu) \cos 2\Phi + \frac{5}{2^{1/2}} \cos 4\Phi\,,$$

where b_{e} is the edge component of the Burgers vector, e_{14} is the 1 to 4 component of the piezoelectric tensor, ν the Poisson modulus, and Φ the polar angle measured from the Burgers vector in a plane perpendicular to the dislocation line.

Taking the Fourier transform of (3.1) introduces a $\delta(k_z)$ function. This implies that no effect will be obtained on the optical absorption for light polarization $\boldsymbol{u}$ parallel to the dislocation line. Considering then light polarization $\boldsymbol{u}$ perpendicular to the dislocation line and making an angle γ with the edge component of the Burgers vector, finally leads after some easy calculations to

$$\alpha_l(\Omega) = \frac{N_l e^4 b_{\text{e}}^2 e_{14}^2 D}{24\pi m_l^{*2} (\varepsilon_0 \varepsilon_{\text{L}})^2\, \hbar \varepsilon_0 n c}\, \frac{A(\gamma)}{\Omega^4}\,, \tag{3.2}$$

where D is the number of dislocation emergences per unit area and

$$A(\gamma) = \frac{27}{4} + 6(1 + 2\nu)^2 - (3\nu + 6) \cos 2\gamma + \frac{3 - 12\nu}{\sqrt{2}} \sin 2\gamma\,. \tag{3.3}$$

$A(\gamma)$ is the anisotropy function calculated in the approximation of an elastic isotropic medium [14]; it shows that for a polarized light beam one expects an absorption about three times higher for $\boldsymbol{u}$ perpendicular to the Burgers vector than for $\boldsymbol{u}$ parallel to the Burgers vector.

Expression (3.3) is valid for any III–V (or II–VI sphalerite) compound and has to be averaged in the case of an unpolarized light beam,

$$\langle A(\gamma)\rangle \approx 11 ,$$

(where Poisson's modulus ν has been set equal to 1/3).

Expression (3.1) shows the specific frequency dependence of $\alpha_l(\Omega) \sim \Omega^{-4}$ which is the same power law we would have obtained considering only the Coulomb potentials of the "core effects" [3].

Note that this specific Ω^{-4} law is somewhat characteristic of dislocation "piezoelectric" or "core effect" scattering potentials since different power laws are obtained, within the same type of calculations [3], for ionized impurities or phonon scattering, respectively, corresponding to power laws of the type $\Omega^{-n/2}$ where n is an odd integer.

Finally, in the case of multi-carrier systems, the total intraband absorption coefficient is given by

$$\alpha_{\mathrm{T}}(\Omega) = \sum_l \alpha_l(\Omega) . \tag{3.4}$$

4. Numerical Calculations and Discussion: The Case of III–V Compounds

4.1 The actual absorption coefficient

Formula (2.10) shows that the intraband absorption coefficient is determined by the square modulus of the total scattering charge distribution. When different scattering centres are simultaneously present (as for instance phonons and dislocations), their potentials are not correlated and it is easy to show that

$$|\varrho_{\mathrm{sc,\,total}}|^2 = \sum_m |\varrho_{\mathrm{sc},\,m}|^2 , \tag{4.1}$$

where m denotes a particular scattering mechanism. Relation (4.1) indicates then that the various contributions to the absorption are simply added to form the total absorption coefficient. Let us, none the less, remark that interference effects between scattering mechanisms may be obtained as soon as they are issuing from the same scattering centres and when the Fourier transform of their scattering charge distribution do not have the same phase. This is particularly the case for "piezoelectric" and "core effect" potentials associated with dislocations.

These interferences being excluded for independent centres, let us denote by α_{ref} the contribution to the intraband absorption of the pre-existing scattering mechanisms of an undeformed (reference) sample, and by α_{disl} the contribution of the dislocation scattering mechanisms. Then the total absorption is given by

$$\alpha_{\mathrm{T}} = \alpha_{\mathrm{ref}} + \alpha_{\mathrm{disl}} \tag{4.2}$$

where α_{ref} has, of course, to be evaluated, taking into account an eventual decrease of the free-carrier density when dislocations have been introduced.

Formulae (2.9, 2.10) show that for any kind of scattering potential, the corresponding absorption coefficient will always remain directly proportional to the free-carrier density. However, the frequency power law is determined by the specific $\boldsymbol{k}$ dependence of the $\varrho_{\mathrm{sc}}(\boldsymbol{k})$ function.

In the extreme temperature range, we are mainly dealing with only one free-carrier family. This is obvious in the case of n-type material, but it represents too a good approximation for p-type material where the density of light holes is negligible when compared to the density of heavy holes (because of the strong difference in their effective masses). So we restrict ourselves to the case of one free-carrier family. The absorption contribution of one scattering mechanism can then be written in the form

$$\alpha = ANX^{-\beta}, \tag{4.3}$$

where $X = \hbar\Omega/E_g$ is a reduced variable introduced for convenience, and β a specific power characterizing the scattering mechanism (equal to 4 in the present case of dislocation scattering).

Introducing (4.3) into (4.2) gives

$$\alpha_T(X) = N\left(\frac{A_{ref}}{X^\beta} + \frac{A_{disl}}{X^4}\right). \tag{4.4}$$

This expression clearly shows the different ways dislocations may influence the total absorption coefficient: (i) by modifying the prominent scattering mechanism ($A_{disl}X^{-4} \gtrsim A_{ref}X^{-\beta}$); (ii) by modifying the free-carrier density (core effects). These two effects are discussed separately in the following sections.

4.2 *The scattering mechanism efficiency*

One discusses, here, the possibility to observe dislocation effects for InSb and GaAs by comparing theoretical results obtained from the present calculation with experimental results found in the literature and describing, in principle, the intraband absorption of "reference" samples.

Table 1

The various coefficients appearing in (4.4)

	A_{ref} (cm²)	β	A_{disl}/D (cm⁴)	β
n-InSb [15]	2×10^{-17}	2	1.5×10^{-26}	4
n-GaAs [16]	4×10^{-20}	3	2×10^{-32}	4

Following [15] and [16], the reference values of Table 1 apply for nondegenerate semiconductors and look rather independent of the free-carrier density. One postulates that they can be applied in the whole frequency range of the forbidden band gap.

Introducing the values of Table 1 into 4.3 shows that:

a) For GaAs, dislocation scattering effects will always be negligible, in the whole frequency range, when compared to the reference scattering effect, even for unrealistic values of the dislocation density as large as 10^{11} cm^{-2}.

b) For InSb, dislocation scattering effects are more important. This results from specific values of the InSb parameters such as its high dielectric constant and its very light electron effective mass. Numerical calculations show that for realistic dislocation densities (10^8 to 10^9 cm^{-2}) dislocation scattering mechanisms would become the prominent ones in the lower part of the energy gap ($X \leq 0.5$ for $D = 5 \times 10^8$ cm^{-2}). In this frequency range one would be, therefore, able to observe the specific Ω^{-4} law.

In the upper part of the energy gap scattering effects of dislocations cannot be observed.

These crude evaluations show that mostly dislocations will not modify the intraband absorption by their scattering effects, i.e. the reference scattering mechanisms (phonons) will remain the prominent effects.

One therefore can practically expect a modification of the absorption coefficient through a variation of the free-carrier density.

4.3 The variation of the intraband absorption

As pointed out in the previous section the variation of the intraband absorption coefficient is, for GaAs as well as for InSb, (at least in the upper part of the energy gap), directly proportional to the variation of the free-carrier density,

$$\frac{\Delta\alpha}{\alpha} = \frac{\Delta N}{N} \,. \tag{4.5}$$

The measurement of $\Delta\alpha/\alpha$ provides then a way to evaluate directly the relative variation of the free-carrier density in the presence of an eventual dislocation trapping.

The calculation of $\Delta N/N$ may be done following standard methods (see for instance [12]), postulating the existence of a dislocation flat band localized at the energy E_D when neutral (E_D is an unknown parameter here) and knowing accurately the density of electrically active dislocations.

Such calculations show that for strongly doped materials (n or $p \geqq 10^{17}$ cm^{-3}), for which the intraband absorption is noticeable, it is necessary to introduce unrealistic dislocation densities in order to observe a variation (decrease) of the free-carrier density and a consequent observable variation of the total absorption coefficient.

On the contrary, the free-carrier density is noticeably reduced by dislocations when it is initially not too large (10^{14} to 10^{16} cm^{-3}). However, for too low free-carrier densities, the intraband absorption is no more observable ($\alpha_T \ll 1$ cm^{-1} for 10^{14} cm^{-3} impurities in InSb).

Therefore, the observability of dislocation effects requires to find a compromise where the initial free-carrier density is neither too high (to avoid unrealistic dislocation densities) nor too low (in order to remain observable).

Such a situation can be found for materials containing 10^{15} to 10^{16} cm^{-3} free carriers and for dislocation densities of the order of 10^8 to 10^9 cm^{-2} (i.e. standard values obtained after some few per cents of plastic deformation).

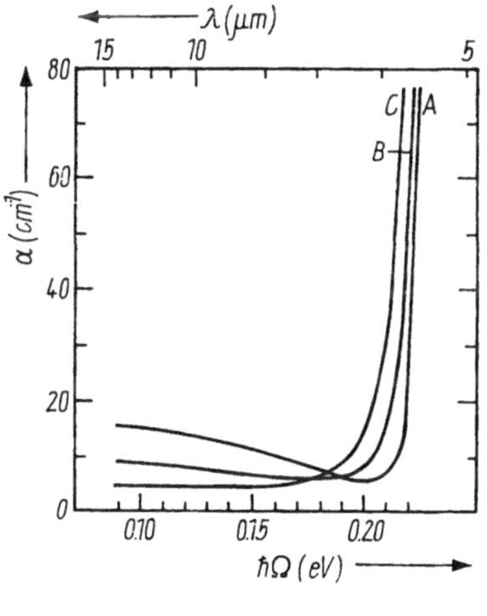

Fig. 1. Absorption spectra at $T = 90$ K in p-type InSb ($N_A - N_D \approx 7 \times 10^{15}$ cm^{-3}) for various uniaxial deformation. A reference sample (undeformed), B and C deformed 3 and 6.3%, respectively

4.4 Experimental case of InSb

Measurements of the intraband absorption have been done on p-type InSb ($N_A \approx \approx 7 \times 10^{15}$ cm^{-3}), on reference and deformed samples [1]. Fig. 1 shows clearly a decrease of the intraband absorption efficiency, parallel to the deformation ε (and therefore the dislocation density). For the most strongly deformed sample ($\varepsilon = 6.3\%$), one just observes a residual absorption continuum, which does not depend anymore on the photon frequency, and the origin of which may be attributed to surface effects.

In the case of sample 2 ($\varepsilon = 3\%$) one observes a relative decrease of α, which may be evaluated to 50% (after having subtracted the residual absorption continuum). This means that a quantity as high as $\approx 3 \times 10^{15}$ cm^{-3} free carriers have been trapped on extrinsic levels introduced by the deformation procedure.

Different origins of these extrinsic levels can be evoked:

(i) either they are connected with any "unwanted" point defects left by the dislocation motion or

(ii) they are actually extrinsic states associated with the dislocations.

Although it is difficult to evaluate the density of point defects left by the dislocation motion, one may have a crude estimate by evaluating the number of intersections of the mobile dislocations with the dislocation forest (with a density D_f) and assimilating this quantity to the order of magnitude of the point defect density. In such a case, one obtains

$$D_{pd} \approx D_f \frac{\varepsilon l}{b}, \tag{4.6}$$

where $\varepsilon l/b$ represents the total quantity of dislocations which escaped from the sample in order to fulfil the total deformation ε.

Evaluating the forest density to 10^9 cm^{-2}, for the deformation $\varepsilon = 3\%$ corresponding to sample 2, (length $l = 1$ cm and $b = 4.6 \times 10^{-8}$ cm) gives $D_{pd} \approx 6 \times 10^{14}$ cm^{-3}. This value is not so far from the one required to justify the experimental decrease of the intraband absorption coefficient. This shows that one cannot eliminate systematically the possibility of an effect which would partly occur because of point defects.

None the less one may evoke some recombinations of these point defects and also argue that each dislocation intersection does not lead systematically to the creation of point defects. In that sense, the upper value of $D_{pd} \approx 6 \times 10^{14}$ cm^{-3} seems to be an overevaluation of the actual density of point defects.

One supposes then that dislocation energy states are only responsible for the observed experimental decrease of the intraband optical absorption. In such an hypothesis, evaluating the dislocation density to 10^9 cm^{-2} (crude estimate by TEM) the variation $\Delta\alpha/\alpha = \Delta N/N \approx 50\%$ can be found for an energy position of the dislocation flat band at 0.16 eV above the valence band.

Unfortunately, this result depends drastically on the actual density of electrically active dislocations which is not, evidently, determined accurately.

5. Conclusion

As soon as some intraband absorption occurs for sufficiently doped materials, dislocations may decrease the absorption coefficient by some free-carrier trapping. This constitutes an indirect evidence of the existence of flat energy bands associated with dislocations which cannot generally be directly observed because of too small optical cross-sections [1, 17].

As for any experiment based on an indirect effect of the presence of these extrastates (Hall effect for instance) it is not possible to determine accurately the position of these energy states without knowing accurately the dislocation density. Moreover, such a determination uses a statistics calculation model which implicitly needs to imagine the spectrum structure associated with the dislocations.

Finally such measurements integrate any possibility to trap the free carriers (as for instance the eventual existence of point defect levels).

None the less, combined with other types of optical measurements (such as photo-conductivity) this kind of observation may help to evidence and localize in the energy band gap, the extrinsic states eventually associated with dislocations.

References

[1] D. Vignaud and J. L. Farvacque, phys. stat. sol. (b) **125**, 785 (1984).
[2] J. L. Farvacque and D. Vignaud, to be published in Phys. Rev. B.
[3] E. Gerlach and P. Grosse, Festkörperprobleme XVII, 157 (1977).
[4] E. Gerlach and J. Mycielski, Verh. DPG. **5**, 746 (1982).
[5] E. Gerlach and M. Rautenberg, phys. stat. sol. (b) **65**, K13 (1974).
[6] J. Lindhard, Kong. Danske Vid. Selsk., mat.-fys. Medd. **28**, No. 8 (1954).
[7] J. L. Farvacque, J. Physique **39**, C2–25 (1978).
[8] F. Düster and R. Labusch, phys. stat. sol. (b) **60**, 161 (1973).
[9] J. C. Doukhan, R. Drope, J. L. Farvacque, E. Gerlach and P. Grosse, phys. stat. sol. (b) **64**, 237 (1974).
[10] J. L. Farvacque and P. Lenglart, phys. stat. sol. (b) **80**, 361 (1977).
[11] J. L. Farvacque and P. Lenglart, phys. stat. sol. (b) **80**, 433 (1977).
[12] R. Masut, C. M. Penchina, and J. L. Farvacque, J. appl. Phys. **53**, 4964 (1982).
[13] C. M. Penchina, J. L. Farvacque, and R. Masut, J. appl. Phys. **53**, 4970 (1982).
[14] D. Vignaud, J. L. Farvacque, and D. Ferre, phys. stat. sol. (b) **110**, 601 (1982).
[15] E. Haga and M. Kimura, J. Phys. Soc. Japan **18**, 777 (1963).
[16] W. G. Spitzer and J. M. Whelan, Phys. Rev. **114**, 59 (1959).
[17] K. Bohm and D. Fischer, J. appl. Phys. **50**, 5453 (1979).

(Received June 27, 1984)

phys. stat. sol. (b) **126**, 361 (1984)

Subject classification: 13.4 and 20.3; 22

Naval Surface Weapons Center, White Oak, Silver Spring[1]) *(a), Department of Physics, Oklahoma State University, Stillwater*[2]) *(b), and Department of Physics, University of Birmingham*[3]) *(c)*

Thermoluminescence Governed by Simultaneous Thermal Stimulation of Electrons and Holes

By

R. CHEN[4]) (a), V. K. MATHUR (a), JOANNE F. RHODES (a), M. D. BROWN (a), S. W. MCKEEVER (b), and R. K. BULL[5]) (c)

The possibility of a simultaneous thermal release of electrons and holes in the thermoluminescence (TL) process was suggested in the early stages of the development of this discipline. However, most investigators utilized a model in which carriers of only one kind are thermally released from their "traps". In this model these carriers then radiatively recombine with carriers of opposite charge in "luminescence centers". This paper presents the results of a theoretical investigation into a more complex model, which allows the thermal release of electrons and holes in the same temperature range. The solution of a set of four simultaneous differential equations is required in this investigation. The equations are not amenable to analytical solution, so the equations are solved numerically for given values of the model parameters. Conclusions are drawn about the shapes of the TL curves, and the expected stability of the corresponding traps at given temperatures, and the apparent effective parameters namely, the activation energy, the frequency factor, and the order of kinetics, are obtained using the "general order" formalism. In all cases examined, there is a good three parameter fit to the TL peak generated by the examined model. This shows that even a rather general TL peak can be described quite accurately by the three parameter model. The activation energy determined using the three parameter analysis is found to agree rather well with the input activation energy of trapped electrons which is used in the original peak generation for the particular cases examined. In cases where the activation energy for holes is small, the glow peaks occur at relatively low temperatures, and are characterized by a very high effective frequency factor in the three parameter analysis. Furthermore, several cases are studied in which the calculated kinetics order is found to be less than 1.0.

La possibilité d'une liberation thermique simultanée des electrons at des trous dans le processus de thermoluminescence avait été suggérée dans les premières étapes du développement de cette discipline; mais la plupart des chercheurs ont préféré utiliser le modèle plus restrictif où un seul type de porteur de charge est libéré thermiquement des pièges pour ensuite se recombiner avec des porteurs de signe opposé dans les "centres". Dans cet article, nous traitons le cas plus complexe où les electrons et les trous peuvent être libérés dans le même domaine de température. Cela appelle la solution d'un système de quatre equations differéntielles simultanées, ce qu'on ne sait bièn faire analytiquement. Au lieu de cela, c'est une methode numériqué que nous utilisons pour résoudre ce système d'equations pour des valeurs données des paramètres. Nous sommes ainsi parvenu à des conclusions, concernant la forme des courbes de TL obtenues, la stabilité des pièges correspondants à attendre à des températures données, et les valeurs effectives des paramètres qu on pourrait déduire d une approche par la methode de "l'ordre général"; savoir, l'energie d'activation, le fac-

[1]) Silver Spring, Maryland 20910, USA.

[2]) Stillwater, Oklahoma 74078, USA.

[3]) P.O. Box 363, Birmingham, B15 2TT, Great Britain.

[4]) Permanent address: Department of Physics and Astronomy, Tel Aviv University.

[5]) Present address: Environmental and Medical Sciences Division, A.E.R.E. Harwell, Didcot, Oxon, Great Britain.

teur de frequence et l'ordre de la cinétique. Dans tous les cas traités numériquement, un très bon accord a été trouve entre la courbe calculée à partir du système différentiel et la courbe a trois paramètres, ce qui nous montre qu'un pic de TL même d'un type fort général se laisse décrire assez exactement par le modèle à trois paramètres. Nous avons trouvé aussi que l'energie d'activation obtenue dans l'analyse a trois paramètres est fort voisine de l'energie d'activation réelle des electrons. Dans les cas où l'energie d'activation des trous a été faible, le pics sont apparu à des temperatures relativement basses, cas caracterisé par un facteur de fréquence effectif de valeur élevée dans l'analyse à trois paramètres. En outre, plusieurs cas ont été appris où l'ordre de la cinétique calculée a été trouvé moins que 1,0.

1. Introduction

The usual model for a single thermoluminescence (TL) peak is that of an electron trapping state situated energetically close enough to the conduction band so that electrons can be released thermally in the temperature range of interest, and subsequently recombine with holes trapped in a "recombination center". The recombination center is usually assumed to be energetically far from the valence band so that no holes are released thermally into the valence band. Thus, only one channel of electron–hole recombination is taken into account. Sometimes, the possibility that thermally excited electrons can retrap into a trapping state of the same kind is included in the analysis. The analogous inverted process in which holes become thermally excited into the valence band and then recombine with electrons in recombination centers, may be as likely to occur as the previous one.

The traffic of carriers between the trapping state, conduction (or valence) band and recombination center was previously shown [1, 2] to be governed by a set of three simultaneous differential equations. The principal conclusion, drawn on the basis of simplifying assumptions, is that the order of kinetics depends on the relative effectiveness of the recombination and retrapping processes [1, 3]. The simplifying assumptions may, however, introduce inaccuracies which can be serious. Therefore, numerical solutions of the set of equations were obtained [4 to 6] without the use of simplifying assumptions, for specific sets of the trapping parameters, and a given heating function. The concentration of carriers in the band was also found as a function of temperature, which has an important bearing on the phenomenon of thermally stimulated conductivity.

In an attempt to describe most of the experimental TL peaks, the "general order kinetics" [7 to 9] was proposed, which is governed by the equation

$$I = -\frac{\mathrm{d}n}{\mathrm{d}t} = s' n^b \exp\left(-\frac{E}{kT}\right), \tag{1}$$

where I is the TL intensity, n the concentration of trapped carriers (cm^{-3}), s' the pre-exponential factor, E the the activation energy (eV), k the Boltzmann constant ($\mathrm{eV\,K^{-1}}$), T the absolute temperature (K), and b the kinetics order. For first order kinetics [10], $b = 1$, for second order, $b = 2$. The solution of (1) was shown [9] to be

$$I = s n_0 \exp\left(-\frac{E}{kT}\right)\left[\frac{(b-1)\,s}{\beta}\int\limits_{T_0}^{T} \exp\left(-\frac{E}{kT'}\right)\mathrm{d}T' + 1\right]^{-b/(b-1)} \tag{2}$$

for $b \neq 1$, where $s = s' n_0^{b-1}$ has dimensions of s^{-1} and n_0 is the initial concentration of trapped carriers.

Rather early in the history of TL, it was suggested by Schön [11, 12] that electrons and holes can be thermally excited to the conduction and valence bands respectively in the same temperature range. This model complicates the physical picture and no less than four differential equations can adequately describe it. Bräunlich and Scharmann [13] studied these equations and gave approximate solutions. They classified the TL into two cases. In one case only electron transitions or only hole transitions contribute to the TL emission, the other being non-radiative or in another part of the spectrum. In the other case both recombinations contribute to the emission. In the former instance, on the basis of assumptions similar to those leading to first order kinetics in the "one channel" classical model, Bräunlich and Scharmann found an expression of the form

$$I(T) = \exp\left(-\frac{E}{kT}\right)\exp\left[\left(-\frac{s}{\beta}\right)\int\limits_{T_0}^{T}\exp\left(-\frac{E_A}{kT'}\right)\mathrm{d}T'\right] \quad (3)$$

which resembles an expression for the first order, except the activation energy for electrons, E, appears in the first exponent and that for holes, E_A, in the second.

The simultaneous thermal release of electrons and holes was further investigated by Gasiot and Fillard [14] and Mathur and Brown [15] in the context of simultaneous measurements of thermoluminescence (TL) and thermally stimulated conductivity (TSC). These authors reported a method for extracting information on hole emission by measuring the ratio of the TSC and TL signals.

In the investigation reported here, the set of four coupled differential equations governing the process of simultaneous electron and hole emission is solved numerically for the first time without simplifying assumptions. This enables the validity of the assumptions and approximations made in previous studies to be checked, and furthermore, thermally stimulated conductivity (TSC) which often accompanies the TL phenomenon can be investigated in this framework.

The present paper, however, is limited to testing the validity of applying the conventional "three-parameter" equation ((1), the three parameters being E, s, b) to describe the glow peaks produced from this model. A TL peak is simulated using the examined model and analyzed by the conventional, somewhat oversimplified, three-parameter equation. The goodness of fit is determined graphically and the effective values of the parameters extracted from the analysis are compared with the known parameters introduced into the original set of equations.

The results obtained indicate that, in each of the cases, the three-parameter equation can be used with confidence to describe the glow peaks produced, and the activation energy so-calculated agrees well with the original value of the electron trap depth. Occasionally unusually high values of the effective frequency factor and values of kinetic order of less than 1.0 were obtained.

2. The Model and Solution

The model considered here includes one electron and one hole trapping state, as shown in Fig. 1. The four inter-related time (or temperature) dependent functions are the following concentrations (in cm^{-3}):

- n concentration of trapped electrons,
- m concentration of trapped holes,
- n_c concentration of electrons in the conduction band,
- p concentration of holes in the valence band.

The parameters involved are N, M concentrations (cm^{-3}) of electrons and holes trapping states, respectively; E_1, E_2 activation energies (eV) for the release of electrons and holes respectively; s_1, s_2 frequency factors (s^{-1}) for the release of electrons and holes, respectively; A_{re}, A_{rh} recombination probabilities ($cm^3\,s^{-1}$) for electrons and

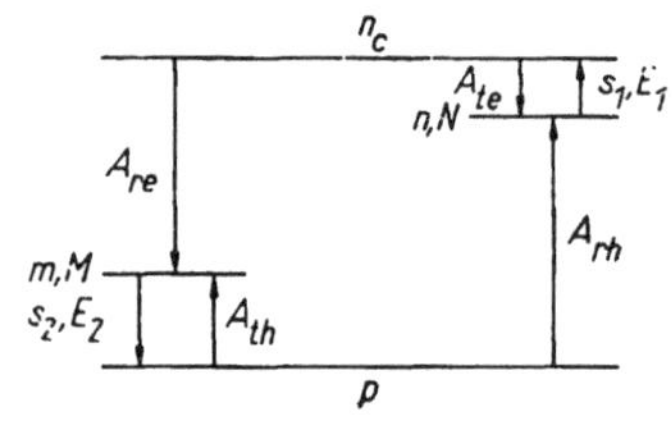

Fig. 1. Energy levels, concentrations and transition probabilities in the model

holes, respectively; A_{te}, A_{th} trapping probabilities ($cm^3 s^{-1}$) for electrons and holes, respectively.

The four simultaneous differential equations as first given by Bräunlich and Scharmann [13] (in a slightly different notation) are:

$$\frac{dn}{dt} = - s_1 \exp\left(-\frac{E_1}{kT}\right) n + A_{te} n_c (N - n) - A_{rh} n p \,, \tag{4}$$

$$\frac{dn_c}{dt} = s_1 \exp\left(-\frac{E_1}{kT}\right) n - A_{te} n_c (N - n) - A_{re} n_c m \,, \tag{5}$$

$$\frac{dp}{dt} = s_2 \exp\left(-\frac{E_2}{kT}\right) m - A_{rh} n p - A_{th} p (M - m) \,, \tag{6}$$

$$\frac{dm}{dt} = -s_2 \exp\left(-\frac{E_2}{kT}\right) m - A_{re} n_c m + A_{th} p (M - m) \,, \tag{7}$$

where k is the Boltzmann constant (eV K^{-1}) and T is the absolute temperature. It is quite obvious that this set of equations incorporates, necessarily, the neutrality condition

$$\frac{dn}{dt} + \frac{dn_c}{dt} = \frac{dm}{dt} + \frac{dp}{dt} \,. \tag{8}$$

The set of differential equations was solved numerically without approximations by utilizing a Runge-Kutta sixth-order predictor–corrector program that had been previously successfully employed [16] for evaluating the dose and dose rate dependences of TL. The main difference is that in the study presented here, two of the coefficients, $\exp(-E_1/kT)$ and $\exp(-E_2/kT)$, in the rate equations are very strongly temperature dependent. It was found that for a heating rate of 1 K/s, a temperature update every 0.1 s (i.e. 0.1 K) gives very good results. The integration interval was smaller, usually 0.001 s. In fact, the program is built in a dynamic manner such that the integration interval varies to ensure a pre-determined accuracy in the results. The program was run on a CDC 6500 computer. A typical run time for producing a glow peak consisting of a 50 K range was about 7 minutes. The resulting curves were analysed using a curve fitting program for the three parameters of (2) [3, 17].

3. Results

The number of possible choices for the sets of parameters that one can assume is, obviously, enormous. In particular, due to the rather preliminary nature of the present work, discussion is limited to a small number of cases chosen to exhibit some features that could not be accounted for by the conventional "one channel" model. Emphasis is placed on the cases where the TL is due to only one of the recombinations (e.g., electrons in the conduction band recombine with holes in centers) the other recombi-

Table 1

Parameters inserted into the simultaneous differential equations simulating TL of electrons recombination in the presence of a thermal release of holes. In all cases, no retrapping of holes is allowed ($A_{th} = 0$) and therefore no particular value is to be assigned to M

$\beta = 1$ K/s	$E_1 = 1$ eV	$s_1 = 10^{12}$ s^{-1}	$s_2 = 10^{12}$ s^{-1}
$A_{re} = 10^{-13}$ cm^3 s^{-1}	$A_{te} = 10^{-12}$ cm^3 s^{-1}	$A_{rh} = 10^{-12}$ cm^3 s^{-1}	
$N = 10^{15}$ cm^{-3}	$m_0 = 10^{14}$ cm^{-3}	$n_{c0} = 10^6$ cm^{-3}	$p_0 = 10^6$ cm^{-3}

$n_0 = 10^{13}$ cm^{-3}

No.	E_2 (eV)	T_m (K)	E (eV)	b	s (s^{-1})
1	0.7	251	0.96	0.63	2.8×10^{19}
2	0.8	285	0.98	0.62	3.2×10^{16}
3	0.9	319	1.01	0.61	9.5×10^{16}
4	1.0	353	1.015	0.59	3.1×10^{13}
5	1.1	386	1.02	0.56	1.6×10^{12}
6	1.2	418	0.98	0.53	5.0×10^{10}
7	1.3	440	0.98	0.67	1.0×10^{10}
8	1.5	448	0.98	1.03	5.3×10^{9}

$n_0 = 10^{14}$ cm^{-3}

No.	E_2 (eV)	T_m (K)	E (eV)	b	s (s^{-1})
9	0.7	269	0.90	1.12	8.9×10^{15}
10	0.8	305	0.94	1.08	3.5×10^{14}
11	0.9	341	0.93	0.98	5.4×10^{12}
12	1.0	376	0.99	1.01	1.7×10^{12}
13	1.1	410	1.00	1.01	1.8×10^{11}
14	1.2	434	0.95	1.18	6.8×10^{9}
15	1.3	443	0.96	1.76	4.4×10^{9}
16	1.5	443	0.99	2.07	9.7×10^{9}

nation is assumed to be either radiationless or not detected. Of course, even in this case, the occurrence of the second recombination is of great importance since it may influence substantially the balance of carriers and therefore, the rate of the recombination which yields TL. In many of these calculations hole retrapping was ignored (i.e. $A_{th} = 0$) and therefore the total number of hole traps, M, was unimpotrant. Thus the number of parameters was reduced by two.

This was done simply to reduce the number of parameters involved and future work will include hole retrapping. In the present work emphasis is put on the apparently dominant parameters which are expected to have maximum impact on the results. In the present situation the electron and hole activation energies E_1 and E_2 are expected to play the dominant role. Of particular importance is the relative magnitude of E_1 and E_2. In the computational work, therefore, $E_1 = 1.0$ eV is kept constant and E_2 is varied.

Table 1 summarizes some of the computational results obtained. The parameters which were not varied are listed in the caption. The table is divided into two groups of results. In the first, the peak follows practically first order kinetics when there is a very low hole emission ($E_2 = 1.5$ eV). In the second, the peak is very close to second order kinetics under the same circumstances. The difference between the two cases

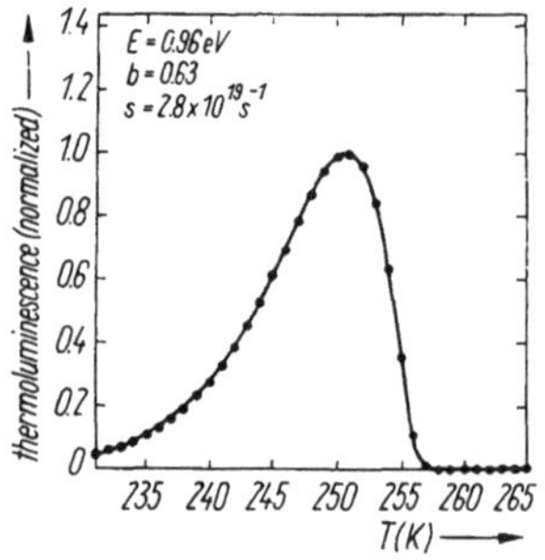

Fig. 2

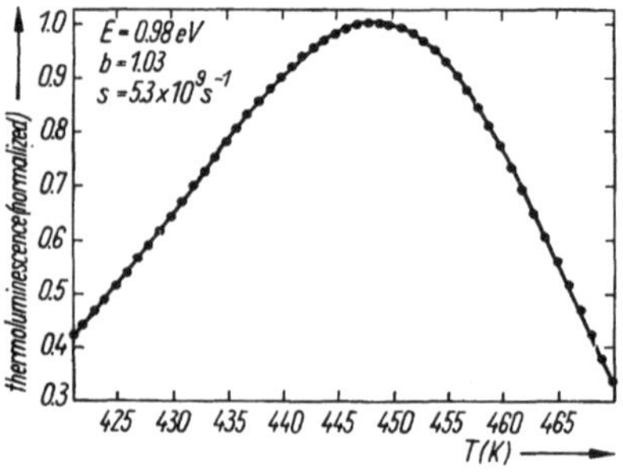

Fig. 3

Fig. 2. First order conditions with strong hole emission; the resulting peak has an effective order of 0.63

Fig. 3. First order conditions with very small hole emission; the resulting peak has an effective order of 1.03

is that for $m_0 = 10^{14}$ cm^{-3}, the first group has $n_0 = 10^{13}$ cm^{-3} whereas in the second, $n_0 = 10^{14}$ cm^{-3}. The table gives the values of E_2 and T_m, the peak temperature, as well as the resulting "three parameters", E, s' and b (order of kinetics) found by curve fitting.

Fig. 2 shows the results for row 1 in Table 1. The dots are the glow peak points simulated by the simultaneous differential equations whereas the solid line is the best fit three parameter curve to the simulated glow peak. The fit is remarkably good. The effective activation energy was found to be 0.96 eV, reasonably close to the value of 1.0 eV inserted into the equations. The quite rapid fall-off of the peak at high temperatures deserves some attention. This is directly related to the resulting order of kinetics. Orders of kinetics smaller than 1 have been reported in other thermally stimulated processes such as thermogravimetry (e.g. see p. 89 in [3]). May and Partridge [8] reported an order of kinetics of 0.7 ± 0.1 for the TL of NaCl, KCl and KBr. Qualitatively, the rapid fall on the high temperature side is related to the thermal exhaustion of holes from the traps. The fact that a peak with relatively high activation energy occurs at such a low temperature is accounted for, in the three parameter presentation, by an unusually high effective frequency factor, 2.8×10^{19} s^{-1}.

Fig. 3 is generated from the data in row 8 of Table 1. Since the activation energy for holes is very high, its influence on the peak shape is minor as compared to the case of no thermal hole emission. The best fit yields values for the effective activation energy of 0.98 eV, order of kinetics of 1.03 and frequency factor of 5.3×10^9 s^{-1}.

Fig. 4 shows the curve for row 9 in the table. Although the situation without hole emission is basically that of second order, with hole emission (and $E_2 = 0.7$ eV) the shape is close to that of first order kinetics, the effective order being 1.12. The effective activation energy in this case is 0.9 eV, 10% below the original electron activation energy. In the cases checked, the initial rise analysis revealed that the increase of the intensity function is dominated by the electron activation energy, 1.0 eV. This is the value which would have been obtained with the initial rise method in the very low temperature range. The effective frequency factor is very high here too, due to the combination of a rather high activation energy with a relatively low temperature of the maximum.

Fig. 5 is generated from the data in row 16. Due to the very high activation energy for the release of holes, 1.5 eV as compared to the 1.0 eV for the electrons, the peak appears to be close to second order (2.07), an order expected with practically no

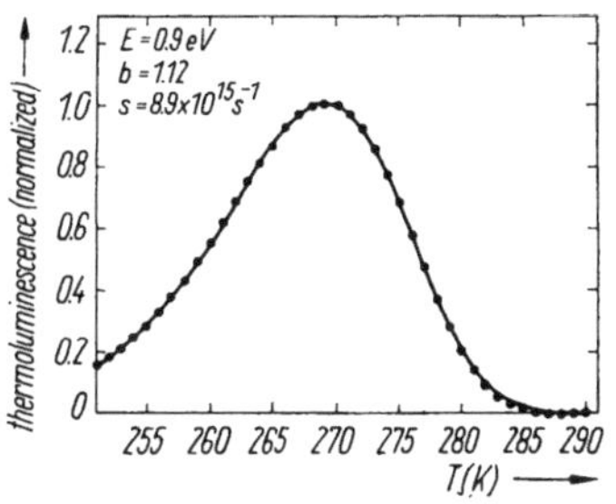

Fig. 4

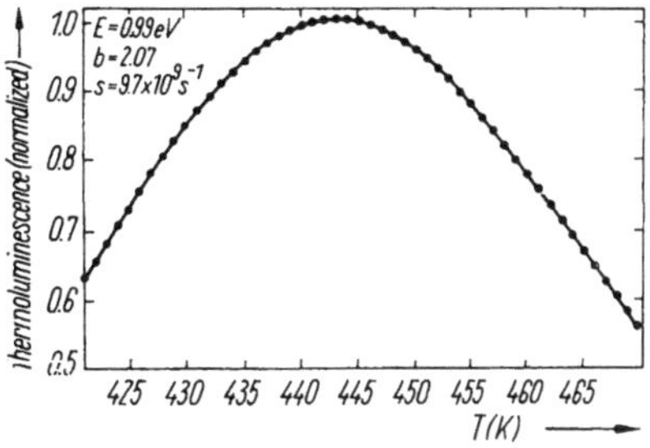

Fig. 5

Fig. 4. Second order conditions with strong hole emission; the resulting peak has an effective order of 1.12

Fig. 5. Second order conditions with very small hole emission; the resulting peak has an effective order of 2.07

thermal release of holes. The effective activation energy was found to be 0.99 eV, in very good agreement with the inserted electron activation energy.

Another interesting simulation can be performed using the same program. This involves the variation of the emitted light and the trapping concentrations of holes with time while at a constant temperature. In the present program, the phosphorescence behavior is studied by setting the heating rate $\beta = 0$.

Fig. 6 depicts the decay of phosphoresence at 450 K for the trapping parameters in row 14. Curve 1 shows the decay of phosphorescence whereas curve 2 shows the decay of the concentration of holes in recombination centers. No simple analytical function is offered for these decay curves.

Fig. 7 gives the decay curve at 270 K with the trapping parameters of row 9 in Table 1. In this case the plot of both the phosphorescence and the concentration $m(t)$ are straight lines on the semi-log scale, indicating an exponential decay, though with

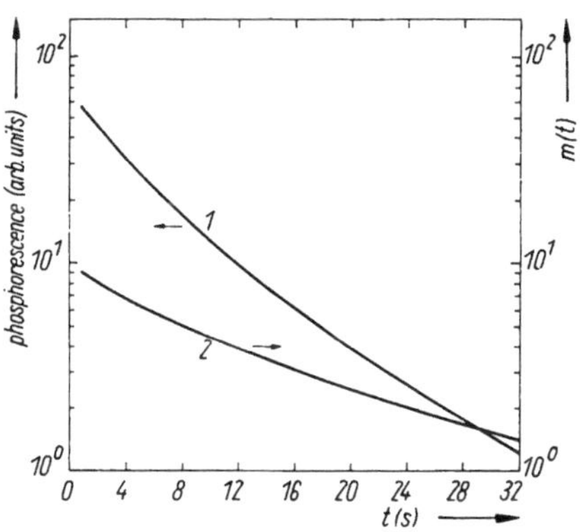

Fig. 6

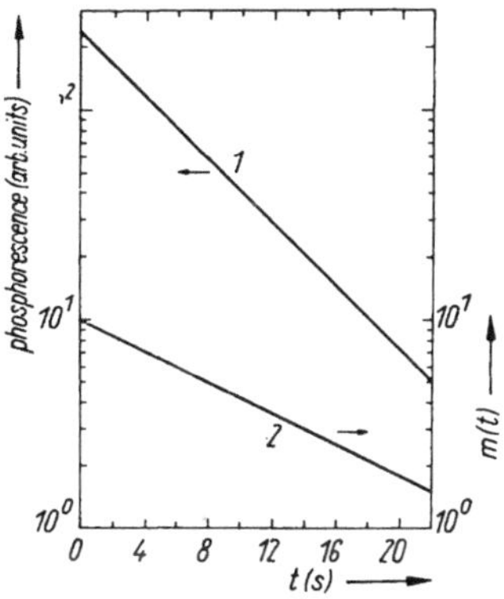

Fig. 7

Fig. 6. Decay curves of phosphorescence (1) and concentration of carriers in centers (2); the parameters are these of row 14 in Table 1, and the decay curve is calculated for 450 K

Fig. 7. Decay curves of phosphorescence (1) and concentration of carriers in centers (2); the parameters are those of row 9 in Table 1, and the decay curve is calculated for 270 K

different slopes in curves 1 and 2. This is opposite to that of the simple first order case in which both the phosphorescence and the concentration of carriers decay at the same rate.

4. Discussion

A model for the thermoluminescence process governed by simultaneous thermal stimulation of electrons and holes is investigated here by numerically solving four coupled rate equations without simplifying assumptions. When the thermal hole emission was substantial, mainly due to low activation energy for holes, the calculated peak occurred at relatively low temperatures, and had an effective cut-off on the high temperature side. In all the cases examined it was possible to fit the calculated glow peak to a three parameter function. This yielded an effective activation energy, frequency factor and order of kinetics. In those peaks which displayed a rapid cut-off on the high temperature side the order of kinetics was much lower than otherwise expected (down to an effective order of 0.53 for the case of row 6 of Table 1). This result suggests an explanation for the experimental result by Partridge and May [8] who reported an effective order of kinetics of 0.7 ± 0.1 for some alkali halides. The present model cannot be utilized in its current form for the case of alkali halides, however. The actual energy band diagram in these materials is almost certainly much more complex than the simple picture described here, and in fact the TL production mechanism may even involve localized transitions between energy levels rather than transitions into and out of the delocalized bands. However, the results do highlight the feasibility of obtaining a kinetic order of less than 1.0.

Regarding the cut-off of TL peaks discribed above it is to be noted that a similar cut-off has been discussed in a theoretical paper by Halperin and Braner [2]. These authors discuss qualitatively the cut-off of different peaks due to a shortage of carriers in the luminescence centers. The shortage they discuss is due, however, to a low initial concentration in the centers rather than to a thermal release of these carriers as discussed in the present work.

Fig. 6 and 7 deserve comment. As seen in Table 1, and discussed above, the effective activation energy turned out to be rather close to the original value of the activation energy for trapped electrons, the largest discrepancy being 10% in the case of row 9 in Table 1. Fig. 7 shows the isothermal decay with time of both $m(t)$ and the phosphorescence intensity for this case. The time constant for the decay of $m(t)$ is governed primarily by the activation energy for the release of holes (0.7 eV) whereas the time constant for the phosphorescence decay is related to the effective activation energy for luminescence (viz. 0.9 eV). Thus, one might expect that the phosphorescence decay would be *slower* than the decay of $m(t)$. However, it can be seen in Fig. 7 that the phosphorescence dacay is *faster* than the decay of $m(t)$. At first sight this might seem "anomalous" in that the luminescence is decaying faster than it should (eg. see Visocekas and Geoffroy [18]). However, it must be remembered that the *effective* frequency factor is unusually high resulting in a TL peak which appears at quite a low temperature (269 K). Thus, the decay rate for phosphorescence is entirely consistent with the TL peak position, and in this respect is not anomalous at all.

For the case of row 14, Table 1, the decay rate for $m(t)$ is expected to be less than that for the phosphorescence, as is observed in Fig. 6.

It is useful to recall that high values for the frequency factor have been reported on several occasions (e.g. [19]). Although it would be unwise at this stage to claim that simultaneous hole and electron release is the cause of these high values (other mechanisms must also be considered (e.g. [20]), it is important to emphasize that high frequency factor values can emerge from applying the simplified three-parameter analysis to this more complicated model. The important question is, can the three-parameter analysis be used to predict the thermal stability of the luminescence despite the existence of a more-complex TL model? The results of Fig. 6 and 7 appear to indicate that it can. Future work is planned on the isothermal decay properties of phosphorescence and the stability of TL peaks in order to test this point further.

Another limitation allowed in this study, besides the neglect of hole retrapping mentioned earlier, was the fact that only the recombination of free electrons with holes in centers was considered to yield TL. This may have excluded interesting results. In fact, it has been suggested [13] in the past that either this recombination, or the recombination of free holes with trapped electrons, or the sum of the two (and possibly some weighted sum) could yield the TL. Finally, it is to be noted that the functions $n_c(t)$ and $p(t)$, namely concentrations of free electrons and holes in the conduction and valence bands respectively, can be obtained from the present computation. The relation of these functions to the measurable thermally stimulated conductivity is to be explored in the future.

References

[1] E. Adirovitch, J. Phys. Radium 17, 705 (1956).
[2] A. Halperin and A. A. Braner, Phys. Rev. 117, 408 (1960).
[3] R. Chen and Y. Kirsh, Analysis of Thermally Stimulated Processes, Pergamon Press, Oxford 1981.
[4] P. Kelly, M. J. Laubitz, and P. Bräunlich, Phys. Rev. B 4, 1960 (1971).
[5] D. Shenker and R. Chen, J. Computer Phys. 10, 272 (1972).
[6] S. Haridoss, J. Computer Phys. 26, 232 (1978).
[7] C. E. May and J. A. Partridge, J. chem. Phys. 40, 1401 (1964).
[8] J. A. Partridge and C. E. May, J. chem. Phys. 42, 797 (1965).
[9] R. Chen, J. Electrochem. Soc. 116, 1254 (1969).
[10] J. T. Randall and M. H. F. Wilkins, Proc. Roy. Soc. A184, 347 (1945).
[11] M. Schön, J. Phys. Radium 17, 689 (1956).
[12] M. Schön, Tech.-wiss. Abh., Osram-Studienges. 7, 175 (1958).
[13] P. Bräunlich and A. Scharmann, phys. stat. sol. 18, 307 (1966).
[14] J. Gasiot and J. P. Fillard, J. appl. Phys. 48, 3171 (1977).
[15] V. K. Mathur and M. D. Brown, J. appl. Phys. 54, 5485 (1983).
[16] R. Chen, S. W. McKeever, and S. A. Durrani, Phys. Rev. B 1, 24, 4931 (1981).
[17] D. Shenker and R. Chen, J. Computer Phys. 10, 272 (1972).
[18] R. Visocekas and A. Geoffroy, phys. stat. sol. (a) 41, 499 (1977).
[19] S. P. Kathuria and C. M. Sunta, J. Phys. D 12, 1573 (1979).
[20] P. D. Townsend, G. C. Taylor, and M. C. Wintersgill, Radiat. Eff. 41, 11 (1979).

(Received June 27, 1984)

phys. stat. sol. (b) **126**, 371 (1984)

Subject classification: 13.4 and 19; 22.8

Institute of Physics, Technical University of Częstochowa[1]) *(a)*
and Institute of Physics, Pedagogical College, Opole[2]) *(b)*

EPR Powder Spectra of Low-Symmetric Mn^{2+} Complexes in Aragonite Lattice[3])

By

J. LECH (a), A. ŚLĘZAK (a), and I. BÓJKO (b)

Electron paramagnetic resonance of Mn^{2+} ions in polycrystalline samples of natural aragonite containing about 100 ppm Mn impurities is studied at X- and K-band. The parameters of the rhombic spin Hamiltonian estimated from the "powder" spectrum are found to be in a good agreement with the parameters obtained with single crystals. It is shown that satisfactorily accurate values of the spin Hamiltonian parameters can be extracted without the use of a simulating method, even in the case of the rather complex "powder" spectrum recorded at X-band.

Die paramagnetische Elektronenresonanz von Mn^{2+}-Ionen in natürlichem polykristallinen Aragonit (ungefähr 100 ppm Manganverunreinigungen) wird im X- und im K-Band untersucht. Es stellt sich heraus, daß die aus dem "Pulver"-Spektrum abgeschätzten Parameter des rhombischen Spin-Hamiltonoperators weitgehend mit den Parametern übereinstimmten, welche sich aus den mit Einkristallen erhaltenen Spektren ergeben. Es wird gezeigt, daß ausreichende Genauigkeit der Parameterwerte des Spin-Hamilton-Operators ohne Simulationsverfahren erreicht werden kann selbst wenn das verhältnismäßig komplizierte „Pulver"-Spektrum im X-Band registriert wird

1. Introduction

The Mn^{2+} ($S = 5/2$, $I = 5/2$) ion is often used as a spin-label to investigate the structure of elementary cells and the distribution of electric fields in polycrystalline solids (see for example [1 to 4]). For this purpose the Mn^{2+} ions originally contained in the sample in the form of a natural admixture may be used as well as those especially introduced in the course of preparation. At the same time the analysis of Mn^{2+} EPR spectra for obtaining information on the lattice structure in a polycrystalline sample is rather difficult. The hyperfine interaction, the zero-field splitting, and the forbidden transitions determine the shape of the spectrum. All these factors, when combined together, make the spectrum extremely complicated. However, using the Mn^{2+} ions in structural studies is advantageous because the EPR spectra of these ions usually reflect the true point symmetry of the lattice [5].

The question of the analysis of Mn^{2+} "powder" EPR spectra in cases of rhombic and lower symmetry is particularly complicated. There are some evidences that different sets of spin Hamiltonian parameters used for computation of the Mn^{2+} "powder" spectra (in a simulation technique adapted for interpreting the experimental spectra) yield very similar patterns [4]. Thus, it is rather important to get an appropriate set of spin Hamiltonian parameters as a starting set in a computer fitting procedure. A simple approach to an extraction of this set from an experimental low-symmetric Mn^{2+} spectrum is presented in this paper. The usefulness and sufficient precision of

---

[1]) 42-200 Częstochowa, Poland.

[2]) 45-060 Opole, Poland.

[3]) This work was supported by the Institute for Low Temperatures and Structure Research, Polish Academy of Sciences, Wrocław.

this approach is shown for the "powder" spectrum of the Mn^{2+} ions in a nine-fold oxygen coordination in the aragonite lattice.

2. EPR of Mn^{2+} Ions in Aragonite Lattice

Aragonite is a rhombic (pseudohexagonal) $CaCO_3$ modification. The EPR of Mn^{2+} ions in natural aragonite single crystals has been studied in detail by the authors of this paper. Some results have been presented earlier [6]. There are two magnetically inequivalent Mn^{2+} sites in the aragonite unit cell which belongs to the space group Pmcn(D_{2h}^{16}). The EPR spectra of both of these sites are well described by a rhombic spin Hamiltonian in the form [6, 7]

$$H = \beta \boldsymbol{B} \cdot \hat{g} \cdot \boldsymbol{S} + \frac{D}{3} O_2^0 + E O_2^2 + \frac{a}{120} (O_4^0 + 5 O_4^4) + \frac{F}{180} O_4^0 + + \boldsymbol{S} \cdot \hat{A} \cdot \boldsymbol{I} - \beta_N \boldsymbol{B} \cdot \hat{g}_N \cdot \boldsymbol{I} + \boldsymbol{I} \cdot \hat{P} \cdot \boldsymbol{I} . \quad (1)$$

The coefficients and operators in (1) have their usual meanings [8] and are considered to be expressed in the system of the principal axes x, y, and z. It is assumed that the principal axes of $\hat{g}$, $\hat{g}_N$, $\hat{A}$, and $\hat{P}$ coincide.

Table 1
Spin Hamiltonian parameters of Mn^{2+} EPR spectra in aragonite lattice (Room temperature, D, E, $a + \frac{2}{3} F$, and A are in 10^{-4} cm^{-1})

	single crystal sample (ref. [6])	polycrystalline and powdered samples			
		X-band (this work)		K-band	Q-band
		method I	method II	(this work)	(ref. [9])
g_x	2.0044				
g_y	1.9910				
g_z	2.0023				
g_{iso}		2.00 ± 0.01	2.00 ± 0.01	2.00 ± 0.002	2.006 ± 0.001
D	218.8	220 ± 8	220 ± 10	220 ± 5	234 ± 10
E	−49.3	50 ± 5	50 ± 10	50 ± 5	39 ± 5
A_x	−88.7				
A_y	−87.6				
A_z	−90.2				
A_{iso}		88 ± 1	88 ± 1	88 ± 1	88 ± 1
$a + \frac{2}{3} F$	4.0				

The spin Hamiltonian parameters are the same for the two sites (Table 1). Thus in a finely powdered aragonite sample an EPR spectrum similar to that of a single Mn^{2+} site should be expected. The spectra of the two sites may be resolved only in samples including large sized and partly ordered microcrystals [9].

3. Experimental

Samples of natural aragonite single crystals were used in this work. They contained about 100 ppm Mn impurities and traces of Fe. Polycrystalline samples were obtained through grinding with an agate mill. Sizes of grains in powdered samples did not exceed 80 μm. The measurements were performed by means of an X-band (9.6 GHz) spectrometer Model SE/X28 and a K-band (24.3 GHz) spectrometer JEOL. The magnetic field strengths were measured with a proton resonance magnetometer. A DPPH probe and the Varian pitch were used as standard g-value markers.

4. Results and Discussion

It has been shown in [6] that the zero-field splitting is rather small in comparison with the Zeeman splitting for the Mn^{2+} ion in the aragonite lattice, even in the X-band. In this case, the line positions of the Mn^{2+} EPR spectra obtained with single crystal samples can be satisfactorily described using the expression obtained for the energy levels by the perturbation method in the usual high-field representation up to third-order terms. For powdered and polycrystalline samples (with local electric fields randomly oriented relatively to an applied magnetic field $\boldsymbol{B}$) all components of the fine structure except that with $M_s = \frac{1}{2} \leftrightarrow -\frac{1}{2}$ (M_s the magnetic quantum number of the electron spin) are broadened to such an extent that they cannot be practically observed if (as usually) the derivative of the absorption is recorded.

Thus, only a resonance signal due to the $|\frac{1}{2}, m\rangle \leftrightarrow |-\frac{1}{2}, m + \Delta m\rangle$ transitions is observed in the experimental spectrum. In the above m is the magnetic quantum number of the nuclear spin I and Δm is the change in m, $\Delta m = 0$ for allowed EPR transitions and $\Delta m \neq 0$ for forbidden ones, $|m + \Delta m| \leq I = 5/2$.

It can be shown that the expression for the resonance line positions due to the mentioned transitions is the following [10]:

$$B = B_0(m, \Delta m) + \frac{16}{B} (|\lambda|^2 - 2\,|\varrho|^2) - (2m + \Delta m)\,\frac{8A}{B^2}\,(9\,|\lambda|^2 - 2\,|\varrho|^2) -$$

$$- (2m + \Delta m)\,(1 + \Delta m)\,\frac{4A^2\sigma}{B^2} - \frac{g_N \beta_N}{g\beta}\,B\,\Delta m + P\,\Delta m(2m + \Delta m)\,, \quad (2)$$

where

$$B_0(m, \Delta m) = B_0 - (2m + \Delta m)\,\frac{A}{2} +$$

$$+ [(2m + \Delta m)^2 + (17 - \Delta m)^2 - 324]\,\frac{A^2}{8B} -$$

$$- [4(m + \Delta m)^3 + 4m^3 - (2m + \Delta m)\,(65 + 134\Delta m)]\,\frac{A^3}{4B^2}\,, \quad (2a)$$

$$B_0 = h\nu/g\beta\,, \quad (2b)$$

$$\sigma = \tfrac{1}{2}\,[D(3\cos^2\theta - 1) + 3E\sin^2\theta\cos 2\varphi]\,, \quad (2c)$$

$$|\lambda|^2 = \sin^2\theta\,[(D - E\cos 2\varphi)^2\cos^2\theta + E^2\sin^2 2\varphi]\,, \quad (2d)$$

$$|\varrho|^2 = \tfrac{1}{16}\,\{[D\sin^2\theta + E(1 + \cos^2\theta)\cos 2\varphi]^2 + 4E^2\cos^2\theta\sin^2 2\varphi\}\,, \quad (2e)$$

$$P = \tfrac{1}{2}\,[P_{||}(3\cos^2\theta - 1) + 3P_{\perp}\sin^2\theta\cos 2\varphi]\,. \quad (2f)$$

θ and φ are the azimuthal and polar angles of the respective magnetic field direction The values g and A for this direction are related to the principal values g_x, g_y, g_z and A_x, A_y, A_z in the known manner [6, 8].

However, when the "powder" spectrum is analyzed the small anisotropy in the hyperfine constant A and g-factor is usually neglected for the Mn^{2+} ion (see for example [4]); i.e. $A = A_{iso}$ and $g = g_{iso}$ within the experimental error. The terms

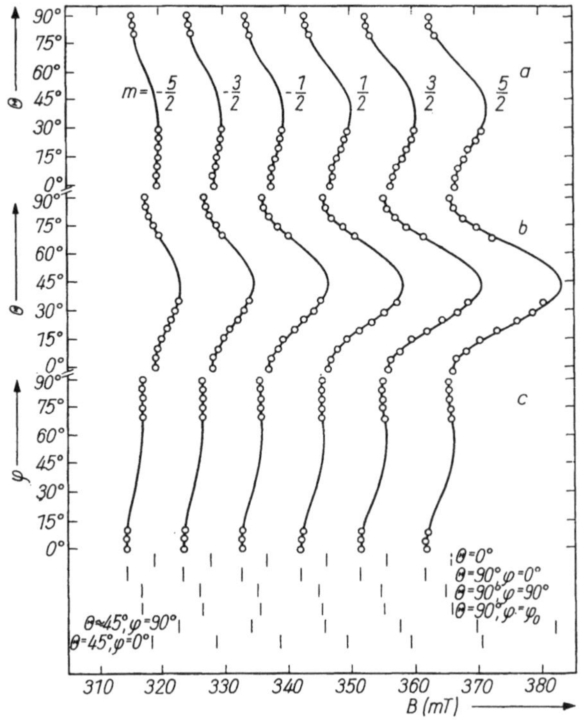

Fig. 1. Angular variations of positions of the allowed $|\frac{1}{2}, m\rangle \leftrightarrow |-\frac{1}{2}, m\rangle$ lines in the X-band EPR spectrum of Mn^{2+} ions in aragonite single crystals and predicted positions of peaks in the "powder" spectrum. The experimental results are denoted with circles. The solid curves show the results of calculations. a), b) and c) are the results for the principal planes (x, z), (y, z), and (x, y), respectively. Short vertical lines at the lower part of the figure indicate the positions of different "powder" peaks

including the parameters a and F cause only very small shifts of the lines under consideration. So these terms are omitted in (2).

As exemplified in Fig. 1, (2) is applicable for a good description of the angular variation of the $|\frac{1}{2}, m\rangle \leftrightarrow |-\frac{1}{2}, m + \Delta m\rangle$ line positions of the Mn^{2+} spectrum in the aragonite single crystal. (Hereafter, the spectrum recorded with the single crystal will be referred to as the "single crystal" spectrum and that obtained with the polycrystalline sample — as the "powder" spectrum.)

There are some singularities (extremal points) in the curves showing these variations in the three principal planes of a paramagnetic complex. In the vicinity of such a singularity a change in the direction of $\boldsymbol{B}$ gives only a small change in the resonance field value; i.e. B_{res} is stationary.

The stationary field values are determined through the conditions [10, 11]

$$\frac{1}{\sin\theta}\frac{\partial B}{\partial \theta} = 0 \quad \text{and} \quad \frac{\partial B}{\partial \varphi} = 0\,. \tag{3}$$

"Powder" peaks occur only at the stationary field values.

There are six possible $\boldsymbol{B}$ field directions (relative to the principal axes of the paramagnetic complex) for which conditions (3) are fulfilled. This means, in turn, that each of the $|\frac{1}{2}, m\rangle \leftrightarrow |-\frac{1}{2}, m + \Delta m\rangle$ hyperfine components of the "single crystal" spectrum can split into six "powder" peaks in the spectrum obtained with a polycrystalline sample (as seen in Fig. 1). Three of these "powder" peaks appear at the $\boldsymbol{B}$ field parallel to each of the three principal axes:

(i) $\boldsymbol{B} \parallel z$ $(\theta = 0°)$, the peak denoted through $B_z \equiv B_1$. (4)

(ii) $$\boldsymbol{B} \parallel \boldsymbol{x}\ (\theta = 90° \text{ and } \varphi = 0°), \text{ the peak } B_x \equiv B_2 \,. \tag{5}$$

(iii) $$\boldsymbol{B} \parallel \boldsymbol{y}\ (\theta = 90° \text{ and } \varphi = 90°), \text{ the peak } B_y \equiv B_3 \,. \tag{6}$$

The other three "powder" peaks correspond to the other extremal points (maxima in Fig. 1) in the curves showing the angular dependences of "single crystal" spectrum line positions obtained for the the principal planes. These peaks are:

(iv) In the (x, y) plane the peak $B_{xy} \equiv B_4$ can be observed at the $\boldsymbol{B}$ direction determined by

$$\theta = 90° \quad \text{and} \quad \cos^2 \varphi_0 = \frac{1}{2}\left(1 + \frac{qD}{E}\right), \tag{7}$$

where

$$q = \frac{2 - (2m + \Delta m)\,\dfrac{A}{B} + 3(2m + \Delta m)\,(1 + \Delta m)\,\dfrac{A^2}{DB}}{18 - 73(2m + \Delta m)\,\dfrac{A}{B}} \,. \tag{7a}$$

This peak is observed when $|q| \leqq |E/D|$.

(v) and (vi): In the planes (x, z) and (y, z) the peaks $B_{xz} \equiv B_5$ and $B_{yz} \equiv B_6$ are observed, when the applied $\boldsymbol{B}$ field is directed at the angles

$$\theta_0 \text{ and } \varphi = 0° \,, \tag{8}$$

and

$$\theta_0 \text{ and } \varphi = 90° \,, \tag{9}$$

respectively.

θ_0 is given by the formula

$$\cos^2 \theta_0 = \frac{10D \mp 6E - (2m + \Delta m)\,(37D \mp 35E)\,\dfrac{A}{B} - 3(2m + \Delta m)\,(1 + \Delta m)\,\dfrac{A^2}{BD}}{\left[18 - 73(2m + \Delta m)\,\dfrac{A}{B}\right](D \mp E)} \,. \tag{10}$$

The different observability of the "powder" peaks is essentially determined by two factors.

The first factor is the number dN of grains including the spins whose directions are closely connected with the values θ and φ determining the stationary field value B_{res}. This number is proportional to the elementary body angle $d\Omega$, i.e.

$$dN \sim d\Omega = 2\pi \sin \theta \, d\theta \, d\varphi \,. \tag{11}$$

Thus, the allowed peaks $B_z \equiv B_1$ $(\theta = 0°)$ are usually very weak and it may be impossible to identify them in the experimental "powder" spectrum.

The probabilities of microwave transitions which determine the relative intensities of the lines in the "single crystal" spectrum are the second factor. The studies carried out in this work have shown that the intensities of the spectrum lines of the Mn^{2+} ions in the aragonite lattice exhibit a strong angular variation when the static magnetic field direction is varied with respect to the principal axes. In order to describe this variation, Bir's theory [12] has been used. This theory gives expressions for the transition probabilities in terms of the spin Hamiltonian parameters.

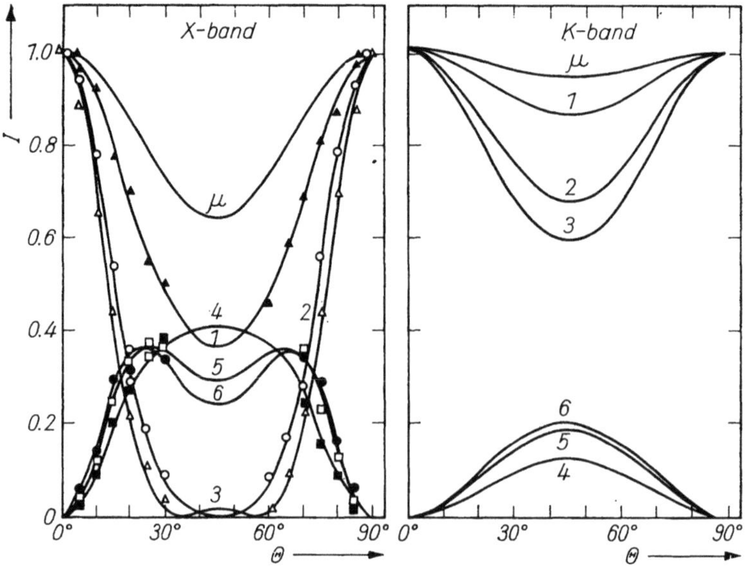

Fig. 2. Angular variations of the relative intensities of the $|\frac{1}{2}, m\rangle \leftrightarrow |-\frac{1}{2}, m'\rangle$ lines in the EPR spectrum of Mn^{2+} ions in the aragonite lattice, principal (y, z) plane. Points denote the experimental results. The solid curves (numbered 1 to 6) are the results of calculations with Bir's theory [12]. μ is the cosine of the angle formed by the effective field directions for the electron states with $M = \frac{1}{2}$ and $M' = -\frac{1}{2}$. $m \leftrightarrow m'$: ▲, (1) $\pm\frac{5}{2} \leftrightarrow \pm\frac{5}{2}$; ○, (2) $\pm\frac{3}{2} \leftrightarrow \pm\frac{3}{2}$; △, (3) $\pm\frac{1}{2} \leftrightarrow \pm\frac{1}{2}$, ■, (4) $\pm\frac{5}{2} \leftrightarrow \pm\frac{3}{2}$; □, (5) $\pm\frac{3}{2} \leftrightarrow \pm\frac{1}{2}$; ●, (6) $\pm\frac{1}{2} \leftrightarrow \mp\frac{1}{2}$

According to Bir the probability P of a transition from the state $|M, m\rangle$ to the state $|M', m'\rangle$ under the action of the microwave field is the product of the electronic transition probability $P_{e,MM'}$ and the nuclear transition probability $P_{n,mm'}$,

$$P = P_{e,MM'} P_{n,mm'} . \tag{12}$$

The $P_{n,mm'}$ are some functions of $\mu_{MM'}$ which is the cosine of the angle between the effective magnetic field directions of the electrons in the initial and the final states. Explicit forms of these functions can be derived for Mn^{2+} from formulae given in Bir's paper [12]. It may be also shown that for the hyperfine transitions $|\frac{1}{2}, m\rangle \leftrightarrow |-\frac{1}{2}, m'\rangle$ (i.e. those which are important in the analysis of the "powder" spectrum under consideration) the μ value is equal to

$$\mu_{1/2,\,-1/2} \equiv \mu = 1 - 512 \frac{|\lambda|^2}{(h\nu)^2} , \tag{13}$$

where $|\lambda|^2$ is defined by (2d).

The relative intensity of the spectrum line is proportional to the relative probability P, i.e.

$$I \sim P . \tag{14}$$

In this work the normalization has been done so that

$$I_{max} = P_{max} = 1 , \tag{15}$$

and the μ's and I's have been evaluated at X-band ($\nu = 9.62$ GHz) and K-band ($\nu = 24.33$ GHz) for the three principal planes of the Mn^{2+} paramagnetic complex in

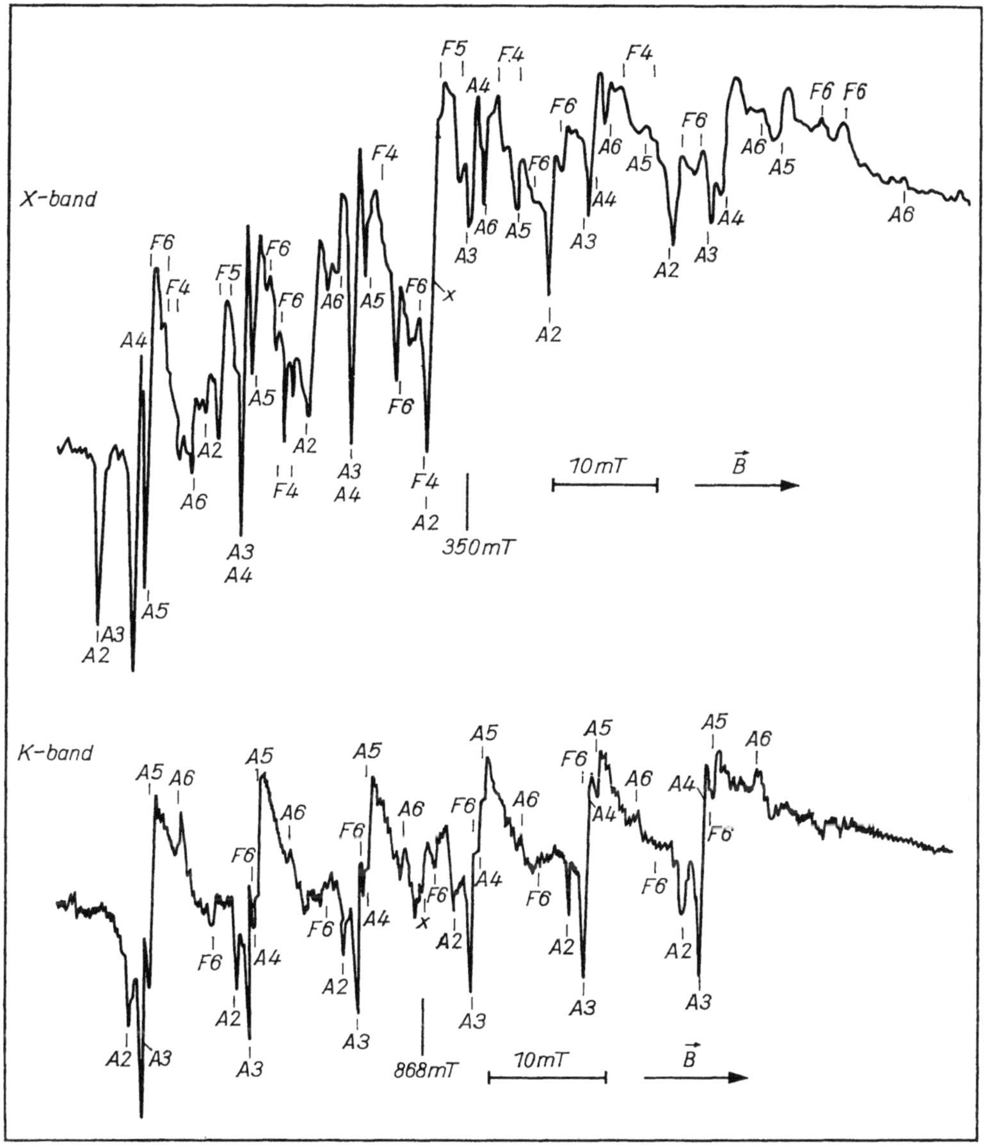

Fig. 3. Experimental EPR spectra of the Mn^{2+} ions in the polycrystalline aragonite sample. The identified allowed peaks are denoted with A, the forbidden ones F. The figures 2 to 6 denote different "powder" peaks defined by (5) to (10) (see the text). X is a signal due to an unidentified absorption centre

the aragonite lattice. Measurements of the line intensities have been performed in detail at the X-band. The calculated and experimental values of the relative intensities of the allowed $|\frac{1}{2}, m\rangle \leftrightarrow |-\frac{1}{2}, m\rangle$ lines and of the forbidden $|\frac{1}{2}, m\rangle \leftrightarrow |-\frac{1}{2}, m \mp 1\rangle$ ones obtained for the (y, z) plane are shown in Fig. 2. The μ variation in this plane is also drawn. As seen in this figure, the observed angular dependence of the $|\frac{1}{2}, m\rangle \leftrightarrow$ $\leftrightarrow |-\frac{1}{2}, m'\rangle$ $(m' - m = \Delta m = 0, \pm 1)$ lines of the "single crystal" spectrum in the

(y, z) plane has been found to agree with the results of Bir's theory, at least at X-band. Similar agreement has been obtained for the other principal planes.

Basing on this fact two conclusions, which seem to be very significant in the interpretation of the Mn^{2+} experimental "powder" spectra, have been drawn. The conclusions are the following:

(i) Bir's theory [12] should be used in the analysis of the Mn^{2+} ion "powder" spectra.

(ii) The curves representing angular variations of the transition probabilities drawn in the principal planes combined with (11) and the curves for the line position anisotropy in the principal planes may be used to predict which of the "powder" peaks determined by (3) to (10) can be observable in the experimental "powder" spectrum.

These conclusions were found to be very useful for an interpretation of the experimental "powder" spectra of the Mn^{2+} ions in aragonite recorded at X- and K-band.

Both spectra are presented in Fig. 3. It can be seen in this figure that the X-band spectrum is more complicated than the K-band one. However, this fact is not surprising as the strong forbidden $\Delta m = \pm 1$ peaks B_{xy}, B_{xz}, and B_{yz} might be predicted at the X-band and the weak ones at the K-band. (These peaks are marked with F4, F5, and F6 in Fig. 3). On the other hand, there were some difficulties with the identification of the allowed peaks B_{xy}, B_{xz}, and B_{yz} (A4, A5, and A6 in Fig. 3) in the X-band spectrum.

Meanwhile, the correct identification of the allowed peaks B_x (A2), B_y (A3), B_{xz} (A5), and B_{yz} (A6) seemed to be necessary for an easy extraction of values of the spin-Hamiltonian parameters D and E. It can be shown for the m-th allowed hyperfine component that

$$\delta_1 = B_y - B_x = B_{A3} - B_{A2} = 8\,\frac{ED}{B_0}\left(1 - \frac{mA}{B_0}\right)\left(1 - 3\,\frac{A}{D}\right) \tag{16}$$

and

$$\begin{aligned}\delta_2 + \delta_3 &= (B_{xz} - B_x) + (B_{yz} - B_y) = (B_{A5} - B_{A2}) + (B_{A6} - B_{A3}) = \\ &= 4t\left[(10 - 9t)\,\frac{D^2}{B_0} + 3\,(2 - 3t)\,\frac{E^2}{B_0}\right] - \\ &\quad - 4mt\left[(74 - 73t)\,\frac{D^2A}{B_0^2} + (70 - 73t)\,\frac{E^2A}{B_0^2} + 6\,\frac{A^2D}{B_0^2}\right].\end{aligned} \tag{17}$$

It is assumed in (16) and (17) that $B = B_0$, g, and A are nearly isotropic, and t is the $\cos^2 \theta_0$ term given by (10).

It has been easy to identify all allowed "powder" peaks (apart from the $B_z \equiv B_1$) in the spectrum recorded at K-band and the peaks A2 and A3 of the low-field hyperfine component of the X-band experimental spectrum. Practically, in dependence on the real, relative signs of D and E, the axes x and y may be exchanged. Thus, the differences δ_1, δ_2, and δ_3 in (16) and (17) should be considered to be absolute values. This means, in turn, that only the absolute D and E values may be extracted from the "powder" spectra. In order to estimate these absolute D and E values, at first an approximate ED absolute product was evaluated using (16) and taking into account only the second term each of $8(ED/B_0)$.

At the K-band this approximate ED value was extracted from the averaged splittings δ_1 measured for every allowed hyperfine component. The value $(120 \pm 4) \times$ $\times\ 10^{-6}\ T^2$ was obtained, in satisfactory agreement with that calculated from the "single crystal" spectrum data, i.e. $124 \times 10^{-6}\ T^2$. It was also quite easy to extract

$A_{iso} \approx A_{\perp} \approx A_x \approx A_y$ from the K-band experimental "powder" spectrum. Then it was possible to estimate the absolute values D and E on the basis of (17) simplified by neglecting the third-order terms and assuming

$$t = \cos^2 \theta_0 \approx \cos^2 45^\circ = \tfrac{1}{2}\,.$$

These D and E values were estimated from each of the six $|\frac{1}{2}, m\rangle \leftrightarrow |-\frac{1}{2}, m\rangle$ components. Then the average values were calculated and used as the starting ones in a numerical iteration procedure to obtain more correct values of t, D, and E. The third-order terms in (16) and (17) were taken into account this time. The calculations were simple and were carried out with a TI-59 calculator. The suitability of using no more than three steps of the iteration procedure was stated. The corrected D and E values and earlier extracted A_{iso} were used for calculating the corrected value of the g_{iso}-factor. The final results are included in the fifth column of Table 1.

As mentioned above, in the case of the X-band experimental "powder" spectrum only the peaks A2 and A3 for the low-field $|\frac{1}{2}, -\frac{1}{2}\rangle \rightarrow |-\frac{1}{2}, -\frac{1}{2}\rangle$ hyperfine component were suitable for identification. It was, however, enough to estimate the approximate ED product, which was found to be $(118 \pm 12) \times 10^{-6}\ T^2$. Further analysis of the X-band "powder" spectrum was performed with numerical techniques. They were used in two ways.

At first, the method of simulating the Mn^{2+} "powder" spectra described in [4] was applied. However, a strong reduction of computing time was possible because only D and values yielding the ED product in the vicinity of the estimated value $118 \times \times 10^{-6}\ T^2$ may be taken as starting ones. The value 9.5 mT was taken as the starting A_{iso} value, as in most experimental evidences for Mn^{2+} ions coordinated by eight and more oxygen ligands, A is close to this value [13]. The value 0.3 mT was used as the starting one for the individual linewidth ΔB_{pp}. The program was written in Fortran and an ODRA 1305 computer was used. Satisfactory agreement between simulated and experimental shapes of the spectrum was achieved for the values included in the third column of Table 1 and for the linewidth $\Delta B_{pp} = 0.35$ mT.

In the second method only the angular variations of the resonance field values and the intensities of the lines $|\frac{1}{2}, m\rangle \rightarrow |-\frac{1}{2}, m + \Delta m\rangle$ ($\Delta m = 0, \pm 1$, and ± 2) of the "single crystal" spectrum were calculated for the three principal planes. Again, the results of the calculations in a graphical form permitted a quick and simple identification of most of the "powder" peaks in the experimental X-band spectrum. This allowed, in turn, to achieve more accurate spin Hamiltonian parameters. The final results obtained in this method are listed in the fourth column of Table 1. They are in clear agreement with those obtained with the simulating method. This means a possibility of a satisfactory interpretation of rather complex "powder" spectra of Mn^{2+} ions without necessity of using the simulating method.

References

[1] D. L. Griscom and R. E. Griscom, J. chem. Phys. **47**, 2711 (1967).
[2] J. S. Shaffer, H. A. Farach, and C. P. Poole, Jr., Phys. Rev. B **13**, 1869 (1976).
[3] V. Beltrán-Lopez and J. Castro-Tello, J. magnetic Resonance **39**, 437 (1980).
[4] J. Kliava and J. J. Purāns, J. magnetic Resonance **40**, 33 (1980).
[5] K. A. Müller, Helv. phys. Acta **31**, 173 (1958).
[6] A. Ślęzak, J. Lech, and I. Bójko, phys. stat. sol. (a) **54**, 755 (1979).
[7] W. D. Posener, J. magnetic Resonance **9**, 128 (1973).
[8] A. Abragam and B. Bleaney, Electron Paramagnetic Resonance of Transition Ions, Clarendon Press, Oxford 1970.
[9] L. K. White, A. Szabo, P. Carkner, and N. D. Chasteen, J. phys. Chem. **81**, 1420 (1977).

[10] J. LECH, Thesis, Silesian University, Katowice 1982.
[11] P. C. TAYLOR, J. F. BAUGHER, and H. M. KRIZ, Chem. Rev. 75, 203 (1975).
[12] G. L. BIR, Fiz. tverd. Tela 5, 2236 (1963).
[13] S. A. ALTSHULER and B. M. KOZYREV, Electron Paramagnetic Resonance of Compounds of Transition Element Groups, Izd. Nauka, Moscow 1971 (in Russian).

(Received July 30, 1984)

phys. stat. sol. (b) **126**, 381 (1984)

Subject classification: 1.2 and 13.4; 18.4; 20.1; 22.8.1

Physikalisches Institut der Universität Karlsruhe[1])

Spectroscopic Study of the Crystal Field Splittings in $TbPO_4$

By

W. Böhm[2]), H. G. Kahle, and W. Wüchner[3])

The optical absorption and emission spectra of Tb^{3+} in $TbPO_4$ in its tetragonal structure above 2.3 K are studied to obtain the energies, symmetry types, and wave functions of mainly the lowest crystal field components of the ground term $^7F_6$. Crystal field parameters are evaluated and used to calculate the anisotropy of the single-ion energy in a semi-classical way. This gives interesting indications for the types of magnetic structures and Jahn-Teller distortion which occur in this substance below 2.3 K.

Die optischen Absorptions- und Emissionsspektren von Tb^{3+} in $TbPO_4$ werden oberhalb 2,3 K untersucht, um die Energien, die Symmetrietypen und die Eigenfunktionen vor allem der tiefsten Kristallfeldkomponenten des Grundterms $^7F_6$ zu bestimmen. Die Kristallfeld-Parameter werden ermittelt und zu einer halbklassischen Berechnung der Einzelionen-Energie benutzt. Hieraus ergeben sich interessante Aufschlüsse darüber, von welcher Art die magnetischen Strukturen und die Jahn-Teller-Verzerrung sein können, die in dieser Substanz unterhalb 2,3 K beobachtet werden.

1. Introduction

The phosphates, vanadates, and arsenates of the rare earths (RE) with the chemical formula $REXO_4$ (X = P, V, As) have obtained substantial attraction in the past because of their interesting low-temperature properties. Many of these compounds undergo a structural phase transition as a consequence of a cooperative Jahn-Teller interaction [1] at a temperature T_D ($<$ 40 K) and/or show a phase transition to a magnetically ordered state at a temperature T_N ($<$ 4 K).

$TbPO_4$ is a member of this series of compounds. The substance has been investigated intensively and repeatedly by different authors with various experimental methods, e.g. specific heat [2 to 5], magnetic measurements [2, 5, 6], optical spectroscopy [7 to 10], and neutron diffraction [11 to 13]. Nevertheless, its properties are less understood than those of all other members since some of the experiments give differing and partly contradictory results.

Free from doubt and common to all investigations is that $TbPO_4$ exhibits a magnetic phase transition at T_N of about 2.2 K to an antiferromagnetic colinear two-sublattice structure with the Tb^{3+} moments in the (110) plane canted off the tetragonal c-axis by an angle θ_0, for which values of $\theta_c = 40°$ [11] and 19.4° [12] have been published.

However, the question if the substance undergoes a structural phase transition above T_N has been open for a long time. Some authors observed a small second anomaly [2, 3] between 3 and 4 K, others did not find any feature in this region. These apparent discrepancies are known meanwhile to be caused by the fact that these substances are very sensitive to external stress (due to sample mounting) and to

[1]) Engesserstr. 7, P.O. Box 6380, D-7500 Karlsruhe 1, FRG.

[2]) Now at SIL, Speyer, FRG.

[3]) Now at BASF, Ludwigshafen, FRG.

internal strain (due to crystal imperfections). The latter seems to be especially important for $TbPO_4$ since it is the first member of the phosphate series crystallizing in the tetragonal zircon structure but still with a tendency to the monoclinic huttonite structure. Therefore, incontestable experimental results can only be obtained with selected crystals which are of good quality, handled with care, and mounted without stress.

A detailed investigation of the linear optical birefringence on carefully prepared crystals showed for the first time a splitting of the phase boundary at T_N [14], which could be confirmed by repeated studies of the specific heat [4, 5]. The transition temperatures are $T_{N,1} = (2.28 \pm 0.02)$ K and $T_{N,2} = (2.15 \pm 0.02)$ K. No indication of a further phase transition at higher temperatures was found.

Recent neutron diffraction measurements [13] demonstrated that the transition at $T_{N,1}$ is purely magnetic with antiferromagnetic ordering of the moments along the tetragonal c-axis (preserving tetragonal lattice symmetry). The second transition at $T_{N,2}$ is due to tilting of the magnetic moments off the c-axis together with a monoclinic distortion of the lattice. Both, tilt angle and monoclinic distortion increase with decreasing temperature. The lowering of the symmetry was also observed via induced infrared activity [10].

For a conclusive theoretical interpretation of the observed magnetic properties and phase transitions, the energies and wave functions of the lowest crystal field components of the Tb^{3+} ion in $TbPO_4$ must be known accurately. By measuring the far-infrared spectra and Zeeman splittings, Lewis and Prinz [9] were able to estimate the energies of the lowest six components, some years ago. Despite of a careful analysis, the distances between the lowest three components, which are especially critical for all calculations, remained somewhat uncertain since the corresponding transitions lay beyond the spectral range of their equipment.

Therefore, we have remeasured the visible absorption and emission spectra of $TbPO_4$ and its Zeeman effect with the aim to win good data for the energies, the wave functions, and the symmetry types of mainly the lowest crystal field components (Section 3). The effective crystal field is handled in the usual way by crystal field parameters (Section 4). With their help, the anisotropy of the energy of the single Tb^{3+} ion in the crystal is calculated in a semi-classical manner, giving instructive suggestions for the types of magnetic ordering and lattice distortion below $T_{N,1}$ and $T_{N,2}$ (Section 5).

2. Experimental Methods

The $TbPO_4$ single crystals used in this investigation were grown by a flux method [15] by which they crystallize in the tetragonal zircon structure $I4_1/amd$ with the point symmetry $\bar{4}m2$ ($\equiv D_{2h}$) for the Tb^{3+} sites. The crystal dimensions were typically about $1 \times 1 \times 3$ mm^3.

The absorption spectrum was taken with high resolution in the known spectroscopic arrangement [16] for temperatures between 2.3 and 20 K and magnetic fields up to 4.0 T. The emission spectrum was recorded on a Jarrell-Ash double monochromator with photomultiplier and photon counter. It was excited with the 488 nm line of an argon laser, which induces direct transitions to the term $^5D_4$ of Tb^{3+}. In this case, the crystal was either immersed into liquid helium ($T = 4.2$ K) or attached to a cold finger in vacuum. The sample temperature was then at about 12 or 100 K, as the cryostat was filled with liquid helium or nitrogen, respectively.

3. Experimental Results

3.1 Absorption spectrum

The optical absorption of Tb^{3+} in $TbPO_4$ shows at temperatures above 2.3 K some peculiarities which complicate the analysis of the spectrum extraordinarily: the lines are unusually broad and generally asymmetric; besides intense lines (doublet–doublet transitions) very weak lines (doublet–singlet transitions) appear; the lines are getting sharper in large magnetic fields, mainly for the perpendicular Zeeman effect.

The transitions $^7F_6$ (ground term) → $^5D_4$ at 20500 cm^{-1} were thoroughly investigated at different temperatures and field directions. With this the excited term $^5D_4$ was completely analysed, at first, and then the energies, the symmetry types, and the Zeeman effect of the lowest components of the ground term $^7F_6$ were determined.

In a crystal field of $\bar{4}m2$ symmetry the term $^7F_6$ is split into three doublets Γ_5 and seven singlets Γ_1, Γ_2, Γ_3, or Γ_4, the term $^5D_4$ is split into two doublets and five singlets. In electric dipole radiation (the only observed type of radiation), the doublet–doublet transitions are visible in π-polarisation (electric field vector $\boldsymbol{E}$ of the light parallel to the crystal c-axis), the doublet–singlet and singlet–doublet transitions in σ-polarisation. Singlet–singlet transitions, only partly not forbidden by selection rules, were not observed.

In the spectrum at $B = 0$, two intense π-transitions and three weak and two very weak σ-transitions are observed at 4 K. Since the lowest ground term component is known to be a doublet Γ_5 [7, 9] this yields the energies of the doublets and singlets of the term $^5D_4$ (see Table 1). The symmetry types of the singlets were derived from the behaviour in magnetic fields of different directions in the following manner.

The Zeeman effect $\boldsymbol{B} \parallel \boldsymbol{c} \parallel [001]$ gives the splittings of the doublets and the shifts of the singlets, which are drawn in Fig. 1 and partially stated in Table 1. To separate a possible asymmetry in the splitting of the lowest ground term doublet, it was presumed that the centre of gravity of the term $^5D_4$ remains unchanged in the magnetic field. The lowest two singlets a and b of the excited term show a strong non-linear Zeeman effect with induced magnetic moments which at large fields asymptotically reach the values $\pm s_{\parallel}\mu_B = \pm (4.8 \pm 0.5)\,\mu_B$ (μ_B Bohr magneton). Since $s_{\parallel}$ is proportional to the Landé factor $g_L = 1.5$ of the term, the wave functions of the singlets a and b must contain about 80% of the states $|M = \pm 4\rangle$ of the free ion and therefore correspond to the symmetry types Γ_1 and Γ_2. Consequently, the singlet e belongs to Γ_1 and the singlets d and g belong to Γ_3 or Γ_4.

Table 1

Crystal field components of the term $^5D_4$ of Tb^{3+} in $TbPO_4$ at 4.2 K (experimental data)

energy (cm^{-1})	symmetry type	symbol in the text	$s_{\parallel}$-factor
20 440.0 ± 0.8	Γ_1	a	−4.8 ± 0.5
20 444.2 ± 0.8	Γ_2	b	+4.8 ± 0.5
20 453.0 ± 0.5	Γ_5	c	3.8 ± 0.3
20 474.8 ± 0.3	Γ_3	d	
20 492.0 ± 0.8	Γ_1	e	
20 498.5 ± 0.5	Γ_5	f	−0.8 ± 0.3*)
20 508.5 ± 0.6	Γ_4	g	

*) The negative sign means that the Zeeman component with crystal quantum number $\bar{\mu} = -1$ is shifted to higher energies.

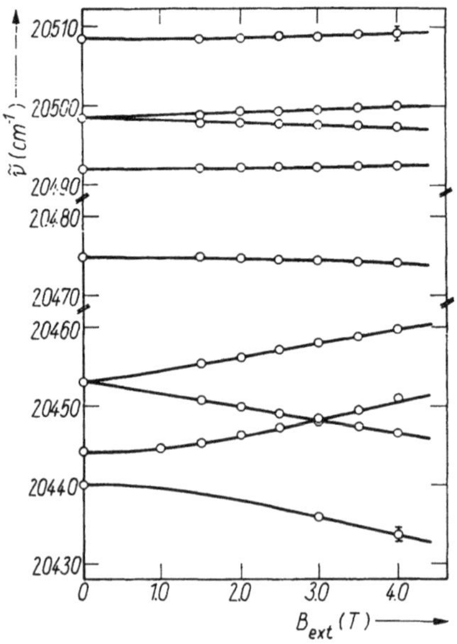

Fig. 1

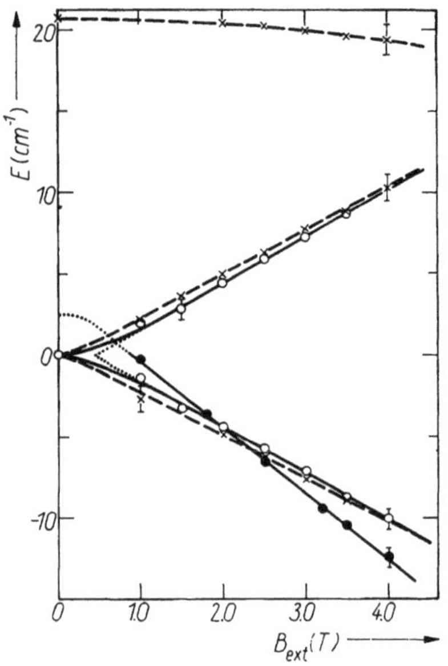

Fig. 2

Fig. 1. Zeeman effect $\boldsymbol{B} \parallel \boldsymbol{c}$ of the term $^5D_4$ at 4.2 K. Some measuring points are missing because of overlap with another intense transition. The drawn curves are fitted to the experimental points

Fig. 2. Zeeman effect $\boldsymbol{B} \parallel \boldsymbol{c}$ of the lowest components of the ground term $^7F_6$ at 4.2 (○) and 20 K (×). The full dots are valid for both temperatures. The drawn curves are fitted to the experimental points

A magnetic field $\boldsymbol{B} \parallel [100]$ reduces the symmetry of the Tb^{3+} sites from $\bar{4}$m2 (D_{2d}) to m (C_s). By this, each doublet is split in two singlets. The correlation of the symmetry types in $\bar{4}$m2 to those in m is known (see, e.g., [17]). From the spectra at $\boldsymbol{B} \parallel [100]$ and the selection rules for electric dipole radiation, valid for the symmetry m, one can easily derive the symmetry types of the components of the term $^5D_4$ in the magnetic field and (with the results of $\boldsymbol{B} \parallel \boldsymbol{c}$) extrapolate them back to those of the high symmetry $\bar{4}$m2 for $B = 0$. They are confirmed by the measurements at $\boldsymbol{B} \parallel [110]$ and are listed in Table 1.

Now that the components of the term $^5D_4$ are known, the Zeeman effect $\boldsymbol{B} \parallel \boldsymbol{c}$ of the lowest ground term components can be derived. The results are shown in Fig. 2. At 4 K, the doublet Γ_5 splits symmetrically with the splitting factor $s_{\parallel} = 6.1 \pm 0.3$. The linear backward-extrapolation of the experimental points to zero splitting renders an internal magnetic field (dipole and exchange) $B_{int} = -(0.45 \pm 0.15)$ T (antiparallel to the external field $\boldsymbol{B}$). At 20 K and in low fields, the doublet splitting is greater than at 4 K, but at large fields it approaches asymptotically the 4 K value. That means, the internal field, in the beginning, is small and can be brought to its full value only by a sufficiently large external field.

At medium magnetic fields $\boldsymbol{B} \parallel \boldsymbol{c}$, additional weak lines appear which correspond to transitions from a singlet II in the ground term (see Table 2). This singlet is very strongly shifted to lower energies by the field. For $B \geqq 2.1$ T it is the lowest ground term component. The measured splitting factor $s_{\parallel} = -8.8 \pm 0.5$ is nearly the maximum possible; therefore, the singlet is an almost pure $|M = -6\rangle$ state in high fields.

Table 2

Crystal field components of the ground term $^7F_6$ of Tb^{3+} in $TbPO_4$

experimental data			calculated data			
energy (cm^{-1})	symmetry type	symbol in the text	energy with J-mixing (cm^{-1})	symmetry	energy without J-mixing (cm^{-1})	wave functions
0.0	Γ_5	I	0.0	Γ_5	0.0	$0.90\|\pm5\rangle + 0.38\|\pm1\rangle + 0.21\|\mp3\rangle$
2.5 ± 1.0	Γ_3	II	2.5	Γ_3	1.9	$0.93\|6\rangle_s + 0.36\|2\rangle_s$
9.5 ± 0.5	Γ_4	III	9.6	Γ_4	10.3	$0.99\|6\rangle_a + 0.13\|2\rangle_a$
20.7 ± 1.0	Γ_1	IV	20.8	Γ_1	25.1	$0.83\|4\rangle_s + 0.56\|0\rangle$
80 ± 5	singlet	V	83.9	Γ_2	95.4	$1.00\|4\rangle_a$
87 ± 5	Γ_5	VI	86.7	Γ_5	96.6	$-0.39\|\pm5\rangle + 0.51\|\pm1\rangle + 0.76\|\mp3\rangle$
?	?	VII	92.7	Γ_3	101.8	$-0.36\|6\rangle_s + 0.93\|2\rangle_s$
254 ± 8	singlet	VIII	251.2	Γ_1	249.1	$-0.56\|4\rangle_s + 0.83\|0\rangle$
262 ± 8	Γ_5	IX	261.2	Γ_5	259.0	$-0.19\|\pm5\rangle + 0.77\|\pm1\rangle - 0.61\|\mp3\rangle$
?	?	X	270.7	Γ_4	267.9	$-0.13\|6\rangle_a + 0.99\|2\rangle_a$

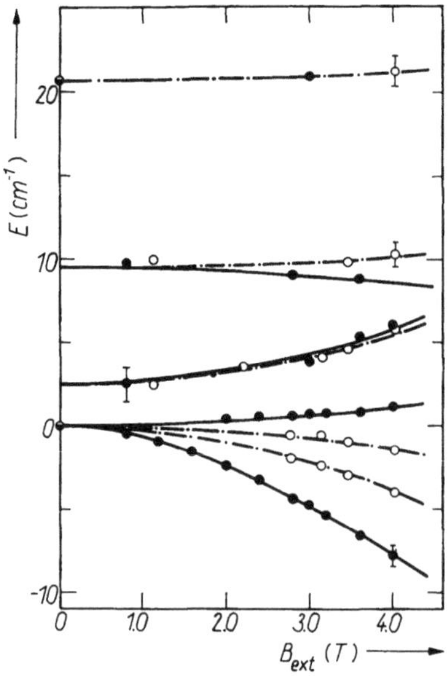

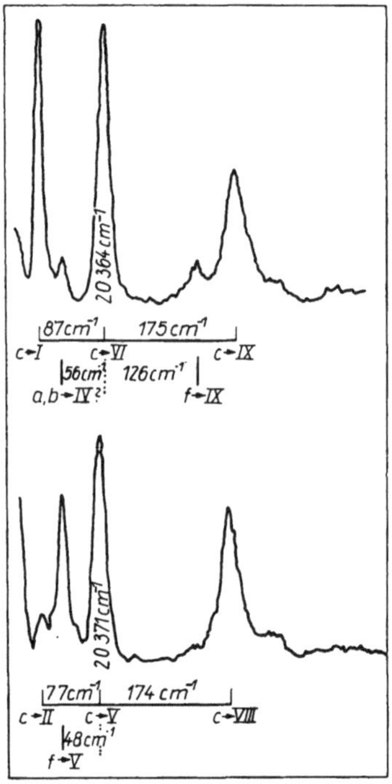

Fig. 3 Fig. 4

Fig. 3. Zeeman effect $\boldsymbol{B} \parallel [100]$ (open circles, dash-dotted curves) and $\boldsymbol{B} \parallel [110]$ (full circles, full curves) of the lowest components of the ground term $^7F_6$ at 4.2 K. The drawn curves are fitted (generally as parabolae $\sim B^2$) to the experimental points

Fig. 4. Emission spectrum of the transitions $^5D_4 \to {}^7F_6$ at 12 K. Top part: π-polarisation, bottom part: σ-polarisation. Wave numbers are increasing to the left. The interpretation of the lines is given below the spectra (see Tables 1 and 2)

The energy $(2.5 \pm 1.0)\ \text{cm}^{-1}$ of this singlet at $B = 0$ follows from the back-extrapolation of the perpendicular Zeeman effect (see Fig. 3).

In strong magnetic fields $\boldsymbol{B} \parallel \boldsymbol{c}$, further lines appear which correspond to transitions from a singlet III at $(9.5 \pm 0.5)\ \text{cm}^{-1}$. At 20 K two weak and sharp lines could be analysed as transitions from a singlet IV at $(20.7 \pm 1.0)\ \text{cm}^{-1}$.

By means of the known symmetry types of the components of the term $^5D_4$ and the selection rules, one gets the symmetry types of the ground term components II, III, and IV without difficulty. They are given in Table 2 and are confirmed by the crystal field calculations (Section 4).

The splitting and shifting of the lowest ground term components for $\boldsymbol{B} \parallel [100]$ and $\boldsymbol{B} \parallel [110]$ are shown in Fig. 3.

3.2 *Emission spectrum*

The spectrum of $TbPO_4$ at room temperature and at 100 K is composed of a large number of broad lines with low intensity which correspond to transitions from nearly all components of the term $^5D_4$. At 12 K, only a few, relatively sharp and intense lines are preserved which almost all originate at the two doublets Γ_5 (c and f). The reason for this is, obviously, that the transition probabilities from the doublets of this term are considerably larger than those from the singlets, as one may also conclude from the absorption spectrum. Therefore, the transitions to the doublets of all terms of the multiplet ^7F appear in π-polarisation, the transitions to the singlets in σ-polarisation.

Here, we restrict ourselves to the transitions to the ground term $^7F_6$, which are shown in Fig. 4 at a temperature of 12 K. The observed lines can easily be correlated to transitions to the already known components as well as to two further groups of excited components of the ground term centred at about 85 and 260 cm^{-1}. In each of the groups, a singlet and a doublet can be determined, their experimental energies are given in Table 2. According to the crystal field calculations (Section 4), each group should contain another singlet. However, their energies cannot be derived from the spectra, because the corresponding transitions either coincide with others or are too weak.

4. Crystal Field Parameters

The splitting of the ground term $^7F_6$ was calculated with full J-mixing within the multiplet ^7F. By this, pure $|LSJM\rangle$ states of the free ion were provided. Deviations from LS-coupling were only in so far considered as the centres of gravity, derived from the emission spectrum, were used for the free ion energies of the terms ^7F_J. The crystal field operator for the site symmetry $\bar{4}2m$ has the form (for relations between the symmetries $\bar{4}2m$ and $\bar{4}m2$, see below)

$$V = A_2^0 \langle r^2\rangle V_2^0 + A_4^0 \langle r^4\rangle V_4^0 + A_4^4 \langle r^4\rangle V_4^4 + A_6^0 \langle r^6\rangle V_6^0 + A_6^4 \langle r^6\rangle V_6^4 \,. \quad (1)$$

The operators V_k^q are defined so that their diagonal elements relating to J are equal to the matrix elements of the Stevens operators $\theta_k(J)\, O_k^q$. The other symbols have their usual meaning. The matrix elements can be calculated as given in [18].

The values of the five unknown crystal field parameters $A_k^q \langle r^k\rangle$ were determined by iterative fitting of the calculated splittings to the experimental energies of the components, especially of the lowest ones which could be measured with greater accuracy. Starting values for the fitting procedure were the parameters of Er^{3+} in YPO_4 [19]. A computer program was used which was developed by Boal et al. [20] for their calculations on terbium aluminium garnet.

The optimum set of parameters determined in this way is given in Table 3 which also contains the results of Lewis and Prinz [9] on $TbPO_4$ and data for other rare

Table 3

Crystal field parameters (in cm^{-1}) for trivalent RE ions in substances with zircon structure

substance	$A_2^0\langle r^2\rangle$	$A_4^0\langle r^4\rangle$	$A_6^0\langle r^6\rangle$	$A_4^4\langle r^4\rangle$	$A_6^4\langle r^6\rangle$	reference
$TbPO_4$	176	14	−50	− 887	+106	this paper
$TbPO_4$	213	25	−43	−1045	−450	[9]
Er^{3+} in YPO_4	141.4	18.1	−40.4	± 837.2	± 88.5	[19]
Er^{3+} in YPO_4	185	18	−40.4	± 800	± 88	[21]
Ho^{3+} in YPO_4	185.3	−1.4	−46.5	− 671	− 33.3	[22]*)
Ho^{3+} in YPO_4	172.1	2.5	−47.4	− 706.6	− 30.1	[23]**)

*) Set No. Ia for the multiplet $^5F'$.

**) With configuration interaction, set for the multiplet $^5F'$ and the terms $^5G_2'$ and $^5S_2'$.

earth phosphates. The calculated energies and symmetry types of all ground term components and the experimental results are compared in Table 2. The agreement is very satisfactory. The wave functions in the last column of Table 2 were calculated with the optimum parameter set in a perturbation without J-mixing. These zero-approximation wave functions are presented as linear combinations of the states $|^7F_6, M\rangle = |M\rangle$ of the free ion. In column 6, the energies calculated in a first-order perturbation without J-mixing are quoted for comparison. Mainly the components V, VI, and VII show distinct deviations from the energies with J-mixing. An important result of the computation has been that the energies of the lowest ground term components react most sensitively to small changes in the crystal field parameters, the wave functions, on the contrary, remain practically unchanged.

By means of the crystal field parameters and wave functions determined by us, the measured Zeeman splittings (see Section 3.1) for $\boldsymbol{B} \parallel \boldsymbol{c}$ and for $\boldsymbol{B} \parallel [100]$ can be described very well. For $\boldsymbol{B} \parallel [110]$ the agreement is worse, so e.g. the splitting of the lowest ground term component I is considerably larger as the calculation shows. A possible explanation is that the magnetic field in this direction — which coincides with the direction of the monoclinic distortion discussed in Section 5 — already above $T_{N,2}$ induces a distortion of the crystal which leads to an additional splitting of the doublets. Such an influence of a magnetic field has been observed e.g. in $DyVO_4$ [24].

Additionally, we have used the calculations with our parameter set for a comparison with the infrared transitions measured by Lewis and Prinz [9] between the ground term components in the experimental range below 5 T and 35 cm^{-1}. Relating to their calculations, we have obtained a distinctly better agreement for $\boldsymbol{B} \parallel [100]$ and a comparable one for $\boldsymbol{B} \parallel \boldsymbol{c}$.

Since for the point symmetry $\bar{4}$m2 of the rare earth sites in the zircon structure repeatedly difficulties and uncertainties have appeared concerning the applied coordinate system, a few clarifying remarks should be added.

The above crystal field calculations, as in most of the publications, were performed for the point symmetry $\bar{4}$2m $\equiv D_{2d}$, and not for the really existing symmetry $\bar{4}$m2. An (x, y, z) coordinate system was used which is rotated around the c-axis $\equiv$ z-axis by the angle $\pi/4$ with respect to the crystallographic (a, b, c) system. For the crystal field parameters, the corresponding transformation leads only to a change in sign for the fourfold parameters $A_k^q\langle r^k\rangle$ with $q = 4$, since here $\exp(iq\pi/4) = -1$. Also the transformation of the basis functions can be given without difficulties. But the nomenclature of the irreducible representations of the symmetry $\bar{4}$m2 and therefore the

Table 4

Basis functions for the symmetries $\bar{4}$2m and $\bar{4}$m2 for even numbered J-values

symmetry $\bar{4}$2m		symmetry $\bar{4}$m2	
representation	basis functions	representation	basis functions
Γ_1	$|0\rangle, |4\rangle_s, |8\rangle_s, \ldots$	Γ_1	$|0\rangle, -|4\rangle_s, |8\rangle_s, \ldots$
Γ_2	$|4\rangle_a, |8\rangle_a, \ldots$	Γ_2	$|4\rangle_a, -|8\rangle_a, \ldots$
Γ_3	$|2\rangle_s, |6\rangle_s, |10\rangle_s, \ldots$	Γ_3	$|2\rangle_a, -|6\rangle_a, |10\rangle_a, \ldots$
Γ_4	$|2\rangle_a, |6\rangle_a, |10\rangle_a, \ldots$	Γ_4	$|2\rangle_s, -|6\rangle_s, |10\rangle_s, \ldots$
Γ_5	$\ldots |\pm5\rangle, |\pm1\rangle, |\mp3\rangle, |\mp7\rangle, \ldots$	Γ_5	$\ldots |\pm5\rangle, -|\pm1\rangle, |\mp3\rangle, -|\mp7\rangle, \ldots$

association of the basis functions with these representations is still unclear. In group-theoretical table books, e.g. of Koster et al. [17], the character table of $\bar{4}2m$, and not of $\bar{4}m2$ is quoted.

It is obvious by going from $\bar{4}2m$ to $\bar{4}m2$ to denote those irreducible representations with the same symbols (e.g. Γ_4) which have the same character in the same class, i.e. which have the same transformation properties (e.g. -1) with respect to the same symmetry elements (e.g. C_2'). This supposition implies that the character table of $\bar{4}m2$ is identical with that of $\bar{4}2m$. We will follow these arguments and stipulate that in going from $\bar{4}2m$ to $\bar{4}m2$ the irreducible representations do not change, e.g. that Γ_4 goes over in Γ_4.

The basis functions of the one-dimensional representations, i.e. the singlets, are the symmetric and anti-symmetric combinations, respectively, of the states $|M\rangle$ of the free ion with even numbered M-values, e.g. $|2\rangle_s = (|2\rangle + |-2\rangle)/\sqrt{2}$, $|2\rangle_a = (|2\rangle - |-2\rangle)/\sqrt{2}$. In Table 4, the correspondence between the representations and the basis functions is shown for even numbered J-values, which are of interest in this paper. The basis functions in the symmetry $\bar{4}2m$ (left-hand side of the table) are transformed to those in the symmetry $\bar{4}m2$ by multiplying each state $|M\rangle$ by a factor $\exp(iM\pi/4)$. For the two-dimensional representations, i.e. the doublets, the basis functions are unchanged by going from $\bar{4}2m$ to $\bar{4}m2$. In the wave functions of the crystal field components which are linear combinations of basis functions, the pre-factor of every second basis function is multiplied by the factor -1. This is indicated by minus signs in front of the corresponding basis functions on the right-hand side of Table 4.

5. Single Ion Anisotropy

The magnetic ordering and the change in crystal symmetry, possibly induced by a cooperative Jahn-Teller effect, which are observed in $TbPO_4$ at low temperatures, are not only determined by the features of the lattice (superexchange, phonon exchange), but decisively depend on the properties of the individual Tb^{3+} ion in the crystal. Therefore, in the following, the single ion anisotropy shall be discussed by the method developed by Pytte and Stevens [25]. In this, one goes from the Hamiltonian of (1) to the classical crystal field energy V_{cl} by replacing the components of the total angular momentum $\boldsymbol{J}$ in the (x, y, z) coordinate system of the point symmetry $\bar{4}2m$ by $J_x = J \sin\theta \cos\varphi$, $J_y = J \sin\theta \sin\varphi$, $J_z = J \cos\theta$. $V_{cl} = V_{cl}(\theta, \varphi)$ gives the energy which is necessary to force the classical angular momentum $\boldsymbol{J}$ of the individual ion in the crystal field in the direction (θ, φ).

Our calculations with the crystal field parameters of the first row of Table 3 result in a very distinct direction dependence of $V_{cl}(\theta, \varphi)$ which is mainly determined by the contributions of the crystal field parts V_2^0 and V_4^4. The three most important sections with the (110), (100), and (001) planes are drawn in Fig. 5 for the first octant. V_{cl} has eight equally deep absolute minima at $\theta = \theta_0, \pi - \theta_0$; $\varphi = 0, \pi/2, \pi, 3\pi/2$ with $\theta_0 = 25°$ which are not very pronounced by the variation in φ-direction. At low temperatures the classical angular momentum will lie in one of the eight minima. In a quantum-mechanical description it will tunnel between the minima, so that the expectation values are $\langle J_x\rangle = \langle J_y\rangle = \langle J_z\rangle = 0$. By applying a magnetic field in z-direction, the four minima at $\theta = \theta_0$ are raised (since angular momentum $\boldsymbol{J}$ and magnetic moment $\boldsymbol{\mu}$ are antiparallel), the four minima at $\theta = \pi - \theta_0$ are lowered. J tunnels only between the four minima mentioned last and one gets $\langle J_z\rangle \neq 0$.

From the pattern of the classical crystal field energy $V_{cl}(\theta, \varphi)$ calculated and presented in Fig. 5 one can understand, in principle, the phase transitions which are observed in $TbPO_4$ at low temperatures.

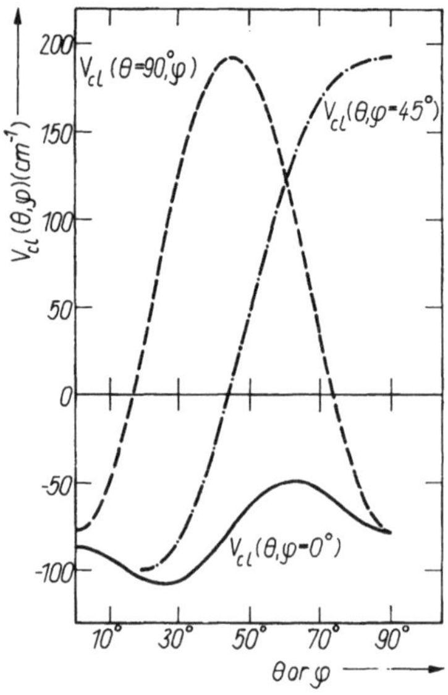

Fig. 5. Anisotropy of the crystal field energy $V_{cl}(\theta, \varphi)$ of the individual ion, calculated with the optimum set of crystal field parameters. Drawn are the sections with the (110) plane $\equiv xz$ plane (θ variable, $\varphi = 0°$, full curve), with the (100) plane (θ variable, $\varphi = 45°$, dash-dotted curve), and with the (001) plane $\equiv xy$ plane. ($\theta = 90°$, φ variable, dotted curve)

1. Magnetic ordering parallel to the crystal c-axis below $T_{N,1}$: As long as the crystal has the high tetragonal symmetry, by magnetic interactions (dipole and exchange) the same four minima are favoured as by an external magnetic field $\boldsymbol{B} \parallel \boldsymbol{c}$, therefore is $\langle J_z \rangle \neq 0$ and $\langle \mu_z \rangle \neq 0$. If the magnetic interactions are getting sufficiently strong with decreasing temperature, the magnetic moments of the individual ions can be ordered parallel to the c-axis. In $TbPO_4$ with a negative B_{int} (see Section 3.1) this ordering really appears with an antiferromagnetic arrangement of the moments below $T_{N,1}$ as neutron diffraction investigations have confirmed [13].

2. Structural phase transitions to monoclinic symmetry with magnetic moments tilted off the c-axis below $T_{N,2}$: The compounds with zircon structure to which $TbPO_4$ belongs, show strong Jahn-Teller interactions which can induce phase transitions with a distortion of the crystal lattice. The type of distortion which is most effective is determined by the high-symmetry crystal field since only those distortions are decisive which alter the minima. The most effective distortion for $TbPO_4$ is obviously such by which only two of the eight minima of the crystal field energy $V_{cl}(\theta, \varphi)$ of Fig. 5 are depressed (or most of all depressed), say those at $\theta = \theta_0$, $\varphi = 0$ and $\theta = \pi - \theta_0$, $\varphi = \pi$. This is a monoclinic distortion.

The relevant lattice angular-momentum interactions contain, in the lowest power in J, the terms $G_1 e_{x^2-y^2}(J_x^2 - J_y^2)$ (orthorhombic distortion) and $G_2 e_{xz}(J_x J_z - J_z J_x)$ (monoclinic distortion), which in the classical approximation can be written as $G_1 e_{x^2-y^2} J^2 \sin^2 \theta \cos 2\varphi$ and $G_2 e_{xz} J^2 \sin 2\theta \cos \varphi$. The first term splits the eight minima of $V_{cl}(\theta, \varphi)$ in two groups of four each, with two different energies. The second term splits every group of four minima in two, two of which, e.g. the above-mentioned two, have the lowest energy.

Due to these considerations it is probable that the crystal, in case it does distort, undergoes a monoclinic distortion in (or near to) the direction of one of the minima of $V_{cl}(\theta, \varphi)$. Then, the principle axis of the single ion μ-tensor is pointing to that direction, too. If exchange and dipole interactions are sufficiently strong and neither of them shows a distinct direction dependence, a magnetic ordering with the moments along or near $\theta = \theta_0$, $\varphi = 0$ can be expected. In $TbPO_4$ the monoclinic distortion and the tilting of the moments in the antiferromagnetic order appear, in reality, below $T_{N,2}$, as a neutron diffraction study has demonstrated [13].

From the single ion anisotropy in the case of $TbPO_4$ one can thus conclude which phase transitions at low temperatures might actually appear. Why at first an antiferromagnetic ordering parallel to c sets in and thereafter the distortion of the lattice, can certainly not be derived alone from these considerations.

6. Discussion

The investigation of the absorption spectrum and its Zeeman effect, especially at high magnetic fields, combined with measurements of the emission spectrum have accomplished a nearly complete analysis of the terms $^7F_6$ and $^5D_4$ of the Tb^{3+} ion in $TbPO_4$ in its tetragonal structure. Above all, it was possible to find relatively certain values for the energies of the lower ground term components which are of particular importance for the calculation and the understanding of the physical properties of this substance at low temperatures.

Our results deviate in parts clearly from the data published up to now, so e.g. in the energy of the ground term component III, in the symmetry types of the components V and VII, and in the energies of the components VIII till X which have not been observed in [9], because of experimental reasons. Consequently our set of crystal field parameters given in the first row of Table 3 differs considerably from that of Lewis and Prinz in the second row of the table. The differences are hereby greater than those to the parameter sets for Er^{3+} and Ho^{3+} in the isostructural YPO_4. The agreement in the values of the crystal field parameters for different compounds with the same structure can at the most be an indication, but it cannot be a proof of the quality of the parameters found by us. In the description of the splitting of the terms of rare earth ions by a small number of parameters, the most important, but by no means all interactions are comprehended; the configuration, spin–spin, and spin–other-orbit interactions remain normally unconsidered. Consequently, this procedure has its limits which were shown clearly by Becker for Ho^{3+} in YPO_4 [22].

In this study, it has been proved that Tb^{3+} in $TbPO_4$ has four crystal field components in the very small energy region from 0 till 21 cm^{-1} of which the lowest is a doublet, so that the properties of the substance are determined even at low temperatures by five crystal field states and therefore its exact theoretical description will be really difficult. A helpful approximation, mainly for qualitative considerations, represents already a model with a doublet and a nearby singlet as the lowest ground term components, as it has been used by Sivardière [26] for a theoretical investigation of the possible structural and magnetic phase transitions in molecular field approximation. The phase transitions which he predicted in the presence of certain interactions are really observed in $TbPO_4$. The possible occurrence of these phase transitions can also be suggested from the single ion anisotropy, as has been done in Section 5 of this paper.

With the present investigation the problem of the low-temperature properties of $TbPO_4$ has been brought closer to an explanation. But its magnetic properties and the phase diagram are not yet completely understood [14]. Further studies for its interpretation are going on in our group.

Acknowledgements

The authors wish to thank Prof. A. Kasten for his interest in this study and stimulating discussions. They also thank Dr. G. Müller-Vogt and Dr. W. Wendl for growing the crystals. One of us, W.B., is deeply indebted to Prof. J. A. Koningstein, Carleton

University, Ottawa (Canada), for his kind hospitality during the stay at his laboratory and the Canadian Council for financial assistance. This work was in part supported by the Deutsche Forschungsgemeinschaft. The computations were performed on the Univac 1108 of the Rechenzentrum der Universität Karlsruhe.

References

[1] G. A. Gehring and K. A. Gehring, Rep. Progr. Phys. **38**, 1 (1975).
[2] H. C. Schopper, Internat. J. Magnetism **3**, 23 (1972).
[3] W. Wüchner, J. Magnetism magnetic Mater. **2**, 203 (1976).
[4] M. Schwab and H. G. Kahle, phys. stat. sol. (b) **84**, 167 (1977).
[5] H. Suzuki and T. Nakajima, J. Phys. Soc. Japan **47**, 1441 (1979).
[6] J. N. Lee, H. W. Moos, and B. W. Mangum, Solid State Commun. **9**, 1139 (1971).
[7] W. Böhm, R. Herb, H. G. Kahle, A. Kasten, J. Laugsch, and W. Wüchner, phys. stat. sol. (b) **54**, 527 (1972).
[8] J. N. Lee and H. W. Moos, Phys. Rev. B **5**, 3645 (1972).
[9] J. F. L. Lewis and G. A. Prinz, Phys. Rev. B **10**, 2892 (1974).
[10] G. A. Prinz and J. F. L. Lewis, J. Magnetism magnetic Mater. **39**, 285 (1983).
[11] S. Spooner, J. N. Lee, and H. W. Moos, Solid State Commun. **9**, 1143 (1971).
[12] J. Coing-Boyat, F. Sayetat, and A. Apostolov, J. Physique **36**, 1165 (1975).
[13] W. Nägele, D. Hohlwein, and G. Domann, Z. Phys. B **39**, 305 (1980).
[14] E. Keller, Diploma Work, Phys. Inst., Univ. Karlsruhe, 1977, to be published.
[15] W. Hintzmann and G. Müller-Vogt, J. Crystal Growth **5**, 274 (1969).
[16] W. Wüchner, W. Böhm, H. G. Kahle, A. Kasten, and J. Laugsch, phys. stat. sol. **54**, 273 (1972).
[17] G. F. Koster, J. O. Dimmock, R. G. Wheeler, and K. Statz, Properties of the Thirty-Two Point Groups, MIT-Press, Cambridge (Mass.) 1963.
[18] G. H. Dieke, Spectra and Energy Levels of Rare Earth Ions in Crystals, Interscience Publishers, 1968 (p. 80).
[19] D. Kuse, Z. Phys. **203**, 49 (1967).
[20] D. Boal, P. Grünberg, and J. A. Koningstein, Phys. Rev. B **7**, 4757 (1973).
[21] Vishwamittar and S. P. Puri, Phys. Rev. B **9**, 4673 (1974).
[22] P. J. Becker, phys. stat. sol. **38**, 379 (1970).
[23] P. J. Becker, G. Domann, and H. G. Kahle, to be published.
[24] C. J. Ellis, K. A. Gehring, M. J. M. Leask, and R. L. White, J. Physique **32**, C1–1024 (1971).
[25] E. Pytte and K. W. H. Stevens, Phys. Rev. Letters **27**, 862 (1971).
[26] J. Sivardière, Phys. Rev. B **8**, 2004 (1973).

(Received August 30, 1984)

phys. stat. sol. (b) **126**, 393 (1984)

Subject classification: 20.3; 2; 13.4; 22

Sektion Physik der Humboldt-Universität zu Berlin[1])

Time-Resolved Luminescence of Charge Carriers in Localized States in Amorphous Semiconductors

By

M. Suhrke and B. Esser

The time-dependent shift and band width of luminescence bands due to radiative transitions between carriers in localized states are calculated. As competitive to the radiative process the escape from the recombination partner over tail states with subsequent non-radiative recombination is assumed. The low temperature case is considered where the hops away over states lower in energy dominate. The temperature dependent corrections due to back hops are found using a self consistent pair approximation for the time decay of the averaged site occupation.

Es wird die Zeitabhängigkeit der Verschiebung und der Breite von Lumineszenzbanden berechnet, die durch strahlende Übergänge zwischen Trägern in lokalisierten Zuständen bedingt sind. Als Konkurrenzprozeß zum strahlenden Übergang wird das Wegspringen vom Rekombinationspartner über Ausläuferzustände mit anschließender nichtstrahlender Rekombination angenommen. Der Tieftemperaturfall wird betrachtet, bei dem das Wegspringen über energetisch tiefere Ausläuferzustände dominiert. Die durch das Rückspringen bedingten temperaturabhängigen Korrekturen werden mittels einer selbstkonsistenten Paar-Approximation für das zeitliche Abklingen der mittleren Besetzung der lokalisierten Zustände gefunden.

1. Introduction

Recombination processes of charge carriers in localized states play an important role in the nonequilibrium behaviour of amorphous semiconductors. Of particular interest are recombination processes involving radiative transitions because of the possibility of performing direct measurements of the change of the luminescence spectrum with time-resolved spectroscopy. Thus in the last few years time-resolved photoluminescence experiments have been used extensively to investigate the electronic states and the dynamics of the recombination processes in amorphous semiconductors [1 to 9]. Most of the experiments deal with the time-resolved characteristics of the luminescence spectrum of amorphous silicon [3, 5 to 9], because this material is of interest for application and the distribution of gap states reflected in such measurements is important for other electronic properties such as photoconductivity [10]. In the case of a-Si the main luminescence band is due to radiative transitions between electrons and holes occupying localized band tail states [11]. One of the important points in the interpretation of the experimental data is the shift of the band to lower energy with increasing time. The mechanism which is responsible for this red shift is the energy dependence of the localization radius of the tail states which results in a smaller overlap and hence a greater lifetime of the nonequilibrium carriers in the deeper tail states [12, 13]. A calculation of the shift of the band and the time dependence of the bandwidth taking explicitly into account the energy dependence of the localization radius was given in [14, 15] for the case of competitive transitions into nonradiative centres.

[1]) Invalidenstr. 42, DDR-1040 Berlin, GDR.

With respect to the recombination in amorphous silicon recently the problems arising from the correlation in the initial distribution of the electrons and holes are intensively discussed, i.e., the question whether the electrons and holes form pairs (geminate model) [1, 12, 13] or are separately trapped (distant pair model) [6, 16, 17] when termalizing after excitation into the tail states. This has important consequences for the dependence of the luminescence intensity on the form of the excitation [6]. On the other hand, the red shift of the luminescence band is a result of the band tailing and hence must be present in the case of the recombination of one type of nonequilibrium carriers occupying a tail state into a defect band as well. This situation which seems to be less investigated is important for time-resolved measurements in a-As_2S_3 [2, 4, 15].

In this paper we consider the mechanism responsible for the shift of the luminescence band in more detail. We assume a recombination model where the escape from the recombination partner occurs by hopping over tail states with subsequent nonradiative recombination, i.e. the tail states "couple" between the initial state and the nonradiative channels (which need not be specified explicitly). Both cases which are of interest for the luminescence involving localized gap states in amorphous semiconductors are investigated: transitions from a band tail into a defect band and transitions between the tail states of two opposite bands. The possibility of back hops is included by using a self-consistent pair approximation in the configurationally averaged occupation function of the localized states. This results in temperature dependent corrections in the band shift and width. The importance of these corrections is, however, sensitive to the form of the density of states. Thus in the low-temperature case, when the escape of the carrier from its recombination partner over states lying lower in energy dominates, it will be seen that the exponential form of the tail manifests itself in a very specific form in the band shift. We complete our considerations by including Coloumb interaction and Stokes shift.

2. Recombination Model and Self-Consistent Approximation for the Time Decay of the Site Occupation

We consider a nonequilibrium distribution of carriers occupying the localized states of a band tail. The carriers are assumed to be created near recombination centres to which radiative recombination is possible. The process competitive to radiative recombination is phonon-assisted hopping away from the recombination centres over the tail states (see Fig. 1). We assume the presence of effective nonradiative recombination channels in the system, so once the carrier has hopped away from its centre the probability of radiative recombination becomes negligible. This recombination model may apply to radiative recombination of electrons in tail states with holes bound

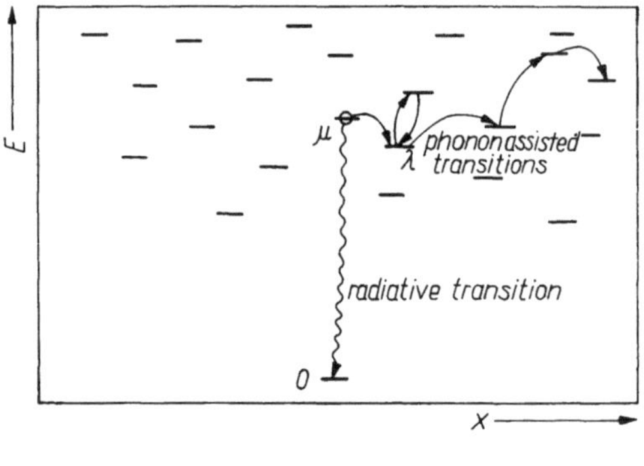

Fig. 1. Recombination model. The carrier in the initial state can recombine radiatively with its nearby recombination centre 0 or escape via phonon-assisted hops over the tail states λ (with included possibility of return to μ)

at D^--centres in chalcogenide glasses [3, 4] or to electron–hole recombination in amorphous silicon if one of the carriers (usual the hole) can be assumed as approximately immobile [12, 13, 18].

The calculation of the rate of radiative recombination in the system considered is part of the general problem of trapping at particular sites for random walks occurring over randomly distributed states. One way for obtaining approximative solutions of this problem are lattice models where one lets the random walk take place over a lattice and mimics the disorder through stepping time distributions [19 to 21] or the approximation of the trapping process by a time-dependent cluster construction using results from the statistics of percolation clusters [22]. Effectively these approaches are restricted to the high-temperature case where the transition probabilities are symmetric. Here we are interested in the low-temperature case with strongly non-symmetric transition probabilities with respect to the energy variables (see (28), (32), and (33)) and another way which uses the equivalence of the master equation with random transition rates to the tight binding model of a disordered electronic system is more appropriate [23]. Using this equivalence the solution of the problem can be formulated in standard terms of a mass operator [23] (or equivalently the T-matrix [24]) which is determined self-consistently. Such self-consistent procedures though differing in details yield good approximations to the time decay of the site occupation [25]. We wish to stress, however, that here we are interested in the calculation of the time evolution of the luminescence spectrum $I(\omega, t)$ which is a more complex quantity than the site occupation $f_\mu(t)$. Hence we derive an expression for the time decay of the site occupation suitable for further calculations according to (24) and containing the simplest nontrivial approximation by treating pair processes self-consistently. Using this approximation we consider both the coordinate and the energy dependences of the phonon-assisted transition probabilities.

We start from the linearized version of the equation for the time-dependent occupation function $f_\lambda(t)$ of the localized states (see e.g. [26]),

$$\frac{\partial f_\lambda}{\partial t} = \sum_{\lambda'} \left[W_{\lambda\lambda'} f_{\lambda'} - W_{\lambda'\lambda} f_\lambda\right] - W_{0\lambda}^{\text{opt}} f_\lambda \,. \tag{1}$$

Here $W_{\lambda\lambda'}(W_{\lambda'\lambda})$ are the phonon-assisted transition rates from state $\lambda'(\lambda)$ to state $\lambda(\lambda')$ and $W_{0\lambda}^{\text{opt}}$ is the rate of radiative recombination from state λ to the state 0 of the recombination centre placed at the origin. We assume the reexcitation from the recombination centre to be negligibly small. Introducing the Laplace transformations $\tilde{f}_\lambda(s)$ and setting $f_\lambda(t = 0) = \delta_{\lambda\mu}$ where μ is the initial state one obtains from (1) the equation for the corresponding Green's function $G_{\lambda\mu}(s) \equiv \tilde{f}_\lambda(s)$,

$$\left[s + \sum_{\lambda'} W_{\lambda'\lambda} + W_{0\lambda}^{\text{opt}}\right] G_{\lambda\mu}(s) = \delta_{\lambda\mu} + \sum_{\lambda'} W_{\lambda\lambda'} G_{\lambda'\mu}(s) \,. \tag{2}$$

Now we assume that the recombination can effectively occur only from the initial state μ nearby the centre 0 due to the rapidly decreasing exponential overlap factor between the states 0 and λ in $W_{0\lambda}^{\text{opt}}$. Hence we can set $W_{0\lambda}^{\text{opt}} \approx \delta_{\lambda\mu} W_{0\mu}^{\text{opt}}$ and in this approximation for the rate of radiative recombination the time dependence of $f_\mu(t)$, i.e. the diagonal part of the Green's function $G_{\mu\mu}(s)$, is of interest only. From (2) $G_{\mu\mu}(s)$ is representable as ($W_{\mu\mu} \equiv 0$)

$$G_{\mu\mu}(s) = \frac{1}{s + \varepsilon_\mu} + \frac{1}{(s + \varepsilon_\mu)^2} \sum_{\lambda'} \frac{W_{\mu\lambda'} W_{\lambda'\mu}}{s + \varepsilon_{\lambda'}} + {}$$
$$+ \frac{1}{(s + \varepsilon_\mu)^2} \sum_{\lambda',\lambda''} \frac{W_{\mu\lambda''} W_{\lambda''\lambda'} W_{\lambda'\mu}}{(s + \varepsilon_{\lambda''})\,(s + \varepsilon_{\lambda'})} + \dots \tag{3}$$

Here ε_λ are the site "energies"

$$\varepsilon_\lambda = \sum_{\lambda'} W_{\lambda'\lambda} + \delta_{\lambda\mu} W_{0\mu}^{\text{opt}} . \tag{4}$$

Equation (3) is analogous to the locator representation for the Green's function of an electron in a tight-binding problem. The representation of the solution of (1) in this form is convenient to derive approximative expressions [23]. In particular the series on the right-hand side of (3) can be represented in the standard form by a mass operator $\Sigma_\mu(s)$, i.e.

$$G_{\mu\mu}(s) = [s + \varepsilon_\mu - \Sigma_\mu(s)]^{-1} , \tag{5}$$

where

$$\Sigma_\mu(s) = \sum_{\lambda'}{}' \frac{W_{\mu\lambda'} W_{\lambda'\mu}}{s + \varepsilon_{\lambda'}} + \sum_{\lambda', \lambda''}{}' \frac{W_{\mu\lambda''} W_{\lambda''\lambda'} W_{\lambda'\mu}}{(s + \varepsilon_{\lambda''}) (s + \varepsilon_{\lambda'})} + \dots . \tag{6}$$

Here $\sum'$ means that the state μ has to be excluded from the sum. In the representation (5), (6) the "powers" and the corresponding mixed terms of all pair, triple, etc. processes are summed up. We set

$$\Sigma_\mu(s) \cong \sum_{\lambda}{}' \frac{W_{\mu\lambda} W_{\lambda\mu}}{s + \varepsilon_\lambda} \tag{7}$$

which corresponds to an approximation taking into account direct back hops from the states λ only. We note that the states μ and λ are coupled to the other states surrounding μ and λ over the transition rates in ε_μ and ε_λ, (4). Hence the carrier has the possibility of hopping away from λ (and μ) or to return by a back hop to μ with a possible subsequent radiative recombination (see Fig. 2).

The function $G_{\mu\mu}(s)$ determines the time evolution of $f_\mu(t)$ at a particular site μ. In what follows we are interested in the luminescence intensity emitted by a macroscopic system, i.e. in a self-averaging quantity in which a summation over the occupation $f_\mu(t)$ for all possible configurations of states 0, μ and the surrounding states λ occurs (see (24)). This average over different state configurations may be divided into two procedures, the average over the configurations of states λ surrounding pairs of states 0 and μ of equal configuration and the average over the configuration of the pair of states 0 and μ involved in the radiative transition itself. For an uncorrelated distribution of states these two averaging procedures decouple. Then in an expression for the luminescence intensity (or in any other expression containing explicitly the states 0 and μ only) the average over the "surrounding states" λ may be performed in $f_\mu(t)$ or $G_{\mu\mu}(s)$ separately. Assuming an uncorrelated distribution of states we now develop a self-consistent procedure for the calculation of the average of $G_{\mu\mu}(s)$ over the states λ, which we denote by $\langle G_{\mu\mu}(s) \rangle_\lambda$. We introduce a new function $g_\mu(s)$ by setting

$$g_\mu(s) = [s + \varepsilon_\mu - \Sigma_\mu^{(0)}(s)]^{-1} , \tag{8}$$

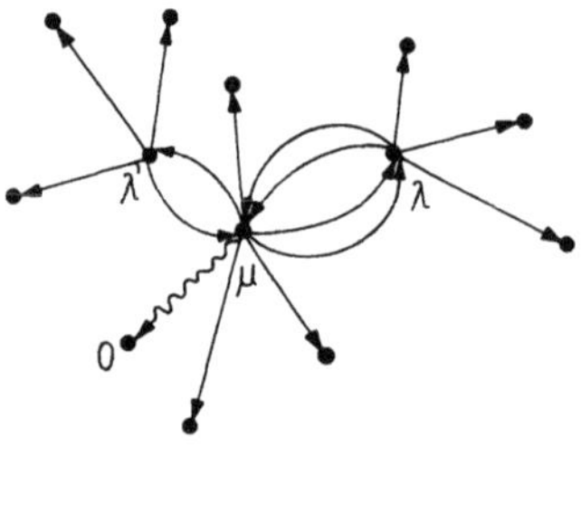

Fig. 2. Graphic representation of the approximation involved in (3), (7). Forth and back hops from (to) the initial state μ, from which radiative recombination can occur, are possible

where

$$\Sigma_\mu^{(0)}(s) = \frac{\sum\limits_\lambda{}' W_{\mu\lambda} W_{\lambda\mu}}{s + \varepsilon(s)} \tag{9}$$

and $\varepsilon(s)$ is an effective medium approximation for the random quantities ε_λ in (7) to be determined self-consistently. Representing $G_{\mu\mu}(s)$ by (8) and (9),

$$G_{\mu\mu}(s) = g_\mu(s) + \sum_{n=1}^{\infty} [g_\mu(s)]^{n+1} [\Sigma_\mu(s) - \Sigma_\mu^{(0)}(s)]^n , \tag{10}$$

we find the function $\varepsilon(s)$ in (9) from the standard condition

$$\langle G_{\mu\mu}(s)\rangle_\lambda = \langle g_\mu(s)\rangle_\lambda , \tag{11}$$

i.e.

$$\langle [g_\mu(s)]^{n+1} [\Sigma_\mu(s) - \Sigma_\mu^{(0)}(s)]^n\rangle_\lambda = 0 . \tag{12}$$

Equation (12) is a self-consistent condition, analogous to the CPA in the alloy problem, which adjusts the mass operator $\Sigma_\mu^{(0)}(s)$ in such a way that (11) is fulfilled. Approximately decoupling the average in (12) into a product of single-site contributions as follows:

$$\langle [g_\mu(s)]^{n+1} [\Sigma_\mu(s) - \Sigma_\mu^{(0)}(s)]^n\rangle_\lambda \cong \langle [g_\mu(s)]^{n+1}\rangle_\lambda \{\langle \Sigma_\mu(s) - \Sigma_\mu^{(0)}(s)\rangle_\lambda\}^n , \tag{13}$$

condition (12) reduces to

$$\langle \Sigma_\mu(s)\rangle_\lambda = \langle \Sigma_\mu^{(0)}(s)\rangle_\lambda . \tag{14}$$

Using (7) and (9) the following equation for $\varepsilon(s)$ is obtained from (14):

$$\left\langle \sum_\lambda{}' \frac{W_{\mu\lambda} W_{\lambda\mu}}{s + \varepsilon_\lambda} \right\rangle_\lambda = \frac{\langle \sum\limits_\lambda{}' W_{\mu\lambda} W_{\lambda\mu}\rangle_\lambda}{s + \varepsilon(s)} . \tag{15}$$

Using the identity

$$\frac{1}{s + \varepsilon_\lambda} = \int_0^\infty \mathrm{d}t\, \mathrm{e}^{-(s+\varepsilon_\lambda)\,t} ,$$

(15) is equivalent to

$$\int_0^\infty \mathrm{d}t\, \mathrm{e}^{-st} \langle \sum_\lambda{}' W_{\mu\lambda}\, \mathrm{e}^{-\varepsilon_\lambda t}\, W_{\lambda\mu}\rangle_\lambda = \frac{\langle \sum\limits_\lambda{}' W_{\mu\lambda} W_{\lambda\mu}\rangle_\lambda}{s + \varepsilon(s)} . \tag{16}$$

Introducing cumulants K_n we represent the average on the left-hand side of (16) in the form

$$\langle \sum_\lambda{}' W_{\mu\lambda}\, \mathrm{e}^{-\varepsilon_\lambda t}\, W_{\lambda\mu}\rangle_\lambda \equiv \langle \sum_\lambda{}' W_{\mu\lambda} W_{\lambda\mu}\rangle_\lambda\, \mathrm{e}^{\sum\limits_n \frac{K_n}{n!}(-t)^n} . \tag{17}$$

Expanding the exponentials in (17) and comparing coefficients for equal powers of t the explicit expressions for K_n are found. Thus the first cumulant is given by

$$K_1 = \frac{\langle \sum\limits_\lambda{}' W_{\mu\lambda}\varepsilon_\lambda W_{\lambda\mu}\rangle_\lambda}{\langle \sum\limits_\lambda{}' W_{\mu\lambda} W_{\lambda\mu}\rangle_\lambda} . \tag{18}$$

To evaluate $f_\mu(t)$ in (24) and explicit expression for $\varepsilon(s)$ is needed. We take into account only the first cumulant K_1 in (17) (the second one would lead to a complex s-dependence of $\varepsilon(s)$ which, however, does not change the t-dependence of $f_\mu(t)$ for long enough times in the considered pair approximation using an expansion of $\varepsilon(s)$ for small s up to linear terms in s because $\varepsilon(s = 0) = K_1/[1 - (K_2/K_1^2)]$ and $K_2/K_1^2 \ll 1$). Thus we obtain

$$\varepsilon(s) \cong K_1 \tag{19}$$

and in this approximation from (9)

$$\Sigma_\mu^{(0)}(s) = \frac{\sigma_\mu}{s + K_1}, \tag{20}$$

where we have introduced the notation

$$\sigma_\mu = \sum_\lambda{}' W_{\mu\lambda} W_{\lambda\mu} . \tag{21}$$

Denoting the inverse Laplace transform by $\mathcal{L}^{-1}\{...\}$ one obtains from (8) and (11) (the configuration average $\langle ... \rangle_\lambda$ and the operation $\mathcal{L}^{-1}\{...\}$ are exchangeable)

$$\langle f_\mu(t) \rangle_\lambda = \mathcal{L}^{-1}\{\langle G_{\mu\mu}(s) \rangle_\lambda\} = \langle \mathcal{L}^{-1}\{g_\mu(s)\} \rangle_\lambda \tag{22}$$

and hence

$$\langle f_\mu(t) \rangle_\lambda = \left\langle e^{-\frac{1}{2}(\varepsilon_\mu + K_1)t} \left\{ \cosh\left[t \sqrt{\sigma_\mu + \frac{1}{4}(\varepsilon_\mu - K_1)^2} \right] - \right.\right.$$
$$\left.\left. - \frac{\varepsilon_\mu - K_1}{2\sqrt{\sigma_\mu + \frac{1}{4}(\varepsilon_\mu - K_1)^2}} \sinh\left[t \sqrt{\sigma_\mu + \frac{1}{4}(\varepsilon_\mu - K_1)^2} \right] \right\} \right\rangle_\lambda . \tag{23}$$

We stress that (23) is valid under the sign $\langle ... \rangle_\lambda$, which in our case is an average over spatial and energetic variables. Dropping in our pair approximation the possibility of radiative recombination and considering processes between μ and only one state λ_1 our procedure would result in the expression $f_\mu(t) = \langle \exp(-W_{\mu\lambda_1} t) \cosh(W_{\mu\lambda_1} t) \rangle_{\lambda_1}$ discussed in [24]. Different from this simple expression (23) includes pair processes between μ and all the other states. We also note that the zero temperature case ($\delta_\mu \equiv 0$) is treated exactly in our approximation, i.e.

$$\langle f_\mu(t) \rangle |_{T \to 0} = \mathcal{L}^{-1} \langle (s + \varepsilon_\mu)^{-1} \rangle_\lambda = \langle e^{-\varepsilon_\mu t} \rangle_\lambda .$$

3. Time Dependence of the Luminescence Spectrum

3.1 General relations

The luminescence intensity emitted by the system is obtained by summing up over all radiative transitions between pairs of states μ (initial state) and 0 (final state), i.e. given by the expression

$$I(\omega, t) = \frac{2ne_0^2(\hbar\omega)^2}{3\pi m_0 c^3} \langle \sum_{0,\mu} W_{0\mu}^{\text{opt}} f_\mu(t) [1 - f_0(t)] \, \delta(\hbar\omega - E_g + E_0 + E_\mu) \rangle . \tag{24}$$

Here e_0 and m_0 are the electron charge and free mass, respectively, n is the refractive index of the network. $E_\mu(E_0)$ is the energy of the initial (final) state referred to the mobility edge $E_c(E_v)$ into the gap, E_g is the gap energy, $E_g = E_c - E_v$. We assume for the asymptotic form of the wave functions of the localized states

$$\psi_l(\boldsymbol{x}) \sim e^{-\gamma_l |\boldsymbol{x} - \boldsymbol{R}_l|} . \tag{25}$$

Here $\boldsymbol{R}_l$ is the position vector of a localized state and γ_l the energy-dependent inverse localization radius. We set

$$\gamma_l = c_l E_l^r \,, \tag{26}$$

where c_l is a constant and r a parameter (for states localized on isolated potential fluctuations $c_l = \sqrt{2m_l}/\hbar$ and $r = 1/2$). The asymptotic form of the matrix elements $W_{0\mu}^{\text{opt}}$ and $W_{\mu\lambda}$ is

$$W_{0\mu}^{\text{opt}} = W_0^{\text{opt}} \exp\{-2\gamma_{0\mu} |\boldsymbol{R}_0 - \boldsymbol{R}_\mu|\} \tag{27}$$

and

$$W_{\mu\lambda} = W_{\text{ph}}(E_\mu, E_\lambda, T) \exp\{-2\gamma_{\mu\lambda} |\boldsymbol{R}_\mu - \boldsymbol{R}_\lambda|\} \,, \tag{28}$$

where

$$\gamma_{\text{if}} = \min\{\gamma_\text{i}, \gamma_\text{f}\} \,.$$

Here $\boldsymbol{R}_0$ and $\boldsymbol{R}_\mu(\boldsymbol{R}_\lambda)$ are the position vectors of the states at the recombination partner 0 and of the $\mu(\lambda)$ tail states. In what follows we assume the localization radius of the state at 0 to be smaller than that of the tail states, i.e.

$$\gamma_{0\mu} = \gamma_\mu \,, \tag{29}$$

in (27).

We consider an uncorrelated distribution of localized states, i.e. the angular brackets denoting the configurational average in (24) factorize into the pair average $\langle ... \rangle_{0,\mu}$ and the average over the states λ surrounding μ, $\langle ... \rangle_\lambda$. As discussed in the preceding section we can average $f_\mu(t)$ over the states λ separately, i.e. (24) is equivalent to

$$I(\omega, t) = \frac{2ne_0^2(\hbar\omega)^2}{3\pi m_0 c^3} \langle \sum_{0,\mu} W_{0\mu}^{\text{opt}} \langle f_\mu(t) \rangle_\lambda \, \delta(\hbar\omega - E_\text{g} + E_0 + E_\mu) \rangle_{0,\mu} \,. \tag{30}$$

In accordance with the linearized version of the kinetic equation (1) here we have set $[1 - f_0(t)] \approx 1$. This is justified if, as assumed in our recombination model, the excited electron can pass to many final states and hence the average occupation of state 0 as one of the final states is small, $f_0(t) \ll 1$. Introducing the density-of-states functions $\varrho_\text{c}(E_\mu)$ and $\varrho_\text{r}(E_0)$ of the initial and final states of the radiative transition, (30) is representable as

$$I(\omega, t) = \frac{2ne_0^2(\hbar\omega)^2}{3\pi m_0 c^3} \int \text{d}E_0 \, \varrho_\text{r}(E_0) \int \text{d}E_\mu \, \varrho_\text{c}(E_\mu) \int \text{d}\boldsymbol{R} \, \times$$
$$\times \, W_0^{\text{opt}} \, \text{e}^{-2\gamma_\mu R} \langle f_\mu(t) \rangle_\lambda \, \delta(\hbar\omega - E_\text{g} + E_0 + E_\mu) \,. \tag{31}$$

The representation (31) contains two important "inputs": the density-of-states model as specified by the functions $\varrho_\text{r}(E_0)$ and $\varrho_\text{c}(E_\mu)$ and the recombination model connected with the function $\langle f_\mu(t) \rangle_\lambda$.

3.2 Average over the surrounding states λ

Now we calculate the average $\langle f_\mu(t) \rangle_\lambda$ explicitly. To this end an approximation for the square root in (23) has to be used. To derive such an approximation we consider the low-temperature case when $W_{\text{ph}}(E_\mu, E_\lambda, T)$ in (28) reduces to

$$W_{\text{ph}}(E_\mu, E_\lambda, T) = W_0 \, \text{e}^{-K_{\mu\lambda}} \,, \tag{32}$$

where

$$K_{\mu\lambda} = \begin{cases} \dfrac{E_\lambda - E_\mu}{kT} & E_\mu < E_\lambda\,, \\ 0 & E_\mu > E_\lambda\,. \end{cases} \tag{33}$$

In this case one obtains for σ_μ given by (21)

$$\sigma_\mu = \sum_\lambda{}' W_0^2\, \mathrm{e}^{-4\gamma_{\mu\lambda}|\boldsymbol{R}_\mu - \boldsymbol{R}_\lambda| - |E_\mu - E_\lambda|/kT}\,, \tag{34}$$

i.e. for T sufficiently low σ_μ decreases according to the factor $\exp\{-|E_\mu - E_\lambda|/kT\}$ (one of the two hops in $W_{\mu\lambda}W_{\lambda\mu}$ is always activated). Hence in the low-temperature case σ_μ becomes arbitrarily small. The further approximations are carried out so that the limit $T \to 0$ is possible in every case. By using an expansion of the square root for the small parameter

$$\delta_\mu = \frac{\sigma_\mu}{(\varepsilon_\mu - K_1)^2}\,, \tag{35}$$

where the difference in the denominator is effectively the radiative transition probability, i.e. temperature independent (see (49)), and taking into account terms up to the first order in δ_μ, we get

$$\langle f_\mu(t)\rangle_\lambda \approx \langle \mathrm{e}^{-\varepsilon_\mu t - \frac{\sigma_\mu}{\varepsilon_\mu - K_1} t} + \delta_\mu\, \mathrm{e}^{-K_1 t + \frac{\sigma_\mu}{\varepsilon_\mu - K_1} t}\rangle_\lambda\,. \tag{36}$$

In the case of not too long times, i.e.

$$t \ll t_\mathrm{c} = \frac{1}{(\varepsilon_\mu - K_1)(1 + 2\delta_\mu)} \ln \frac{1}{\delta_\mu} \approx \frac{1}{\varepsilon_\mu - K_1} \ln \frac{1}{\delta_\mu}\,, \tag{37}$$

($t_\mathrm{c} \to \infty$ if $T \to 0$, in the low-temperature case the t-interval becomes effectively unrestricted) we can neglect the second term on the right-hand side of (36). In the first term we use the following approximation:

$$\frac{1}{\varepsilon_\mu - K_1} \approx \frac{1}{\langle\varepsilon_\mu\rangle_\lambda - K_1} \equiv a_\mu\,. \tag{38}$$

Then for an uncorrelated distribution of states λ the average

$$\langle f_\mu(t)\rangle_\lambda = \langle \mathrm{e}^{-\varepsilon_\mu t - a_\mu \sigma_\mu t}\rangle_\lambda \tag{39}$$

factorizes into contributions from different states λ. One obtains

$$\langle f_\mu(t)\rangle_\lambda = \mathrm{e}^{-W_{0\mu}^{\mathrm{opt}} t - \alpha_\mu(t)}\,, \tag{40}$$

where

$$\alpha_\mu(t) = \int \mathrm{d}E_\lambda\, \varrho_\mathrm{c}(E_\lambda) \int \mathrm{d}\boldsymbol{R}_\lambda\, [1 - \mathrm{e}^{-W_{\lambda\mu} t - a_\mu W_{\lambda\mu} W_{\mu\lambda} t}]\,. \tag{41}$$

Here $\varrho_\mathrm{c}(E_\lambda)$ is the density of tail states. We assume for $\varrho_\mathrm{c}(E)$

$$\varrho_\mathrm{c}(E) = N_\mathrm{c}\, \mathrm{e}^{-(E/\Delta_\mathrm{c})^p}\,; \qquad 0 \leqq E \leqq +\infty\,, \tag{42}$$

where N_c is a constant, Δ_c characterizes the decrease of $\varrho_\mathrm{c}(E)$, and p is a parameter. Considering the case

$$\frac{kT}{\Delta_\mathrm{c}} \ll 1 \tag{43}$$

and taking into account terms up to the first order in kT/Δ_c, $\alpha_\mu(t)$ reduces to

$$\alpha_\mu(t) = \frac{n_c^<(E_\mu)}{\gamma_\mu^3}\,\beta_{T=0}(t) + \frac{kT\,\varrho_c(E_\mu)}{\gamma_\mu^3}\,\beta_{T>0}(a_\mu, t)\,. \tag{44}$$

The quantity $n_c^<(E_\mu)$ is the concentration of tail states below the energy E_μ

$$n_c^<(E_\mu) = \int_{E_\mu} dE\,\varrho_c(E) \tag{45}$$

and the first term in (44) connected with $n_c^<(E_\mu)$ contains all processes in which the electron can hop away from μ by entering energetically lower states $\lambda(E_\lambda > E_\mu)$. $\beta_{T=0}(t)$ is a well-known function [27, 28],

$$\beta_{T=0}(t) = \frac{\pi}{6}\int_0^{W_0t} dx\,[\ln W_0t - \ln x]^3\,e^{-x} = -\pi\sum_{n=1}^{\infty}\frac{(-W_0t)^n}{n!\,n^3}\,. \tag{46}$$

Its asymptotic expression for long times ($W_0t \gg 1$) is

$$\beta_{T=0}(t) \approx \frac{\pi}{6}\left[(C + \ln W_0t)^3 + \frac{\pi^2}{2}(C + \ln W_0t) + 2\zeta(3)\right], \tag{47}$$

where C is Euler's constant and $\zeta(x) = \sum_{n=1}^{\infty} 1/n^x$.

The second term on the right-hand side of (44) is due to forth and back hops, i.e. proportional to the number of states within the energy range kT around E_μ contained in a volume of the extension of the localization radius γ_μ^{-1} (the contribution of this term vanishes for $T \to 0$). Its time dependence is connected with the function $\beta_{T>0}(a_\mu, t)$ which can be written as

$$\beta_{T>0}(a_\mu, t) = \frac{\pi}{6}\int_0^{W_0t} dx\,[\ln W_0t - \ln x]^3 \left\{\frac{1 + 2\frac{a_\mu}{t}x}{x + \frac{a_\mu}{t}x^2}\left[1 - e^{-x-\frac{a_\mu}{t}x^2}\right] + \right.$$

$$\left. + 2\frac{e^{-x}}{x}\left[1 - e^{-\frac{a_\mu}{t}x^2}\right] - e^{-x}\left[C + \ln\left(\frac{a_\mu}{t}x^2\right) - \mathrm{Ei}\left(-\frac{a_\mu}{t}x^2\right)\right]\right\}, \tag{48}$$

where

$$-\mathrm{Ei}(-x) = \int_x^{\infty} dt\,\frac{e^{-t}}{t}\,.$$

The asymptotic expression for $\beta_{T>0}(a_\mu, t)$ in the case $W_0t \gg 1$ is given below. Using (4) and (18) one obtains from (38) in the low-temperature case for a_μ,

$$a_\mu \approx \frac{1}{W_{0\mu}^{\mathrm{opt}}}\,. \tag{49}$$

3.3 Time-resolved shift of the band

Now we consider the time-dependent shift of the luminescence band for radiative transitions between tail states and recombination centres characterized by a density of states $\varrho_r(E)$ (see Fig. 3) as it takes place in chalcogenide glasses [4, 29, 30]. This is the simplest case for which the relaxation process of the nonequilibrium electrons filling the conduction-band tail most directly manifests itself in the shift of the luminescence band. To this end we also assume that there is no Coulomb interaction between the electrons and the recombination centres [30]. In the next section we will consider the additional effects of the Coulomb interaction and of Stokes shifts.

We evaluate the multiple integral in (31) by performing series expansions in the exponent of the integrand in the regions of the main contributions to the E- and R-integrals and using the integral representation of the δ-function in (31). With (40), (44) we have

$$I(\omega, t) = \frac{n e_0^2 \hbar\omega^2}{3\pi^2 m_0 c^3} \int_{-\infty}^{\infty} d\tau \, e^{\frac{i\tau}{\hbar}(\hbar\omega - E_g)} \int dE_0 \, \varrho_r(E_0) \, e^{\frac{i\tau}{\hbar} E_0} \int dE_\mu \, \varrho_c(E_\mu) \times$$
$$\times \, e^{\frac{i\tau}{\hbar} E_\mu - \frac{n_c^{<}(E_\mu)}{\gamma_\mu^3} \beta_{T=0}(t)} \times$$
$$\times \int d\boldsymbol{R} \, W_0^{\text{opt}} \exp\left(-2\gamma_\mu R - W_0^{\text{opt}} t \, e^{-2\gamma_\mu R} - \frac{kT}{\gamma_\mu^3} \varrho_c(E_\mu) \, \beta_{T>0}(a_\mu, t)\right). \tag{50}$$

The R-integral is calculated around the maximum of the integrand, which is the electron–centre separation of those pairs of states 0, μ giving the main contribution to the radiative recombination at time t [13],

$$R_0(E_\mu, t) = \frac{1}{2\gamma_\mu} \ln W_0^{\text{opt}} t \,, \tag{51}$$

by expanding the exponent at R_0 up to terms of second order in $R - R_0$ and setting $R = R_0$ in the other terms. Then $a_\mu(R_0) = t$ and we obtain the asymptotic expression

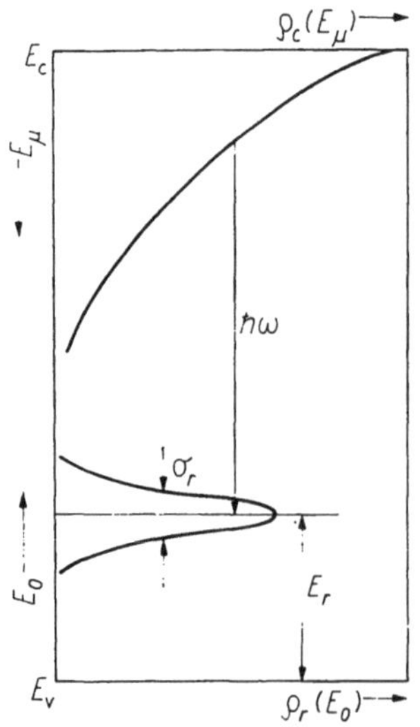

Fig. 3. Density-of-states model for tail-to-peak recombination

for $\beta_{T>0}(a_\mu(R_0), t)$ in the time region $W_0 t \gg 1$,

$$\beta_{T>0}(a_\mu(R_0), t) \equiv \beta_{T>0}(t) \approx \frac{\pi}{12} \ln^4 W_0 t + 0.328 \ln^3 W_0 t + 4.135 \ln^2 W_0 t + 0.554 \ln W_0 t + 6.883 . \tag{52}$$

The E_μ-integral is performed generalizing the procedure given in [14] i.e.

$$I_{E_\mu} = N_c \int \frac{dE_\mu}{\gamma_\mu^3} e^{-G(E_\mu) + \frac{i\tau}{\hbar} E_\mu} , \tag{53}$$

where $G(E_\mu)$ is with (26)

$$G(E_\mu) = \left(\frac{E_\mu}{\Delta_c}\right)^p + c_\mu^{-3} E_\mu^{-3r} [n_c^<(E_\mu) \beta_{T=0}(t) + kT \varrho_c(E_\mu) \beta_{T>0}(t)] . \tag{54}$$

The minimum of $G(E_\mu)$ can be found from

$$pkx_0^{p+3r} = 3r \left[\Delta_c^{-1} n_c^<(x_0) \beta_{T=0}(t) + \frac{kT}{\Delta_c} \varrho_c(x_0) \beta_{T>0}(t)\right] + x_0 \varrho_c(x_0) \left[\beta_{T=0}(t) + p \frac{kT}{\Delta_c} x_0^{p-1} \beta_{T>0}(t)\right]$$

with $k = c_\mu^3 \Delta_c^{3r-1}$ and $x_0(t) = E_{\mu_0}(t)/\Delta_c$. In the case $p = 1$, $r = 1/2$, $c_\mu = \sqrt{2m_\mu}/\hbar$, this reduces to

$$f(x_0) \equiv \frac{x_0^{5/2} e^{x_0}}{x_0 + 3/2} = k^* \beta_T(t) . \tag{55}$$

Here the following notations were introduced:

$$k^* = \frac{N_c}{k} = \frac{\hbar^3 n_c}{(2m_\mu \Delta_c)^{3/2}} , \qquad \beta_T(t) = \beta_{T=0}(t) + \frac{kT}{\Delta_c} \beta_{T>0}(t) , \tag{56}$$

where n_c is the concentration of the states in the conduction-band tail. An approximative solution of (55) for $W_0 t \gg 1$ is

$$x_0(t) \approx \ln k^* \beta_T(t) - \tfrac{3}{2} \ln \ln k^* \beta_T(t) . \tag{57}$$

Expanding $G(E_\mu)$ at E_{μ_0} up to quadratic terms in $E_\mu - E_{\mu_0}$ we find for the E_μ-integral

$$I_{E_\mu} \approx \frac{k^*}{x_0} \sqrt{\frac{2\pi}{x_0 + 3/2}} e^{\frac{i\tau}{\hbar} \Delta_c x_0 - x_0 - \frac{\tau^2 \Delta_c^2 x_0}{2\hbar^2 (x_0 + 3/2)}} . \tag{58}$$

For the energy distribution of the recombination centres we assume

$$\varrho_r(E_0) = N_v e^{-\frac{(E_0 - E_r)^2}{\sigma_r^2}} \tag{59}$$

with the peak maximum at E_r and the peak width σ_r (see Fig. 3). Then for a peak deep enough in the tail, i.e. $E_r/\sigma_r \gg 1$, one obtains

$$I_{E_0} = \int dE_0 \, \varrho_r(E_0) e^{\frac{i\tau}{\hbar} E_0} \approx \sqrt{\pi} N_v \sigma_r e^{\frac{i\tau}{\hbar} E_r - \frac{\tau^2}{4\hbar^2} \sigma_r^2} . \tag{60}$$

Finally we have

$$I(\omega, t) \approx \frac{\pi^{5/2} N_v \sigma_r^2 n e_0^2 \hbar \omega^2 k^*}{3 e m_0 c^3} \frac{\ln^2 W_0^{\text{opt}} t}{t} \frac{e^{-x_0}}{x_0 \sqrt{x_0 + 3/2}} \times$$

$$\times \int_{-\infty}^{\infty} d\tau \exp\left\{\frac{i\tau}{\hbar} (\hbar\omega - E_g + E_r + \Delta_c x_0) - \frac{\tau^2}{2\hbar^2} \left(\frac{\Delta_c^2 x_0}{x_0 + 3/2} + \frac{\sigma_r^2}{2}\right)\right\}. \tag{61}$$

Now, the time-dependent moments of the luminescence band can be found by using the standard relations for the moments. For the central frequency of the band ω_c we obtain

$$\hbar\omega_c(t) = E_g - E_r - E_{\mu_0}(t) \tag{62}$$

and for the band-width

$$\sigma^2(t) = \frac{\sigma_r^2}{2} + \frac{\Delta_c E_{\mu_0}(t)}{E_{\mu_0}(t)/\Delta_c + 3/2} \tag{63}$$

where with (47), (52), (56), and (57), for $W_0 t \gg 1$,

$$E_{\mu_0}(t) = \Delta_c x_0(t) \approx \Delta_c \ln \left\{\frac{\pi k^*}{6} \left[1 + \frac{kT}{2\Delta_c} \ln W_0 t\right] \ln^3 W_0 t\right\}. \tag{64}$$

The meaning of the energy $E_{\mu_0}(t)$ is seen by neglecting the τ^2-term in (61). Then, the τ-integral can be written as

$$I_\tau \sim \delta\big(\hbar\omega - E_g + E_r + E_{\mu_0}(t)\big) . \tag{65}$$

From (65) it is obvious that at the time t the main part of the recombining electrons comes from the energy region around $E_{\mu_0}(t)$ in the conduction-band tail. After initially increasing with time the width of this region becomes nearly constant for long times as is seen from (63).

In (64) the double logarithmic time-dependence in $E_{\mu_0}(t)$ is important, i.e. according to (62) the shift becomes very slow when the region of deep tail states is entered. This is a result of the logarithmic dependences in the functions $\beta_{T=0}(t)$ and $\beta_{T>0}(t)$ (47) and (52) and of the exponential form of the density-of-states tail, as is seen from (55), (57). The mechanism behind this slow dependence is that in the low-temperature case the escape from the recombination partner occurs mainly over tail states lying lower in energy than the initial state. In case of an exponential tail the concentration of such states, $n_c^<(E_\mu)$, is a decreasing function of the energy itself and hence the decay of the site occupation (40) becomes slower for deeper tail states. The corrections due to forth and back hops give rise to the temperature dependent term of (64). It is seen that such hops result in a larger shift compared to the $T = 0$ case. According to the logarithmic dependence this effect is, however, weak. Summarizing we can conclude that the exponential form of the tail results in a specific, slow time dependence of both the band shift and the linewidth.

4. Influence of Coulomb Interaction and Stokes Shift

For a-Si the case of radiative transitions between electrons and holes localized in conduction and valence band tail states and interacting via Coulomb interaction is important [11]. Assuming the recombination model of Section 2, i.e., that the electron can escape from the hole by hopping over tail states with subsequent nonradiative recombination, the calculations of the preceding section are easily extended to this case. We note that in a-Si an efficient nonradiative channel is connected with nonradiative centres correlated to the ESR spin density [1]. Hence with respect to a-Si the recombination model of this paper corresponds to an indirect coupling of the excited electron to such centres over the tail states, whereas in [14] the shift and the band-width were calculated for direct competitive transitions to these centres (which of the models should be applied depends on the centre concentration). For the valence-band tail we assume

$$\varrho_{\rm v}(E_0) = N_{\rm v}\, {\rm e}^{\frac{-E_0}{\Delta_{\rm v}}} . \tag{66}$$

The Coulomb interaction is included by substituting[2])

$$\delta(\hbar\omega - E_{\rm g} + E_0 + E_\mu) \to \delta\left(\hbar\omega - E_{\rm g} + E_0 + E_\mu + \frac{e_0^2}{\varepsilon\,|\boldsymbol{R}_{\rm c} - \boldsymbol{R}_\mu|}\right) \tag{67}$$

in (31). Here ε is the dielectric constant of the network and $\boldsymbol{R}_\mu(\boldsymbol{R}_0)$ are the position vectors of the localized electron (hole) in the conduction (valence) band tail. Performing the calculations along the lines of the previous section we get for the τ-integral in (62)

$$I_\tau = \int\limits_{-\infty}^{\infty} {\rm d}\tau \frac{1}{1 - (i\tau/\hbar)\,\Delta_{\rm v}} \exp\left\{\frac{i\tau}{\hbar}\left[\hbar\omega - E_{\rm g} + E_{\mu_0}(t) + \frac{e_0^2}{\varepsilon R_0\big(E_{\mu_0}(t), t\big)}\right] - \right.$$
$$\left. - \frac{\tau^2}{2\hbar^2}\left[\frac{\Delta_{\rm c} E_{\mu_0}(t)}{E_{\mu_0}(t)/\Delta_{\rm c} + 3/2} + \left(\frac{e_0^2}{2\varepsilon\gamma\big(E_{\mu_0}(t)\big)\, R_0^2\big(E_{\mu_0}(t), t\big)}\right)^2\right]\right\}, \tag{68}$$

where $R_0\big(E_{\mu_0}(t), t\big) \equiv R_0(t)$ and $E_{\mu_0}(t)$ are given by (51) and (64), respectively. For the time dependence of the centre of the band and the dispersion one obtains

$$\hbar\omega_{\rm c}(t) = E_{\rm g} - E_{\mu_0}(t) - \Delta_{\rm v} - \frac{e_0^2}{\varepsilon R_0(t)} \tag{69}$$

and

$$\sigma^2(t) = \frac{\Delta_{\rm c} E_{\mu_0}(t)}{E_{\mu_0}(t)/\Delta_{\rm c} + 3/2} + \Delta_{\rm v}^2 + \left[\frac{e_0^2}{2\varepsilon\gamma\big(E_{\mu_0}(t)\big)\, R_0^2(t)}\right]^2 . \tag{70}$$

The last terms in (69), (70) are the contributions due to the Coulomb interaction.

The importance of multiphonon processes accompanying the optical transition and resulting in a Stokes shift is discussed for both a-Si and the chalcogenide glasses [1, 4]. Proceeding as in [31] we can include the Stokes shift by substituting the δ-function of energy conversation by

$$C_{\rm phon}(E) = \frac{1}{2\pi\hbar} \int\limits_{-\infty}^{\infty} {\rm d}\tau\, {\rm e}^{\frac{i\tau}{\hbar} E + \chi(\tau)} , \tag{71}$$

where $\chi(\tau)$ includes the multiphonon processes. The approximation involved in using (71) is to consider the same coupling strength for all tail states (for the derivation see

---

[2]) We wish to stress that we take into account the Coulomb interaction with respect to the energy conservation for the radiative transition, but not in the solution of the kinetic equation $\langle f_\mu(t)\rangle_\lambda$ i.e. the influence of the Coulomb interaction in the dynamic correlation between electron and hole is neglected.

[31]). Proceeding again as in the previous section and considering tail-to-tail transitions one obtains additive contributions both to $\hbar\omega_c(t)$ and to $\sigma^2(t)$,

$$\hbar\omega_c(t) = E_g - E_{\mu_0}(t) - \Delta_v - \frac{e_0^2}{\varepsilon R_0(t)} - W_p \tag{72}$$

and

$$\sigma^2(t) = \frac{\Delta_c E_{\mu_0}(t)}{E_{\mu_0}(t)/\Delta_c + 3/2} + \Delta_v^2 + \left[\frac{e_0^2}{2\varepsilon\gamma(E_{\mu_0}(t))\, R_0^2(t)}\right]^2 + \\ + \hbar\omega_{\text{phon}} W_p \coth \frac{\hbar\omega_{\text{phon}}}{2kT}\,. \tag{73}$$

Here W_p is the deformation energy and ω_{phon} a characteristic phonon frequency. Considering multiphonon processes accompanying tail-to-peak transitions one obtains contributions of the same form in (62) and (63), respectively.

References

[1] R. A. STREET, Adv. Phys. **30**, 593 (1981).
[2] G. S. HIGASHI and M. KASTNER, Phys. Rev. B **24**, 2295 (1981).
[3] W. CZAJA and S. KURITA, Helv. phys. Acta **53**, 620 (1980).
[4] J. SHAH, Phys. Rev. B **21**, 4751 (1980).
[5] M. A. PAESLER and W. PAUL, Phil. Mag. B **41**, 393 (1981).
[6] D. J. DUNSTAN, Phil. Mag. B **46**, 579 (1982).
[7] P. K. BHAT, A. J. RHODES, T. M. SEARLE, I. G. AUSTIN, and J. ALLISON, Phil. Mag. B **47**, L 99 (1983).
[8] B. A. WILSON, P. HU, T. M. JEDJU, and J. P. HARBISON, Phys. Rev. B **28**, 5901 (1983).
[9] H. BERGNER, V. BRÜCKNER, D. DIETRICH, and W. NOWICK, phys. stat. sol. (b) **120**, 655 (1983).
[10] F. EVANGELISTI, P. FIORINI, G. FORTUNATO, A. FROVA, C. GIOVANNELLA, and R. PERUZZI, J. non-cryst. Solids **55**, 91 (1983).
[11] R. FISCHER, Festkörperprobleme **17**, 85 (1977).
[12] R. A. STREET and J. C. KNIGHTS, Phil. Mag. B **42**, 551 (1980).
[13] R. A. STREET, Solid State Commun. **34**, 157 (1980).
[14] B. ESSER, phys. stat. sol. (b) **114**, 175 (1982).
[15] M. SUHRKE, phys. stat. sol. (b) **118**, K79 (1983).
[16] D. J. DUNSTAN, Phys. Rev. B **28**, 2252 (1983).
[17] S. P. DEPINNA, Phys. Rev. B **28**, 5327 (1983).
[18] S. KIVELSON and C. D. GELATT, Phys. Rev. B **26**, 4646 (1982).
[19] E. W. MONTROLL and G. H. WEISS, J. math. Phys. (New York) **6**, 167 (1965).
[20] H. SCHER and M. LAX, Phys. Rev. B **7**, 4491, 4502 (1973).
[21] G. ZUMOFEN, J. KLAFTER, and A. BLUMEN, J. chem. Phys. **79**, 5131 (1983).
[22] B. ESSER and I. P. ZVYAGIN, phys. stat. sol. (b) **104**, 523 (1981).
[23] B. MOVAGHAR, J. Phys. C **13**, 4915 (1980).
[24] D. L. HUBER, in: Laser Spectroscopy of Solids, Ed. W. M. YEN and P. M. SELZER, Springer-Verlag Berlin/Heidelberg/New York 1981.
[25] R. F. LORING, H. C. ANDERSEN, and M. D. FAYER, J. chem. Phys. **76**, 2015 (1982).
[26] V. L. BONCH-BRUEVICH, R. ENDERLEIN, B. ESSER, R. KEIPER, A. G. MIRONOW, and I. P. ZVYAGIN, Einführung in die Elektronentheorie ungeordneter Halbleiter, VEB Deutscher Verlag der Wissenschaften, Berlin 1984.
[27] D. G. THOMAS, J. J. HOPFIELD, and W. M. AUGUSTYNIAK, Phys. Rev. **140**, A 202 (1965).
[28] A. BLUMEN, J. chem. Phys. **72**, 2632 (1980).
[29] C. TSANG and R. A. STREET, Phys. Rev. B **19**, 3027 (1979).
[30] R. A. STREET and N. F. MOTT, Phys. Rev. Letters **35**, 1293 (1975).
[31] B. ESSER and D. SPRENGER, phys. stat. sol. (b) **94**, 583 (1979).

(Received September 3, 1984)

phys. stat. sol. (b) **126**, 407 (1984)

Subject classification: 13.4 and 20.3; 22.5.4

Institute of Physics, Academy of Sciences of the Estonian SSR, Tartu[1])

Decay Kinetics and Polarization of the A_T Emission of Pb^{2+} Centres of Different Structure in $KBr:PbBr_2$

By

ZH. EGEMBERDIEV, V. NAGIRNYI, T. SOOVIK, and S. ZAZUBOVICH

Three types of $Pb^{2+}v_c^-$ dipole and interstitial bromine atom associates ($H_z(Pb^{2+})$ centres) as well as a Pb^{2+} ion associated with the next-nearest-neighbouring cation vacancy ($Pb^{2+}v_c^-$ [001] centres) are found in a $KBr:PbBr_2$ crystal. The luminescence of all these centres is caused by electronic transitions from tetragonal minima of the adiabatic potential energy surface (APES) of the triplet excited state of the Pb^{2+} ion. A comparison of the spectral, polarization and kinetic characteristics of their emission is performed. The position and structure of the defect associated with the Pb^{2+} ion are shown to have strong influence on the luminescence properties of $KBr:PbBr_2$ crystal.

Обнаружены три типа ассоциатов междоузельных атомов брома с диполями $Pb^{2+}v_c^-$($H_z(Pb^{2+})$-центров) и ассоциаты иона Pb^{2+} с далекой катионной вакансией ($Pb^{2+}v_c^-$[001]-центры) в кристалле KBr–$PbBr_2$. Люминесценция всех этих центров обусловлена электронными переходами из тетрагональных минимумов триплетного возбужденного состояния иона Pb^{2+}. Проведено сопоставление спектральных, поляризационных и кинетических характеристик их излучения. Показано, что положение и структура дефекта, ассоциированного с ионом Pb^{2+}, существенно влияет на люминесцентные свойства KBr–$PbBr_2$.

1. Introduction

Earlier two types of $Pb^{2+}v_c$ dipole and interstitial chlorine atom (Cl_i^0) associates have been found and investigated in X-irradiated $KCl:PbCl_2$ crystals [1 to 3]. They differ by the position of the cation vacancy v_c^- (nn or nnn) with respect to the Pb^{2+} ion, and are labelled in [3] as $H_z^{[110]}(Pb^{2+})$ and $H_z^{[001]}(Pb^{2+})$ centres, respectively. The structure of these centres is determined by the ESR method [2]. At 80 K $H_z^{[110]}(Pb^{2+})$ centres are mainly created as a result of the trapping of mobile H centres by $Pb^{2+}v_c^-$ [110] dipoles, while $H_z^{[001]}(Pb^{2+})$ centres are formed thermally from $H_z^{[110]}(Pb^{2+})$ owing to the jumping of v_c^- from (nn) to (nnn) position with respect to Pb^{2+}.

The luminescence of $H_z(Pb^{2+})$ centres is caused by electronic transitions from tetragonal minima on the APES of the triplet excited state of the Pb^{2+} ion perturbed by v_c^- and an interstitial chlorine atom [1, 3]. The latter in form of a Cl_2^- molecular ion occupies an anion site and a cation vacancy.

In [4] photostimulated reorientations of $H_z^{[001]}(Pb^{2+})$ centres in the triplet excited state of the Pb^{2+} ion at $T > 110$ K have been found. This phenomenon is explained by the dissociation of the two-component (v_c^-, Cl_i^0) defect near the excited Pb^{2+} ion and the following jump of the interstitial chlorine atom onto one of the ⟨100⟩ directions perpendicular to the Pb^{2+}–v_c^- axis. The cation vacancy remains at its site, which guarantees the memorizing of the original state.

Thanks to the availability of several types of lead centres with known geometrical structure in $KCl:PbCl_2$, a comparison of their luminescence characteristics in [5]

[1]) 142 Riia, 202400 Tartu, USSR.

allowed to obtain important information about the effect of the position and structure of defects associated with the Pb^{2+} ion on the energy structure of the triplet excited state of Pb^{2+} centres and the processes occurring in it as well as to elucidate the character and relative contribution of different interactions in these centres.

So far only $Pb^{2+}v_c^-$ [110] centres have been investigated in KBr:$PbBr_2$ [6 to 13]. Their luminescence properties have a number of peculiarities whose comprehension needs a comparison of the characteristics of Pb^{2+} centres of different structure. Pb^{2+} ions are characterized by a rather strong spin–orbit interaction comparable with the Jahn-Teller one. For this reason the clarification of the question about the adequacy of the model proposed in [14] for describing the Pb^{2+} centres is of great interest. In the model the interaction of impurity optical electrons with non-totally symmetric vibrations of the crystal lattice is considered as the main one, while spin–orbit and other interactions, as perturbations. On the other hand, due to the availability of the intra-centre emission, $H_z(Pb^{2+})$ centres in alkali halides may be used as rather sensitive and selective luminescent detectors for investigating the different processes with the participation of H centres. For these reasons a detailed study of $H_z(Pb^{2+})$ luminescence properties is necessary.

With this aim in view we have undertaken a search for and study of the luminescence of $H_z(Pb^{2+})$ centres in KBr:$PbBr_2$ and compared the characteristics of Pb^{2+} ions with different nearest surroundings.

2. Experimental Procedure

A KBr:$PbBr_2$ single crystal was grown by the Stockbarger method from zone-refined KBr salt. The Pb^{2+} concentration in the sample was about 10^{17} cm^{-3}. The samples newly quenched from 700 °C were X-irradiated (tube with Cu anode, 50 kV, 20 mA) through a beryllium window of a cryostat for 4 h at 80 K, further they were heated up to 200 K to remove H_A and V_K centres. At this operation Pb^{3+} and some quantity of $H_z^{[110]}(Pb^{2+})$ centres were created. Then at 80 to 150 K F, Pb^+, and Pb^{3+} centres were destroyed optically; this resulted in the growth of the $H_z^{[110]}(Pb^{2+})$ number by an order. On heating the crystal with the created $H_z^{[110]}(Pb^+)^2$ centres up to 215 K, $H_z^{[001]}(Pb^{2+})$ centres arise.

The method of investigation of the spectral, kinetic, and polarization characteristics of emission was analogous to that described in [3 to 5].

3. Experimental Results

3.1 $H_z^{[110]}(Pb^{2+})$ centres

In non-irradiated newly-quenched KBr:$PbBr_2$ crystals the associates of the Pb^{2+} ion with the cation vacancy — $Pb^{2+}v_c^-$ [110] centres — prevail [10]. Their characteristics according to data in [13] are depicted in Fig. 1a.

On X-irradiation at 80 K a small number of centres with 3.07 eV emission is created. Their quantity grows considerably at optical destruction of both electronic (F, Pb^+) and hole (Pb^{3+}) centres at 80 K. It is known that at the decay of excitons created during the electron or hole recombination process mobile H centres arise. The number of centres with 3.07 eV emission decreases sharply above 160 K (Fig. 2, curve 1).

The optical spectra of 3.07 eV emission centres are shown in Fig. 1b. A comparison of Fig. 1a and 1b shows that the absorption and emission spectra of the centres investigated are only slightly shifted to the low-energy side relative to the corresponding spectra of $Pb^{2+}v_c^-$ [110] centres. The half-widths and Stokes shifts are comparable.

The decay kinetics of 3.07 eV emission is analogous to that of $Pb^{2+}v_c^-$ [110]. It consists of two components — fast (FC) and slow (SC), the light sum of FC being con-

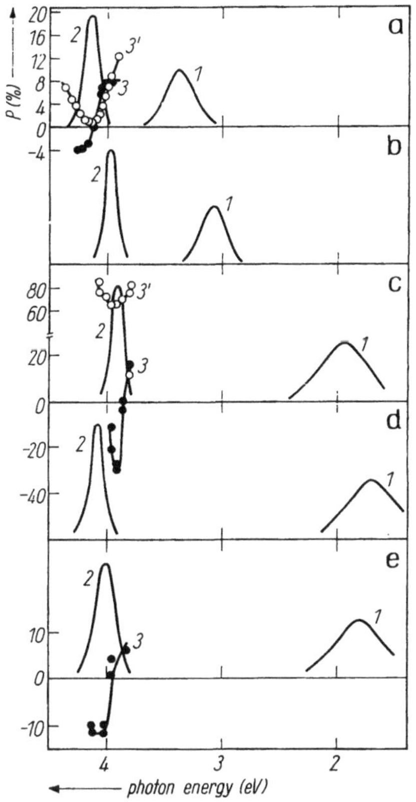

Fig. 1. Emission (1) and excitation (2) spectra of the stationary A_T emission and polarization spectra of the slow (3) and fast (3′) decay components of the A_T emission of a) $Pb^{2+}v_c^-$ [110], b) $H_z^{[110]}(Pb^{2+})$, c) $H_z^{[001]}(Pb^{2+})_{ter}$, d) $Pb^{2+}v_c^-[001]$, and e) $H_z^{[001]}(Pb^{2+})_{opt}$ centres at 80 K

siderably larger than that of SC. The $\tau_{SC}(T)$ dependences of the centres investigated and of $Pb^{2+}v_c^-$ [110] centres (Fig. 3, curve 1) are similar. The polarization spectra of the separate decay components are also similar to the corresponding spectra of the 3.4 eV emission of $Pb^{2+}v_c^-$ [110] (Fig. 1a, curves 3, 3′). Namely, for both centres a strong dependence of the polarization degree of FC on the energy of exciting photons is observed (curve 3′). The polarization degree of SC changes its sign in the region of the absorption band and its maximum value in the negative region is half that in the positive region. This fact testifies [15] that X and Y minima of the triplet excited state of these centres are the lowest ones (v_c^- is in the XY-plane). The emission of both centres is polarized along the C_4 axis of the crystal. The thermal quenching of 3.07 and 3.4 eV emissions takes place in approximately the same temperature interval.

An analogous resemblance of characteristics has been observed for $Pb^{2+}v_c^-$ [110] and $H_z^{[110]}(Pb^{2+})$ centres in KCl:$PbCl_2$ [3, 5]. The conditions of the creation of centres with 3.07 eV emission in KBr:$PbBr_2$ and $H_z^{[110]}(Pb^{2+})$ in KCl:$PbCl_2$ are also similar. It may be concluded that the centres with 3.07 eV emission represent $H_z^{[110]}(Pb^{2+})$ centres in KBr:$PbBr_2$ crystals.

3.2 $H_z^{[001]}(Pb^{2+})$ centres

In the temperature region from 160 to 210 K $H_z^{[110]}(Pb^{2+})$ centres transform into new ones with 1.9 eV emission (Fig. 2, curve 2). The number of the latter reaches its maximum at 215 K and then decreases. At 270 K they are practically destroyed.

The excitation spectrum of the 1.9 eV emission (Fig. 1c) is shifted slightly to the low-energy side with respect to the excitation spectra of $Pb^{2+}v_c^-$ [110] and $H_z^{[110]}(Pb^{2+})$ emission, while the half-width of the emission band and Stokes shift of the new centres are considerably larger than those of [110] centres (see Table 1).

The kinetics of the 1.9 eV emission consists of two components, slow and fast, but their light sums S_{SC} and S_{FC} differ noticeably less than in the case of $Pb^{2+}v_c^-$ [110] and $H_z^{[110]}(Pb^{2+})$. The decay time τ_{SC} is equal to 4×10^{-3} s at 1.7 K and decreases

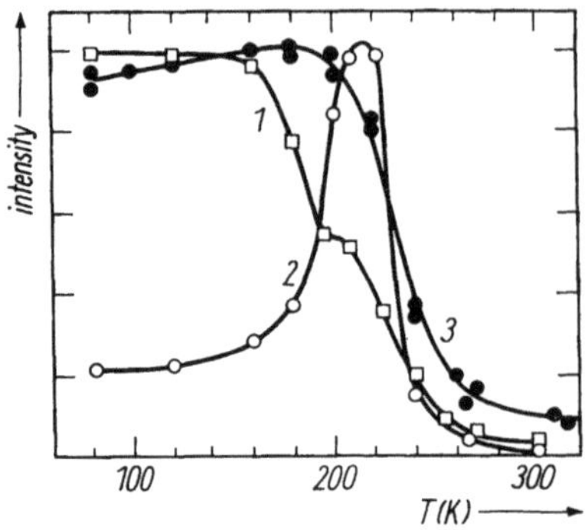

Fig. 2

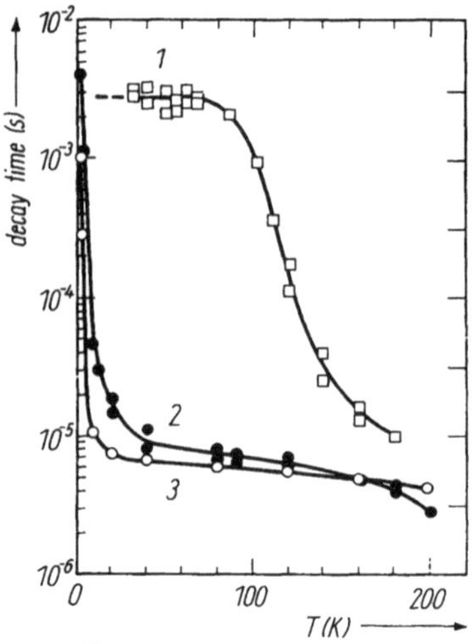

Fig. 3

Fig. 2. Dependences on pulse annealing temperature measured at 80 K for the intensity of the emission of $H_z^{[110]}(Pb^{2+})$ (1), $H_z^{[001]}(Pb^{2+})_{ter}$ (2), and $H_z^{[001]}(Pb^{2+})_{opt}$ (3) centres

Fig. 3. Temperature dependences of decay time of the slow component of $Pb^{2+}v_c^-$ [110] (1) and $H_z^{[001]}(Pb^{2+})_{ter}$ (2) emission, and of the light sum of the slow decay component of $H_z^{[001]}(Pb^{2+})_{ter}$ emission (3)

abruptly to 1.5×10^{-5} s at 20 K (Fig. 3, curve 2). At further temperature increase only a slight shortening of τ_{SC} is observed. The light sum S_{SC} also decreases drastically from 1.7 to 20 K (curve 3). It means that the decrease of τ_{SC} is caused by some non-radiative process.

The polarization degree of the fast component is positive. It reaches the value of 80% on excitation at the shoulders of the absorption band and decreases to 60% when approaching the band maximum (Fig. 1c, curve 3'). The polarization degree of the slow component changes its sign within the absorption band and, besides, near 80 K its absolute value in the negative region is twice that in the positive region (curve 3, see also Fig. 4).

The optical characteristics and the formation conditions for centres with 1.9 eV emission are similar to those of $H_z^{[001]}(Pb^{2+})$ centres in $KCl:PbCl_2$ [3]. An analogous

Table 1

Maxima of the absorption ($E_m^{\varkappa}$) and emission (E_m^I) bands as well as half-widths of the emission bands (δ^I) and Stokes shifts (S) for the Pb^{2+} centres of different types in $KBr:PbBr_2$ crystals at $T = 80$ K

centres	$E_m^{\varkappa}$ (eV)	E_m^I (eV)	δ^I (eV)	S (eV)
$Pb^{2+}v_c^-$ [110]	4.14	3.4	0.22	0.74
$H_z^{[110]}(Pb^{2+})$	3.96	3.07	0.26	0.89
$H_z^{[001]}(Pb^{2+})_{ter}$	3.9	1.9	0.48	2.0
$H_z^{[001]}(Pb^{2+})_{opt}$	4.0	1.8	0.45	2.2
$Pb^{2+}v_c^-$ [001]	4.07	1.7	0.48	2.37

structure is most probable. Let us label these centres with 1.9 eV emission as $H_z^{[001]}(Pb^{2+})_{ter}$.

3.3 $Pb^{2+}v_c^-[001]$ and $H_z^{[001]}(Pb^{2+})_{opt}$ centres

In non-irradiated newly-quenched crystals there exist some centres with emission at about 1.7 eV. Their spectra are shown in Fig. 1d.

On X-irradiation at 80 K centres with 1.8 eV emission arise beside $H_z^{[001]}(Pb^{2+})$ centres. Their number grows essentially at optical destruction of electron and hole centres at 80 K. It increases slightly on heating the crystal up to 180 K (Fig. 2, curve 3). These centres are thermally destroyed in the same temperature region as $H_z^{[110]}(Pb^{2+})_{ter}$. The characteristics of these centres are shown on Fig. 1e.

Comparing Fig. 1c, d, and e, one can see that the spectral and polarization characteristics of the emission of these three types of centres are similar. The absorption and emission bands are close. Their half-widths and Stokes shifts are comparable (see Table 1). The polarization spectra of SC are of the same type: the absolute value of the negative polarization degree is twice that of the positive one. The emission is polarized along the C_4 axis and is practically not quenched at least until 210 K. The ratios of the light sums of SC and FC are comparable. It is worth noting that the temperature dependences of τ_{SC} and S_{SC} of all these centres are qualitatively analogous (see, e.g., Fig. 3, curves 2, 3). Consequently, these three types of centres must have a similar geometrical structure. By analogy with $KCl:PbCl_2$ [5] it may be concluded that in all these cases the cation vacancy is in (nnn) position with respect to Pb^{2+} ion ([001] centres).

The centres with 1.7 eV emission existing in small quantity in a non-irradiated crystal may be considered as associates of the Pb^{2+} ion and v_c^- in remote (nnn) position, i.e. $Pb^{2+}v_c^-$ [001] centres. Their number increases slightly after X-irradiation of the crystal at 80 K and heating up to 200 K. Possibly they are created as a result of the recombination of mobile V_K centres with electronic $Pb^+v_c^-$ [001] which are formed thermally from $Pb^+v_c^-$ [110] centres (an analogous process was studied in detail for $Sn^+v_c^-$ centres in [16]).

The number of $Pb^{2+}v_c^-$ [001] centres increases also at the optical destruction of $Pb^+v_c^-$ [001] centres and decreases if F centres are optically destroyed.

The centres with 1.8 eV emission arise at 80 K probably as a result of the trapping of mobile H centres generated on exciton decay by $Pb^{2+}v_c^-$ [001] centres. The excitons are created by X-irradiation or optically in recombination processes. Thus, the 1.8 eV emission centres represent, contrary to 1.9 eV emission centres, rather optically created than thermally created $H_z^{[001]}(Pb^{2+})$ centres. Let us denote them as $H_z^{[001]}(Pb^{2+})_{opt}$ centres. Some differences in the characteristics of $H_z^{[001]}(Pb^{2+})_{ter}$ and $H_z^{[001]}(Pb^{2+})_{opt}$ indicate some difference in their structure, that is caused probably by a different position of the molecular ion $[Cl_2^-]_{ac}$ with respect to Pb^{2+}. One of the possible models of $H_z^{[001]}(Pb^{2+})_{opt}$ is depicted in Fig. 6e.

The form of the polarization spectra of SC of all these [001] centres testifies [15] that the Z minima of the APES of the triplet excited state are the lowest ones (Z is the direction ([001]) from Pb^{2+} to v_c^-).

The triplet relaxed excited state of the Pb^{2+} ion has a sufficiently large energy difference between the radiative doublet and the metastable minimum, therefore an abrupt shortening of τ_{SC} to 20 K cannot be accounted for by the non-radiative transitions between them. It is not unlikely that this temperature dependence of τ_{SC} may be caused by the transitions between the lowest (Z) minimum of the A_T state and that of the A_X state, thermal equilibrium being achieved at ≈ 20 K.

3.4 Photostimulated reorientation of $H_z^{[001]}(Pb^{2+})_{ter}$ centres

It is shown in [4] that photostimulated reorientations of $H_z^{[001]}(Pb^{2+})$ centres in KCl:$PbCl_2$ occur after the thermal equilibrium between the radiative and metastable minima of the triplet excited state of Pb^{2+} ion has been achieved. Therefore, they reveal themselves in the polarization characteristics of SC only. The polarization of FC is temperature independent at $T < 210$ K. As SC is dominating in the emission of $H_z^{[001]}(Pb^{2+})_{ter}$ centres in KCl:$PbCl_2$ at $T > 80$ K, the reorientations reveal themselves as well in the inversion of the polarization spectrum of the stationary emission. Due to the relatively large light sum of FC and the high positive degree of its polarization, only a somewhat unusual alteration of the polarization degree of the stationary $H_z^{[001]}(Pb^{2+})$ emission is observed in KBr:$PbBr_2$ when the temperature rises (P decreases at temperatures from 77 K to 150 K, then grows again). The inversion of the polarization spectrum analogous to that in KCl:$PbCl_2$ is observed only for the slow decay component of this emission in the temperature region from 100 to 200 K (Fig. 4 and 5). As in the case of KCl:$PbCl_2$, the inversion is accompanied by the shortening of the emission decay time (Fig. 3, curve 3). However, it is difficult to single out a new decay component caused by the reorientation process.

Thus, the similarity of the processes of photostimulated reorientations of $H_z^{[001]}(Pb^{2+})_{ter}$ centres in KCl:$PbCl_2$ and KBr:$PbBr_2$ is evident. For details, see [4].

4. Summary

Basing on the data of [3 to 5] and of the present work, let us compare the characteristics of five types of Pb^{2+} centres in KCl and KBr, namely, of Pb^{2+} ions associated

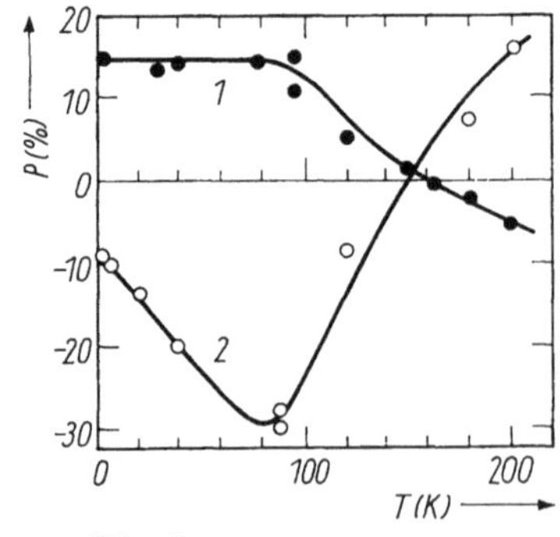

Fig. 4

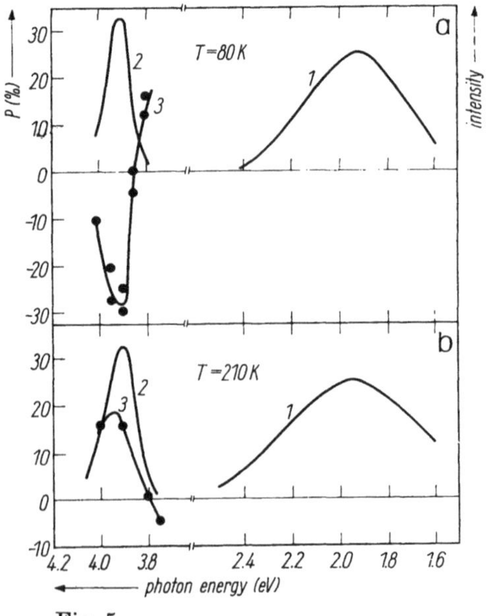

Fig. 5

Fig. 4. Temperature dependences of polarization degree of the slow decay component of $H_z^{[001]}(Pb^{2+})_{ter}$ emission at excitation in the low-energy (1) and high-energy (2) tails of the A absorption band

Fig. 5. Emission (1), excitation (2), and polarization (3) spectra of the $H_z^{[001]}Pb^{2+})_{ter}$ emission measured at a) $T = 80$ K and b) 210 K

with v_c^- in (nn) position (a), with v_c^- in (nn) position and interstitial halide atom (b), with v_c^- in (nnn) position (c), with v_c^- in (nnn) position and interstitial halide atom (d, e), and of Pb^{2+} ions, not associated with any defect (f) (Fig. 6).

The common properties of Pb^{2+} centres investigated are the following: (i) their absorption spectra are close; (ii) two components of the emission decay are observable indicating the existence of a metastable minimum below the radiative one of the APES of the triplet excited state, the energy difference δ between them is of the same order ($\approx 10^{-2}$ eV); (iii) the polarization spectra of anisotropic Pb^{2+} centres change their sign within the absorption band testifying the splitting of X, Y, Z minima due to the influence of the defect; (iv) the emission from metastable minima is polarized along the C_4 axis for Pb^{2+} centres of different structure.

On the basis of these data it may be concluded that the luminescence of all Pb^{2+} centres investigated is caused by electronic transitions from tetragonal minima of the triplet excited state of Pb^{2+} ions and its properties may be described adequately in terms of the model described in [14]. Consequently, in spite of the comparatively large spin–orbit interaction in the Pb^{2+} ion, the symmetry and the structure of the excited states of Pb^{2+} centres are determined by the interaction of optical impurity electrons with non-totally symmetric lattice vibrations (Jahn-Teller effect). The hyperfine interaction also reveals itself in the decay kinetics of the emission from the metastable minima [17]. However, the combined effect of spin–orbit and vibronic interactions gives the main contribution to the radiative decay of these minima of Pb^{2+} centres (nuclear spin is 1/2).

The cation vacancy usually associated with the Pb^{2+} ion disturbs the energy equivalence of the Jahn-Teller minima of the same symmetry but of different orientations (X, Y, Z), and is considered in the theory [14] as a perturbation. The splitting energy is $\varphi \approx 10^{-2}$ eV. Nevertheless, v_c^- affects noticeably the luminescence characteristics of Pb^{2+} centres. First of all, if v_c^- compensating surplus charge of Pb^{2+}, is removed from the (nn) position, the electron–phonon interaction in the excited state of the Pb^{2+} centre increases considerably. This leads to the increase of the Stokes shift by about 2.5 times, of emission band half-widths by about 1.7 to 2 times, and of the energy barriers for the transition between X, Y, Z minima by about an order. Moreover, the ratio of light sums of fast and slow components changes. Namely, the light sum of SC increases at the expense of FC. The increase of the prob-

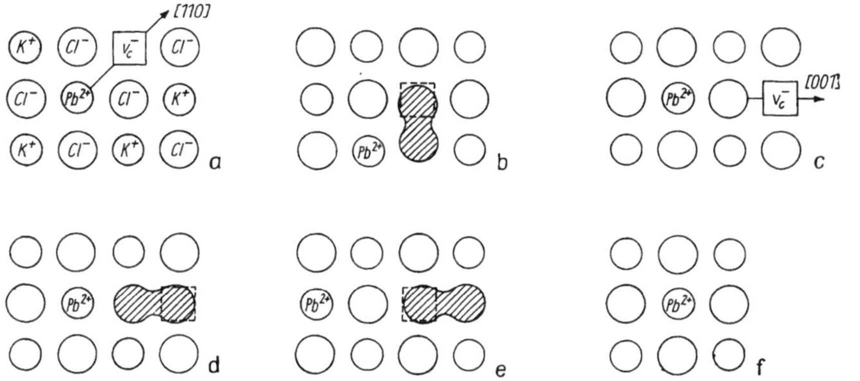

Fig. 6. Schematic models of a) $Pb^{2+}v_c^-$ [110], b) $H_z^{[110]}(Pb^{2+})$, c) $Pb^{2+}v_c^-$ [001], d) $H_z^{[001]}(Pb^{2+})_{ter}$, e) $H_z^{[001]}(Pb^{2+})_{opt}$, and f) $Pb^{2+}[O_h]$ centres in $KCl:PbCl_2$ and $KBr:PbBr_2$ crystals. Indices [110] and [001] point to the orientation of v_c^- with respect to the Pb^{2+} ion

ability of non-radiative transitions from radiative to metastable minima as a result of strengthening the electron–phonon interaction may account for this effect. It is not excluded, nevertheless, that also the ratio of the initial populations of radiative and metastable minima alters.

Secondly, polarization properties of the emission depend essentially on the availability of defects near Pb^{2+} and their disposition. An analysis of the polarization spectra of Pb^{2+} centres of different structure shows [15] that the APES minima whose orbital wave functions have a projection in the direction from Pb^{2+} to v_c^-, i.e. X, Y minima in case of [110] centres and Z minimum in case of [001] centres, are the lowest ones in the triplet excited state. This indicates that non-electrostatic interactions of the optical electrons of Pb^{2+} with its nearest neighbourhood play a dominant role [5].

From the comparison of the characteristics of $Pb^{2+}v_c^-$ and $H_z(Pb^{2+})$ centres with the same orientation of the Pb^{2+}–v_c^- dipole (Fig. 1) one can see that the interstitial atom near Pb^{2+} causes a small shift of the absorption and emission spectra of $Pb^{2+}v_c^-$ centres and some alteration of the energy parameters of their excited state.

However, the most interesting manifestation of the two-component (X_i^0, v_c^-) structure of the defect perturbing the Pb^{2+} ion is observed in the polarization properties of the emission of $H_z^{[001]}(Pb^{2+})$ centres. On absorption of linearly polarized light by the Pb^{2+} ion a dissociation of the complicated defect occurs with the subsequent reorientation of the centre. The mechanism of this phenomenon is discussed in detail in [4]. It should be emphasized that the reorientation takes place only after the thermal equilibrium between radiative and metastable minima has been achieved and thus only the slow decay component displays the corresponding peculiarities. These reorientations lead neither to a depolarization of the emission (as it is observed in case of $Sn^{2+}v_c^-$ centres [18]) nor to an "alignment" of $H_z^{[001]}(Pb^{2+})$ centres in the crystal (as is in case of F_A or M centres (see, e.g., [19]).

References

[1] Zh. Egemberdiev, Trudy IF Akad. Nauk Est. SSR **51**, 163 (1980).
[2] W. Van Puymbroeck and D. Schoemaker, Phys. Rev. B **23**, 167 (1981).
[3] Zh. Egemberdiev, S. Zazubovich, and V. Nagirnyi, phys. stat. sol. (b) **109**, 173 (1982).
[4] V. Nagirnyi, T. Soovik, and S. Zazubovich, phys. stat. sol. (b) **125**, 139 (1984).
[5] V. Nagirnyi, Trudy IF Akad. Nauk Est. SSR **55**, 198 (1984).
[6] S. Zazubovich and T. Kuketaev, Trudy IFA Akad. Nauk Est. SSR **31**, 190 (1966).
[7] A. Fukuda, S. Makishima, T. Mabuchi, and R. Onaka, J. Phys. Chem. Solids **28**, 1763 (1967).
[8] L. Mărculescu and C. Ghiţă, phys. stat. sol. (b) **75**, 375 (1976).
[9] J. L. Pascual, L. Arizmendi, F. Jaque, and F. Aguiló-López, J. Lum. **17**, 325 (1978).
[10] M. F. Trinkler and L. E. Trinkler, Optika i Spektroskopiya **46**, 91 (1979).
[11] K. Schmitt, V. S. Sivasankar, and P. W. M. Jacobs, J. Lum. **27**, 313 (1982).
[12] D. J. Simkin, J. G. Kang, T. F. Bellivean, P. W. M. Jacobs, A. Scacco, M. J. Stillman, and V. S. Sivasankar, Radiat. Eff. **73**, 81 (1983).
[13] Zh. Egemberdiev, V. Nagirnyi, T. Soovik, and S. Zazubovich, phys. stat. sol. (b), to be published.
[14] V. V. Hizhnyakov, Preprint FI-36, Tartu 1975.
[15] V. Hizhnyakov and S. Zazubovich, phys. stat. sol. (b) **86**, 733 (1978).
[16] C. J. Delbecq, R. Hartford, D. Schoemaker, and P. H. Yuster, Phys. Rev. B **13**, 3631 (1976).
[17] V. Hizhnyakov, T. Soovik, and S. Zazubovich, phys. stat. sol. (b) **106**, 581 (1981).
[18] V. Hizhnyakov, S. Zazubovich, and T. Soovik, phys. stat. sol. (b) **66**, 727 (1974).
[19] F. Lüty, Physics of Color Centres, Ed. W. B. Fowler, Academic Press, Inc., New York/London 1968 (p. 209).

(Received July 26, 1984)

phys. stat. sol. (b) **126**, 415 (1984)

Subject classification: 14.3.4

Institute of Physics, Academy of Sciences of the Belorussian SSR, Minsk[1])

On the Theory of Crystal Excitation by Electric Discharges

By

S. V. Voitikov and V. P. Gribkovskii

The problem is considered of semiconductor excitation by the streamer discharge electric field front in the case where the carrier generation processes do not reduce to the quasiclassical Zener tunneling because of high discharge velocities V and a steep front of the field. The streamer-excited interband transitions, unlike previously known ones, are shown to be indirect even in the first-order processes taking place without phonon assistance. The first-order transition slope is equal to $\hbar V$. The calculated interband transition probabilities ω_{cv} and the carrier concentrations N_e, N_h directly depend on the velocity of discharge, the form of its electric field, and the polarity: ω_{cv}, N_e, and N_h are the greater, the higher the discharge velocity, the thinner the field front, and the faster the field intensity drop outside the discharge.

Рассмотрена задача о возбуждении полупроводников фронтом электрического поля стримерных разрядов, когда вследствие больших скоростей распространения V и из-за наличия крутого фронта поля в них процессы генерации носителей не сводятся к квазиклассическому зинеровскому туннелированию. Показано, что при стримерном возбуждении межзонные переходы, в отличие от известных ранее, являются непрямыми уже в процессах первого порядка, т. е. происходящих без участия фононов. Тангенс угла наклона безфононных переходов равен $\hbar V$. Рассчитанные вероятности межзонных переходов ω_{cv} и концентрации создаваемых носителей N_e, N_h непосредственно зависят от скорости разряда, формы его электрического поля, полярности разряда: ω_{cv}, N_e и N_h тем больше, чем больше скорость разряда, уже фронт поля и чем быстрее спадает напряженность поля вдали от разряда.

1. Introduction

Excitation of high-resistivity direct-gap semiconductors by electric field pulses results in the appearance on the surface and in the bulk of a crystal of streamers, a kind of incomplete electric breakdown, which is not attended with crystal destruction [1, 2] (see also references cited in [2]). The streamers propagate at a rate of 10^8 to 5×10^9 cm/s, in their channels non-equilibrium carriers with a concentration of 10^{18} to 10^{20} cm^{-3} are formed [1 to 3] (see also references cited in [3]) and under certain conditions high-power ($\approx 10^4$ W) ns and ps laser radiation is observed [3]. Such discharges may be an effective means of bulk excitation of solids.

In [1], streamer propagation through a crystal is described on the basis of the solution of microscopic electrodynamics and balance equations for charge carriers. The processes of carrier generation — impact and photoionization, field excitation — are looked upon as quasi-classical. The rate of carrier field generation and the impact and photoionization coefficients were taken from the solutions of stationary problems of exciting crystals by homogeneous fields and do not depend on the polarity and velocity of discharge and the shape of its electrical field. A model of streamer formation in the directions of acoustic phonon focusing is proposed, but further travel of the discharge through the crystal is not discussed [4].

[1]) Minsk 220602, USSR.

Since discharges propagate at velocities which exceed the sound velocity and the drift velocities of carriers and have a steep electric field front, the field inhomogeneity and the instability of the phenomenon must be taken into account in the description of carrier generation itself.

Non-equilibrium electrons and holes generated by the streamer field can be formed outside the discharge, in front of it, and at either side of the field front. The wave functions of such excited states of the crystal are eigenfunctions of different Hamiltonians. The quasi-classical consideration of carrier generation does not allow these differences to be taken into account.

In the present paper, we calculate probabilities of electron–hole pair generation and concentrations of non-equilibrium carriers created by the electric field of the discharge that has already passed the stage of formation and is propagating through the crystal at the velocity $\boldsymbol{V}$.

2. Probabilities of Interband Transitions

The wave function of the crystal electron system in its final state, i.e. in the presence of an electron with the quasi-momentum $\hbar \boldsymbol{k}$ in the c-band and a hole with a quasi-momentum $\hbar \boldsymbol{q}$ in the v-band, is the Slater determinant

$$|\mathrm{f}\rangle = A\{\Psi_{\mathrm{v}\boldsymbol{q}_1}(\boldsymbol{r}_1) \dots \Psi_{\mathrm{c}\boldsymbol{k}}(\boldsymbol{r}_n) \dots \Psi_{\mathrm{v}\boldsymbol{q}_N}(\boldsymbol{r}_N)\}\,, \tag{1}$$

where A is the antisymmetrization operator, N is the number of crystal atoms, enclosed in braces is the product of the v-band Bloch functions except for $\Psi_{\mathrm{v}\boldsymbol{q}_n}(\boldsymbol{r}_n)$ in place of which $\Psi_{\mathrm{c}\boldsymbol{k}}(\boldsymbol{r}_n)$ is given. The wave function of the initial state $|\mathrm{i}\rangle$ is the v-band Slater determinant.

The operators of the electron potential energy in the discharge field $W(\boldsymbol{r} \cdot \boldsymbol{v} - Vt) \equiv W(r_{\parallel} - Vt)$ where $\boldsymbol{v} = \boldsymbol{V}/V$ or the holes $-W(r_{\parallel} - Vt)$ which are equal to zero prior to the discharge pass ($t = -\infty$) are "stepped" and at $t = +\infty$ they take the values W_0 and $-W_0$ for electrons and holes generated behind the discharge, respectively, and remain equal to zero when the quasi-particles are formed in front of the discharge. As can easily be shown,

$$W(r_{\parallel} - Vt) = \exp\left(-it\,\frac{\boldsymbol{V}\cdot\hat{\boldsymbol{p}}}{\hbar}\right) W(r_{\parallel}) \exp\left(it\,\frac{\boldsymbol{V}\cdot\hat{\boldsymbol{p}}}{\hbar}\right), \tag{2}$$

where $\hat{\boldsymbol{p}}$ is the momentum operator.

The final states of the crystal $|\mathrm{f}\rangle$ are eigenstates of different Hamiltonians H_f since they depend on the electron and hole velocities. Therefore, the following four types of transitions take place:

(i) The electron and the hole are generated behind the discharge; $H_\mathrm{f} = H_0 + (W_0 - W_0) = H_0$.

(ii) The electron and the hole are generated in front of the discharge; $H_\mathrm{f} = H_0 + (0 - 0) = H_0$.

(iii) The electron is generated in front and the hole behind the discharge; $H_\mathrm{f} = H_0 + (W_0 - 0) = H_0 + W_0$.

(iv) The electron is generated behind and the hole in front of the discharge; $H_\mathrm{f} = H_0 + (0 - W_0) = H_0 - W_0$.

The amplitude of the pair generation probability is

$$A_{\mathrm{cv}} = \lim_{\substack{t\to\infty \\ t_0\to-\infty}} \left\langle \mathrm{f} \left| \exp\left(\frac{iH_\mathrm{f}t}{\hbar}\right) U(t, t_0) \exp\left(-\frac{iH_0t_0}{\hbar}\right) \right| \mathrm{i} \right\rangle, \tag{3}$$

where $U(t, t_0)$ is the evolution operator. The initial and final Bloch states being infinite, the procedure of excitation adiabatical switching on and off is essential. Such procedures are developed in detail in the theory of scattering and consist in introducing the cut-off factor $\exp(-\beta|t|)$, $\beta \to +0$ into the equation for $U(t, t_0)$. Here the Sunakawa method [5] is used,

$$U(t, t_0) = 1 + \lim_{\beta\to+0} \frac{1}{i\hbar} \int_{t_0}^{t} \mathrm{d}t' \exp(-\beta|t'|)\, U(t, t')\, \hat{W}(t');$$

$$\hat{W}(t') = \sum_j W(r_{j\|} - Vt') . \tag{4}$$

In the first order of perturbation

$$A_{\mathrm{cv}}^{(\nu)} = \lim_{\beta\to+0} \frac{1}{i\hbar} \int_{-\infty}^{\infty} \mathrm{d}t \exp(-\beta|t|) \left\langle \mathrm{f} \left| \exp\left(\frac{iH_{\mathrm{f}}t}{\hbar}\right) \hat{W} \exp\left(-\frac{iH_0 t}{\hbar}\right) \right| \mathrm{i} \right\rangle =$$

$$= \lim_{\beta\to+0} \frac{1}{i\hbar} \int_{-\infty}^{\infty} \mathrm{d}t \exp\left(-\beta|t| + it \frac{\mathscr{E}_{\mathrm{g}}^{(\nu)} + \mathscr{E}_{\mathrm{c}\boldsymbol{k}} + \mathscr{E}_{\mathrm{v}\boldsymbol{q}}}{\hbar}\right) \langle \mathrm{f}|\, \hat{W}\, |\mathrm{i}\rangle , \tag{5}$$

where ν marks the type of transitions, $\mathscr{E}_{\mathrm{g}}^{(1)} = \mathscr{E}_{\mathrm{g}}^{(2)} = \mathscr{E}_{\mathrm{g}}$, $\mathscr{E}_{\mathrm{g}}^{(3)} = \mathscr{E}_{\mathrm{g}} + W_0$, $\mathscr{E}_{\mathrm{g}}^{(4)} = \mathscr{E}_{\mathrm{g}} - W_0$, $\mathscr{E}_{\mathrm{g}}$ is the bandgap, $\mathscr{E}_{\mathrm{c}\boldsymbol{k}}$ the electron energy in the c-band, $\mathscr{E}_{\mathrm{v}\boldsymbol{q}}$ the energy of the hole in the v-band. The matrix element of the operator which is a sum of one-particle operators is expressed by a one-particle operator [6],

$$\langle \mathrm{f}|\, \hat{W}\, |\mathrm{i}\rangle = \langle \mathrm{c}\boldsymbol{k}|\, W(r_\| - Vt)\, |\mathrm{v}\boldsymbol{q}\rangle =$$

$$= \left\langle \mathrm{c}\boldsymbol{k} \left| \exp\left(-it \frac{\boldsymbol{V}\cdot\hat{\boldsymbol{p}}}{\hbar}\right) W(r_\|) \exp\left(it \frac{\boldsymbol{V}\cdot\hat{\boldsymbol{p}}}{\hbar}\right) \right| \mathrm{v}\boldsymbol{q} \right\rangle . \tag{6}$$

Since the displacement operator $\exp(\pm it\boldsymbol{V}\cdot\hat{\boldsymbol{p}}/\hbar)$ is diagonal in $\boldsymbol{p}$-representation, then

$$\langle \mathrm{c}\boldsymbol{k}|\, W(r_\| - Vt)\, |\mathrm{v}\boldsymbol{q}\rangle =$$

$$= \int \mathrm{d}^3\boldsymbol{p}\, \mathrm{d}^3\boldsymbol{p}'\, \langle \mathrm{c}\boldsymbol{k} \mid \boldsymbol{p}\rangle \exp\left(-it \frac{\boldsymbol{V}\cdot\hat{\boldsymbol{p}}}{\hbar}\right) \langle \boldsymbol{p}|\, W(r_\|)\, |\boldsymbol{p}'\rangle \times$$

$$\times \exp\left(it \frac{\boldsymbol{V}\cdot\boldsymbol{p}'}{\hbar}\right) \langle \boldsymbol{p}' \mid \mathrm{v}\boldsymbol{q}\rangle = \sum_{\boldsymbol{G}} \exp\left[-it\boldsymbol{V}\cdot(\boldsymbol{k} - \boldsymbol{q} + \boldsymbol{G})\right] \times$$

$$\times \frac{1}{(2\pi\hbar)^3} W(\boldsymbol{k} - \boldsymbol{q} + \boldsymbol{G}) \int_{\Omega} u_{\mathrm{c}\boldsymbol{k}}^* \exp(-i\boldsymbol{G}\cdot\boldsymbol{r})\, u_{\mathrm{v}\boldsymbol{q}}\, \mathrm{d}^3\boldsymbol{r} , \tag{7}$$

where $\boldsymbol{G}$ is the reciprocal lattice vector, $u_{\mathrm{c}\boldsymbol{k}}$ and $u_{\mathrm{v}\boldsymbol{q}}$ are periodical parts of the Bloch wave functions $\Psi_{\mathrm{c}\boldsymbol{k}}$ and $\Psi_{\mathrm{v}\boldsymbol{q}}$, respectively,

$$W(\boldsymbol{k} - \boldsymbol{q} + \boldsymbol{G}) = \int \mathrm{d}^3\boldsymbol{r} \exp\left[-i\boldsymbol{r}\cdot(\boldsymbol{k} - \boldsymbol{q} + \boldsymbol{G})\right] W(r_\|) =$$

$$= (2\pi)^2\, \delta(\boldsymbol{k}_\perp - \boldsymbol{q}_\perp + \boldsymbol{G}_\perp) \int_{-\infty}^{\infty} \mathrm{d}r_\| \times$$

$$\times \exp\left[-ir_\|(k_\| - q_\| + G_\|)\right] W(r_\|) , \tag{8}$$

Ω is the volume of a unit cell. Let us introduce the following notations:

$$\int_{\Omega} u_{\mathrm{c}\boldsymbol{k}}^* \exp(-i\boldsymbol{G}\cdot\boldsymbol{r})\, u_{\mathrm{v}\boldsymbol{q}}\, \mathrm{d}^3\boldsymbol{r} \equiv (\mathrm{c}\boldsymbol{k}|\, \exp(-i\boldsymbol{G}\cdot\boldsymbol{r})\, |\mathrm{v}\boldsymbol{q}) . \tag{9}$$

It is shown below that the Fourier transform $W(\alpha)$ for the stepped potentials is calculated as an analytical continuation from the lower half-plane $\operatorname{Im}\alpha < 0$ and has a first-order pole at the origin $\alpha = 0$. The contour of integration with respect to α (in (7) $\alpha = (p_{\|} - p'_{\|})/\hbar$) should pass under the point $\alpha = 0$ [7], since on the real axis the integral is indefinite. Since the Bloch wave functions are analytical functions of the quasi-wave vector components in some non-zero strips containing a real axis as an interior of a set [7, 8], the contour deformation does not affect the analyticity of integrands in (8) and makes possible the calculation of the transition amplitude by straightforward integration,

$$A_{\mathrm{cv}}^{(\nu)} = -i \sum_{\boldsymbol{G}} \delta(\boldsymbol{k}_{\perp} - \boldsymbol{q}_{\perp} + \boldsymbol{G}_{\perp})\, \delta[\mathscr{E}_{\mathrm{g}}^{(\nu)} + \mathscr{E}_{\mathrm{c}\boldsymbol{k}} + \mathscr{E}_{\mathrm{v}\boldsymbol{q}} - \hbar V(k_{\|} - q_{\|} + G_{\|})] \times$$
$$\times\, W(k_{\|} - q_{\|} + G_{\|})\, (\mathrm{c}\boldsymbol{k}|\exp(-i\boldsymbol{G}\cdot\boldsymbol{r})\,|\mathrm{v}\boldsymbol{q}) \,. \tag{10}$$

As one would expect, the perpendicular part of the electron–hole pair quasi-wave vector is equal to zero to an accuracy of the perpendicular part of the reciprocal lattice vector. The real roots of the equation

$$\mathscr{E}_{\mathrm{g}}^{(\nu)} + \mathscr{E}_{\mathrm{c}\boldsymbol{k}} + \mathscr{E}_{\mathrm{v}\boldsymbol{q}} - \hbar V(k_{\|} - q_{\|} + G_{\|}) = 0 \tag{11}$$

determine the longitudinal part of the quasi-wave vectors. If $\boldsymbol{G} \neq 0$, electron–hole pair generation belongs to the class of umklapp processes. The reciprocal lattice vector should be chosen such that the roots of (11) are real.

According to (11), which is the law of energy conservation, under the action of the rapidly moving electric field front in the crystal indirect interband transitions must occur, including first-order processes, i.e. without the assistance of phonons or other quasi-particles which would change the electron and hole quasi-momenta. The slope of the transition in the band picture is $\hbar V$ (see Fig. 2). The second-order phonon-assisted processes are characterized by a tilted fan of transitions.

The rate of non-equilibrium carrier pair generation is expressed through the amplitude A_{cv} by the known relation [5, 9],

$$P_{\mathrm{cv}}^{(\nu)} = \frac{2\pi}{\hbar}\, \delta[\mathscr{E}_{\mathrm{g}}^{(\nu)} + \mathscr{E}_{\mathrm{c}\boldsymbol{k}} + \mathscr{E}_{\mathrm{v}\boldsymbol{q}} - \hbar V(k_{\|} - q_{\|} + G_{\|})] \times$$
$$\times\, |W(k_{\|} - q_{\|} + G_{\|})\, (\mathrm{c}\boldsymbol{k}|\exp(-i\boldsymbol{G}\cdot\boldsymbol{r})\,|\mathrm{v}\boldsymbol{q})|^2 \,. \tag{12}$$

To calculate the pair formation probability by the discharge field, we formally consider this process in terms of the scattering theory [9] as one-dimensional inelastic discharge scattering by the crystal with electron–hole pair generation in the latter. The final discharge velocity V_{f} will be other than V due to the law of conservation of energy and momentum, but this difference is extremely small, since the discharge "mass" is evidently much greater than the electron and hole masses, $M \gg m_{\mathrm{c}}, m_{\mathrm{v}}$. The probability of inelastic scattering per unit time is given by (12) to whose δ-function argument containing the difference in internal energy of the crystal final and initial states, the difference should be added in kinetic energy of the discharge before and after collision with the crystal, $\hbar^2 K_{\mathrm{f}}^2/2M - \hbar^2 K^2/2M$ where $K_{\mathrm{f}} = MV_{\mathrm{f}}/\hbar$ and $K = MV/\hbar$. The discharge "flow" after the collision followed by excitation of the non-equilibrium carrier pair in the crystal is

$$J_{\mathrm{f}} = 2\pi \int \mathrm{d}K_{\mathrm{f}} P_{\mathrm{cv}} = \frac{(2\pi)^2}{\hbar^2 V}\, |W(k_{\|} - q_{\|} + G_{\|})\, (\mathrm{c}\boldsymbol{k}|\exp(-i\boldsymbol{G}\cdot\boldsymbol{r})\,|\mathrm{v}\boldsymbol{q})|^2 \,. \tag{13}$$

The discharge "flow" before the collision is $J_i = V$. The sought probability of the electron–hole generation is equal to the "flow" relation J_f/J_i,

$$w_{cv} = \left(\frac{2\pi}{\hbar V}\right)^2 |W(k_{||} - q_{||} + G_{||}) (c\boldsymbol{k}| \exp(-i\boldsymbol{G}\cdot\boldsymbol{r}) |v\boldsymbol{q})|^2 , \tag{14}$$

where the electron and hole quasi-wave vectors are determined by the energy conservation law (11). The probability w_{cv} directly depends on the discharge velocity, the function type $W = W(r_{||})$, and the discharge polarity (the sign of W_0). Formula (14) holds for different band structures.

The Fourier transform $W(\alpha)$ is determined by the shape of the electric discharge field. At $r_{||} \to \pm\infty$ the intensity $E(r_{||})$ falls to zero and, since the discharge velocity exceeds the drift velocities of carriers, it falls faster than for a statically screened field at least, i.e. faster than exp $(-\text{const}\,|r_{||}|)$. This condition is obviously satisfied, since the discharge electric field front is compressed by a focused phonon jet [4]. In this case the Fourier transform $W(\alpha)$ has a finite value and is found as an analytical continuation from the lower half-plane Im $\alpha < 0$,

$$W(\alpha) = \int_{-\infty}^{\infty} dx \exp(-i\alpha x) \int_{-\infty}^{x} dy\, eE(y) = -\frac{1}{i\alpha} \exp(-i\alpha x)\, eE(x) \Big|_{-\infty}^{\infty} +$$

$$+ \frac{1}{i\alpha} \int_{-\infty}^{\infty} dr_{||} \exp(-i\alpha r_{||})\, eE(r_{||}) = \frac{W_0}{i\alpha} \varphi(\alpha) . \tag{15}$$

Here, $\varphi(\alpha)$ is the function which decreases fast on the real axis at $|\alpha|\, l > 2$, $\varphi(0) = 1$, $W_0 = eEl$ is the height of a potential energy step (see Fig. 1), e the electron charge, l the average field front thickness, E the mean field intensity. It can be seen from (15) that $W(\alpha)$ has a first-order pole at $\alpha = 0$ and is analytically continued to the upper half-plane. As shown by example of the existing theories of tunneling p–n junctions [7, 10], even results corresponding to the step $W(\alpha)$ in the form of a bias plane

$$E_1(r_{||}) = E\theta\left(\frac{l}{2} - |r_{||}|\right), \qquad W_1(\alpha) = \frac{W_0}{i\alpha} \frac{\sin(\alpha l/2)}{\alpha l/2}, \tag{16}$$

where θ is a step-function, are not beyond the first approximation accuracy in the amplitude calculation.

In order to determine the influence of the field shape on w_{cv}, we represent Fourier transforms $W(\alpha)$ for two more dependences $E(r_{||})$ differently decreasing at $r_{||} \to +\infty$,

$$E_2(r_{||}) = E \cos^2\left(\frac{\pi r_{||}}{2l}\right) \theta\left(\frac{l}{2} - |r_{||}|\right), \qquad W_2 = \frac{W_0}{i\alpha} \frac{\cos(\alpha l/2)}{1 + \alpha^2 l^2/\pi}, \tag{17}$$

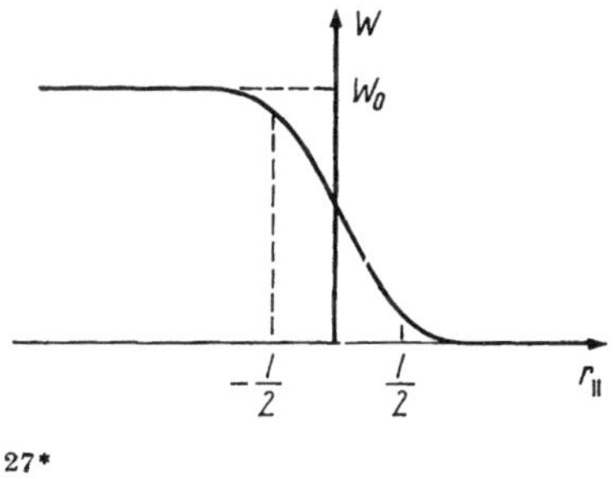

Fig. 1. An approximate form of the carrier potential energy in the streamer discharge field

$$E_3(r_{||}) = E\left(2\frac{r_{||}^2}{l^2}+1\right)^{-1/2} K_1\left(\sqrt{2\frac{r_{||}^2}{l^2}+1}\right),$$

$$W_3 = \frac{W_0}{i\alpha}\exp\left(1-\sqrt{1+\frac{\alpha^2 l^2}{2}}\right), \tag{18}$$

where K_1 is a modified Bessel function. At $r_{||} \to \pm\infty$, $E_3(r_{||}) \approx E(l/\sqrt{2}\,r_{||}) \times \exp(-\sqrt{2}\,|r_{||}|/l)$. It may be concluded, by comparing $W_1(\alpha)$, $W_2(\alpha)$, $W_3(\alpha)$ that the faster the $E(r_{||})$ drop to zero at $r_{||} \to \pm\infty$ and the smaller the front thickness l, the greater the probability w_{cv} of pair generation.

Due to the fast decrease in $\varphi(k_{||} - q_{||} + G_{||})$ the overwhelming majority of transitions occur in the vicinity of the band extrema. These are first-type transitions (Fig. 2 a, b). When the band extrema are situated at the same point $\boldsymbol{k}_0$ (Fig. 2a), the matrix element $(\mathrm{c}\boldsymbol{k} \mid \mathrm{v}\boldsymbol{q})$ calculated by the $\boldsymbol{k}\cdot\boldsymbol{p}$ perturbation approach is

$$(\mathrm{c}\boldsymbol{k} \mid \mathrm{v}\boldsymbol{q}) = \frac{\hbar^2}{m_0^2}\frac{(\boldsymbol{k}-\boldsymbol{q})\cdot\boldsymbol{p}_{cv}}{\mathscr{E}_g + \mathscr{E}_{c\boldsymbol{k}} + \mathscr{E}_{v\boldsymbol{q}}} \tag{19}$$

and

$$w_{cv} = 4\pi^2\frac{W_0^2\,|\boldsymbol{p}_{cv}|^2}{m_0^2}\,B_{cv}\,, \tag{20}$$

$$B_{cv} = \frac{1}{V^2}\frac{\varphi^2(k_{||}-q_{||})}{(\mathscr{E}_g + \mathscr{E}_{c\boldsymbol{k}} + \mathscr{E}_{v\boldsymbol{q}})^2}, \tag{21}$$

where m_0 is the real electron mass,

$$\boldsymbol{p}_{cv} = \int_\Omega u_{c\boldsymbol{k}_0}^* \hat{\boldsymbol{p}} u_{v\boldsymbol{k}_0}\,\mathrm{d}^3\boldsymbol{r}\,. \tag{22}$$

For parabolic bands

$$k_{||} - q_{||} = \frac{m_{c||}+m_{v||}}{\hbar}V - \left(q_{||} + \frac{m_{v||}}{\hbar}V\right) - \left\{m_{c||}\left[\frac{m_{c||}+m_{v||}}{\hbar^2}V^2 - \frac{2\mathscr{E}_g}{\hbar^2} - \left(\frac{1}{m_{c\perp}}+\frac{1}{m_{v\perp}}\right)q_\perp^2 - \frac{1}{m_{v||}}\left(q_{||}+\frac{m_{v||}}{\hbar}V\right)^2\right]\right\}^{1/2}, \tag{23}$$

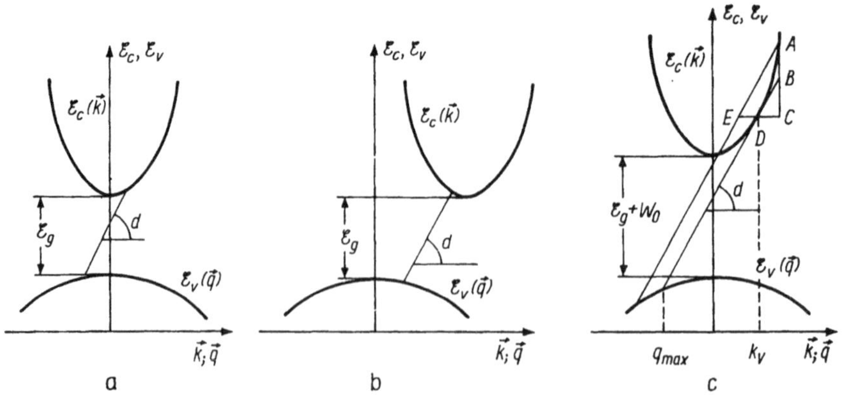

Fig. 2. Interband transitions excited by the streamer discharge field. a), b) First type of transitions (electrons and holes appear behind the discharge front); c) third type of transitions (the velocities of the electrons appearing in the c-band exceed the discharge velocity V); $\tan d = \hbar V$

where

$$m_{c,v\|}^{-1} = \sum_j m_{c,vj}^{-1} \cos \gamma_j , \qquad m_{c,v\perp}^{-1} = \sum_j m_{c,vj}^{-1} \sin \gamma_j ,$$

γ_j is an angle between the vector $\boldsymbol{V}$ and the j-axis of the system of coordinates where the effective mass tensors are diagonal.

The surface $B_{cv}(q_\|, q_\perp)$ has the form of a hill (Fig. 3) whose horizontal sections are half-ellipses. The equation of the ellipse is

$$\left(\frac{1}{m_{c\|}} + \frac{1}{m_{v\|}}\right)\left(q_\| + \frac{m_{v\|}}{m_{c\|} + m_{v\|}} \alpha\right)^2 + \left(\frac{1}{m_{c\perp}} + \frac{1}{m_{v\perp}}\right) q_\perp^2 = R^2 , \tag{24}$$

$$R^2 = \frac{m_{c\|} + m_{v\|}}{\hbar^2} V^2 - \frac{2\mathcal{E}_g}{\hbar^2} - \frac{1}{m_{c\|} + m_{v\|}} \left(\alpha - \frac{m_{c\|} + m_{v\|}}{\hbar} V\right)^2, \tag{25}$$

where α is the difference $k_\| - q_\|$ corresponding to the section height. The greater the discharge velocity V, the higher and wider the surface $B_{cv}(q_\|, q_\perp)$, i.e. the greater the probabilities of carrier pair generation.

3. Non-Equilibrium Carrier Concentrations

The non-equilibrium electron and hole concentrations are determined by the integral

$$N_e = N_h = \frac{2}{(2\pi)^3} \int d^3\boldsymbol{q}\, w_{cv} = \frac{1}{2\pi^2} \int dq_\|\, dq_\perp\, q_\perp w_{cv} , \tag{26}$$

which is reduced to

$$N_e = N_h = \frac{2}{3} \frac{W_0^2 |p_{cv}|^2}{\hbar^2} \left(\frac{1}{m_{c\|}} + \frac{1}{m_{v\|}}\right)^{-1/2} \left(\frac{1}{m_{v\perp}} + \frac{1}{m_{c\perp}}\right)^{-1} \int_{\alpha_1}^{\alpha_2} d\alpha\, R^3(\alpha) \left(-\frac{\partial B_{cv}}{\partial \alpha}\right), \tag{27}$$

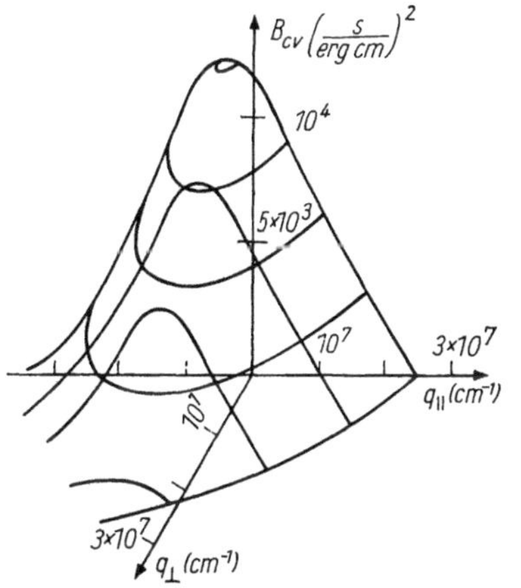

Fig. 3

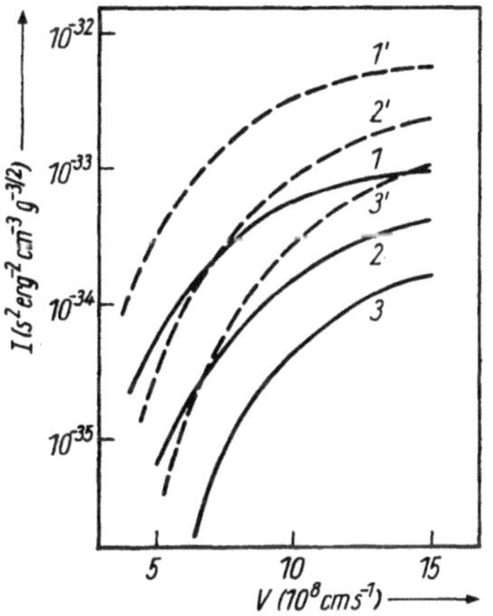

Fig. 4

Fig. 3. The function $B_{cv}(q_\|, q_\perp)$ (21) for hexagonal CdS. Discharge propagates along the c-axis; $W = W_3(\alpha)$ (18); $V = 6 \times 10^8$ cms^{-1}; $l = 8 \times 10^{-7}$ cm

Fig. 4. $I(V)$ dependence (29) for hexagonal CdS. Solid and dashed lines show discharges propagating perpendicular to and along the c-axis, respectively; (1), (1') $l = 6 \times 10^{-7}$, (2), (2') 8×10^{-7}, (3), (3') 10^{-6} cm

where α_1 is a minimum possible value α,

$$\alpha_1 = \frac{m_{c\parallel} + m_{v\parallel}}{\hbar} V \left[1 - \sqrt{1 - \frac{2\mathcal{E}_g}{V^2(m_{c\parallel} + m_{v\parallel})}}\right], \tag{28}$$

$\alpha_2 = (m_{c\parallel} + m_{v\parallel}) V/\hbar$. Computations show that the integral which enters into (27),

$$I(V) = \int_{\alpha_1}^{\alpha_2} d\alpha\, R^3(\alpha) \left(-\frac{\partial B_{cv}}{\partial \alpha}\right), \tag{29}$$

is a monotonically increasing function of the discharge velocity V (Fig. 4). Unlike the known inelastic processes whose probabilities are proportional to V^{-2} [9], w_{cv} and N_e, N_h for the interband transitions under consideration increase as V grows though they have a factor V^{-2}, since in this case the quasi-momenta $\hbar(k_\parallel - q_\parallel)$ of electron–hole pairs and, therefore, the difference in energy of excited and initial states of the crystal decrease sufficiently fast.

Transitions between bands whose extrema are situated at different points $\boldsymbol{k}_0$ and $\boldsymbol{q}_0$ of the Brillouin zone (Fig. 2b) do not radically differ from those described above, but the calculation of transition probabilities and carrier concentrations presents difficulties, since the matrix elements $(c\boldsymbol{k} \mid v\boldsymbol{q})$ are unknown.

Consider now a third type of transitions between the c- and v-branches of the direct-gap spectrum when the velocity of electrons generated in the c-band exceeds the discharge velocity, $k_\parallel = k_V + \Delta k_\parallel > k_V$, and the electrons are situated in the spectrum above the point of tangency D (Fig. 2c). The geometrical proportions of the cuts AB:BC = ED:DC make it possible to write down

$$\frac{\mathcal{E}_{c\boldsymbol{k}_\perp, k_V+\Delta k_\parallel} - \mathcal{E}_{c\boldsymbol{k}_\perp, k_V} - \hbar V \Delta k_\parallel}{\hbar V k_\parallel} = \frac{\Delta q_\parallel}{\Delta k_\parallel} \cdot \frac{\hbar V}{\hbar V + (\partial \mathcal{E}_{v\boldsymbol{q}}/\partial q_\parallel)}, \tag{30}$$

hence

$$\Delta^3 \boldsymbol{q} = \Delta^2 \boldsymbol{q}_\perp\, \Delta q_\parallel = \Delta^2 \boldsymbol{k}_\perp\, \Delta k_\parallel \frac{k_\parallel - k_V}{2} \frac{\partial^2 \mathcal{E}_{c\boldsymbol{k}}}{\partial k_\parallel^2} \frac{1}{\hbar V + \partial \mathcal{E}_{v\boldsymbol{q}}/\partial q_\parallel}, \tag{31}$$

$\partial^2 \mathcal{E}_{c\boldsymbol{k}}/\partial k_\parallel^2$ and $\partial \mathcal{E}_{v\boldsymbol{q}}/\partial q_\parallel$ being calculated at $k_\parallel = k_V$ and $q_\parallel = q_{\parallel,\max}$, respectively. From the equality of the number of electrons with $k_\parallel > k_V$ in the cell $\Delta^3 \boldsymbol{k}$ to the number of holes generated in the cell $\Delta^3 \boldsymbol{q}$,

$$n(\boldsymbol{k})\, \Delta^3 \boldsymbol{k} = \frac{2}{(2\pi)^3} w_{cv}\, \Delta^3 \boldsymbol{q} = \frac{1}{(2\pi)^3} w_{cv} \frac{\partial^2 \mathcal{E}_{c\boldsymbol{k}}}{\partial k_\parallel^2} \frac{k_\parallel - k_V}{\hbar V + (\partial \mathcal{E}_{v\boldsymbol{q}}/\partial q_\parallel)} \Delta^3 \boldsymbol{k}, \tag{32}$$

we obtain a distribution function of unscattering electrons out-stripping the discharge

$$n(\boldsymbol{k}) = \frac{1}{(2\pi)^3} w_{cv} \frac{\partial^2 \mathcal{E}_{c\boldsymbol{k}}}{\partial k_\parallel} (k_\parallel - k_V) \left(\hbar V + \frac{\partial \mathcal{E}_{v\boldsymbol{q}}}{\partial q_\parallel}\right)^{-1}. \tag{33}$$

Such a dependence is qualitatively similar to a running electron distribution [11]. Perhaps it is the running electrons that are responsible for the high-speed propagation of electric discharges [12].

4. Conclusion

Thus, under the action of the streamer discharge electric field indirect interband transitions (Fig. 2) must occur in crystals according to (11) even without other quasi-particle participation. In contrast to the quasi-classical interpretation of carrier generation as Zener tunneling in an inhomogeneous field, the probabilities of interband transitions w_{cv} (14) and the non-equilibrium electron and hole concentrations

N_e, N_h (27) directly depend on the velocity of discharge, the shape and intensity of its electric field, the discharge polarity. They are the greater, the higher the discharge velocity, the thinner the field front, and the faster the field intensity drops to zero outside the discharge.

References

[1] N. G. Basov, A. G. Molchanov, A. S. Nasibov, A. Z. Obidin, A. N. Pechenov, and Yu. M. Popov, Zh. eksper. teor. Fiz. **70**, 1751 (1976).
[2] V. P. Gribkovskii, A. A. Gladyshchuk, V. V. Zubritskii, V. V. Parashchuk, and G. P. Yablonskii, phys. stat. sol. (a) **77**, 765 (1983).
[3] A. Z. Obidin, A. N. Pechenov, Yu. M. Popov, and V. A. Frolov, Kvantovaya Elektronika **10**, 1165 (1983).
[4] L. A. Chernozatonskii, Zh. eksper. teor. Fiz., Pisma **38**, 225 (1983).
[5] S. Sunakawa, Quantum Theory of Scattering, Izd. Mir, Moscow 1979 (in Russian).
[6] F. Bassani and G. Pastori Parravichini, Electron States and Optical Transitions in Solids, Izd. Nauka, Moscow 1982.
[7] R. T. Shuey, Phys. Rev. **137**, A1268 (1965).
[8] W. Kohn, Phys. Rev. **115**, 809 (1959).
[9] L. D. Landau and E. M. Lifshits, Quantum Mechanics, Izd. Nauka, Moscow 1974 (in Russian).
[10] D. R. Fredkin and G. H. Wannier, Phys. Rev. **128**, 2054 (1962).
Y. Takeuti and H. Funada, J. Phys. Soc. Japan **20**, 1954 (1965).
[11] A. B. Levinson, Fiz. tverd. Tela **6**, 2113 (1964).
[12] L. P. Babich, Dokl. Akad. Nauk SSSR **263**, 76 (1982).

(Received June 13, 1984)

phys. stat. sol. (b) **126**, 425 (1984)

Subject classification: 2 and 14.1; 21

Fakultät für Physik der Universität Konstanz[1])

Electrical Resistivity of Amorphous SnMn

By

D. Korn

Amorphous alloys of SnMn are produced by quench condensation from the vapour phase onto a substrate held at 5 K. The Mn concentration of the alloys varies between 5 and 60 at%. The electrical resistivity is measured in situ in the temperature range 1.8 to 390 K. The electrical resistivity of all amorphous SnMn alloys increases with decreasing temperature without approaching a constant value at low temperature. To explain the almost constant negative temperature coefficient the assumption is made that in amorphous Sn the orbital angular momenta of the p-electron interact via spin–orbit coupling with the Mn ion spin.

Amorphe SnMn-Legierungen wurden durch abschreckende Kondensation auf einer Unterlage von 5 K hergestellt. Die Mn-Konzentration der Legierungen beträgt 5 bis 60 at%. Der elektrische Widerstand wird sofort nach dem Aufdampfen im Temperaturbereich von 1.8 bis 390 K gemessen. Der elektrische Widerstand von amorphem SnMn steigt mit fallender Temperatur fast linear an, ohne bei tiefen Temperaturen gegen einen konstanten Wert zu gehen. Zur Erklärung des negativen Temperaturkoeffizienten wird angenommen, daß die p-Elektronen des amorphen Sn am Ort des Mn-Ionenspins mittels Spin–Bahn-Kopplung einen temperaturabhängigen Streuquerschnitt bewirken.

1. Introduction

The electrical resistivity of amorphous metals and alloys has approximately the same magnitude as the resistivity of liquid alloys. The temperature coefficient of the resistivity has positive and negative values. A negative temperature coefficient of the resistivity is remarkable because the density of the conduction electrons is near to the value of the free electron gas [1]. This is valid for amorphous alloys consisting of noble metals and non-transition metals with unfilled p-shells. The negative temperature coefficient of these alloys decreases at low temperature [2].

On the other hand, there exist amorphous alloys with a temperature coefficient which does not decrease at low temperature. These alloys contain a component with an unfilled d-shell. An example is $Cu_{60}Zr_{40}$ [3]. Amorphous $Cu_{60}Zr_{40}$ has no localized magnetic moment [4]. Some amorphous alloys have a localized magnetic moment (CrPdSi, MnPdSi, [5, 6]) and a resistivity behaviour described with the Kondo theory. The magnetic moment is that of e.g. Mn and the amorphous state is due to PdSi. In contrast to these ternary alloys such a separation concerning the magnetic moment and the amorphous state is not easy with respect to the binary system SnMn. The electrical resistivity of SnMn increases at low temperature similar to CrPdSi [5] and CuZr [3]. But opposite to CuZr the Mn in Sn has a magnetic moment. That is valid for the crystalline, the amorphous [7], and the liquid phase [8, 9]. The electrical resistivity of liquid SnMn is investigated by Gasser and Kleim [10]. Crystalline SnMn shows a resistance anomaly at low temperature (Kondo effect as reported by Buckel et al. [11]). The resistance anomaly of SnMn is supported by disorder of the Sn lattice [12]. In the concentration range 1.5 to 5 at% Mn resistance maxima exist in the crystalline

[1]) PF 5560, D-7750 Konstanz, FRG.

Table 1
Transition temperature amorphous–crystalline T_t, annealing temperature T_a, residual resistivity ϱ_0, negative slope $-d\varrho/dT$ taken at 5 K

SnMn (at% Mn)	T_t (K)	T_a (K)	ϱ_0 (μΩ cm)	$-d\varrho/dT$ (nΩ cm/K)
5	35	25	53.9	10.1
		40	36.3	−13.1
		80	34.5	−11.6
		150	29.0	− 8.9
		220	21.5	− 4.6
		290	6.6	− 2.1
6	50	25	62.1	19.6
		40	57.8	11.7
		80	40.9	− 5.7
		150	34.5	− 6.2
		220	24.3	− 2.7
		290	6.8	0
7	90	25	74.9	33.2
		80	71.0	
		150	46.4	
		290	8.7	
8	127	25	78.7	38.8
		40	78.3	38.8
		80	77.7	37.3
		150	54.0	21.4
		220	45.4	17.0
		290	8.6	− 2.9
8 (another film)	125	25	77.0	36.8
		40	76.6	36.3
		80	76.3	36.8
		150	54.7	20.5
		220	47.0	9.9
		290	9.0	− 2.9
15	323	25	133	64
		40	133	64
		80	133	65
		150	133	63
		290	151	75
		390	10	
25	≈ 270	25	161	73
		40	161	77
		80	162	78
		150	164	81
		290	106	10
		390	34	
36	259	40	177	92
		80	178	97
		150	180	104
		290	179	48
		390	99	
60	> 390	25	282	240
		40	282	
		80	278	230
		390	310	

state [12]. The crystalline and the amorphous phases differ only little with regard to the magnetic properties [7] but the electrical resistance behaves very different in the two phases [12].

2. Experimental Essentials

Mn is insoluble in Sn within the concentration range 5 to 60 at% Mn [13] except the intermetallic compound Sn_2Mn. Therefore, the samples are prepared by quenching from the vapour phase without fractionation onto a cold crystalline quartz plate. The method has already been described [14, 12]. The films produced by this manner have thicknesses between 60 and 120 nm. The thickness is determined optically by an interferometric method. The resistance is measured in situ by a conventional four-electrode method with a resolution of 10^{-5}. The resistivity has an absolute error of 10% mainly due to the thickness. After condensation at 5 K the film is stepwise annealed to temperatures T_a given in Table 1. The film thickness exceeds the mean free path of the electrons which has the order of magnitude of 1 nm in amorphous alloys. Thus, bulk properties are measured. This has also been proved experimentally with amorphous SnAu films [15]. So the films serve only as an implement to investigate amorphous SnMn because no other method exists until now to bring this system into the amorphous state.

3. Results

Fig. 1 shows as one example the resistivity of a SnMn film as a function of temperature. By condensation at 5 K the film grows in the amorphous state. The resistivity has a negative temperature coefficient. Temperature coefficient and residual resistivity ϱ_0 do not change by annealing to $T_a = 80$ K. Even no appreciable change in resistivity behaviour is observed for $T_a = 150$ K. Annealing to $T_a = 290$ K increases the residual resistivity as often observed for amorphous alloys with negative temperature coefficient. At higher temperature a steep decrease marks the transition from the amorphous to the crystalline state. The connection between the decrease of the resistivity and the transition amorphous–crystalline is already known by X-ray and electron diffraction investigation from the systems SnCu [16, 17] and SnAu [18]. In the crystalline state the normal positive temperature coefficient of the resistivity is observed (Fig. 1).

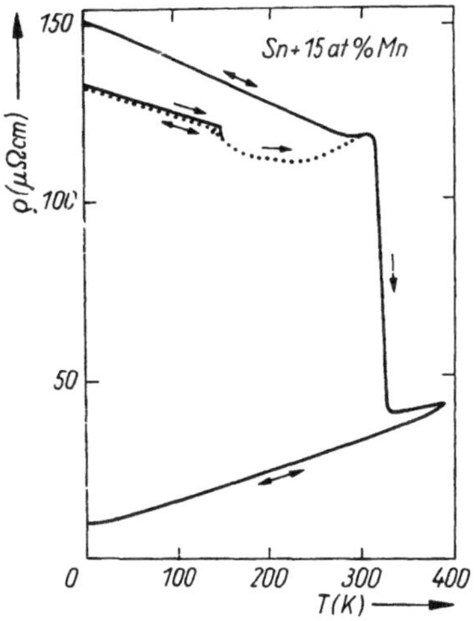

Fig. 1. Electrical resistivity of Sn + 15 at% Mn as a function of temperature. Reversible branches of the curves are marked with double arrows

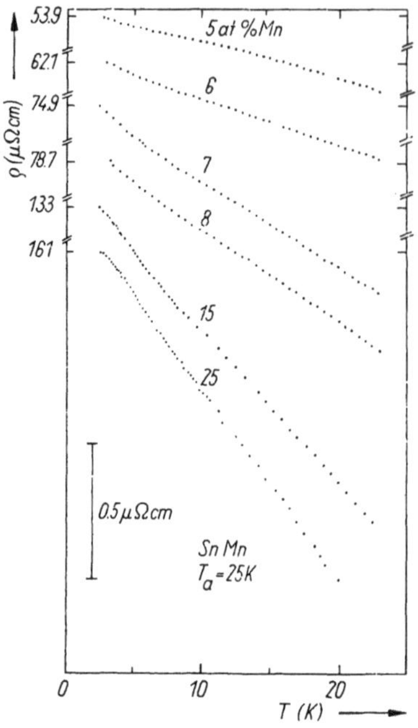

Fig. 2

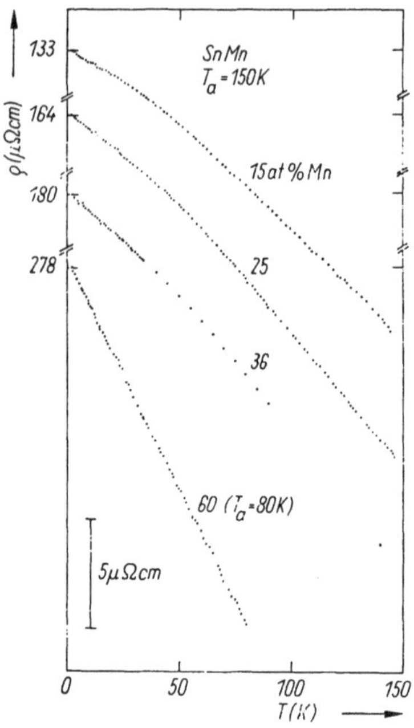

Fig. 3

Fig. 2. Electrical resistivity of amorphous SnMn alloys annealed at the temperature $T_a = 25$ K. Ordinate zero suppressed. Scale (0.5 μΩ cm) valid for all curves

Fig. 3. Electrical resistivity of amorphous SnMn alloys as a function of temperature. Temperature of annealing $T_a = 150$ K except Sn + 60 at% Mn: $T_a = 80$ K

In the following the resistivity ϱ of the amorphous phase is only considered at temperatures T which are lower than the annealing temperature T_a (reversible $\varrho(T)$). The electrical resistivity of amorphous SnMn alloys annealed at $T_a = 25$ K is plotted in Fig. 2. The electrical resistivity increases almost linearly with falling temperature without approaching a constant value at low temperature. The same holds for SnMn alloys annealed at $T_a = 150$ K (Fig. 3). The negative slope of the resistivity increases with concentration within the investigated range from 5 to 60 at% Mn in Sn (Fig. 2 and 3). Additional values concerning the negative slope and further data are given in Table 1. The residual resistivity ϱ_0 is taken at 2 K. Not extrapolating to $T = 0$ means an error of 10^{-3} which is neglected. The residual resistivity increases with Mn concentration in amorphous SnMn (Table 1). The temperature coefficient α of the electrical resistivity is shown in Fig. 4. The value of α is constant between 8 and 36 at% Mn in amorphous Sn and independent of the annealing temperature,

$$\alpha = \frac{\mathrm{d}\varrho}{\varrho_0\,\mathrm{d}T} = (-0.5 \pm 0.05) \times 10^{-3}\ \mathrm{K}^{-1}\,.$$

(α is taken at 5 K. It does not change much with temperature since the resistivity varies almost linearly with temperature (Fig. 2 and 3).)

The amorphous alloys exist only below the transition temperature T_t. The transition temperature rises with concentration (Table 1). At 36% Mn a minimum of T_t is observed due to the intermetallic compound in the crystalline state (generally a stable

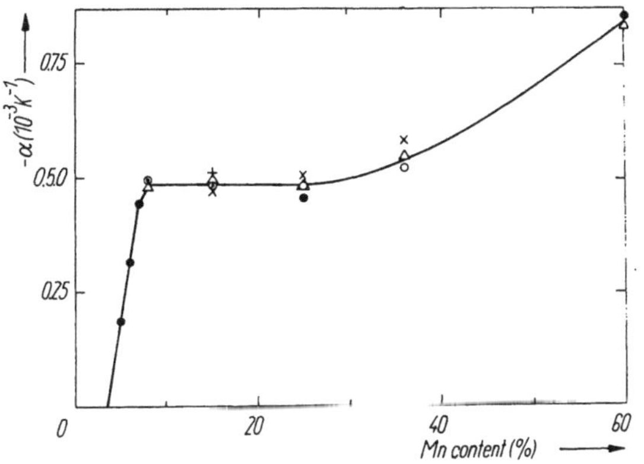

Fig. 4

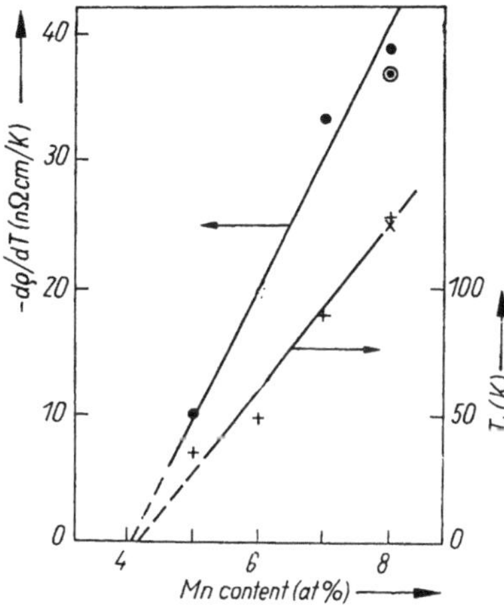

Fig. 5

Fig. 4. Negative temperature coefficient ($-\alpha$) of the electrical resistivity of amorphous SnMn alloys. Different annealing temperatures. ● $T_a = 25$, ○ 40, △ 80, × 150, + 290 K

Fig. 5. Transition temperature T_t amorphous–crystalline and negative slope ($-\mathrm{d}\varrho/\mathrm{d}T$) against Mn concentration in amorphous Sn. Negative slope taken at 5 K and valid for $T_a = 25$ K. Symbols as in Fig. 4

crystalline phase reduces T_t, e.g. amorphous $Au_{50}Sn_{50}$ [15]). The transition temperature vanishes at 4 at% Mn in Sn (Fig. 5). Hence SnMn cannot be amorphous below that concentration. Furtheron the negative slope $-\mathrm{d}\varrho/\mathrm{d}T$ of the amorphous SnMn alloys is plotted in Fig. 5. As may be seen from this figure the negative slope vanishes at the same concentration of 4 at% Mn as the amorphous phase.

4. Discussion

The electrical resistivity of amorphous SnMn is similar to that of AlMn [19]. The resistivity of AlMn decreases also nearly linearly with rising temperature. Opposite to SnMn the magnetic moment of AlMn quenched from the liquid is due to Mn clusters (e.g. triples) [20, 21] at low temperature. The question is whether the negative temperature coefficient of amorphous SnMn is caused by the magnetic moment of Mn or has to be ascribed to the amorphous state of the system. The strong difference of the electrical resistivity in the crystalline and the amorphous phase (Section 1) means that the negative temperature coefficient of amorphous SnMn is not the Kondo effect. Crystalline Sn with 5 at% Mn has a resistance maximum [12]. This maximum does not exist in the amorphous phase (Fig. 2). The disappearance of the maximum in the amorphous phase cannot be explained as a mean-free-path effect because the residual resistivities of the crystalline and the amorphous phase have the same order of magnitude [12].

Fig. 5 shows that the negative slope of the electrical resistivity and the transition temperature amorphous–crystalline vanish at the same concentration. That means the negative slope of Sn with concentrations larger than 4 at% is due to the amorphous state of SnMn.

The number of phonons in the solid state (both for crystalline and amorphous) decreases at low temperature. Therefore, the electrical resistivity should approach a constant value at low temperature if the resistivity is connected with a phonon-dependent structure factor. That is not valid for amorphous SnMn since the resistivity increases almost linearly with falling temperature. Consequently the temperature dependence of the electrical resistivity is of electronic origin. On the other hand, the negative slope is connected with the amorphous phase. Traditionally solid-state physics is based on the assumption that the atoms or ions, respectively, which build up the solid state, do not change their electronic configuration with variation of temperature. This assumption is not absolutely fulfilled in the case of amorphous alloys. NMR measurements concerning $(Ni_yPt_{1-y})_{75}P_{25}$ yield that the Knight shift changes strongly with temperature at the Pt atom [22, 23]. A charge transfer to the Pt atom occurs resulting in a temperature-dependent electronic cross-section of the d-atom. The change of the cross-section of the d-atom with temperature causes a temperature-dependent contribution to the electrical resistivity. The resistivity of $(Ni_yPt_{1-y})_{75}P_{25}$ has mainly a negative temperature coefficient [24].

The question arises if such a mechanism may be possible in SnMn. Mössbauer measurements concerning amorphous SnCu give evidence for a s^2p^2 configuration of amorphous Sn [25]. That means the Sn atoms change the electronic configuration by transition from the amorphous to the crystalline state (that implies also a change of the atomic scattering factor and the pseudo-potential). In the amorphous phase the Sn atoms have the electronic configuration of free Sn atoms with two p-electrons. Investigations of the Hall effect with respect to amorphous Sn alloys result in an electron density of about four electrons per Sn atom [1]. Hence the p-electrons in Sn are free electrons.

Normally the orbital angular momentum of the p- and d-electrons is quenched in the solid state. However, there is experimental evidence that under certain conditions an angular momentum exists. For example, in CuMn the spin of the Mn is fixed by spin–orbit interaction with Au or Pt atoms (added to the alloy) into certain directions [26, 27]. That is only possible by the existence of a z-component of the orbital angular momentum. With XPS the spin–orbit interaction of the 6p-electrons (conduction electrons) of Pb metal can be observed. The measured value of the spin–orbit interaction is the same as that of the free Pb atom [28, 29]. That means in solid metals the z-component of the orbital angular momentum of electrons can exist for a certain time even if these electrons are involved in the chemical bond or contribute to the electrical conductivity.

Let us now consider amorphous Sn with free p-electrons. The spin of the Mn atom interacts by exchange with the spins of the p-electrons. The orbital angular momenta of the p-electrons ($l = 1$) or d-electrons ($l = 2$) interact by three mechanisms: a) interaction of the orbital angular momenta as in the atom according to Hund's second rule, b) quadrupole interaction of electrons with $l \neq 0$, c) magnetic dipole interaction of electrons with $l \neq 0$. If in amorphous Sn the interactions a, b, c) are larger than the spin–orbit coupling of the p-electrons, then the spin–orbit coupling of the p-electrons is broken and a p-electron has a free orbital angular momentum. It can orient to the direction of a magnetic field. The density of the p-electrons is one per half atomic volume that means about one p-electron orients with respect to one Mn spin. According to the interactions a, b, c) between the p-electrons the energy states are not discrete but a continuum results. The ion spin S of the Mn has a magnetic moment

$$\mu_s = 2\sqrt{S(S+1)}\,\mu_B\,,$$

μ_B being Bohr's magneton number. This magnetic moment produces a magnetic field B at the distance a_0,

$$B = \frac{\mu_0}{4\pi}\frac{\mu_s}{a_0^3}.$$

The p-electron with the magnetic moment $\mu = \sqrt{l(l+1)}\, \mu_B$ orients to the field direction at the temperature T. The component of the magnetic moment in the direction of the magnetic field is given by the Langevin function $L(x)$ in the case of continuous energy states,

$$\langle \cos \vartheta \rangle = L(x) = \coth x - \frac{1}{x},$$

with

$$x = \frac{\mu B}{kT}.$$

$E_{lS} = \mu B$ means the spin–orbit interaction between p-electron and Mn spin. Numerical values are $E_{lS} = 0.05$ eV [30] and $kT = 10^{-3}$ eV for $T = 10$ K.

$$x = \frac{E_{lS}}{kT} = 50 \gg 1$$

yields

$$L = 1 - \frac{kT}{E_{lS}}.$$

The result is a temperature-dependent contribution of the orbital angular momentum of the p-electron to the Mn spin. The consequence is a temperature-dependent change of the p-orbital wave function at the place of the Mn spin and hence a change in the cross-section for conduction electrons,

$$Q = Q_0 \left(1 - \frac{kT}{E_{lS}}\right).$$

The cross-section gives a contribution to the electrical resistivity

$$\varrho = \frac{n_1}{n} \frac{m v_F Q}{e^2}$$

(n_1 density of the Mn ion spins; n density of electrons, one Sn atom with four electrons; m, e electron mass and charge, respectively; v_F Fermi velocity),

$$\frac{\varrho}{\varrho_0} = 1 - \frac{kT}{E_{lS}}$$

with

$$\varrho_0 = \frac{m v_F}{4e^2} Q_0 c .$$

The last two formulae contain the following points which roughly agree with the experimental results of Section 3:

(i) The electrical resistivity falls linearly with increasing temperature (compare with Fig. 2 and 3).

(ii) The negative temperature coefficient amounts to (with $E_{lS} = 0.05$ eV)

$$\alpha = \frac{d\varrho}{\varrho_0\, dT} = -2 \times 10^{-3}\ \mathrm{K}^{-1} .$$

(iii) The residual resistivity grows proportionally to Mn concentration (Table 1).

(iv) Taking for the diameter of the p-electrons the distance of the Sn atoms yields $Q_0 = 10$ Å^2 and $\varrho_0/c = 2\ \mu\Omega$ cm/%. This value is lower than the experimental one

(Table 1: 5 to 10 $\mu\Omega$ cm/%). But we have to keep in mind that additional mechanisms to that mentioned contribute to the residual resistivity. The experimental value of α (Section 3) is smaller than the estimated α above. That is also due to the large experimental ϱ_0.

5. Conclusion

SnMn can be amorphous at concentrations above 4 at% Mn. The electrical resistivity of the amorphous SnMn alloys increases without approaching a constant value by lowering the temperature. The negative temperature coefficient vanishes at the same concentration of 4 at% Mn as the transition temperature amorphous–crystalline demonstrating that the negative temperature coefficient is connected with the amorphous phase. With respect to the amorphous state we assume as a model that the angular momenta of the atoms in the amorphous solid are not quenched. This leads in the case of amorphous Sn alloys to a temperature-dependent contribution of the orbital angular momentum to the ion spin of Mn which explains the behaviour of the electrical resistivity.

References

[1] P. HÄUSSLER and F. BAUMANN, Z. Phys. **B49**, 303 (1983).
[2] D. KORN, H. PFEIFLE, and G. ZIBOLD, Z. Phys. **270**, 195 (1974).
[3] T. MURATA, S. TOMIZAWA, T. FUKASE, and T. MASUMOTO, Scripta metall. **10**, 181 (1976).
[4] Z. ALTOUNIAN, TU GUO-HUA, and J. O. STROM-OLSEN, Solid State Commun. **40**, 221 (1981).
[5] R. HASEGAWA and C. C. TSUEI, Phys. Rev. B **2**, 1631 (1970).
[6] R. HASEGAWA, J. Phys. Chem. Solids **32**, 2487 (1971).
[7] U. HENGER and D. KORN, Verh. DPG (VI) **18**, 706 (1983).
[8] E. WACHTEL and R. ULRICH, Z. Metallk. **59**, 227 (1968).
[9] E. W. COLLINGS, Solid State Commun. **8**, 381 (1970).
[10] J. G. GASSER and R. KLEIM, Inst. Phys. Conf. Ser. No. 30, Chapter 2, Part 2, p. 352 (1977).
[11] W. BUCKEL, M. DIETRICH, and J. KESSLER, Z. Phys. **245**, 283 (1971).
[12] D. KORN and K. MAURER, J. Physique **44**, 827 (1983).
[13] M. HANSEN, Constitution of Binary Alloys, McGraw-Hill Publ. Co., London 1958 (p. 955).
[14] H. HALLER, D. KORN, and G. ZIBOLD, phys. stat. sol. (a) **68**, 135 (1981).
[15] E. BLASBERG, D. KORN, and H. PFEIFLE, J. Phys. F **9**, 1821 (1979).
[16] W. RÜHL, Z. Phys. **138**, 121 (1954).
[17] W. BUCKEL, Z. Phys. **138**, 136 (1954).
[18] H. LEITZ, Z. Phys. **B40**, 65 (1980).
[19] E. BABIĆ, R. KRSNIK, B. LEONTIĆ, Z. VUČIĆ, and I. ZORIĆ, Phys. Rev. Letters **27**, 805 (1971).
[20] M. MILYAK and J. R. COOPER, Physica (Utrecht) **86/88B**, 476 (1977).
[21] J. R. COOPER and M. MILYAK, J. Phys. F **6**, 2151 (1976).
[22] W. A. HINES, K. GLOVER, W. G. CLARK, L. T. KABACOFF, C. U. MODZELEWSKI, R. HASEGAWA, and P. DUWEZ, Phys. Rev. B **21**, 3771 (1980).
[23] H. R. KHAN and K. LÜDERS, phys. stat. sol. (b) **108**, 9 (1981).
[24] A. K. SINHA, Phys. Rev. B **1**, 4541 (1970).
[25] J. BOLZ and F. POBELL, Z. Phys. **B20**, 95 (1975).
[26] P. M. LEVY, C. MORGAN-POND, and A. FERT, J. appl. Phys. **53**, 2168 (1982).
[27] J. J. PREJEAU, M. J. JOLICLERC, and P. MONOD, J. Physique **41**, 427 (1980).
[28] L. LEY, R. POLLAK, S. KOWALCZYK, and D. A. SHIRLEY, Phys. Letters A **41**, 429 (1972).
[29] L. LEY, S. P. KOWALCZYK, F. R. MCFEELY, and D. A. SHIRLEY, Phys. Rev. B **10**, 4881 (1974).
[30] C. E. MOORE, Atomic Energy Levels II, Nat. Bur. Standards NSRDS-NBS 35, Washington, D.C. 1971 (p. 27).

(Received December 8, 1983)

phys. stat. sol. (b) 126, K1 (1984)
Subject classification: 2; 21.7

Laboratorio de Defectos en Sólidos, Facultad de Física,
Pontificia Universidad Católica de Chile, Santiago[1]

Liquid Structure Factor of Sb: Application of the Canessa-Mariani-Vignolo Model

By

E. CANESSA, D.F. MARIANI, and J. VIGNOLO

Introduction In a recent paper /1/ (to be referred to as I), we have presented an expression for the static structure factor or interference function a(Q) of non-simple liquid metals, i.e. Hg, Ga, Sn, Bi, Si, Ge, and Sb, which shows a subsidiary shoulder on the right-hand side of the main peak. In I, we have interpreted the shoulder in a(Q) for these metals as a dual structure in the liquid, representing the two substructures by two types of clusters, and applied our a(Q) expression to liquid Hg, Ga, Sn, and Bi. The present interest is to apply this model approach to calculate a(Q) in liquid Sb, see I for details.

Results and discussion The observed a(Q) for simple liquid metals, e.g. monovalent /2, 3/ and heavy metals /4, 5/, are described by applying the analytic solution of the Percus-Yevick integral equation for hard sphere (HS) fluids in which the ion-ion repulsive forces are idealized /6/. Nevertheless, for non-simple liquid metals, the HS model potential has been treated as a reference potential, being the difference between the effective interatomic potential and the HS potential considered as a weak perturbation /4 to 5, 7/; e.g., Bretonnet et al. (BGBK) /8/ have used a parabolic well potential as a perturbation to deduce a(Q) within the framework of the mean spherical approximation /9/. On the other hand, also for a(Q) of non-simple liquid metals, we have presented /1/ a new approach based on the concept of a double structure in the liquid, which is represented by two types of the non-interacting HS. In our case, the structures have been approximated by the superposition of two modified HS interference functions by introducing both parameters, the correlation amplitude

---

1) Casilla 114-D, Santiago, Chile.

Table 1

Parameters used in a(Q) calculations of liquid Sb. The number density used is 0.03203 atoms/$(10^{-10}$ m$)^3$. For details of notation see the original papers

	HS model /6/	CMV's model /1/						BGBK's model /8/		
	η	η_1	ε_1	δ_1	η_2	ε_2	δ_2	η	ε/k_BT	λ
Fig. 1a	0.41	0.34	0.749	1.848	0.67	1.981	0.114	0.394	1.069	2.405
Fig. 1b	0.41	0.34	0.749	1.792	0.67	1.213	0.211	0.418	1.054	2.392

(δ_1 and δ_2) and the correction factor (ε_1 and ε_2) for the fraction of the total volume occupied by the HS (η_1 and η_2). We note that the double HS model proposed by Orton /10 to 14/ uses also the superposition of two HS interference functions, but they are not modified.

We now apply our (CMV) model to liquid Sb. The results of the calculations are shown in Fig. 1a and b. Furthermore, the experimental data /15/ and the values obtained using the HS and BGBK models are also shown. The marks in the figures indicate the value of twice the Fermi momentum. The parameters used and the values of the calculated a(0) are compiled in Table 1 and 2, respectively.

From Fig. 1a it can be seen that the CMV model agrees with experiment excellently and gives better results than the HS model, for $Q \lessapprox 2\,k_F$, and that the BGBK model shows a small hump at low Q as well as a strong dip near the top of the second peak, models being both fitted on the first peak. On the other hand, it can be seen from Fig. 1b that the BGBK model yields the same features as shown in Fig. 1a, and that the CMV model also agrees with data excellently for $Q \gtrapprox 2\,k_F$, both models being fitted over all Q. Finally, these

Table 2

Calculated a(0) values

	expt. /15/	HS model	CMV's model	BGBK's model
Fig. 1a	0.019	0.037	0.028	0.020
Fig. 1b	0.019	0.037	0.024	0.018

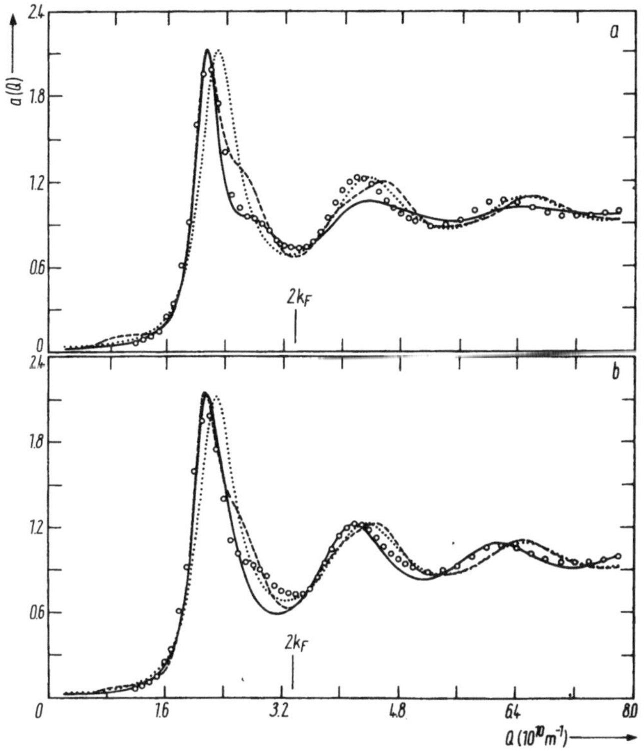

Fig. 1. a) Calculated structure factor a(Q) of liquid Sb at 933 K in comparison with experimental results o from /15/;(...) HS model. CMV's model (solid curve) and BGBK's model (dashed curve) are fitted on the main peak, b) symbols and curves as in a) but the CMV and BGBK models are now fitted over all Q (momentum transfer)

results tell us that our analytic form reproduces a(Q) of liquid Sb (including the value of a(0)) better than the HS model for Q below 2 k_F, and that our results are comparable to those of BGBK for Q above 2 k_F; then, as a first approximation, a double structure could also be assumed in liquid Sb.

The authors wish to acknowledge the support of Dirección de Investigación of the Pontificia Universidad Católica de Chile, DIUC.

References

/1/ E. CANESSA, D.F. MARIANI, and J. VIGNOLO, phys. stat. sol. (b) 124, 465 (1984).

/2/ A.M. BRATKOVSKY, V.G. VAKS, S.P. KRAVCHUK, and A.V. TRELIFOV, J. Phys. F 12, 1293 (1982).

/3/ M.S. ZEI, Phys. Rev. B 24, 4509 (1981).
/4/ O.J. EDER, E. ERDPRESSER, B. KUNSCH, H. STILLER, M. SUDA, and P. WEINZIERL, J. Phys. F 9, 1215 (1979).
/5/ O.J. EDER, E. ERDPRESSER, B. KUNSCH, H. STILLER, and M. SUDA, J. Phys. F 10, 183 (1980).
/6/ N.W. ASHCROFT and J. LEKNER, Phys. Rev. 145, 83 (1966).
/7/ M. SILBERT and W.H. YOUNG, Phys. Letters A 58, 469 (1976).
/8/ J.L. BRETONNET, J.G. GASSER, A. BATH, and R. KLEIM, phys. stat. sol. (b) 114, 243 (1982).
/9/ J.L. LEBOWITZ and J.K. PERCUS, Phys. Rev. 144, 251 (1966).
/10/ B.R. ORTON, Z. Naturf. 30a, 1500 (1975).
/11/ B.R. ORTON, Z. Naturf. 31a, 397 (1976).
/12/ B.R. ORTON, Z. Naturf. 32a, 332 (1977).
/13/ B.R. ORTON, J. Physique, Suppl. 41, C8-280 (1980).
/14/ J.P. GABATHULER and S. STEEB, Z. Naturf. 34a, 1314 (1979).
/15/ Y. WASEDA and K. SUZUKI, phys. stat. sol. (b) 47, 581 (1971).

(Received July 31, 1984)

phys. stat. sol. (b) 126, K5 (1984)

Subject classification: 6 and 8; 12.1; 21.2

Department of Applied Physics and Mathematics, Mining College, Akita University[1]

Compression Effect on Mode Grüneisen Parameters of Alkali Metals

By

T. SOMA and H.-MATSUO KAGAYA

Previously, we /1/ have presented the local Heine-Abarenkov-type model potential of alkali metals in the perturbational scheme, and applied to the calculation of phonon frequencies /2/. Recently, we /3/ have studied the pressure effect on the phonon dispersion curves, and then, the mode Grüneisen parameters under atmospheric pressure and the temperature dependence of the thermal expansion coefficient /4/. In the present note, we calculate the compression effect on the mode Grüneisen parameters of alkali metals in relation to our recent study /3/ about the lattice dynamics of alkali metals under pressure.

The mode Grüneisen parameter $\gamma_i(\vec{q})$ for the i-th mode with phonon frequency $\nu_i(\vec{q})$ is a measure of the volume Ω-dependence and defined by

$$\gamma_i(\vec{q}) = -\frac{d[\ln \nu_i(\vec{q})]}{d[\ln \Omega]} . \qquad (1)$$

Recently, we /4/ obtained $\gamma_i(\vec{q})$ of alkali metals under atmospheric pressure along the symmetry directions [001], [111], and [011]. Using the same treatment as our previous ones /1 to 4/, we show the compression effect on the mode Grüneisen parameters at the representative points H([001]), P([111] /2), N([011] /2), in Fig. 1a to e. In these figures, the numerical results with Vashishta-Singwi screening function /1/ are given, and the numerical curves of $\gamma_i(\vec{q})$ using other screening functions /1/ are qualitatively the same as those in Fig. 1. The maximum variation width of $\gamma_i(\vec{q})$ amounted to 0.2 to 0.3 under atmospheric pressure, and increased somewhat from 0.2 to 0.5 at the compressed volume $\Omega/\Omega_0 = 0.6$. From Fig. 1, we see that the characteristic mode Grüneisen parameters at P and N_L are largely increased as the crystal volume is compressed.

---

1) Akita 010, Japan.

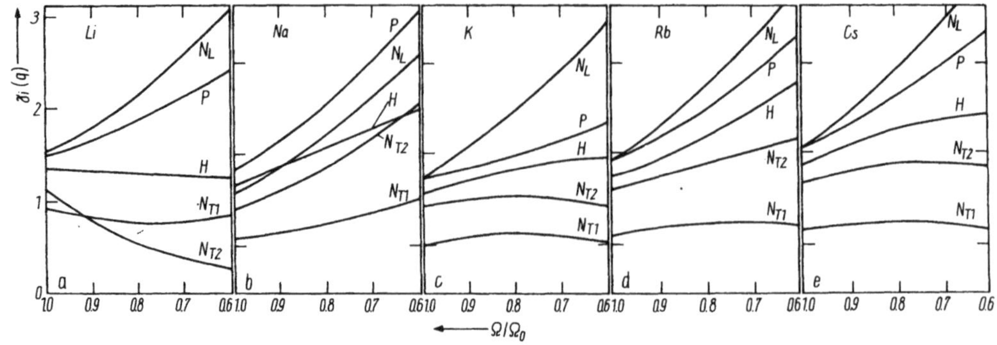

Fig. 1. Volume dependence of the mode Grüneisen parameters $\gamma_i(\vec{q})$ at the representative points H([001]), P([111]/2), and N([011]/2) for a) Li, b) Na, c) K, d) Rb, and e) Cs

Over the long-wavelength region $\vec{q} \to 0$, we cannot determine the corresponding mode Grüneisen parameter $\gamma_i(\vec{q} \to 0)$ because of the very small shift of phonon frequency in (1). In the long-wavelength limit, we obtain the corresponding mode Grüneisen parameters $\gamma_i(\vec{q} \to 0)$ from the volume dependence of the elastic stiffness constants $dC_j/d\Omega$. For example, in the main symmetry directions such as [001], [111], and [011], we obtain

$$\gamma_i(\vec{q} \to 0) = -\frac{1}{6} - \Omega \sum_j b_j \frac{dC_j}{d\Omega} \Big/ \left(2 \sum_j b_j C_j\right) , \tag{2}$$

where b_j and C_j are the linear coefficients and elastic constants, respectively,

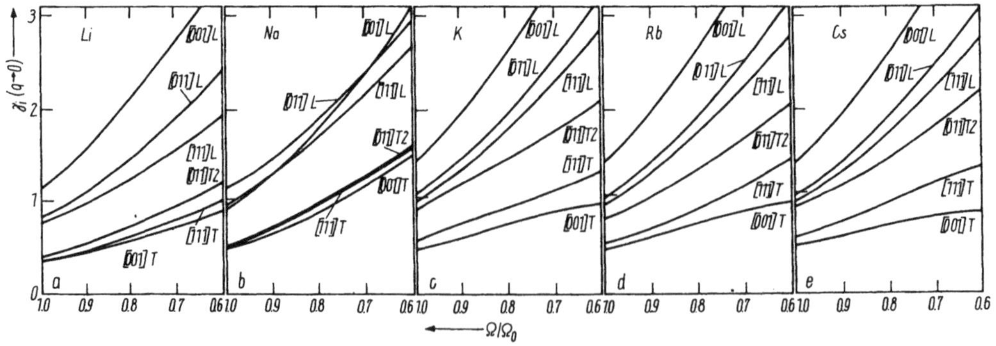

Fig. 2. Volume dependence of the long-wave mode Grüneisen parameters $\gamma_i(\vec{q} \to 0)$ along [001], [111], and [011] directions for a) Li, b) Na, c) K, d) Rb, and e) Cs

for the i-th mode in question. The compression effect on the long-wave mode Grüneisen parameter $\gamma_i(\vec{q} \to 0)$ along [001], [111], [011] directions for alkali metals with Vashishta-Singwi screening is shown in Fig. 2a to e. The increase of $\gamma_i(\vec{q} \to 0)$ for the longitudinal mode is remarkably in common with all alkali metals. We studied theoretically the volume dependence of the mode Grüneisen parameters, and our obtained data are useful in studies of the anharmonicity of alkali metals and alkali alloys.

References

/1/ T. SOMA, Physica (Utrecht) 97B, 76 (1979).
/2/ T. SOMA, phys. stat. sol. (b) 99, 195 (1980).
/3/ T. SOMA, Y. KIMURA, and H.-MATSUO KAGAYA, phys. stat. sol. (b) 112, 151 (1982).
/4/ T. SOMA, H.-MATSUO KAGAYA, and Y. KIMURA, phys. stat. sol. (b) 116, 57 (1983).

(Received September 5, 1984)

phys. stat. sol. (b) 126, K9 (1984)

Subject classification: 12.2; 6; 22.2; 22.4

Department of Applied Physics and Mathematics, Mining College, Akita University[1]

Dynamical Elastic Constants of Covalent Compounds

By

H.-MATSUO KAGAYA and T. SOMA

Recently we /1, 2/ have proposed the extended binding force, which included partially ionic contributions essential to the optical phonon frequency splitting in the long wavelength limit, and calculated the phonon dispersion curves of III-V covalent crystals using the Ashcroft empty core potential. Then, we /3/ have presented the local Heine-Abarenkov model potential for III-V and II-VI tetrahedral compounds, and applied to the calculation of the phonon curves for these compounds /4/. By introducing the local Heine-Abarenkov potential, the agreement with the observed neutron scattering data was remarkable as compared to Ashcroft's potential, mainly due to the good fitting of TA modes at the zone boundaries.

In the dynamical treatment, elastic stiffness constants are determined from the long-wave phonon velocity of the acoustical modes in the lattice vibrations and the procedures are as follows:

$$D^{\alpha\beta}(\vec{q}) \xrightarrow{\vec{q}\to 0} \Omega(\alpha\beta,\gamma\delta)\, q_\gamma q_\delta \quad , \tag{1}$$

$$C_{\alpha\beta\gamma\delta} = (\alpha\gamma,\beta\delta) + (\gamma\beta,\alpha\delta) - (\gamma\delta,\alpha\beta) \quad , \tag{2}$$

$$B_D = (C_{11} + 2C_{12})/3 = (C_{xxxx} + 2C_{xxyy})/3 \quad , \tag{3}$$

$$C_{44} = C_{xyxy} \quad , \tag{4}$$

and

$$C' = (C_{11} - C_{12})/2 = (C_{xxxx} - C_{xxyy})/2 \quad . \tag{5}$$

The local Heine-Abarenkov type potential is representative of the self-consistent one with two parameters. The important difference between the local Heine-Abarenkov potential and the simplified Ashcroft one is that the latter shows oscillations with large amplitude in the region of large $q(>2\,k_F)$ com-

1) Akita 010, Japan.

T a b l e 1

Dynamical elastic constants C_{11}, C_{12}, C_{44}, C', and B_D (in units of 10^{-1} N) of GaP, GaAs, GaSb, InP, InAs, and InSb

compound	potential	C_{11}	C_{12}	C_{44}	C'	B_D
GaP	LHA 1	1.296 to 1.401	0.593 to 0.804	0.504 to 0.655	0.261 to 0.395	0.875 to 0.978
	LHA 2	1.273 to 1.394	0.595 to 0.799	0.518 to 0.666	0.255 to 0.394	0.862 to 0.974
	A	1.576 to 1.685	1.098 to 1.283	0.379 to 0.507	0.174 to 0.263	1.274 to 1.399
	obs. /5/	1.412	0.625	0.705	0.393	0.887
GaAs	LHA 1	1.090 to 1.172	0.496 to 0.689	0.422 to 0.502	0.247 to 0.313	0.704 to 0.849
	LHA 2	1.012 to 1.213	0.505 to 0.667	0.448 to 0.527	0.249 to 0.349	0.711 to 0.842
	A	1.171 to 1.272	0.828 to 0.961	0.253 to 0.285	0.155 to 0.207	0.968 to 1.050
	obs. /6/	1.188	0.537	0.594	0.325	0.754
GaSb	LHA 1	0.876 to 0.924	0.290 to 0.406	0.442 to 0.494	0.214 to 0.286	0.485 to 0.578
	LHA 2	0.862 to 0.914	0.268 to 0.381	0.456 to 0.511	0.233 to 0.294	0.465 to 0.557
	A	0.790 to 0.882	0.681 to 0.748	0.097 to 0.218	0.073 to 0.096	0.722 to 0.756
	obs. /7/	0.884	0.403	0.432	0.240	0.563

Table 1 - Continued

InP	LHA 1	0.933 to 1.047	0.574 to 0.700	0.493 to 0.547	0.172 to 0.224	0.746 to 0.799
	LHA 2	1.024 to 1.077	0.564 to 0.688	0.508 to 0.563	0.189 to 0.253	0.750 to 0.787
	A	0.942 to 0.984	0.778 to 0.834	0.263 to 0.300	0.054 to 0.103	0.847 to 0.870
	obs. /8/	1.011	0.561	0.456	0.225	0.711
InAs	LHA 1	0.829 to 0.866	0.411 to 0.499	0.402 to 0.422	0.166 to 0.219	0.583 to 0.621
	LHA 2	0.844 to 0.887	0.419 to 0.508	0.399 to 0.421	0.178 to 0.226	0.593 to 0.633
	A	0.801 to 0.895	0.629 to 0.712	0.165 to 0.191	0.069 to 0.112	0.714 to 0.748
	obs. /9/	0.833	0.453	0.396	0.190	0.579
InSb	LHA 1	0.601 to 0.636	0.297 to 0.358	0.297 to 0.303	0.126 to 0.154	0.398 to 0.444
	LHA 2	0.611 to 0.645	0.300 to 0.361	0.294 to 0.301	0.131 to 0.161	0.404 to 0.453
	A	0.560 to 0.646	0.478 to 0.539	0.134 to 0.166	0.036 to 0.073	0.515 to 0.548
	obs. /10/	0.667	0.364	0.302	0.151	0.465

Table 2

Dynamical elastic constants C_{11}, C_{12}, C_{44}, C', and B_D (in units of 10^{-1} N) of ZnS, ZnSe, ZnTe, and CdTe

compound	potential	C_{11}	C_{12}	C_{44}	C'	B_D
ZnS	LHA 1	0.916 to 1.072	0.581 to 0.793	0.461 to 0.543	0.169 to 0.240	0.722 to 0.844
	LHA 2	0.940 to 1.045	0.583 to 0.798	0.539 to 0.619	0.145 to 0.223	0.734 to 0.833
	A	1.472 to 1.563	0.808 to 1.063	0.360 to 0.491	0.222 to 0.355	0.930 to 1.101
	obs. /11/	1.046	0.653	0.461	0.196	0.784
ZnSe	LHA 1	0.842 to 0.936	0.522 to 0.693	0.438 to 0.500	0.150 to 0.199	0.691 to 0.737
	LHA 2	0.801 to 0.907	0.520 to 0.713	0.435 to 0.512	0.153 to 0.183	0.663 to 0.727
	A	1.350 to 1.474	0.976 to 1.090	0.234 to 0.308	0.150 to 0.225	1.005 to 1.126
	obs. /11/	0.810	0.488	0.441	0.161	0.595
ZnTe	LHA 1	0.745 to 0.772	0.413 to 0.494	0.342 to 0.360	0.132 to 0.167	0.527 to 0.564
	LHA 2	0.720 to 0.748	0.391 to 0.471	0.356 to 0.374	0.129 to 0.175	0.534 to 0.562
	A	0.899 to 0.924	0.693 to 0.749	0.157 to 0.185	0.082 to 0.104	0.760 to 0.804
	obs. /12/	0.711	0.407	0.313	0.152	0.508
CdTe	LHA 1	0.585 to 0.621	0.343 to 0.403	0.201 to 0.215	0.092 to 0.121	0.424 to 0.474
	LHA 2	0.583 to 0.636	0.349 to 0.420	0.208 to 0.215	0.083 to 0.121	0.427 to 0.491
	A	0.657 to 0.681	0.545 to 0.591	0.108 to 0.144	0.042 to 0.056	0.582 to 0.621
	obs. /13/	0.535	0.368	0.199	0.083	0.424

pared with the former, and that the deviation from the pure Coulomb potential in the region of small q is self-consistently treated for the former. This self-consistency in the region of small q is essential to the calculation of the dynamical elastic constants obtained from the long-wave phonon velocity.

In the present work, we study in detail effects of both model potential and electronic screening on the dynamical elastic constants of III-V and II-VI tetrahedral compounds. We show the resulting data of C_{11}, C_{12}, C_{44}, C', and B_D in Table 1 and 2 for III-V and II-VI compounds, respectively. In these tables, the variation width of dynamical elastic constants obtained using the different forms of both the model potential and the screening function /1 to 4/ is shown. We see that the agreement of the dynamical elastic constants, i.e., the acoustic modes in the long-wavelength region with the observed data /5 to 13/ is good using the local Heine-Abarenkov model potential, contrary to the Ashcroft type. We note also that the good agreement of the dynamical bulk modulus B_D with the observed data is superficial, because the dynamical bulk modulus does not coincide with the static bulk modulus obtained from the homogeneous deformation method because of the so-called compressibility problem /14, 15/.

References

/1/ T. SOMA and H.-MATSUO KAGAYA, phys. stat. sol. (b) 118, 245 (1983).
/2/ T. SOMA and H.-MATSUO KAGAYA, phys. stat. sol. (b) 119, K41 (1983).
/3/ T. SOMA and H.-MATSUO KAGAYA, phys. stat. sol. (b) 119, 547 (1983).
/4/ H.-MATSUO KAGAYA and T. SOMA, phys. stat. sol. (b) 124, 37 (1984).
/5/ R. WEIL and W.O. GROVES, J. appl. Phys. 39, 4049 (1968).
/6/ H.J. MCSKIMIN, A. JAYARAMAN, and P. ANDREATCH, J. appl. Phys 38, 2362 (1967).
/7/ H.J. MCSKIMIN, A. JAYARAMAN, P. ANDREATCH, and T.B. BATEMAN, J. appl. Phys. 39, 4127 (1968).
/8/ D.N. NICHOLS, D.S. RIMAI, and R.J. SLADEK, Solid State Commun. 36, 667 (1980).
/9/ D. GERLICH, J. appl. Phys. 34, 2915 (1963).
/10/ L.J. SLUTSKY and C.W. GARLAND, Phys. Rev. 113, 167 (1959).
/11/ D. BERLINCOURT, H. JAFFE, and L.R. SHIOZAWA, Phys. Rev. 129, 1009 (1963).

/12/ B.H. LEE, J. appl. Phys. 41, 2988 (1970).

/13/ H.J. MCSKIMIN and D.G. THOMAS, J. appl. Phys. 33, 56 (1962).

/14/ E.G. BROVMAN, YU. KAGAN, and A. KHOLAS, Soviet Phys. - J. exper. theor. Phys. 30, 883 (1970).

/15/ T. SOMA, J. Phys. Soc. Japan 36, 1292 (1974).

(Received May 30, 1984)

phys. stat. sol. (b) 126, K15 (1984)

Subject classification: 4 and 6; 8; 23

Condensed Matter Research Centre,
Department of Postgraduate Studies and Research in Physics,
Rani Durgawati University, Jabalpur

Mean Square Displacements of Rare Gas Solids

By

R.K. SINGH and D.K. NEB

During the last few years the descriptions of properties of rare gas solids (RGS) in terms of interatomic potentials have made considerable progress /1 to 4/. Most of these investigations, however, deal with the usual two body Lennard-Jones (6 to 12) or "exp-6" potential or its variants. A critical assessment of these studies reveals that various critical phenomena, such as Cauchy violation ($C_{12} \neq C_{44}$), phonon dispersion along ($\vec{q}$, $\vec{q}$, $\vec{q}$) directions, and structure of rare gas solids could not be explained by these two-body potentials /5/. This is mainly because these properties are considerably influenced by the three-body interactions (TBI) /6/. The TBI proposed by Axilrod and Teller /7/ is successful in predicting various crystal properties, such as elastic, phonon dispersion, etc. of RGS. Recently, Singh and Neb /8/ have proposed another type of TBI which arise from variable induced dipoles when RGS atoms overlap. The superiority of the TBI model proposed by Singh and Neb /8/ is that it has only two parameters and the three-body interactions are accounted for realistically through the overlap integrals. This model has explained satisfactorily the harmonic and anharmonic elastic and thermodynamical properties of perfect and mixed /8 to 10/ RGS and the dynamical properties /11/ of RGS.

The present note aims to report our theoretical results on mean-squared displacements (MSD) of RGS, calculated from the proposed TBI model. This property of RGS has already been investigated by Goldman /13/ and Kushwaha /14/, using the two-body potentials. Their studies, however, involve too many parameters fitted to experimental results on elastic and phonon properties. Thus, these models (and their parameters) are of limited physical significance. The present model is, however, free from such limitations as indicated in our earlier papers /8/.

---

1) Jabalpur 482001, India.

The study of MSD is of great importance for understanding the interactions of lattice with the electromagnetic waves, Mössbauer effect, etc.

The expression for MSD of the k-th particle at a temperature T can be written as /14/

$$\langle u^2 \rangle_k = \frac{\hbar}{N m_k} \sum_{\vec{q}} \sum_j \frac{|e(k/\vec{q}, j)|^2 \left[(n(w_j(\vec{q})) + \frac{1}{2}\right]}{w_j(\vec{q})} , \qquad (1)$$

where $w_j(\vec{q})$ is the eigenfrequency for the j-th phonon branch corresponding to the wave vector $\vec{q}$; $e(k/\vec{q}, j)$ is the corresponding eigenvector of the k-th particle with mass m_k. N is the division of $\vec{q}$-points in the first Brillouin zone. $n(w(\vec{q}, j))$ is the well known phonon occupation number at temperature T.

In the present calculations, we have divided the translational vector of the reciprocal space in 512 equally spaced points in the first Brillouin zone.

The values of parameters used to calculate the eigenfrequencies and eigenvectors for all the four RGS have been reported in our earlier papers /8, 12/. As already mentioned, the present model has effectively two parameters and explains various elastic and dynamical properties of these solids satisfactorily /8, 12/. The MSD of RGS at various temperatures (starting from 0 K to respective melting points) have been calculated using (1) and presented in Fig. 1a to d. The theoretical results of Goldman /13/ and Kushwaha /14/ have also been included in the figures for a comparison.

It is revealed from Fig. 1a to d that our calculated results on MSD vary with temperature systematically and the trends are similar to those obtained

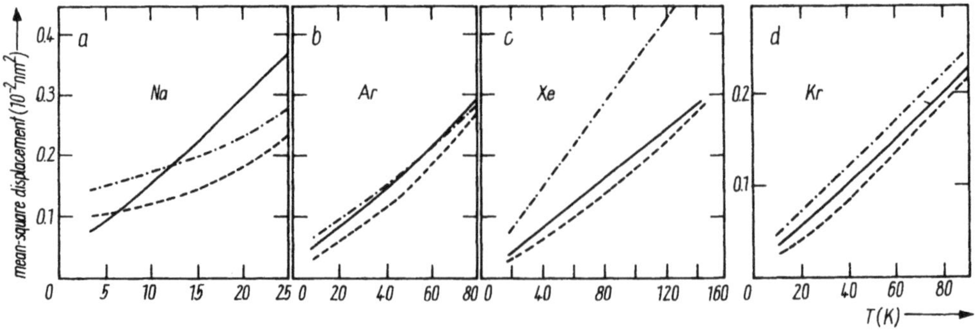

Fig. 1. Mean square displacements versus temperature for a) Ne, b) Ar, c) Xe, and d) Kr; —— present work, – – – /13/; —·—·— /14/

Table 1

Validity of Lindemann's melting criterion

solids	melting point (K)	MSD $\langle u^2 \rangle$ (10^{-2} nm^2)	RMSD $\langle u^2 \rangle^{1/2}$ (10^{-1} nm)	nearest-neighbour distance (NND) (10^{-1} nm)	RMSD/NND
Ne	25	0.405	0.636	3.155	0.202
Ar	84	0.323	0.568	3.756	0.151
Kr	117	0.323	0.668	3.991	0.142
Xe	161	0.378	0.615	4.335	0.142

by Goldman /13/ and Kushwaha /14/ although these results differ in respect of magnitude. This difference is expected because of different models used for calculations and also for different degrees of accuracy in predicting the eigen-frequencies of the phonon states. In the present calculations, the Brillouin zone (BZ) has been divided in 512 equal parts. It is naturally expected that for a better accuracy, particularly in the low temperature region, the BZ may be further divided. Moreover, the present model is based on harmonic approximation and hence the values of $\langle u^2 \rangle_k$ do not contain any anharmonic contribution. In the absence of any experimental results a judgement of success of any model in respect of prediction of MSD cannot be made at this point.

The Lindeman criterion for melting points of RGS is also satisfied by our calculated MSD, according to which the ratio of nearest neighbour distance to the root MSD should be a constant. In Table 1 we present these results showing the validity of this criterion. A similar result has also been reported by Kushwaha /14/.

The authors are thankful to Council of Scientific and Industrial Research for financial assistance and to Dr. S. P. Sanyal for discussions.

References

/1/ J.A. BARKER and A. POMPE, Austral. J. Chem. 21, 1683 (1968).

/2/ M.V. BOBETIC and J.A. BARKER, Phys. Rev. B 2, 4169 (1970).

/3/ M.L. KLEIN and R.D. MURPHY, Phys. Rev. B 6, 2433 (1972).

/4/ M.L. KLEIN and T.R. KOEHLER, in: Rare Gas Solids, Ed. M.L. ELEIN and J.A. VENEBLES, Academic Press, New York 1976.

/5/ M.L. KLEIN and J.A. VENABLES, (Ed.), Rare Gas Solids, Academic Press, New York 1976.

/6/ S. LEHRI and M.P. VERMA, phys. stat. sol. (b) 92, 363 (1979).

/7/ B.M. AXILROD and E. TELLER, J. chem. Phys. 11, 229 (1943); 17, 1349 (1949).

/8/ R.K. SINGH and D.K. NEB, phys. stat. sol. (b) 107, K47 (1981); 112, 735 (1982).

/9/ R.K. SINGH, S.P. SANYAL, and D.K. NEB, to be published.

/10/ R.K. SINGH, D.K. NEB, and S.P. SANYAL, J. Phys. C 16, 3409 (1983); phys. stat. sol. (b) 116, 289 (1983).

/11/ D.K. NEB, Ph.D. Thesis, R.D. University, Jabalpur 1983.

/12/ R.K. SINGH and D.K. NEB, to be published.

/13/ V.V. GOLDMAN, Phys. Rev. 174, 1041 (1968).

/14/ M.S. KUSHWAHA, phys. stat. sol. (b) 96, 301 (1979).

(Received August 13, 1984)

phys. stat. sol. (b) 126, K19 (1984)
Subject classification: 1.2 and 10

G. Nadjakov Institute of Solid State Physics, Bulgarian Academy of Sciences, Sofia[1)]

On the Long-Range Random Critical Behaviour

By
E.R. KORUTCHEVA and D.I. UZUNOV

The scaling properties of isotropic impure systems with quenched point impurities are well known in the case of short-range spatial correlations /1 to 4/. The case in point is the disorder of type "random critical temperature". Recently, Weinrib and Halperin /5/ have generalized the treatment of this quenched disorder including into consideration long-range spatial Gaussian correlations between the impurities. They have performed their analysis to first order in $\varepsilon = 4 - d$, using the isotropic Landau-Ginzburg-Wilson Hamiltonian

$$\frac{H}{k_B T} = \frac{1}{2}\int d^d x \left\{ r_o \varphi^2 + (\nabla\varphi)^2 + \varphi_r \varphi^2 + \frac{u}{2}\varphi^4 \right\} \quad (1)$$

in d-dimensional space, where the fluctuation order parameter $\varphi(\vec{x})$ is considered as an n-component vector, $\left\{\varphi = \varphi_\alpha, \ \alpha = 1, 2, \ldots, n\right\}$, $r_o \sim (T - T_c^o)$, T_c^o is the bare critical temperature, and $\varphi_r(x)$ is a random function intended to describe the quenched effects. The random correlations between the impurities are given in the momentum space by $g(k) = \left\langle |\varphi_r(\vec{k})|^2 \right\rangle_{av}$. They are discussed /5/ in the form

$$g(k) = \Delta_o + \Delta_1 k^{a-d} \ , \quad (2)$$

where Δ_o, Δ_1, and a are parameters such that $g(k) \geqq 0$ for any momentum in the domain $0 < k < \Lambda$ ($|\vec{k}| \equiv k$, Λ is the cut-off). The parameter a describes the power law for the correlations in the position space: $\bar{g}(R) \sim R^{-a}$, where $\bar{g}(R)$ with $R = |x - x'|$ is the Fourier transform of g(k). Note, that in the case of $\Delta_1 = 0$, the problem reduces to the description of short-range correlations only /1 to 4/. We define the quantity $\delta = 4 - a$ /5/. Qualitatively new results are obtained if one assumes that δ is of order ε^1. In this way a new fixed point of the renormalisation-group equations has been obtained /5/. This fixed point is ex-

1) blvd. Lenin 72, 1184 Sofia, Bulgaria.

pected to describe an essentially new long-range random critical behaviour. There is a region on the plane (y, n), $y = \varepsilon/\delta$, where the new fixed point is stable and, hence, the corresponding critical behaviour is accessible. For example, if n = 2, the stability holds if $-0.5 < y < 0.7$.

In this note, using calculations to second order in ε, we present the critical exponents describing the long-range random critical behaviour. For this aim we shall essentially use the results to first order in ε /5/. It has been also demonstrated /5/ that the result $\nu = 2/a$ for the correlation length exponent ν is exact, i.e., it should be valid to any order in ε. Therefore, to obtain the exponents to second order in ε we need the calculation of one exponent to this order in ε and the scaling relations. Most convenient is to calculate Fisher's exponent η.

Using a standard mass renormalisation /4/, the bare Green function is given in the form

$$G_o(k, r) = \left\{k^2 + r\right\}^{-1} , \tag{3}$$

where $r \sim (T - T_c)$ is the renormalised value of r_o (T_c being the renormalised critical temperature). The full Green function after the averaging over the impurities $G(k) = \left\langle G(k; \varphi_r(k))\right\rangle_{av}$ is obtained from the Dyson equation

$$G^{-1}(k, r) = G_o^{-1}(k, r) - \Sigma(k, r) , \tag{4}$$

where $\Sigma(k, r)$ is the self-energy function. G(k) is expected in the form $G^{-1}(k) \sim k^{2-\eta} + r$. On the critical line (r = 0), the exponent is given by /4/

$$\eta = \frac{\Delta\Sigma(k)}{k^2 \ln k} , \tag{5}$$

where

$$\Delta\Sigma(k) = \Sigma(k, 0) - \Sigma(0, 0) . \tag{6}$$

Taking in mind that u, Δ_o, Δ_1, and δ are of order ε^1 and accounting for the k-dependent contributions to the self-energy in order ε^2 (see Fig. 1), we obtain

$$\eta = \frac{1}{256\,\pi^4}\left\{2(n+2)u^{*2} - 2(n+2)(\Delta_o^* + \Delta_1^*)u^* + (\Delta_o^* + \Delta_1^*)^2 + 8\pi^2(\delta - \varepsilon)\Delta_1^*\right\} , \tag{7}$$

where u^*, Δ_o^*, and Δ_1^* are the values at the long-range fixed point of the parameters u, Δ_o, and Δ_1, respectively. For these values we use the results in

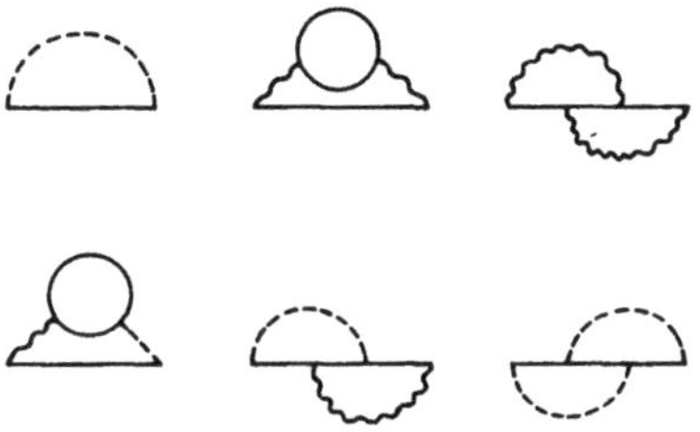

Fig. 1. k-dependent contributions to Σ. The solid lines represent G_o, the dashed lines represent g(k), the wavy lines stand for the interaction u

/5/. In our notations,

$$u^* = 8\pi^2 \frac{3\delta - \varepsilon}{5n + 4} ,$$

$$\Delta_o^* = 4\pi^2 \frac{[2(n+2)\varepsilon - (n+8)\delta]^2}{(\delta - \varepsilon)(5n + 4)^2} ,$$

and

$$\Delta_1^* = 4\pi^2 \frac{6n(n+2)\varepsilon^2 - (11 n^2 + 32 n - 16)\varepsilon\delta + 4(n+8)(n-1)\delta^2}{(\delta - \varepsilon)(5n + 4)^2} . \tag{8}$$

Equations (7) and (8) yield

$$\eta = \frac{1}{16(5n + 4)^2} \left\{ 8n(n + 2)\varepsilon^2 + (n - 4)^2 \delta(2\varepsilon - 3\delta) \right\} . \tag{9}$$

The other exponents describing the long-range random behaviour are determined by the scaling relations /4/ using $\nu = 2/a$, $a = 4 - \delta$, and the result (9). Expanding ν in powers of δ, we find

$$\nu = \frac{1}{2} + \frac{\delta}{8} + \frac{\delta^2}{32} + O(\delta^3) . \tag{10}$$

Thus, the specific heat exponent is

$$\alpha = \frac{\varepsilon - \delta}{2} + \frac{\delta(\varepsilon - \delta)}{8} , \tag{11}$$

the exponent describing the susceptibility is

$$\gamma = 1 + \frac{\delta}{4} + \frac{1}{32(5n + 4)^2} \left\{ (53n^2 + 56n + 80)\delta^2 - 2(n - 4)^2 \varepsilon\delta - 8n(n+2)\varepsilon^2 \right\} , \tag{12}$$

the order parameter exponent is

$$\beta = \frac{1}{2} + \frac{\delta - 2\varepsilon}{8} + \frac{1}{64(5n+4)^2}\Big\{(47n^2 - 104n + 16)\delta^2 - 2(49n^2 + 88n + 16)\varepsilon\delta +$$
$$+ 8n(n+2)\varepsilon^2\Big\}\,, \tag{13}$$

and

$$\tilde{\delta} = 3 + \varepsilon + \frac{1}{8(5n+4)^2}\Big\{4(23n^2 + 36n + 16)\varepsilon^2 - (n-4)^2\delta(2\varepsilon - 3\delta)\Big\}, \tag{14}$$

where $\tilde{\delta}$ is the usual exponent /4/ defined by $\langle\varphi\rangle \sim h^{1/\tilde{\delta}}$ (h is the external field conjugated to the order parameter $\langle\varphi\rangle$).

References

/1/ T.C. LUBENSKY, Phys. Rev. B 11, 3573 (1975).
/2/ D.E. KHMELNITSKII, Zh. eksper. teor. Fiz. 68, 1960 (1975); Soviet Phys. - J. exper. theor. Phys. 41, 981 (1976).
/3/ G. GRINSTEIN and A. LUTTER, Phys. Rev. B 13, 1329 (1976).
/4/ S. MA, Modern Theory of Critical Phenomena, Benjamin, London 1976.
/5/ A. WEINRIB and B.I. HALPERIN, Phys. Rev. B 27, 413 (1983).

(Received June 26, 1984)

phys. stat. sol. (b) 126, K23 (1984)
Subject classification: 18.2; 1.2

Sektion Physik der Martin-Luther-Universität, Halle[1)]

Anisotropic Interactions in 2 + ε Dimensions

By
S. TRIMPER

Isotropic magnetic systems with a continuous rotational symmetry have some peculiarities compared with Ising-like systems below the phase transition temperature T_c. There is no spontaneous symmetry breaking in $d \leqq 2$ dimensions /1/. The transverse susceptibility is infinite everywhere on the coexistence curve.

Real magnetic systems do not have complete rotational invariance. In this note we consider the case of such perturbations to an isotropic system. The low temperature phase is investigated using renormalization group arguments. The classical system under consideration is described by an n component order parameter field. It is divided into cells, the linear size of which is greater than any atomic scale. In the low temperature phase the modul of the order parameter is different from zero and nearly equal to its maximal value (setting unity in this note). The orientation of the vector field expressed by the n-component unit vector $\vec{n}(\vec{x})$, however, is changed slowly from cell to cell. Therefore we consider the orientational behaviour of the system. The Hamiltonian of the fixed length system can be written in the form /2, 3/

$$H = \int d^d x \left\{ \alpha_{\mu\nu} (\partial_\mu n_\alpha)(\partial_\nu n_\alpha) \right\} , \qquad \vec{n}^2 = 1 . \tag{1}$$

The symmetric tensor $\alpha_{\mu\nu}$ represents the exchange interactions /3/ which we chose for simplicity as diagonal $\alpha_{\mu\nu} = J\delta_{\mu\nu}$. The magnetic anisotropy can be expressed by the following perturbations to (1):

$$H_v = \int d^d x \, v_{\alpha\beta\gamma\delta} n_\alpha n_\beta n_\gamma n_\delta \tag{2}$$

and

$$H_w = \int d^d x \, w_{\alpha\beta\gamma\delta\varrho\nu} \, n_\alpha n_\beta n_\gamma n_\delta n_\varrho n_\nu . \tag{3}$$

---

1) Friedemann-Bach-Platz, DDR-4020 Halle, GDR.

They break the O(n) symmetry of (1). For instance, in the case of tetragonal and rhomboedrical perturbations one finds for an n = 3 component system

$$H_v = v\int d^dx \left[n_x^2 n_y^2\right] , \quad H_v = v\int d^dx \left[n_z(n_x^3 - 3 n_x n_y^2)\right] ,$$

respectively, whereas in the case of hexagonal symmetry a term

$$H_w = w\int d^dx \left[n_x n_y(n_x^4 - (10/3)n_x^2 n_y^2 + n_y^4)\right]$$

occurs.

Following the renormalization group method proposed by Polyakov /2/ we decompose $\vec{n}(\vec{x})$ with Fourier components $0 \leqq |\vec{k}| \leqq \Lambda$ into short- and long-wavelength parts as

$$\vec{n}(\vec{x}) = \vec{m}(\vec{x})(1 - \varphi^2(\vec{x}))^{1/2} + \vec{e}_i(\vec{x})\varphi_i(\vec{x}), \quad \varphi^2 = \sum_{i=1}^{n-1} \varphi_i^2 , \tag{4}$$

where $\vec{m}(\vec{x})$ is a slowly varying unit vector with Fourier components $0 \leqq |\vec{k}| \leqq \Lambda' < \Lambda$. The vectors $\{\vec{m}, \vec{e}_i \ (i = 1, \ldots, n-1)\}$ form an orthogonal basis. The rapid variations of the fluctuations are in the fields $\varphi_i(\vec{x})$ which have Fourier components $\Lambda' < |\vec{k}| < \Lambda$. If we take $\Lambda' = \Lambda - d\Lambda$ there result the renormalization group equations in $2 + \varepsilon$ dimensions,

$$\frac{dT}{dl} = -\varepsilon T + \frac{n-2}{2\pi} T^2 , \tag{5a}$$

$$\frac{dv}{dl} = 2v - \frac{n+6}{2\pi} vT , \tag{5b}$$

$$\frac{dw}{dl} = 2w - \frac{n+7}{\pi} wT , \quad l = \ln\frac{\Lambda_o}{\Lambda} , \quad T \triangleq \frac{T}{J} . \tag{5c}$$

Λ_o is a reference cut-off. For l = 0 the renormalized parameters T(l), v(l), and w(l) correspond to the bare parameters. Since the resulting equations (5) are valid for an arbitrary tensor structure of $v_{\alpha\beta\gamma\delta}$ and $w_{\alpha\beta\gamma\delta\varrho\nu}$ we have omitted the indices. Equation (5a) is in accord with /2/ and also found with the field theoretical renormalization method /4/. The investigation of the fourth-order perturbation is already performed using a different approach /5/. The eigenvalues describing the decay of the perturbation can be calculated near the fixed point ($T^* = 2\pi\varepsilon/(n-2)$, $v^* = w^* = 0$)

$$\lambda_v = 2 - \varepsilon\Delta_v , \quad \Delta_v = (n+6)/(n-2) \; ; \tag{6a}$$

$$\lambda_w = 2 - \varepsilon\Delta_w , \quad \Delta_w = 2(n+7)/(n-2) \tag{6b}$$

which indicate that the anisotropic interactions are strongly relevant near d = 2 dimensions.

To calculate the equation of states we have to consider the influence of an external field which couples linear to the vector field $\vec{n}(\vec{x})$. Using the same procedure like before we find

$$\frac{dh}{dl} = 2h + \frac{n-3}{4\pi} h \; T \; . \tag{7a}$$

The corresponding exponent of the stability matrix is

$$\lambda_h = 2 - \varepsilon\Delta_h , \quad \Delta_h = -(n-3)/2(n-2) \; . \tag{7b}$$

In the same way we obtain the equation for the magnetization on the scale l

$$\frac{dm}{dl} = - \frac{n-1}{4\pi} m \; T \; . \tag{8}$$

The differential equations (5), (7a), and (8) can be solved in a closed form. For instance, it follows

$$w(l) - w(0)\exp(\lambda_w l)\left[t + \tau \exp(\varepsilon l)\right]^{\Delta_w} , \tag{9a}$$

$$h(l) = h(0) \exp(\lambda_h l) \left[t + \tau \exp(\varepsilon l)\right]^{\Delta_h} \tag{9b}$$

with $\tau = 1 - t = 1 - T/T_c$, $T_c = 2\pi - \varepsilon J \Lambda_0^{-\varepsilon}/(n-2)$. (The fixed point is identified with the phase transition temperature /2/.) Since through the equation of states the magnetization is given in terms of the applied field and the anisotropy v or w, respectively, we expect also a relation between the renormalized parameters. Using the equations for those parameters a straightforward calculation yields

$$m(l)^{\omega-\delta} v(l)/h(l) = \text{const} \quad \text{with} \quad \delta = \lambda_h/\varepsilon\beta, \quad \omega = \lambda_v/\varepsilon\beta \tag{10a}$$

and

$$m(l)^{\Delta-\delta} w(l)/h(l) = \text{const} \quad \text{with} \quad \Delta = \lambda_w/\varepsilon\beta, \quad \beta = (n-1)/2(n-2) \; . \tag{10b}$$

From (9) one concludes an increasing of the parameters h(l) and w(l), respectively, with growing up of the scaling factor l. This fact is in contradiction with the requirement that the anisotropy and the external field are considered as small perturbations. The anomalous increase of h(l) and w(l) is stopped at a certain l^*. An appropriate choice of l^* fulfils the relation $h(l^*) = 1 - w(l^*)$ /5, 6/. Using (10b) and the solution of (8) we find

$$h\,m^{-\delta} + w\,m^{-\Delta} = t^{-2/\varepsilon}\,\Lambda_0^2\left[1 - \left(\frac{\tau^{\beta}}{m}\right)^{1/\beta}\right]^{2/\varepsilon} . \tag{11}$$

The same procedure applied to the fourth-order anisotropy leads to the equation of states

$$h\,m^{-\delta} + v\,m^{-\omega} = t^{-2/\varepsilon}\,\Lambda_0^2\left[1 - \left(\frac{\tau^{\beta}}{m}\right)^{1/\beta}\right]^{2/\varepsilon}$$

which is in accordance with /5/.

From (11) we can also find the density of the free energy f to

$$f = \int h\,dm = m^{(2-\alpha)/\beta}\left\{-\frac{\beta\,w\,m^{-\Delta}}{2-\alpha-\Delta\beta} + \Lambda_0^2\,t^{\alpha-1}\,\frac{\beta}{2-\alpha}\left[1 - \left(\frac{\tau^{\beta}}{m}\right)^{1/\beta}\right]^{2-\alpha}\right\} \tag{12}$$

The critical exponents $\alpha = 1 - 2/\varepsilon$, β, and δ fulfil the scaling laws. Further conclusions can be drawn following /5/. In the special case of d = 2 dimensions ($\varepsilon = 0$) it follows $T_c(w = 0) = 0$.

The equation of states has for zero field the form

$$m = \left[1 + \frac{T(n-2)}{4\pi}\ln\frac{w}{\Lambda_0^2}\right]^{\beta} .$$

From here we obtain the phase transition temperature

$$T_c(w \neq 0) = \frac{4\pi}{n-2}\Big/\ln\frac{\Lambda_0^2}{w} , \quad (n > 2) .$$

References

/1/ A. Z. PATASINSKII and V. L. POKROVSKII, Fluctuation Theory of Phase Transitions, Nauka, Moscow 1982 (in Russian).

/2/ A.M. POLYAKOV, Phys. Letters B 59, 79 (1975).

/3/ L.D. LANDAU and E.M. LIFSCHITZ, Elektrodynamik der Kontinua, Akademie-Verlag, Berlin 1967.

/4/ E. BREZIN and J. ZINN-JUSTIN, Phys. Rev. B 14, 3110 (1976).

/5/ T. NATTERMANN, phys. stat. sol. (b) 90, 105 (1978).

/6/ S. TRIMPER, to be published.

(Received August 13, 1984)

phys. stat. sol. (b) 126, K29 (1984)
Subject classification: 18

Departamento de Física, Universidade Estadual de Maringá[1)]

Random-Field Ising Model with a Zero Temperature Multicritical Point

By
L.R. EVANGELISTA and V.K. SAXENA

The existence of a tricritical point in the phase diagram of a random-field Ising model has been a point of much discussion /1 to 4/. Aharony /1/ suggested that at low temperature the associated phase transition should be of first order (second order) when the symmetric distribution function of the random fields P(H) has a minimum (maximum) at zero field. Andelman /3/ extended the discussion of Aharony to the case of distributions with a maximum at zero field and argued that the low temperature phase transition in such a case can be of second order as well as of first order depending on the form of the distribution function. Recently /4/, the question of the existence of a low temperature first-order phase transition and consequently a tricritical point was further discussed for the case of the distribution with a maximum at zero field. It has been shown /4/ that for such a distribution there may exist a low temperature first-order phase transition depending on certain inequalities satisfied by the higher derivatives of the distribution at zero field.

In the present note we present the study of a random-field Ising model, with a distribution for which the system has a second-order phase transition at non-zero finite temperatures (T) and exhibits an infinite order multicritical point at zero temperature.

We consider the following model:

$$\mathcal{H} = - \sum_{(ij)} J S_i S_j - \sum_i H_i S_i \quad , \tag{1}$$

where $S_i = \pm 1$ are the Ising spins. J is the nearest-neighbour interaction. H_i are the random local fields having a distribution

$$P(H) = \frac{1}{2H_0} \quad \text{for} \quad |H| \leq H_0 \; ,$$
$$P(H) = 0 \quad \text{for} \quad |H| > H_0 \; . \tag{2}$$

1) Caixa Postal 331, 87100 Maringá, Paraná, Brazil.

As is usual for such models /1, 2/ we use the mean-field approximation and write the mean-field Hamiltonian as

$$\mathcal{H}_{MFA} = \frac{1}{2} NJzM^2 - \sum_i (JzM + H_i)S_i \quad , \tag{3}$$

N being the number of spins. z is the number of nearest neighbours and M is the averaged order parameter (magnetization). Within the mean-field approximation the averaged free energy F can be written as

$$F/N = \frac{1}{2} JzM^2 - \frac{1}{2\beta H_o} \int_{-H_o}^{H_o} dH \ln\{2 \cosh\beta(JzM + H)\} \tag{4}$$

with $\beta = 1/kT$, k being the Boltzmann constant. The order parameter M is given, by minimizing F, as

$$M = \frac{1}{2H_o} \int_{-H_o}^{H_o} dH \tanh \beta(JzM + H) \ . \tag{5}$$

In order to investigate the second-order phase transition, the transition line can be obtained in the usual way simply by linearizing (5), which gives

$$T_c = 2h_o/\ln\{(1 + h_o)/(1 - h_o)\}, \tag{6}$$

where $T_c = kT/Jz$ and $h_o = H_o/Jz$. Analysing (6) we find that the system undergoes a second-order phase transition at temperatures $0 < T_c \leqq 1$, given by the solution of (6) for the values $1 > h_o \geqq 0$. As can be seen, the point $h_o = 1$ is a singular point of (6) and therefore the point $h_o = 1$, $T_c = 0$ is not a second-order transition point of the model considered.

Analyzing (4) and (5) at T = 0, we find that at T = 0 and for $h_o < 1$, M = 1 is a stable phase of the system whereas for $h_o \geqq 1$, M = 0 is the only stable phase. Therefore, at T = 0 there is a strong jump in the value of M from M = 1 to M = 0, while for $T \approx 0$ but $T \neq 0$ the transition is continuous. In order to investigate the behaviour of the system at T = 0, making the Landau expansion

$$F = aM^2 + bM^4 + cM^6 + dM^8 + \dots \tag{7}$$

of the free energy, (4), for T = 0, we observe that all the higher order coefficients, b, c, d, ... are zero. Thus, in terms of the Landau theory of phase transitions, the point T = 0, $h_o = 1$ can be looked at as a kind of infinite order multicritical point.

As is obvious, this result is completely different from the other theoretical results involving other types of random-field distribution functions where a simple tricritical point has been indicated at some temperatures generally different than zero. In conclusion, we feel that the existence of a first-order phase transition, a tricritical point or a multicritical point is strongly dependent on the type of the random-field distribution function considered and the behaviour of a system observed with a particular distribution may not be considered as universal.

References

/1/ A. AHARONY, Phys. Rev. B 18, 3318 (1978).
/2/ T. SCHNEIDER and E. PYTTE, Phys. Rev. B 15, 1519 (1977).
/3/ D. ANDELMAN, Phys. Rev. B 27, 3079 (1983).
/4/ S. GALAM and J.L. BIRMAN, Phys. Rev. B 28, 5322 (1983).

(Received August 31, 1984)

phys. stat. sol. (b) 126, K33 (1984)
Subject classification: 18.4

Institute of Mathematics and Physics, Titograd[1] (a) and
Faculty of Physics, Skopje[2] (b)

On the Ground-State Properties of the Ising Antiferromagnet with General Spin on the Honeycomb Lattice

By
D. HAJDUKOVIĆ (a) and M. MIJATOVIĆ (b)

The model considered in this note is the Ising model on the honeycomb lattice described by the Hamiltonian

$$\mathcal{H} = J \sum_{\langle ij \rangle} S_i S_j - H \sum_i S_i \ , \tag{1}$$

where S_i is the z projection of the spin at site i ($S_i = S, S-1, \ldots, -S$), J is the antiferromagnetic interaction ($J > 0$), the applied field is positive ($H > 0$) and parallel to the z axis, and $\langle ij \rangle$ stipulates that the summation is restricted over those pairs of lattice sites which are the nearest neighbours.

It is well established /1 to 4/ that such an Ising antiferromagnet may have nonzero ground-state entropy in a critical magnetic field equal to

$$H_c = 6JS \ . \tag{2}$$

We are going to estimate the value of this ground-state entropy in the case of general spin S. To this end we strictly use a recently proposed and very accurate approximate method /5, 6/.

A honeycomb lattice with $N = L \times L$ spins can be conceived as a set of L parallel chains as shown in Fig. 1. (Only two neighbouring chains, the p-th and (p + 1)-th chain are shown with dashed lines.) In the first step let us consider an arbitrary p-th chain ($p = 1, 2, \ldots, L$). There are L/2 bonds between the spins of this chain (thick segments on dashed lines in Fig. 1) and, because we are interested in ground-state configurations only, it is not possible to have a bond with both ends in a spin state different from S. Thus, if on the p-th chain there are n spins in spin state different from S the number of configurations is

1) 81000 Titograd, Yugoslavia.

2) P.O. Box 162, Skopje, Yugoslavia.

$$P(n) = 2^n(2S)^n \binom{L/2}{n} . \tag{3}$$

In the second step let us consider two coupled chains, e.g. the p-th and (p+1)-th chain. These two coupled chains really form a one-dimensional Ising lattice with 2L sites. For a given number 2n of spins which are in spin states different from S there are

$$P(2n) = (2S)^{2n} \binom{2L - 2n}{2n} \tag{4}$$

configurations of these two coupled chains /7/.

We can consider (4) as a product of two factors, the factor (3) for the isolated p-th chain and a factor

$$C(p + 1) = S^n \frac{\binom{2L - 2n}{2n}}{\binom{L/2}{n}} \tag{5}$$

for the (p+1)-th chain which takes into account the influence of the p-th chain. Then in agreement with the proposed approximation /5, 6/ in order to find the concomitant ground-state entropy per site, in the thermodynamic limit, we have to find the maximum of (5). It occurs for

$$2n = xL , \tag{6}$$

where x is the smallest positive root in the interval (0, 1) of the equation

$$16S(1 - x)^3 = x(2 - x)^2 . \tag{7}$$

If we know the solution of (7) we can obtain the maximum $[C(p+1)]_{max}$ of (5) and thereby the ground-state entropy per site

$$\sigma_{H_c} = \lim_{L \to \infty} \frac{1}{L} \ln [C(p + 1)]_{max} = \ln \frac{(2 - x)^2}{4(1 - x)^{3/2}} . \tag{8}$$

The calculated ground-state entropies for S = 1/2, 1, 3/2, are

$$\sigma_{H_c}(1/2) = 0.4354, \quad \sigma_{H_c}(1) = 0.6202, \quad \sigma_{H_c}(3/2) = 0.7436.$$

In the cases of triangular and square lattices /5, 6/ our approximation is in perfect agreement with the previous computerized analysis. So, we hope it is a good approximation for the honeycomb lattice as well. Unfortunately, up to

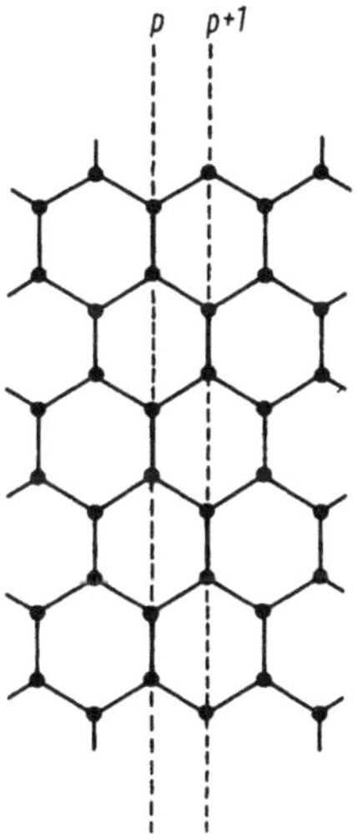

Fig. 1. Honeycomb lattice with N = L x L spins

our best knowledge there is no exact result or computerized analysis on the ground-state entropy of the honeycomb lattice. Thus, the question stays open how good our approximation is in this case. We can say only that the result $\sigma_{H_c}(1/2) = 0.4354$ is in good agreement with known bounds /4/ of ground-state entropy.

At the end let us note that beside the ground-state entropy the ground-state magnetization can be estimated, too. Thus, for the S = 1/2 model, from (7) we obtain x = 0.4822. Within our approach the magnetization is simply given by m = 1 - x = 0.5178.

References

/1/ J.E. BROOKS and C. DOMB, Proc. Roy. Soc. (London) A207, 343 (1952).
/2/ C. DOMB, Adv. Phys. 9, 149 (1960).
/3/ J.C. BONNER and M.E. FISHER, Proc. Phys. Soc. (London) 80, 508 (1963).
/4/ D. HAJDUKOVIĆ and S. MILOŠEVIĆ, J. Phys. A 15, 3561 (1982).
/5/ D. HAJDUKOVIĆ and S. MILOŠEVIĆ, Phys. Letters A 95, 249 (1983).
/6/ D. HAJDUKOVIĆ, I.M. SAVIĆ, and LJ. DAMJANOVIĆ, to be published in Z. Phys. B.
/7/ D. HAJDUKOVIĆ and S. MILOŠEVIĆ, Phys. Letters A 95, 57 (1983).

(Received August 17, 1984)

phys. stat. sol. (b) 126, K37 (1984)

Subject classification: 18.4

Institute of Mathematics and Physics, Titograd[1] (a) and
Department of Physics and Meteorology,
Faculty of Natural and Mathematical Sciences, Belgrade[2] (b)

On the Critical Properties of the Potts Antiferromagnets on the Bond-Decorated B.C.C. and F.C.C. Lattice

By

D. HAJDUKOVIĆ (a) and I.M. SAVIĆ[3] (b)

The antiferromagnetic Potts model has attracted increasing attention after a rescaling prediction, by Berker and Kadanoff /1/, that a distinctive low-temperature phase in which correlations decay algebraically with distance can exist in these systems. For the q-state antiferromagnetic Potts model this behaviour is permitted when q is less than a cut-off value q_o. This cut-off value and the critical point depend on the topological characteristics of the lattice and are not known for models in $d \geqq 3$ dimensions. In the present note we are going to obtain approximate singular trajectories (i.e. the functions $e^{K_c^*} = f(q)$, $K_c^* < 0$ being the critical coupling) for the antiferromagnetic Potts model on the decorated b.c.c. and f.c.c. lattice. We are interested in these decorated models because we expect a similar behaviour for the usual (i.e. undecorated) model on b.c.c. and f.c.c. lattice. Such a similarity in the behaviour is known for example in the case of bond-decorated /2/ and undecorated square lattices /3/.

To be definite let us consider the q-state Potts model on the decorated b.c.c. (or f.c.c.) lattice described by the Hamiltonian

$$\mathcal{H} = -\varepsilon \sum_{\langle ij \rangle} \delta_{Kr}(\sigma_i, \sigma_j) , \qquad \varepsilon < 0 . \tag{1}$$

Here $\sigma_i = 1, 2, \ldots, q$ specifies the spin state at the i-th site, $\delta_{Kr}(\alpha, \beta)$ is the Kronecker delta, and the sum is taken over the nearest-neighbour sites on the lattice. Taking the partial traces over the bond-decorating sites of a lattice

---

1) 81000 Titograd, Yugoslavia.

2) P.O. Box 550, 11001 Belgrade, Yugoslavia.

3) To whom all correspondence should be addressed.

with pure antiferromagnetic ($K^* < 0$) interactions leads to an effective b.c.c. (or f.c.c.) lattice. This b.c.c. (f.c.c.) lattice has ferromagnetic interactions ($K > 0$) determined by

$$e^{K} = (e^{2K^*} + q - 1)/(2e^{K^*} + q - 2) \tag{2}$$

which gives

$$e^{K^*} = e^{K} - \left[(e^{K} - 1)(e^{K} + q - 1)\right]^{1/2} . \tag{3}$$

On the other hand, there is a good approximate critical condition /4/ for the ferromagnetic Potts model on the b.c.c. and f.c.c. lattice

$$e^{\gamma K_c/2} - 2^{1/(\gamma/2-1)} e^{\gamma K_c/4} = q - 1 , \tag{4}$$

where γ is the coordination number ($\gamma = 8$ for b.c.c. and $\gamma = 12$ for f.c.c. lattice) and the condition (4) is valid for $q > 1$ (including the case when q is a continuous variable). If we introduce the critical value e^{K_c} determined by (4), equation (3) gives the critical point $e^{K_c^*}$ of the Potts model on the decorated b.c.c. (or f.c.c.) lattice. The corresponding singular trajectories are shown in Fig. 1.

The cut-off value q_o for the antiferromagnetic model follows immediately from (2) by setting $e^{K^*} = 0$ and using (4). It is a solution of

$$\left[(q-1)/(q-2)\right]^{\gamma/2} - 2^{1/(\gamma/2-1)} \left[(q-1)/(q-2)\right]^{\gamma/4} = q - 1 . \tag{5}$$

The cut-off values for the bond-decorated b.c.c. ($\gamma = 8$) and f.c.c. ($\gamma = 12$) lattice are

$$q_o^{\text{b.c.c.}} = 3.801, \quad q_o^{\text{f.c.c.}} = 4.692 . \tag{6}$$

There are some indications /5, 6/ that the cut-off value of the bond-decorated lattice is a lower bound for the cut-off

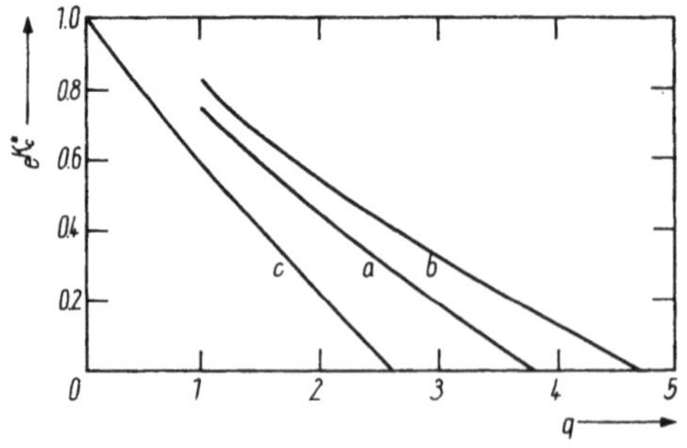

Fig. 1. Singularity trajectories for the antiferromagnetic Potts model on the bond-decorated b.c.c. (a) and f.c.c. (b) lattice. The exact trajectory for the decorated square lattice (c) is given for comparison

value of the corresponding undecorated lattice and we expect it is also true in this case for values (6).

References

/1/ A.N. BERKER and L.P. KADANOFF, J. Phys. A 13, L259 (1980).
/2/ F.Y. WU, J. stat. Phys. 23, 773 (1980).
/3/ R.J. BAXTER, Critical Antiferromagnetic Square Lattice Potts Model, Proc. Roy. Soc. A (1982).
/4/ I.M. SAVIĆ and D. HAJDUKOVIĆ, phys. stat. sol. (b) 123, K35 (1984).
/5/ F.Y. WU, Rev. mod. Phys. 54, 235 (1982).
/6/ D. HAJDUKOVIĆ, J. Phys. A 16, 2881 (1983).

(Received August 20, 1984)

phys. stat. sol. (b) 126, K41 (1984)
Subject classification: 13 and 18; 22.2.1

Department of Pure Physics,
University College of Science and Technology, Calcutta[1)]

Magnetic Susceptibility of Electrons in Ultrathin Films of Semiconductors in the Presence of a Quantizing Magnetic Field

By
M. MONDAL and K. P. GHATAK

In recent years there has been considerable interest in the study of various physical properties of 2D-electron gases in ultrathin films and multilayer heterostructures of semiconductors /1 to 10/. In such films, the restriction of the motion of the carriers in the direction normal to the film of the layer (say, the z-direction) may be viewed as carrier confinement in a one-dimensional potential well leading to quantization of the wave-vector of carriers in the z-direction. In the presence of a strong quantizing magnetic field along the z-direction, the free motion parallel to the film is also quantized forming Landau levels and leading to diamagnetism. Moreover, the spinning motion of the electrons leads to paramagnetism due to spin splitting of the Landau levels. In the present communication, an attempt is made to study the doping dependence of the dia- and paramagnetic susceptibilities of electrons in ultrathin semiconducting films taking degenerate n-GaAs as an example.

In an ultrathin film of a semiconductor like GaAs (having a parabolic $\bar{E}(k)$ relation) with the z-direction normal to the film, the modified electron energy spectrum in the presence of a quantizing magnetic field B along the z-direction can be expressed as

$$E_{nl} = (n + \frac{1}{2})\hbar\omega_0 + E_l \ , \tag{1}$$

where E_{nl} is the energy in the presence of magneto-size quantization as measured from the edge of the conduction band in the absence of any quantization, n (= 0, 1, 2, ...) is the Landau quantum number, $\hbar = h/2\pi$, h is the Planck constant, $\omega_0 \equiv eB/m^*$, e is the electron charge, m^* is the mass of the electrons, $E_l \equiv \left[\hbar^2/2m^*\right] (l\pi/d_0)^2$, $l(= 1, 2, 3, \ldots)$ is the size quantum number, and d_0 is the thickness of the semiconducting film. Using (1) the density-of-states function is given by

1) 92, Acharya P.C. Road, Calcutta 700009, India.

$$N(E) = \frac{eB}{d_o \pi \hbar} \sum_{n=0}^{n_{max}} \sum_{l=1}^{l_{max}} \delta(E - E_{nl}) , \qquad (2)$$

where δ is the Dirac delta function. Incidentally, it may be stated in this context that the density-of-states function in ultrathin semiconducting films in the presence of magnetic quantization is independent of energy, which is in direct contrast to that observed in bulk specimens of the same semiconductor and is the basis of the various peculiar phenomena associated with the magnetic quantization of 2D electrons. Using (2), we can express the free energy of the electrons as

$$u = n_o E_F - \frac{eBk_B T}{d_o \pi \hbar} \sum_{n=0}^{n_{max}} \sum_{l=1}^{l_{max}} F_o(\eta_{nl}) , \qquad (3)$$

where n_o is the electron concentration, E_F is the Fermi energy in the presence of magneto-size quantization as measured from the edge of the conduction band in the absence of any quantization, k_B is the Boltzmann constant, T is the temperature, $\eta_{nl} \equiv (E_F - E_{nl})/k_B T$, and $F_j(\eta_{nl})$ is the Fermi-Dirac integral of order j as defined by Blakemore /11/. Using (3), the diamagnetic part of the susceptibility of the electrons can be written as

$$\chi_d = -\mu_o \frac{\partial^2 u}{\partial B^2} = -\mu_o \left(\frac{2}{B^2} A_{11} - \frac{1}{k_B T B^2} A_{22} + P - Q \right) , \qquad (4)$$

where μ_o is the permeability of the free space,

$$A_{ij} \equiv \frac{eB}{d_o \pi \hbar} \sum_{n=0}^{n_{max}} \sum_{l=1}^{l_{max}} (E_{nl})^i F_{-j}(\eta_{nl}) ,$$ i is a set of real numbers,

$$P \equiv (n_o - A_{o2}) E_F'' , \qquad E_F'' \equiv \frac{\partial^2 E_F}{\partial B^2} ,$$

$$Q \equiv \left(\frac{1}{k_B T} A_{o2} \right) (E_F')^2 + \left(\frac{2}{B} A_{o1} - \frac{2}{B k_B T} A_{12} \right) E_F' ,$$

and $E_F' \equiv \partial E_F / \partial B$. It appears then that the computation of χ_d from (4) requires expressions for E_F' and E_F'', respectively. Incidentally these can be obtained from the expression for the electron concentration which can in turn be written, using (2), as

$$n_o = \frac{eB}{d_o \pi \hbar} \sum_{n=0}^{n_{max}} \sum_{l=1}^{l_{max}} F_{-1}(\eta_{nl}) \; . \qquad (5)$$

Since the electron concentration is not affected by magnetic quantization, we can get the expressions for E_F' and E_F'' from (5), respectively, as

$$E_F' \equiv (A_{02})^{-1} \left(\frac{1}{B} A_{12} - \frac{n_o k_B T}{B} \right) \qquad (6)$$

and

$$E_F'' \equiv (A_{02})^{-1} \left[\frac{2 n_o k_B T}{B^2} - (k_B T)^{-1} \left\{ A_{23} B^{-2} - 2 A_{13} B^{-1} E_F' + A_{03} (E_F')^2 \right\} \right] \qquad (7)$$

When spin-splitting is considered, the magnetic moment can be expressed as

$$M = \frac{eB\beta g_o}{4\pi\hbar d_o} \sum_{n=0}^{n_{max}} \sum_{l=1}^{l_{max}} \left[F_{-1}(\varphi_{nl,+}) - F_{-1}(\varphi_{nl,-}) \right] , \qquad (8)$$

where g_o is the Landé g-factor, β is the Bohr magneton, $\varphi_{nl,\pm} \equiv (k_B T)^{-1} (E_f - E_{nl} \pm \frac{1}{2} g_o B\beta)$, and E_f is the Fermi energy when occupations of the spin-split levels are considered.

Thus, we get the paramagnetic part of the susceptibility of the electrons from (8) as

$$\chi_p = \mu_o \frac{\partial M}{\partial B} = \mu_o \left[\frac{M}{B} + \frac{g_o \beta}{k_B T} \left\{ D_{02,+} - D_{02,-} \right\} \frac{\partial E_f}{\partial B} + \frac{g_o \beta}{k_B T} \left\{ D_{12,+} - D_{12,-} \right\} \right] , \qquad (9)$$

where

$$D_{iJ,\pm} \equiv \frac{eB}{2\pi\hbar d_o} \sum_{n=0}^{n_{max}} \sum_{l=1}^{l_{max}} \left[\frac{-E_{nl}}{B} \pm \frac{1}{2} g_o \beta \right]^i F_J(\varphi_{nl,\pm}) \; .$$

Incidentally, following the method as given above the expressions for n_o and $\partial E_f / \partial B$ in the presence of spin-splitting can respectively be expressed

$$n_o = D_{01,+} + D_{01,-} \qquad (10)$$

and

$$\frac{\partial E_f}{\partial B} = - (D_{02,+} + D_{02,-})^{-1} \left(\frac{n_o k_B T}{B} + D_{12,+} + D_{12,-} \right) . \qquad (11)$$

Using the appropriate equations we have computed the doping dependence of the dia- and paramagnetic susceptibilities of electrons under the electric quantum-limit ($l = 1$) in ultrathin films of n-GaAs at 2 K in the presence of a quantizing magnetic field as shown in Fig. 1. The parameter values used in these com-

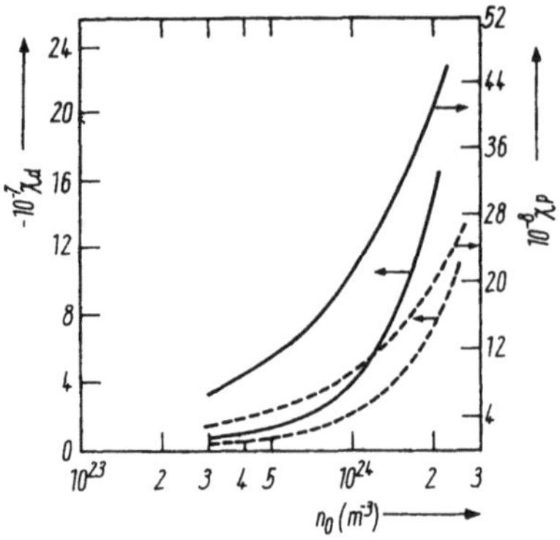

Fig. 1. Plots of the diamagnetic and paramagnetic susceptibilities of electrons in ultrathin films of n-GaAs at 2 K corresponding to two different values of a quantizing magnetic field; B = 15 T (——), 30 T (- - -)

putations are as follows: $m^* = 0.07\ m_0$, $d_0 = 45$ nm, B = 15 and 30 T. The value of g_0 is taken to be 2, because it is observed /12/ that at low temperatures, when the magnetic field is of the order of 10 T or higher g_0 tends to that for bulk electrons. It appears from the figure that the values of χ_d and χ_p are both higher than their bulk values in GaAs and increase with increasing electron concentration. It has also been observed that $\chi_d = -\lambda\chi_p$, where the factor λ decreases with increasing electron concentration in the presence of a constant magnetic field (the value of λ typically varying between 1 to 1/3). Thus at relatively high values of the electron concentration, we have approximately the relation $\chi_d = -(1/3)\chi_p$ which is in fact the relation first derived by Landau /13/ for 3D-electron gases by considering the magnetic quantum number to be fairly large. It appears then that a strong quantizing magnetic field would not affect the validity of Landau's basic relation for electron gases if these are two-dimensional. This follows from the fact that the electron concentration increases very much with the consequent rise of the Fermi level with respect to the bottom of the lowest Landau level resulting in the occupation of a large number of Landau levels even under strong magnetic quantization. Thus, the magnetic quantum number becomes large for large values of the electron concentration leading to the picture assumed in the Landau theory. It may be noted that the electron concentration in ultrathin semiconducting film has been expressed per unit volume for the convenience of comparison with the results of calculation for the bulk semiconductor. Furthermore, the effect of collision broadening has not been considered here. Though the effects of broadening are usually small at low temperatures, the sharpness of the amplitudes in the figure would somewhat be reduced by its influence. Nevertheless our present analysis would remain valid since the effects of collision broadening can usually be taken into account by an effective increase in temperature. Besides, the effects of electron-electron interactions which become increasingly important with increasing electron concentration have not been considered in the present derivation. Finally it may be remarked that, though in a more rigorous treatment the many-body effects should be considered along with a self-consistent procedure, the simplified analysis exhibits the major features of the magnetic susceptibility of the electrons in ultrathin semiconducting films with reasonable accuracy.

References

/1/ A.A. COTTEY, J. Phys. C 9, 2365 (1976).
/2/ W.L. FREEMAN, Ph.D. Thesis, Clemson University (USA) 1976 and references cited therein.
/3/ M. BARKOWSKI, M. JOLOCHOWSKI, Z. KORCZAK, M. SUBOTOWICZ, and M. ZATUZNY, in: Physics of Semiconductors, Ed. B.L.H. WILSON, Institute of Physics, Bristol 1978 (p. 745) and references cited therein.
/4/ A.A. COTTEY, phys. stat. sol. (b) 102, 577 (1980).
/5/ T. ANDO, A.H. FOWLER, and F. STERN, Rev. mod. Phys. 54, 437 (1982) and references cited therein.
/6/ K.P. GHATAK, K.K. GHOSH, H.K. MUKHERJEE, and A.N. CHAKRAVARTI, phys. stat. sol. (b) 110, 323 (1982).
/7/ A.N. CHAKRAVARTI, K.P. GHATAK, K.K. GHOSH, S. GHOSH, and A. DHAR, Z. Phys. B 17, 149 (1982).
/8/ A.N. CHAKRAVARTI, K.P. GHATAK, D. DHAR, and K.K. GHOSH, Czech. J. Phys. B 33, 65 (1983).
/9/ M. MONDAL and K.P. GHATAK, phys. stat. sol. (b) 120, K63 (1983).
/10/ N.B. MUSTAFAEV and M.G. SHAKHTAKHTINSKII, phys. stat. sol. (b) 120, K23 (1983).
/11/ J.S. BLAKEMORE, Semiconductor Statistics, Pergamon Press, London 1962 (p. 79).
/12/ K.F. KOMATSUBARA, K. NARITA, Y. KATAYAMA, and N. KOTERA, J. Phys. Chem. Solids 35, 723 (1974).
/13/ L.D. LANDAU, Z. Phys. 64, 629 (1930).

(Received July 26, 1984)

phys. stat. sol. (b) 126, K47 (1984)
Subject classification: 13.1; 22

Department of Pure Physics,
University College of Science and Technology, Calcutta[1)]

On the Effective Mass in Tetragonal Semiconductors in the Presence of an Arbitrarily Oriented Quantizing Magnetic Field

By
M. MONDAL and K. P. GHATAK

It is well-known that the effective mass of carriers in semiconductors is a very important parameter since it influences almost all the transport properties and has been studied in /1 to 6/ under various physical conditions. Nevertheless, a generalized expression of the effective mass in tetragonal semiconductors in the presence of arbitrary magnetic quantization has yet to be derived by considering the generalized dispersion relation of the conduction electrons allowing all types of anisotropies in the energy spectrum instead of the simplified derivation which occurs from the assumption of isotropic spin-orbit splitting /5, 6/, the more so, since the above class of semiconductors is being increasingly investigated /4 to 9/ for their peculiar physical characteristics. In what follows, we shall make such an investigation by taking n-Cd_3As_2 as an example. However, we shall consider the effective mass of the electrons only at the Fermi surface since at low temperatures the electrons at the Fermi surface are the major participants in electron transport.

The generalized dispersion relation for the conduction electrons in tetragonal semiconductors in the presence of arbitrary magnetic quantization can be expressed, according to Chuiko /9/, as

$$A(E) = a_{\pm}(E) + b(E)(k_z^{\prime\prime})^2 \quad , \tag{1}$$

where

$$A(E) \equiv \gamma(E) + E(E - E_g)\left[\frac{2}{9}(\Delta^2 - \Delta_\perp^2)\right] ,$$

$$\gamma(E) \equiv \left[E(E - E_g)\left\{E(E + \Delta) + \delta\left(E + \frac{2}{3}\Delta\right)\right\}\right],$$

$$a_{\pm}(E) \equiv \frac{eB}{\hbar}(2n + 1)\left[F_1(E)\left\{F_1(E)\cos^2\Theta + f_2(E)\sin^2\Theta\right\}\right]^{1/2} \pm$$
$$\pm\, eB\, P_\perp (3\hbar)^{-1}\left\{\left(E + \delta + \frac{\Delta^2 - \Delta_\perp^2}{3\Delta}\right)^2 \Delta^2\cos^2\Theta\, P_\perp^2 + E^2 P^2 \Delta_\perp^2 \sin^2\Theta\right\}^{1/2},$$

---

1) 92, Acharya P.C. Road, Calcutta 700009, India.

$$F_1(E) \equiv f_1(E) + \left[P_\perp^2(\Delta^2 - \Delta_\perp^2)\frac{1}{9}\right] ,$$

$$f_1(E) \equiv P_\perp^2\left[E(E + \frac{2}{3}\Delta) + \delta(E + \frac{\Delta}{3})\right] ,$$

$$f_2(E) \equiv P^2\left[E(E + \frac{2}{3}\Delta)\right] ,$$

$$b(E) \equiv \left[F_1(E)\cos^2\Theta + f_2(E)\sin^2\Theta\right]\left[F_1(E)f_2(E)\right] ,$$

$$k_z^{\shortparallel} \equiv (k_x \sin\Theta + k_z \cos\Theta) ,$$

E is the energy as measured from the edge of the conduction band in the absence of any quantization, E_g is the band gap, e is the electron charge, $\hbar \equiv h/2\pi$, h is the Planck constant, n is the Landau quantum number, B is the quantizing magnetic field which makes an angle Θ with the k_z direction and lies in the $k_x k_z$ plane, Δ is the spin-orbit splitting parameter along a four-fold axis of the tetragonal crystal, $\Delta_\perp$ is the corresponding parameter in a plane perpendicular to this axis, δ is the splitting of the p-bands by the tetragonal lattice fields, and $P_\perp$ and P are the anisotropic momentum matrix elements along directions perpendicular and parallel to the four-fold axis, respectively.

Therefore, the effective electron mass at the Fermi level E_F corresponding to the n-th magnetic subband can be written as

$$m^*_{n\pm}(E_F) = (\hbar^2/2)\left[\left\{A'(E_F) - a'_\pm(E_F)\right\}\left\{b(E_F)\right\}^{-1} - \right.$$
$$\left. - \left\{A(E_F) - a_\pm(E_F)\right\} b^{-2}(E_F) b'(E_F)\right] , \qquad (2)$$

where

$$A'(E_F) \equiv \left[\gamma'(E_F) + \frac{2}{9}(2E_F - E_g)(\Delta^2 - \Delta_\perp^2)\right] ,$$

$$\gamma'(E_F) \equiv \gamma(E_F)\left[E_F^{-1} + (E_F - E_g)^{-1}\right] + E_F(E_F - E_g)(2E_F + \Delta + \delta) ,$$

$$a'_\pm(E_F) \equiv \frac{eB}{\hbar}(n + \frac{1}{2})\left\{2F_1(E_F)f_1'(E_F)\cos^2\Theta + f_1'(E_F)f_2(E_F)\sin^2\Theta + \right.$$
$$\left. + F_1(E_F)f_2'(E_F)\sin^2\Theta\right\}\left[F_1(E_F)\left\{F_1(E_F)\cos^2\Theta + f_2(E_F)\sin^2\Theta\right\}\right]^{-1/2} \pm$$
$$\pm P_\perp eB(3\hbar)^{-1}\left\{\left(E + \delta + \frac{\Delta^2 - \Delta_\perp^2}{3\Delta}\right)\Delta^2 P_\perp^2 \cos^2\Theta + E P^2 \Delta_\perp^2 \sin^2\Theta\right\} \times$$
$$\times \left\{\left(E + \delta + \frac{\Delta^2 - \Delta_\perp^2}{3\Delta}\right)^2 \Delta^2 P_\perp^2 \cos^2\Theta + E^2 P^2 \Delta_\perp^2 \sin^2\Theta\right\}^{-1/2} ,$$

$$f_1'(E_F) \equiv P_\perp^2(2E_F + \frac{2}{3}\Delta + \delta),$$

$$b'(E_F) \equiv b(E_F)\Big[f_1'(E_F)\{F_1(E_F)\}^{-1} + f_2'(E_F)\{f_2(E_F)\}^{-1} - b(E_F)\{F_1(E_F)\cos^2\Theta + f_2(E_F)\sin^2\Theta\}^{-1}\{f_1'(E_F)\cos^2\Theta + f_2'(E_F)\sin^2\Theta\}\Big].$$

For $\Delta_\perp = \Delta$, (2) reduces to the same simplified expression as given for the first time by Wallace /5/. Incidentally, under the substitutions $\delta = 0$ and $\Delta_\perp = \Delta$, (2) takes the form

$$m^*_{n\pm}(E_F) = \hbar^2 [2L]^{-1}\Big[\alpha(E_F) \pm \beta(E_F)\Big], \tag{3}$$

$$L \equiv (P P_\perp)^2 \{P_\perp^2 \cos^2\Theta + P^2 \sin^2\Theta\}^{-1},$$

$$\alpha(E_F) \equiv \varphi(E_F)\Big[E_F^{-1} + (E_F - E_g)^{-1} + (E_F + \Delta)^{-1} - (E_F + \frac{2}{3}\Delta)^{-1}\Big],$$

where

$$\varphi(E_F) \equiv \Big[E_F(E_F - E_g)(E_F + \Delta)\Big]\Big[E_F + \frac{2}{3}\Delta\Big]^{-1},$$

$$\beta(E_F) \equiv S(E_F + \frac{2}{3}\Delta),$$

and

$$S \equiv \frac{eB\Delta P_\perp^2}{3\hbar} P[L]^{-1/2}.$$

Thus for $\delta = 0$ and $\Delta_\perp = \Delta$, the effective Fermi level mass is independent of the magnetic quantum number. Further, from (2) we observe that the difference between the effective Fermi level masses corresponding to the n-th and (n + 1)-th subbands becomes independent of n for large values of n. It appears that the evaluation of the effective mass at the Fermi level requires a relation between the electron concentration and the Fermi energy under magnetic quantization. Using (1), the electron concentration can be expressed as

$$n_o = \frac{eB}{2\pi^2\hbar}\sum_{n=0}^{n_{max}} \int_{E_{n,\pm}}^{\infty} \Big[\{A(E) - a_\pm(E)\}\{b(E)\}^{-1}\Big]^{1/2}\Big[-\frac{\partial f(E)}{\partial E}\Big] dE, \tag{4}$$

where $E_{n,\pm}$ can be determined from the equation $a_\pm(E_{n,\pm}) = A(E_{n,\pm})$ and $f(E)$ is the occupation probability factor.

For n-Cd_3As_2, using the appropriate equations (neglecting spin-splitting effects) together with the parameters /7, 8/:

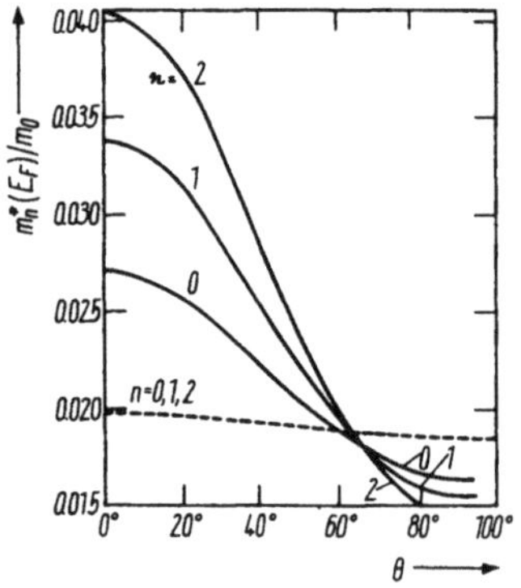

Fig. 1. Plot of the normalized effective Fermi level masses as a function of Θ corresponding to n = 0, 1, and 2 in degenerate n-Cd_3As_2 for B = 2T, n_o = $= 10^{23}\ m^{-3}$ and T = 4.2 K. The dashed plot corresponds to $\delta = 0$ and $\Delta_\perp = \Delta$

$$E_g = -0.095\ eV, \Delta = 0.27\ eV, \quad P = 7.21\times10^{-10}\ eV\ m\ ,$$

$$P_\perp = 7.43\times10^{-10}\ eV\ m\ , \qquad \delta = 0.085\ eV\ , \Delta_\perp = 0.29\ eV,$$

and T = 4.2 K we have plotted $m_n(E_F)/m_o$ corresponding to the first three subbands with orientation of the magnetic field with respect to the k_z-axis (Θ) and also with the electron concentration (n_o) as shown in Fig. 1 and 2, respectively. From these figures and the above discussion, the following features follow:

i) the dependence of the effective Fermi level mass on n is due to the combined effects of δ and the anisotropic spin-orbit splittings,

ii) the effective Fermi level masses corresponding to different n closely approach each other for relatively large values of n_o,

iii) in the limit of large n, the different effective Fermi level masses closely approach each other and become independent of n, and

(iv) the dependence of $m_n^*(E_F)$ on n is significantly influenced by the values of Θ.

Incidentally, it may be noted that though the effects of collision broadening are usually small at low temperatures the sharpness of the amplitudes of the plots would somewhat be reduced by the effects of collision broadening. Besides

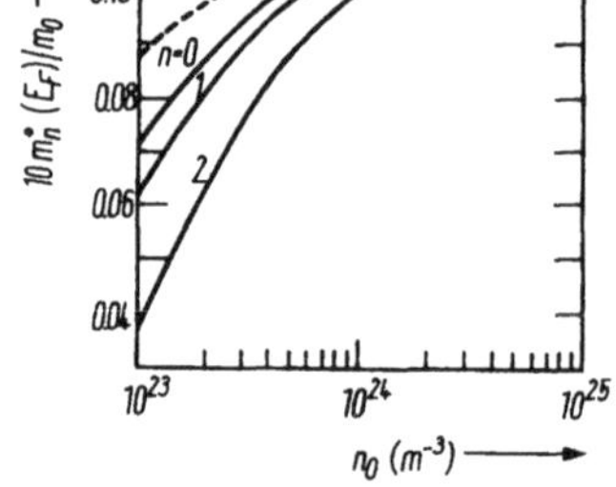

Fig. 2. Plot of the normalized effective Fermi level masses corresponding to n = 0, 1, and 2 as a function of n_o in degenerate n-Cd_3As_2 for B = 2T, Θ = 60° and T = 4.2 K. The dashed plot corresponds to $\delta = 0$ and $\Delta = \Delta_\perp$.

the effects of electron-electron interactions which become increasingly important with increasing electron concentration have not been considered here. Though in a more rigorous treatment the many-body effects should be considered along with a self-consistent procedure, this simplified analysis exhibits the basic qualitative features of the effective mass in tetragonal semiconductors in the presence of arbitrary magnetic quantization with reasonable accuracy. It may finally be noted that the SdH oscillations which occur in degenerate semiconductors would further be influenced by the index dependent effective mass in tetragonal semiconductors having negative band gap and the contribution of the oscillatory effective mass to the oscillatory mobility would be important.

References

/1/ U. ROSSLER, Solid State Commun. 49, 945 (1984).

/2/ A.N. CHAKRAVARTI, A.K. CHOWDHURY, K.P. GHATAK, S. GHOSH, and A. DHAR, Appl. Phys. 25, 105 (1981) and references cited therein.

/3/ A.N. CHAKRAVARTI, K.P. GHATAK, A. DHAR, and K.K. GHOSH, Czech. J. Phys. B33, 65 (1983).

/4/ A.N. CHAKRAVARTI, K.P. GHATAK, K.K. GHOSH, S. GHOSH, and A. DHAR, Z. Phys. B 17, 149 (1982).

/5/ P.R. WALLACE, phys. stat. sol. (b) 92, 49 (1979).

/6/ M. SINGH and P.R. WALLACE, J. Phys. C 16, 3877 (1983) and references cited therein.

/7/ F.A.P. BLOM, J.W. CREMERS, J.J. NEVE, and M.J. GELTEN, Solid State Commun. 33, 69 (1980).

/8/ G.P. CHUIKO and I.A. TOPLINSKAYA, Soviet Phys. - Semicond. 17, 706 (1983).

/9/ G.P. CHUIKO, Soviet Phys. - Semicond. 17, 301 (1983) and references cited therein.

(Received July 26, 1984)

phys. stat. sol. (b) 126, K53 (1984)

Subject classification: 13.3 and 13.5.1

Heat Physics Department, Academy of Sciences of the Uzbek SSR, Tashkent[1)]

The Localization Energy of Wannier-Mott Excitons at the Interface

By

I.M. NEPOMNYASHI and V.N. NISHANOV

There is a number of papers /1 to 3/ devoted to the observation and calculation of the Wannier exciton energy spectra. In /1/ a criterion of existence of localized exciton states induced by the electrostatic image forces has been obtained. Besides, there are some experimental indications on the existence of such states. To identify the excitons localized near the surface it is necessary to know the localization energy of excitons induced by the electrostatic image forces. With this purpose in this note the values of the ground state surface exciton energy for a number of semiconductors (Ge, PbTe, GaAs, CdAs) and the radius of these localized states have been obtained, besides a diagram of existence of the states mentioned as a function of the dielectric and band characteristics of semiconductor and substrate has been constructed.

We suppose that the necessary condition of localization is satisfied, i.e. $\varepsilon_2 > \varepsilon_1$ (where ε_1, ε_2 are the dielectric permittivities of the semiconductor and the substrate, respectively). As it was done in /1/, we shall consider semiconductors with strong band anisotropy, so the effective masses of quasiparticles satisfy the proper condition $m = m_x = m_y \ll m_z$ (the direction of the 0Z-axis is perpendicular to the interface). Using the adiabatic approximation we obtain the equation the solution of which describes the spectra of the surface levels.

In this case in the adiabatic approximation the spectra of the surface exciton levels may be obtained from the solutions of the Schrödinger equation

$$-\frac{\hbar^2}{2M}\frac{d^2\eta}{dz^2} + W(z)\eta(z) = E\,\eta(z) \qquad (1)$$

with the proper boundary condition

$$\eta(0) = 0 \;, \qquad (2)$$

---

1) 700100 Tashkent, USSR.

where $M = m_{ez} + m_{hz}$, W(z) is the excitonic term which plays the role of the potential well caused by the image forces /1/

$$W(z) = -\frac{e^2 \gamma}{2\varepsilon_1 z}\left\{1 - x\left[\frac{x\pi}{4}(H_1(x) - N_1(x) - x(H_0(x) - N_0(x))) - \frac{1}{2\gamma}\right]\right\}. \quad (3)$$

The relation between z and x is given by

$$z = \frac{\varepsilon_1 \hbar^2 x}{4e^2 \mu \gamma} \frac{1}{\frac{2}{\gamma} + 4x - x\pi\left(H_1(x) - N_1(x) + x\left[H_0(x) - N_0(x)\right]\right)} ,$$

$\gamma = \varepsilon_2 - \varepsilon_1/\varepsilon_2 + \varepsilon_1$, H_n, N_n are the Struve and Neumann functions, respectively. Equation (1) is solved using the variational procedure with the proper trial function η;

$$\eta = 2\alpha^{3/2}\, z\, e^{-\alpha z} .$$

The ground state energy E_0 is found from the set of equations:

$$V(\beta) = \beta^2 - \gamma\beta - \frac{\tilde{m}^2}{2\pi\gamma}\beta^3 K(\beta;\gamma) , \quad (4)$$

$$\frac{dV}{d\beta} = 0 . \quad (5)$$

Here

$$V = E_0 \frac{2\hbar^2 \varepsilon_1^2}{M e^4} , \quad \beta = \alpha \frac{\hbar^2 \varepsilon_1^2}{M e^2} , \quad \tilde{m} = \frac{M}{4\mu} , \quad \mu = \frac{m_e m_h}{m_e + m_h} ,$$

$$K(\beta,\gamma) = \int_0^\infty dx\, x^2 \left\{H_1(x) - N_1(x) - x\left[H_0(x) - N_0(x)\right] - \frac{2}{\pi\gamma x}\right\} \times$$

$$\times \frac{(x^2+1)\frac{H_1(x) - N_1(x)}{2x} - \left[H_0(x) - N_0(x)\right] - \frac{x}{\pi} - \frac{1}{\pi x^2 \gamma}}{\left\{\frac{2}{\pi\gamma x} - \left[H_1(x) - N_1(x)\right] - x\left[H_0(x) - N_0(x)\right] + \frac{4}{\pi}\right\}^3} \times$$

$$\times \exp\left\{-\frac{2\beta\tilde{m}}{-\left\{H_1(x) - N_1(x) - x\left[{}_0(x) - H_0(x)\right] + \frac{4}{\pi}\right\}\gamma\pi + \frac{2}{x}}\right\} . \quad (6)$$

Since the surface exciton levels are counted from the two-dimensional excitonic Rydberg, the localization exciton energy $E_1 = V + 1/\tilde{m}$, and the criterion of existence of the bound state on the surface will be the inequality $E_1 < 0$.

Using the asymptotic behaviour of the Struve and Neumann functions for $x < 1$ and $x > 1$ we can get a qualitative estimate of the exciton binding energy on the surface

$$E_1 = \begin{cases} \frac{1}{\tilde{m}} + \beta^2 - \gamma\beta - e^{-\beta\tilde{m}}\left[\beta^2\tilde{m}\left(\frac{1}{2} - \gamma\right) + \beta(1 - \gamma) + \frac{1}{\tilde{m}}\right] + \\ + \exp(-\beta\tilde{m}/(1-\gamma))\left[\frac{\beta^2\tilde{m}}{2} + \beta(1-\gamma) + \frac{1}{\tilde{m}}(1-\gamma)^2\right] - \frac{(1-\gamma)^2}{\tilde{m}}, & \gamma \neq 1 , \\ \frac{1}{\tilde{m}} + \beta^2 - \gamma\beta - \frac{\beta^3\tilde{m}^2}{150}\exp(-\beta\tilde{m}/2) + c^{-\beta\tilde{m}}\left[\frac{\beta^2\tilde{m}}{2} - \frac{1}{\tilde{m}}\right], & \gamma = 1 . \end{cases} \tag{7}$$

In the case of strong anisotropy $\tilde{m} \gg 1$ it is easy to obtain from (7) and (5) the asymptotic value of the localization energy

$$E_1^{as} = \frac{\gamma^2}{4} , \tag{8}$$

i.e. the minimum value of the energy is

$$E_1^{as} = -\frac{1}{4} ,$$

and the proper localization radius is

$$\varrho_0 = \frac{3}{\gamma} , \tag{9}$$

and the minimum value $\varrho_0 = 3$.

To find the exact values of the ground state energy (4) and (5) were solved numerically, the results obtained are shown in Fig. 1. Here the values of the surface excitonic levels are presented for a number of anisotropic semiconductors (Ge, PbTe, GaAs, CdAs) as a function of γ. As it can be seen from Fig. 1, with increasing γ (the maximum value for γ is unity) the absolute value of the exciton localization energy increases, too, in the case of highly anisotropic semiconductors (Ge, PbTe) the values of this energy as a function of γ are described by formula (8). This is also illustrated by Fig. 2, where the E_1-dependence on $\tilde{m}$ is shown; it is seen that for $\tilde{m} > 15$ the values of the localization energy are described by formula (8).

The radius of the exciton localization appropriate for the states mentioned above increases with decreasing γ (see Fig. 3). It is due to the fact that the width of the potential well induced by the electrostatic image forces is directly proportional to γ (see (3)). For highly anisotropic semiconductors (Ge, PbTe) ϱ_0 as a function of γ is described by the expression (9).

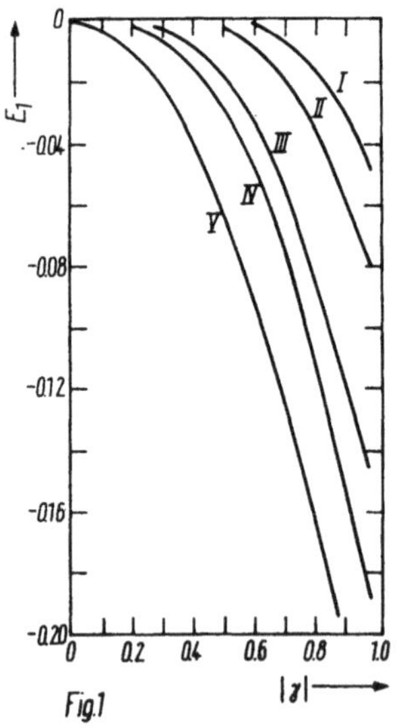

Fig.1

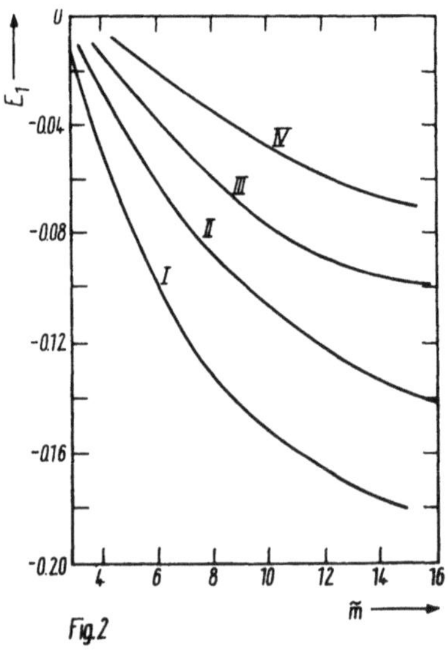

Fig.2

Fig. 1. The ground state energy of surface excitons; (I) $\tilde{m}$ = 4.01, (II) $\tilde{m}$ = 4.85, (III) $\tilde{m}$ = 10.4, (IV) $\tilde{m}$ = 14.6, (V) $\tilde{m}$ = ∞

Fig. 2. The ground state energy of surface excitons; (I) $|\gamma| = 1$, (II) $|\gamma| = 0.9$, (III) $|\gamma| = 0.8$, (IV) $|\gamma| = 0.7$

In this connection it should be noted that non-adiabatic corrections may be considered negligible, since the exciton localization radii for all values of γ are much larger than the effective Bohr radius /1/. (As it is seen in /1/ the adiabatic approximation is valid within distances less than the two-dimensional Bohr radius of the exciton, ($r_o = \hbar^2 \varepsilon_1 / 4\mu e^2$).)

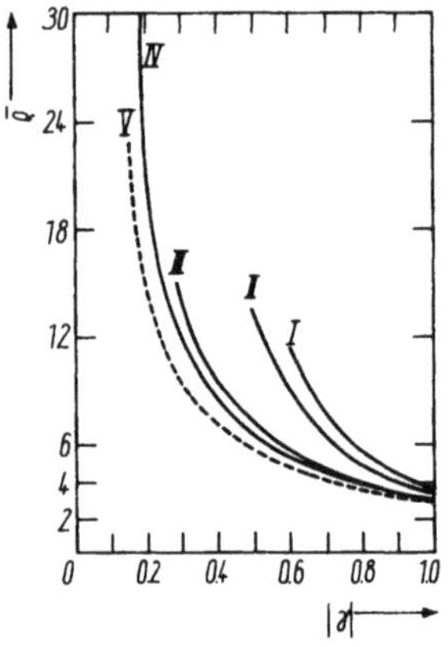

Fig. 3. Radius of exciton localization at the interface; (I) $\tilde{m}$ = 4.01, (II) $\tilde{m}$ = 4.85, (III) $\tilde{m}$ = 10.4, (IV) $\tilde{m}$ = = 14.6, (V) $\tilde{m}$ = ∞

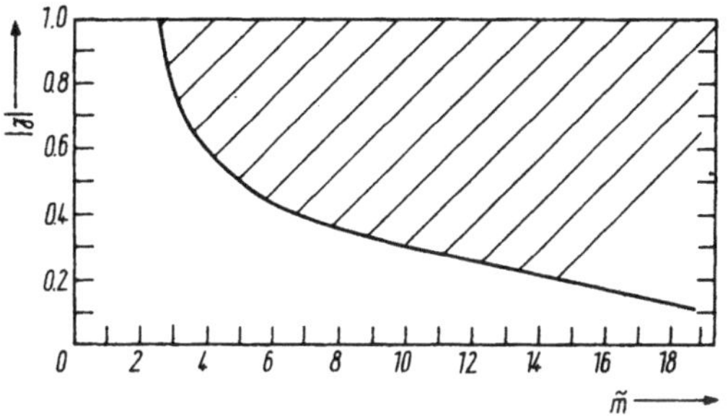

Fig. 4. Diagram of existence of localized exciton states

The diagram in Fig. 4 shows the region of localized states for various dielectric and band characteristics of the system under consideration. The dashed area indicates the region where localized states exist. It can be seen from the diagram that localized states exist for certain values of γ and $\tilde{m}$. Indeed, in semiconductors with $\tilde{m} < 3$ there are no surface levels for any γ. Note, that the region of localized states shown in Fig. 4 is in accordance with the criterion of exciton localization on the surface, $\gamma > \mu/M$, obtained in /1/.

In conclusion we note that the results obtained in this study may be applicable to the problem of small impurity levels in semiconductors.

The authors are indebted to Yu.E. Lozovik for discussions.

References

/1/ YU.E. LOZOVIK and V.N. NISHANOV, Fiz. tverd. Tela 18, 3267 (1976); Preprint ISAN 12/76, 1976.

/2/ V.A. KISELEV, Fiz. tverd. Tela 20, 1191 (1978).

/3/ V.G. LITOVCHENKO, Principles of the Physics of Semiconductor Sandwich Systems, Naukova Dumka, Kiev 1980 (in Russian).

(Received July 23, 1984)

phys. stat. sol. (b) 126, K59 (1984)

Subject classification: 13.5.1 and 20.1; 22.3

Institute of Applied Physics, Academy of Sciences of the Moldavian SSR, Kishinev[1)]

Free Exciton and Exciton-Impurity Complexes of Zinc Diphosphide and Zinc Diarsenide

By

V.V. SOBOLEV and A.I. KOZLOV

This note reports on reflection and transmission measurements of ZnP_2 and $ZnAs_2$ single crystals near E_g in the range 4.2 to 293 K. All the investigated specimens had p-type conductivity with carrier concentrations of 10^{15} to 10^{17} cm^{-3}. The measurements were carried out on cleavages or natural mirror-like platelets. X-ray examination showed the $\vec{c}$ and $\vec{b}$ axes and more seldom the $\vec{c}$ and $\vec{a}$ axes to lie in the plane of the samples. The diffraction patterns showed the melt grown crystals to be monoclinic (C_{2h}^5). The vapour grown crystals had twinning of the monoclinic lattice.

Fully polarized structures C_1, C_2 with maxima at 1.038 and 1.046 eV for $ZnAs_2$, 1.562 and 1.590 eV for ZnP_2 have been observed in the reflection spectra at 4.2 K and with light polarized parallel to the $\vec{c}$ axis (Fig. 1). These structures correspond to the free exciton states with n = 1 and n = 2. The states with n = 3, 4, 5 have been detected also in ZnP_2 at 1.596, 1.598, and 1.599 eV. The main exciton maxima have high values of reflectivity (up to 70% for ZnP_2 and 40% for $ZnAs_2$) and enormous oscillator strength /1/. This structure is shifted with the average temperature coefficient of about $-(1.0$ to $1.2)\times10^{-4}$ eV K^{-1} over the range 4.2 to 80 K and $-(3.2$ to $3.3)\times10^{-4}$ eV K^{-1} over the range 80 to 293 K. The states with n = 2 for $ZnAs_2$ and those with n = 4, 5 for ZnP_2 are fully washed out at 80 K. Assuming a hydrogenic series, the exciton binding energy G^c is estimated as 12 meV for $ZnAs_2$ and 47 meV for ZnP_2. The direct exciton gap E_g^C is determined to be 1.050 eV for $ZnAs_2$ and 1.601 eV for ZnP_2 at 4.2 K. The results obtained do not depend on the growing methods of both crystals.

The reflectivity spectra were structureless for $\vec{E} \parallel \vec{b}$ and $\vec{E} \parallel \vec{a}$. However, as seen in Fig. 2, the transmission lines B_1, B_2 at 1.038 and 1.048 eV for

1) Grosula 5, 277028 Kishinev, USSR.

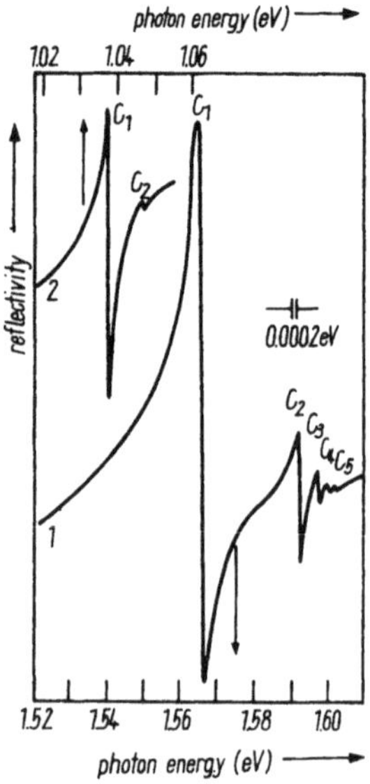

Fig. 1. Reflection spectra of (1) cleaved ZnP_2 and (2) $ZnAs_2$ at 4.2 K and $\vec{E} \parallel \vec{c}$

$ZnAs_2$ and B_1, B_2, B_3 at 1.558, 1.591, and 1.598 eV for ZnP_2 have been observed at $\vec{E} \parallel \vec{b}$ and 4.2 K. It is possible that these weak lines are due to band fine structure. The lines A_1, A_2, A_3 at 1.574, 1.594, and 1.599 eV have been detected in ZnP_2 for $\vec{E} \parallel \vec{a}$ simultaneouly with the series B. These two hydrogen-like series beginning with n = 1 have nearly the same energy gaps (E_g^A = 1.602 eV and E_g^B = 1.603 eV). At the same time the exciton binding energies of these series are different (G^A = 35 meV and G^B = 49 meV). This indicates the dielectric constant to be anisotropic. The calculations of E_g and G for ZnP_2 were carried out from the states with n = 2 and n = 3. The experimental structures with n = 1 proved to be shifted by 7 meV for series B and C and by 3 meV for series A as compared with calculated data. A detailed analysis is difficult to perform due to the absence of an exciton theory in strongly anisotropic monoclinic crystals.

Thus it is important to note, that the intensive series of free excitons in the reflection spectra and the two weak series in transmission represent another excitonic structure than the three allowed series of hexagonal $A^{II}B^{VI}$ crystals. Our data indicate the unusual influence of the crystal field and spin-orbit interaction on the upper valence band structure of the monoclinic zinc diphosphide and zinc diarsenide. It is interesting that a similar situation has been found in the reflection spectra of orthorhomic InBr /2/.

In addition to free-exciton structures, we have observed in the transmission spectra of ZnP_2 and $ZnAs_2$ exciton-impurity complexes (EIC) in the long-wave region in the vicinity of the exciton state with n = 1 (Fig. 2). The EIC structures have been obtained also in the absorption /3/ and photoluminescence /4/ spectra of ZnP_2 at 4.2 K. We have observed for the first time the EIC in the reflection spectra of natural faces of ZnP_2 and $ZnAs_2$ crystals grown from the vapour phase, at 4.2 and 80 K /1, 5/. Some complexes have been detected even at 293 K. The free exciton and EIC peaks have different temperature co-

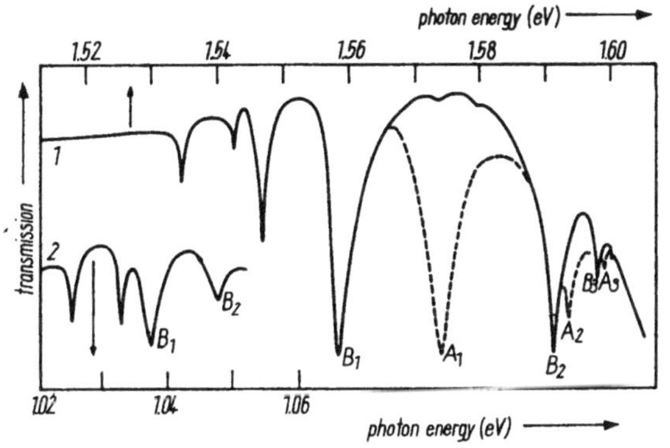

Fig. 2. Transmission spectra of ZnP_2 (1) and $ZnAs_2$ (2) at 4.2 K and $\vec{E} \parallel \vec{b}$ (solid curve), $\vec{E} \parallel \vec{a}$ (dashed curve)

efficients. The intensity of some complexes may exceed that of the free exciton with n = 1. This experimental result confirms the theoretical prediction of the gigantic oscillator strength of EIC in semiconductors /6/. These structures for both cleaved crystals have not been usually observed in reflection (Fig. 1). The lack of correlation between the reflectivity and transmission for some ZnP_2 and $ZnAs_2$ samples indicates that the concentration of impurities on which the exciton is localised, may be strongly different for volume and surface. The variation of the number, position, and intensity of the EIC reflectivity and transmission lines from sample to sample means that they do not form a single series.

The authors wish to thank L. A. Bityutskaya, E. N. Bormontov, S.F. Marenkin, P. A. Romanyk, E.M. Smolyarenko, K. A. Sokolovskii, and I.I. Tychina for the samples given.

References

/1/ V.V. SOBOLEV, A.I. KOZLOV, I.I. TYCHINA, P.A. ROMANYK, and E.M. SMOLYARENKO, Zh. eksper. teor. Fiz., Pisma 34, 115 (1981).

/2/ M. YOSHIDA, N. OHNO, K. NAKAMURA, and Y. NAKAI, phys. stat. sol. (b) 109, 503 (1982).

/3/ I.S. GORBAN, V.V. LUGOVSKII, I.I. TYCHINA, and A.V. FEDOTOVSKII, Zh. eksper. teor. Fiz., Pisma 17, 193 (1973).

/4/ R.S. BERG and P.Y. YU, Solid State Commun. 46, 101 (1983).
/5/ V.V. SOBOLEV and A.I. KOZLOV, Tezisy dokl. Vses. sovesh. Eksitony v poluprovodnikakh, Izd. Akad. Nauk SSSR, Leningrad 1982 (p. 42).
/6/ E.I. RASHBA, Fiz. Tekh. Poluprov. 8, 1241 (1974).

(Received July 23, 1984)

phys. stat. sol. (b) 126, K63 (1984)
Subject classification: 13.4 and 20.3; 22.5.2

Institute of Physics, Academy of Sciences of the Estonian SSR, Tartu[1)]

Slow Decay Component of the A_T Emission in KBr:Pb Crystals

By
ZH. EGEMBERDIEV[2)], V. NAGIRNYI, T. SOOVIK, and S. ZAZUBOVICH

The luminescence of KBr:Pb crystals has been studied in several papers /1 to 7/, but the obtained results are often contradictory. It may be caused by very active aggregation processes leading to a strong dependence of the KBr:Pb luminescence characteristics on lead concentration as well as on the history of the sample. It has been established /4/ that in well-quenched crystals containing a small amount of lead the dominating centres are $Pb^{2+}v_c^-$ dipoles. The 3.4 eV emission of these centres is caused by electronic transitions from the tetragonal minima of the adiabatic potential energy surface (APES) of the triplet excited state /1, 2/ and labeled as A_T emission. Due to some peculiarities, the A_T emission of KBr:Pb differs essentially as compared with the other alkali halides doped with mercury-like ions. In particular, the polarization spectrum of the A_T emission /2/ is not of such "sign-changing" form as it is in the case of the other anisotropic centres, and its decay time is very short (10^{-8} to 10^{-9} s /6/). Recently /7/ the slow component has been found in the KBr:Pb emission decay but its properties are still not reported.

In the present note we have extracted and studied the slow (equilibrium) component in the A_T emission decay and established the qualitative analogy of the A_T emission characteristics both in KBr:Pb and KCl:Pb crystals.

A KBr:$PbBr_2$ single crystal was grown by the Stockbarger method from zone-refined KBr. The concentration of lead was about 10^{17} cm^{-3}. Before the measurements the samples were heated for 20 min at 700 °C and then rapidly cooled down to room temperature. The emission was excited within the A absorption band (from 4.4 to 3.9 eV) by a spark pulse in air through a quartz

---

1) Riia 142, 202400 Tartu, Estonian SSR, USSR.

2) Present address: Pedagogical Institute, ul. Sverdlova 331, 714010 Osh, Kirgisian SSR, USSR.

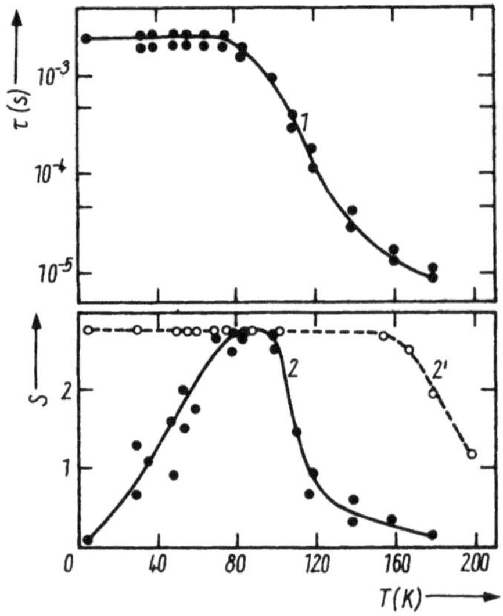

Fig. 1. Temperature dependences of the decay time (1) and light sums (2, 2') of the slow (1, 2) and fast (2') components of the A_T emission decay in KBr: Pb crystal. Curves 2 and 2' are normalized to the maximal S values for both curves, whereas $S_{max}(2) \approx 10^{-2} S_{max}(2')$

monochromator (ZMR-3) and polaroid ($\vec{E}_{exc} \parallel [001]$). The 3.4 eV emission was observed through optical filters (UFS-6+ +BS-6) and another polaroid in the [010] direction and detected by an EMI 9813 QB photomultiplier. The decay curves of the emission polarized along the [001] and [100] directions ($I_{\parallel}(t)$ and $I_{\perp}(t)$, respectively) were registered by the synchronous photon counting method using a multichannel analyser (type NTA 1024 EMG, Hungary) with minimum channel time 5 μs.

The decay time of the slow component of the A_T emission (τ_s) at T < 80 K is temperature independent and equal to $(2.5 \pm 0.5) \times 10^{-3}$ s (Fig. 1, curve 1). As the temperature rises from 80 to 140 K, τ_s decreases sharply by two orders. This shortening of τ_s may arise not only from the participation of the radiative state in the luminescence but also from thermal quenching of the A_T emission (curve 2). Therefore the $\tau_s(T)$ dependence may not be used for calculation of the energetic distance between the radiative and metastable minima of the A_T state.

The light sum of the slow component (S_s) at 70 to 100 K is at least by two orders smaller than that of the fast component (S_f). S_s decreases as the temperature lowers (curve 2), and at 4.2 K it is not measurable. It shows that the probability of the radiationless transitions from the upper radiative to the lower metastable minima in KBr: Pb is very small and remarkably temperature dependent.

The thermal quenching of the slow component takes place at a considerably lower temperature (≈100 K) than that of the fast component (160 K) (curves 2, 2'). Probably this may be explained by the more effective non-radiative deexcitation

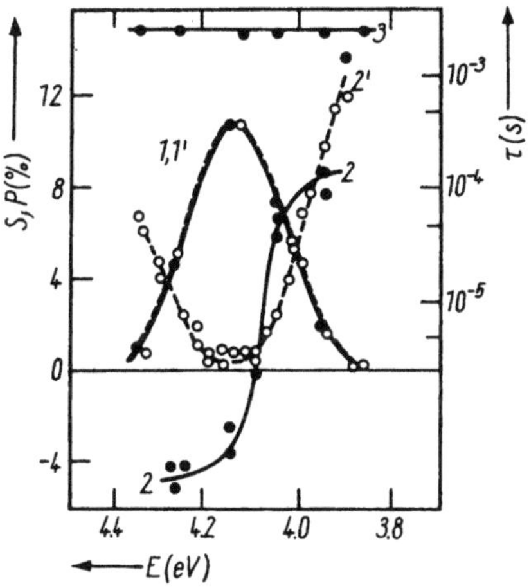

Fig. 2. Excitation (1, 1'), polarization (2, 2'), and decay time (3) spectra of the slow (1 to 3) and fast (1', 2') components of the A_T emission of KBr:Pb at 77 K

of the metastable state, taking into account the large difference between the lifetimes of the metastable and the radiative states (τ_s is by five orders longer than τ_f).

The excitation spectra of the low and fast components coincide completely (Fig. 2, curves 1, 1').

The polarization spectrum of the slow component $P_s(h\nu_{exc})$ has a "sign-changing" shape which is characteristic of the anisotropic mercury-like centres and is caused by the presence of v_c^- near Pb^{2+} /8/ (Fig. 2, curve 2). Just as in the case of KCl:Pb /1/, the absolute value of the negative polarization degree in KBr:Pb is by about two times smaller than that of the positive one.

The small value of P_f and the shape of the polarization spectrum $P_f(h\nu_{exc})$ may be connected with the peculiarities in the relaxation processes in the excited state leading to almost equal populations of the emitting minima of various orientations (especially at excitation near the A absorption band maximum).

The spectral, polarization, and kinetic characteristics of the principal (fast) component in the A_T emission decay obtained by us (Fig. 1 and 2, curves 1', 2')are in a complete agreement with those reported previously on $Pb^{2+}v_c^-$ centres in KBr:Pb /2, 4, 6/. Therefore, in spite of the rather large lead concentration in our samples needed for careful examination of the weak slow component, we are sure that the slow component considered here belongs really to $Pb^{2+}v_c^-$ centres. This conclusion is also confirmed by the coincidence of the excitation spectra for the fast and slow components (Fig. 2, curves 1, 1') and by the constant τ_s value at excitation in the whole region of the A absorption band of the $Pb^{2+}v_c^-$ centres (curve 3).

These results illustrate that the main properties of the A_T emission in both KBr:Pb and KCl:Pb crystals - presence of the slow and fast decay components, the behaviour of $\tau(T)$, S(T) as well as the shape of the polarization spectra - are very similar. It means that the structure of the APES for the A_T state of $Pb^{2+}v_c^-$ centres in both crystals must be essentially the same. Namely, the metastable minima are situated below the radiative ones. At $T < 80$ K the slow component of the A_T emission is caused by electronic transitions from the metastable minima ($\tau_s(T)$ = const). As this emission is polarized, the low (tetragonal) symmetry of the metastable minima can be concluded. The shape of the polarization spectrum ($|-P_s| \approx 2|+P_s|$) means /8/ that in KBr:Pb as well as in KCl:Pb crystals the X and Y minima of the relaxed A_T state are the lowest ones. If the Pb^{2+} ion is associated with the nearest-neighbouring cation vacancy v_c^- /5/ (situated in the X, Y plane), then the non-electrostatic interactions dominate in $Pb^{2+}v_c^-$ centres.

Thus, in spite of a rather strong spin-orbit interaction (ξ = 0.78 eV in KBr:Pb and 0.95 eV in KCl:Pb /9/) comparable with the Jahn-Teller one, the properties of the A_T emission and the structure of the APES of the triplet excited state of $Pb^{2+}v_c^-$ centres can be adequately described in terms of the model /8/ which considers the Jahn-Teller interaction as the main one for the s^2-type impurity ions in alkali halides.

References

/1/ S. ZAZUBOVICH and T. KUKETAEV, Trudy Inst. Fiz. i Astron. Akad. Nauk Eston. SSR 31, 190 (1966).

/2/ A. FUKUDA, S. MAKISHIMA, T. MABUCHI, and R. ONAKA, J. Phys. Chem. Solids 28, 1763 (1967).

/3/ L. MĂRCULESCU and C. GHITA, phys. stat. sol. (b) 75, 375 (1976).

/4/ J.L. PASCUAL, L. ARIZMENDI, F. JAQUE, and F. AGULLÓ-LÓPEZ, J. Lum. 17, 325 (1978).

/5/ M. TRINKLER and L. TRINKLER, Optika i Spektroskopiya 46, 91 (1979).

/6/ K. SCHMITT, V.S. SIVASANKAR, and P.W.M. JACOBS, J. Lum. 27, 313 (1982).

/7/ D.J. SIMKIN, J.G. KANG, T.F. BELLIVEAN, P.W.M. JACOBS, A. SCACCO, M.J. STILLMAN, and V.S. SIVASANKAR, Radiat. Eff. 73, 81 (1983).

/8/ V. HIZHNYAKOV, Preprint FI-36, Tartu 1975.
V. HIZHNYAKOV and S. ZAZUBOVICH, phys. stat. sol. (b) 86, 733 (1978).

/9/ R.E. CHANEY, P.W.M. JACOBS, and T. TSUBOI, Canad. J. Phys. 51, 2242 (1973).

(Received July 4, 1984)

phys. stat. sol. (b) 126, K67 (1984)
Subject classification: 13.3; 22.1

Physics Department, Wright State University, Dayton[1]

C_5H_9 Cluster as a Model for the (111) 1x1 Diamond Surface

By
G. BAMBAKIDIS and W.T. MC KEOWN[2]

The diamond (111) surface has been studied for several years, both for its intrinsic interest and also as a prototype of the tetrahedrally coordinated semiconductors so important as electronic materials.

Intrinsic diamond is an insulator with an indirect band gap of 5.4 eV /1/. For the clean, unrelaxed, unreconstructed (111) 1x1 surface, self-consistent pseudopotential calculations by Ihm et al. /2/ and LCAO calculations by Vanderbilt and Louie /3/ indicate the existence of a narrow band of partially occupied surface states lying within the gap. We report here the results of a self-consistent cluster calculation for this surface using the scattered-wave self-consistent-field method with local exchange (SCF-Xα-SW method) and a "muffin-tin" potential /4/. The results of this calculation agree well with the Xα-pseudopotential and LCAO results.

Fig. 1 shows the cluster model used in the calculation. The cluster in Fig. 1a consists of a carbon atom at the center of the tetrahedron formed by its four nearest neighbours in the diamond lattice. This is the smallest cluster having the full tetrahedral point-group symmetry T_d of diamond. The 12 dangling bonds which normally terminate on second-nearest neighbors are here tied off by hydrogen atoms centered at the second-nearest neighbor sites in the lattice. The electronic structure of the resulting C_5H_{12} cluster has been calculated by Watkins and Messmer /5/ using the muffin-tin SCF-Xα-SW method, and we have taken their results as representative of the diamond crystal. We model the (111) 1x1 surface by breaking the three C-H bonds perpendicular to a face of the tetrahedron, as shown in Fig. 1b, where the bottom face is chosen to represent the surface. The symmetry of the resulting C_5H_9 cluster is lowered to C_{3v}.

---

1) Dayton, Ohio 45435, USA.

2) Present address: AFWAL/AADO-2, WPAFB, Ohio 45433, USA.

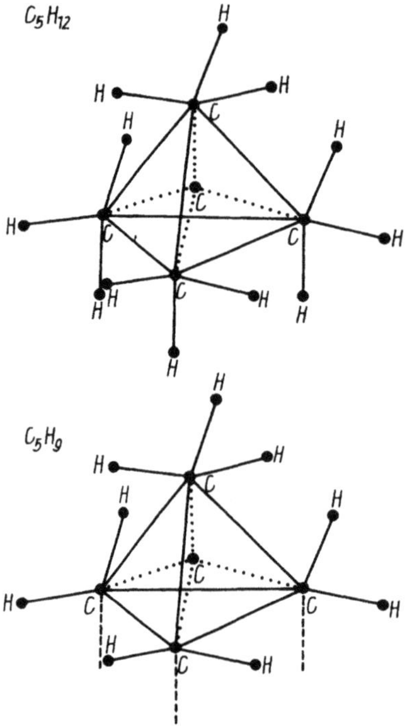

Fig.1

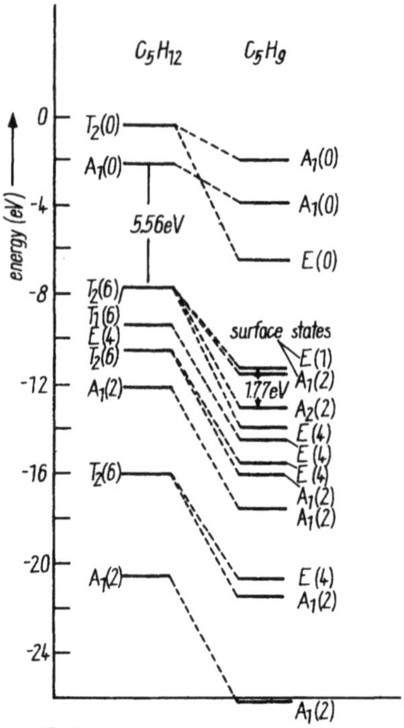

Fig.2

Fig. 1. The two clusters considered in this calculation. a) C_5H_{12}, representative of bulk diamond, b) C_5H_9, representative of the unreconstructed diamond (111) surface. The dashed lines indicate dangling bonds

Fig. 2. Energy level diagram for C_5H_{12} (from /5/) and C_5H_9, including representations and occupancies

We have calculated the electronic structure of the C_5H_9 cluster, using touching spheres and an exchange parameter α of 0.75982 for C and 0.97804 for H. Identical radii were used for all atomic spheres, based on the C-C nearest neighbor distance of 1.54 Å in diamond. No spin polarization was assumed. The results are shown on an energy level diagram in Fig. 2, together with the C_5H_{12} results from /5/. In the figure, the symmetry and occpancy are given for each level. It is evident that the cluster C_5H_{12}, corresponding to bulk diamond, has a well defined energy gap of 5.56 eV between occupied and empty states. This agrees well with the experimental gap energy. The six-fold degeneracy (three-fold orbital times two-fold spin) of the T_1 and T_2 states at the top of the valence band also agree with band-structure calculations and with experiment /6/. (Spin-orbit coupling has not been included in this calculation, but for each cluster it would produce a small splitting of the T-type states into two-fold degenerate A-type states and four-fold degenerate E states.) Relative to C_5H_{12}, the C_5H_9 levels are shifted downward in energy, and due to the lowered structural symmetry there is a splitting of the T-type states according to

$$T_1 \rightarrow A_2 + E \ ,$$
$$T_2 \rightarrow A_1 + E \ . \tag{1}$$

This splitting is based on the correlation table connecting the representations of the T_d group with its C_{3v} subgroup /7/. In particular, the T states at the top of the valence band split away and form a narrow group of partially occupied surface states of width 0.24 eV located in the lower half of the gap. The "Fermi level" in our cluster model, which is at -11.35 eV, lies 1.77 eV from the top of the valence band. These results agree quite well with the calculations of /2/, in which a band of partially filled surface states approximately 0.2 eV wide and 1.7 eV above the valence band is found, and with the results of /3/, in which a band 0.22 eV wide centered at 1.88 eV above the valence band is obtained. In /2/ a value of the work function Φ of about 7 eV for the clean 1x1 surface is also deduced, whereas our model would predict $\Phi \approx 11$ eV.

Experimentally, surface states on the (111) 1x1 diamond surface have not been observed. This surface is chemically and structurally unstable, the dangling bonds tending to be terminated by absorbed impurity atoms such as hydrogen /8/, while at high temperatures surface reconstruction occurs /9/. Nevertheless, the agreement of our cluster model with the pseudopotential and LCAO calculations for the infinite surface indicates that cluster models are capable of giving information relating to electronic structure, to the same level of reliability as models which treat an extended surface. The effects of surface relaxation may, of course, be significant, but have not been considered in any of the calculations mentioned here.

The authors acknowledge several useful discussions with Dr. S.V. Pepper and G.C. Abell.

References

/1/ C. KITTEL, Introduction to Solid State Physics, 4th ed., Wiley, New York 1971 (p. 364).

/2/ J. IHM, S.G. LOUIE, and M.L. COHEN, Phys. Rev. B 17, 769 (1978).

/3/ D. VANDERBILT and S.G. LOUIE, J. Vacuum Sci. Technol. B 1, 723 (1983).

/4/ J.C. SLATER and K.H. JOHNSON, Phys. Rev. B 5, 844 (1972).

/5/ G.D. WATKINS and R.P. MESSMER, Phys. Rev. Letters 32, 1244 (1974).

/6/ C. KITTEL, Quantum Theory of Solids, 1st ed., Wiley, New York 1971 (Chap. 14).

/7/ E.B. WILSON, Jr., J.C. DECIUS, and P.C. CROSS, Molecular Vibrations, McGraw-Hill, New York 1955 (p. 340).

/8/ G. VIDALI, M.W. COLE, and W.H. WEINBERG, Phys. Rev. Letters 51, 118 (1983).

/9/ S.V. PEPPER, Surface Sci. 123, 47 (1982).

(Received August 1, 1984)

phys. stat. sol. (b) 126, K71 (1984)
Subject classification: 13.4; 10.2; 22.6

Sektion Physik der Wilhelm-Pieck-Universität Rostock, Wissenschaftsbereich Isolatorphysik[1])

Cluster-Bethe-Lattice Approach to the Electronic Structure of a Twofold Coordinated Silicon Atom as a New Intrinsic Defect in v-SiO_2

By
V. BETHKENHAGEN and K. HÜBNER

Recently, Skuja et al. /1/ have related the 5.03 eV optical absorption band and the 4.3 eV photoluminescence to an intrinsic defect of vitreous (v) SiO_2 which has been discussed before only as a theoretical possibility /2/: the twofold coordinated silicon atom. Until now, the 4.3 eV luminescence has been supposed to be correlated with the 7.6 eV absorption /3/ or the 5.9 eV absorption /4/ and has been interpreted to be due to an oxygen vacancy /4/ or valence-alternation pair defects /3, 5/. The aim of this note is to investigate the electronic structure of a twofold coordinated silicon and to show that it is a good model to explain the correlation of the 5.03 eV absorption and the 4.3 eV luminescence as pointed out by Skuja et al. /1/. The approach used is the cluster-Bethe-lattice method /6/.

Fig. 1 shows the cluster which contains the defect, embedded into a Bethe lattice representing the non-crystalline SiO_2. The O-Si-O angle in the cluster is assumed to be the same as for an ideal SiO_4 tetrahedron, i.e. 109.5°. The local electronic densities of states (LDOS) of the cluster atoms and of the constituents of the SiO_2 Bethe lattice (Si-O-Si angle 144°) were calculated with the help of the Green

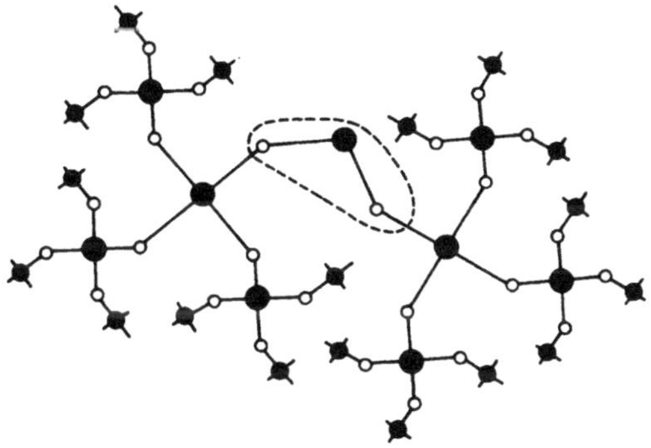

Fig. 1. Bethe-lattice representation of a twofold coordinated Si atom in v-SiO_2. The dashed curve encloses the defect cluster; ● silicon, o oxygen

1) Universitätsplatz 3, DDR-2500 Rostock, GDR.

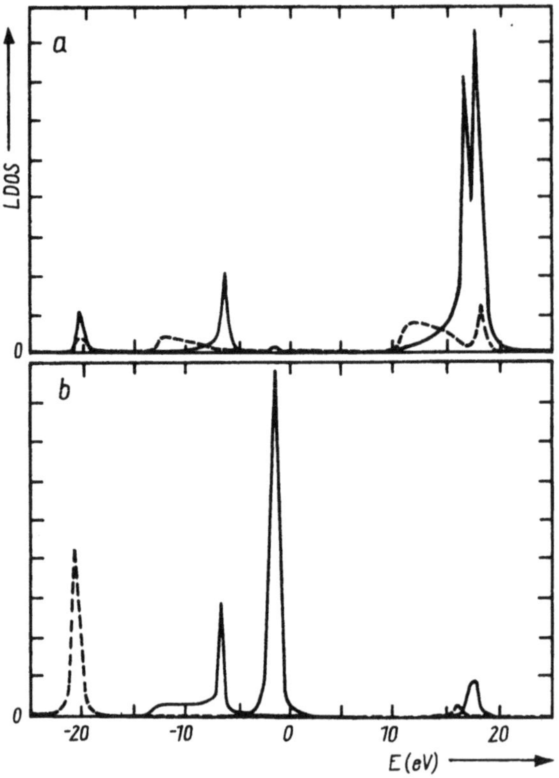

Fig. 2. Local electronic density of states (LDOS) of an a) Si atom in the SiO_2 Bethe lattice and b) O atom in the SiO_2 Bethe lattice (all in arbitrary units). The dashed and solid lines show the s and p parts of the LDOS, respectively

function technique, using a nearest-neighbour tight-binding Hamiltonian with Si3s, Si3p, O2s, and O2p atomic wave functions as a basis set. The tight-binding parameters for the cluster are assumed to be the same as for the Bethe lattice and have the following values (in eV): $(pp\sigma) = 5.9$, $(pp\pi) = -1.7$, $(sp\sigma) = -5.4$, $(ps\sigma) = -7.0$, and $(ss\sigma) = -3.5$; the orbital energies in eV are $\langle Os|H|Os \rangle = -16.4$, $\langle Op|H|Op \rangle = -1.3$, $\langle Sis|H|Sis \rangle = 5.9$, $\langle Sip|H|Sip \rangle = 12.2$ /7/.

The resulting electronic density of states of the SiO_2 Bethe lattice (Fig. 2) reveals a gap of about 10 eV between the oxygen related lone-pair band as the highest valence band and a Si3s like band as the deepest conduction band. The

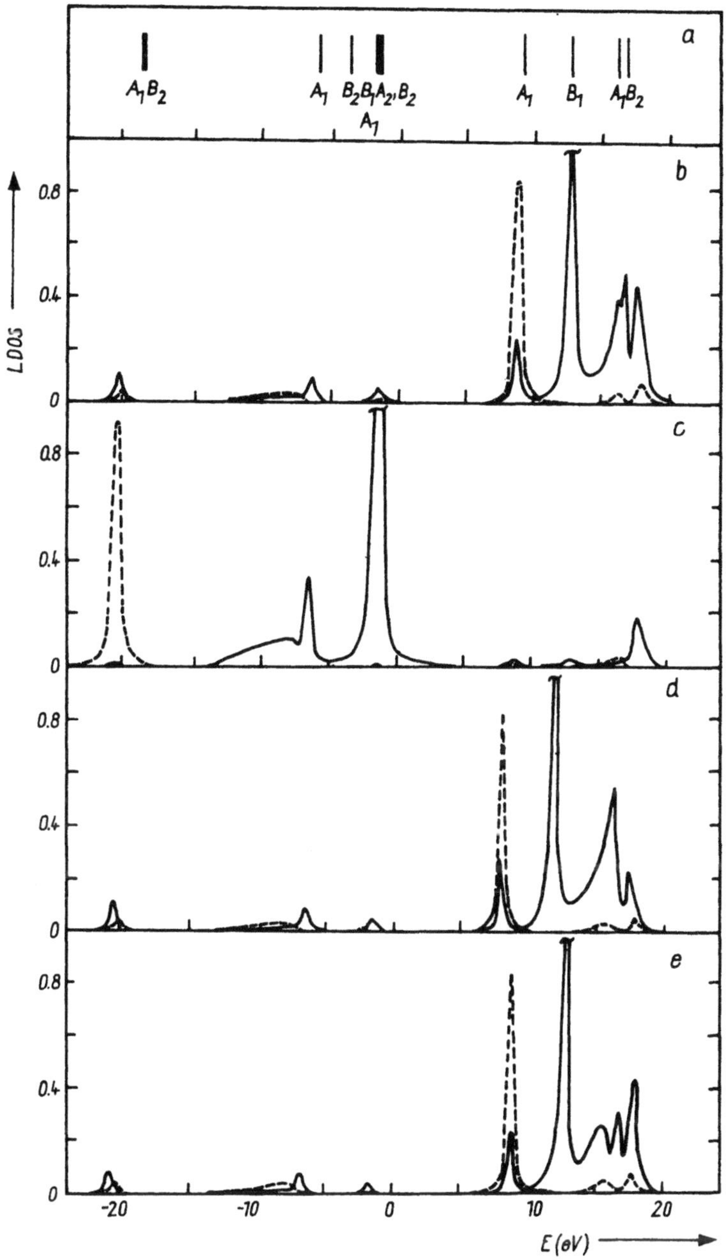

Fig. 3. a) Eigenvalues of an isolated SiO_2 molecule with symmetry classification (see text) and local electronic density of states (LDOS) of b) a twofold coordinated Si atom, c) two oxygen neighbours of the twofold coordinated Si atom, d) a twofold coordinated Si atom with a downward shift of the orbital energy of 1.0 eV, e) a twofold coordinated Si atom with an O-Si-O bond angle variation of 40° (all in arbitrary units). The dashed and solid lines show the s and p parts of the LDOS, respectively

LDOS of the twofold coordinated Si atom shows a sharp peak at 8.6 eV in the gap (Fig. 3b) with mainly Sis contributions and a Sip like peak at 12.6 eV in the conduction-band region of the SiO_2 Bethe lattice. The LDOS of its two equivalent oxygen neighbours (Fig. 3c) has weak resonance states at the energetic positions of these defect peaks.

Ciraci and Ercoc /2/ have published comparable results for a so-called oxygen divacancy in the β-cristobalite structure of SiO_2. They modelled the divacancy by a 48-atom supercell with periodic boundary conditions where oxygen atoms were removed from the defect sites. The electronic structure of such a supercell was calculated using a tight-binding Hamiltonian including also oxygen-oxygen interactions. Ciraci and Ercoc have found three states below or near above the conduction band edge. One of these states (the second) is localized at two threefold coordinated silicon atoms in the neighbourhood of the twofold coordinated silicon and, therefore, it does not appear in the LDOS of the defect cluster discussed in the present paper.

To interpret our results we have solved the eigenvalue problem of an SiO_2 molecule (which has C_{2v} symmetry) using the tight-binding parameters given above. We have not considered a cluster charge, since it does not essentially alter the symmetry properties of the cluster. The eigenvalues near the positions of the two defect peaks of the LDOS of the twofold coordinated Si atom correspond to eigenvectors which are of A_1 symmetry for the gap state and B_1 symmetry for the conduction band state (see Fig. 3a, b). This suggests to correlate these states with the optical absorption of 5.03 eV which was shown to be an A_1-to-B_1 transition /1/. It should be noted that the designation has been done following Skuja et al. /1/ assuming that a Si3p orbital perpendicular to the O-Si-O plane belongs to the B_1 representation. Due to the approximative character of Bethe-lattice calculations the discrepancy of the energetic distances of the calculated states to the experimentally observed energy of the optical transition can be considered to be not significant.

The charge transfer between the cluster and the surrounding Bethe lattice leads to a cluster population of $(SiO_2)^{-0.91}$, i.e. the cluster contains about 17 valence electrons. By filling up the discrete energy levels of the isolated SiO_2 molecule with 17 electrons we get an unpaired electron at the A_1 level near the gap state of the cluster LDOS. Therefore, in difference to Ciraci and Ercoc /2/, we conclude that this defect peak could be half filled in the ground state. This result allows us to explain the 5.03 eV excitation of the B_1-to-A_1 luminescence transition /1/ as an excitation from the defect state in the gap to its counterpart in the conduction band. Due to our investigations, it seems to be incorrect to assume an SiO_2^{2-} unit as a quasi-molecule representation of the twofold coordinated silicon (see /1/). Indeed, the calculated charge of the oxygen neighbours of the twofold coordinated silicon is 1.16 electrons per oxygen, but the silicon charge should be taken into account. We get for the silicon defect atom a deficit of 1.41 electrons.

In correspondence with LDOS calculations of other intrinsic defects of v-SiO_2 /5, 7/ we have modified the orbital energies of the Si defect atom. A downward shift of the orbital energy of 1.0 eV leads to an A_1 peak at 7.7 eV and a B_1 peak at 11.6 eV (Fig. 3d). We have also investigated variations of the O-Si-O angle of the cluster. Their influence on the LDOS of the defect atom is very small as shown in Fig. 3e for an angle of 149.5^o. Such a small influence of bond-angle variations on the LDOS was already pointed out by the authors also for other intrinsic defects of v-SiO_2 /7/.

In conclusion it was shown that the twofold coordinated Si atom in v-SiO_2 preduces a defect peak of A_1 symmetry in the gap and a peak of B_1 symmetry in the conduction band which could give rise for known optical transitions. Therefore, our calculations support the suggestions of Skuja et al. /1/ about the existence of twofold coordinated Si atoms as intrinsic defects of v-SiO_2. More direct experimental evidence is needed, however, to estimate also the concentration of these defects.

This work was performed with the support of a contract between the Wilhelm-Pieck University Rostock and the Latvian State University Riga. The authors are indebted to Dr. L.N. Skuja, Institute of Solid State Physics of the Latvian State University, for valuable discussions.

References

/1/ L.N. SKUJA, A.N. STRELETSKY, and A.B. PAKOVICH, Solid State Commun. 50, 1069 (1984).

/2/ S. CIRACI and S. ERCOC, Solid State Commun. 40, 801 (1981).

/3/ C.M. GEE and M. KASTNER, Phys. Rev. Letters 42, 1765 (1979).

/4/ C.E. JONES and D. EMBREE, J. appl. Phys. 47, 5365 (1976).

/5/ E.P. O'REILLY and J. ROBERTSON, Phys. Rev. B 27, 3780 (1983).
/6/ J.D. JOANNOPOULOS, Phys. Rev. B 16, 2764 (1977).
/7/ V. BETHKENHAGEN and K. HÜBNER, phys. stat. sol. (b) 125, K79 (1984).

(Received September 21, 1984)

phys. stat. sol. (b) 126, K77 (1984)
Subject classification: 13.4 and 19; 22.5.3

Institute of Physics, Technical University of Szczecin[1)]

Dynamic Relativistic Crystal Field Contributions to the Zero-Field Splitting of S-State $4f^7$ Ions

By
W. PASTUSIAK

In our previous papers /1, 2, 4 to 6/ we have considered static effects linear and quadratic in the crystal field based on the concept of the relativistic crystal field /3/. The comparison of the theoretical results with experimental data /2, 5, 6/ shows that the relativistic spin-lattice coupling is one of the most important mechanisms responsible for the zero-field splitting.

So far, we have not accounted for the fact that the molecular complex is not stationary, but exhibits vibrations around some average position. In this note we shall consider some aspects of dynamic effects in the zero-field splitting of the $4f^7$ ions.

As is well known, the operator describing the potential energy of an ion in a crystal field V_c can be expanded in a Taylor series about the equilibrium positions of the ligands. This series is normally written as a sum of terms with the first term (denoted by $V^{(0)}$) representing the crystal field operator corresponding to the equilibrium position and the next terms are denoted by $V^{(1)}$, $V^{(2)}$, etc. Consequently, the Hamiltonian of the paramagnetic ion in a crystal field may be written as follows:

$$H = H_o + V^{(0)} + V^{(1)} + V^{(2)} + \dots , \quad (1)$$

where H_o is the "free ion" Hamiltonian. Treating the interaction with the crystal field as a perturbation of intra-ionic interaction we obtain two mechanisms up to the second order of perturbation theory which lead to a temperature dependence. In the first of these one-phonon terms $V^{(1)}$ are taken in second order and can schematically be written as

$$\langle {}^8\mathcal{S}_{7/2}\, M | V^{(1)} | A\mathcal{S}\mathcal{L} J\, M' \rangle \langle A\mathcal{S}\mathcal{L} J\, M' | V^{(1)} | {}^8\mathcal{S}_{7/2}\, M'' \rangle , \quad (2)$$

where the ground $|{}^8\mathcal{S}_{7/2} M\rangle$ and excited $|A\mathcal{S}\mathcal{L} J M'\rangle$ states can be written in the form

---

1) al. Piastów 17, 70-310 Szczecin, Poland.

$$\left|{}^8\mathcal{S}_{7/2}\,M\right\rangle = \sum_{\bar{\alpha},\bar{S},\bar{L}} \mu({}^8\mathcal{S}_{7/2}\,\bar{\alpha}\bar{S}\,\bar{L})\left|\bar{\alpha}\bar{S}\,\bar{L}\,\tfrac{7}{2}\,M\right\rangle, \tag{3}$$

$$\left|A\mathcal{S}\ell J M'\right\rangle = \sum_{\alpha,S,L} \mu(A\mathcal{S}\ell J|\,\alpha S\,L)\left|\alpha S\,L\,J\,M'\right\rangle, \tag{4}$$

where $\mu({}^8\mathcal{S}_{7/2}|\bar{\alpha}\bar{S}\,\bar{L})$ and $\mu(A\mathcal{S}\ell J\,|\,\alpha S\,L)$ determine the contribution of the states with the well defined quantum numbers $\bar{L}$, $\bar{S}$ (L, S) to the ground and excited states, respectively; and A, $\bar{\alpha}$, α denote a set of additional quantum numbers indespensable for specifying a given state.

The second mechanism is a two-phonon first-order one:

$$\left\langle {}^8\mathcal{S}_{7/2}\,M|V^{(2)}|\,{}^8\mathcal{S}_{7/2}\,M'\right\rangle. \tag{5}$$

Because the second-order mechanism with respect to Q/R (Q vibrational amplitude and R is the metal-ligand distance) is of the same order as the first-order one and the first excited state lies over 30 000 cm^{-1} above the ground one, the one-phonon second-order mechanism seems to be negligible compared to the two-phonon first-order one. So far only the one-phonon second-order mechanism has been microscopically investigated /7, 8/. This mechanism yielded, however, negligible contributions. Therefore, let us discuss the two-phonon first-order mechanism.

Obviously, $V^{(2)}$ must transform according to the identity representation of a point symmetry group of the paramagnetic ion in the crystal field. The group theoretical analysis of $V^{(2)}$ leads to the conclusion that $V^{(2)}$ may be written as

$$V^{(2)} = \sum_{K=2,4,6}\;\sum_{\alpha,\beta,\Gamma,\gamma} A^{K}_{\beta\Gamma\gamma}\,C^{(K)}_{\alpha\Gamma\gamma}\,(Q_i\cdot Q_j)_{\beta\Gamma\gamma}\,, \tag{6}$$

where $A^{K}_{\beta\Gamma\gamma}$ is the K-th order two-phonon orbit-lattice coupling parameter, $C^{(K)}_{\alpha\Gamma\gamma}$ is a symmetry adapted linear combination of spherical harmonics normalized like the Legendre polynominals and transforming according to the γ-th component of the irreducible representation Γ, α classifies the repeating representation Γ in the decomposition of the 2K + 1 representation $D^{(K)}$ of the O_3 group in given point group, $(Q_i\cdot Q_j)_{\beta\Gamma\gamma}$ is a linear combination of products of normal coordinates transforming like $\Gamma\gamma$, and β is a set of symbols distinguishing different linear combinations of $Q_i'\cdot Q_j$ with the same transformation properties. We have assumed here that all irreducible representation are real.

Following Wybourne /3/ we can write the second-order orbit-lattice interaction $V^{(2)}$ in the relativistic approach:

$$V_{rel}^{(2)} = \sum_{k'=0,1} \sum_{k} \sum_{K=2,4,6} \sum_{\alpha,\beta,\Gamma,\gamma} A_{\alpha\beta\Gamma}^{K}(k', k) W_{\alpha\Gamma\gamma}^{(k'k)K}(Q_i \cdot Q_j)_{\beta\Gamma\gamma} \qquad (7)$$

where $A_{\alpha\beta\Gamma}^{K}(k', k)$ is the K-th order parameter describing two-phonon relativistic orbit-lattice coupling and can be expressed through the parameter $A_{\alpha\beta\Gamma}^{K}$ and the relativistic integrals $b_K(k', k)$ given by Wybourne /3/. $W_{\alpha\Gamma\gamma}^{(k'k)K}$ is a symmetry adapted linear combination of the components of the Judd tensor operators $W^{(k'k)K}$ of rank k' with respect to the spin angular momentum, of the rank k with respect to the orbital angular momentum, and rank K with respect to the total angular momentum.

The temperature dependence of the zero-field splitting can now be obtained if one writes the ground state as a product of both electronic and vibrational parts and then performs the thermal average procedure /9/.

On the other hand, the temperature dependence of the spin-Hamiltonian parameters can be described as

$$b_Q^K(T) = b_Q^K(RL) + C_Q^K f(T) \quad , \qquad (8)$$

where $b_Q^K(T)$ is the K-th rank spin-Hamiltonian parameter, $b_Q^K(RL)$ is the same parameter corresponding to the "rigid lattice" /10/, and C_Q^K is a temperature independent parameter describing the magnitude of the orbit-lattice interaction. f(T) determines the temperature dependence of the spin-Hamiltonian parameters.

The contribution resulting from the two-phonon first-order mechanism to the C_Q^K values can be achieved in the way presented in our previous papers. The calculation of this contribution reduces itself to the comparison of the matrix elements of the spin-Hamiltonian in the effective spin equal to 7/2 with the matrix elements of operators (6) or (7).

An analysis of these latter matrix elements leads to the conclusion that relativistic crystal field contributions cannot be neglected.

As an example let us consider the zero-field splitting of $4f^7$ ions in a cubic crystal field for eightfold coordination. We shall focus ourselves on doubly degenerate vibrations connected with the normal coordinates $Q_{E\theta}$ and $Q_{E\varepsilon}$ (irreducible representation and their rows are denoted like in Griffith /11/). The direct product of two representations $E_g \times E_g$ decomposes into the irreducible representations $A_{1g} + A_{2g} + E_g$. Considering only the second-order two-phonon spin-lattice interaction (K = 2) and $\Gamma = E_g$ we obtain the following forms for $V^{(2)}$: (i) in the non-relativistic approach

$$V^{(2)}(E_g \times E_g) = A^2_{E_g}\left[C^{(2)}_{E_g\Theta}(Q_{E_g i}\cdot Q_{E_g j})_{E_g\Theta} - C^{(2)}_{E_g\varepsilon}(Q_{E_g i}\cdot Q_{E_g j})_{E_g\varepsilon}\right], \quad (9)$$

where

$$A^2_{E_g} = e\cdot e_{eff}\langle r^2\rangle/18\sqrt{3}\, a^5 ,$$

$$C^{(2)}_{E_g\Theta} = C^{(2)}_0 , \quad (10)$$

$$C^{(2)}_{E_g\varepsilon} = (C^2_2 + C^2_{-2})/\sqrt{2} , \quad (11)$$

$$(Q_{E_g i}\cdot Q_{E_g j})_{E_g\Theta} = (Q_{E_g\Theta})^2 + (Q_{E_g\varepsilon})^2 , \quad (12)$$

$$(Q_{E_g i}\cdot Q_{E_g j})_{E_g\varepsilon} = Q_{E_g\Theta}\cdot Q_{E_g\varepsilon} + Q_{E_g\varepsilon}\cdot Q_{E_g\Theta} ; \quad (13)$$

(ii) in the relativistic approach

$$V^{(2)}_{rel}(E_g \times E_g) = A^2_{E_g\Theta}(1, 1)\Big[W^{(11)2}_{E_g\Theta} \times$$

$$\times (Q_{E_g i}\cdot Q_{E_g j})_{E_g\Theta} - W^{(11)2}_{E_g\varepsilon}(Q_{E_g i}\cdot Q_{E_g j})_{E_g\varepsilon}\Big], \quad (14)$$

where

$$A^2_{E_g}(1, 1) = e\cdot e_{eff} b_2(1, 1)/18\sqrt{3}\, a^5 ,$$

whereas $W^{(11)2}_{E_g i}$ ($i = \Theta, \varepsilon$) can be obtained by replacing $C^{(2)}_Q$ by appropriate $W^{(11)2}_Q$ in (10), (11).

We neglected here the term with spin independent operator $W^{(02)2}$.

Because the $W^{(11)2}$ operator has nonzero matrix elements between the $^8S_{7/2}$ and $^6P_{7/2}$ states and $\mu(^8\mathcal{S}_{7/2}|^6P_{7/2})\cdot\mu(^8\mathcal{S}_{7/2}|^8S_{7/2}) \approx 10^{-1}$, whereas the $C^{(2)}$ operator has nonzero matrix elements between $^6P_{7/2}$ and $^6D_{7/2}$ and $\mu(^8\mathcal{S}_{7/2}|^6P_{7/2})\mu(^8\mathcal{S}_{7/2}|^6D_{7/2}) \approx 10^{-3}$, we conclude that in order to compare the contributions resulting from relativistic and non-relativistic mechanisms it is sufficient to compare the quantities $10^{-1}\, e\cdot e_{eff}\, b_2(1, 1)/18\sqrt{3}\, a^5$ and

$10^{-3} e \cdot e_{ff} \langle r^2 \rangle / 18\sqrt{3}\, a^5$. Because $b_2(1, 1) \approx 10^{-2} \langle r^2 \rangle$ the contribution resulting from the relativistic mechanism cannot be neglected in the temperature dependence of the $4f^7(^8S_{7/2})$ ion electrons coupling with the crystal lattice.

Finally, we would like to point out that the temperature dependence of the spin-Hamiltonian parameters as interpreted microscopically in this note is predominantly caused by the two-phonon first-order mechanism with relativistic contributions being quite important.

I thank Dr. J. Kuriata for suggestion of the problem and for helpful discussions concerning this work.

References

/1/ W. PASTUSIAK and J. KURIATA, phys. stat. sol. (b) 111, K147 (1982).
/2/ W. PASTUSIAK, phys. stat. sol. (b) 116, 63 (1983).
/3/ B.G. WYBOURNE, J. chem. Phys. 43, 4506 (1965).
/4/ W. PASTUSIAK, phys. stat. sol. (b) 120, K151 (1983).
/5/ W. PASTUSIAK and J. KURIATA, Acta Phys. Polon., to be published.
/6/ J. KURIATA and W. PASTUSIAK, Acta Phys. Polon., to be published.
/7/ T.J. MENNE, J. Phys. Chem. Solids 28, 1629 (1967).
/8/ CHAO-YUAN HUANG, Phys. Rev. 159, 683 (1967).
/9/ G. PFISTER, W. DREYBRODT, and W. ASSMUS, phys. stat. sol. 36, 351 (1969).
/10/ S.B. OSEROFF and R. CALVO, Phys. Rev. B 5, 2474 (1972).
/11/ J.S. GRIFFITH, The Theory of Transition-Metal Ions, University Press, Cambridge 1961.

(Received May 30, 1984)

phys. stat. sol. (b) 126, K83 (1984)
Subject classification: 10.2 and 14.3; 22.1.1

Institute of Solid State Physics, Academy of Sciences of the USSR, Chernogolovka[1)]

Electron Scattering by a Charged Dislocation

By
R. A. VARDANIAN and G.G. KIRAKOSIAN

It is well known that edge dislocations in an electronic semiconductor show acceptor action and have a negative charge. The electrostatic potential in cylindrical coordinates (ϱ, φ, z) with the z-axis directed along the dislocation has for high and low temperatures T the form /1/

$$U(\varrho) = T \begin{cases} 2\alpha K_0(\varrho/\varrho_D) \; ; & \alpha \ll 1 \\ 1 - \frac{1}{4}\frac{R^2 - \varrho^2}{\varrho_D^2} - 2\alpha \ln(\varrho/R) \; ; & a \leqq \varrho \leqq R \quad \alpha \gg 1 \\ (R/\varrho)^{1/2} \exp\left[(R - \varrho)/\varrho_D\right] \; ; & \varrho \geqq R \end{cases} \quad (1)$$

where $\alpha = e^2 f(T)/\varkappa a T$, $\varkappa$ is the dielectric constant, a is the distance between the dangling bonds, f(T) is the occupation coefficient of the dislocation by electrons, $K_0(x)$ is the Macdonald function, $R = \left[f(T)/\pi a n_d\right]^{1/2}$ is the radius of the Read cylinder, and $\varrho_D = (\varkappa T/4\pi n_d e^2)^{1/2}$ is the Debye screening radius defined by the donor concentration n_d (we put $k_B = \hbar = 1$). Equations (1) are valid under the assumption that T is high enough so that the donors are completely ionized.

The dislocation influence on the electron mobility in n-type semiconductors was theoretically investigated in /2 to 4/ using the Born approximation because the scattering problem with potential (1) cannot be solved exactly. The condition for the applicability of the Born approximation is /2/

$$\delta \equiv \left| 2\pi^2 \mu \int_0^\infty H_0^{(1)}(k\varrho) I_0(k\varrho) U(\varrho) \varrho \, d\varrho \right| \ll 1 , \quad (2)$$

where $H_0^{(1)}(x)$ is the Hankel function of the first kind, $I_0(x)$ is the Bessel function, $k = (2\mu\varepsilon)^{1/2}$ is the modulus of the two-dimensional wave vector in the plane (ϱ, φ), and μ is the electron effective mass. It is easy to show by

---

1) 142 432 Chernogolovka, Moscow District, USSR.

substituting (1) into (2) that practically within the whole temperature range where the acceptor action of the dislocation is important, $f(T) \approx 10^{-2}$ to 10^{-1} /5/, the scattering has a quasi-classical character since the value

$$\delta = \pi k \frac{\alpha T}{\varepsilon} \begin{cases} \varrho_D ; & \alpha \ll 1 \\ R ; & \alpha \gg 1 \end{cases} \tag{3}$$

turns out to be sufficiently larger than 1.

For not very large dislocation concentrations when their contributions to the scattering are additive, the transport scattering time is given by

$$\tau^{-1} = N_D \langle V \rangle \varrho_t , \tag{4}$$

where $\langle V \rangle$ is the two-dimensional mean velocity of the carrier in the band, N_D is the dislocation concentration, and

$$\varrho_t = \int_0^{2\pi} (1 - \cos\varphi) \, |F(\varphi)|^2 \, d\varphi \tag{5}$$

is the scattering radius. The electron scattering amplitude by the angle φ for a two-dimensional problem is given by /6/

$$F(\varphi) = (2\pi i k)^{-1/2} \sum_{m=0}^{\infty} \varepsilon_m \left[\exp(2i\delta_m) - 1\right] \cos(m\varphi) ,$$
$$\varepsilon_0 = 1 ; \quad \varepsilon_m = 2 , \quad m = 1, 2, \ldots , \tag{6}$$

where m is the magnetic quantum number and δ_m is the scattering phase. One can obtain the expression for the scattering radius, substituting (6) into (5) and integrating over φ, in the form

$$\varrho_t = \frac{4}{k} \sum_{m=0}^{\infty} \sin^2(\delta_{m+1} - \delta_m) . \tag{7}$$

Because in the quasi-classical approximation only large values of m are of vital importance one can use for the scattering phase /7/

$$\delta_m = -\frac{\mu}{k} \int_{m/k}^{\infty} \frac{U(\varrho) \, d\varrho}{(1 - m^2/k^2\varrho^2)^{1/2}} \tag{8}$$

and pass in (7) from summation to integration over m putting $\delta_{m+1} - \delta_m \approx$ $\approx d\delta_m/dm$

$$\varrho_t = \frac{4}{k} \int_0^{\infty} \sin^2(d\delta_m/dm) dm \tag{9}$$

substituting (1) into (8) and performing simple but lengthy calculations we get

$$\left|\delta_m\right| = \begin{cases} \frac{\alpha T}{\varepsilon} k R \left[(1 - m^2/k^2 R^2)^{1/2} - \frac{m}{kR} \arccos\left(\frac{m}{kR}\right)\right]; & m < kR \\ & \alpha \gg 1 \\ \frac{\sqrt{\pi}}{4} \frac{T}{\varepsilon} \alpha^{-1/4} kR \exp\left[(kR - m)/k\varrho_D\right]; & m > kR \\ \frac{\pi}{2} \frac{\alpha T}{\varepsilon} k\varrho_D \exp(-m/k\varrho_D) ; & \alpha \ll 1 . \end{cases} \quad (10)$$

After integrating (9) over m and averaging the result obtained with the equilibrium Boltzmann distribution function for band electrons we obtain finally for the transport scattering time

$$\tau^{-1} = N_D \langle v \rangle \begin{cases} 2 R(1 + \ln\alpha/8\sqrt{\alpha}) ; \alpha \gg 1 , \\ \frac{\pi}{2} \varrho_D \alpha \quad ; \alpha \ll 1 . \end{cases} \quad (11)$$

The last expression shows that the scattering radius is a monotonous function in the whole temperature range, at low temperatures determined by the Read radius and weakly depending on temperature $\sim\sqrt{f(T)}$. At high temperatures τ^{-1} depends on the screening radius ϱ_D, the dislocation occupation, and the carrier energy in the band.

The scattering rate by acoustic phonons is

$$\tau_{ph}^{-1} \approx \frac{E_c^2 \mu^{3/2} T^{3/2}}{\varrho_L s^2} , \quad (12)$$

where E_c is the deformation potential constant, s is the sound velocity, and ϱ_L the crystal density. The comparison of the two latter equations shows that for different semiconductor parameters one or another scattering mechanism is predominant. For example in germanium with $\mu \approx 10^{-28}$ g, $E_c \approx 10$ eV, $\varkappa = 16$, $s = 5\text{x}10^5$ cm/s, $\varrho_L \approx 10$ g/cm^3, $n_d \approx 10^{14}$ cm^{-3}, even at low concentrations of dislocations ($N_D \approx 10^7$ cm^{-2}) the contributions of these two channels are comparable already at room temperatures when $f(T) \approx 10^{-2}$. At lower temperatures scattering by the dislocation electrostatic potential begins to predominate.

The authors are grateful to V. Ya. Kravchenko, Yu. A. Ossipyan, and S. A. Shevchenko for useful discussion of the results and the experimental situations.

References

/1/ R. A. VARDANIAN, Zh. eksper. teor. Fiz. 73, 2313 (1977).

/2/ E.M. KUZNETSOVA, Fiz. tverd. Tela 3, 1987 (1961).

/3/ B. PÖDÖR, phys. stat. sol. 16, K137 (1966).

/4/ B. PÖDÖR, Acta Phys. Acad. Sci. Hungaria 23, 393 (1967).

/5/ A.I. KOLYUBAKIN, YU.A. OSSIPYAN, and S. A. SHEVCHENKO, Zh. eksper. teor. Fiz. 77, 975 (1979).

/6/ V.V. BABIKOV, Metod fazovikh funktsii v kvantovoi mekhanike, Izd. Nauka, Moscow 1976.

/7/ L.D. LANDAU and E.M. LIFSHITS, Quantum Mechanics, Non-Relativistic Theory, 2nd ed., Pergamon, Oxford 1965.

(Received July 19, 1984)

phys. stat. sol. (b) 126, K87 (1984)
Subject classification: 7 and 14.3; 14.1; 22.7

Institute of Physics, Czechoslovak Ademy of Sciences, Prague[1)]

Semiconductor-Semimetal Transition in Bi-Sb Alloys Observed in Ultrasound Attenuation

By
J. DOMINEC

The semimetallic nature of bismuth is caused by the small energy overlap between the states of the fifth and the sixth Brillouin zones. This overlap decreases gradually by adding antimony to the alloy and vanishes at the concentration of 7% of antimony in the alloy /1, 2/. The alloys containing 7 to 22% of antimony are semiconductive.

In a magnetic field, the electrons occupy discrete energy levels. The energy of an electron is /1/

$$E = \frac{|e|\hbar B}{2}\left(\frac{2n+1}{m_o} \pm \frac{1}{m_s}\right) + \frac{p^2}{2m_{\parallel}} , \qquad (1)$$

where B is the magnetic field strength, m_o and m_s are the orbital and the spin cyclotron masses, p is the momentum along the magnetic field, $m_{\parallel}$ is the longitudinal orbital mass along the magnetic field. The relative motion of the levels of different zones in a magnetic field depends on the relative magnitudes m_o and m_s of these zones. We will denote the states of the fifth zone by the index h, the ones of the sixth zone by the index e. If

$$A = 1/m_o^e - 1/m_s^e + 1/m_o^h - 1/m_s^h < 0 , \qquad (2)$$

then the gap between the fifth and the sixth zones decreases /1/ in increasing magnetic field. This situation occurs in semiconductive Bi-Sb alloys when the magnetic field is parallel to the trigonal axis. The highest level of the fifth zone and the lowest level of the sixth zone become close and eventually cross /1/. This cemiconductor-metal transition causes changes of the ultrasound attenuation /4/.

The samples were cut from rods grown by the zone melting method. The concentration of antimony (determined from their specific weight) was 7.5, 8.5, and 12.8%.

---

1) Na Slovance 2, 18040 Praha 8, Czechoslovakia.

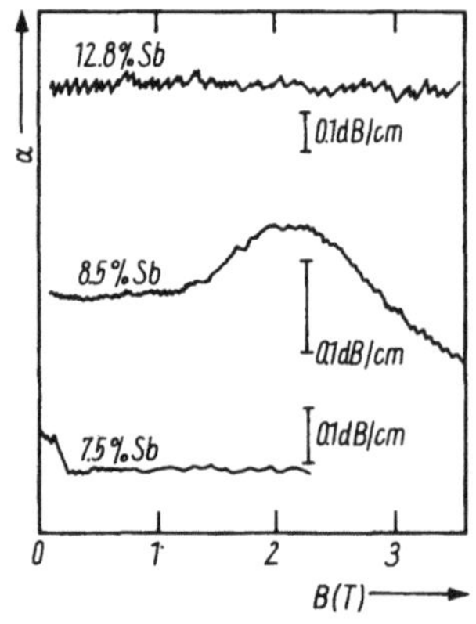

Fig. 1. Ultrasound attenuation in semiconductive Bi-Sb alloys. Frequency of ultrasound 190 MHz, temperature 4.2 K

The magnetic field was parallel to the trigonal axis. The measurement was performed at liquid helium temperature.

The ultrasound attenuation was measured by the pulse method. As the basal attenuation of the samples was high, $LiNbO_3$ transducers were used to obtain a strong enough acoustic signal.

To understand the measured attenuation in the three samples (Fig. 1), let us have a look, how various quantities depend on the magnetic field strength - the schematic view is depicted in Fig. 2. We have drawn the energy of sound phonons ① (having frequency 190 to 510 MHz), the energy of thermal phonons ② (for temperature 2 to 4.2 K), the difference between levels with $\Delta n = 1$ but with the same spin ③, and the energy gap between the states of the fifth and the sixth zones ④. The area ③ has been calculated with the help of data /3/ for the levels of the sixth zone (the fifth zone data are not available), the low limit applies to 12.8% Sb, the high limit to 7.5% Sb. The area ④ has been estimated with the help of the figure of Brandt et al. /1, 2/. It applies to 7 to 8% Sb. The broadening of the levels due to various defects is expected to be of the same order as the influence of thermal phonons (a guess based on the similarity with semimetallic Bi-Sb samples - the results will be published elsewhere). It is probable that there exist also impurity levels in the samples, but their behaviour in magnetic field is not known enough.

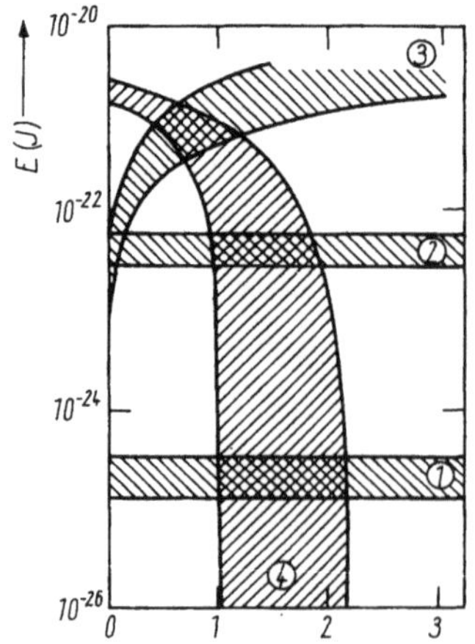

Fig. 2. The dependence of various quantities on magnetic field strength under the condition of the experiment: ① energy of ultrasound phonons, ② energy of thermal phonons, ③ difference between neighbour Landau levels, ④ gap between the valence and the conduction band; for details see text

The changes of the attenuation (Fig. 1), dependent on magnetic field are caused by the interaction between ultrasound phonons and electrons. The excitation of electrons is liable to the law of conservation of energy; therefore the excitation of an electron into another Landau level is improbable. The excitations occur within the highest occupied level. The attenuation goes up as the number of the "accessible" states increases. Therefore it will increase around the point of the transition to the metallic state.

The measured dependence of the sound attenuation shows a hump near 0.2 T (for 7.5% Sb) respectively 2.3 T (for 8.5% Sb). The attenuation change in the sample containing 7.5% Sb can be explained either by the proposed semiconductor-metal transition or by quantum oscillations, analogous to the Shubnikov-de Haas ones. The attenuation in the sample with 8.5% Sb can be explained only by the transition, as the Subnikov-de Haas oscillations are periodic in the reverse magnetic field. Therefore, it seems likely, that also the changes in the sample with 7.5% Sb are caused by the semiconductor-metal transition.

The attenuation in the sample with 12.8% Sb does not change in the investigated field range - it reflects the fact that the transition point lies outside.

The frequency dependence of the attenuation in the sample containing 8.5% Sb is shown in Fig. 3. The hump near 2.3 T grows bigger on increasing the frequency of ultrasound. It reflects the fact that the higher the frequency of phonons, the more electrons are excited by the sound wave.

The observed dependence of the attenuation is in agreement with the theory /4/ and with the work of Buga et al. /5/, in which the transition is achieved via the change of stress.

The author wishes to thank Dr. W. Kraak from Humboldt University Berlin for providing the samples.

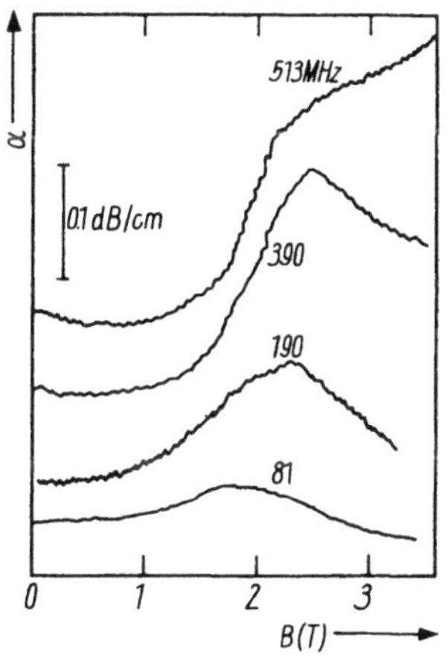

Fig. 3. Ultrasound attenuation in a sample containing 8.5% Sb; temperature 4.2 K

References

/1/ N.B. BRANDT and E.A. SVISTOVA, J. low.-Temp. Phys. 2, 1 (1970).

/2/ N.B. BRANDT, S.M. CHUDINOV, and V.G. KARABAEV, Zh. eksper. teor. Fiz. 70, 2296 (1976).

/3/ G. OELGART and R. HERRMANN, phys. stat. sol. (b) 75, 189 (1976).
H. KRÜGER, R. HERRMANN, B. FELLMUTH, and W. KRAAK, phys. stat. sol. (b) 93, 661 (1979).
N.A. REDKO, V.I. POLSHIN, V.V. KOSAREV, and G.A. IVANOV, Fiz. tverd. Tela 25, 3138 (1983).

/4/ V.N. DAVYDOV and M.I. KAGANOV, Zh. eksper. teor. Fiz. 67, 1491 (1974).

/5/ S.G. BUGA, L.K. ZAREMBO, A.I. KOROBOV, V.A. KULBACHINSKII, and V.V. RZHEVSKII, Fiz. tverd. Tela 26, 293 (1984).

(Received July 20, 1984)

phys. stat. sol. (b) 126, K91 (1984)

Subject classification: 14.1 and 15; 21

Sektion Physik der Technischen Universität Dresden,
Wissenschaftsbereich Theoretische Physik[1)]

The Influence of Electron-Phonon Interaction on the Thermoelectric Power at High Temperatures

By

E. CHILLA and W. SCHILLER

The temperature dependence of the electrical resistivity ϱ of some transition metals and A-15 compounds shows a remarkable deviation from the usual Bloch-Grüneisen law ($\varrho \sim T$) at high temperatures with the tendency to saturation /1, 2/. The anomalous temperature dependence of the resistivity gives rise to the question about the temperature dependence of other transport parameters. The aim of this note is to discuss the consequences of deviations from linearity (DFL) in the electrical resistivity (saturation phenomenon) on the thermoelectric power (TEP).

We start from a resistivity theory based on force-force correlation functions, which describe the deviations from linearity

i) via multiple scattering corrections (MS) /3/,

ii) via a self-consistent calculation of the electrical resistivity (SC) /4/. In both cases we use the following single-band Hamiltonian:

$$H = \sum_{\vec{k}} \varepsilon_{\vec{k}} c^+_{\vec{k}} c_{\vec{k}} + \sum_{\vec{q}} \hbar\omega_{\vec{q}} b^+_{\vec{q}} b_{\vec{q}} + \frac{2c_{ep}}{\sqrt{\Omega}} \sum_{\vec{k},\vec{q}} F_{\vec{q}} c^+_{\vec{k}+\vec{q}} c_{\vec{k}} (b_{\vec{q}} - b^+_{-\vec{q}}) , \qquad (1)$$

where all symbols have their usual meaning (cf. /3/).

In the case (i) the high temperature resistivity ϱ_{MS} ($T > \Theta_D$, Θ_D Debye temperature) can be written as /3/

$$\varrho_{MS} = \tilde{A} \frac{T}{\Theta_D} \int_0^{q_D} dq \int_{-1}^{+1} dz \frac{q^4}{k_F^5} \frac{\mathrm{Im}\Sigma_{MS}}{\left[\frac{\hbar^2 q}{m^*}(z k_F + \frac{q}{2}) - \mathrm{Re}\Sigma_{MS}\right]^2 + \left[\mathrm{Im}\Sigma_{MS}\right]^2} \qquad (2)$$

where $\tilde{A} = (3c_{ep}^2 m^* q_D)/(2e^2\hbar)$, $z = \cos\vartheta$ (ϑ angle between $\vec{k}_F$ and $\vec{q}$), q_D and k_F are the Debye and the Fermi wave numbers. The self-energy $\Sigma_{MS}(k_F, q, z)$ due to electron-phonon interaction is temperature dependent and describes the influence of multiple scattering corrections yielding DFL in the resistivity.

---

1) Mommsenstr. 13, DDR-8027 Dresden, GDR.

In /3/ the quasi-elastic approximation is used for the evaluation of the self-energy, which consists in the neglect of the phonon energies $\hbar\omega_q$ in the energy differences of the scattering processes. Furthermore, in this approximation the scattering takes place on the Fermi surface only and for a spherical one there holds $|\vec{k}_F + \vec{q}| \approx k_F$, where q is the momentum of the scattered phonon. In this approach the self-energy (cf. /3/) has the following simple structure:

$$\mathrm{Re}\Sigma_{MS} = 0 \;,$$

$$\mathrm{Im}\Sigma_{MS} = AT\left\{\left[\left(\frac{2k_F}{q_o}\right)^2 - 1\right]\Theta(z_o - z) + 1\right\}, \tag{3}$$

where

$$z_o = \frac{q_D^2 - 2k_F q_D}{2k_F q} - \frac{q}{2k_F} \;, \qquad A = \frac{c_{ep}^2 m^* q_D^3}{2\pi\hbar^2 \Theta_D k_F} \;,$$

and

$$\Theta(z_o - z) = \begin{cases} 1 & \text{for } z_o \geqq z \\ 0 & \phantom{\text{for }} z_o < z \end{cases} .$$

In the case (ii) the high temperature resistivity ϱ_{SC} has the following form /4/:

$$\varrho_{SC} = \frac{aT}{1 + bT} \tag{4}$$

with

$$a = \frac{3\pi m^{*2} c_{ep}^2 q_D^4 k_B}{8 e^2 \hbar^4 k_F^6 V_o}$$

and

$$b = \frac{4m^{*2} c_{ep}^2 k_B}{\pi^2 \hbar^5 k_F V_o} \left| P\left(\frac{q_D}{2k_F}\right)\right| \;, \qquad q_D < 2k_F \;,$$

where $P(x) = 2x + \ln|(1 + x)/(1 - x)|$. In deriving (4), the imaginary part of the self-energy at the Fermi energy is determined self-consistently

$$\mathrm{Im}\Sigma_{SC} = \frac{e^2 \hbar k_F^3}{6 m^* \pi^2} \frac{aT}{1 + bT} \;. \tag{5}$$

The real part of the self energy does not influence the electrical resistivity.

The thermopower can be calculated by using Mott's formula /5/, the validity of which is shown for independent electrons interacting with adiabatic phonons for arbitrary interaction strength /6/. In both cases the self-energies

(3), (5) are calculated in the limit $\omega = i\eta \to 0$ (cf. /3, 4/), corresponding to the static or adiabatic phonon approximation /7/, so the application of Mott's formula is possible. In the case of the free electron dispersion relation this formula has the following structure:

$$S = \frac{\pi^2 k_B^2 T}{6 |e| \varepsilon_F} \frac{\partial \varrho}{\partial \ln k_F} . \tag{6}$$

In the case (i) we find for the thermoelectric power

$$S_{MS} = -\frac{\pi^2}{3} \frac{k_B^2 T}{|e| \varepsilon_F} \left\{ 5 + \frac{\tilde{A}}{k_F^4 \varrho_{MS}} \frac{T}{\Theta_D} \int_0^{q_D} dq \int_{-1}^{+1} dz\, q^4 I_{MS} \right\} \tag{7}$$

with

$$I_{MS} = L \left\{ 2L \frac{d}{dk_F} \mathrm{Im}\Sigma_{MS} - \frac{d}{dk_F} \ln \mathrm{Im}\Sigma_{MS} + \frac{2L}{\mathrm{Im}\Sigma_{MS}} \left(\frac{\hbar^2 q}{m^*} \right)^2 z k_F \left(z + \frac{q}{2k_F} \right) \right\} \tag{8}$$

$$L = \frac{\mathrm{Im}\Sigma_{MS}}{\left[\frac{\hbar^2 q}{m^*} (z k_F + \frac{q}{2}) \right]^2 + \left[\mathrm{Im}\Sigma_{MS} \right]^2} .$$

The derivative of the imaginary part has the following form:

$$\frac{\partial}{\partial k_F} \mathrm{Im}\Sigma_{MS} = A \frac{T}{\varepsilon_F} \left\{ -1 + \left[\left(\frac{2k_F}{q_D} \right)^2 + 1 \right] \Theta(z_o - z) + \left[\left(\frac{2k_F}{q_D} \right)^2 - 1 \right] \times \right.$$
$$\left. \times \frac{q^2 - q_D^2}{2 q k_F} \delta(z - z_o) \right\} . \tag{9}$$

In the case (ii) we find for the thermoelectric power

$$S_{SC} = -\frac{\pi^2 k_B^2}{6 |e| \varepsilon_F} T \left\{ \frac{6 + b \left[5 - M \left(\frac{q_D}{2k_F} \right) \right]}{1 + bT} \right\} \tag{10}$$

with $M(x) = 2H(x)/ |P(x)|$ and $H(x) = x^3/(1 - x^2)$.

In both cases the resistivity formula contains the following parameters: $\alpha = (c_{ep} m^*/c_{ep}^o m)^2$ describing the strength of the electron-phonon coupling with $(c_{ep}^o)^2 = 1.5 \times 10^{-76}$ $(Ws)^2 m^4$ /8/, q_D, k_F, and the sound velocity v_o. Furthermore, the thermoelectric power depends on the effective mass m^*. Evaluating resistivity and thermoelectric power we have assumed one electron per lattice site (cf. /3/). The sign of TEP is dependent on the sign of the charge

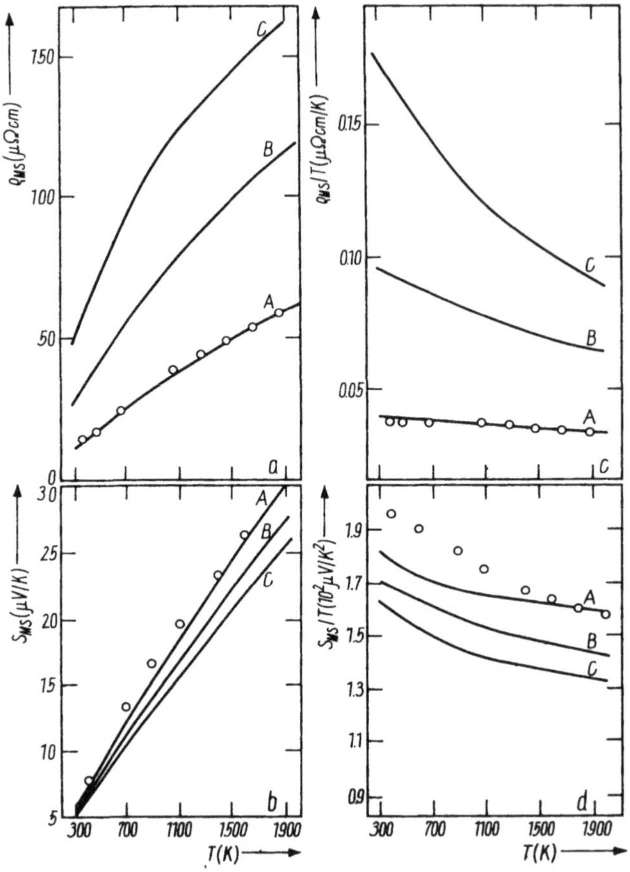

Fig. 1. a) Resistivity ϱ_{MS}, b) thermoelectric power S_{MS}, c) resistivity divided by temperature ϱ_{MS}/T, and d) thermoelectric power divided by temperature S_{MS}/T in dependence on temperature T; Θ_D = 240 K, m^* = 1.5 m, v_o = = 2000 m/s, and with α = 0.09 (A), 0.225 (B), 0.45 (C). Circles are measurements for Pt /10, 11/

carriers (electrons or holes). By variation of α we can study the influence of the electron-phonon scattering strength on the transport quantities.

In the weak scattering limit the expressions for the resistivity (2), (4) and the thermoelectric power (7), (10) yield the expected results well-known from Boltzmann transport theory

$$\varrho = \frac{3\pi c_{ep}^2 q_D^5 m^{*2}}{8 e^2 k_F^6 \hbar^3 \Theta_D} T \quad , \qquad S = -\frac{\pi^2 k_B^2}{|e| \varepsilon_F} T \quad , \tag{11}$$

where both quantities show a linear temperature dependence for $T > \Theta_D$. In Fig. 1 resistivity (2) and TEP (7) are calculated for Θ_D = 240 K, v_o = = 2 km/s, m^* = 1.5, and three different values for the electron-phonon coupling strength α. Both resistivity (Fig. 1a) and TEP (Fig. 1b) show deviations from the linear temperature dependence usually expected, but for the resistivity the deviations are much stronger than for the TEP (cf. Fig. 1c, d). In Fig. 1a, b the parameters are chosen in order to try a comparison with experimental results for systems, where the resistivity shows the DFL and also the thermopower is known from experiments. The calculated values (curves A in Fig. 1a, b) agree qualitatively with measurements for Pt (Θ_D = 240 K /12/) /10, 11/, (an analogous picture can be found for Pd /9, 11/ and ordered Ni-P alloys /5/). Increasing the coupling strength α the DFL are enhanced. Furthermore, with increasing α the magnitude of the resistivity increases also, but the magnitude of the TEP decreases (cf.

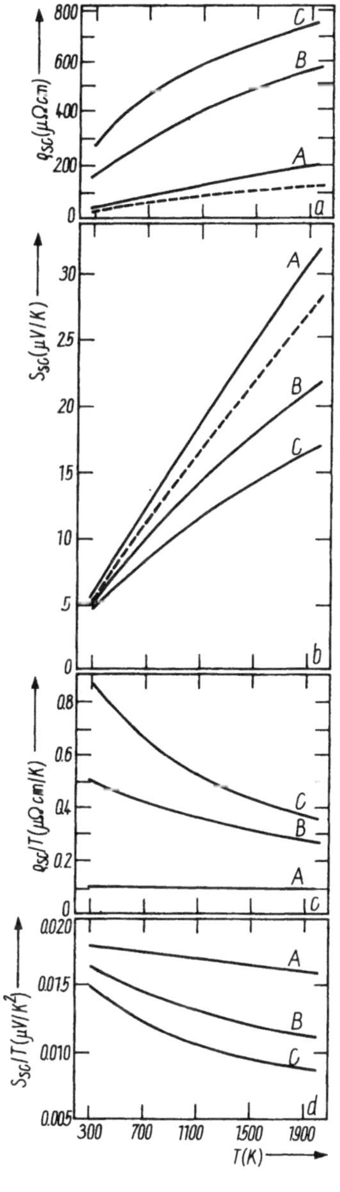

Fig. 2. a) Resistivity ϱ_{SC}, b) thermoelectric power S_{SC}, c) resistivity divided by temperature ϱ_{SC}/T, and d) thermoelectric power divided by temperature S_{SC}/T in dependence on temperature T; Θ_D = 240 K, m^* = 1.5 m, v_o = = 2000 m/s, and with α = 0.225 (A), 1.125 (B), 2.25 (C). Dashed lines correspond to curves B of Fig. 1a and Fig. 1b, respectively

Fig. 1a, b). The same qualitative picture can be found by using the self-consistent theory for the description of the DFL in the electrical resistivity (5) and for the TEP (11) (Fig. 2a, b). Especially, for the same set of parameters (cf. curve A in Fig. 2) the results for the transport quantities calculated by the multiple scattering theory are qualitatively in agreement with the self-consistent theory. The self-energy (5) of the self-consistent theory allows to take into account higher values for the electron-phonon interaction strength in contradiction to the self-energy (3), which is calculated only in the lowest order of the interaction strength. Increasing α the effect of DFL becomes stronger and the magnitude of the resistivity is also increased (cf. curves B, C in Fig. 2a, c). The TEP shows smaller deviations from linearity in comparison to the resistivity (cf. curves B, C in Fig. 2d) and with increasing α the TEP is decreased for a given temperature (cf. curves B, C in Fig. 2b). In discussing the consequences of DFL (saturation phenomenon in the electrical resistivity we have found the effect of DFL also in the thermoelectric power.

The authors are indebted to Dr. V. Christoph and Prof. K. Elk for helpful discussions, Dr. J. Schreiber for helpful discussions and reading of the manuscript, and Prof. P. Ziesche for his support of this work.

References

/1/ Z. FISK and G.W. WEBB, Phys. Rev. Letters 36, 1084 (1976).
/2/ P.B. ALLEN, in: Superconductivity in d- and f-Band Metals, Ed. H. SUHL, and M.B. MAPLE, Academic Press, New York 1980 (p. 291).
/3/ W. SCHILLER and G. LANGOWSKI, J. Phys. F 12, 449 (1982).
/4/ V. CHRISTOPH and W. SCHILLER, J. Phys. F 14, 1173 (1984).
/5/ P.J. COTE and L.V. MEISEL, Phys. Rev. B 20, 3030 (1979).
/6/ M. JONSON and G.D. MAHAN, Phys. Rev. B 21, 4223 (1980).
/7/ S.M. GIRVIN and M. JONSON, Phys. Rev. B 22, 3583 (1980).
/8/ V. CHRISTOPH, phys. stat. sol. (b) 91, 593 (1979).
/9/ F.Y. FRADIN, Phys. Rev. Letters 33, 158 (1974).
/10/ F.J. BLATT, P.A. SCHROEDER, C.L. FOILES, and D. GREIG, Termoelektrodvizushchaya sila metallov, Metallurgisdat, Moscow 1980.
/11/ C.D. HODGMAN, Handbook of Chemistry and Physics, Cleveland 1956.
/12/ C. KITTEL, Einführung in die Festkörperphysik, Akademische Verlagsgesellschaft Geest & Portig K.-G., Leipzig 1973.

(Received August 15, 1984)

phys. stat. sol. (b) 126, K97 (1984)

Subject classification: 14.3.4

Institute of Experimental Physics, Attila Jozsef University, Szeged[1)]

On the Poole-Frenkel Constant

By

M.I. TÖRÖK

The Frenkel theory /1/ of field enhanced emission from Coulombic donor-like centers has been successfully used to explain the non-ohmic feature of many amorphous materials /2, 3/ and to interpret some recent experiments on crystalline semiconductors, too /4 to 6/

The original theorem is based on the following assumptions: (i) the donor-like centers are neutral, when occupied by electrons, (ii) the potential distribution around the donor can be described by a simple Coulombic potential, (iii) the effect of surrounding atoms can be taken into account by the high-frequency dielectric constant, (iv) the concentration of filled donors is practically constant, and (v) the field enhanced excitation from the donor centers and the recapture of electrons by the same kind of centers is in equilibrium. We note that it is sufficient when the assumptions (ii) and (iii) hold for the long-range part of the potential only since the maximum of the potential lies in a distance of more than 10 lattice spacings from the site in the field direction. This approximation holds for deep levels, too /7, 8/. The one-dimensional treatment yields the well-known result:

$$\sigma(F) = A \exp\left(-\frac{\Delta E_{eff}}{2kT}\right) = A \exp\left(-\frac{\Delta E_d - \beta F^{1/2}}{2kT}\right) , \qquad (1)$$

where $A = e\,\mu(N_c N_d/2)^{1/2}$, ΔE_d is the ionisation energy of the donor in absence of external field, and $\beta = (e^3/\varepsilon\varepsilon_{rel}\pi)^{1/2}$ is the Poole-Frenkel constant.

With the Poole-Frenkel plot (i.e. plotting ln σ versus $F^{1/2}$) (1) yields a straight line with a slope of $\beta/2kT$. Experimentally, however, it is often found that the slope of the lines differs from that predicted by (1). Several theorems of different character have been proposed to explain this discrepancy /9 to 11/. Simmons /12/ reasoned the presence of shallow traps.

Yeargan and Taylor /13/ pointed out that two limiting cases exist for an n-type material containing both donor and acceptor impurities: (i) for $n < N_a$

1) H-6720 Szeged, Hungary.

the exponent in (1) will be $\Delta E_d/kT$, (ii) for $N_a < n \ll N_d$ it becomes $\Delta E_d/2kT$. They proposed that the experimental results could be explained by supposing the existence of an "intermediate case" with an exponent $\Delta E_d/skT$ where $1 \leqq s \leqq 2$, and this supposition has been widely used /14, 15/. It is easy to show that this interpretation is incorrect. The concentration of electrons for compensated semiconductors with $N_d < N_a$ can be written as follows /16/:

$$n = (N_a + N_{cd}) \left\{ \left[\frac{1}{4} + \frac{N_{cd}(N_d - N_a)}{(N_{cd} + N_a)^2} \right]^{1/2} - \frac{1}{2} \right\} , \qquad (2)$$

where $N_{cd} = N_c \exp(-\Delta E_d/kT)$. The slope in question obeys the relation $s^{-1} = \ln n/\ln N_{cd}$. The density of electrons can be approximated by $n = N_{cd}(N_d - N_a)/2N_a$ for $N_{cd} < N_a^2/N_d$ and by $n = \left[N_{cd}(N_d - N_a)\right]^{1/2}$ for $N_{cd} > N_a$. The intermediate case corresponds to that range of N_{cd}, where of the approximations is valid. In this range, however, the slope of the ln n versus ln N_{cd} plot continuously decreases, i.e. the Poole-Frenkel plot of experimental data yields a curve, and not a straight line. Moreover, the concentration of electrons in this range increases by not more than one order; however, when the Poole-Frenkel effect is observed, the conductivity varies usually through several orders. For this reason the whole curve cannot fall in the intermediate range.

A very obvious case has been disregarded in the literature of the Poole-Frenkel theory: the effect of multiple ionisation. A simple algebra shows that the ionisation step

$$D^{(z-1)+} \longrightarrow D^{z+} + e \qquad (3)$$

in high field has the ionisation energy of

Table 1

material	$\sqrt{z}$	references
amorphoss Ge	1.32 to 1.41	/14/
amorphous Ge	1.34	/17/
amorphous SiO	1.21 to 1.51	/18/
Si_3N_4 electronic	1.36	/19/
Si_3N_4 electronic	1.32	/13/

$$\Delta E_z(F) = \Delta E_z - z^{1/2} \beta F^{1/2} \quad . \qquad (4)$$

The slope will be $\beta z^{1/2}/kT$ or $\beta z^{1/2}/2kT$, depending on the degree of compensation, i.e. the slope of the Poole-Frenkel plot will vary if the charge of the ionised donor is other than 1. For illustration some earlier experimental results which can be interpreted by z = 2 and s = 2 are summarised in Table 1.

References

/1/ J. FRENKEL, Phys. Rev. 54, 647 (1938).

/2/ A.K. JONSCHER, J. Electrochem. Soc. 116C, 217 (1967).

/3/ J.J. O'DWYER, The Theory of Electrical Conduction and Breakdown in Solid Dielectrics, Clarendon Press, Oxford 1973.

/4/ G. OTTAVIANI, C. CANALI, C. JACOBINI, A. ALBERIGI QUARANTA, and Z. ZAINO, J. appl. Phys. 44, 360 (1973).

/5/ R.D. HARRIS, J.L. NEWTON, and G.D. WATKINS, Phys. Rev. Letters 10, 1271 (1982).

/6/ L.C. KIMMERLING and J.L. BENTON, Appl. Phys. Letters 39, 410 (1981).

/7/ M. ALTARELLI and F. BASSANI, Impurity States: Theoretical, in: Handbook of Semiconductors, Ed. W. PAUL, North Holland Publ. Co., Amsterdam 1983 (p. 280).

/8/ M. JAROS, Adv. Phys. 29, 409 (1980).

/9/ R.M. HILL, Thin Solid Films 8, R21 (1971).

/10/ M.R. BOON, Thin Solid Films 8, R5 (1971).

/11/ R.B. HALL, Thin Solid Films 8, 263 (1971).

/12/ J.G. SIMMONS, Phys. Rev. 155, 657 (1967).

/13/ J.R. YEARGAN and H.L. TAYLOR, J. appl. Phys. 39, 5600 (1968).

/14/ G.A.N. CONNEL, D.L. CAMPHAUSEN, and W. PAUL, Phil. Mag. 26, 541 (1972).

/15/ L. FRITZSCH and W. BOBE, phys. stat. sol. (b) 58, K49 (1973).

/16/ R.A. SMITH, Semiconductors, Univ. Press, Cambridge 1961 (Chap. 4.3).

/17/ N. COITROU and L. VESCAN, Thin Solid Films 3, 269 (1969).

/18/ M. DEERY, J.G. PERKINS, and K.G. STEPHENS, Thin Solid Films 8, R16 (1971).

/19/ S.M. HU, D.R. KERR, and L.V. GREGOR, Appl. Phys. Letters 10, 97 (1967).

(Received July 26, 1984)

phys. stat. sol. (b) 126, K101 (1984)

Subject classification: 14.1 and 19; 21.1

Institut für Atom- und Festkörperphysik der Freien Universität Berlin, Berlin (West)[1]

Electrical Resistivity and NMR Investigations of the V-Co System at High Temperatures

By

G. HÜBER and D. PLOUMBIDIS

Studies of the electrical and magnetic properties of transition metal systems are of considerable interest. This is not only because of their technological importance (e.g. production of electrical resistors with definite characteristics) but also because of their particular electronic and magnetic structures.

Electrical resistivity ρ and Knight shift K measurements concerning the alloys V_3Co and equiatomic V-Co have been presented in our previous work /1/. In the present study we report on the first high temperature measurements of ρ and K for the alloys V-38 at% Co and V-55 at% Co. For both, the resistivity and the NMR measurements we used samples which were prepared starting from the same alloy bulk material. The alloys were produced by repeated melting of the appropriate quantities of V and Co (with a purity of 99.995%) in an argon furnace. Further experimental details are described in our earlier reports /1, 2/.

The temperature dependence of resistivity of the alloys V-38 at% Co and V-55 at% Co is shown in Fig. 1a and b, respectively. The temperature coefficient β of ρ for the V-38 at% Co system is positive between 290 and 600 K; but for higher temperatures, up to 1300 K, it is negative. For the V-55 at% Co system β is negative over the entire temperature range, as is the case for the V-50 at% Co system about which we had reported previously /1/. The variation of the resistivity with temperature is reversible in the range 290 to 860 K (the measuring points for this range are labelled with closed squares). For higher temperatures $\rho(T)$ shows a pronounced hysteresis. The cooling procedure (Fig. 1a and b, open circles) from about 1300 K down to 290 K takes a very long time (30 days) until the resistivity reaches its initial value. The measurements have been repeated several times and the ρ values have been

---

1) Arnimallee 14, D-1000 Berlin 33, Berlin (West).

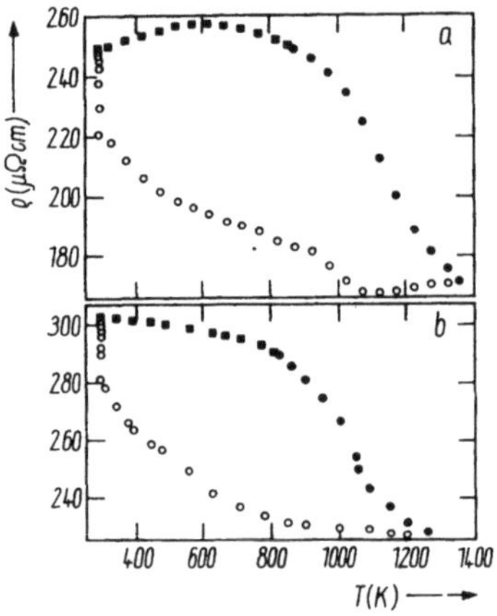

Fig. 1. a) The temperature dependence of the electrical resistivity ϱ of the V-38 at% Co alloy. Closed circles denote increasing, open circles decreasing temperature. The variation of ϱ with temperature is reversible in the range 290 to 860 K (closed squares). b) The temperature dependence of the electrical resistivity ϱ of the V-55 at% Co alloy. (Labelling as in a))

found to be reproducible within ±2%. So the temperature coefficient of the electrical resistivity in the σ-phase of the V-Co system, as these three alloys show, is negative in contrast to the A-15 phase (V_3Co).

The V_3Co alloy shows a positive β over the entire temperature range (290 to 1300 K), and β decreases continuously with increasing temperature /1/. That is, V_3Co exhibits "resistivity saturation" like other A-15 structure systems (e.g. Nb_3Sn) /3/. This temperature behavior of ϱ of the V_3Co alloy has been described phenomenologically by the "shunt resistor model" where the electron-phonon scattering processes are considered to be dominant /1/. Now negative temperature coefficients of ϱ have been found for a number of transition metal alloys /4 to 8/. There are different phenomenological descriptions concerning the negative β, but no satisfactory quantitative theoretical approximation has been given up to now.

It is recognized that metallic alloys with a relatively high resistivity (150 to 300 μΩcm) and a negative β are characterized by a high amount of disorder. There are various types of disorder which influence ϱ and β, such as disorder of topological, compositional, thermal, or magnetic origin. For a deeper insight into the structure of this complex matter and for finding out which interactions mainly cause this temperature behavior of ϱ, it is important to add to the resistivity experiments the results of experiments using other techniques, e.g. Hall effect, Knight shift, magnetoresistance, and magnetic susceptibility experiments. In this sense we measured, additionally to the resistivity, the ^{51}V Knight shift in dependence on the temperature for the alloys V-38 at% Co and V-55 at% Co. The results are shown in Fig. 2a and b. The Knight shift increases with increasing temperature for both alloys in the high temperature range. This behavior of K(T) can be interpreted in the frame of a model, which

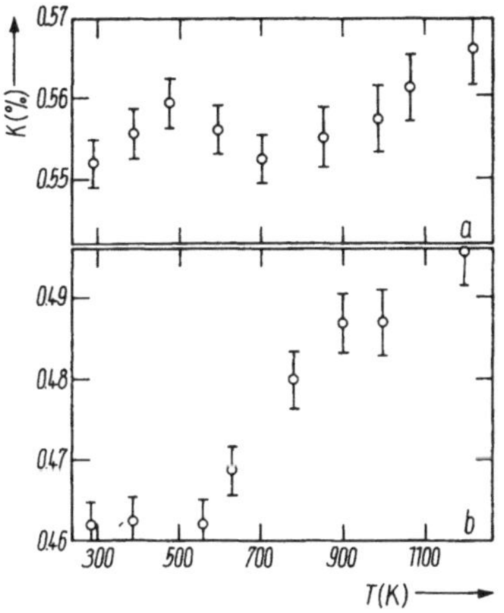

Fig. 2. a) The temperature dependence of the ^{51}V Knight shift in the V-38 at% Co alloy. b) The temperature dependence of the ^{51}V Knight shift in the V-55 at% Co alloy

in the past has been successfully applied in order to explain Knight shift results in transition metal systems /9 to 12/. According to this model, the Knight shift part corresponding to the d-spin susceptibility is negative and decreases with increasing temperature. Thus, if the negative contribution to the Knight shift decreases, the total Knight shift has to increase with increasing temperature. This is in agreement with our experimental results shown in Fig. 2.

At low temperatures the d-spin susceptibility has higher values and therefore the total Knight shift is smaller. If the Co concentration increases the d-spin susceptibility, as expected, increases, and the Knight shift has to decrease. This is what our measurements show (Fig. 2). On the other hand, the resistivity of both alloys decreases with increasing temperature in the high temperature range and ϱ increases with increasing Co concentration (Fig.1).

Now this correlation on the base of our present and previous experimental results concerning ϱ and K of the V-Co system supports strongly the assumption that the relatively high electrical resistivity and its negative temperature coefficient of the alloys V-38 at% Co, V-50 at% Co, and V-55 at% Co is mainly caused by a scattering process of magnetic origin.

Further systematic experiments concerning Hall effect and magnetic susceptibility of these systems will help to give a more detailed analysis of their electronic and magnetic properties.

This work was supported by the "Deutsche Forschungsgemeinschaft".

References

/1/ D. PLOUMBIDIS and G. HÜBER, Z. Naturf. 37a, 319 (1982).

/2/ D. PLOUMBIDIS, Exper. Tech. Phys. 30, 323 (1982).

/3/ P.B. ALLEN, Proc. Physics of Transition Metals 1980, The Institute of Physics, Bristol/London 1981 (p. 425).

/4/ P. CHEVENARD, J. Inst. Metals 36, 39 (1926).

/5/ H. THOMAS, Z. Phys. 129, 219 (1951).

/6/ R.G. LOASBY, Proc. Phys. Soc. 72, 425 (1958).

/7/ J.H. MOOIJ and M. DE JONG, J. Vacuum Sci. Technol. 9, 446 (1972).

/8/ J.H. MOOIJ, phys. stat. sol. (a) 17, 521 (1973).

/9/ D. PLOUMBIDIS, Z. Phys. B 28, 61 (1977).

/10/ R. RÜNGER and D. PLOUMBIDIS, phys. stat. sol. (b) 122, K49 (1984).

/11/ A.M. CLOGSTON, V. JACCARINO, and Y. YAFET, Phys. Rev. 134, A650 (1964).

/12/ I. SENEL and D. PLOUMBIDIS, Z. Naturf. 39a, 203 (1984).

(Received August 31, 1984)

phys. stat. sol. (b) 126, K105 (1984)
Subject classification: 6 and 16; 22.2.1

V.I. Lenin State University, Kishinev[1)]

Magnetic Oscillations of the Acoustoelectrical Current of Photoelectrons in p-GaAs Type Semiconductors

By
G.M. SHMELEV, NGUYEN HONG SHON, and G.I. TSURKAN

In the present note we shall consider the acoustoelectrical (AE) current, which arises owing the phonon drag of photoelectrons; the latter occurs in the conduction band under optical transitions, induced by a linear polarized electromagnetic wave, in p-GaAs type semiconductors. The sample is placed in a non-quantized magnetic field: $\omega_c \ll \varepsilon_o$ (ω_c is the cyclotron frequency; the initial energy of excited electrons $\varepsilon_o \gg T$ (T is the lattice temperature), $\hbar = 1$). The length of the sound wave is considered to be small in comparison with the length of the free path of electrons, the screening radius, and the Larmor radius. It is known /1/ that only carriers with electron crystal momentum $p \geqq q/2$ (or energy $\varepsilon > \varepsilon_1 = q^2/8m$, m is the effective carrier mass, q is the phonon wave number) can interact with the monochromatic sound (according to the laws of conservation), therefore under the condition $\varepsilon_1 \gg T$ the thermalized carriers need not be considered. In p-GaAs (with hole concentration $N_h \gtrapprox 2\times10^{17}\ \text{cm}^{-3}$) the main mechanism of energy and momentum relaxations is the interaction of hot photoelectrons with heavy holes with the transition of the latter into the light hole band /2, 3/. The characteristic peculiarities of this effective mechanism are the following:

1) by neglecting the weak logarithmic dependences of the relaxation time of energy (and $m = m_l$, the mass of the light hole) $\tau_\varepsilon(\varepsilon) = \tau_p(\varepsilon) = \tau_1(\varepsilon) \sim \varepsilon^{3/2}$ ($\tau_\varepsilon(\varepsilon)$, $\tau_1(\varepsilon)$, and $\tau_p(\varepsilon)$ relaxation times for symmetrical $f^{(0)}$ and antisymmetrical $\vec{f}^{(1)}$ parts of the distribution function (DF) and momentum relaxation time, respectively);

2) the ratio

$$\xi = \frac{D_1(\varepsilon)\tau_1(\varepsilon)}{\varepsilon^2} \sim \left|\ln(p/\sqrt{2m_hT})\right|^{-1} \ll 1 \qquad (1)$$

when $p \gg \sqrt{2m_hT}$. In inequality (1)

---

1) Sadovaya 60, 277003 Kishinev, USSR.

$$D_1(\varepsilon) = \int (\varepsilon - \varepsilon')^2 W_1(\varepsilon, \varepsilon') g(\varepsilon') d\varepsilon' ,$$

where

$$W_1(\varepsilon, \varepsilon') = \frac{1}{2} \int_0^\pi W_{\vec{p} \to \vec{p}'} \cos\Theta \sin\Theta d\Theta ,$$

$g(\varepsilon)$ is the density of states, $W_{\vec{p} \to \vec{p}'}$ is the transition probability $\vec{p} \to \vec{p}'$, and Θ is the angle included between $\vec{p}$ and $\vec{p}'$. Condition (1) permits to neglect the summand $\sim D_1(\varepsilon)$ in the kinetic equation written in the diffusion approximation (see (5));

3) it dominates over other mechanisms in some energy region $\varepsilon_- < \varepsilon < \varepsilon_+$, where $\varepsilon_- \sim T(m_h/m_l)$ /3/. Outside this region, if electrons interact with the acoustic lattice vibrations, then $\varepsilon_+ \sim T(\tau_1^{ac}(T)/\tau_1(T))^{1/2}$. Further we shall show that these peculiarities lead to an oscillatory dependence of the AE current on the magnetic field strength.

The kinetic equation for the DF of photoelectrons is

$$(\omega_c [\vec{p}, \vec{h}] , \frac{\partial f}{\partial \vec{p}}) = St[f] + St^{es}[f] + G(\vec{p}) , \quad (\vec{h} = \vec{H}/H) , \tag{2}$$

where the first two summands in the right part of (2) are collision integrals with scatterers and sound, the latter summand is the generation rate of photoelectrons:

$$G(\vec{p}) = \left\{1 + \alpha_1 \vec{c} - \omega_c [\vec{h}, \vec{c}] \frac{\partial}{\partial \Omega} \alpha_2 \right\} G_o \delta(\varepsilon - \varepsilon_o) , \ (\varepsilon_o = \frac{\mu}{m_c} (\Omega - \varepsilon_g)) , \tag{3}$$

where the first term in (3) describes the isotropic generation, the next term is responsible for the ballistic photogalvanic current (PGC) /4, 5/, and the latter term determines the magneto-induced contribution in PGC /6/; $G_o g(\varepsilon_o)$ is the number of electrons which enter the conduction band in 1s per cm^3, $c_i = \varepsilon_{ikl} e_k e_l$, $\vec{e}$ is the unit vector of the light polarization, ε_{ikl} is the unit antisymmetric tensor of third order, Ω is the light frequency, μ is the equivalent mass of electron and hole, ε_g is the width of the band gap. In p-GaAs type semiconductors at transitions from the valence band Γ_{15} into the s-state of the conduction band /5, 6/ $\alpha_1 = 4/m_e \mathcal{P} a_B$ and $\alpha_2 = p_o m_c/m_e m_h \mathcal{P}$ ($\mathcal{P}^2 = 3\varepsilon_g/4m_c$, m_e is the free electron mass, m_c is the effective electron mass in c-band, a_B is the Bohr radius of the exciton).

We consider that the initial energy of photoelectrons satisfies the inequalities

$$\varepsilon_- < \varepsilon_1 < \varepsilon_o < \varepsilon_+ , \tag{4}$$

then the equation of the antisymmetric part of the DF $f^{(1)}$ under condition (1) is the following:

$$\omega_c\left[\vec{h}, \vec{f}^{(1)}\right] = -\frac{\vec{f}^{(1)}}{\tau_p(\varepsilon)} + \frac{1}{g(\varepsilon)}\frac{\partial}{\partial\varepsilon}\left\{\frac{g(\varepsilon)}{\tau_1(\varepsilon)}\vec{f}^{(1)}\right\} -$$

$$-\frac{3\pi}{2}\left|C_{\vec{q}}\right|^2 N_{\vec{q}}\,\omega_q\Theta(\varepsilon-\varepsilon_1)\frac{1}{\varepsilon}\frac{\partial f^{(0)}}{\partial\varepsilon}\vec{\chi} +$$

$$+\left\{\alpha_1\vec{c} - \omega_c\left[\vec{h}, \vec{c}\right]\frac{\partial}{\partial\Omega}\alpha_2\right\}G_o\delta(\varepsilon-\varepsilon_o)\ , \qquad (5)$$

where $C_{\vec{q}}$ is the matrix element of interaction of the electron with the sound, $N_{\vec{q}}$ is the number of external phonons in 1 cm^3, $\vec{\chi} = \vec{q}/q$, $\omega_q = qv_s$, v_s is the speed of sound. In the region $\varepsilon > \varepsilon_1$ under condition (4) the DF $f^{(0)}$ is the following /7/:

$$f^{(o)}(\varepsilon) \approx \frac{G_o\tau_\varepsilon(\varepsilon_o)}{\varepsilon}\Theta(\varepsilon_o-\varepsilon)\ . \qquad (6)$$

For $\vec{\chi} \parallel \vec{OX}$, $\vec{H} \parallel \vec{OZ}$, from (5) and (6) we obtain

$$f_x^{(1)} + if_y^{(1)} = \frac{3\pi}{2}\left|C_{\vec{q}}\right|^2 N_{\vec{q}}\,\omega_q f^{(o)}(\varepsilon_1)\frac{\tau_1(\varepsilon)}{\varepsilon_o^{1/2}\varepsilon^{3/2}}(\chi_x + i\chi_y)\Theta(\varepsilon_o-\varepsilon)\times$$

$$\times\Theta(\varepsilon-\varepsilon_1)\exp\left\{-\int_\varepsilon^{\varepsilon_o}\frac{\tau_1(\varepsilon')}{\varepsilon'\tau_p(\varepsilon')}\left[1 + i\omega_c\tau_p(\varepsilon')\right]d\varepsilon'\right\}+$$

$$+\left(\alpha_1 - i\omega_c\frac{\partial}{\partial\Omega}\alpha_2\right)G_o\tau_1(\varepsilon)\frac{\varepsilon_o^{1/2}}{\varepsilon^{3/2}}c_x\Theta(\varepsilon_o-\varepsilon)\times$$

$$\times\exp\left\{-\int_\varepsilon^{\varepsilon_o}\frac{\tau_1(\varepsilon')}{\varepsilon'\tau_p(\varepsilon')}\left[1 + i\omega_c\tau_p(\varepsilon')\right]d\varepsilon'\right\}. \qquad (7)$$

In (7) we used the initial condition in the form $\vec{f}^{(1)}(\infty, \vec{H}) = 0$.

The total current in the sample is given by the formula

$$\vec{j}_{tot} = \vec{j}^{AE} + \vec{j}^{PG} = \frac{e}{3m}\sum_{\vec{p}} p\vec{f}^{(1)}\ , \qquad (8)$$

where $\vec{j}^{AE}$ is the AE current, which is determined by the first summand in (7)

$$j_x^{AE} + i\,j_y^{AE} = \frac{j_o}{\varphi_o^2}\left\{\left[1-\cos(\varphi_o-\varphi_1)\right] - i\left[(\varphi_o-\varphi_1) - \sin(\varphi_o-\varphi_1)\right]\right\}, \qquad (9)$$

here $\varphi_{o(1)} = \frac{2}{3}\omega_c\tau_1(\varepsilon_{o(1)})$, $j_o = \frac{2}{3}\frac{e^2 n}{m}\tau_1(\varepsilon_o)E_w$ (E_w is the Weinreich field, n is the carrier concentration). The PGC $\vec{j}^{PG}$ is studied in /8/.

The main peculiarity of (9) is the periodical dependence of $\vec{j}^{AE}$ on H (for a given ε_0 and ε_1): the magnetic oscillation of the AE current. The appearance in the argument of sin and cos of the difference $\tau \equiv \frac{2}{3}\left[\tau_1(\varepsilon_0) - \tau_1(\varepsilon_1)\right]$ is not accidental. In reality this difference is the time of descent of an "average" electron from energy ε_0 to ε_1, while the momentum rotates by an angle $\omega_c\tau$. Note, that there is an oscillatory dependence of the AE current on the wave vector of sound. Apart from the conditions for correctness of formula (9), which has been stated above, we add the condition of weak damping of magnetic oscillations: $\xi\varphi_0^2 \ll 1$.

Note, that the magnetic oscillations of the AE current arise under non-polarized illumination, too (this follows from (9)).

For example, let us give a numerical estimate. For $\varepsilon_g = 1.4$ eV, $m = 0.065\, m_e$, $v_s = 3.5\times10^5$ cm/s, $\varrho = 5.3$ g/cm^3, $\Lambda \approx 10$ eV, $\tau_\varepsilon(\varepsilon_0) = \tau_1(\varepsilon_0) = \tau_p(\varepsilon_0) = 5\times10^{-13}$ s and a source of light with power $\approx 10^6$ W/cm^2, $\hbar\Omega = 2$ eV, the sound intensity $W \approx 10^{-2}$ W/cm^2, $\omega_q \approx 10^{10}$ s^{-1} having $j_0 \approx 10^{-4}$ A/cm^2 and the magnetic oscillation should be observed under $\varphi_0 \gtrsim 1$, which corresponds to fields $H \gtrsim 10^4$ Oe.

References

/1/ E.M. EPSHTEIN, Dokl. Akad. Nauk SSSR 173, 320 (1967).

/2/ M.I. DYAKONOV, V.I. PEREL, and I.I. YASSIEVICH, Fiz. Tekh. Poluprov, 11, 1364 (1977).

/3/ B.P. ZAKHARCHENYA, D.N. MIRLIN, V.I. PEREL, and I.I. RESHINA, Uspekhi fiz. Nauk 136, 459 (1982).

/4/ V.I. BELINICHER and V.N. NOVIKOV, Fiz. Tekh. Poluprov. 15, 1957 (1981).

/5/ E.L. IVCHENKO and G.E. PIKUS, in: Problemi sovremenoi fiziki, Nauka, Leningrad 1980 (p. 275 to 293).

/6/ E.L. IVCHENKO, YU.B. LIANDA-GELLER, G.E. PIKUS, and R.YA. RASULOV, Fiz. Tekh. Poluprov. 18, 93 (1984).

/7/ V.D. DIMNIKOV and V.I. PEREL, Fiz. Tekh. Poluprov. 13, 707 (1979).

/8/ NGUYEN HONG SHON and G.M. SHMELEV, Fiz. tverd. Tela 26, 1863 (1984).

(Rec ived May 7, 1984)

Pre-Printed Titles

Papers to be published
in "physica status solidi (b) - basic research"
Vol. 126, No. 2

Original Papers

F. BECHSTEDT, R. ENDERLEIN, and O. HEINRICH, Berlin
The Effect of Many-Electron Correlation on Photothresholds of Semiconductors and Valence Band Discontinuities at Heterojunctions

A.M. EL-DIB and H.F. HASSAN, Nasr City, Cairo
Temperature Dependence of the Electronic Energy Bands of $NaNO_2$ Crystal from 20 to 245 ^{o}C

V.N. RUDKO and V.I. SUGAKOV, Kiev
Superconductivity of Metals with a Superlattice of Voids

R.P. SINGH, J.P.S. RANA, A. DAUD, and G.B. VARSHNEY, Aligarh
The Role of the Zero-Point Vibration towards Crystal Properties of Rubidium Halide Crystals

A.O.O. MEKHRABOV and M. DOYAMA, Tokyo
Electronic Theory of Atomic Short-Range Order for Ternary Alloys Using the Pseudopotential Approximation and Its Comparison with Experiments

V.V. NEMOSHKALENKO and A.N. TIMOSHEVSKII, Kiev
The Peculiarities of the Electronic Structure of $BaTiO_3$ in $ATiO_3$ (A = Ca, Sr, Ba) Series

M. JIRSA and V. KAMBERSKY, Prague
Angular Dependence of Spin Wave Resonance in Thin Films with Asymmetrical Boundary Conditions

E.V. ZAROCHENTSEV, K.B. TOLPYGO, and E.P. TROITSKAYA, Donetsk
Pressure Effect on Band Structure of Neon

K. MASUDA-JINDO, Yokohama
Tight-Binding Electronic Theory for Lattice Defects in Transition Metals

I.G. LANG, S.T. PAVLOV, A.V. PROKAZNIKOV, and A.V. GOLTSEV, Leningrad
General Theory of Secondary Radiation in Condensed Matter

V. NAGIRNYI, T. SOOVIK, and S. ZAZUBOVICH, Tartu
Decay Kinetics and Polarization of the A_X Emission of KI:In Crystal

M.J. MARCINKOWSKI, Maryland
The Role of Surface Dislocations in the Continuum Theory of Lattice Defects

L.E. KHOLODENKOV, A.G. MAKHANEK, and A.A. KAMINSKII, Minsk and Moscow
One- and Two-Photon Luminescence Excitation of Eu^{3+} in a LaF_3 Monocrystal

PREM CHAND and M. UMAR, Kanpur and Aligarh
Electron Parmagnetic Resonance Study of Copper (II) Ions in Single Crystal of Strontium Formate Dihydrate

U. PIETSCH, Leipzig
Simulation of Dynamical Bond Charge Transfer Properties in Elemental Semiconductors by Means of a Simple Quantum-Mechanical Model

YU.A. ĖKMANIS and I.S. RADCHENKO, Riga and Krivoi Rog
Light Ėxtinction by Ċolloidal Thallium in Ionic Crystals

G.I. PILIPENKO, D.V. OPARIN, O.I. TYUTYUNNIK, and F.F. GAVRILOV, Sverdlovsk
Photostimulated Transformations of Complex Colour Centres in Lithium Hydride Single Crystals

C. TRALLERO GINER and O. SOTOLONGO COSTA, Havana
One-Phonon Resonance Raman Scattering

JUH TZENG LUE, Taiwan
Time-Resolved Reflectivity Measurements on RF Sputtered a-Si:H Alloy

S. JALAJ, I. SINGH, and S.K. JOSHI, Roorkee and Berlin (West)
Tight-Binding Calculations for the Angle-Resolved Local Density of States of InP(110) Surfaces

O. P. HANSEN and I. F. I. MIKHAIL, Copenhagen and Cairo
Galvanomagnetic Effects of Bismuth in Non-Quantising Magnetic Fields

M. ASCHE and O. G. SARBEI, Berlin
The Role of Electron-Hole Interaction in the Cooling Process of Highly Excited Carriers

PAN JIN-SHENG, Chungchun
The Surface or Interface Polaron in Polar Crystals (I)

D. SEN and S. K. SEN, Jadavpur
Transport in Simple Metals in the Framework of a Unified Pseudopotential Study

G. THÖRNER and G. BORSTEL, Osnabrück
Relativistic Theory of Photoemission from Solids (I)

C. TUNCAY and M. TOMAK, Ankara and Trieste
Ideal Vacancy States in Si

K. R. ALLAKHVERDIEV, U. A. ALESHENKO, N. A. BAKHYSHOV, L. K. VODOPYANOV, F. M. GASHIMZADE, R. M. SARDARLY, and V. YA. SHTEINSHRAIBER, Baku
Vibrational Spectrum of TlS_xSe_{1-x} Solid Solutions

M. S. JAHAN and D. W. COOKE, Memphis, Tennessee and Los Alamos, New Mexico
Thermal Evolution of Low-Temperature Manganese Centers in X-Irradiated CaF_2:Mn

K. IWAHANA, Wien
Magnetic Field Dependence of the Spin Polarization of the F Center

L. E. NAGLI, Riga
F-Centres in KCl-Tl Crystals

I. G. RATISHVILI, Tbilisi
Annihilation of Frenkel Pairs in B. C. C. Metals (α-Iron) (II)

V. M. YARTSEV, Chelyabinsk
Charge Transfer and Electron-Molecular Vibration Coupling in Tetramerized Quasi-1D Semiconductors

M.A. ALAM and M. TOMAK, Trieste
Electrical Resistivity of Liquid Noble Metal Alloys

B. HOLLAND, H.S. GREENSIDE, and M. SCHLÜTER, Murray Hill, New Jersey
Cohesive Properties of Ge, Si, and Diamond Calculated with Minimum Basis Sets

V.I. TORGASHEV, V. DVOŘÁK, and F. SMUTNÝ, Prague
On Phase Transitions in $LiNH_4SO_4$

E. KADEN and H. L. GÜNTER, Hannover
The Thermoelectric Power of p-Ge at Low Temperatures

M. RÖSLER and W. BRAUER, Berlin
On the Theory of Ion-Induced Electron Emission

B.G. DICK, Salt Lake City, Utah
Electric Field-Induced First Order Raman Scattering in Alkali Halides

U.H. GLÄSER and H. LÖFFLER, Halle
Effective Medium Approach for the Residual Resistivity of Alloys Containing Guinier-Preston Zones

K. KAMBAS, A. ANAGNOSTOPOULOS, S. VES, and J. SPYRIDELIS, Thessaloniki
Optical Absorption Edge Investigation of $CdIn_2S_4$ an β-In_2S_3 Compounds

Papers to be published
in "physica status solidi (a) - applied research"
Vol. 86, No. 1

Review Article

C. BOULESTEIX, Marseille
A Survey of Domains and Domain Walls by Crystallographic Phase Transitions Causing a Change of the Lattice

Original Papers

N.E. ALEKSEEVSKII, A.V. MITIN, E.P. KHLYBOV, CZ. BAZAN, B. GREN, and A. GILEWSKI, Moscow, Troitsk, and Wrocław
The Influence of Alloying on the Properties of Superconducting Molybdenum Sulphides

V.V. KIRSANOV, S.B. KISLITSIN, and E.M. KISLITSINA, Alma-Ata
Interstitial Migration in a Stress Gradient

I. TALE and J. ROSA, Riga and Prague
Fractional Glow Technique Spectroscopy of Traps in Heavily Doped AlN:O

S.C. SABHARWAL, S.P. KATHURIA, and B. GHOSH, Bombay
Kinetics of Thermoluminescence in γ-Irradiated NaI(Tl)

A.P. TYUTNEV, V.S. SAENKO, A.V. VANNIKOV, and E.D. POZHIDAEV, Moscow
Concerning the Radiation-Induced Surface Conductivity in Polymers

H.M. EISSA and M.M. EL-OKER, Kuwait and Cairo
The Interaction of Straight Dislocations with Dislocation Loops in an Anisotropic Material (Zinc)

A.A. WAFIK and S.A. MAZEN, Cairo and Zagazig
Barkhousen Jumps and Resistivity of Cu-Cd Ferrites

M. TAKEDA and H. HASHIMOTO, Osaka
A High Resolution Electron Microscopy Study of Domain Structure in CuAuI

W. TORBICZ, D. SOBCZYNSKA, A. OLSZYNA, G. FORTUNATO, and A. D'AMICO, Warsaw
A Stable $Pd/BN/Si_3N_4/SiO_2/Si$ FET for Hydrogen Detection

E. HARTMANN, L. KOVÁCS, and J. PAITZ, Budapest
Electrical Conductivity of Gadolinium-Gallium Garnet (GGG) Crystals

W. BRAUNE, M. SCHNÜRER, N. KUBICKI, and R. HERRMANN, Berlin
C-U Measurements of Anodized InSb MOS Structures at 77 and 4.2 K

V.V. ARISTOV, L.S. KOKHANCHIK, and YU.I. VORONOVSKII, Chernogolovka
Potential Contrast of Ferroelectric Domains of Lithium Niobate in SEM

S.B. SEVASTIANOV, N.N. GERASIMENKO, and N.V. VERSHININA, Novosibirsk
Inhomogeneities in Irradiated SiO_2-Si Structures

CH. LUSHCHIK, J. KOLK, A. LUSHCHIK, and N. LUSHCHIK, Tartu
Radiational Creation of Frenkel Defects in KCl-Tl

H.-R. HÖCHE and J. SCHREIBER, Halle (Saale)
Anisotropic Deformation Behaviour of GaAs

M.B. SHAH and M.S.C. BOSE, Bombay
Magnetic NDT Technique to Evaluate Fatigue Damage

P.C. MORAIS, A.L. TRONCONI, and K. SKEFF NETO, Brasilia
Photoacoustic Signal Behavior near the Ferromagnetic Resonance Field

R. CAPELLETTI and M. MANFREDI, Parma
The Role of Eu^{++} Aggregation on Optical Absorption, Emission, Luminescence Time Decay, and ITC Plots in $NaCl:Eu^{++}$

G.P. KRAMAR and YA.I. PANOVA, Leningrad
The Permeability Dispersion in Ferrites at Different Levels of Direct Current Magnetic Biasing

P.A. ALEKSANDROV, A.M. AFANASEV, and S.A. STEPANOV, Moscow
Bragg-Laue Diffraction in Inclined Geometry

M. HIDAKA, H. FUJII, B.J. GARRAD, and B.M. WANKLYN, Fukuoka and Oxford
Structural Phase Transitions of $RbVF_4$

A. A. KAMINSKII, E. L. BELOKONEVA, B. B. MILL, YU.V. PISAREVSKII, S. E. SARKISOV, I. M. SILVESTROVA, A. V. BUTASHIN, and G. G. KHODZHABAGYAN, Moscow
Pure and Nd^{3+}-Doped $Ca_3Ga_2Ge_4O_{14}$ and $Sr_3Ga_2Ge_4O_{14}$ Single Crystals, Their Structure, Optical, Spectral-Luminescence, Electromechanical Properties, and Stimulated Emission

M. F. ZHU, F. C. T. SO, E. T.-S. PAN, and M-A. NICOLET, Pasadena, California
Investigation of Amorphous $W_{60}Zr_{40}$ Film as a Diffusion Barrier in Metallization Schemes

P. MALÉTRAS et P. DUVAL, Mont-Sain-Aignan
Evolution comparée d'alliages Cu-Al trempés revenus avant et après amincissement

J. W. TOMM, KARIN HERRMANN, C. BARTHEL, and U. BARTHEL, Berlin
On the Dispersion of the Refractive Index in Active Layers of Lead-Salt Injection Lasers

U. BRAATZ and D. ZAPPE, Dresden
Investigation of Space Charge Limited, Poole-Frenkel, and Isothermal Current on HgI_2 Single Crystals

N. V. RUMAK, V. V. KHATKO, and V. N. PLOTNIKOV, Minsk
Structure and Properties of Silicon Dioxide Thermal Films (I)

M. G. BLANCHIN and L. A. BURSILL, Villeurbanne, and Parkville, Victoria
Extended versus Small Defect Equilibria in Non-Stoichiometric Rutile (I)

G. SORGE, J. JANICH, L. A. SHUVALOV, and N. M. SHCHAGINA, Wittenberg and Moscow
The Influence of an Electric DC Field on the New Intermediate Phase between the α- and β-phases of $NaH_3(SeO_3)_2$ Crystals

M. TOPIĆ, A. MOGUŠ-MILANKOVIĆ, and Z. KATOVIĆ, Zagreb
Thermally Stimulated Depolarization Current Study in Novolac Phenol-Formaldehyde Resin

H. YAMAGUCHI, T. TSUCHIDA, I. HASHIMOTO, Y. OHTANI, and S. OHTAKI, Tokyo
Stage II Recovery of Defects in Cold-Worked Aluminium

H. BENDER, Antwerpen
Investigation of the Oxygen-Related Lattice Defects in Czochralski Silicon by Means of Electron Microscopy Techniques

D. GIGNOUX, J.C. GOMEZ-SAL, and J. RODRIGUEZ FERNANDEZ, Grenoble and Santander
Magnetic Structure of the Er_2Pt Antiferromagnet

E. ZSCHECH, W. BLAU, H. VEGA, K. KLEINSTÜCK, S. MAGER, M.A. KOZLOV, and M.A. SHEROMOV, Dresden and Novosibirsk
EXAFS and X-Ray Diffractional Investigation of the Heusler-Type Alloys Co_2MnSi and $Fe_{2.4}Mn_{0.6}Al$

A. KOZANECKI and H. RZEWUSKI, Warsaw
Distant DAP Bands in Electron-Irradiated and Ion-Implanted CdSe

A.J. MUHAMMAD, D.V. MORGAN, and A.E. GUILE, Leeds
The Effect of Overlap of Ohmic Contacts on the Characteristics of a MIM Structure Having a Single Discrete Shallow Trap Level

DANG TRAN QUAN, Rennes
Electrical Properties and Optical Absorption of SnSe Evaporated Thin Films

D. FASOLD, K. HEHL, and S. JETSCHKE, Jena
Optical Reflection and Transmission through a Flat System with a Thick Fluctuating Layer

T. OHKAWA and H. HASHIMOTO, Kanagawa and Osaka
An X-Ray Wave Theory for Heavily Distorted Crystals (I)

M. TAKAHASHI, Y. SASAKI, and S. ISHIO, Sendai
Magnetic Freezing in Solution-Quenched Dilute Fe-Cu Alloys

Classification Scheme — Continued

19. Magnetic Resonance

20. Optical Properties
    - 20.1 Spectra. Optical Constants (X-Ray Spectra See 20)
    - 20.2 Lasers
    - 20.3 Luminescence (X-Ray Spectra See 20)

21. Metals. Alloys
    - 21.1 Transition Metals (Group Sc to Zn) and Their Alloys
        - 21.1.1 Fe and Fe Alloys
    - 21.2 Alkali Metals
    - 21.3 Alkali-Earth Metals
    - 21.4 Rare-Earth Metals
    - 21.5 Actinides
    - 21.6 Noble Metals
    - 21.7 Semi-Metals (Semiconducting Alloys 22.7)

22. Semiconductors and Ionic Crystals
    - 22.1 Elements
        - 22.1.1 Germanium
        - 22.1.2 Silicon
        - 22.1.3 Group VI Elements
    - 22.2 III–V Compounds
        - 22.2.1 Arsenides
        - 22.2.2 Phosphides
        - 22.2.3 Antimonides
        - 22.2.4 Mixed Crystals
    - 22.3 IV–IV, II–V, and III–VI Compounds
        - 22.3.1 Mixed Crystals
    - 22.4 II–VI Compounds
        - 22.4.1 Sulphides
        - 22.4.2 Selenides
        - 22.4.3 Tellurides
        - 22.4.4 Mixed Crystals
    - 22.5 Halides
        - 22.5.1 Silver Halides
        - 22.5.2 Alkali Halides
        - 22.5.3 Alkali-Earth Halides
        - 22.5.4 Mixed Crystals
    - 22.6 Simple Oxides (CdO, ZnO See 22.4)
        - 22.6.1 Mixed Crystals
    - 22.7 Semiconducting Intermetallic Compounds
        - 22.7.1 Mixed Crystals
    - 22.8 Tertiary and Higher Compounds
        - 22.8.1 Salts of Oxy-Acids (Carbonates, Phosphates, Silicates Including Mica etc.)
        - 22.8.2 Spinels, Garnets, Ferrites
    - 22.9 Organic Semiconductors

23. Solidified Gases

Attention

The preceding classification scheme is closely akin to the scheme used by Physics Abstracts to classify the various fields of solid state physics. A paper will either be classified under one or more of the subsections or, if none of these are relevant, it will just appear under the general section heading (i. e. sections and subsections are equivalent in this sense).

The *subject classification* given at the head of each paper begins with the main subject matter of the paper, and this is followed, if necessary, by figures referring to subsidiary subject matter and the type of solids investigated.

The papers published in this issue are listed in the order of their subject classification at the end of the table of contents.

Manuscripts and letters for
physica status solidi (b) — basic research and
physica status solidi (a) — applied research

should be sent to

DDR-1086 Berlin, Leipziger Straße 3–4, P. B. 1233

or

Internationaler Buchversand GmbH, 1000 Berlin 19, Eschenallee 21.

The object of "physica status solidi (b)" and "physica status solidi (a)" is to publish rapidly important and topical results concerning fundamental research in the field of solid state physics and aspects of application and experimental techniques of fundamental solid state research, respectively. The issues of "physica status solidi (b)" and "physica status solidi (a)" will be published on the 1st and the 16th day of each month, respectively. Either series will contain the following categories:

1. Review articles written by arrangement with the editors. They will be remunerated.
2. Original papers selected from those which have not been submitted for publication elsewhere, after being reviewed. (The latest date for acceptance of contributions is the 8th and the 23rd day of each month for "physica status solidi (b)" and "physica status solidi (a)", respectively, 50 days before the delivery of the journal.)
3. Short notes published in the off-set part of the journal. They are, if necessary after being reviewed, selected by editorial staff. (The latest date for acceptance of manuscripts for this part is the 12th and the 27th day of each month for "physica status solidi (b)" and "physica status solidi (a)", respectively, 16 days before the delivery of the journal.)
4. Pre-printed titles of papers to be published in the next issue of both series.

Editorial Information

Manuscripts should be submitted preferentially in English. Besides, papers in German, French, and Russian are admitted. The final decision on the incorporation of the papers in the series (b) or (a) is taken by the editorial staff after consultation of the referee. Authors are asked to observe the following rules:

1. Manuscripts have to be written as concisely as possible. As most, 20 and 4 pages (28 lines per page; abstracts, tables, references, figures, and captions of figures included, where 3 figures are counted as about one page) are admitted for original papers and short notes, respectively. All manuscripts should be type-written on one side only. Two copies are required. Badly arranged manuscripts are returned.
2. The author's name, initials, and institution should be given.
3. One abstract of about ten lines in English and one in German, French, or Russian are required for original papers.
4. Because of lack of space conclusions (if any) must not have the character of a summary (i. e. substantially a repetition of the contents of the abstract).
5. Formulae should not be written too small and not with pencil. Separate lines for formulae are desirable. Indistinct writing will lead to a return of the manuscript.
6. Drawings should be enclosed on separate sheets of paper, not exceeding the maximum and minimum sizes of 20×30 and 10×15 cm^2, respectively. Photographs should be submitted, on white paper, only for review articles and original papers. Duplicates are required.
7. Captions for the figures should be submitted on separate sheets of paper, and special instructions with regard to the size of illustrations should be given on these.
8. References should be numbered in order of appearance and should be given, on a separate sheet, in the form used in physica status solidi.
9. SI-units should be used.
10. Each paper should contain a proposal for the subject classification according to the classification scheme appearing on the second and third cover of each issue.

The authors will be sent galley proofs of review articles and original papers. Proof reading of short notes and pre-printed titles will be done by the editorial staff.

For each original paper and short note a total of 75 reprints is available free of charge. The number of reprints free of charge per review article is 25. Requests for additional reprints (against payment) should be given together with the submitted manuscripts. **The deadline for orders of additional reprints is the 15th and the 30th of each month** for "physica status solidi (b)" and "physica status solidi (a)", respectively, two weeks before the delivery of the journal.

Bei Fragen zur Produktsicherheit wenden Sie sich bitte an:
If you have any questions regarding product safety,
please contact:

Walter de Gruyter GmbH
Genthiner Straße 13
10785 Berlin
productsafety@degruyterbrill.com